Probability and its Applications

A Series of the Applied Probability Trust

Editors: J. Gani, C.C. Heyde, T.G. Kurtz

Springer
New York
Berlin
Heidelberg
Hong Kong
London
Milan
Paris
Tokyo

Probability and its Applications

Anderson: Continuous-Time Markov Chains.
Azencott/Dacunha-Castelle: Series of Irregular Observations.
Bass: Diffusions and Elliptic Operators.
Bass: Probabilistic Techniques in Analysis.
Choi: ARMA Model Identification.
Daley/Vere-Jones: An Introduction to the Theory of Point Processes.
  Volume I: Elementary Theory and Methods, Second Edition.
de la Peña/Giné: Decoupling: From Dependence to Independence.
Del Moral: Feynman-Kac Formula: Genealogical and Interacting Particle Systems
  with Applications
Durrett: Probability Models for DNA Sequence Evolution.
Galambos/Simonelli: Bonferroni-type Inequalities with Applications.
Gani (Editor): The Craft of Probabilistic Modelling.
Grandell: Aspects of Risk Theory.
Gut: Stopped Random Walks.
Guyon: Random Fields on a Network.
Kallenberg: Foundations of Modern Probability, Second Edition.
Last/Brandt: Marked Point Processes on the Real Line.
Leadbetter/Lindgren/Rootzén: Extremes and Related Properties of Random Sequences
  and Processes.
Nualart: The Malliavin Calculus and Related Topics.
Rachev/Rüschendorf: Mass Transportation Problems. Volume I: Theory.
Rachev/Rüschendorf: Mass Transportation Problems. Volume II: Applications.
Resnick: Extreme Values, Regular Variation and Point Processes.
Shedler: Regeneration and Networks of Queues.
Silvestrov: Limit Theorems for Randomly Stopped Stochastic Processes
Thorisson: Coupling, Stationarity, and Regeneration.
Todorovic: An Introduction to Stochastic Processes and Their Applications.

Pierre Del Moral

Feynman-Kac Formulae

Genealogical and Interacting
Particle Systems with Applications

With 15 Illustrations

 Springer

Pierre Del Moral
Laboratoire de Statistique et Probabilités
Université Paul Sabatier
118, Route de Narbonne
31062 Toulouse, Cedex 4
France
delmoral@cict.fr

Series Editors

J. Gani
Stochastic Analysis
  Group, CMA
Australian National
  University
Canberra, ACT 0200
Australia

C.C. Heyde
Stochastic Analysis
  Group, CMA
Australian National
  University
Canberra, ACT 0200
Australia

T.G. Kurtz
Department of
  Mathematics
University of Wisconsin
480 Lincoln Drive
Madison, WI 53706
USA

Library of Congress Cataloging-in-Publication Data
Del Moral, Pierre.
 Feynman-Kac formulae : genealogical and interacting particle systems with applications /
Pierre Del Moral.
  p. cm. — (Probability and its applications)
 Includes bibliographical references and index.
 ISBN 0-387-20268-4 (alk. paper)
 1. Path integrals. 2. Evolution equations. 3. Quantum theory. 4. Vector spaces. I. Title.
II. Springer series in statistics. Probability and its applications
QC174.17.P27D45 2004
530.12—dc22   2003063340

ISBN 0-387-20268-4   Printed on acid-free paper.

© 2004 Springer-Verlag New York, LLC
All rights reserved. This work may not be translated or copied in whole or in part without the written permission of the publisher (Springer-Verlag New York, Inc., 175 Fifth Avenue, New York, NY 10010, USA), except for brief excerpts in connection with reviews or scholarly analysis. Use in connection with any form of information storage and retrieval, electronic adaptation, computer software, or by similar or dissimilar methodology now known or hereafter developed is forbidden.
The use in this publication of trade names, trademarks, service marks, and similar terms, even if they are not identified as such, is not to be taken as an expression of opinion as to whether or not they are subject to proprietary rights.

Printed in the United States of America. (MVY)

9 8 7 6 5 4 3 2 1   SPIN 10942918

Springer-Verlag is a part of *Springer Science+Business Media*

springeronline.com

To Laurence, Tiffany and Timothée

Preface

The central theme of this book concerns Feynman-Kac path distributions, interacting particle systems, and genealogical tree based models. This recent theory has been stimulated from different directions including biology, physics, probability, and statistics, as well as from many branches in engineering science, such as signal processing, telecommunications, and network analysis. Over the last decade, this subject has matured in ways that make it more complete and beautiful to learn and to use. The objective of this book is to provide a detailed and self-contained discussion on these connections and the different aspects of this subject. Although particle methods and Feynman-Kac models owe their origins to physics and statistical mechanics, particularly to the kinetic theory of fluid and gases, this book can be read without any specific knowledge in these fields. I have tried to make this book accessible for senior undergraduate students having some familiarity with the theory of stochastic processes to advanced postgraduate students as well as researchers and engineers in mathematics, statistics, physics, biology and engineering.

I have also tried to give an "exposé" of the modern mathematical theory that is useful for the analysis of the asymptotic behavior of Feynman-Kac and particle models. Researchers and applied mathematicians will find a collection of modern techniques from various branches of probability and stochastic analysis, including convergence of empirical processes, fluctuation and large-deviation analysis, semigroup and martingale techniques and propagation of chaos, as well as the asymptotic stability and the concentration of measure-valued processes, functional inequalities, ergodic coefficients, and contractions of Markov operators and nonlinear semigroups.

Besides the mathematical analysis of Feynman-Kac distribution flows and interacting particle models, I have developed a rather large class of applications to specific models from various scientific disciplines. The practitioner will find a source of useful convergence estimates as well as a detailed list of concrete examples of particle approximations for real models, including restricted Markov chain simulations, random motions in absorbing media, spectral analysis of Schrödinger operators and Feynman-Kac semigroups, rare event analysis, Dirichlet boundary problems, nonlinear filtering problems, interacting Kalman-Bucy filters, directed polymer simulations, and interacting Metropolis type algorithms. While this diversity of application model areas is part of the charm of the particle theory of Feynman-Kac models the list topics above is not exhaustive and actually only reflects the tastes and interests of the author.

One objective in writing this book was to throw some new light on some interesting links between sometimes too disconnected physical, engineering, and mathematical domains. In this connection, I would like to thanks Springer-Verlag and the editorial board for the invitation to write a book on this theme. I undertook this project for two main reasons. First I felt that there was no accessible treatment on Feynman-Kac path models and their interacting particle approximation schemes. Second, the abstract concepts and the probability theory are now at a point where they provide a natural and unifying mathematical basis for a large class of heuristic-like Monte Carlo algorithms currently used in Bayesian statistics, engineering science, and in physics and biology since the beginning of 1950s. I also hope that practitioners as well as graduate students from Bayesian schools will find a great advantage in using the abstract Feynman-Kac and particle theory developed in this book, provided they overcome the fear of seeing integral operators rather than summations or integrals with respect to some density function. Besides its mathematical elegance, the abstract formulation is less "notationally consuming" and of great practical value. It gives a powerful applied tool to be used in modeling nonlinear estimation problems as well as in studying and developing interacting particle approximation models. I hope that these ideas will fruitfully serve the further development of the field and that their propagation will influence other new application areas.

The material in this book can serve as a basis for different types of advanced courses on probability. The first type, geared towards pure applications of particle methods, could be centered around the Feynman-Kac modeling techniques and their application model areas discussed in Chapters 2, 3, 11 and 12. To aid more detailed studies these lectures could be completed either with the presentation of one of the more application-related articles selected from the list of references or with a new application model. More theoretical types of courses could cover the material in Chapters 4 through 10. A semester-long course would cover the stability and the annealed properties of Feynman-Kac semigroups derived in Chapters 4, 5, and 6 (possibly excluding Chapter 5). There is also enough material in the book to support three other sequences of courses. These lectures would cover, respectively, propagation of chaos (Chapters 7 and 8), central limit theorems (Chapters 7 and 9) and large-deviation principles (Chapter 10).

A part of the material presented in this book is based on a series of lectures I delivered at the 24th Finnish Summer School on Probability Theory in Lahti in spring 2002 and that was arranged by the Finnish Graduate School in Stochastics and the Rolf Nevanlinna Institute. It is also partly based on a second-year graduate course on particle methods and nonlinear filtering I gave at the Operations Research and Financial Engineering department of Princeton University in fall 2001. An overview was presented in three one-hour lectures for the Symposium on Numerical Stochastics (April 1999) at the Fields Institute for Research in Mathematical Sciences (Toronto) and at the same time was presented at the University of Alberta, Edmonton, with the support of the Canadian Mathematics of Information Technology and Complex Systems project.

Some of the material developed in this book results from various fruitful collaborations with Frédéric Cérou, Dan Crisan, Donald Dawson, Arnaud Doucet, François Le Gland, Michael Kouritzin, Pascal Lézaud, Michel Ledoux, Terry Lyons, Philip Protter, Samuel Tindel, Frederi Viens, Tim Zajic, and particularly with Alice Guionnet, Jean Jacod, and Laurent Miclo. Most of the text also proposes many new contributions to the subject. The reader will find a series of new deeper studies of topics such as contraction properties of nonlinear semigroups, functional entropy inequalities, uniform and precise increasing propagation-of-chaos estimates, central limit and Berry-Esseen type theorems, large-deviation principles for strong topologies on path-distribution spaces, new branching and genealogical particle models, advanced Feynman-Kac modeling techniques, and a fairly new class of application modeling areas.

While continuous time models and their applications in physics and engineering sciences are discussed in Chapters 1 and 12, I have not hesitated to concentrate the exposition on discrete Feynman-Kac and particle models. The reasons are twofold:

First, the analysis of discrete time models only requires a small prerequisite on Markov chains, while the study of continuous time models would have required different and specific knowledge on stochastic analysis, particularly on interacting jump models and Markov processes taking values in path spaces. Since one of the objectives was to prepare a text that was as self-contained as possible, a full presentation of both classes of models would have been too much digression.

The second reason is that apart from some mathematical technicalities the asymptotic analysis of continuous time models is in some sense more sophisticated but generally follows the same intuitions and the same line of argument as in the discrete time case. On the other hand, to my knowledge, various convergence theorems such as the uniform and increasing propagation of chaos, the Berry-Esseen estimates, and the strong large-deviation principles, presented respectively in Chapters 8, 9, and 10, remain nowadays open problems for continuous time models.

The reader interested in continuous time models can complete the study of this book following the four articles [96, 97, 98, 102]. For an introduction to interacting particle interpretation of continuous time Feynman-Kac models we recommend the review article on genetic type models [96] as well as [98] and [97, 102]. The last two referenced articles discuss both discrete and continuous time with applications to Schödinger generators, filtering problems and fixed points of integro-differential equations. I recommend [102] as a start for someone who has never studied the subject before. The article [97] provides a series of advanced lectures on simple central limit theorems and exponential estimates. Strong propagation-of-chaos results using coupling and semigroup techniques can be found in [95]. The series of articles above can also be completed by some studies on particle approximation of stochastic Feynman-Kac flows [63, 66, 64, 65] and their applications in the numerical solution of stochastic partial differential equations with non-linear potential [296]. I hope the former volume and the list of contributions to continuous time models above will guide the reader to understand the current state of the art on the topic and contribute to many interesting open problems.

I end this preface with a few words of advice to readers who are anxious about spending too much time on unnecessary immersions. The introduction of this book is an important step in entering into Feynman-Kac modeling and interacting-particle methods. The applications described in this opening section as well as in Chapters 11 and 12 should help the reader to find a concrete basis for going further through the mathematical aspects of this theory. A complete description on how the theory is applied in each application model area would of course require separate volumes with precise computer simulations and comparisons with different types of particle models and other existing algorithms. I have chosen to treat each subject in a rather short but self-contained way. Some applications are nowadays routine, and in this case I provide precise pointers to existing more application-related articles in the literature. Most applications also provide new insight on the theoretical and potential applications of Feynman-Kac and particle methods to statistical physics and engineering science. In this case the programming of these new particle algorithms is left to the reader. One natural path of "easy reading" will probably be to choose a familiar or attractive application area and to explore some selected parts of the book in terms of this choice. Nevertheless, this advice must not be taken too literally. To see the impact of genealogical tree-based particle methods, it is essential to understand the full force of Feynman-Kac modeling techniques on various research domains. Upon doing so, the reader will have a powerful weapon for the discovery of new Feynman-Kac interpretations and related particle numerical models. The principal challenge is to understand the theory and the branching particle models well enough to reduce them to practice.

I did not try to avoid repetition and each chapter starts with an introduction connecting the results developed in earlier parts with the current analysis. With a few exceptions, this book is self-contained and each chapter can be read independently of the other ones. To get up to speed on Chapter 12 on applications, the reader is recommended to start with Chapters 2 and 3, which contain the main concepts on Feynman-Kac modeling and interacting processes. Chapter 11 should also not be skipped since it contains a series of recipes on particle models to be combined with one another and applied in each application model area. In general, I did not give references in the text but in the introduction at the beginning of each chapter. I already apologize for possible errors or for references that have been omitted due to the lack of accurate information.

Finally, I would like to express my gratitude to the Centre National de la Recherche Scientifique (CNRS) which gave me the freedom and the opportunity to undertake this project and the Université Paul Sabatier of Toulouse. I am also grateful to the University of Melbourne and Purdue University as well as to Princeton University, where part of this project was developed. Last but not least, I would like to extend my thanks to John Kimmel for his precious editorial assistance as well as for his encouragements during these last two years.

Toulouse, France
September, 2003

Pierre Del Moral

Contents

1 Introduction   **1**
- 1.1 On the Origins of Feynman-Kac and Particle Models 1
- 1.2 Notation and Conventions 7
- 1.3 Feynman-Kac Path Models 11
  - 1.3.1 Path-Space and Marginal Models 11
  - 1.3.2 Nonlinear Equations 13
- 1.4 Motivating Examples 14
  - 1.4.1 Engineering Science 14
  - 1.4.2 Bayesian Methodology 21
  - 1.4.3 Particle and Statistical Physics 22
  - 1.4.4 Biology 25
  - 1.4.5 Applied Probability and Statistics 28
- 1.5 Interacting Particle Systems 29
  - 1.5.1 Discrete Time Models 30
  - 1.5.2 Continuous Time Models 34
- 1.6 Sequential Monte Carlo Methodology 37
- 1.7 Particle Interpretations 39
- 1.8 A Contents Guide for the Reader 41

2 Feynman-Kac Formulae   **47**
- 2.1 Introduction 47
- 2.2 An Introduction to Markov Chains 48
  - 2.2.1 Canonical Probability Spaces 49
  - 2.2.2 Path-Space Markov Models 51

		2.2.3	Stopped Markov chains	52

```
        2.2.3  Stopped Markov chains ................  52
        2.2.4  Examples ...........................  55
   2.3  Description of the Models ....................  58
   2.4  Structural Stability Properties ...............  61
        2.4.1  Path Space and Marginal Models ......  62
        2.4.2  Change of Reference Probability Measures  63
        2.4.3  Updated and Prediction Flow Models ..  65
   2.5  Distribution Flows Models ....................  68
        2.5.1  Killing Interpretation ...............  71
        2.5.2  Interacting Process Interpretation ...  73
        2.5.3  McKean Models ........................  76
        2.5.4  Kalman-Bucy filters ..................  79
   2.6  Feynman-Kac Models in Random Media ...........  81
        2.6.1  Quenched and Annealed Feynman-Kac Flows  83
        2.6.2  Feynman-Kac Models in Distribution Space  85
   2.7  Feynman-Kac Semigroups .......................  87
        2.7.1  Prediction Semigroups ................  88
        2.7.2  Updated Semigroups ...................  91

3  Genealogical and Interacting Particle Models        95
   3.1  Introduction ................................  95
   3.2  Interacting Particle Interpretations .........  96
   3.3  Particle models with Degenerate Potential ....  99
   3.4  Historical and Genealogical Tree Models ..... 103
        3.4.1  Introduction ........................ 103
        3.4.2  A Rigorous Approach
               and Related Transport Problems ...... 105
        3.4.3  Complete Genealogical Tree Models ... 108
   3.5  Particle Approximation Measures ............. 109
        3.5.1  Some Convergence Results ............ 112
        3.5.2  Regularity Conditions ............... 115

4  Stability of Feynman-Kac Semigroups                121
   4.1  Introduction ................................ 121
   4.2  Contraction Properties of Markov Kernels .... 122
        4.2.1  $h$-relative Entropy ................ 122
        4.2.2  Lipschitz Contractions .............. 127
   4.3  Contraction Properties of Feynman-Kac Semigroups  132
        4.3.1  Functional Entropy Inequalities ..... 134
        4.3.2  Contraction Coefficients ............ 138
        4.3.3  Strong Contraction Estimates ........ 142
        4.3.4  Weak Regularity Properties .......... 144
   4.4  Updated Feynman-Kac Models .................. 146
   4.5  A Class of Stochastic Semigroups ............ 152
```

5 Invariant Measures and Related Topics — 157
- 5.1 Introduction — 157
- 5.2 Existence and Uniqueness — 160
- 5.3 Invariant Measures and Feynman-Kac Modeling — 161
- 5.4 Feynman-Kac and Metropolis-Hastings Models — 164
- 5.5 Feynman-Kac-Metropolis Models — 166
  - 5.5.1 Introduction — 166
  - 5.5.2 The Genealogical Metropolis Particle Model — 170
  - 5.5.3 Path Space Models and Restricted Markov Chains — 172
  - 5.5.4 Stability Properties — 179

6 Annealing Properties — 187
- 6.1 Introduction — 187
- 6.2 Feynman-Kac-Metropolis Models — 189
  - 6.2.1 Description of the Model — 189
  - 6.2.2 Regularity Properties — 191
  - 6.2.3 Asymptotic Behavior — 193
- 6.3 Feynman-Kac Trapping Models — 197
  - 6.3.1 Description of the Model — 197
  - 6.3.2 Regularity Properties — 198
  - 6.3.3 Asymptotic Behavior — 201
  - 6.3.4 Large-Deviation Analysis — 204
  - 6.3.5 Concentration Levels — 208

7 Asymptotic Behavior — 215
- 7.1 Introduction — 215
- 7.2 Some Preliminaries — 217
  - 7.2.1 McKean Interpretations — 218
  - 7.2.2 Vanishing Potentials — 219
- 7.3 Inequalities for Independent Random Variables — 221
  - 7.3.1 $\mathbb{L}_p$ and Exponential Inequalities — 222
  - 7.3.2 Empirical Processes — 227
- 7.4 Strong Law of Large Numbers — 231
  - 7.4.1 Extinction Probabilities — 231
  - 7.4.2 Convergence of Empirical Processes — 236
  - 7.4.3 Time-Uniform Estimates — 244

8 Propagation of Chaos — 253
- 8.1 Introduction — 253
- 8.2 Some Preliminaries — 255
- 8.3 Outline of Results — 258
- 8.4 Weak Propagation of Chaos — 261
- 8.5 Relative Entropy Estimates — 262
- 8.6 A Combinatorial Transport Equation — 267
- 8.7 Asymptotic Properties of Boltzmann-Gibbs Distributions — 271

xvi   Contents

    8.8 Feynman-Kac Semigroups 277
        8.8.1 Marginal Models 278
        8.8.2 Path-Space Models 280
    8.9 Total Variation Estimates 282

9 Central Limit Theorems — 291
    9.1 Introduction 291
    9.2 Some Preliminaries 293
    9.3 Some Local Fluctuation Results 295
    9.4 Particle Density Profiles 300
        9.4.1 Unnormalized Measures 300
        9.4.2 Normalized Measures 301
        9.4.3 Killing Interpretations and Related Comparisons . . 303
    9.5 A Berry-Esseen Type Theorem 306
    9.6 A Donsker Type Theorem 318
    9.7 Path-Space Models 322
    9.8 Covariance Functions 327

10 Large-Deviation Principles — 331
    10.1 Introduction 333
    10.2 Some Preliminary Results 339
        10.2.1 Topological Properties 339
        10.2.2 Idempotent Analysis 340
        10.2.3 Some Regularity Properties 344
    10.3 Crámer's Method 347
    10.4 Laplace-Varadhan's Integral Techniques 351
    10.5 Dawson-Gärtner Projective Limits Techniques 359
    10.6 Sanov's Theorem 363
        10.6.1 Introduction 363
        10.6.2 Topological Preliminaries 364
        10.6.3 Sanov's Theorem in the τ-Topology 370
    10.7 Path-Space and Interacting Particle Models 374
        10.7.1 Proof of Theorem 10.1.1 374
        10.7.2 Sufficient Conditions 376
    10.8 Particle Density Profile Models 377
        10.8.1 Introduction 377
        10.8.2 Strong Large-Deviation Principles 379

11 Feynman-Kac and Interacting Particle Recipes — 387
    11.1 Introduction 387
    11.2 Interacting Metropolis Models 389
        11.2.1 Introduction 389
        11.2.2 Feynman-Kac-Metropolis and Particle Models . . . 390
        11.2.3 Interacting Metropolis and Gibbs Samplers 393
    11.3 An Overview of some General Principles 394

- 11.4 Descendant and Ancestral Genealogies 396
- 11.5 Conditional Explorations 400
- 11.6 State-Space Enlargements and Path-Particle Models 402
- 11.7 Conditional Excursion Particle Models 404
- 11.8 Branching Selection Variants 405
  - 11.8.1 Introduction 405
  - 11.8.2 Description of the Models 408
  - 11.8.3 Some Branching Selection Rules 409
  - 11.8.4 Some $\mathbb{L}_2$-mean Error Estimates 411
  - 11.8.5 Long Time Behavior 417
  - 11.8.6 Conditional Branching Models 419
- 11.9 Exercises 420

12 Applications   427
- 12.1 Introduction 427
- 12.2 Random Excursion Models 429
  - 12.2.1 Introduction 429
  - 12.2.2 Dirichlet Problems with Boundary Conditions 431
  - 12.2.3 Multilevel Feynman-Kac Formulae 436
  - 12.2.4 Dirichlet Problems with Hard Boundary Conditions 440
  - 12.2.5 Rare Event Analysis 444
  - 12.2.6 Asymptotic Particle Analysis of Rare Events 447
  - 12.2.7 Fluctuation Results and Some Comparisons 450
  - 12.2.8 Exercises 453
- 12.3 Change of Reference Measures 459
  - 12.3.1 Introduction 459
  - 12.3.2 Importance Sampling 460
  - 12.3.3 Sequential Analysis of Probability Ratio Tests 462
  - 12.3.4 A Multisplitting Particle Approach 463
  - 12.3.5 Exercises 465
- 12.4 Spectral Analysis of Feynman-Kac-Schrödinger Semigroups 469
  - 12.4.1 Lyapunov Exponents and Spectral Radii 470
  - 12.4.2 Feynman-Kac Asymptotic Models 471
  - 12.4.3 Particle Lyapunov Exponents 473
  - 12.4.4 Hard, Soft and Repulsive Obstacles 475
  - 12.4.5 Related Spectral Quantities 477
  - 12.4.6 Exercises 479
- 12.5 Directed Polymers Simulation 484
  - 12.5.1 Feynman-Kac and Boltzmann-Gibbs Models 484
  - 12.5.2 Evolutionary Particle Simulation Methods 487
  - 12.5.3 Repulsive Interaction and Self-Avoiding
           Markov Chains 488
  - 12.5.4 Attractive Interaction and Reinforced Markov Chains 490
  - 12.5.5 Particle Polymerization Techniques 490
  - 12.5.6 Exercises 495

12.6 Filtering/Smoothing and Path estimation 497
    12.6.1 Introduction 497
    12.6.2 Motivating Examples 500
    12.6.3 Feynman-Kac Representations 505
    12.6.4 Stability Properties of the Filtering Equations . . . 508
    12.6.5 Asymptotic Properties of Log-likelihood Functions . 510
    12.6.6 Particle Approximation Measures 512
    12.6.7 A Partially Linear/Gaussian Filtering Model 513
    12.6.8 Exercises 520

References   **523**

Index   **549**

1
Introduction

1.1 On the Origins of Feynman-Kac and Particle Models

The field of Feynman-Kac and particle models is one of the most active contact points between probability, engineering, and the natural sciences. It is hard to know where to start in describing its early contributions.

The origins of Feynman-Kac formulae certainly started with the work of R.P. Feynman who in his doctoral dissertation (Princeton, 1942) provides a heuristic connection between the Schrödinger equation and N. Wiener path integral theory. These lines of investigations were pursued and amplified by M. Kac in the early 1950s (see for instance [194]). The idea was to express the semigroup of a quantum particle evolving in a potential in terms of a functional path-integral formula. Intuitively speaking, Feynman-Kac measures enter the effects of the potential in the distribution of the paths of the particles. This "change of probability" on path space associated with a given potential function has considerably influenced several research directions in mathematical physics, stochastic processes, and other scientific disciplines. One of the fascinations of these models today is their use to model a rather large class of physical, biological, and engineering problems. From the point of view of physics, they represent for instance the path distribution of a single particle evolving in absorbing and disordered media (see for instance [295]). In this interpretation, the potential function represents a "killing or creation" rate related to the absorbing nature of the medium.

More generally, they can be regarded as the Boltzmann-Gibbs distribution of certain physical or biological quantities, such as directed polymers in physical chemistry or genetic infinite-population models (see [92, 209, 280] and references therein). In this context, the potential function can be regarded as a Hamiltonian or an energy function related to internal interactions or to the selection pressure of the environment. From the perspective of the engineer or the applied statistician, it usually represents a conditional distribution of a certain unknown quantity with respect to some observation process. This interpretation is currently used in advanced signal processing, particularly in filtering estimation and Bayesian analysis (see [97, 125] and references therein). In these settings, the potential is rather regarded as a likelihood function of the states with respect to some observation process or some reference path.

Stochastic particle algorithms belong to the class of Monte Carlo methods. Their sources may be found in the foundations of probability theory with the pioneering work of J. Bernoulli (*Ars Conjectandi* published in 1716), who introduced the concept of the probability of an event as the ratio of favorable outcomes with respect to the number of all possible independent outcomes. A decisive step in the modern development of probability theory was the introduction in the 1920s by A.A. Markov (*Calculus of Probabilities*, 3rd ed., St. Petersburg, 1913) of a theory of stochastic processes that studies sequences of random variables evolving with time. This new branch of probability led to rather intense activity in various scientific disciplines. The theory of Markov processes provides natural probabilistic interpretations of various evolution models arising in engineering and the natural sciences. One critical aspect of particle methods as opposed to any other numerical method is that it provides a "microscopic particle interpretation" of the physical or engineering evolution equation at hand. Another advantage of these probabilistic techniques is that they do not use any regularity information on the coefficients of the models and they apply to large scale models. The increasing fascination of these particle methods today is their use to solve numerically nonlinear equations in distribution space. The nonlinear structure of these distribution models induces a natural interaction or a branching mechanism in the evolution of the particle approximation model. This rather recent aspect of particle methods takes its origins from the 1960s with the development of fluid mechanisms and statistical physics. We refer the reader to the pioneering works of McKean [243, 244] (see also the more recent treatments [27, 28, 245, 273, 294] and references therein).

The use of interacting particle methods in engineering science and more particularly in advanced signal processing is more recent. The first rigorous study in this field seems to be the article [75] published in 1996 on the applications of particle methods to nonlinear estimation problems. This article provides the first convergence result for a new class of interacting particle models originally presented as heuristic schemes in the beginning of the 1990s in three independent chains of articles [164, 163], [204], and [43, 205, 104, 105]. These studies were followed by four other articles [76, 83, 84, 86] revealing the generality and the impact of these particle methods in solving numerically a rather large class of discrete generation and abstract nonlinear measure-valued processes. In the same period two other independent works [64, 65] proposed another class of particle branching variants for solving continuous-time filtering problems. Incidentally, and as we noticed in a more recent work [99], all of these mathematical techniques also apply directly without further work to analyze the asymptotic behavior of a class of genealogical tree particle models currently used in nonlinear smoothing and path estimation problems.

Although precise mathematical statements and various detailed applications are provided in the further development of this introductory chapter (see for instance Section 1.4 pp. 14-29), to motivate this introduction we already present one particular example from engineering science and more particularly from advanced signal processing, that gives some insights as to what this book is about. The example we have chosen is taken from [104, 105]. It is also known as "the Singer model" and is often used as a simplified radar model. We consider a three-dimensional Markov chain $X_n = (X_n^{(1)}, X_n^{(2)}, X_n^{(3)})$, $n \in \mathbb{N}$. The three coordinates represent respectively the acceleration, the speed, and the position of an abstract target evolving in the real line according to the dynamical equation

$$\begin{cases} X_n^{(1)} &= X_{n-1}^{(1)} + \epsilon_n \, W_n \\ X_n^{(2)} &= (1 - \alpha \, \Delta) \, X_{n-1}^{(2)} + \beta \, \Delta \, X_n^{(1)} \\ X_n^{(3)} &= X_{n-1}^{(3)} + \Delta \, X_n^{(2)} \end{cases} \tag{1.1}$$

where (α, β) is a pair of constant parameters and $\Delta \in (0, 1)$ corresponds to the radar sampling period. The initial random variable X_0 represents the unknown random location of the target. The random changes of the acceleration coordinates may be modeled by a sequence of independent Bernoulli random variables $\varepsilon_n (\in \{0, 1\})$ and a sequence of independent uniform random variables W_n ($\in [0, a]$ for some $a \in \mathbb{R}$). The target X_n is partially observed by the radar measurements. The observations delivered at each time $n \geq 0$ by the radar have the form

$$Y_n = X_n^{(3)} + \Delta \, V_n$$

The random perturbations in radar measurements (induced for instance by the thermic noise in complex electronic devices) are often modeled by choosing a sequence of independent Gaussian random variables V_n with zero mean and say, unit variance. In Figure 1.1, we have represented three consecutive radar measurements of the evolving target X_n, X_{n+1}, and X_{n+2}.

The nonlinear filtering problem consists in estimating the conditional distribution of the random path $(X_0, \ldots, X_n)$ of the target given the ob-

4   1. Introduction

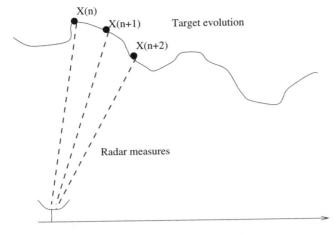

FIGURE 1.1. Radar processing

servations Y_p delivered by the radar up to time n

$$\text{Law}(X_0, \ldots, X_n | Y_0, \ldots, Y_n) \tag{1.2}$$

The particle approximation model of these path distributions is constructed as follows. First, at time $t_0 = 0$, we sample N independent locations of the target, say $(X_0^i)_{1 \leq i \leq N}$, with the initial acceleration/speed/position random components $X_0^i = (X_0^{(1),i}, X_0^{(2),i}, X_0^{(3),i})$. Then, we evolve randomly these initial points according to the dynamical equation (1.1) up to some fixed time, say $t_1 (> t_0)$. In other words, we sample N independent copies $X_{t_0,t_1}^i = (X_0^i, \ldots, X_{t_1}^i)$ of the target from the origin t_0 up to time t_1. The likelihood W_{t_0,t_1}^i of each of these paths is defined by the formula

$$W_{t_0,t_1}^i = \exp\left(-\frac{1}{2} \sum_{t_0 \leq p < t_1} [Y_p - X_p^{(3),i}]^2\right)$$

Loosely speaking, these $[0,1]$-valued random exponential parameters measure the adequacy of the sampled targets with respect to the observation sequence. The way to update the initial configuration with respect to the radar observations is not unique. For instance, we can select randomly N conditionally independent paths $\widehat{X}_{t_0,t_1}^i = (\widehat{X}_0^i, \ldots, \widehat{X}_{t_1}^i)_{1 \leq i \leq N}$ with respective distributions

$$W_{t_0,t_1}^i \, \delta_{X_{t_0,t_1}^i} + (1 - W_{t_0,t_1}^i) \sum_{j=1}^N \frac{W_{t_0,t_1}^j}{\sum_{k=1}^N W_{t_0,t_1}^k} \, \delta_{X_{t_0,t_1}^j}$$

In other words, with a probability W_{t_0,t_1}^i, we keep the path X_{t_0,t_1}^i; otherwise we replace it by a new one X_{t_0,t_1}^j randomly chosen in the current

1.1 On the Origins of Feynman-Kac and Particle Models

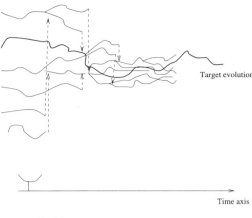

FIGURE 1.2. Particle radar processing

configuration with a probability proportional to its likelihood W_{t_0,t_1}^j. After this updating stage, we again sample N independent copies $X_{t_1,t_2}^i = (X_{t_1}^i, \ldots, X_{t_2}^i)$ of the target from t_1 up to another fixed time, say t_2, and starting at $X_{t_1}^i = \widehat{X}_{t_1}^i$. We update this new configuration with respect to the next sequence of observations from t_1 to t_2 as above by replacing (t_0, t_1) by (t_1, t_2), and so on.

In Figure 1.2, we have represented the genetic type evolution of $N = 5$ particles. The radar measurements give some information to each exploring particle on the evolution parameter of the target. The jumps correspond to the selection transition where a particle with poor likelihood prefers to select a new site.

The rationale behind this heuristic-like algorithm is to construct a stochastic and adaptive grid with a refined degree of precision on the regions with high conditional probability mass. This simple example discussed above leads inevitably to the following questions: Is this evolutionary stochastic grid model well-founded? Is it possible to calibrate the "speed" of convergence when the precision parameter N tends to infinity? If so, then what can we say about the long time behavior of these algorithms? Can we extend these ideas to more general optimization and simulation problems? If we interpret this genetic type particle scheme as a birth and death model, what can we say about the corresponding genealogical trees? The main difficulty in the asymptotic analysis of these algorithms comes from the interacting selection mechanism. As we mentioned earlier, the first well-founded proof of these particle algorithms can be found in [75]. The central idea was to connect the desired conditional distributions with a new class of discrete generation and Feynman-Kac particle approximation models. Surprisingly enough, we shall see in the further development of this book that the occupation measures of the genealogical tree model associated with the genetic type particle algorithm above converge to the desired conditional distribution on path space (1.2). We shall also prove that the ancestral lines of each current individual can be regarded with some respect as a collection of approximating independent samples of the complete path of the target given the observations.

The idea of duplicating in a dynamical way better-fitted individuals and moving them one step forward to explore state-space regions is the basis of various stochastic search algorithms. In Section 1.7, we shall provide a rather detailed catalog of models arising in engineering sciences, physics, and biology built on this natural exploration strategy. In this connection, we mention that these heuristic ideas seem to have emerged in biology in the beginning of the 1950s with the article of M.N. Rosenbluth and A.W. Rosenbluth on macromolecular simulations [280] as well as in physics with the article of Kahn and Harris [177].

A more systematic and recent study on particle methods and abstract Feynman-Kac models in general metric spaces was initiated in the chain of articles [91, 93, 95, 97, 98, 99]. The range of applications of these particle techniques is attested by the number of articles in engineering and applied statistics and particularly in Bayesian literature. For instance, the book [125] provides a detailed panorama of recent Bayesian applications in seemingly disconnected areas such as target tracking, computer vision, and financial mathematics, as well as in biology and in directed polymer simulations. Unfortunately, these lines of research seem to be developing in a blind way with at least no visible connections with the physical and the mathematical sides of this field.

All these developments also revealed and tied up strong and fruitful connections with classical genetic algorithms. These models were introduced by J.H. Holland in [183] in 1975. During the last thirty years, these powerful stochastic search algorithms have been used with success in the numerical solution of a wide range of global optimization problems. We refer the reader to the chain of articles [5, 191, 192, 258, 285, 299, 306, 309] and references therein. The first well-founded proof of the convergence of genetic algorithms towards a set of global minima of a potential function on a finite state space was due to R. Cerf in 1994 in a chain of articles [45, 46, 47, 48]. This line of research was extended and simplified in [94]. These last referenced articles provide respectively a large-deviation analysis and a semigroups approach combined with log-Sobolev inequalities to study the concentration properties of genetic algorithms with fixed population size as the time tends to infinity.

One of the main objectives of this book is to provide a unifying treatment on Feynman-Kac and particle methods. Most of the book is concerned with abstract mathematical models in general measurable state spaces. Precise applications will be discussed in full detail in a separate chapter. In each application model area we consider, we will provide a specific interpretation of these abstract Feynman-Kac and particle models with a detailed list of contributions and references. An important part of the book is concerned with the asymptotic behavior of particle models as the size of the systems tends to infinity. Special attention is paid to the delicate and probably the most important problems of the long time behavior of particle algorithms.

In the further development of this preliminary chapter, we provide a detailed introduction on discrete and continuous time Feynman-Kac models and genealogical and interacting particle methods. We underline the fundamental concepts and the general mathematical structure of these models leaving aside precise constructions with "unnecessary" technical assumptions. We motivate the forthcoming development of the book with illustrations of these abstract mathematical models on several concrete examples from advanced signal processing, microstatistical mechanics, polymer chemistry, and applied statistics. We also provide several particle interpretations in connection with the application model areas we consider. We finally discuss the connections between these particle models and a class of existing Monte Carlo algorithms currently used in Bayesian statistics, quantum physics, and operations research. We end the chapter with a detailed guide to the contents of this book.

1.2 Notation and Conventions

In this preliminary section, we have collected some basic notation and conventions that we have tried to keep consistently throughout the book.

We denote respectively by $\mathbb{N}$, $\mathbb{Z}$, and $\mathbb{R}$ the fields of all positive integers, the set of all integers, and the field of all real numbers.

We denote by $\mathcal{M}(E)$ the set of bounded and signed measures on a given measurable space $(E, \mathcal{E})$. By $\mathcal{M}_0(E)$ and $\mathcal{M}_+(E) \subset \mathcal{M}(E)$, we denote respectively the subset of measures with null total mass and the subset of positive measures. Finally, $\mathcal{P}(E)$ and $\mathcal{B}_b(E)$ denote respectively the set of probability measures and bounded measurable functions on a given measurable space $(E, \mathcal{E})$. As usual $\mathcal{B}_b(E)$ is regarded as a Banach space with the supremum norm

$$\|f\| = \sup_{x \in E} |f(x)|$$

We shall slightly abuse the notation and denote by 0 and 1 the zero and the unit elements in the semirings $(\mathbb{R}, +, \times)$ and $(\mathcal{B}_b(E), +, \times)$. We always assume implicitly that $\{x\} \in \mathcal{E}$ for any $x \in E$ and we write δ_x, the Dirac measure at x. Unless otherwise stated, the set $\mathcal{M}(E)$ is endowed with the σ-algebra generated by $\mathcal{B}_b(E)$; that is, the coarsest σ-algebra on $\mathcal{M}(E)$ such that the linear functionals

$$\mu \in \mathcal{M}(E) \longrightarrow \mu(f) = \int f(x)\, \mu(dx) \ \in \mathbb{R}$$

are measurable. We also denote $\mathrm{Osc}_1(E)$, the convex set of $\mathcal{E}$-measurable functions f with oscillations less than one; that is,

$$\mathrm{osc}(f) = \sup\{|f(x) - f(y)|\ ;\ x, y \in E\} \leq 1$$

We also use the notation, for any $f \in \mathcal{B}_b(E)$,

$$\|f\|_{\mathrm{osc}} = \|f\| + \mathrm{osc}(f)$$

$(.)^+$, $(.)^-$, and $\lfloor . \rfloor$ denote respectively the positive, negative, and integer part functions. The maximum and minimum operations are denoted respectively by $\vee$ and $\wedge$:

$$\begin{aligned} a \vee b &= \max(a, b), & a^+ &= a \vee 0 \\ a \wedge b &= \min(a, b), & -a^- &= a \wedge 0 \end{aligned}$$

We extend the preceding operations on the set of functions $\mathcal{B}_b(E)$. For instance, for a given $f \in \mathcal{B}_b(E)$, we denote by f^+ and $-f^-$ its positive and negative parts

$$f^+(x) = f(x) \vee 0 \quad \text{and} \quad -f^-(x) = f(x) \wedge 0$$

For any pair of signed measures $\mu, \eta \in \mathcal{M}(E)$, we say that μ is absolutely continuous with respect to η and we write $\mu \ll \eta$ if $\eta(A) = 0$ whenever $\mu(A) = 0$, $A \in \mathcal{E}$. When $\mu \ll \eta$, sometimes we say that η is the dominating measure of μ. We recall that the Radon-Nikodym derivative of μ with respect to a dominating measure η is the unique function $x \in E \longrightarrow \frac{d\mu}{d\eta}(x)$ (up to sets of η-measure zero) such that for any $A \in \mathcal{E}$

$$\eta(A) = \int_A \frac{d\mu}{d\eta}(x)\ \eta(dx)$$

The relative entropy $\mathrm{Ent}(\mu_1|\mu_2)$ and the total variation distance $\|\mu_1 - \mu_2\|_{\mathrm{tv}}$ between probability measures $\mu_1, \mu_2 \in \mathcal{P}(E)$ are defined by

$$\mathrm{Ent}(\mu_1|\mu_2) = \int \log \frac{d\mu_1}{d\mu_2}\ d\mu_1$$

if $\mu_1 \ll \mu_2$ and ∞ otherwise, and

$$\begin{aligned} \|\mu_1 - \mu_2\|_{\mathrm{tv}} &= \sup_{A \in \mathcal{E}} |\mu_1(A) - \mu_2(A)| \\ &= \frac{1}{2} \sup\{|\mu_1(f) - \mu_2(f)|\ ;\ f \in \mathcal{B}_b(E)\ ;\ \|f\| \leq 1\} \end{aligned}$$

For a distribution $\mu \in \mathcal{P}(E)$ and $p \geq 1$, we also write $\|.\|_{p,\mu}$, the $\mathbb{L}_p(\mu)$-norm

$$\|f\|_{p,\mu} = \left(\int |f|^p\ d\mu\right)^{\frac{1}{p}}$$

A (bounded) integral operator from a measurable space $(E_0, \mathcal{E}_0)$ into another measurable space $(E_1, \mathcal{E}_1)$ is an integral kernel $M(x_0, dx_1)$ such that for any $(x_0, A_1) \in (E_0 \times \mathcal{E}_1)$ we have

- $M(x_0, .) \in \mathcal{M}(E_1)$ and $\sup_{y_0 \in E_0} |M(y_0, E_1)| < \infty$.
- The mapping $y_0 \in E_0 \to M(y_0, A)$ is a $\mathcal{E}_0$-measurable function.

We say that an integral operator M is a Markov kernel from E_0 into E_1 when we have $M(x_0, .) \in \mathcal{P}(E_1)$ for any $x_0 \in E_0$.

We also recall that any integral transition $M_1(x_0, dx_1)$ from a measurable space $(E_0, \mathcal{E}_0)$ into another measurable space $(E_1, \mathcal{E}_1)$ generates two operators, one acting on bounded $\mathcal{E}_1$-measurable functions $f_1 \in \mathcal{B}_b(E_1)$ and taking values in $\mathcal{B}_b(E_0)$

$$\forall (x_0, f_1) \in (E_0 \times \mathcal{B}_b(E_1)), \quad (M_1 f_1)(x_0) = \int_{E_1} M_1(x_0, dx_1) \; f_1(x_1)$$

and the other one acting on measures $\mu_0 \in \mathcal{M}(E_0)$ and taking values in $\mathcal{M}(E_1)$

$$\forall (\mu_0, A_1) \in (\mathcal{P}(E_0) \times \mathcal{E}_1), \quad (\mu_0 M_1)(A_1) = \int_{E_0} \mu_0(dx_0) \; M_1(x_0, A_1)$$

Finally, if $M_2(x_1, dx_2)$ is a Markov transition from $(E_1, \mathcal{E}_1)$ into another measurable space $(E_2, \mathcal{E}_2)$, then we denote by $M_1 M_2$ the composite operator

$$(M_1 M_2)(x_0, dx_2) = \int_{E_1} M_1(x_0, dx_1) \; M_2(x_1, dx_2)$$

For any $\mathbb{R}^d$-valued function $f = (f^i)_{1 \leq i \leq d} \in \mathcal{B}_b(E_1)^d$, any integral operator M from E_0 into E_1, and any $\mu \in \mathcal{M}(E_0)$ and $(x^1, \ldots, x^d) \in E_0^d$, we will slightly abuse the notation, and we write $M(f)$ and $\mu(f)$ the $\mathbb{R}^d$-valued function and the point in $\mathbb{R}^d$ given by

$$M(f) \stackrel{\text{def.}}{=} (M(f^1), \ldots, M(f^d)) \quad \text{and} \quad \mu(f) \stackrel{\text{def.}}{=} (\mu(f^1), \ldots, \mu(f^d))$$

For any $x \in E_0$ and $\varphi^1, \varphi^2 \in \mathcal{B}_b(E_1)$, we also simplify the notation and we write

$$M[(\varphi^1 - M\varphi^1)(\varphi^2 - M\varphi^2)](x)$$

instead of

$$M[(\varphi^1 - M(\varphi^1)(x))(\varphi^2 - M(\varphi^2)(x))](x)$$
$$= M(\varphi^1 \varphi^2)(x) - M(\varphi^1)(x) \; M(\varphi^2)(x)$$

Throughout this book, we will consider various collections of Markov kernels $K_{\eta_0}(x_0, dx_1)$ from a measurable space $(E_0, \mathcal{E}_0)$ into another measurable space $(E_1, \mathcal{E}_1)$ and indexed by the set of all probability measures

$\eta_0 \in \mathcal{P}(E_0)$. To avoid repetition, we will always implicitly suppose that for any $A_1 \in \mathcal{E}_1$ the mappings

$$(x_0, \eta_0) \in \mathcal{P}(E_0) \longrightarrow K_{\eta_0}(x_0, A_1) \in [0,1]$$

are measurable. In other words, $K_{\eta_0}(x_0, dx_1)$ can be regarded as a single Markov kernel from $(E_0 \times \mathcal{P}(E_0))$ into E_1.

Let $(E_n, \mathcal{E}_n)$, $n \in \mathbb{N}$, be a collection of measurable spaces, and let M_n be a sequence of (bounded) integral operators from E_{n-1} into E_n. We use the notation $M_{p,n}$ to denote the integral operator from E_p into E_n defined by

$$M_{p,n} = M_{p+1} M_{p+2} \ldots M_n \quad \text{with} \quad M_{n,n} = Id$$

For time-homogeneous models, we simplify the notation and we denote by M^n the n iterates of the operator M

$$M^n = M \, M^{n-1} \quad \text{with} \quad M^0 = Id$$

We examine in this book discrete and continuous time mathematical models. To distinguish the discrete and continuous time indices, we use as traditionally the letters $n, p, q \in \mathbb{N}$ for discrete time parameters and the letters $r, s, t \in \mathbb{R}_+ = [0, \infty)$ for continuous time parameters.

Given a sequence of measurable spaces $(E_n, \mathcal{E}_n)$ we often use the letters x_n, y_n, z_n to denote the points in E_n. By f_n and μ_n we denote respectively a given bounded measurable function and a probability measure on E_n. To simplify the presentation sometimes we also use the notation

$$\forall 0 \leq p \leq n \quad E_{[p,n]} = \prod_{q=p}^{n} E_q = (E_p \times \ldots \times E_n) \quad \text{and} \quad E_{(p,n]} = E_{[p+1,n]}$$

We shall often use the letter c to denote any nonnegative universal constant whose values may vary from line to line but do not depend on the time parameter nor on the Feynman-Kac model. We shall denote by $a = (a(p))_{p \geq 0}$ or $b = (b(p))_{p \geq 0}$ a sequence of nonnegative universal constants whose values may also vary from line to line. We shall use the letter a when the sequence does not depend on the Feynman-Kac model and the letter b in the opposite situation.

We also use the conventions

$$\sum_{\emptyset} = 0, \quad \prod_{\emptyset} = 1, \quad \sup_{\emptyset} = -\infty, \quad \inf_{\emptyset} = +\infty$$

1.3 Feynman-Kac Path Models

The main object of this section is to present the Feynman-Kac path distribution models discussed in this book. For the time being, these models are described in a somewhat heuristic way. The full mathematical description will be given in the forthcoming development of Chapter 2.

1.3.1 Path-Space and Marginal Models

In the discrete time situation, Feynman-Kac path measures are traditionally defined by the following formulae:

$$\mathbb{Q}_n(d(x_0,\ldots,x_n)) = \frac{1}{\mathcal{Z}_n} \left\{ \prod_{p=0}^{n-1} G_p(x_p) \right\} \mathbb{P}_n(d(x_0,\ldots,x_n)) \qquad (1.3)$$

The measure $\mathbb{P}_n$ represents the probability distribution of the path sequence $(X_0,\ldots,X_n)$ of a Markov chain X taking values in some measurable space $(E,\mathcal{E})$, and the potential functions G_n are $\mathcal{E}$-measurable nonnegative functions such that the normalizing constants are well-defined; that is, for any $n \in \mathbb{N}$,

$$\mathcal{Z}_n = \int_{E^{(n+1)}} \prod_{p=0}^{n-1} G_p(x_p) \; \mathbb{P}_n(d(x_0,\ldots,x_n)) \quad \in (0,\infty)$$

When the potential functions $G_n = \exp V_n$ are related to some energy function V_n, the measures $\mathbb{Q}_n$ can be written in the more familiar form

$$\mathbb{Q}_n(d(x_0,\ldots,x_n)) = \frac{1}{\mathcal{Z}_n} \exp\left\{ \sum_{p=0}^{n-1} V_p(x_p) \right\} \mathbb{P}_n(d(x_0,\ldots,x_n))$$

The continuous time models are defined similarly by the Feynman-Kac path measures

$$d\mathbb{Q}_t = \frac{1}{\mathcal{Z}_t} \exp\left\{ \int_0^t V_s(X_s)\,ds \right\} d\mathbb{P} \qquad (1.4)$$

where $\mathbb{P}$ is the distribution of a canonical Markov process X_t taking values in some measurable space $(E,\mathcal{E})$ and V_t, $t \in \mathbb{R}_+$, is a collection of measurable functions such that the normalizing constants are well-defined in the sense that

$$\mathcal{Z}_t = \mathbb{E}\left(\exp\left\{ \int_0^t V_s(X_s)\,ds \right\}\right) \quad \in (0,\infty)$$

Even if they look innocent, these Feynman-Kac path measures are very complex mathematical objects. To get some feeling of their complexity, we note that the continuous normalizing constants $\mathcal{Z}_t$ are expressed in terms of functional integrals on path spaces. In the same way, solving the discrete time constants $\mathcal{Z}_n$ requires the computation of n integrations over $(E, \mathcal{E})$.

To get one step further in our discussion, it is convenient to introduce the terminal time marginals of these path distributions. In the discrete time case, the corresponding distribution flow η_n, $n \in \mathbb{N}$, is defined for any bounded $\mathcal{E}$-measurable function f by the Feynman-Kac formulae

$$\eta_n(f) = \gamma_n(f)/\gamma_n(1) \quad \text{with} \quad \gamma_n(f) = \mathbb{E}\left(f(X_n) \prod_{p=0}^{n-1} G_p(X_p)\right) \quad (1.5)$$

The continuous time marginal distribution flow η_t, $t \in \mathbb{R}_+$, is defined in the same way by the Feynman-Kac formulae

$$\eta_t(f) = \gamma_t(f)/\gamma_t(1) \quad \text{with} \quad \gamma_t(f) = \mathbb{E}\left(f(X_t) \exp\left\{\int_0^t V_s(X_s) ds\right\}\right)$$

A simple calculation shows that the normalizing constants can be expressed in terms of the normalized distribution flow with the formulae

$$\mathcal{Z}_n = \gamma_n(1) = \prod_{p=0}^{n-1} \eta_p(G_p) \quad \text{and} \quad \mathcal{Z}_t = \gamma_t(1) = \exp \int_0^t \eta_s(V_s) ds$$

This shows that the unnormalized flow can be computed in terms of the normalized distributions. More precisely, we easily deduce the following representations from the display above:

$$\gamma_n(f) = \eta_n(f) \prod_{p=0}^{n-1} \eta_p(G_p) \quad \text{and} \quad \gamma_t(f) = \eta_t(f) \exp \int_0^t \eta_s(V_s) ds \quad (1.6)$$

The second important observation is that the marginal distributions have the same mathematical structure as the path measures. To make precise this fundamental stability property, we suppose we are given an auxiliary Markov chain X'_n taking values in some measurable space $(E', \mathcal{E}')$. We associate with X'_n the sequence of n-stopped processes

$$X_n = (X'_{p \wedge n})_{p \geq 0} \in E = (E')^{\mathbb{N}}$$

It can be easily verified that X_n is again a Markov chain taking values in the set of E'-valued countable sequences. We further suppose that the potential functions G_n only depend on the nth time value of the stopped chain; that is, we have that

$$G_n(X_n) = G_n((X'_{p \wedge n})_{p \geq 0}) = G'_n(X'_n)$$

for some $\mathcal{E}'$-measurable function G'_n. In this situation, η_n are distributions on $(E')^{\mathbb{N}}$, and their marginals $\mathbb{Q}'_n \in \mathcal{P}((E')^{n+1})$ on the first $(n+1)$ coordinates coincide with the Feynman-Kac path measure defined as in (1.3) by replacing the pair (X_n, G_n) by (X'_n, G'_n). It also follows that the marginal distributions $\eta'_n \in \mathcal{P}(E')$ of $\mathbb{Q}'_n$ with respect to the terminal time coincide with the Feynman-Kac measure defined as in (1.5) by replacing the pair (X_n, G_n) by (X'_n, G'_n). We can summarize the structural property above with the following synthetic diagram

$$\mathbb{Q}_n \xleftarrow{\text{path measure}} \eta_n \xrightarrow{n-\text{time marginal}} \eta'_n$$

Similar arguments apply to continuous time models, but the construction of the stopped path process is technically more involved (see for instance [72] and references therein).

These two apparently innocent observations will in fact be essential in the development of this book. The first pair of formulae (1.6) lead to a natural way to define a particle-unbiased estimate of unnormalized Feynman-Kac flows. It is also the basis for a semigroup and martingale methodology for studying the asymptotic behavior of particle models. The second observation on the path Markov chain is essential in the construction of genealogical tree-based models. It also allows direct transfer of several mathematical results on the time marginals to path-space models.

1.3.2 Nonlinear Equations

In some research areas, such as nonlinear filtering and nonlinear differential equations literature, the Feynman-Kac distribution flow models are alternatively defined as a solution of a nonlinear and measure-valued equation.

To better connect our abstract models with these subjects, we give next a brief description of these alternative representations. One drawback of this modeling technique is that it often requires more regularity conditions on the test functions as well as on the underlying Markov process and potential functions. On the other hand, this measure-valued process approach gives a strong basis for constructing particle approximation models. For all of these reasons, we have chosen to devote a brief introduction on the dynamical structure of Feynman-Kac distributions.

In the discrete time situation, we denote by M_n the Markov kernel of the chain X_n and denote by Ψ_n the Boltzmann-Gibbs transformation from the set of probability measures η on E into itself defined by

$$\Psi_n(\eta)(dx) = \frac{1}{\eta(G_n)} G_n(x) \, \eta(dx)$$

To simplify the presentation, we assume that the potential functions are strictly positive so that the mapping Ψ_n is well-defined on the whole set of distributions. Using the Markov property and the multiplicative nature of the Feynman-Kac models, we check that the distribution flows γ_n and η_n satisfy the recursive equations

$$\gamma_{n+1} = \gamma_n Q_{n+1} \quad \text{and} \quad \eta_{n+1} = \Psi_n(\eta_n) M_{n+1} \tag{1.7}$$

with

$$Q_{n+1}(f)(x) = G_n(x) \, M_{n+1}(f)(x)$$

and

$$M_{n+1}(f)(x) = \int_E M_{n+1}(x, dy) \, f(y)$$

As we already mentioned, the continuous time situation is technically more involved. To keep things as simple as possible, let us assume that E is a Polish space and X_t is an E-valued Markov process with time-inhomogeneous infinitesimal generator L_t. Under appropriate regularity conditions, the distribution flows γ_t and η_t satisfy for sufficiently regular test functions f the nonlinear equations

$$\frac{d}{dt}\gamma_t(f) = \gamma_t(L_t(f)) + \gamma_t(fV_t)$$

and

$$\frac{d}{dt}\eta_t(f) = \eta_t(L_t(f)) + \eta_t(f \, (V_t - \eta_t(V_t))) \tag{1.8}$$

Even if they look innocent, the equations (1.7) and (1.8) can rarely be solved analytically, and their solution requires extensive calculations.

1.4 Motivating Examples

The abstract Feynman-Kac models presented in Section 1.3 are at the corner of diverse disciplines. We give next a nonexhaustive list of the applications discussed in this book together with some comments concerning the Feynman-Kac interpretation of nonlinear estimation problems. In each of the application model areas, we provide several motivating and illuminating examples for the forthcoming particle algorithms developed in this book. The reader who wishes to know more details about a more specific application is recommended to consult Chapters 11 and 12, entirely devoted to applications of Feynman-Kac and particle methods.

1.4.1 *Engineering Science*

Feynman-Kac path measures are currently used in engineering science and particularly in financial mathematics, signal processing, and nonlinear filtering problems. They often go by various names, such as the Bayesian posterior, the conditional distributions of signal, or the Boltzmann-Gibbs measure, depending on the model areas. In rare event analysis, they represent the distribution of a Markov process in the rare event regime (see Section 12.2.5). They also appear naturally in the mathematical description of certain statistical methods. For instance, they represent a change of reference probability measure in importance sampling techniques (see Section 2.4.2). A full description of all of these models would of course be too much digression. Some of them will be discussed in the further development of Chapter 12. Because of their importance in practice, we have chosen in this introduction to concentrate on application models in advanced signal processing, particularly in nonlinear filtering problems.

Nonlinear Filtering

We recall that the filtering problem consists in computing the conditional distribution of a path signal process given its noisy and partial observations. In the discrete time situation, the state signal is an $\mathbb{R}^d$-valued Markov chain, usually defined through a recursion of the form

$$X_n = F_n(X_{n-1}, W_n)$$

where X_0 and W_n are independent and $\mathbb{R}^d$-valued random variables and F_n is a collection of Borel functions from $\mathbb{R}^{d+d}$ into $\mathbb{R}^d$. The recursive equation above may represent the random evolution of a target in tracking problems (see [253]), the evolution of an aircraft in radar processing (see [103]), or inertial navigation errors in GPS signal processing (see [43]). The noise sequence W_n has different possible interpretations. First, it represents the uncertainties in the choice of the stochastic mathematical model. More interestingly, it models some unknown quantities we want to estimate. For instance, in tracking problems, W_n corresponds to the unknown control laws of a noncooperative target.

The state of the signal is not directly observed. The observation process is traditionally defined in terms of a sequence of $\mathbb{R}^{d'}$-valued random variables given by

$$Y_n = H_n(X_n, V_n)$$

The perturbation sequence V_n consists of $\mathbb{R}^{d'}$-valued independent random variables independent of the signal X. H_n is a measurable function from $\mathbb{R}^{d+d'}$ into $\mathbb{R}^{d'}$. We further suppose that the distributions of the random variables $H_n(x, V_n)$ have the form

$$\text{Proba}\,(H_n(x, V_n) \in dy) = g_n(x,y)\; q_n(dy)$$

where q_n is a given positive measure on $\mathbb{R}^{d'}$ and $g_n(x,.)$ a density function. The statistical nature of the perturbation sequence V_n depends on the form of the sensors. For instance, they may represent thermic noises resulting from electronic devices, the uncertainties in the sensor model, or unknown quantities such as the atmospheric propagation delays or clock bias in global positioning system (GPS) processing. For more details, we refer the reader to the set of referenced articles.

Under our statistical assumptions, it is convenient to note that the observation variables $(Y_0, \ldots, Y_n)$ are independent conditionally on the signal path $(X_0, \ldots, X_n)$. That is, we have in a symbolic form

$$\text{Proba}\left((Y_0, \ldots, Y_n) \in d(y_0, \ldots, y_n) \mid (X_0, \ldots, X_n)\right) \quad (1.9)$$
$$= \prod_{p=0}^{n} g_p(X_p, y_p) \, q_p(dy_p)$$

If we take the nonhomogeneous potential functions

$$G_n(x) = g_n(x, y_n)$$

in the Feynman-Kac path model (1.3), then from Bayes' rule it becomes intuitively clear that

$$\mathbb{Q}_n = \text{Law}\left((X_0, \ldots, X_n) \mid (Y_0, \ldots, Y_{n-1}) = (y_0, \ldots, y_{n-1})\right)$$

Continuous time problems are defined in terms of a pair signal/observation Markov process (S_t, Y_t) taking values $\mathbb{R}^{d+d'}$. It is the solution of a pair of Itô's stochastic differential equations

$$dS_t = A(t, S_t) \, dt + B(t, S_t) \, dW_t + \int_{\mathbb{R}^m} C(t, S_{t-}, u) \, (\mu(dt, du) - \nu(dt, du))$$

and

$$dY_t = H_t(S_t) \, dt + \sigma \, dV_t$$

(V, W) is a $(d_v + d_w)$-dimensional standard Wiener process, σ is a strictly positive parameter, μ is a Poisson random measure on $\mathbb{R}_+ \times \mathbb{R}^{d_\mu}$ with intensity measure $\nu(dt, du) = dt \otimes F(du)$, and F is a positive σ-finite measure on $\mathbb{R}^{d_\mu}$. The mappings $A : \mathbb{R}_+ \times \mathbb{R}^d \to \mathbb{R}^d$, $B : \mathbb{R}_+ \times \mathbb{R}^d \to \mathbb{R}^d \otimes \mathbb{R}^{d_w}$, $C : \mathbb{R}_+ \times \mathbb{R}^d \times \mathbb{R}^{d_\mu} \to \mathbb{R}^d$, and $H : \mathbb{R}_+ \times \mathbb{R}^d \to \mathbb{R}^{d'}$ are Borel functions, S_0 is a random variable independent of (V, W, μ), and $Y_0 = 0$. Here again the first equation represents the evolution laws of the physical signal process at hand. For instance, the Poisson random measure μ may represent jump variations of a moving and noncooperative target (see for instance [105]).

The traditional nonlinear filtering problem in continuous time is to estimate the conditional distribution of the signal S_t given the observations Y_s from the origin $s = 0$ up to time t. The Kallianpur-Striebel formula (see for instance [195, 262]) states that there exists a reference probability measure $\mathbb{P}_0$ under which the signal and the observations are independent. In addition, for any measurable function f_t on the space $D([0, t], \mathbb{R}^d)$ of $\mathbb{R}^d$-valued càdlàg paths from 0 to t, we have that

$$\mathbb{E}(f_t((S_s)_{s \leq t}) \mid \mathcal{Y}_t) = \frac{\mathbb{E}_0(f_t((S_s)_{s \leq t}) \, Z_t(S, Y) \mid \mathcal{Y}_t)}{\mathbb{E}_0(Z_t(S, Y) \mid \mathcal{Y}_t)}$$

where $\mathcal{Y}_t = \sigma(Y_s, s \leq t)$ represents the sigma-field generated by the observation process and

$$\log Z_t(S, Y) = \int_0^t H_s(S_s) \, dY_s - \int_0^t H_s^\star(S_s) H_s(S_s) \, ds$$

This filtering problem is also related to the numerical solution of some nonlinear stochastic partial differential equations. To be more precise, we introduce the end time marginals of the preceding conditional distributions defined for any bounded Borel function f on $\mathbb{R}^d$ by

$$\widehat{\eta}_t(f) = \mathbb{E}(f(S_t) \mid \mathcal{Y}_t)$$

For sufficiently regular test functions, we can prove that the optimal filter $\widehat{\eta}_t$ satisfies the Kushner-Stratonovitch equation

$$d\widehat{\eta}_t(f) = \widehat{\eta}_t(L_t(f)) \, dt + \widehat{\eta}_t((H - \widehat{\eta}_t(H))^\star \, f) \, (dY_t - \widehat{\eta}_t(H) \, dt)$$

where L_t represents the infinitesimal generator of S_t. Next we examine three situations:

- *Discrete time formulation:*
  Let t_n, $n \geq 0$, be a given time mesh with $t_0 = 0$ and $t_n \leq t_{n+1}$. Also let X_n be the sequence of random variables defined by

$$X_n = S_{[t_n, t_{n+1}]}$$

  By construction, X_n is a nonhomogeneous Markov chain taking values at each time n in the space $E_n = D([t_n, t_{n+1}], \mathbb{R}^d)$. From previous considerations, the observation process Y_t can be regarded as a random environment. Given the observation path, we define the "random" potential functions G_n on $E_n = D([t_n, t_{n+1}], \mathbb{R}^d)$ by setting for any $x_n = (x_n(s))_{t_n \leq s \leq t_{n+1}} \in E_n$

$$G_n(x_n) = \exp\left(\int_{t_n}^{t_{n+1}} H_s(x_n(s)) \, dY_s - \int_{t_n}^{t_n} H_s^\star(x_n(s)) H_s(x_n(s)) \, ds\right)$$

  By construction, we can check that the quenched Feynman-Kac path measures (1.3) associated with the pair (X_n, G_n) coincide with the Kallianpur-Striebel representation and we have

$$\mathbb{Q}_n = \text{Law}\left(S_{[t_0, t_1]}, \ldots, S_{[t_n, t_{n+1}]} \mid \mathcal{Y}_{t_n}\right)$$

- *Discrete time observations:*
  Suppose that the observations are only delivered by the sensors at some fixed times t_n, $n \geq 0$, with $t_n \leq t_{n+1}$. Also suppose that we are interested in computing the conditional distributions

$$\text{Law}\left(S_{[t_0, t_1]}, \ldots, S_{[t_n, t_{n+1}]} \mid Y_{t_1}, \ldots, Y_{t_{n+1}}\right)$$

18   1. Introduction

Notice that

$$(Y_{t_{n+1}} - Y_{t_n}) = \int_{t_n}^{t_{n+1}} H_s(S_s)\,ds + \sigma\,(V_{t_{n+1}} - V_{t_n})$$

and $\sigma\,(V_{t_{n+1}} - V_{t_n})$ are independent and random variables with Gaussian density g_n. Arguing as in the discrete time formulation and using the same notation as there, we introduce the "random" potential functions G_n defined for any $x_n = (x_n(s))_{t_n \leq s \leq t_{n+1}} \in E_n$ by

$$G_n(x_n) = g_n\left(Y_{t_{n+1}} - Y_{t_n} - \int_{t_n}^{t_{n+1}} H_s(x_n(s))\,ds\right)$$

By construction, the quenched Feynman-Kac path measures (1.3) associated with the pair (X_n, G_n) now have the following interpretation

$$\mathbb{Q}_n = \mathrm{Law}\left(S_{[t_0,t_1]}, \ldots, S_{[t_n,t_{n+1}]} \mid Y_{t_1}, \ldots, Y_{t_n}\right)$$

- Robust equation:
The precise description of the robust equation is technically more involved than in the discrete time interpretation. The idea is to remove the stochastic integrals in the exponential terms to work with a robust pathwise version of the conditional distributions. More precisely, using the Girsanov formula, we can construct a novel reference probability measure $\widehat{\mathbb{P}}$ so that

$$\mathbb{E}(f_t((S_s)_{s \leq t}) \mid \mathcal{Y}_t) = \frac{\widehat{\mathbb{E}}(f_t((S_s)_{s \leq t})\,\widehat{Z}_t(S,Y) \mid \mathcal{Y}_t)}{\widehat{\mathbb{E}}(\widehat{Z}_t(S,Y) \mid \mathcal{Y}_t)}$$

with

$$\log \widehat{Z}_t(S,Y) = H_s^\star(S_t) Y_t + \int_0^t U_s(S_s, Y_s)\,ds$$

where U_s is a given collection of measurable functions on $\mathbb{R}^{d+d'}$. In contrast to the previous change of reference measure under $\widehat{\mathbb{P}}$, the canonical Markov process S_t now depends on the observations. This robust description is clearly related to the continuous time Feynman-Kac path measures (1.4). For instance, the robust version $\widehat{\eta}_{y,t}$ of the optimal filter is now given for any continuous observation path $y = (y_t)_{t \geq 0}$ by the formula

$$\widehat{\eta}_{y,t}(f) = \frac{\eta_{y,t}(e^{H_s^\star(\cdot) y_t} f)}{\eta_{y,t}(e^{H_s^\star(\cdot) y_t})}$$

The distribution flow $\eta_{y,t}$ is defined by the Feynman-Kac measures

$$\eta_{y,t}(f) = \gamma_{y,t}(f)/\gamma_{y,t}(1)$$

and
$$\gamma_{y,t}(f) = \mathbf{E}\left(f(X_t^y)\,\exp\left\{\int_0^t U_s(X_s^y, y_s)ds\right\}\right)$$

where X_t^y is a Markov process with the same law as the signal process under $\widehat{\mathbb{P}}$. The robustness property ensures that the pathwise version of the optimal filter is a continuous function with respect to the observation process. In practice, this robustness property is a fundamental requirement of any filter, as small variations of the observation sequence should not influence drastically the solution.

Examples

Nonlinear filtering problems have become increasingly important in engineering and operations research literature. In order to provide a concrete basis for the further development of this book, we propose hereafter some examples of application areas as well as some discrete time filtering problems in their simplest form. For a more thorough discussion, we refer the reader to Section 12.6. Next, we examine three estimation problems corresponding respectively to tracking analysis, stochastic volatility estimation, and speech recognition. In each situation, we describe the choice of the potential function for which the corresponding Feynman-Kac measures are versions of the desired conditional distributions. In all situations, the underlying Markov model coincides with the signal process.

1. One traditional situation in *tracking problems* is to estimate the location of a moving target X_n in the quarter plane $E = \mathbb{R}_+^2$ with a fixed observer at the origin $(0,0)$ taking angular measurements Y_n. One of the simplest models for the signal is the Markov chain $X_n = (X_n^1, X_n^2) \in \mathbb{R}_+^2$ defined by the recursive equations
$$\begin{cases} X_n^1 = X_{n-1}^1 + W_n^1 \\ X_n^2 = X_{n-1}^2 + W_n^2 \end{cases}$$

The velocities W_n^1, W_n^2 and the initial conditions X_0^1, X_0^2 are independent and identically distributed random variables. The noisy angular positions delivered by the sensor are given by the equation
$$Y_n = \arctan(X_n^2/X_n^1) + V_n$$

The perturbations V_n are assumed to be independent, identically distributed, and centered Gaussian random variables with unit variance. In this example, we clearly have
$$\text{Proba}\left((\arctan(x^2/x^1) + V_n) \in dy\right) = \frac{1}{\sqrt{2\pi}}\,e^{-\frac{1}{2}(y - \arctan(x^2/x^1))^2}\,dy$$

and we can alternatively choose the potential functions

$$G_n((x^1, x^2)) = \exp\left(-(y_n - \arctan(x^2/x^1))^2/2\right)$$

or

$$G_n((x^1, x^2)) = \exp\left(-(\arctan(x^2/x^1))^2/2 + y_n \ \arctan(x^2/x^1)\right)$$

2. In *financial engineering and economics*, one recent area of research is the estimation of the stochastic volatility of the price of a given asset. In this context, the signal X_n represents the random evolution of the logarithmic volatility and the observation sequence Y_n represents the change of amplitude of the return series. The simplest filtering model associated with this volatility estimation problem is described inductively as

$$\begin{cases} X_n &= a\ X_{n-1} + b\ W_n \\ Y_n &= e^{(X_n+c)/2}\ V_n \end{cases}$$

where W_n, V_n, and X_0 are independent random sequences, $a, b, c \in \mathbb{R}$, and V_n is a centered Gaussian distribution with unit variance. In this example, we have

$$\text{Proba}\left((x + c + \log V_n^2) \in dy\right)$$

$$= \tfrac{1}{\sqrt{2\pi}}\ e^{-\frac{1}{2}[e^{(y-(x+c))} - (y-(x+c))]}\ 1_{\mathbb{R}_+}(y)\ dy$$

and we can take

$$G_n(x) = \exp\left(-\frac{1}{2}[e^{(\log y_n^2 - (x+c))} - (\log y_n^2 - (x+c))]\right)$$

3. In *speech separation analysis*, we want to recover mutually independent signals while observing some noisy mixture of them. This estimation problem is also called the *blind source separation problem*. It is traditionally defined by taking a system of d sources transmitting a d-dimensional signal $X_n = (X_n^i)_{1 \le i \le d}$ satisfying a system of recursive formulae of the form

$$\begin{cases} X_n^i &=\ F_n^i(X_{n-1}^i, W_n^i) \\ i &=\ 1, \ldots, d \end{cases}$$

where X_0^i and W_n^i are independent and real-valued random variables and F_n^i are measurable functions from $\mathbb{R}^2$ into $\mathbb{R}$. We observe a mixture of these signals usually defined in matrix form as follows

$$Y_n = A_n X_n + V_n$$

where A_n are $(d' \times d)$ matrices. The perturbation sequence V_n is a collection of independent random variables with a density distribution $g_n(v)$ with respect to the Lebesgue measure on $\mathbb{R}^{d'}$. Arguing as before, we can take the potential functions

$$G_n(x) = g_n(y_n - A_n x)$$

4. *Data assimilation* methodologies often refer to high-dimensional estimation problems such as those arising in forecast prediction analysis. In this application area, the signal process X_n may represent an ocean dynamic model such as the Miami Isopycnic Coordinate Ocean Model [33], an Indian Ocean model [140, 141], the classical Lorentz attractor [266], or a barotropic ocean model [198]. In this context, the measurements Y_n represent acoustic tomography data (see [198]) or sea level anomalies and sea surface temperature data (see [140]). Here again, the problem is to estimate the conditional distributions of the signal given its noisy and partial observations. In forecast and oceanographic model literature, these classical filtering models are also called sequential data assimilation.

1.4.2 Bayesian Methodology

Many estimation problems arising in operations research can be thought of as filtering problems. This point of view is at the heart of Bayesian methodology. In this branch of statistics, the conditional distributions (1.9) are the so-called "Bayesian posterior" and the path distribution of the signal is called the "prior". Although most of the research articles in this field are often written in a somewhat heuristic way, the Bayesian literature abounds with applications of nonlinear filtering models in many research areas, including neural networks, robot localizations, time series estimation and target recognition. We refer the interested reader to the set of articles in the collective book [125]. To better connect the Feynman-Kac measures (1.3) with the "Bayesian language," we observe that

$$\mathbb{Q}_n(d(x_0, \ldots, x_n))$$
$$= \mathbb{Q}_{n-1}(d(x_0, \ldots, x_{n-1})) \; \frac{\mathcal{Z}_{n-1}}{\mathcal{Z}_n} \; G_{n-1}(x_{n-1}) \; M_n(x_{n-1}, dx_n)$$

where M_n stands for the Markov transition of the chain X_n. To get rid of the normalizing constants, we use the proportional sign $\propto$ and we rewrite the expression above as

$$\mathbb{Q}_n(d(x_0, \ldots, x_n)) \propto \mathbb{Q}_{n-1}(d(x_0, \ldots, x_{n-1})) \; Q_n(x_{n-1}, dx_n) \qquad (1.10)$$

with the positive integral operator

$$Q_n(x_{n-1}, dx_n) = G_{n-1}(x_{n-1}) \; M_n(x_{n-1}, dx_n)$$

Also notice that any positive integral operator Q_n can be written as above. To prove this elementary observation, we simply take

$$G_{n-1}(x_{n-1}) = Q_n(1)(x_{n-1}) \quad \text{and} \quad M_n(x_{n-1}, dx_n) = \frac{Q_n(x_{n-1}, dx_n)}{Q_n(1)(x_{n-1})}$$

This alternative representation of the Feynman-Kac path measures (1.3) is commonly used in Sequential Monte Carlo literature. Since various authors in this field seem to be reluctant to use measure theory, the formula displayed above is more often written in terms of distribution density functions. In the further development of this book, we will see that (1.10) is in fact equivalent to writing a measure-valued dynamical equation. As a result, most of the algorithms discussed in Sequential Monte Carlo literature coincide with the particle approximation model of these nonlinear equations in distribution space.

1.4.3 Particle and Statistical Physics

In physics, Feynman-Kac formulae occur in a variety of topics, such as trapping problems, Schrödinger equations, quantum physics, and micro-statistical mechanics.

Trapping Analysis

In the discrete time situation, we can formally model a killed particle motion in an absorbing medium by "adding" in the random evolution of a Markov chain X_n a trapping mechanism

$$X_n \xrightarrow{\text{trapping}} \widehat{X}_n \xrightarrow{\text{exploration}} X_{n+1}$$

The trapping transition consists in killing the particle at site X_n with a probability $(1 - G_n(X_n))$, where G_n is a $[0,1]$-valued potential function. Notice that the particle is not trapped when visiting regions where the potential function G_n is equal to 1. The opposite regions, where G_n is equal to 0, are called hard obstacles. "Soft obstacles" correspond to regions in the medium where $G_n \in (0,1)$. During the exploration phase $\widehat{X}_n \longrightarrow X_{n+1}$, the particle $\widehat{X}_n$ simply evolves in the medium to a new location X_{n+1} randomly chosen according to the distribution $M_{n+1}(\widehat{X}_n, \cdot)$. Let T denote the lifetime of the particle. By construction, the normalizing constants $\mathcal{Z}_n$ in (1.3) represent the probability that the particle X_n is still alive at time n and

$$\mathbb{Q}_n = \text{Law}((X_0, \ldots, X_n) \mid T \geq n)$$

The continuous time Feynman-Kac path formula (1.4) can also be interpreted as the distribution of a trapped Markov motion. In this context, the killed Markov process evolves randomly in the medium according to an

L-motion where L is an infinitesimal generator. It is killed at rate $U(X_t)$, where U is a nonnegative potential function on the medium. By taking $V_t = -U$ in (1.4) and $\mathbb{P}$ as the distribution of the Markov L-motion, we end up formally with

$$\mathbb{Q}_t = \mathrm{Law}((X_s)_{s \leq t} \mid T > t)$$

where T stands for the lifetime of the killed process.

Schrödinger Operators

The Feynman-Kac path distributions (1.4) are also closely related to Schrödinger equations. To describe these interesting links, we further suppose that $E = \mathbb{R}^d$ and X_t is a time-homogeneous Markov process with infinitesimal generator L. We also suppose that the potential function is time-homogeneous with $V_t = V$ and we denote by V^+ and $-V^-$ its positive and negative parts. In quantum physics and microstatistical mechanics, the potentials V^+ and V^- are respectively called the "creation" and "killing" potentials (see [254]). When V^+ is null, we have $V = -V^-$. In this situation, the Feynman-Kac positive measures

$$\gamma_t(f) = \mathbb{E}\left(f(X_t) \, \exp\left\{ -\int_0^t V^-(X_s) ds \right\} \right)$$

satisfy for sufficiently regular test functions f the linear equation

$$\frac{d}{dt} \gamma_t(f) = \gamma_t(L^v(f))$$

with the Schrödinger operator $L^v = L - V^-$. In this case, we can argue as above and interpret γ_t as the distribution of a single Markov particle evolving in a random medium with killing rate V^-. This interpretation does not hold true when the potential V has a "creation" component. To understand the role of V^+, it is convenient to work with the normalized distribution flow $\eta_t(f) = \gamma_t(f)/\gamma_t(1)$. We recall that η_t satisfy for sufficiently regular test functions f the nonlinear equation

$$\frac{d}{dt} \eta_t(f) = \eta_t(L(f)) + \eta_t(f(V - \eta_t(V))) \tag{1.11}$$

As traditionally, we can interpret η_t as the law of a nonhomogeneous Markov process $\overline{X}_t$ with a (nonunique) nonhomogeneous infinitesimal generator L_{η_t} satisfying the compatibility condition

$$L_\eta(f) = \eta(L(f)) + \eta(f(V - \eta(V)))$$

for any distribution η on $\mathbb{R}^d$. We can choose for instance

$$L_\eta = L + \widehat{L}_\eta^- + \widehat{L}_\eta^-$$

with the jump type generators

$$\widehat{L}_\eta^-(f)(x) = V^-(x) \int (f(y) - f(x))\, \eta(dy)$$

$$\widehat{L}_\eta^+(f)(x) = \int (f(y) - f(x))\, V^+(y)\, \eta(dy)$$

The corresponding Markov process $\overline{X}_t$ is a jump type nonlinear Markov process. Between the jumps, it evolves randomly according to an L-motion. At rate V^-, it is killed and instantly jumps to a new site randomly chosen according to the current distribution η_t. The "creation" term V^+ induces an auxiliary nonhomogeneous killing rate $\eta_t(V^+)$. At this rate, the particle dies and instantly a particle randomly chosen with a distribution proportional to $V^+(x)\, \eta_t(dx)$ splits into two offsprings.

The interpretation above is connected to the construction of the Schrödinger process using renormalizing techniques. We refer the interested reader to the book [254].

We also mention that for sufficiently regular L-motions the top eigenvalue $\lambda(V)$ of the Schrödinger operator $L^v = L + V$ can be formulated in terms of the long time behavior of the nonlinear distribution flow model (see Section 12.4 and the article [102]). For instance, for sufficiently stable L-motions, we will see that

$$\lambda(V) = \lim_{t \to \infty} \frac{1}{t} \int_0^t \eta_s(V)\, ds$$

In the reversible situation, we can also prove that $\lambda(V)$ coincides with the top of the spectrum of Schrödinger operator L^v and the density of the stationary distribution of the flow η_t (with respect to the reversible measure) is proportional to the corresponding eigenvector.

Interacting Jump and Boltzmann Type Models

Nonlinear equations of type (1.11) can be interpreted as evolutionary type interacting jump models. The quadratic structure of the evolution can be extended to the situation where the potential function U acts on the transition space $E \times E$. The corresponding equation is now given by

$$\frac{d}{dt}\eta_t(f) = \eta_t(L(f)) + \int_{E \times E} (f(x) - f(y))\, U(x,y)\, \eta_t(dx)\eta_t(dy)$$

Here again, we have a Feynman-Kac interpretation. Simply note that the "implicit Feynman-Kac" flow defined by

$$\eta_t(f) = \mathbb{E}\left(f(X_t)\, \exp\left\{\int_0^t \int_E [U(X_s, x) - U(x, X_s)]\eta_s(dx)ds\right\}\right)$$

satisfies the desired equation. In addition, using the same arguments as in [98], one can check that the flow is the unique solution of the integral equation

$$\eta_t(f)$$

$$= \eta_0(P_{0,t}(f)) + \int_0^t \int_{E \times E} [P_{s,t}(f)(x) - P_{s,t}(f)(y)] \; U(x,y) \; \eta_s(dx) \; \eta_s(dy) ds$$

where $P_{s,t}$ stands for the semigroup of X.

These continuous time Feynman-Kac models can also be interpreted as particular examples of generalized and spatially homogeneous Boltzmann models introduced by S. Méléard in [245] and further developed in a chain of articles [165, 166, 246, 247, 248]. Whenever it exists, the Feynman-Kac representation often provides natural semigroup and martingale techniques to analyze the long time behavior of the flow and/or the asymptotic behavior of the interacting jump approximation models. We refer the interested reader to the articles [95, 97, 98, 102].

1.4.4 Biology

In biology, Feynman-Kac formulae also occur in a variety of topics, such as chemical polymerizations, and genetic and genealogical population models.

Directed Polymers

In this application model area, the path distributions (1.3) represent the Boltzmann-Gibbs measures associated with a random and directed polymer chain. In this context, the underlying Markov chain X_n represents random chemical polymerizations in a given solvent E. To simplify the presentation, we further suppose that X_n is a nonhomogeneous Markov chain

$$X_n = X'_{[0,n]} =_{\text{def.}} (X'_0, \ldots, X'_n) \in E_n = \underbrace{E \times \ldots \times E}_{(n+1)\text{times}}$$

The elementary random variables X'_p, with $p \leq n$, represent the monomers in a directed chain X_n with a polymerization degree n. In this situation the function G_n on E_n reflects the intermolecular attraction or repulsive potential interactions between the monomers. The precise structure of these intermolecular interactions is usually unknown. We often need to represent these chemical reactions by a simplified model. It is often assumed that the "free chemical construction" of a directed polymer is Markovian. In other words, the monomer sequence X'_n is an E-valued Markov chain. Under this Markovian hypothesis, random polymerizations with strong repulsions are closely related to nonintersecting Markov chains. Indeed, if we choose in

(1.3) the potential functions

$$G_n(x_0, \ldots, x_n) = 1_{E - \{x_0, \ldots, x_{n-1}\}}(x_n)$$

then the corresponding Feynman-Kac path measures represent the distributions of the path of a nonintersecting Markov chain

$$\mathbb{Q}_n = \mathrm{Law}((X_0, \ldots, X_n) \mid X'_p \neq X'_q \, ; \, \forall 0 \leq p < q < n)$$

Note that the normalizing constants are well-defined as soon as for any $n \in \mathbb{N}$ we have

$$\mathcal{Z}_n = \mathrm{Proba}(X'_p \neq X'_q \, ; \, \forall 0 \leq p < q < n) > 0$$

More regular intermolecular interactions are represented by a Hamiltonian energy function H_n and a cooling temperature parameter T_n. These models are again described by Boltzmann-Gibbs measures (1.3) with path potential functions

$$G_n(x_0, \ldots, x_n) = \exp\left(-\frac{1}{T_n} H_n(x_0, \ldots, x_n)\right)$$

Branching Population Models

Feynman-Kac measures are also related to genetic evolution models. This connection will become transparent when we describe in Section 1.5 the particle interpretations of Feynman-Kac distribution flows. In particular, we will see that the genealogical tree occupation measure associated with a "simple genetic" model converges as the size of the population tends to infinity to a Feynman-Kac path measure. Genetic models can be thought of as the evolution of a population on a given state space whose individuals reproduce or die subject to a natural evolution interaction mechanism. The members of the population are often called particles in reference to physics application model areas.

These genetic processes differ from branching models in which the particles do not interfere with one another such as the generalized Galton-Watson models developed in the book by T.E. Harris [176]. For a more thorough study of branching measure-valued process from the point of view of the general theory of Markov processes, we recommend the reader to the book by E.B. Dynkin [128].

Our next objective is to better connect these two classes of branching models and underline their kinship with the Feynman-Kac models discussed in this book.

We denote by $S = \cup_{p \geq 0} E^p$ the state space of the branching population model whose members take values in some measurable space $(E, \mathcal{E})$. The integer parameter p represents the size of the population, and for $p = 0$ we use the convention $E^0 = \{c\}$, where c represents a cemetery or coffin state. We consider a sequence of Markov transitions $M_n(x, dy)$ on E and a collection of $\mathbb{N}$-valued random variables $(g_n^i(x))_{i\geq 1, x\in E}$ with uniformly finite first and second moments. We also assume that, for each $x \in E$, $(g_n^i(x))_{i\geq 1}$ are identically distributed, and we set $\mathbb{E}(g_n^1(x)) = G_n(x)$.

Our branching process $(\chi_n)_{n\geq 0}$ consists in a Markov chain taking values in S. Its initial state χ_0 is assumed to be a single E-valued random particle with a given distribution $\eta_0 \in \mathcal{P}(E)$, and its elementary transitions are defined in terms of a two-step branching/exploration mechanism:

$$\chi_n \in E^{p_n} \xrightarrow{\text{branching}} \widehat{\chi}_n \in E^{\widehat{p}_n} \xrightarrow{\text{exploration}} \chi_{n+1} \in E^{p_{n+1}}$$

During the branching stage, each individual χ_n^i gives birth to a random number of offsprings $g_n^i(\chi_n^i)$, with $1 \leq i \leq p_n$. At the end of the branching transition, we have a population $\widehat{\chi}_n = (\widehat{\chi}_n^1, \ldots, \widehat{\chi}_n^{\widehat{p}_n})$ with $\widehat{p}_n = \sum_{i=1}^{p_n} g_n^i(\chi_n^i)$ particles. Whenever the system dies at a given time n, we have $(\widehat{p}_n, \widehat{\chi}_n) = (0, c)$, and we set $(\widehat{p}_k, \widehat{\chi}_k) = (p_{k+1}, \chi_{k+1}) = (0, c)$ for any $k \geq n$. Note that the reproduction ability of each individual is not affected by the rest of the population. During the exploration transition, each particle evolves randomly according to the elementary Markov transition M_{n+1}.

By construction, we have $p_{n+1} = \widehat{p}_n$, and $\chi_{n+1} = (\chi_{n+1}^i)_{1\leq i \leq p_{n+1}}$ forms a sequence of independent random variables with respective distributions

$$(M_{n+1}(\widehat{\chi}_n^i, \cdot))_{1\leq i \leq p_{n+1}}$$

We introduce the point distributions

$$s(\chi_n) = \sum_{j=1}^{p_n} \delta_{\chi_n^j} \quad \text{and} \quad s(\widehat{\chi}_n) = \sum_{j=1}^{\widehat{p}_n} \delta_{\widehat{\chi}_n^j} = \sum_{j=1}^{p_n} g_n^j(\chi_n^j)\, \delta_{\chi_n^j}$$

with the convention $s(c) = \sum_\emptyset = 0$, the null measure on E. For any $f \in \mathcal{B}_b(E)$, we easily check that

$$\mathbb{E}(s(\chi_{n+1})(f) | \widehat{\chi}_n) = \sum_{j=1}^{p_n} g_n^j(\chi_n^j) M_{n+1}(f)(\chi_n^j)$$

$$\mathbb{E}(s(\chi_{n+1})(f) | \chi_n) = \sum_{j=1}^{p_n} G_n(\chi_n^j)\, M_{n+1}(f)(\chi_n^j) = s(\chi_n) Q_{n+1}(f)$$

with the bounded positive semigroup Q_{n+1} on $\mathcal{B}_b(E)$ defined by

$$Q_{n+1}(f)(x) = G_n(x)\, M_{n+1}(f)(x)$$

Using a simple induction, we check that the Feynman-Kac flow associated with the pair potential/transition (G_n, M_n) corresponds to the first moments of the branching point distributions. That is, we have that

$$\mathbb{E}(s(\chi_n)(f)) = \gamma_n(f) = \mathbb{E}\left(f(X_n) \prod_{q=0}^{n-1} G_q(X_q)\right)$$

28   1. Introduction

where X_n represents an E-valued Markov chain with elementary transitions M_n and initial transition $\eta_0 = \text{Law}(\chi_0)$. This representation of birth and death models in terms of a point distribution is due to Moyal [251, 252] and it has been further developed by Harris [175, 176].

From previous considerations, we observe that an elementary branching approximation algorithm of the unnormalized Feynman-Kac flow γ_n could consist in sampling N independent branching models of the previous form.

We finally mention that these simplified models are currently used in statistical physics to study the transport and the multiplication of neutrons (see for instance [176] and references therein). They also have some interesting features of more sophisticated population evolution models arising in molecular biology [203]. For a more recent account on branching population models, we refer the reader to the books [14, 128, 180] and references therein. Related continuous time branching interpretation models of Feynman-Kac formulae can also be found in [96].

1.4.5 Applied Probability and Statistics

There is a remarkable advantage in formulating a nonlinear estimation problem in terms of a Feynman-Kac path measure. From the theoretical point of view, Feynman-Kac models represent a change of probability measures. The change of probability mass expresses in some way the relative information between a complex distribution and a simpler reference one. The frustration practitioners may get when developing these abstract models will be released since the various particle interpretations of these measures will give instantly a collection of powerful simulation tools for the numerical solution of the nonlinear problem at hand.

These Feynman-Kac modeling techniques offer nice and natural probabilistic interpretations of Dirichlet problems with boundary conditions (see Section 12.2.2 and Section 12.2.4) and of fixed points of non linear integral operators (see Chapter 5 and Section 12.4). They also appear in the sequential analysis of probability test ratio (see Section 12.3.3) and in Monte Carlo Markov chain problems (see Chapters 5 and 6, and Section 11.2). To illustrate these comments, we give next one original way to describe the conditional distribution of a Markov chain restricted to its terminal value in terms of a Feynman-Kac path measure (see [80, 81] and Section 5.5).

Let π and L be respectively a probability measure and a Markov kernel on a measurable state space $(S, \mathcal{S})$. We associate with the pair (π, L) the S-valued Markov chain $(\Omega, F, Y, \mathbb{P}_\pi^L)$ with initial distribution π and Markov transitions L. Let K be an auxiliary Markov kernel on S such that the pair measures on $E = (S \times S)$ defined by

$$\begin{aligned}(\pi \times K)_1(d(y, y')) &= \pi(dy)\ K(y, dy') \\ (\pi \times L)_2(d(y, y')) &= \pi(dy')\ L(y', dy)\end{aligned}$$

are mutually absolutely continuous. We further require that the triplet (π, K, L) be chosen so that the Radon-Nikodym derivative

$$G = \frac{d(\pi \times L)_2}{d(\pi \times K)_1}$$

is a bounded and strictly positive function on E. Using a time reversal technique (see Section 5.5.3), we can prove that for any bounded measurable function f_n on S^{n+1} we have

$$\mathbb{E}_\pi^L(f_n(Y_n, Y_{n-1}\ldots, Y_0) \mid Y_{n+1} = y)$$
$$= \frac{\mathbb{E}_y^K(f_n(Y_1\ldots, Y_{n+1}) \prod_{p=0}^n G(Y_p, Y_{p+1}))}{\mathbb{E}_y^K(\prod_{p=0}^n G(Y_p, Y_{p+1}))} \qquad (1.12)$$

In the display above, $\mathbb{E}_\pi^L$ represents the expectations with respect to $\mathbb{P}_\pi^L$. Similarly, $\mathbb{E}_y^K$ stands for the expectation with respect to the distribution $\mathbb{P}_y^K$ of a (canonical) Markov chain with transitions K and starting at y.

This description of a chain restricted to its terminal value can be expressed as a Feynman-Kac path measure (1.3) through a state-space enlargement and a clear time reversal.

When the Markov transition L is sufficiently regular, we also notice that in some sense

$$\mathbb{P}_\pi^L(Y_0 \in dy' \mid Y_{n+1} = y) = \pi(dy') \frac{dL^{n+1}(y', \cdot)}{d\pi L^{n+1}}(y) \xrightarrow[n \to \infty]{} \pi(dy')$$

From this property, we will prove that the marginal Feynman-Kac distribution flow converges to π as the time parameter tends to infinity. In Section 5.5, we will also use these observations to construct a genealogical path particle simulation technique for sampling restricted Markov models. As the form of the potential function already indicates, we will see that the corresponding particle model behaves as a sequence of interacting Metropolis models. We already mention that these Feynman-Kac-Metropolis models have "better" asymptotic properties than a traditional Metropolis-Hastings model.

1.5 Interacting Particle Systems

From the pure mathematical point of view, particle methods can be viewed as a kind of *stochastic linearization technique* for solving nonlinear equations in distribution space. The idea is to associate to a given nonlinear dynamical structure a sequence of E^N-valued Markov processes such that the N-empirical measures of the configurations converge as $N \to \infty$ to the desired distribution. The parameter N represents the precision parameter and the size of the systems. The state components of the E^N-valued Markov process are called particles.

The choice of the particle model is far from being unique. Loosely speaking, it depends on the "Markov interpretation" of the nonlinear equation at hand. The objective of this section is to design a general strategy to construct a class of particle approximation models for Feynman-Kac distribution flows. Before entering into a precise description of these models, we provide hereafter a brief reminder on the dynamical structure of the discrete and continuous time Feynman-Kac flows. We first recall that these models are respectively defined by

$$\eta_n(f) = \gamma_n(f)/\gamma_n(1) \quad \text{and} \quad \eta_t(f) = \gamma_t(f)/\gamma_t(1)$$

with

$$\gamma_n(f) = \mathbb{E}\left(f(X_n) \prod_{p=0}^{n-1} G_p(X_p)\right)$$

and

$$\gamma_t(f) = \mathbb{E}\left(f(X_t) \exp\left\{\int_0^t V_s(X_s)ds\right\}\right)$$

The discrete and continuous time stochastic processes X_n and X_t are non-homogeneous Markov processes taking values in a measurable space $(E, \mathcal{E})$. X_n is a Markov chain with elementary transitions M_n, and X_t is a Markov process with infinitesimal generators L_t. The potential functions G_n and V_t are bounded $\mathcal{E}$-measurable functions on E such that the normalizing constants are well-defined. To simplify the presentation, we further assume that G_n is strictly positive. In Section 1.3, we have seen that these distributions satisfy respectively the nonlinear equations

$$\eta_{n+1} = \Psi_n(\eta_n)M_{n+1} \quad \text{and} \quad \frac{d}{dt}\eta_t(f) = \eta_t(L_t(f)) + \eta_t(f\ (V_t - \eta_t(V_t)))$$

with the Boltzmann-Gibbs transformation Ψ_n from the set of probability measures η on E into itself defined by

$$\Psi_n(\eta)(dx) = \frac{1}{\eta(G_n)}\ G_n(x)\ \eta(dx)$$

1.5.1 Discrete Time Models

In the discrete time situation, we observe that there exists a nonunique collection of Markov transitions $K_{n,\eta}$ indexed by the set of probability measures $\eta \in \mathcal{P}(E)$ and the time parameters $n \in \mathbb{N}$ and such that

$$\eta_{n+1} = \eta_n K_{n+1,\eta_n}$$

For instance, we can choose

$$K_{n+1,\eta_n}(x,dz) = S_{n,\eta_n} M_{n+1}(x,dz) = \int_E S_{n,\eta_n}(x,dy) M_{n+1}(y,dz) \quad (1.13)$$

with the collection of selection type transitions

$$S_{n,\eta_n}(x,dy) = \epsilon_n \, G_n(x) \, \delta_x(dy) + (1 - \epsilon_n \, G_n(x)) \, \Psi_n(\eta_n)(dy)$$

In the display above the nonnegative parameters $\epsilon_n \geq 0$ are chosen such that $\epsilon_n G_n \leq 1$. The N-particle model associated with a given collection of Markov transitions $K_{n,\eta}$ satisfying the compatibility condition

$$\eta K_{n,\eta} = \Psi_{n-1}(\eta) M_n$$

is an E^N-valued Markov chain

$$\xi_n^{(N)} = \left(\xi_n^{(N,1)}, \xi_n^{(N,2)}, \ldots, \xi_n^{(N,N)}\right)$$

Its elementary transitions are defined as

$$\mathrm{Proba}\left(\xi_{n+1}^{(N)} \in d(x^1,\ldots,x^N) | \xi_n^{(N)}\right) = \prod_{i=1}^N K_{n,\frac{1}{N}\sum_{j=1}^N \delta_{\xi_n^{(N,j)}}}(\xi_n^{(N,i)}, dx^i)$$

The initial system $\xi_0^{(N)}$ consists of N independent and identically distributed random variables with common law η_0. For sufficiently regular transitions, we prove that for any time horizon n and as $N \to \infty$

$$\eta_n^N = \frac{1}{N}\sum_{i=1}^N \delta_{\xi_n^{(N,i)}} \longrightarrow \eta_n$$

The convergence above can be understood in various ways. These asymptotic properties will be discussed in full detail in Chapter 7. Mimicking formula (1.6), we also construct an unbiased estimate for the unnormalized model and prove that as $N \to \infty$

$$\gamma_n^N(.) = \eta_n^N(.) \prod_{p=0}^{n-1} \eta_p^N(G_p) \longrightarrow \gamma_n(.) = \eta_n(.) \prod_{p=0}^{n-1} \eta_p(G_p)$$

To get an intuitive feel of the motion of the particle models associated with Feynman-Kac flows, we provide a brief description of the particle model associated with the choice of transitions $K_{n,\eta}$ defined in (1.13). To clarify the presentation, we suppress the index parameter N and write ξ_n and ξ_n^i instead of $\xi_n^{(N)}$ and $\xi_n^{(N,i)}$. In this simplified notation, we first observe the selection transitions take the form

$$S_{n,\frac{1}{N}\sum_{j=1}^N \delta_{\xi_n^j}}(\xi_n^i, .)$$

$$= \epsilon_n \, G_n(\xi_n^i) \, \delta_{\xi_n^i}(.) + (1 - \epsilon_n \, G_n(\xi_n^i)) \, \Psi_n\left(\tfrac{1}{N}\sum_{j=1}^N \delta_{\xi_n^j}\right)(.) \quad (1.14)$$

with the discrete Boltzmann-Gibbs distribution given by

$$\Psi_n\left(\frac{1}{N}\sum_{j=1}^{N}\delta_{\xi_n^j}\right) = \sum_{j=1}^{N}\frac{G_n(\xi_n^j)}{\sum_{k=1}^{N}G_n(\xi_n^k)}\,\delta_{\xi_n^j} \qquad (1.15)$$

From this observation, we see that the two-step transitions of the nonlinear distribution flow

$$\eta_n \xrightarrow{S_{\eta_n}} \Psi_n(\eta_n) = \eta_n S_{n,\eta_n} \xrightarrow{M_{n+1}} \eta_{n+1} = \Psi_n(\eta_n) M_{n+1}$$

are approximated by a two-step and E^N-valued Markov chain

$$\xi_n \in E^N \xrightarrow{\text{selection}} \widehat{\xi}_n \in E^N \xrightarrow{\text{mutation}} \xi_{n+1} \in E^N$$

In other words, during the selection transition, each particle evolves randomly as $\xi_n^i \longrightarrow \widehat{\xi}_n^i$ according to the transition (1.14). With a probability $\epsilon_n G_n(\xi_n^i)$, it stays in the same location and we set $\widehat{\xi}_n^i = \xi_n^i$; otherwise, we select randomly a new particle $\widetilde{\xi}_n^i$ in the current configuration with the Boltzmann-Gibbs distribution (1.15) and we set $\widehat{\xi}_n^i = \widetilde{\xi}_n^i$. During the mutation stage, each particle evolves randomly as $\widehat{\xi}_n^i \longrightarrow \xi_{n+1}^i$ according to the mutation transition M_{n+1}. Note that the selection transitions $\xi_n^i \to \widehat{\xi}_n^i$ can alternatively be described by the following acceptance/rejection mechanism

$$\widehat{\xi}_n^i = g_n^i\,\xi_n^i + (1 - g_n^i)\,\widetilde{\xi}_n^i$$

where g_n^i is a collection of conditionally independent Bernoulli random variables with respective distributions

$$\mathbb{P}(g_n^i = 1 \mid \xi_n) = 1 - \mathbb{P}(g_n^i = 0 \mid \xi_n) = \varepsilon_n G_n(\xi_n^i) \qquad (1.16)$$

If we let τ^i be the first time we have $g_n^i = 0$, then it is readily checked that for any $n \geq 0$

$$\mathbb{P}(\tau^i \geq n) = \mathbb{E}\left(\prod_{p=0}^{n-1}\varepsilon_p G_p(X_p)\right)$$

and therefore

$$\text{Law}(\xi_0^i, \ldots, \xi_n^i \mid \tau^i \geq n) = \mathbb{Q}_n$$

More generally, on the intersection of events $\cap_{i=1}^{q}(\tau^i \geq n)$, the path particles $(\xi_0^i, \ldots, \xi_n^i)_{1 \leq i \leq q}$ are independent and identically distributed with common law $\mathbb{Q}_n$. Loosely speaking, up to the first interaction time, the interacting particle model produces independent samples according to the desired distribution.

In Figure 1.3, we have presented the schematic selection/mutation transitions of eight particles. The particles with low potential are killed, while the one with high potential duplicates into several offsprings.

1.5 Interacting Particle Systems   33

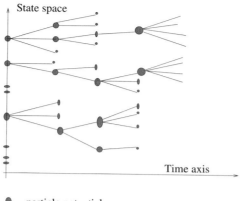

● =particle potential
N=8 particles

FIGURE 1.3. Genetic particle model

In the time homogeneous situation and under appropriate regularity conditions these particle models also provide a natural particle approximation of the fixed point (whenever it exists) of the nonlinear transformations

$$\mu \in \mathcal{P}(E) \longrightarrow \Phi(\mu) = \Psi(\mu)M \in \mathcal{P}(E)$$

We have in some sense, and as n and N tend to infinity

$$\eta_n^N \longrightarrow \eta_\infty = \Phi(\eta_\infty)$$

We will make precise this convergence with several uniform $\mathbb{L}_p$ and strong-propagation-of-chaos estimates. In this connection, we mention that the increasing propagation-of-chaos properties developed in this book also provide precise informations on the asymptotic behavior of the invariant measure of the interacting process ξ_n as the size of the systems tends to infinity.

If we interpret the selection transition as a birth and death process, then arises the important notion of the ancestral line of a current individual. More precisely, when a particle $\widehat{\xi}_{n-1}^i \longrightarrow \xi_n^i$ evolves to a new location ξ_n^i, we can interpret $\widehat{\xi}_{n-1}^i$ as the parent of ξ_n^i. Looking backwards in time and recalling that the particle $\widehat{\xi}_{n-1}^i$ has selected a site ξ_{n-1}^j in the configuration at time $(n-1)$, we can interpret this site ξ_{n-1}^j as the parent of $\widehat{\xi}_{n-1}^i$ and therefore as the ancestor $\xi_{n-1,n}^i$ at level $(n-1)$ of ξ_n^i. Running back in time we trace mentally the whole ancestral line

$$\xi_{0,n}^i \longleftarrow \xi_{1,n}^i \longleftarrow \ldots \longleftarrow \xi_{n-1,n}^i \longleftarrow \xi_{n,n}^i = \xi_n^i$$

of each current individual. More interestingly, the occupation measures of the corresponding N-genealogical tree model converge as $N \to \infty$ to the

34   1. Introduction

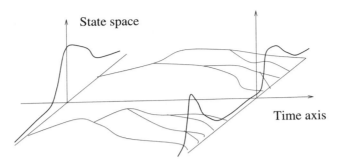

FIGURE 1.4. Genealogical tree model

Feynman-Kac path measures. In a sense to be given, we have the convergence, as $N \to \infty$,

$$\frac{1}{N}\sum_{i=1}^{N} \delta_{(\xi^i_{0,n},\xi^i_{1,n},\ldots,\xi^i_{n,n})} \longrightarrow \mathbb{Q}_n$$

Figure 1.4 represents the genealogical tree model associated with ten interacting particles. The initial probability density gives an idea of the initial configuration of the system, while the concentration of current individuals provides an approximation of the current probability density. Note that the ancestors concentration has to be related to the marginal at time 0 of $\mathbb{Q}_n$.

We finally recall that the Feynman-Kac measures $\mathbb{Q}_n$ have the same dynamical structure as their end time marginals η_n. From this stability property, we will see that the genealogical tree-based model is nothing but the N-particle model associated with a nonlinear distribution flow in path space.

1.5.2 Continuous Time Models

Using the same line of argument as in the discrete time situation, the continuous time evolution equation (1.8) can be rewritten in the form

$$\frac{d}{dt}\eta_t(f) = \eta_t(L_{t,\eta_t}(f))$$

where $L_{t,\eta}$ is a nonunique collection of infinitesimal generators satisfying the compatibility condition

$$\eta_t(L_{t,\eta_t}(f)) = \eta_t(L_t(f)) + \eta_t(f(V_t - \eta_t(V_t))) \qquad (1.17)$$

To construct simple examples of generators satisfying this condition, we proceed as on page 24. We let V_t^+ and $(-V_t^-)$ be the positive and negative parts of the potential function V_t and we denote by $\widehat{L}_\eta^-$ and $\widehat{L}_\eta^-$ the jump type generators

$$\widehat{L}^-_{t,\eta}(f)(x) = V_t^-(x) \int (f(y) - f(x)) \, \eta(dy)$$

$$\widehat{L}^+_{t,\eta}(f)(x) = \int (f(y) - f(x)) \, V_t^+(y) \, \eta(dy)$$

Arguing as before, we check that the class of generators given by

$$L_{t,\eta} = L_t + \widehat{L}_{t,\eta} \quad \text{with} \quad \widehat{L}_{t,\eta} = \widehat{L}^-_{t,\eta} + \widehat{L}^+_{t,\eta}$$

satisfies the desired compatibility condition (1.17). Note that we can alternatively use the potential functions $(V^+_{t,\eta_t}, V^-_{t,\eta_t})$ associated with the positive/negative part decomposition

$$(V_t - \eta_t(V_t)) = V^+_{t,\eta_t} - V^-_{t,\eta_t}$$

The N-particle model associated with a compatible class of generators $L_{t,\eta}$ is a nonhomogeneous Markov process

$$\xi^{(N)}_t = (\xi^{(N,1)}_t, \xi^{(N,2)}_t, \ldots, \xi^{(N,N)}_t)$$

taking values in E^N. Its infinitesimal generator $\mathcal{L}^N_t$ is defined for a sufficiently regular test function φ on E^N by the formula

$$\mathcal{L}^N_t(\varphi)(x^1, \ldots, x^N) = \sum_{i=1}^N L^{(i)}_{t, \frac{1}{N} \sum_{j=1}^N \delta_{x^j}}(\varphi)(x^1, \ldots, x^i, \ldots, x^N)$$

We have used the notation $L^{(i)}_{t,\eta}$ instead of $L_{t,\eta}$ when the generator acts on the ith component of the test function.

Loosely speaking, the motion of the N-particle model associated with the previous example is decomposed into three mechanisms. As in the discrete time case, we simplify the presentation, suppressing the index parameter N, and we write ξ_t and ξ^i_t instead of $\xi^{(N)}_t$ and ξ^i_t.

Between the interacting jumps, each particle ξ^i_t evolves randomly according to an L_t-motion. During its exploration of the state space, it is killed at rate $V^+(\xi^i_t)$ and instantly a randomly chosen particle in the configuration splits into two offsprings. In addition, at rate $\eta^N_t(V^-_t) = \frac{1}{N} \sum_{j=1}^N V^-_t(\xi^j_t)$, a randomly chosen particle ξ^i_t is replaced by a new one ξ^j_t randomly chosen with a probability proportional to its adaptation $V^-_t(\xi^j_t)$. The initial system $\xi_0 = (\xi^i_0)_{1 \leq i \leq N}$ consists of N independent and identically distributed random particles with common law η_0.

In Figure 1.5, we have presented the schematic evolution of five particles. Note that, in contrast to the discrete time case, each particle interacts and jumps to a new selected site at random exponential times.

36   1. Introduction

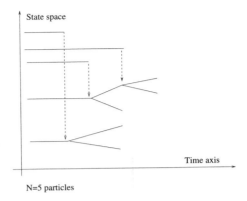

FIGURE 1.5. Genetic particle model (continuous time)

For sufficiently regular generators $L_{t,\eta}$, we prove that, for any fixed time horizon t, as $N \to \infty$ and in a sense to be given

$$\eta_t^N = \frac{1}{N} \sum_{i=1}^{N} \delta_{\xi_t^i} \longrightarrow \eta_t$$

As for discrete time models mimicking formula (1.6), we construct an unbiased estimate for the unnormalized distribution flow. We prove that as N tends to infinity

$$\gamma_t^N(.) = \eta_t^N(.) \ \exp \int_0^t \eta_s^N(V_s)ds \ \longrightarrow \ \gamma_t(.) = \eta_t(.) \ \exp \int_0^t \eta_s(V_s)ds$$

As in the discrete time situation for sufficiently regular and time homogeneous Feynman-Kac models, the particle approximation measures η_t^N also provide a natural particle approximation of the fixed point η_∞ (whenever it exists) of the integro-differential equation (1.17) and we have in some sense and as t and N tend to infinity

$$\eta_t^N \longrightarrow \eta_\infty$$

We can also interpret interacting jumps as birth and death mechanisms and define in this way the ancestral lines $(\xi_{s,t}^i)_{s \leq t}$ of each current individual $\xi_{t,t}^i = \xi_t^i$. Using the structural stability properties of Feynman-Kac flows, we also prove that the genealogical tree particle model is nothing but the N-particle model associated with the path-distribution flows $\mathbb{Q}_t$. We use this observation to check that the occupation measures of the corresponding N-genealogical model converge as $N \to \infty$ to the Feynman-Kac path measures. That is, in a sense to be given, we have the convergence, as N tends to infinity,

$$\frac{1}{N} \sum_{i=1}^{N} \delta_{(\xi_{s,t}^i)_{s \leq t}} \longrightarrow \mathbb{Q}_t$$

1.6 Sequential Monte Carlo Methodology

One central question arising in scientific computing and applied mathematics is to find numerical approximations of a given nonnegative measure π on some measurable space $(E, \mathcal{E})$. In other words, the objective is to evaluate for any bounded measurable function f the integrals $\pi(f) = \int \pi(dx)\, f(x)$ or more generally the value $F(\pi)$ of some functional $F : \mathcal{P}(E) \to \mathbb{R}$. One important and rather generic class of measures arising in various scientific disciplines is the Boltzmann-Gibbs distributions defined by

$$\Psi(\eta)(dx) = \frac{1}{\eta(G)}\, G(x)\, \eta(dx) \qquad (1.18)$$

where G represents a potential function on some measurable space $(E, \mathcal{E})$ and η a given (probability) measure on E such that $\eta(G) > 0$. We first notice that any Feynman-Kac distribution and therefore all the examples we have presented so far are particular Boltzmann-Gibbs measures (for more details we refer the reader to Section 11.9 and to Exercise 11.9.1). In physics, the complex interaction between atoms in ferromagnetic spin models is often expressed in terms of Boltzmann-Gibbs measures (see Section 11.2). In biology, these distributions are also used as a simplified model for DNA sequences and protein chains [215, 227, 234, 256]. In engineering science, optimization and regulation problems consist in computing the global minima of a given numerical cost function V on a given state space E. For instance, the cost function can measure the performance of a given quality test schedule performed at each successive stage of an industrial production chain. In some image reconstruction problems, the function V measures the resemblance of a brightness pattern of pixel configurations to a given reference image [18, 113, 154]. In both cases, the desired minima can be interpreted as the modes of a Boltzmann-Gibbs measure with potential function $G = e^{-V}$.

More generally, in the Bayesian perspective, "any estimation problem" can be interpreted as a Boltzmann-Gibbs measure of the form (1.18). In this context, the reference measure ν and the function G are the so-called *prior/posterior* distributions. These Bayesian models are often described by a pair (X, Y) of random $(d+d')$-dimensional random variables having a joint density probability of the form $p_{X,Y}(x, y) = p_{Y|X}(y|x)\, p_X(x)$, where p_X is the marginal of $p_{X,Y}$ with respect to the first component X, and $p_{Y|X}(\cdot|x)$ is the conditional density of Y given $X = x$. The random variable Y represents the partial and noisy observation of some random state X. To estimate X given the observation $Y = y$, we need to compute the conditional density

$$p_{X|Y}(x|y) = \frac{p_{Y|X}(y|x)}{p_Y(y)}\, p_X(x)$$

with the normalizing constant $p_Y(y) = \int p_{Y|X}(y|x) p_X(x) dx$. If we fix the observation data $Y = y$ and if we set $\nu(dx) = p_X(x)\, dx$ and $G(x) = p_{Y|X}(y|x)$, then these conditional distributions coincide with the Boltzmann-Gibbs distributions defined in (1.18). In this context, we also notice that the normalizing constant $p_Y(y)$ coincides with the corresponding partition function. We emphasize that the Bayesian methodology is often more directly related to nonlinear filtering and Feynman-Kac models. In this interpretation, it is often more judicious to take advantage of the dynamical or the multiplicative structure of the desired conditional distributions to define a working interacting particle algorithm. In statistical literature, these filtering models are rather called hidden Markov chain models. This situation is often related to distribution densities $p_{X,Y} = p_{\theta,X,Y}$, whose values depend on some unknown random variable θ. Traditionally, θ represents the unknown distribution of X and we want to estimate the latter using the observations of the state sequence X. If we set $\mathcal{X} = (\theta, X)$, we see that this situation reduces to the first one by a simple state-space enlargement. We refer the reader to any textbook on hidden Markov chain and Bayesian statistics (see for instance [132, 199, 206, 223, 275, 237] and references therein).

The idea of classical Monte Carlo Markov chain methods (often abbreviated MCMC) is to find a judicious Markov kernel K_η with invariant measure $\eta = \eta K_\eta$. When the Markov chain with elementary transition kernel K_η is sufficiently mixing, one expects after some long and spaced runs to generate approximate samples of the limiting distribution $\Psi(\eta)$. There exist several popular algorithms, including the Gibbs-Sampler and the Metropolis-Hastings model. It is of course out of the scope of this book to review the modeling, the convergence, and the applications of these models. For more details, we refer the reader to Chapter 5 and references therein. We underline that these "linear type" algorithms suffer from two main numerical problems. First of all, their convergence to the desired equilibrium Boltzmann-Gibbs measure strongly depends on the oscillations of the potential function G. In addition, they are not recursive with respect to the time parameter. For instance, in every filtering problem, the target distribution is the conditional distribution of a signal process given the observations delivered at each time by some sensor. In this context, it is hopeless to run a different Markov chain with transition between each two consecutive observations. More generally, suppose we are given a collection of target distributions η_n on some measurable spaces $(E_n, \mathcal{E}_n)$. Further assume that they satisfy a recursive equation of the form

$$\eta_{n+1} = \eta_n K_{n+1,\eta_n}$$

for some collection of Markov transitions $K_{n+1,\eta}$ from some measurable space $(E_n, \mathcal{E}_n)$ into $(E_{n+1}, \mathcal{E}_{n+1})$. Here again, to approximate these measure-valued equations or generate samples according to each η_n, it is not really judicious to run a different Markov chain model with invariant distribution

η_n. The advantages in applying the interacting particle interpretations of these nonlinear models are twofold. First, they are recursive with respect to the time parameter. On the other hand, even when the target distribution η is not related to some of these recursions, we can use the time reversal Feynman-Kac representation (1.12) presented in Section 1.4.5 and its particle approximation model to improve the convergence to equilibrium of the algorithm.

For various statistical engineers or physicists, any Monte Carlo Markov chain algorithm is pinched up between a Metropolis-Hastings model and a Gibbs sampler. This rather simplistic idea is partly true if one believes that any kind of stochastic algorithm is a simple Markov chain. We emphasize here that evolutionary type models such as genetic type branching algorithms and the Feynman-Kac-Metropolis algorithms described in Section 5.4 are nonlinear type models and do not fit in the same framework as MCMC methods. These modern strategies rather belong to mean field interacting particle systems and nonlinear measure-valued processes. In connection with the traditional Metropolis acceptance/rejection transition, the idea here is simply to duplicate better-fitted proposals. Loosely speaking, in these branching type algorithms, the time precision parameter of traditional MCMC methods is replaced by a population size parameter. Various authors have tried to find the exact origins of this natural splitting idea (some recent attempts can be found in [125, 139, 157, 167, 227]). In our opinion, these evolutionary techniques of duplicating better-fitted individuals are rather natural and arise in so many human endeavors that it is more interesting to trace back in time their use in different application areas.

1.7 Particle Interpretations

The particle models described above can be sought in many different ways, depending on the Feynman-Kac application model areas we consider. Apart from the pure mathematical interpretation as a stochastic linearization technique, three additional interpretations can be underlined.

From the point of view of physics, statistical mechanics, biology, and industrial chemistry, these particle models provide a concrete microscopic interpretation of the evolution of some physical quantity. For instance, in the trapping problems examined in Section 1.4.3, the particle model may represent the evolution of a collection of particles in an absorbing medium. When a particle enters an obstacle and dies, a particle with better adaptation splits into two offsprings. These interacting jump mechanisms are also related to the microstatistical interpretation of Schrödinger equations discussed in Section 1.4.3. In this context, the particle model can be seen as the evolution of relativistic quantum particles with creation and killing transitions (see also [254]). We also mention that the continuous time Feynman-Kac model and its particle approximation model can be seen as a simple generalized Boltzmann equation (as defined in [245]) with the corresponding Nanbu particle interpretation. In this situation, the particle model represents the evolution of colliding molecules in a rarefied gas model. In this context, genealogical models represent the historical process of splitting particles in trapping models or the history of collisions in rarefied gas equations. The genetic type particle models introduced in Section 1.5.1 and Section 1.5.2 are clearly related to natural evolution models. They can be used to model random mutation/selection transitions in gene analysis. In this connection, the genealogical models can be seen as the random evolution of the ancestral lines of species. In polymer analysis, the Feynman-Kac path models represent the Boltzmann-Gibbs measures of a polymerization sequence (see Section 1.4.4 and Section 12.5). In this context, the particle motion can be regarded as the chemical construction of monomers with intermolecular interaction in a given solvent. The corresponding genealogical models represent the chemical construction of a sequence of flexible and directed polymers.

From the point of view of evolutionary mathematics and engineering sciences, these probabilistic methods can be viewed as global and adaptive stochastic searches of a state space. The motion and the adaptation of individuals are related to natural evolution mechanisms such as gene mutations and/or type selection. This interpretation also enters physical and biological intuitions in engineering problems. We can relate in this way learning and adaptation mechanisms with evolutionary and microscopic particle motions. In this connection, particle search methods can also be regarded as a stochastic grid approximation. They also complement more traditional deterministic grid techniques. Their main advantage is to refine the precision of the grid in accordance with the mass variation dictated by the distribution equation. In this interpretation, genealogical models represent the adaptation history of the grid.

From a statistical point of view, the genealogical and interacting particle models presented in this book can also be regarded as a new approximation simulation technique for sampling conditional distributions and more generally for sampling according to Boltzmann-Gibbs type measures on path spaces. They complement classical Monte Carlo strategies such as the Gibbs sampling and the Metropolis-Hastings algorithm. In contrast to the latter, particle simulation methods are recursive with respect to the time parameter. One of the most illuminating examples that illustrates this point of view is the Feynman-Kac-Metropolis interpretation of a restricted Markov chain model described in the introductory Section 1.4.5 and further developed in Section 5.5. The genealogical approximation models provide a natural simulation technique for drawing path samples according to the distribution of a Markov chain restricted to its terminal values. Furthermore, the underlying genetic type model can be interpreted as a sequence of interacting

Metropolis models. These particle models are not of pure mathematical interest. We will see that they have "better asymptotic properties" than traditional Monte Carlo Markov chain methods.

During the last few decades, the application of particle methods has grown, establishing unexpected connections with a number of other fields, including biology [227, 231, 232, 233, 234, 235, 280], electromagnetics [191, 299], neural networks [151], computer networks and telecommunication analysis [274], traffic control [5], image processing [297], financial mathematics [23, 285, 302], global optimization problems [108, 160, 161, 162, 192, 193, 205, 306], statistics [8, 9, 124], and signal processing [7, 10, 122, 123, 143].

In each discipline, these particle methods have taken different names but often have a common theoretical basis. For instance, in applied probability and engineering literature, there exists at least a dozen different names for the same particle algorithm: branching and interacting particle systems [75, 97, 156], Monte Carlo optimal filters [39, 185, 204], genetic algorithms [181, 182], bootstrap or particle filters [123, 160, 239, 274], sampling-importance-resampling [57, 164], condensation filters [144], population Monte Carlo algorithms and Sequential Monte Carlo methods [51, 81, 122, 125, 207, 228], spawning filters [145], switching algorithm [224], sampled stochastic processes [238], auxiliary particle filters [269], matrix reconfigurations [179], quantum and diffusion Monte Carlo methods [12, 242], go with the winner [4, 263, 264], multisplitting and restart method [49, 261], interacting Metropolis algorithm [80], and ensemble Kalman filters [135].

All of these particle models are built on the same paradigm: When exploring a state space with many particles, we duplicate better fitted individuals at the expense of having light particles with poor fitness die. In this connection, we also mention that this selection mechanism has been associated in biology and engineering science with the following list of botanical names, to name a few: branching selections, bootstrap, adaptive dynamics, switching, prune enrichments, cloning, reconfiguration, stratification, resampling, rejuvenation, acceptance/rejection, spawning, and "go with the winner".

1.8 A Contents Guide for the Reader

The book is divided into eight main parts devoted respectively to Markov path processes, to the precise description of the Feynman-Kac models and their structural properties, to genealogical and particle models, to stability properties of Feynman-Kac semigroups, to invariant measures, to annealed properties, to the asymptotic behavior of particle models, and to applications in statistics, physics, and engineering sciences.

In Chapter 2, we give a detailed and rigorous mathematical description of Feynman-Kac models with an emphasis on path-valued processes and unnormalized models. To make the book self-contained, we have provided in the introductory Section 2.2 a brief discussion on Markov chains. We give three different presentations arising in engineering science, applied probability, and Bayesian statistical literature. We also discuss the coordinate method of constructing a Markov chain on a canonical probability space. We conclude with the construction of path-space Markov models. Then we list three important structural stability properties connected respectively to path/marginal models, change of reference probability, and updated/prediction flow models. These properties show the potential of Feynman-Kac modeling techniques. They exhibit several degrees of freedom in the Feynman-Kac interpretations of a given nonlinear estimation problem. To illustrate the impact of these results, we already mention that the structural stability property related to path and marginal models will lead to a natural construction of genealogical particle models. After this round of modeling and structural analysis, we propose two probabilistic interpretations of Feynman-Kac distribution flow models. The first one concerns the traditional particle-trapping interpretation arising in physics literature. The second one is related to interacting and measure-valued processes theory. We provide a collection of mean field and McKean type interpretations of these flows. In a third part, we discuss Feynman-Kac models in random media. We connect the annealed and quenched flows in terms of a Feynman-Kac model in distribution space. The impact of this modeling technique will be illustrated in nonlinear filtering problems with the construction of interacting Kalman-Bucy filters. Finally, we describe the semigroup structure of discrete time Feynman-Kac models. These updated and prediction semigroups will be of constant use in the development of this book.

Chapter 3 is concerned with genealogical and interacting particle models. It is itself essentially decomposed into three main parts. We first give a rigorous mathematical construction of the particle model associated with a McKean interpretation of a nonlinear and measure-valued process. This particle method is presented here as a stochastic linearization technique of a nonlinear equation in distribution space. Then we illustrate this abstract formulation with a detailed discussion on the particle interpretations of Feynman-Kac models. We also show that genealogical particle tree models are natural particle interpretations of Feynman-Kac models in path space. Finally, to prepare the applications developed in the next part of the book, we review the various particle measures introduced so far with some motivating convergence results. We also end this part with a brief discussion on the main regularity conditions used in this book.

Chapters 4, 5, and 6 are devoted to qualitative properties of Feynman-Kac semigroups. Chapter 4 focuses on the stability of Feynman-Kac semigroups. In a first part, we analyze the contraction properties of a general

Markov kernel with respect to an abstract h-relative entropy. We exhibit several functional inequalities in terms of the Dobrushin ergodic coefficient. Then we use these results to study the stability properties of nonlinear Feynman-Kac semigroups. We derive precise contraction estimates with respect to a class of h-relative entropies, including the traditional Boltzmann entropy or the total variation distance as well as the $\mathbb{L}_2$ or Havrda-Charvat entropies and the Hellinger integrals. We finally apply these results to study the stability properties of several classes of models including updated Feynman-Kac semigroups and stochastic models arising in filtering problems.

In Chapter 5, we discuss the existence and the uniqueness of invariant measures. We also design a class of Metropolis type Feynman-Kac models admitting a given distribution as an invariant measure. We connect this model with restricted Markov chains with respect to their terminal values. We compare the path-particle algorithms associated with these Feynman-Kac models with the traditional Metropolis Markov chain. Finally we give a detailed study on the stability properties of this particular class of models.

In Chapter 6, we examine the long time behavior of a Feynman-Kac distribution flow model associated with a cooling schedule. We first examine the annealed properties of the Feynman-Kac-Metropolis model introduced in the preceding chapter. We will see that this model can also be regarded as a nonlinear simulated annealing algorithm. We also discuss the annealed properties of a class of Feynman-Kac models arising in trapping analysis. Special attention is paid to the limiting concentration regions. We propose a set of sufficient conditions under which the algorithm converges as the time parameter tends to infinity to the global minima of a potential function.

Chapters 7, 8, 9, and 10 are concerned with the asymptotic behavior of particle methods when the size of the systems tends to infinity. We have chosen to present these convergence results with respect to the traditional increasing degrees of refinement. A good deal of the theory consists of the study of various limit theorems. We provide a detailed analysis going from the simple and traditional law of large numbers to more sophisticated empirical process theorems, uniform estimates with respect to the time parameter, weak and strong propagation of chaos, central limit theorems, Berry-Esseen inequalities, Donsker type theorems, and large-deviation principles. For the convenience of the reader, we have tried to present an "exposé" of the mathematical theory we shall be using on each of these different subjects in a self-contained treatment. Each chapter often starts with a preliminary discussion on some more or less traditional results. These introductory sections also provide several new and complementary results to some well-known theorems such as a Berry-Esseen inequality for martingale sequences as well as a complement to the Laplace-Varadhan integral transfer lemma in the context of large deviations.

Chapters 11 and 12 are concerned with the applications of the Feynman-Kac modeling and the particle methodology developed in this book. Chapter 11 should not be skipped. It provides a series of Feynman-Kac and interacting particle recipes to be used in different application model areas. In Chapter 12, we describe in some detail applications to restricted Markov chain sampling, spectral analysis of Feynman-Kac-Schrödinger semigroups, fixed point approximations of nonlinear equations in distribution space, rare events estimation, Dirichlet problems with boundary conditions, flexible polymer chains, filtering, and path estimations of a signal. The emphasis is given to the Feynman-Kac modeling of the various mathematical quantities we want to estimate in each of these disciplines:

- In statistical simulation problems, we provide a Feynman-Kac formulation of the laws of restricted Markov models. First we examine a class of Markov chains restricted to a given space-time tube. In another context, we construct a Feynman-Kac semigroup admitting a given distribution as an invariant measure. This construction is based on a judicious choice of Metropolis type potential function and on a time reversal technique. In this situation the resulting particle model will behave as a sequence of interacting Metropolis algorithms. Furthermore, the corresponding genealogical tree-based model can be regarded as a path particle simulation technique for drawing samples according to the law of Markov path models restricted to their terminal values.

- In the spectral analysis of Feynman-Kac-Schrödinger semigroups, we describe the Lyapunov exponent and related spectral quantities in terms of the invariant measure of a Feynman-Kac model. The particle models associated with these exponents provide a natural microscopic particle interpretation of these physical quantities. For instance, in trapping analysis, these exponents represent the cost of performing long crossing in an absorbing medium without being absorbed. In this context, the particle coefficients represent the mean averaged energy of a flow of interacting particle systems.

- In rare events analysis, we propose a multilevel Feynman-Kac path model to describe the conditional distribution of a stochastic motion in a rare event regime. The multilevel decomposition can be interpreted as the different successive steps the process needs to pass to enter into the upper rare level. In this context, the genealogical particle model can be seen as the random tree of different possible trajectories leading to this level.

- In biology and chemistry, flexible polymer chains are often described in terms of a Boltzmann-Gibbs distribution. We connect this formulation with Feynman-Kac models on path spaces. In this connection each random path can be seen as a sequence of reacting monomers in a given solvent. The potential function represents the attractive or repulsive interaction between the elementary monomers of a polymer chain. As an aside, we mention that self-avoiding random walks are simple polymerization models with repulsive interactions. The path particle algorithms associated with these models can be interpreted as a particle simulation method for sampling flexible polymer chains.

- In signal processing, we underline the role of Feynman-Kac path measures in the description of path estimation and nonlinear smoothing problems. We discuss four different ways to introduce a nonlinear filtering model: the engineering description in terms of a signal/sensor equation, the probabilistic interpretation with the pair (signal, observation) Markov model, the traditional change of reference probability measure technique, and the Bayesian posterior/prior interpretation. In this context, we propose a new model of quenched and annealed formulae to describe a partial linear/Gaussian filtering problem. As mentioned above, the particle interpretation of the annealed model consists of a sequence of interacting Kalman-Bucy filters.

2
Feynman-Kac Formulae

2.1 Introduction

In this chapter, we introduce the mathematical structure of Feynman-Kac models.

In a preliminary section, Section 2.2, we provide a brief introduction to Markov chain models in abstract path spaces.

In Section 2.3, we present the abstract construction of Feynman-Kac distribution models on general nonhomogeneous state spaces. We also fix some notation and terminology currently used in this book.

Section 2.4 provides some structural properties of Feynman-Kac models. We connect path-space models with their time marginals as well as connecting updated models with prediction ones. We also examine the stability of Feynman-Kac formulae through a change of reference probability measure. The stability properties developed in this section provide a natural and simple tool to transfer results from marginal or prediction models to path-space models or updated ones.

Section 2.5 is concerned with two physical interpretations of Feynman-Kac distribution flow models. The first one is the traditional interpretation in terms of a single particle motion in an absorbing medium. The second interpretation comes from nonlinear and measure-valued processes theory. We provide an alternative physical interpretation of Feynman-Kac models in terms of interacting jump type Markov models.

In Section 2.6, we discuss Feynman-Kac models in random media. We provide an original description of the annealed distribution flows in terms of a Feynman-Kac model associated with a measure-valued Markov chain. These modeling techniques will be used in Section 12.6.7 to define in a natural way a sequence of interacting Kalman-Bucy filters. To better connect our work with related literature on the subject, we mention that continuous time Feynman-Kac formulae with random potentials have been used in a different context as explicit examples of PDEs in a random environment. In discrete space, Carmona and Molchanov gave in [40] a detailed treatment of the so-called parabolic Anderson model. Large time asymptotics and related Lyapunov exponent estimations for such stochastic Feynman-Kac flows can also be found in [13, 29, 31, 41, 42, 152].

Section 2.7 is concerned with the fine semigroup structure of Feynman-Kac distribution flows.

Section 2.4 and Section 2.6 can be skipped at a first reading. They may be used from time to time as a reference in some precise situations. Sections 2.5 and 2.7 are the core of this chapter. They provide a detailed discussion on the Feynman-Kac models currently used in the further development of this book.

2.2 An Introduction to Markov Chains

A Markov chain is a sequence of random variables X_n with a time index $n \in \mathbb{N}$ and taking values at each time n in some measurable state space $(E_n, \mathcal{E}_n)$. In contrast to any class of random sequences the future of a Markov evolution model is independent of the past when the present is given. Loosely speaking, a Markov model is a random and discrete generation model whose elementary transitions only depend on the more recent information on its past history.

To motivate this abstract Markov dependence property and to introduce the various application areas we have ahead, we have chosen to present next three traditional formulations.

The first one can be regarded as a stochastic version of a control system. the random component of these models usually consists of a sequence of independent random variables U_n, $n \in \mathbb{N}$, taking values in some measurable control space $(C_n, \mathcal{C}_n)$. Formally, they are defined inductively by a recursive equation of the form

$$X_n = F_n(X_{n-1}, U_n) \tag{2.1}$$

The drift function F_n is a given measurable function from $E_{n-1} \times C_n$ into E_n, and X_0 is an arbitrary random initial condition with distribution $\mu \in \mathcal{P}(E_0)$. The probabilistic interpretation of the random sequence U_n depends on the problem at hand. For more details, we refer the reader to the discussion on filtering models given in Section 1.4.1.

Another more probabilistic way to specify a Markov chain is to consider its elementary transitions

$$\text{Proba}(X_n \in dx_n \mid X_{n-1} = x_{n-1}) = M_n(x_{n-1}, dx_n)$$

The Markov kernel $M_n(x_{n-1}, dx_n)$ denotes the probability that the chain $X_{n-1} = x_{n-1} \in E_{n-1}$ at time $(n-1)$ will be at time n "in an infinitesimal neighborhood" dx_n of the point $x_n \in E_n$. To connect this formulation with the previous one, we observe that the Markov kernel associated with the Markov chain (2.1) is simply given by

$$M_n(x_{n-1}, dx_n) = \text{Proba}(F_n(x_{n-1}, U_n) \in dx_n)$$

From the practitioner's point of view, there may be very little difference between these two formulations. The first interpretation is often used in engineering science when the Markov chain represents the evolution of a physical quantity with a precise structure.

The second formulation is more tractable when the Markov model is defined by complicated rules with no precise dynamical structure.

Finally, in Bayesian literature, the elementary transitions of a chain are often described in terms of a hypothetical density function

$$\text{Proba}(X_n \in dx_n \mid X_{n-1} = x_{n-1}) = p_n(x_n|x_{n-1})\,dx_n$$

Some authors also use the synthetic notation

$$X_n \mid X_{n-1} = x_{n-1} \sim p_n(x_n|x_{n-1})\,dx_n$$

The term $p_n(x_n|x_{n-1})$ represents the density of the Markov transition with respect to some reference probability measure dx_n.

2.2.1 Canonical Probability Spaces

The rigorous mathematical formulation in terms of Markov kernels is traditionally used in applied probability literature. Aside from inherent mathematical interest, this measure-theoretic formulation also provides a natural semigroup formulation for the asymptotic analysis of Markov chains. Furthermore, the forthcoming Markov chain models are traditionally described on a canonical probability space, and their "dynamical structure" only depends on the underlying probability distribution. Loosely speaking, this presentation is often less "notationally consuming":

The same canonical process may have different probabilistic interpretations depending on the different probability distributions we consider on the canonical space.

For all these reasons, we will use as often as we can this modern probabilistic and canonical formulation in the modeling and the analysis of Feynman-Kac semigroups.

Since one of the goals of this section is to guide the reader to abstract Markov chains on general nonhomogeneous state spaces, we give next a brief presentation of the rigorous mathematical construction of a canonical Markov chain. This construction is sometimes called the coordinate method of constructing the Markov chain. These models will be of constant use in the further development of this chapter.

To start with, we use the Markov dependence property to check that

$$\text{Proba}((X_0, \ldots, X_n) \in d(x_0, \ldots, x_n))$$

$$= M_n(x_{n-1}, dx_n)\text{Proba}((X_0, \ldots, X_{n-1}) \in d(x_0, \ldots, x_{n-1}))$$

$$= \mu(dx_0)\, M_1(x_0, dx_1) \ldots M_n(x_{n-1}, dx_n)$$

From this display, we define the distribution $\mathbb{P}_{\mu,n}$ of the sequence

$$(X_0, \ldots, X_n)$$

on $\Omega_n = E_{[0,n]}(= \prod_{p=0}^n E_p)$ equipped with the product σ-algebra $\mathcal{F}_n = \prod_{p=0}^n \mathcal{E}_p$ by setting in the integral sense

$$\mathbb{P}_{\mu,n}(d(x_0, \ldots, x_n)) = \mu(dx_0)\, M_1(x_0, dx_1) \ldots M_n(x_{n-1}, dx_n)$$

By the consistency of the collection $\mathbb{P}_{\mu,n}$, $n \in \mathbb{N}$, there exists an overall distribution $\mathbb{P}_\mu$ on $\Omega = \prod_{n \geq 0} E_n$ for which $\mathbb{P}_{\mu,n}$ are finite-dimensional distributions (see Ionescu Tulcea's theorem, p. 249 in [287]). That is, for any $A_n \in \mathcal{E}_n$, $n \geq 0$, and any cylinder set of the form

$$C_n(A_0, \ldots, A_n) = \{\omega = (\omega_n)_{n \geq 0} \in \Omega\,;\, \forall 0 \leq p \leq n \quad \omega_p \in A_p\}$$

we have that

$$\mathbb{P}_\mu(C_n(A_0, \ldots, A_n)) = \mathbb{P}_{\mu,n}(A_0 \times \ldots \times A_n)$$

If we denote by X_n, $n \in \mathbb{N}$, the sequence of canonical mappings

$$X_n : \omega = (\omega_n)_{n \geq 0} \in \Omega \longrightarrow X_n(\omega) = \omega_n \in E_n$$

then we find that

$$C_n(A_0, \ldots, A_n) = \{\omega = (\omega_n)_{n \geq 0} \in \Omega\,;\, \forall 0 \leq p \leq n \quad X_p(\omega) \in A_p\}$$

This readily yields that

$$\mathbb{P}_\mu((X_0, \ldots, X_0) \in (A_1 \times \ldots \times A_n))$$
$$= \int_{A_0 \ldots \times A_n} \mu(dx_0)\, M_1(x_0, dx_1) \ldots M_n(x_{n-1}, dx_n)$$

The canonical probability space defined in this way

$$\left(\Omega = \prod_{n \geq 0} E_n, \ \mathcal{F} = (\mathcal{F}_n)_{n \in \mathbb{N}}, \ X = (X_n)_{n \in \mathbb{N}}, \ \mathbb{P}_\mu\right)$$

is called the canonical Markov chain with transitions M_n and initial distribution μ. The probability measure $\mathbb{P}_\mu$ on the canonical space $(\Omega, \mathcal{F}, \mathcal{F}_\infty, X)$, with $\mathcal{F}_\infty = \vee_{n \geq 0} \mathcal{F}_n$, is called the distribution or the law of the (canonical) Markov chain.

2.2.2 Path-Space Markov Models

The abstract mathematical modeling presented in the preceding section is particularly useful to describe Markov motions on path spaces. We now discuss some of these models. Let $(E'_n, \mathcal{E}'_n)$ be an auxiliary collection of measurable spaces, and let X'_n be a nonanticipative sequence of E'_n-valued random variables in the sense that the distribution of X'_{n+1} on E'_{n+1} only depends on the random states $(X'_0, \ldots, X'_n)$. By direct inspection, we notice that under some appropriate measurability conditions the path sequence

$$X_n = X'_{[0,n]} = (X'_0, \ldots, X'_n) \tag{2.2}$$

forms a nonhomogeneous Markov chain taking values in the product spaces

$$E_n = E'_{[0,n]} = (E'_0 \times \ldots \times E'_n)$$

In this situation, each point $x_n = (x'_0, \ldots, x'_n) \in E_n$ has to be thought of as a path from the origin up to time n.

The archetype of such Markov path models is the situation where X'_n is an E'_n-valued Markov chain with not necessarily homogeneous transitions $M'_n(x_{n-1}, dx_n)$ from E'_{n-1} into E'_n. In this context, the Markov transitions M_n of the path chain (2.2) and M'_n are connected by the formula

$$M_{n+1}((x_0, \ldots, x_n), d(y_0, \ldots, y_n, y_{n+1}))$$
$$= \delta_{(x_0, \ldots, x_n)}(d(y_0, \ldots, y_n)) \ M'_{n+1}(y_n, dy_{n+1}) \tag{2.3}$$

Of course, the time parameter can be added to the state space as an additional deterministic variable. As an aside, if we consider the sequence $\overline{X}_n = (n, X_n)$ on the state space $E = \cup_{p \geq 0}(\{p\} \times E_p)$, we do get a time-homogeneous Markov chain with elementary transitions

$$M'((n, x), d(p, y)) = \delta_{n+1}(p) \ M_p(x, dy) \tag{2.4}$$

Nevertheless, the Markov kernels (2.4) contain a Dirac measure, and various regularity conditions needed later are not preserved by this state-space enlargement. On the other hand, the time parameter here is often interpreted as the length of a path. It seems therefore notationally more transparent to consider nonhomogeneous path spaces.

Definition 2.2.1 *The Markov chain in path space*

$$X_n = X'_{[0,n]} \in E_n = E'_{[0,n]}$$

associated with an E'_n-valued Markov chain X'_n is called the historical process or the path process of the chain X'_n.

The motion of the historical process X_n simply consists of extending each path of X'_n with an elementary M'_n-transition. We have the synthetic diagram

$$X_{n-1} = X'_{[0,n-1]} \longrightarrow X_n = X'_{[0,n]} = (X'_{[0,n-1]}, X'_n)$$

with $X'_n \sim M'_{n+1}(X'_{n-1}, \cdot)$.

2.2.3 Stopped Markov chains

We consider a Markov chain X'_n taking values in some measurable spaces $(E'_n, \mathcal{E}'_n)$, with elementary transitions M'_n from E'_{n-1} into E'_n. We further assume that X' is defined on the canonical space

$$\left(\Omega = \prod_{n \geq 0} E'_n, \ \mathcal{F} = (\mathcal{F}_n)_{n \in \mathbb{N}}, \ X' = (X'_n)_{n \in \mathbb{N}}, \ (\mathbb{P}_x)_{x \in E'_0} \right)$$

where $\mathcal{F}_n$ is the natural filtration generated by the random variables X'_p, with $p \leq n$.

A finite stopping time T with respect to $\mathcal{F}$ is a random variable taking values in $\mathbb{N}$ and such that for any $x \in E'_0$ we have

$$\{T = n\} \in \mathcal{F}_n \quad \text{and} \quad \mathbb{P}_x(T < \infty) = 1$$

As an aside, let us check that the chain X'_n always satisfies the strong Markov property with respect to T. We recall that the σ-field associated to T is given by

$$\mathcal{F}_T = \{ A \in \mathcal{F}_\infty \ : \ A \cap \{T \leq n\} \in \mathcal{F}_n, \quad \forall \, n \geq 0 \}$$

For any $A \in \mathcal{F}_T$, any collection of subsets $B_n \in \mathcal{F}_n$, and any $p \geq 0$ we have

$$\mathbb{P}_x(A \cap (X'_{T+1} \in B_{T+1}, \ldots, X'_{T+p} \in B_{T+p}))$$

$$= \sum_{n \geq 0} \mathbb{E}_x(1_{A \cap \{T=n\}}$$

$$\times \int_{B_{n+1} \times \ldots \times B_{n+p}} M'_{n+1}(X'_n, dx_1) \ldots M'_{n+p}(x_{p-1}, dx_p))$$

$$= \mathbb{E}_x(1_A \int_{B_{T+1} \times \ldots \times B_{T+p}} M'_{T+1}(X'_T, dx_1) \ldots M'_{T+p}(x_{p-1}, dx_p))$$

We conclude the strong Markov property
$$\mathbb{P}_x(X'_{T+1} \in B_{T+1}, \ldots, X'_{T+p} \in B_{T+p} \mid \mathcal{F}_T)$$
$$= \int_{B_{T+1} \times \ldots \times B_{T+p}} M'_{T+1}(X'_T, dx_1) \ldots M'_{T+p}(x_{p-1}, dx_p)$$

We let X_n be the stopped path-valued process defined by
$$X_n = (n \wedge T, X'_{[0, n \wedge T]}) \in E_n = \cup_{p=0}^n (\{p\} \times E'_{[0,p]})$$

with
$$X'_{[0, n \wedge T]} = (X'_0, \ldots, X'_{n \wedge T})$$
$$= \sum_{p=0}^{n-1} X'_{[0,p]} 1_{T=p} + X'_{[0,n]} 1_{T \geq n}$$

To describe precisely its elementary transitions we need a few observations. Since we have
$$\{T \geq n\} = \{\omega = (x_n)_{n \geq 0} : T(\omega) \geq n\} \in \mathcal{F}_{n-1} = \sigma(X'_p, \ p < n)$$

there exists a measurable set $A_n \subset E'_{[0, n-1]}$ such that
$$\{T \geq n\} = \{X'_{[0, n-1]} \in A_n\} \tag{2.5}$$

Note that
$$\{T \geq n\} = \{T \wedge n = n\} \quad \text{and} \quad \{T < n+1\} = \{T \leq n\} = \{X'_{[0,n]} \in A_{n+1}^c\}$$

and therefore
$$\{T \wedge n = n\} \cap \{X'_{[0,n]} \in A_{n+1}^c\} = \{T = n\}$$

Using this set-realization of the stopping time we find the decomposition
$$X_{n+1} = \left(1_{T \wedge n < n} + 1_{T \wedge n = n} 1_{A_{n+1}^c}(X'_{[0,n]})\right) X_n$$
$$+ 1_{T \wedge n = n} 1_{A_{n+1}}(X'_{[0,n]}) \ (n+1, X'_{[0, n+1]})$$

The elementary transitions M_n of X_n are now given for any $p \leq n$ and $(x_0, \ldots, x_p) \in E'_{[0,p]}$ by the formula

$M_{n+1}((p, (x_0, \ldots, x_p)), d(q, (y_0, \ldots, y_q)))$

$= \left(1_{p<n} + 1_{p=n} 1_{A_{n+1}^c}(x_0, \ldots, x_n)\right) \delta_{(p,(x_0, \ldots, x_p))}(d(q, (y_0, \ldots, y_q)))$

$+ 1_{p=n} 1_{A_{n+1}}(x_0, \ldots, x_n)$

$\qquad \times \delta_{(n+1, (x_0, \ldots, x_n))}(d(q, (y_0, \ldots, y_n))) \ M'_{n+1}(dy_n, dy_{n+1})$

54   2. Feynman-Kac Formulae

The archetype of such stopped processes is the situation where T represents the exit time of a time-homogeneous Markov chain X' from a given measurable set $A \subset E'$. In this case we have

$$\{T \geq n\} = \{X'_{[0,n-1]} \in A^n\}$$

Note that, in this situation, the stopped process $X'_{n \wedge T}$ is itself a Markov chain with elementary transitions

$$K(x, dy) = 1_{A^c}(x)\, \delta_x(dy) + 1_A(x)\, M'(x, dy)$$

We finally mention that the Markov chain X_n introduced above converges almost surely to the excursion valued random variable X_T as n tends to infinity. Indeed, if we set $\Omega_X = \{\lim_{n \to \infty} X_n = X_T\}$, then we have $\mathbb{P}_x(\Omega_X \cap \{T \leq n\}) = 1$, for every $n \geq 0$. By the monotone convergence theorem, we conclude that $\mathbb{P}_x(\Omega_X) = 1$. Also note that, for any $f \in \mathcal{B}_b(\cup_{n \geq 0}(\{n\} \times E'_{[0,n]}))$, we have

$$|\mathbb{E}(f(X_n)) - \mathbb{E}(f(X_T))| = |\mathbb{E}([f(X_n) - f(X_T)]1_{T>n})| \leq 2\|f\|\mathbb{P}(T > n) \to 0$$

If T is not almost surely finite, we observe that the above analysis remains valid on the event $1_{T<\infty}$. In this case, we have

$$|\mathbb{E}(f(X_n)1_{T<\infty}) - \mathbb{E}(f(X_T)1_{T<\infty})| \leq 2\|f\|\mathbb{P}(n < T < \infty) \to 0$$

Random excursion and stopped processes appear to be useful in the Feynman-Kac modeling of Dirichlet boundary problems (see Section 12.2.2 and Section 12.2.4). To illustrate this observation we return to the time-homogeneous model stopped when it exists the set A described above. We also consider the bounded linear mapping

$$D : \varphi \in \mathcal{B}_b(E') \longrightarrow D(\varphi) \in \mathcal{B}_b(E')$$

defined by

$$D(\varphi)(x) = \mathbb{E}_x(\varphi(X_T))$$

It is easy to check that $D(f)$ satisfies the Dirichlet problem

$$\begin{cases} M'(D(\varphi))(x) = D(\varphi)(x) & \text{for } x \in A \\ D(\varphi)(x) = \varphi(x) & \text{for } x \in A^c \end{cases}$$

In much the same way, if we choose the function

$$f(T, X'_{[0,T]}) = \varphi(X'_T) + \sum_{0 \leq p < T} g(X'_p)$$

for some $\varphi, g \in \mathcal{B}_b(E')$, then we readily check that the function

$$h(x) = \mathbb{E}_x \left(\varphi(X'_T) + \sum_{0 \leq p < T} g(X'_p) \right)$$

satisfies the (generalized) Poisson problem

$$\begin{cases} h(x) = M'(h)(x) + g(x) & \text{for} \quad x \in A \\ h(x) = \varphi(x) & \text{for} \quad x \in A^c \end{cases}$$

In this situation we have

$$|\mathbb{E}_x(f(X_n)) - \mathbb{E}_x(f(X_T))| \leq (\operatorname{osc}(\varphi) + \|g\|) \, \mathbb{E}_x(T \, 1_{T>n}) \to 0$$

for any $n \geq 1$, and as soon as $\mathbb{E}_x(T) < \infty$.

2.2.4 Examples

Example 2.2.1 (Simple random walks) *A well know example of Markov chain is the simple random walk on the lattice $E = \mathbb{Z}$ and defined by*

$$X_n = X_0 + \sum_{i=1}^n \varepsilon_i = X_{n-1} + \varepsilon_n \qquad (2.6)$$

where $(e_i)_{i \geq 1}$ is a sequence of independent and identically distributed random variables with common law $\mathbb{P}(\varepsilon_1 = +1) = p$ and $\mathbb{P}(\varepsilon_1 = -1) = q$, with $p, q \in (0,1)$ and $p + q = 1$. The above Markov chain is sometimes used to model a gambler's random ruin process starting with X_0 euros and loosing or wining 1 euros with probability p or q. More generally a simple random walk on the d-dimensional lattice $E = \mathbb{Z}^d$ is the Markov chain X_n with elementary transitions $M(x, .) = \sum_{|e|=1} p(e) \, \delta_{x+e}$ for some sequence $(p(e))_{|e|=1} \in [0,1]^{2d}$ with $\sum_{|e|=1} p(e) = 1$.

Example 2.2.2 (Birth and death model) *The birth and death model is a simple Markov chain X_n on $E = \mathbb{N}$ where the state 0 is an absorbing barrier. Its elementary transitions are given for any $x > 0$ by*

$$M(x, dy) = p(x) \, \delta_{x+1}(dy) + q(x) \, \delta_{x-1}(dy)$$

where for any $x > 0$ we have $p(x), q(x) \in [0,1]$ with $p(x) + q(x) = 1$ and the absorbing condition is given by $M(0, \{0\}) = 1$.

Example 2.2.3 (Storage and dam model) *Various storage processes arising in engineering and financial sciences have a Markovian representation. A typical toy example is the water flow reservoir. In this context, X_n represents the storage level and the amount of water in the dam. The inflows, demands, and the percentage loss due to evaporation are represented y a non negative random variables I_n, D_n and ε_n. Assuming that $(I_n, D_n, \varepsilon_n)$ are independent variables, the Markov chain X_n evolution is given by the recursion*

$$X_n = ((1 - \varepsilon_n)X_{n-1} + (I_n - D_n))^+$$

Example 2.2.4 (Auto-regressive models) *Consider the $\mathbb{R}$-valued and random recurrence relation*

$$X'_n = \sum_{k=1}^{p} a(k) X'_{n-k} + W_n$$

where $a = [a(1), \ldots, a(p)] \in \mathbb{R}^p$ is a deterministic vector, and W_n is a collection of independent and real-valued random variables. If we consider the p-length vector

$$X_n = \begin{bmatrix} X'_n \\ X'_{n-1} \\ \vdots \\ X'_{n-p+1} \end{bmatrix}$$

then we find that

$$X'_n = a\ X_{n-1} + W_n$$

from which we easily conclude that X_n is a p-dimensional Markov chain.

Example 2.2.5 (Queueing model) *Consider a single-line service queue in which one customer is served per unit of time. The random number of new arrivals at time n is specified by a distribution μ_n on $\mathbb{N}$. Assuming that these arrival numbers are independent, the Markov transitions of the queue length X_n at time n are given for any $j \geq 0$ and $i \geq 1$ by the formulae*

$$\mathbb{P}(X_{n+1} = (i-1) + j \mid X_n = i) = \mu_{n+1}(j) = \mathbb{P}(X_{n+1} = j \mid X_n = 0)$$

Example 2.2.6 (Urn model) *We consider n urn with black and white balls. At each time a ball is randomly chosen, and returned to the urn with an extra-ball of the same color. Let $X_n = (B_n, W_n)$ be the random numbers of black and white balls. It is easily checked that X_n is a Markov chain taking values in $\mathbb{N}^2$, and its transitions are given for any $b + w \geq 1$ by*

$$M((b, w), \cdot) = \frac{b}{b+w} \delta_{(b+1, w)} + \frac{w}{b+w} \delta_{(b, w+1)}$$

Example 2.2.7 (Branching model) *Consider an elementary population branching model in which each ith individual member produces at time n a random number of offsprings g_n^i. We assume that $(g_n^i)_{i \geq 1, n \geq 0}$ and independent and identically distributed random variables taking values in $\mathbb{N}$, and we let $\mathbb{E}(g_n^i) = g$ be the expected number of offsprings. The number of individuals at time n is a Markov chain X_n starting with a single individual $X_0 = 1$, taking values in $\mathbb{N}$, and it has the representation $X_n = \sum_{i=1}^{X_{n-1}} g_n^i$. Note that if $g < 1$, then we have*

$$\mathbb{E}(\sum_{n \geq 0} X_n) = 1/(1-g) \quad \text{and} \quad \mathbb{P}(\sum_{n \geq 0} X_n < \infty) = 1$$

In this case, X_n tends to 0 almost surely as n tends to infinity. When $m = 1$, X_n is a martingale that converge almost surely to the only possible value, namely 0 (∞ being clearly excluded). If $m > 1$, recalling that $\mathbb{P}(\lim_{p\to\infty} X_p = 0 \mid X_1 = j) = \mathbb{P}(\lim_{p\to\infty} X_p = 0 \mid X_0 = 1)^j$, the sequence of random variables

$$M_n = \mathbb{P}(\lim_{p\to\infty} X_p = 0)^{X_n}$$

is a martingale, and we have

$$\mathbb{P}(\lim_{p\to\infty} M_p = 1) = \mathbb{P}(\lim_{p\to\infty} X_p = 0), \quad \mathbb{P}(\lim_{p\to\infty} M_p = 0) = \mathbb{P}(\lim_{p\to\infty} X_p = \infty)$$

Example 2.2.8 (Independent path sequences) Let U_n, $n \in \mathbb{N}$, be a collection of independent random variables taking values in some measurable state spaces $(C_n, \mathcal{C}_n)$. The sequence of random variables defined by $X_n = (U_0, \ldots, U_n) \in E_n = (C_0 \times \ldots \times C_n)$ forms a Markov chain.

Example 2.2.9 (Excursion valued Markov chains) Let Y_n be a Markov chain taking values at each time $n \in \mathbb{N}$ on some measurable space $(S_n, \mathcal{S}_n)$. Also let T_n, $n \in \mathbb{N}$, be a collection of nondecreasing stopping times (with respect to the filtration $F_n = \sigma(Y_0, \ldots, Y_n)$ associated with Y_n) such that the pair sequence (T_n, Y_{T_n}) is a Markov chain on $E = \cup_{n\geq 0}(\{n\} \times S_n)$. For any $0 \leq p \leq n$ we write

$$Y_{[p,n]} = (Y_q)_{p \leq q \leq n} \in S_{[p,n]} = (S_p \times \ldots \times S_n)$$

the excursion of Y_q from time p up to time n. We can check that the following sequences are Markov chains:

$$(T_n, Y_{[T_{n-1}, T_n]}), \quad (T_n - T_{n-1}, Y_{[T_{n-1}, T_n]}) \quad \text{and} \quad (T_n - T_{n-1}, Y_{T_n})$$

Let $p \geq 1$ be a fixed integer parameter. We easily check that the random sequence

$$X_n^{(p)} = (Y_{pn}, Y_{pn+1}, \ldots, Y_{p(n+1)}) \in E_n^{(p)} = S_{[pn, p(n+1)]}$$

is a Markov chain. For any nondecreasing sequence of integers t_n, $n \in \mathbb{N}$, the random sequence $X_n = Y_{[t_n, t_{n+1}]} \in E_n = S_{[t_n, t_{n+1}]}$ also forms a Markov chain.

Example 2.2.10 (Restricted Markov chains) Let Y_n be a time-homogeneous Markov chain on some measurable space $(E, \mathcal{E})$. We further assume that $A \in \mathcal{E}$ is a recurrent set (in the sense that Y_n visits A infinitely often) and $Y_0 \in A$. We define the nondecreasing sequence of returns times to A

$$T_n = \inf\{p > T_{n-1} : Y_p \in A\} \quad \text{with} \quad T_0 = 0$$

The sequence $X_n = Y_{T_n}$ forms a time-homogeneous Markov chain taking values in A and its elementary transition is defined by

$$M(x, dy) = \mathbb{P}_x(Y_{T_1} \in dy)$$

2.3 Description of the Models

From the discussion given in the introduction of the book, we see that the Feynman-Kac path measures can be sought in many different ways. If we want to capture the full force of these models, it is therefore necessary to undertake their analysis in an abstract and nonhomogeneous setting. In this section, we introduce an abstract class of Feynman-Kac models in general nonhomogeneous state spaces. These models are built with two main ingredients: a Markov chain associated with a reference probability measure and a sequence of potential functions related to the mass repartition of the Feynman-Kac measures. Our first task is to introduce these two mathematical objects.

Let $(E_n, \mathcal{E}_n)$, $n \in \mathbb{N}$, be a collection of measurable spaces. We consider a collection of Markov transitions $M_n(x_{n-1}, dx_n)$ from E_{n-1} into E_n and a given probability measure $\mu \in \mathcal{P}(E_0)$. We associate with the latter a nonhomogeneous Markov chain

$$\left(\Omega = \prod_{n \geq 0} E_n, \ \mathcal{F} = (\mathcal{F}_n)_{n \in \mathbb{N}}, \ X = (X_n)_{n \in \mathbb{N}}, \ \mathbb{P}_\mu \right)$$

taking values at each time n on E_n, with elementary transitions M_n and initial distribution μ. When the initial distribution $\mu = \delta_x$ is concentrated at a single point $x \in E_0$, we simplify notation and we write $\mathbb{P}_x$ instead of $\mathbb{P}_{\delta_x}$. We use the notation $\mathbb{E}_\mu(.)$ and $\mathbb{E}_x(.)$ for the expectations with respect to $\mathbb{P}_\mu$ and $\mathbb{P}_x$. In this simplified notation, we notice that

$$\mathbb{P}_\mu(.) = \int_{E_0} \mu(dx) \ \mathbb{P}_x(.)$$

and for any $F_n \in \mathcal{B}_b(E_{[0,n]})$ we have

$$\mathbb{E}_\mu(F_n(X_0, \ldots, X_n)) = \int_{E_{[0,n]}} F_n(x_0, \ldots, x_n) \ \mathbb{P}_{\mu,n}(d(x_0, \ldots, x_n))$$

with the distribution $\mathbb{P}_{\mu,n}$ on $E_{[0,n]}$ given by

$$\mathbb{P}_{\mu,n}(d(x_0, \ldots, x_n)) = \mu(dx_0) \ M_1(x_0, dx_1) \ldots M_n(x_{n-1}, dx_n)$$

Let $G_n : E_n \to [0, \infty)$ be a given collection of bounded and $\mathcal{E}_n$-measurable nonnegative functions such that for any $n \in \mathbb{N}$

$$\mathbb{E}_\mu \left(\prod_{p=0}^{n} G_p(X_p) \right) > 0$$

Next we present the definitions of the Feynman-Kac models associated with the pair potential/kernel (G_n, M_n). We start with the traditional description of a path-space model. In reference to filtering literature, we adopt the following terminology.

2.3 Description of the Models

Definition 2.3.1 *The Feynman-Kac prediction and updated path models associated with the pair (G_n, M_n) (and the initial distribution μ) are the sequence of path measures defined respectively by*

$$\mathbb{Q}_{\mu,n}(d(x_0,\ldots,x_n)) = \frac{1}{\mathcal{Z}_n} \left\{ \prod_{p=0}^{n-1} G_p(x_p) \right\} \mathbb{P}_{\mu,n}(d(x_0,\ldots,x_n))$$

$$\widehat{\mathbb{Q}}_{\mu,n}(d(x_0,\ldots,x_n)) = \frac{1}{\widehat{\mathcal{Z}}_n} \left\{ \prod_{p=0}^{n} G_p(x_p) \right\} \mathbb{P}_{\mu,n}(d(x_0,\ldots,x_n))$$

(2.7)

for any $n \in \mathbb{N}$. The normalizing constants

$$\mathcal{Z}_n = \mathbb{E}_\mu \left(\prod_{p=0}^{n-1} G_p(X_p) \right) \quad \text{and} \quad \widehat{\mathcal{Z}}_n = \mathcal{Z}_{n+1} = \mathbb{E}_\mu \left(\prod_{p=0}^{n} G_p(X_p) \right)$$

are also often called the partition functions.

The measures $\mathbb{Q}_{\mu,n}$ and $\widehat{\mathbb{Q}}_{\mu,n}$ are alternatively defined for any test function $F_n \in \mathcal{B}_b(E_{[0,n]})$ by the formulae

$$\mathbb{Q}_{\mu,n}(F_n) = \frac{1}{\mathcal{Z}_n} \mathbb{E}_\mu \left(F_n(X_0,\ldots,X_n) \prod_{p=0}^{n-1} G_p(X_p) \right)$$

$$\widehat{\mathbb{Q}}_{\mu,n}(F_n) = \frac{1}{\widehat{\mathcal{Z}}_n} \mathbb{E}_\mu \left(F_n(X_0,\ldots,X_n) \prod_{p=0}^{n} G_p(X_p) \right)$$

This "weak" description of the measures in terms of the expectation $\mathbb{E}_\mu(.)$ with respect to the law of a reference Markov chain is more tractable than the previous one. Definition 2.3.1 shows the correspondence between Boltzmann-Gibbs and Feynman-Kac models. The difference between these two models concerns the role of the time parameter. In contrast to Boltzmann-Gibbs measures, the Feynman-Kac models have a particular dynamic structure. To get one step further in this discussion, it is convenient to introduce the flow of the time marginals.

Definition 2.3.2 *The sequence of bounded nonnegative measures γ_n and $\widehat{\gamma}_n$ on E_n defined for any $f_n \in \mathcal{B}_b(E_n)$ by*

$$\gamma_n(f_n) = \mathbb{E}_\mu \left(f_n(X_n) \prod_{p=0}^{n-1} G_p(X_p) \right)$$

and

$$\widehat{\gamma}_n(f_n) = \mathbb{E}_\mu \left(f_n(X_n) \prod_{p=0}^{n} G_p(X_p) \right)$$

are respectively called the unnormalized prediction and the updated Feynman-Kac model associated with the pair (G_n, M_n). The sequence of distributions η_n and $\widehat{\eta}_n$ on E_n defined for any $f_n \in \mathcal{B}_b(E_n)$ as

$$\eta_n(f_n) = \gamma_n(f_n)/\gamma_n(1) \quad \text{and} \quad \widehat{\eta}_n(f_n) = \widehat{\gamma}_n(f_n)/\widehat{\gamma}_n(1) \qquad (2.8)$$

are respectively called the normalized prediction and updated Feynman-Kac model associated with the pair (G_n, M_n).

There exist several ways to extend these formulae to a more general potential. The interested reader is referred to the book by A.S. Sznitman [295], and Section X.11 in M. Reeds and B. Simon [277], or the book by M. Nagasawa [254].

To better connect these objects, it is convenient to make a couple of remarks. First we observe that for $n = 0$ we have $\eta_0 = \gamma_0 = \mu \in \mathcal{P}(E_0)$. On the other hand, we have for any $n \in \mathbb{N}$

$$\widehat{\mathcal{Z}}_n = \mathcal{Z}_{n+1} = \widehat{\gamma}_n(1) = \gamma_{n+1}(1) = \mathbb{E}_\mu\left(\prod_{p=0}^n G_p(X_p)\right)$$

In this connection, we also observe that $\gamma_n(1) \leq \|G_n\|\, \gamma_{n-1}(1)$, from which we conclude that

$$\mathcal{Z}_n = \gamma_n(1) > 0 \iff \forall 0 \leq p \leq n \quad \gamma_p(1) > 0$$

We end this section with an important formula that relates the "unnormalized" models γ_n and $\widehat{\gamma}_n$ with the Feynman-Kac distribution flow η_p, $p \leq n$. We start by noting that

$$\gamma_n(f_n G_n) = \mathbb{E}_\mu\left(f_n(X_n)\, G_n(X_n) \prod_{p=0}^{n-1} G_p(X_p)\right) = \widehat{\gamma}_n(f_n)$$

By direct inspection, this yields

$$\widehat{\eta}_n(f_n) = \frac{\gamma_n(f_n G_n)}{\gamma_n(G_n)} = \frac{\gamma_n(f_n G_n)/\gamma_n(1)}{\gamma_n(G_n)/\gamma_n(1)} = \frac{\eta_n(f_n G_n)}{\eta_n(G_n)}$$

Thus, we are led to introduce the following transformation.

Definition 2.3.3 *The Boltzmann-Gibbs transformation associated with a potential function G_n on $(E_n, \mathcal{E}_n)$ is the mapping*

$$\Psi_n : \eta \in \mathcal{P}_n(E_n) \longrightarrow \Psi_n(\eta) \in \mathcal{P}_n(E_n)$$

from the subset $\mathcal{P}_n(E_n) = \{\eta \in \mathcal{P}(E_n)\,;\, \eta(G_n) > 0\}$ into itself and defined for any $\eta \in \mathcal{P}(E_n)$ by the Boltzmann-Gibbs measure

$$\Psi_n(\eta)(dx_n) = \frac{1}{\eta(G_n)}\, G_n(x_n)\, \eta(dx_n)$$

Note that the Boltzmann-Gibbs transformation is well-defined as soon as G_n is not the null function. In this notation, we see that

$$\widehat{\eta}_n = \Psi_n(\eta_n) \qquad (2.9)$$

In the reverse angle, the distribution flow η_n is connected to $\widehat{\eta}_{n-1}$ by the formula

$$\eta_n = \widehat{\eta}_{n-1} M_n \qquad (2.10)$$

To see this claim we simply use the Markov property of X_n in the definition of η_n to check that

$$\begin{aligned}\gamma_n(f_n) &= \mathbb{E}_\mu\left(f_n(X_n) \prod_{p=0}^{n-1} G_p(X_p)\right) \\ &= \mathbb{E}_\mu\left(M_n(f_n)(X_{n-1}) \prod_{p=0}^{n-1} G_p(X_p)\right) = \widehat{\gamma}_{n-1}(M_n(f_n))\end{aligned}$$

From this we find that

$$\begin{aligned}\eta_n(f_n) &= \gamma_n(f_n)/\gamma_n(1) \\ &= \widehat{\gamma}_{n-1}(M_n(f_n))/\widehat{\gamma}_{n-1}(1) = \widehat{\eta}_{n-1} M_n(f_n)\end{aligned}$$

We are now in a position to state the announced formula, whose proof is left as a simple exercise.

Proposition 2.3.1 *For any $n \in \mathbb{N}$ and for any $f_n \in \mathcal{B}_b(E_n)$, we have*

$$\gamma_n(f_n) = \eta_n(f_n) \prod_{p=0}^{n-1} \eta_p(G_p) \quad \text{and} \quad \widehat{\gamma}_n(f_n) = \widehat{\eta}_n(f_n) \prod_{p=0}^{n} \eta_p(G_p)$$

2.4 Structural Stability Properties

Feynman-Kac models have various structural stability properties. In Section 2.4.1, we will see that the path measures and their time marginals have the same algebraic structure, provided the potential energy of a path only depends on the current state. We will use this property to define genealogical tree-based approximations of Feynman-Kac path measures. In Section 2.4.2, we describe the class of Feynman-Kac flows that is connected to a given reference model by a change of probability measure. Finally, Section 2.4.3 connects updated and prediction models.

These three structural stability properties are not of pure mathematical interest. They confer to these formulae a stable and rich algebraic structure that allows direct transfer of many known results on marginal or prediction models to path-space formulae or updated flows.

2.4.1 Path Space and Marginal Models

Let X_n be a nonhomogeneous Markov chain with Markov transitions M_{n+1} from E_n into E_{n+1} and initial distribution $\mu \in \mathcal{P}(E_0)$. Also let G_n be a given collection of bounded measurable nonnegative functions on E_n such that for any $n \in \mathbb{N}$, $\mathbb{E}_\mu(\prod_{p=0}^n G_p(X_p)) > 0$, where $\mathbb{E}_\mu(.)$ represents the expectation with respect to the distribution $\mathbb{P}_\mu$ of X_n. We consider the Feynman-Kac path measures $\mathbb{Q}_{\mu,n}$ associated with the pair (G_n, M_n) and defined by

$$\mathbb{Q}_{\mu,n}(d(x_0,\ldots,x_n)) = \frac{1}{\mathcal{Z}_n} \left\{ \prod_{p=0}^{n-1} G_p(x_p) \right\} \mathbb{P}_{\mu,n}(d(x_0,\ldots,x_n))$$

where $\mathbb{P}_{\mu,n}$ denotes the probability measure of the path $(X_0,\ldots,X_n)$

$$\mathbb{P}_{\mu,n}(d(x_0,\ldots,x_n)) = \mu(dx_0)\, M_1(x_0, dx_1)\ldots M_n(x_{n-1}, dx_n)$$

We further suppose that

$$X_n = X'_{[0,n]}(= (X'_0,\ldots,X'_n)) \in E_n = E'_{[0,n]}(= (E'_0 \times \ldots \times E'_n))$$

represents the path process associated with an E'_n-valued Markov chain X'_n with Markov transitions M'_{n+1} from E'_n into E'_{n+1}. By construction, we notice that the initial random variable $X'_0 = X_0$ is distributed according to $\mu \in \mathcal{P}(E_0)(= \mathcal{P}(E'_0))$. In this situation, the unnormalized prediction measures γ_n on E_n have the form

$$\gamma_n(f_n) = \mathbb{E}_\mu \left(f_n(X'_{[0,n]}) \prod_{p=0}^{n-1} G_p(X'_{[0,p]}) \right)$$

As a result, the corresponding normalized distributions η_n are given by

$$\eta_n(d(x'_0,\ldots,x'_n)) = \frac{1}{\gamma_n(1)} \left\{ \prod_{p=0}^{n-1} G_p(x'_0,\ldots,x'_p) \right\} \mathbb{P}'_{\mu,n}(d(x'_0,\ldots,x'_n))$$

where $\mathbb{P}'_{\mu,n} = \mathbb{P}_\mu \circ (X'_{[0,n]})^{-1}$ stands for the distribution of the path of X' from the origin up to time n. In other words

$$\mathbb{P}'_{\mu,n}(d(x'_0,\ldots,x'_n)) = \mu(dx'_0) M'_1(x'_0, dx'_1)\ldots M'_n(x'_{n-1}, dx'_n)$$

Next we examine the situation where the potential functions G_n only depend on the terminal point of the path. That is, we have that

$$G_n : x_n = (x'_0,\ldots,x'_n) \in E_n \to G_n(x'_0,\ldots,x'_n) = G'_n(x'_n)$$

In this case, we readily check that the n-time marginal distribution η_n of the path measure $\mathbb{Q}_{\mu,n}$ coincides with the Feynman-Kac path measure $\mathbb{Q}'_{\mu,n}$

2.4 Structural Stability Properties 63

associated with the pair (G'_n, M'_n). More precisely, we have that

$$\begin{aligned}
\eta_n(d(x'_0, \ldots, x'_n)) &= \mathbb{Q}'_{\mu,n}(d(x'_0, \ldots, x'_n)) \\
&= \frac{1}{\mathcal{Z}'_n} \left\{ \prod_{p=0}^{n-1} G'_p(x'_p) \right\} \mathbb{P}'_{\mu,n}(d(x'_0, \ldots, x'_n))
\end{aligned}$$
(2.11)

with the same partition functions $\mathcal{Z}'_n = \mathcal{Z}_n = \gamma_n(1) \; (> 0)$.

Moreover, their nth time marginals η'_n are again defined for any test function $f'_n \in \mathcal{B}_b(E'_n)$ by the Feynman-Kac formulae

$$\eta'_n(f'_n) = \gamma'_n(f'_n)/\gamma'_n(1) \quad \text{with} \quad \gamma'_n(f'_n) = \mathbb{E}_\mu \left(f'_n(X'_n) \prod_{p=0}^{n-1} G'_p(X'_p) \right)$$

We will use these structural properties of Feynman-Kac models in several places in this book. Given a reference Feynman-Kac distribution model η_n, we will use the notation $\mathbb{Q}_n$ to represent the corresponding path-distribution model, and whenever η_n is already a path measure we will denote by η'_n the marginal distribution flow. We summarize the preceding discussion with the following synthetic diagram:

$$\mathbb{Q}_n \quad \xleftarrow{\text{path measure}} \quad \eta_n \quad \xrightarrow{\text{marginal measure}} \quad \eta'_n$$

2.4.2 Change of Reference Probability Measures

In this section, we describe a class of Feynman-Kac models that are equivalent by a change of reference measure on the canonical space

$$\left(\Omega = \prod_{n \geq 0} E'_n, \; \mathcal{F}' = (\mathcal{F}'_n)_{n \in \mathbb{N}}, \; X' = (X'_n)_{n \in \mathbb{N}} \right)$$

where $(E'_n, \mathcal{E}'_n)$, $n \in \mathbb{N}$, is a given collection of measurable spaces. Let μ and $\overline{\mu}$ be two distributions on E_0 with $\mu \ll \overline{\mu}$. Also let $M'_n(x_{n-1}, dx_n)$ and $\overline{M}'_n(x_{n-1}, dx_n)$ be collections of Markov transitions from E'_{n-1} into E'_n such that for any $x_{n-1} \in E'_{n-1}$

$$M'_n(x_{n-1}, dx_n) \ll \overline{M}'_n(x_{n-1}, dx_n)$$

As in the beginning of Section 2.3, we associate with the pairs (μ, M'_n) and $(\overline{\mu}, \overline{M}'_n)$ the laws $\mathbb{P}_\mu$ and $\overline{\mathbb{P}}_{\overline{\mu}}$ on the canonical space of two nonhomogeneous Markov chains with the n-time marginals

$$\mathbb{P}'_{\mu,n} = \mathbb{P}'_\mu \circ (X'_{[0,n]})^{-1} \quad \text{and} \quad \overline{\mathbb{P}}'_{\overline{\mu},n} = \overline{\mathbb{P}}'_{\overline{\mu}} \circ (X'_{[0,n]})^{-1}$$

given by

$$\mathbb{P}'_{\mu,n}(d(x_0,\ldots,x_n)) = \mu(dx_0) M'_1(x_0, dx_1) \ldots M'_n(x_{n-1}, dx_n)$$
$$\overline{\mathbb{P}}'_{\overline{\mu},n}(d(x_0,\ldots,x_n)) = \overline{\mu}(dx_0) \overline{M}'_1(x_0, dx_1) \ldots \overline{M}'_n(x_{n-1}, dx_n)$$

Under our assumptions, the measure $\mathbb{P}'_\mu$ is locally absolutely continuous with respect to $\overline{\mathbb{P}}'_\mu$ and for each $n \geq 0$ and for $\overline{\mathbb{P}}'_{\mu,n}$ almost every sequence $(x_0,\ldots,x_n) \in E'_{[0,n]}$ we have

$$\frac{d\mathbb{P}'_{\mu,n}}{d\overline{\mathbb{P}}'_{\mu,n}}(x_0,\ldots,x_n) = \frac{d\mu}{d\overline{\mu}}(x_0) \prod_{p=1}^{n} \frac{dM'_p(x_{p-1},\cdot)}{d\overline{M}'_p(x_{p-1},\cdot)}(x_p)$$

We use the notation $\mathbb{E}_\mu(.)$ and $\overline{\mathbb{E}}_{\overline{\mu}}(.)$ for the expectation with respect to $\mathbb{P}_\mu$ and $\overline{\mathbb{P}}_{\overline{\mu}}$. We denote by

$$X_n = X'_{[0,n]} \in E_n = E'_{[0,n]}$$

the historical process of the chain X'_n and by M_n and $\overline{M}_n$ the corresponding Markov transitions under the reference measures $\mathbb{P}_\mu$ and $\overline{\mathbb{P}}_{\overline{\mu}}$. Finally, let G_n be a given collection of potential functions

$$G_n : (x_0,\ldots,x_n) \in E_n = E'_{[0,n]} \longrightarrow G_n(x_0,\ldots,x_n) \in [0,\infty)$$

such that for any $n \geq 0$

$$\widehat{\gamma}_n(1) = \mathbb{E}_\mu \left(\prod_{p=0}^{n} G_p(X_p) \right) = \mathbb{E}_\mu \left(\prod_{p=0}^{n} G_p(X'_{[0,p]}) \right) > 0$$

We recall that the updated Feynman-Kac distribution flow model associated with the pair (G_n, M_n) is given for any $f_n \in \mathcal{B}_b(E_n)$ by

$$\widehat{\eta}_n(f_n) = \widehat{\gamma}_n(f_n)/\widehat{\gamma}_n(1)$$

with

$$\widehat{\gamma}_n(f_n) = \mathbb{E}_\mu \left(f_n(X'_{[0,n]}) \prod_{p=0}^{n} G_p(X'_{[0,p]}) \right)$$

By construction, we have that

$$\widehat{\gamma}_n(f_n) = \overline{\mathbb{E}}_{\overline{\mu}} \left(f_n(X'_{[0,n]}) \prod_{p=0}^{n} \overline{G}_p(X'_{[0,p]}) \right)$$

with the potential function $\overline{G}_n$ on $E'_0 \times \ldots \times E'_n$ defined by

$$\overline{G}_n(x_0,\ldots,x_n) = G_n(x_0,\ldots,x_n) \times \frac{dM'_n(x_{n-1},\cdot)}{d\overline{M}'_n(x_{n-1},\cdot)}(x_n)$$

and the convention for $n = 0$, $M'_0(x_{-1}, .) = \mu$, and $\overline{M}'_0(x_{-1}, .) = \overline{\mu}$ so that

$$\overline{G}_0(x_0) = G_0(x_0) \, \frac{d\mu}{d\overline{\mu}}(x_0)$$

From these observations, we prove easily the following proposition.

Proposition 2.4.1 *The prediction and updated Feynman-Kac models associated with the pairs (G_n, M_n) and $(\overline{G}_n, \overline{M}_n)$ coincide.*

Each representation of a Feynman-Kac model in terms of a pair (G_n, M_n) will correspond to a different physical interpretation and will lead to different particle approximation models. As we shall see in the forthcoming sections, M_n correspond to the elementary transitions of a random Markov particle evolving in an environment with absorbing potential G_n.

We end this section we an elementary Feynman-Kac formula which allows to change the potential functions without changing the underlying Markov chain. Let G_n and $\overline{G}_n$ be a sequence of positive potential functions on the state spaces $E'_{[0,n]}$. We further assume that the ratio function $G_n/\overline{G}_n$ is a well defined bounded function. We associate to the pair $(G_n, \overline{G}_n)$ the transformations $\pi^G_{n,\overline{G}}$ from $\mathcal{B}_b(E'_{[0,n]})$ into itself defined by

$$\pi^G_{n,\overline{G}}(f_n)(x_0, \ldots, x_n) = f_n(x_0, \ldots, x_n) \prod_{p=0}^{n} \frac{G_p(x_0, \ldots, x_p)}{\overline{G}_p(x_0, \ldots, x_p)}$$

It is now readily checked that for any $f \in \mathcal{B}_b(E'_{[0,n]})$ we have the equivalent formulations

$$\widehat{\gamma}^G_n(f_n) =_{\text{def.}} \mathbb{E}_\mu \left(f_n(X'_{[0,n]}) \prod_{p=0}^{n} G_p(X'_{[0,p]}) \right) = \widehat{\gamma}^{\overline{G}}_n(\pi^G_{n,\overline{G}}(f_n)) \quad (2.12)$$

From the numerical point of view, the choice of the pair (G_n, M_n) is related to some physical knowledge of the probability mass evolution of the Feynman-Kac models. In some instances, a judicious choice of Markov transitions will drive the particles in the regions with high probability mass. Proposition 2.4.1 gives the way to change the potential functions accordingly. In other instances, the choice of the absorbing potential functions may induce a too crude selection transition. In this case formula (2.12) gives the way to choose the potential function with changing the test function.

2.4.3 Updated and Prediction Flow Models

In this short section, we discuss the connections between the updated and prediction flow models introduced in Section 2.3. We further require that the pairs (G_n, M_n) satisfy for any $x_n \in E_n$ and $n \in \mathbb{N}$ the following condition:

$$\widehat{G}_n(x_n) = M_{n+1}(G_{n+1})(x_n) \in (0, \infty) \tag{2.13}$$

66   2. Feynman-Kac Formulae

In this situation, the integral operators

$$\widehat{M}_n(x_{n-1}, dx_n) = \frac{M_n(x_{n-1}, dx_n)\, G_n(x_n)}{M_n(G_n)(x_{n-1})} \qquad (2.14)$$

are well-defined Markov kernels from E_{n-1} into E_n. Also notice that in this case $\widehat{\eta}_0 = \Psi_0(\eta_0)$ is a well-defined distribution on E_0 provided $\eta_0(G_0) > 0$.

We associate with $\widehat{\eta}_0$ and $\widehat{M}_n$ the probability distribution $\widehat{\mathbb{P}}_{\widehat{\eta}_0}$ of a canonical Markov chain with initial distribution $\widehat{\eta}_0$ and elementary transitions $\widehat{M}_n$. By construction, the nth time marginals $\widehat{\mathbb{P}}_{\widehat{\eta}_0,n}$ of $\widehat{\mathbb{P}}_{\widehat{\eta}_0}$ are given by

$$\widehat{\mathbb{P}}_{\widehat{\eta}_0,n}(d(x_0, \ldots, x_n)) = \widehat{\eta}_0(dx_0)\widehat{M}_1(x_0, dx_1) \ldots \widehat{M}_n(x_{n-1}, dx_n) \qquad (2.15)$$

The Feynman-Kac path measures associated with these new pairs of potential/kernel $(\widehat{G}_n, \widehat{M}_n)$ are now defined by

$$\widehat{\mathbb{Q}}_{\eta_0,n}(d(x_0, \ldots, x_n)) = \frac{1}{\widetilde{Z}_n} \left\{ \prod_{p=0}^{n-1} \widehat{G}_p(x_p) \right\} \widehat{\mathbb{P}}_{\widehat{\eta}_0,n}(d(x_0, \ldots, x_n))$$

Under condition (2.13), we prove that the normalizing constants $\widetilde{Z}_n$ are always strictly positive. It is also easily proved that $\widehat{\mathbb{Q}}_{\eta_0,n}$ can alternatively be written in terms of the original pairs (G_n, M_n) with the formula

$$\widehat{\mathbb{Q}}_{\eta_0,n}(d(x_0, \ldots, x_n)) = \frac{1}{\widehat{Z}_n} \left\{ \prod_{p=0}^{n} G_p(x_p) \right\} \mathbb{P}_{\eta_0,n}(d(x_0, \ldots, x_n))$$

and the normalizing constant $\widehat{Z}_n = \eta_0(G_0)\, \widetilde{Z}_n > 0$. From this simple observation, we prove the following proposition.

Proposition 2.4.2 *The updated Feynman-Kac models $\widehat{\mathbb{Q}}_{\eta_0,n}$ and $\widehat{\eta}_n$ associated with the pair (G_n, M_n) coincide with the prediction Feynman-Kac models associated with the pair $(\widehat{G}_n, \widehat{M}_n)$ and starting at $\widehat{\eta}_0 = \Psi_0(\eta_0)$. We have for any $n \in \mathbb{N}$, $f_n \in \mathcal{B}_b(E_n)$, and $F_n \in \mathcal{B}_b(E_{[0,n]})$*

$$\widehat{\mathbb{Q}}_{\eta_0,n}(F_n) = \frac{\widehat{\mathbb{E}}_{\widehat{\eta}_0}(F_n(X_0, \ldots, X_n)\, \prod_{p=0}^{n-1} \widehat{G}_p(X_p))}{\widehat{\mathbb{E}}_{\widehat{\eta}_0}(\prod_{p=0}^{n-1} \widehat{G}_p(X_p))}$$

and $\widehat{\eta}_n(f_n) = \widehat{\mathbb{E}}_{\widehat{\eta}_0}(f_n(X_n)\, \prod_{p=0}^{n-1} \widehat{G}_p(X_p))/\widehat{\mathbb{E}}_{\widehat{\eta}_0}(\prod_{p=0}^{n-1} \widehat{G}_p(X_p))$.

This stability property indicates one way to transfer results between updated and prediction models. In this connection, we already mentioned that in some instances it is more judicious to interpret $\widehat{\eta}_n$ as the prediction flow associated with the pair $(\widehat{G}_n, \widehat{M}_n)$ and starting at $\widehat{\eta}_0$. To illustrate this assertion, we examine hereafter the situation where the potential function G_n may take some null values and we set

$$\widehat{E}_n = \{x_n \in E_n\,;\, G_n(x_n) > 0\}$$

It may happen that the set $\widehat{E}_n$ is not M_n-accessible from any point in E_{n-1}. In this case, we may have $M_n(x_{n-1}, \widehat{E}_n) = 0$ for some $x_{n-1} \in E_{n-1}$ and therefore $M_n(G_n)(x_{n-1}) = 0$. In this situation, the Markov kernel $\widehat{M}_n$ introduced in (2.14) is not well-defined on the whole set E_{n-1}. This irregularity property creates some technical difficulties in defining properly the dynamical structure of the corresponding Feynman-Kac models. Next we weaken (2.13) and we consider the condition

$$(\mathcal{A}) \quad \forall\, x_n \in \widehat{E}_n \quad M_{n+1}(x_n, \widehat{E}_{n+1}) > 0 \quad \text{and} \quad \eta_0(\widehat{E}_0) > 0 \qquad (2.16)$$

In words, this condition says that the set $\widehat{E}_{n+1}$ is accessible from any point in $\widehat{E}_n$. In this case, we readily check that condition (2.13) is only met for any $x_n \in \widehat{E}_n$. More precisely, we have for any $n \in \mathbb{N}$ and $x_n \in \widehat{E}_n$

$$\widehat{G}_n(x_n) = M_{n+1}(G_{n+1})(x_n) \in (0, \infty)$$

More interestingly, in this situation the integral operators $\widehat{M}_n$ defined for any $x_{n-1} \in \widehat{E}_{n-1}$ by

$$\widehat{M}_n(x_{n-1}, dx_n) = \frac{M_n(x_{n-1}, dx_n)\, G_n(x_n)}{M_n(G_n)(x_{n-1})}$$

are well-defined Markov kernels from $\widehat{E}_{n-1}$ into $\widehat{E}_n$. Finally we note that, for any $\eta_0 \in \mathcal{P}(E_0)$ with $\eta_0(\widehat{E}_0) > 0$, the updated measure $\widehat{\eta}_0 = \Psi_0(\eta_0)$ is such that $\widehat{\eta}_0(\widehat{E}_0) = 1$. From the preceding discussion we see that the distributions $\widehat{\mathbb{P}}_{\widehat{\eta}_0}$ defined in (2.15) are such that $\widehat{\mathbb{P}}_{\widehat{\eta}_0}(\widehat{E}_{[0,n]}) = 1$. In addition, the distribution on $\widehat{\Omega}_n = \widehat{E}_{[0,n]}$ defined by

$$\widehat{\mathbb{P}}_{\widehat{\eta}_0, n}(d(x_0, \ldots, x_n)) = \widehat{\eta}_0(dx_0) \widehat{M}_1(x_0, dx_1) \ldots \widehat{M}_n(x_{n-1}, dx_n)$$

can be extended by consistency arguments to the whole canonical space

$$\left(\widehat{\Omega} = \prod_{n \geq 0} \widehat{E}_n,\ \widehat{\mathcal{F}} = (\widehat{\mathcal{F}}_n)_{n \in \mathbb{N}},\ \widehat{X} = (\widehat{X}_n)_{n \in \mathbb{N}} \right)$$

Under $\widehat{\mathbb{P}}_{\widehat{\eta}_0}$, the canonical process is a Markov chain $\widehat{X}_n$ with initial distribution $\widehat{\eta}_0$ and elementary transitions $\widehat{M}_n$ from $\widehat{E}_{n-1}$ into $\widehat{E}_n$. Summarizing the discussion above, we get the following proposition.

Proposition 2.4.3 *When the accessibility condition* $(\mathcal{A})$ *is met, the updated Feynman-Kac measures* $\widehat{\mathbb{Q}}_{\eta_0, n} \in \mathcal{P}(\widehat{E}_{[0,n]})$ *and* $\widehat{\eta}_n \in \mathcal{P}(\widehat{E}_n)$ *can be interpreted as the prediction models associated with the pair potential/kernel* $(\widehat{G}_n, \widehat{M}_n)$ *on the restricted state spaces* $(\widehat{E}_n, \widehat{\mathcal{E}}_n)$.

For instance, if we choose $G_n = 1_{\widehat{E}_n}$, then under $\widehat{\mathbb{P}}_{\widehat{\eta}_0}$ the canonical process is the "local" restriction of the original chain with transition M_n to the sets $\widehat{E}_n$. That is, we have for any $x_{n-1} \in \widehat{E}_{n-1}$

$$\widehat{M}_n(x_{n-1}, dx_n) = \frac{M_n(x_{n-1}, dx_n)\, 1_{\widehat{E}_n}(x_n)}{M_n(x_{n-1}, \widehat{E}_n)}$$

2.5 Distribution Flows Models

In previous sections, we have introduced a variety of Feynman-Kac modeling techniques. We have underlined several structural stability properties and the interplay between these distributions. In the present section, we provide two different physical interpretations. The first one is the traditional trapping interpretation. The leading idea consists in turning a sub-Markov property into the Markov situation by adding a cemetery or coffin state to the state spaces. In this context, the Feynman-Kac models represent the conditional probabilities of a nonabsorbed Markov particle evolving in an environment with obstacles. One drawback of this physical interpretation is that the resulting killed Markov model is defined on a different canonical probability space and the particle motion is instantly stopped as soon as it is trapped by an obstacle.

In the second part of this section, we adopt a different point of view. This alternative physical interpretation is based on measure-valued and interacting processes ideas. Loosely speaking, instead of adding an auxiliary cemetery point to the state space, when the particle dies then instantly a new particle is created at a site randomly chosen according to the current distribution of the model. This interacting jump process is defined on the same original canonical space by changing the reference probability measure by an appropriate McKean measure. Under the latter, the Feynman-Kac measures represent the distribution laws of a birth and death process. This interacting process interpretation will be the stepping stone of our construction of particle and genealogical approximation models.

To clarify the presentation, we will assume that the potential functions are strictly positive; that is, we have for any $x_n \in E_n$, $G_n(x_n) > 0$. As noted in Section 2.4.3, when the potential functions G_n are not strictly positive, we have to carry out a more careful analysis. In this situation, it is often more convenient and natural to work on the state spaces

$$\widehat{E}_n = \{x_n \in E_n \;;\; G_n(x_n) > 0\}$$

with the potential functions

$$x_n \in \widehat{E}_n \longrightarrow \widehat{G}_n(x_n) = M_{n+1}(G_{n+1})(x_n)$$

This strategy works when the accessibility condition $(\mathcal{A})$ presented in (2.16) is met. In the final part of this section, we will discuss the difficulties that can arise when ($\mathcal{A}$) is not met. The main advantage of the preceding condition is that the updated model $\widehat{\eta}_n$ can be regarded as a distribution flow on $\mathcal{P}(\widehat{E}_n)$. More precisely, it can be interpreted as a prediction model with transitions $\widehat{M}_n$ from $\widehat{E}_{n-1}$ into $\widehat{E}_n$ and potential functions $\widehat{G}_n$ on $\widehat{E}_n$ such that for any $x_n \in \widehat{E}_n$, $\widehat{G}_n(x_n) > 0$. On the other hand, since the potential functions G_n are assumed to be bounded, we can replace in the definition of the normalized measures $\eta_n, \widehat{\eta}_n$ the functions G_n by $G_n/\|G_n\|$ without altering their nature. From all these observations, we see that there is no great loss of generality in considering potential functions G_n in the half unit ball. Unless otherwise stated, in this section we will assume that $0 < G_n(x_n) \leq 1$.

Definition 2.5.1 *We identify the potential functions G_n with the Boltzmann multiplicative operator $\mathcal{G}_n$ on $\mathcal{B}_b(E_n)$ defined for any $f_n \in \mathcal{B}_b(E_n)$ and $x_n \in E_n$ by the equation*

$$\mathcal{G}_n(f_n)(x_n) = G_n(x_n) \, f_n(x_n)$$

We can alternatively see $\mathcal{G}_n$ as the integral operator on E_n defined by

$$\mathcal{G}_n(x_n, dy_n) = G_n(x_n) \, \delta_{x_n}(dy_n)$$

In this connection, we see that $\mathcal{G}_n$ is a sub-Markov kernel

$$\mathcal{G}_n(x_n, E_n) = G_n(x_n) \leq 1$$

We continue our program, noting that the positive measures γ_n satisfy a linear equation of the form

$$\gamma_n = \gamma_{n-1} Q_n \qquad (2.17)$$

where Q_n are the bounded nonnegative operators on $\mathcal{B}_b(E_n)$ defined by $Q_n = \mathcal{G}_{n-1} M_n$. To emphasize the role of each quantity we also note that

$$Q_n(f_n) = G_{n-1} \, M_n(f_n) \quad \text{and} \quad 0 < Q_n(1) = G_{n-1} \leq 1$$

The right-hand side in the last display shows that the sub-Markov property of $\mathcal{G}_n$ is transferred to Q_n. In the next two propositions, we have collected some structural properties of Feynman-Kac flows.

Proposition 2.5.1 *The unnormalized prediction and updated Feynman-Kac measures γ_n and $\widehat{\gamma}_n$ associated with the pair (G_n, M_n) satisfy the linear recursive equations $\gamma_n = \gamma_{n-1} Q_n$ and $\widehat{\gamma}_n = \widehat{\gamma}_{n-1} \widehat{Q}_n$ with the bounded nonnegative operators defined*

$$Q_n = \mathcal{G}_{n-1} M_n \quad \text{and} \quad \widehat{Q}_n = M_n \mathcal{G}_n$$

Proposition 2.5.2 *The normalized prediction and updated Feynman-Kac distributions η_n and $\widehat{\eta}_n$ associated with the pair (G_n, M_n) satisfy the nonlinear recursive equations $\eta_n = \Phi_n(\eta_{n-1})$ and $\widehat{\eta}_n = \widehat{\Phi}_n(\widehat{\eta}_{n-1})$ with the mappings Φ_n and $\widehat{\Phi}_n$ from $\mathcal{P}(E_{n-1})$ into $\mathcal{P}(E_n)$ defined for any $\eta \in \mathcal{P}(E_{n-1})$ by*

$$\Phi_n(\eta) = \Psi_{n-1}(\eta) M_n \quad \text{and} \quad \widehat{\Phi}_n(\eta) = \Psi_n(\eta M_n) = \widehat{\Psi}_{n-1}(\eta) \widehat{M}_n$$

In the last display, Ψ_n and $\widehat{\Psi}_n$ denote the Boltzmann-Gibbs transformations on $\mathcal{P}(E_n)$ given by

$$\Psi_n(\mu)(dx_n) = \frac{G_n(x_n)}{\mu(G_n)} \mu(dx_n) \quad \text{and} \quad \widehat{\Psi}_n(\mu)(dx_n) = \frac{\widehat{G}_n(x_n)}{\mu(\widehat{G}_n)} \mu(dx_n)$$

The pair potentials/kernels $(\widehat{G}_n, \widehat{M}_n)$ are defined for any $f_n \in \mathcal{B}_b(E_n)$ by the formulae

$$\widehat{G}_n = M_{n+1}(G_{n+1}) \quad \text{and} \quad \widehat{M}_n(f_n) = M_n(f_n G_n)/M_n(G_n)$$

We end this section with a discussion of the case where G_n may take null values. First, we recall that the accessibility condition $(\mathcal{A})$ ensures that the Feynman-Kac normalizing constants are always well-defined. To prove this assertion, we can use the interpretation of the flow $\widehat{\eta}_n$ as the prediction flow associated with the pair $(\widehat{G}_n, \widehat{M}_n)$ and check that

$$\mathbb{E}_{\eta_0}\left(f_n(X_n) \prod_{p=0}^{n} G_p(X_p)\right) = \eta_0(G_0)\, \widehat{\mathbb{E}}_{\widehat{\eta}_0}\left(f_n(X_n) \prod_{p=0}^{n-1} \widehat{G}_p(X_p)\right) > 0$$

This shows in particular that for any $n \in \mathbb{N}$ we have

$$\eta_n \in \mathcal{P}_n(E_n) = \{\eta \in \mathcal{P}(E_n)\,;\, \eta(G_n) > 0\}$$

and the Feynman-Kac flow is a well-defined two-step updating/prediction model

$$\eta_n \in \mathcal{P}_n(E_n) \xrightarrow{\text{updating}} \widehat{\eta}_n \in \mathcal{P}(\widehat{E}_n) \xrightarrow{\text{prediction}} \eta_{n+1} \in \mathcal{P}_{n+1}(E_{n+1})$$

When the accessibility condition $(\mathcal{A})$ is not met, then the set $\widehat{E}_{n+1}$ is not M_{n+1}-accessible from any point in $\widehat{E}_n$, and it may happen that

$$\widehat{\eta}_n M_{n+1}(G_{n+1}) = \eta_{n+1}(G_{n+1}) = 0$$

In this situation, the Feynman-Kac flow η_n is well-defined up to the first time τ we have $\eta_\tau(G_\tau) = 0$. At time τ, the measure η_τ cannot be updated anymore. Recalling that $\eta_\tau(G_\tau) = \gamma_{\tau+1}(1)/\gamma_\tau(1)$, we also see that τ coincides with the first time the Feynman-Kac normalizing constants become null, that is

$$\gamma_{\tau+1}(1) = \mathbb{E}_{\eta_0}\left(\prod_{p=0}^{\tau} G_p(X_p)\right) = 0$$

2.5.1 Killing Interpretation

The first way to turn the sub-Markovian kernels $\mathcal{G}_n$ into the Markov case consists in adding a common cemetery point c to the state spaces E_n and in extending the various quantities as follows.

- The test functions $f_n \in \mathcal{B}_b(E_n)$ and the potential functions G_n are first extended to $E_n^c = E_n \cup \{c\}$ by setting $f_n(c) = 0 = G_n(c)$.

- The Markov transitions M_{n+1} from E_n into E_{n+1} are extended to transitions M_{n+1}^c from E_n^c into E_{n+1}^c by setting $M_{n+1}^c(c,.) = \delta_c$ and for each $x_n \in E_n$ $M_{n+1}^c(x_n, dx_{n+1}) = M_{n+1}(x_n, dx_{n+1})$.

- Finally, the Markov extension $\mathcal{G}_n^c$ of $\mathcal{G}_n$ on $E_n \cup \{c\}$ is given by

$$\mathcal{G}_n^c(x_n, dy_n) = G_n(x_n)\,\delta_{x_n}(dy_n) + (1 - G_n(x_n))\,\delta_c(dy_n) \qquad (2.18)$$

Note that for any $x_{n-1} \in E_{n-1}$ and $A_n \in \mathcal{E}_n$ we have

$$Q_n^c(x_{n-1}, A_n) = G_{n-1}(x_{n-1})\,M_n(x_{n-1}, A_n)$$

The corresponding Markov chain

$$\left(\Omega^c = \prod_{n\geq 0} E_n^c, \mathcal{F}^c = (\mathcal{F}_n^c)_{n\geq 0}, X = (X_n)_{n\geq 0}, \mathbb{P}_\mu^c\right)$$

with initial distribution $\mu \in \mathcal{P}(E_0)$ and elementary transitions

$$Q_{n+1}^c = \mathcal{G}_n^c M_{n+1}^c \qquad (2.19)$$

can be regarded as a Markov particle evolving in an environment with absorbing obstacles related to potential functions G_n. In view of (2.19), we see that the motion is decomposed into two separate killing/exploration transitions:

$$X_n \xrightarrow{\text{killing}} \widehat{X}_n \xrightarrow{\text{exploration}} X_{n+1}$$

This killing/exploration mechanism represents the overlapping of the two elementary transitions $\mathcal{G}_n^c$ and M_n^c. They are defined as follows:

- **Killing:** If $X_n = c$, we set $\widehat{X}_n = c$. Otherwise the particle X_n is still alive. In this case, with a probability $G_n(X_n)$ it remains in the same site so that $\widehat{X}_n = X_n$, and with a probability $1 - G_n(X_n)$ it is killed and we set $\widehat{X}_n = c$.

- **Exploration:** First, since there is probably no life after death when the particle has been killed, we have $\widehat{X}_n = c$ and we set $\widehat{X}_p = X_p = c$ for any $p > n$. Otherwise the particle $\widehat{X}_n \in E_n$ evolves to a new location X_{n+1} in E_{n+1} randomly chosen according to the distribution $M_{n+1}(X_n, .)$.

In this physical interpretation, the Feynman-Kac distribution flows $\widehat{\eta}_n$ and η_n represent the conditional distributions of a nonabsorbed Markov particle. To see this claim, we denote by T the time at which the particle has been killed
$$T = \inf\{n \geq 0\,;\, \widehat{X}_n = c\}$$
By construction, we have

$\mathbb{P}^c_\mu(T > n)$

$= \mathbb{P}^c_\mu(\widehat{X}_0 \in E_0, \ldots, \widehat{X}_n \in E_n)$

$= \displaystyle\int_{E_0 \times \ldots \times E_n} \mu(dx_0)\, G_0(x_0)\, M_1(x_0, dx_1) \ldots M_n(x_{n-1}, dx_n) G_n(x_n)$

$= \mathbb{E}_\mu \left(\displaystyle\prod_{p=0}^n G_p(X_p) \right)$

This also shows that the normalizing constants of $\widehat{\eta}_n$ and η_n represent respectively the probability for the particle to be killed at a time strictly greater than or at least equal to n. In other words, we have that
$$\widehat{\gamma}_n(1) = \mathbb{P}^c_\mu(T > n) \quad \text{and} \quad \gamma_n(1) = \mathbb{P}^c_\mu(T \geq n)$$

Similar arguments yield that
$$\widehat{\gamma}_n(f_n) = \mathbb{E}^c_\mu(f_n(X_n)\, 1_{T>n}) \quad \text{and} \quad \gamma_n(f_n) = \mathbb{E}^c_\mu(f_n(X_n)\, 1_{T \geq n})$$
where $\mathbb{E}^c_\mu(.)$ is the expectation with respect to $\mathbb{P}^c_\mu(.)$. From the observations above, we conclude that
$$\eta_n(f_n) = \mathbb{E}^c_\mu(f_n(X_n) \mid T \geq n) \quad \text{and} \quad \widehat{\eta}_n(f_n) = \mathbb{E}^c_\mu(f_n(X_n) \mid T > n)$$

To get one step further in our discussion, it is convenient to introduce some additional terminology.

Definition 2.5.2 *The subsets $G_n^{-1}((0,1))$ and $G_n^{-1}(0)$ are called respectively the sets of soft and hard obstacles (at time n).*

By construction, a particle entering into a hard obstacle is instantly killed. When it enters into a soft obstacle, its lifetime decreases.

Let $\widehat{E}_n = E_n - G_n^{-1}(0)$, and suppose the accessibility condition $(\mathcal{A})$ introduced on page 67 is met. Let $(\widehat{G}_n, \widehat{M}_n)$ be the restrictions to the state spaces $\widehat{E}_n$ of the pair potentials/kernels defined in Proposition 2.5.2. From the discussion given in Section 2.4.3, the updated Feynman-Kac model associated with the pair (G_n, M_n) with initial distribution η_0 coincides with the prediction model associated with the pair $(\widehat{G}_n, \widehat{M}_n)$ with initial distribution $\widehat{\eta}_0$. Furthermore, if we replace in the preceding construction the mathematical objects (η_0, E_n, G_n, M_n) by $(\widehat{\eta}_0, \widehat{E}_n, \widehat{G}_n, \widehat{M}_n)$, we define a particle motion in an absorbing medium with no hard obstacles. Loosely speaking, this strategy consists in replacing the hard obstacles by repulsive obstacles. It is instructive to examine the situation where $G_n = 1_{\widehat{E}_n}$. In this case, the Feynman-Kac model associated with (η_0, G_n, M_n) corresponds to a particle motion in an absorbing medium with pure hard obstacle sets $\widehat{E}_n$, while the Feynman-Kac model associated with $(\widehat{\eta}_0, \widehat{G}_n, \widehat{M}_n)$ corresponds to a particle motion in an absorbing medium with only soft obstacles related to the potential functions defined for any $x_n \in \widehat{E}_n$ by

$$\widehat{G}_n(x_n) = M_{n+1}(G_{n+1})(x_n) = \mathbb{P}^c_{\widehat{\eta}_0}(X_{n+1} \in \widehat{E}_{n+1} \mid X_n = x_n)$$

Note that the less chances we have to enter in $\widehat{E}_{n+1}$ from some region the more stringent is the obstacle.

2.5.2 Interacting Process Interpretation

In interacting process literature, Feynman-Kac flows are alternatively seen as a nonlinear measure-valued process. For instance, the distribution sequence η_n defined in (2.8) is regarded as a solution of nonlinear recursive equations of the form

$$\eta_{n+1} = \eta_n K_{n+1,\eta_n} \qquad (2.20)$$

with the initial distribution $\eta_0 = \mu \in \mathcal{P}(E_0)$ and a collection of Markov kernels $K_{n+1,\mu}$ from E_n into E_{n+1}. As mentioned in the introduction the choice of $K_{n+1,\mu}$ is far from being unique. From (2.9) and (2.10), we easily see that we can choose

$$K_{n+1,\eta} = S_{n,\eta} M_{n+1} \qquad (2.21)$$

with the Markov kernels $S_{n,\eta}$ on E_n defined by

$$S_{n,\eta}(x_n, dy_n) = G_n(x_n)\, \delta_{x_n}(dy_n) + (1 - G_n(x_n))\, \Psi_n(\eta)(dy_n) \qquad (2.22)$$

Note that the evolution equation corresponding to this choice of kernels is decomposed into two separate transitions

$$\eta_n \xrightarrow{S_{n,\eta_n}} \widehat{\eta}_n = \eta_n S_{n,\eta_n} \xrightarrow{M_{n+1}} \eta_{n+1} = \widehat{\eta}_n M_{n+1} \qquad (2.23)$$

In contrast to the first killing interpretation given in Section 2.5.1, we have here turned the sub-Markovian kernel $\mathcal{G}_n$ into the Markov case in a nonlinear way by replacing the Dirac measure on the cemetery point c by the Gibbs-Boltzmann distribution $\Psi_n(\eta_n)$.

The nonlinear measure-valued process (2.23) can be interpreted as the evolution of the laws of a nonhomogeneous Markov chain with a two-step transition $S_{n,\eta_n} M_{n+1}$ that depends on the distribution η_n of the current value of the chain. The precise mathematical description of this Markovian interpretation is simply based on the construction of a judicious probability measure on the canonical space. These mathematical objects are currently used in the literature on mean field interacting processes. For continuous time models, the desired distribution on the canonical space cannot be described explicitly. We usually need to resort to some "fixed point argument" to ensure the existence and uniqueness of these measures. In the discrete time situation, they can be described in a very simple and natural way in terms of the transitions $K_{n,\eta}$. We give next an abstract formulation of these probability measures, and we check that they satisfy all the requirements.

Definition 2.5.3 *The* **McKean measure** *associated with a collection of Markov kernels* $(K_{n+1,\eta})_{\eta \in \mathcal{P}(E_n), n \in \mathbb{N}}$ *with initial distribution* $\eta_0 \in \mathcal{P}(E_0)$ *is a probability measure* $\mathbb{K}_{\eta_0}$ *on the canonical space*

$$\left(\Omega = \prod_{n \geq 0} E_n, \ \mathcal{F} = (\mathcal{F}_n)_{n \in \mathbb{N}}, \ X = (X_n)_{n \in \mathbb{N}} \right)$$

Its n-time marginals $\mathbb{K}_{\eta_0,n} = \mathbb{K}_{\eta_0} \circ (X_0, \ldots, X_n)^{-1}$ *are given by*

$$\mathbb{K}_{\eta_0,n}(d(x_0, \ldots, x_n)) = \eta_0(dx_0) \ K_{1,\eta_0}(x_0, dx_1) \ \ldots \ K_{n,\eta_{n-1}}(x_{n-1}, dx_n) \tag{2.24}$$

where $\eta_n \in \mathcal{P}(E_n)$ *is the solution of the recursive equation*

$$\eta_{n+1} = \eta_n K_{n+1,\eta_n}$$

with the initial distribution η_0.

By construction, we have for any $n \in \mathbb{N}$ in a synthetic integral form

$$\mathbb{K}_{\eta_0}((X_0, \ldots, X_n) \in d(x_0, \ldots, x_n))$$

$$= \eta_0(dx_0) \ K_{1,\eta_0}(x_0, dx_1) \ \ldots \ K_{n,\eta_{n-1}}(x_{n-1}, dx_n)$$

This clearly implies that under $\mathbb{K}_{\eta_0}$ the canonical Markov chain X_n has elementary transitions $K_{n,\eta_{n-1}}$ and initial distribution η_0. We denote by $\overline{\mathbb{E}}_{\eta_0}(.)$ the expectation with respect to $\mathbb{K}_{\eta_0}(.)$. To prove that η_n is the law

2.5 Distribution Flows Models

of X_n under $\mathbb{K}_{\eta_0}$, we simply check that for any test function $f_n \in \mathcal{B}_b(E_n)$

$$\overline{\mathbb{E}}_{\eta_0}(f_n(X_n))$$

$$= \int_{E_0 \times \ldots \times E_n} f_n(x_n) \, \eta_0(dx_0) \, K_{1,\eta_0}(x_0, dx_1) \ldots K_{n,\eta_{n-1}}(x_{n-1}, dx_n)$$

$$= \int_{E_n} f_n(x_n) \, \eta_{n-1} K_{n,\eta_{n-1}}(dx_n) = \eta_n(f_n).$$

Nonlinear measure-valued equations are usually not attached to a particular McKean measure. The choice of the latter depends on the physical interpretation of the model. To distinguish these possibly different choices of models, we will adopt the following terminology.

Definition 2.5.4 *Suppose $\eta_n \in \mathcal{P}(E_n)$ is a sequence of distributions satisfying a recursive equation $\eta_{n+1} = \Phi_{n+1}(\eta_n)$ for some measurable mappings $\Phi_{n+1} : \mathcal{P}(E_n) \to \mathcal{P}(E_{n+1})$, $n \in \mathbb{N}$. A collection of Markov kernels $K_{n+1,\eta}$, $\eta \in \mathcal{P}(E_n)$, $n \in \mathbb{N}$, satisfying the compatibility condition*

$$\Phi_{n+1}(\eta) = \eta K_{n+1,\eta}$$

for any $\eta \in \mathcal{P}(E_n)$ and $n \in \mathbb{N}$ is called the McKean interpretation of the flow η_n.

In comparison with (2.19), under $\mathbb{K}_{\eta_0}$ the motion of the canonical model $X_n \to X_{n+1}$ is the overlapping of an interacting jump and an exploration transition

$$X_n \xrightarrow{\text{interacting jump}} \widehat{X}_n \xrightarrow{\text{exploration}} X_{n+1} \qquad (2.25)$$

These mechanisms are defined as follows:

- **Interacting jump:** Given the position and the $\mathbb{K}_{\eta_0}$-distribution η_n at time n of the particle X_n, it performs a jump to a new site randomly chosen according to the distribution

$$S_{n,\eta_n}(X_n, .) = G_n(X_n) \, \delta_{X_n} + (1 - G_n(X_n)) \, \Psi_n(\eta_n)$$

In other words, with a probability $G_n(X_n)$ the particle remains in the same site, and we set $\widehat{X}_n = X_n$. Otherwise it jumps to a new location randomly chosen according to the Boltzmann-Gibbs distribution

$$\Psi_n(\eta_n)(dx_n) = \frac{1}{\eta_n(G_n)} \, G_n(x_n) \, \eta_n(dx_n)$$

Notice that during this transition the particle is attracted by regions with high potential in accordance with the updating transformation of the model.

- **Exploration:** The exploration transition coincides with that of the killed particle model. During this stage, the particle $\widehat{X}_n$ evolves to a new site X_{n+1} randomly chosen according to $M_{n+1}(\widehat{X}_n, .)$.

2.5.3 McKean Models

In this section, we discuss in some detail the nonuniqueness of McKean interpretations of Feynman-Kac models. In Section 2.5.2, we have already seen that different choices of Markov kernels $K_{n,\eta}$ satisfying the compatibility condition $\eta K_{n+1,\eta} = \Phi_{n+1}(\eta)$ $(= \Psi_n(\eta) M_{n+1})$ correspond to different McKean interpretations. One natural strategy to construct compatible kernels is to find a collection of selection transitions $S_{n,\eta}$ on E_n such that

$$\eta S_{n,\eta} = \Psi_n(\eta)$$

If we set $K_{n+1,\eta} = S_{n,\eta} M_{n+1}$, then we clearly obtain the desired compatible transitions. Our immediate objective is to compare the McKean models associated with the two choices

1) $\quad S_{n,\eta}(x_n, .) = \Psi_n(\eta)$
2) $\quad S_{n,\eta_n}(x_n, .) = G_n(x_n)\, \delta_{x_n} + (1 - G_n(x_n))\, \Psi_n(\eta_n)$

In the first case, the selection transition $S_{n,\eta}(x_n, dy_n)$ does not depend on the current location x_n. As a result, the particle selects more often a new location even if it fits with the potential function. In this sense, this model contains more randomness than the second one. Also notice that the corresponding McKean measure $\mathbb{K}_{\eta_0}$ is now a tensor product measure

$$\mathbb{K}_{\eta_0} = \otimes_{n \geq 0}\, \eta_n = (\eta_0 \otimes \eta_1 \otimes \ldots)$$

Under this new reference measure, the canonical process is again an interacting jump model of the form (2.25). The mutation remains the same, but the jump transition is a little more simple. Here the particle X_n selects randomly a new site $\widehat{X}_n$ with the Boltzmann-Gibbs distribution $\Psi_n(\eta_n)$ associated with the law η_n of current state X_n.

Next we examine the somehow degenerate situation where the potential functions are constant, $G_n = 1$. In this case, the Feynman-Kac flow (2.20) represents the distributions of the random states X_n of the chain with Markov transitions M_n and we have $\eta_{n+1} = \eta_n M_{n+1}$. In the second model, the jump transition disappears and we have

$$S_{n,\eta}(x_n, dy_n) = \delta_{x_n}(dy_n) \quad \text{and} \quad K_{n+1,\eta} = M_{n+1}$$

The corresponding McKean measure is simply the distribution of the chain with Markov transitions M_{n+1}

$$\mathbb{K}_{\eta_0, n}(d(x_0, \ldots, x_n)) = \eta_0(dx_0)\, M_1(x_0, dx_1)\, \ldots\, M_n(x_{n-1}, dx_n)$$

On the other hand, in the first case model we have $S_{n,\eta}(x_n, .) = \eta_n$ and $K_{n+1,\eta}(x_n, .) = \eta_{n+1}$. The McKean measure is now the tensor product of the distribution laws of the chain with transitions M_n

$$\mathbb{K}_{\eta_0, n}(d(x_0, \ldots, x_n)) = \eta_0(dx_0)\, \eta_1(dx_1)\, \ldots\, \eta_n(dx_n)$$

As we shall see in the further development of Chapter 3, these two models will have a similar interacting particle interpretation. We already mentioned that the particle interpretation of the first model coincides with the traditional selection/mutation genetic algorithm. The particle model associated with the second McKean interpretation is numerically "more stable". We will make this assertion precise in Chapter 9 with a comparison of the variances in the central limit theorems associated with these two models. This discussion seems to indicate that it is preferable to use the second McKean interpretations but this is always possible. To clarify this comment, we recall that we have made the noninnocent assumption that G_n takes values in $(0, 1)$. This condition is crucial to define the second model but it is not essential in the first interpretation. Indeed, the Boltzmann-Gibbs distributions $\Psi_n(\eta_n)$ are well-defined for any strictly positive functions G_n. Of course this situation can be embedded in the first one by replacing G_n by $G_n/\|G_n\|$. But if we do so, then running the corresponding particle algorithm we shall need to compute at each time the supremum norms $\|G_n\|$ of the current potential function (at least on the current configuration).

Next we relax condition $G_n \leq 1$ but still suppose that G_n is strictly positive. Since the potential functions are bounded, we can always find a nonnegative number $\epsilon_n \geq 0$ such that $\epsilon_n G_n \leq 1$. Arguing as usual, we check that for any $\eta \in \mathcal{P}(E_n)$ we have $\Psi_n(\eta) = \eta S_{n,\eta}$ with the Markov transition $S_{n,\eta}$ from E_n into itself defined by

$$S_{n,\eta}(x_n, .) = \epsilon_n G_n(x_n)\, \delta_{x_n} + (1 - \epsilon_n G_n(x_n))\, \Psi_n(\eta) \qquad (2.26)$$

It is interesting to note that this formulation contains the two cases examined above. The case 1) corresponds to the situation $\epsilon_n = 0$ and, whenever $G_n \leq 1$, case 2) corresponds to the choice $\epsilon_n = 1$. We also note that we can choose a parameter $\epsilon_n = \epsilon_n(\eta_n) \geq 0$ that depends on the current distribution η. For instance, we can choose

$$1/\epsilon_n(\eta) = \eta - \text{ess} - \sup G_n$$

so that (2.26) reads

$$S_{n,\eta}(x_n, .) = \frac{G_n(x_n)}{\eta - \text{ess sup}\, G_n} \delta_{x_n} + \left(1 - \frac{G_n(x_n)}{\eta - \text{ess sup}\, G_n}\right) \Psi_n(\eta) \qquad (2.27)$$

We end this section with two alternative McKean interpretations. In the first one, we use the decomposition $G_n = G_n^+ + G_n^-$ with

$$G_n^+ = G_n\, 1_{G_n \geq 1} \quad \text{and} \quad G_n^- = G_n\, 1_{G_n \leq 1}$$

For a given distribution η on E_n, three situations may occur:

1. If $\eta(G_n \geq 1) = 0$, then $G_n = G_n^- \leq 1$ η-almost surely. In particular, we have $\eta(G_n^-) > 0$, and the Boltzmann-Gibbs distribution $\Psi_n^-(\eta)$ on

E_n associated with the potential function G_n^- is well-defined. Moreover we have $\Psi_n(\eta) = \Psi_n^-(\eta) = \eta S_{n,\eta}^-$ with the Markov transition

$$S_{n,\eta}^-(x_n, .) = G_n^-(x_n)\, \delta_{x_n} + (1 - G_n^-(x_n))\, \Psi_n^-(\eta)$$

2. If $\eta(G_n \leq 1) = 0$, then $G_n = G_n^+ \geq 1$ η-almost surely. In particular, we have $\eta(G_n^+) \geq 1$, and the Boltzmann-Gibbs distribution $\Psi_n^+(\eta)$ on E_n associated with the potential function G_n^+ is well-defined. Furthermore, we prove easily that $\Psi_n(\eta) = \Psi_n^+(\eta) = \eta S_{n,\eta}^+$ with the Markov transition

$$S_{n,\eta}^+(x_n, .) = \frac{1}{\eta(G_n^+)}\, \delta_{x_n} + \left(1 - \frac{1}{\eta(G_n^+)}\right) \Psi_n^+(\eta)$$

3. Finally, if $\eta(G_n \leq 1) \wedge \eta(G_n \geq 1) > 0$, we can use for instance the decomposition

$$\Psi_n(\eta) = \frac{\eta(G_n^+)}{\eta(G_n^+ + G_n^-)}\, \Psi_n^+(\eta) + \frac{\eta(G_n^-)}{\eta(G_n^+ + G_n^-)}\, \Psi_n^-(\eta) = \eta S_{n,\eta}$$

with the Markov transition defined by

$$S_{n,\eta} = \frac{\eta(G_n^+)}{\eta(G_n^+ + G_n^-)}\, S_{n,\eta}^+ + \frac{\eta(G_n^-)}{\eta(G_n^+ + G_n^-)}\, S_{n,\eta}^-$$

When the potential $G_n = \exp V_n$ is related to some (bounded) energy function V_n, we can use the decomposition

$$G_n = G_n^+\, G_n^- \quad \text{with} \quad G_n^+ = e^{V_n^+} \geq 1 \quad \text{and} \quad G_n^- = e^{-V_n^-} \leq 1$$

where V_n^+ and $(-V_n^-)$ are the positive and negative parts of V_n. By the multiplicative form of the decomposition above, we clearly have the formulae

$$\Psi_n = \Psi_n^+ \circ \Psi_n^- = \Psi_n^- \circ \Psi_n^+$$

where Ψ_n^+ and Ψ_n^- denote the Boltzmann-Gibbs transformations associated with the potential functions G_n^+ and G_n^-. From these observations, we can decompose the mapping $\eta \to \Psi_n(\eta)$ in two different ways. We can use for instance the decomposition

$$\eta \xrightarrow{S_{n,\eta}^+} \eta_n^+ = \eta S_{n,\eta}^+ \xrightarrow{S_{n,\eta_n^+}^-} \eta_n^+ S_{n,\eta_n^+}^- = \Psi_n(\eta)$$

2.5.4 Kalman-Bucy filters

The measure-valued equations presented in Proposition 2.5.2 can rarely be solved analytically and recursively in time, except in some particular situations. When the state spaces are finite, with a reasonably small cardinality, the integral operators reduce to finite sums, and the solution reduces to simple algebraic computations. Another generic situation where an explicit solution exists is known in filtering literature as the linear/Gaussian filtering problem. Rather than rederiving rigorously the optimal Kalman-Bucy equations from the start, we provide in this section a short and informal way to obtain these explicit solutions. For a more detailed discussion of linear filtering problems, the reader is referred to the pioneering articles of R.E. Kalman and R.S. Bucy [196, 197]. A rigorous derivation of extended Kalman-Bucy solutions, and more recent developments, can be found in textbook by A.N. Shiryaev [287].

We consider a $\mathbb{R}^{p+q}$-valued Markov chain (X_n, Y_n) defined by the recursive relations

$$\begin{cases} X_n = A_n X_{n-1} + a_n + B_n W_n, & n \geq 1 \\ Y_n = C_n X_n + c_n + D_n V_n, & n \geq 0 \end{cases} \quad (2.28)$$

for some $\mathbb{R}^{d_w}$ and $\mathbb{R}^{d_v}$-valued independent random sequences W_n and V_n, independent of X_0, some matrices A_n, B_n, C_n, D_n with appropriate dimensions and finally some $(p+q)$-dimensional vector (a_n, c_n). We further assume that W_n and V_n centered Gaussian random sequences with covariance matrices R_n^v, R_n^w and X_0 is a Gaussian random variable in $\mathbb{R}^p$ with a mean and covariance matrix denoted by

$$\widehat{X}_0^- = \mathbb{E}(X_0) \quad \text{and} \quad \widehat{P}_0^- = \mathbb{E}((X_0 - \mathbb{E}(X_0))(X_0 - \mathbb{E}(X_0))')$$

In the further development of this section we shall denote by $\mathcal{N}(m, R)$ a Gaussian distribution a d-dimensional space $\mathbb{R}^d$ with mean vector $m \in \mathbb{R}^d$ and covariance matrix $R \in \mathbb{R}^{d \times d}$

$$\mathcal{N}(m, R)(dx) = \frac{1}{(2\pi)^{d/2}\sqrt{|R|}} \exp\left[-2^{-1}(x-m)R^{-1}(x-m)'\right] dx$$

We also fix a sequence of observations $Y = y$ and we introduce the non homogeneous potential/transitions (G_n, M_n) defined as follows:

- We let $G_n : \mathbb{R}^p \to (0, \infty)$ be the defined by the Radon-Nykodim derivative

$$G_n(x_n) = \frac{d\mathcal{N}(C_n x_n + c_n, D_n R_n^v D_n')}{d\mathcal{N}(0, D_n R_n^v D_n')}(y_n)$$

- We let M_{n+1} be the Gaussian transition on $\mathbb{R}^p$ defined by

$$M_{n+1}(x_n, dx_{n+1}) = \mathcal{N}(A_{n+1} x_n + a_{n+1}, B_{n+1} R_{n+1}^w B_{n+1}')(dx_{n+1})$$

2. Feynman-Kac Formulae

To have a well-defined pair of potentials/kernels (G_n, M_n), we have implicitly assumed that the covariance matrices are non degenerate. The distribution flow defined for any $f \in \mathcal{B}_b(\mathbb{R}^p)$ by the Feynman-Kac formulae

$$\eta_n(f) = \gamma_n(f)/\gamma_n(1) \quad \text{with} \quad \gamma_n(f) = \mathbb{E}(f(X_n) \prod_{p=0}^{n-1} G_p(X_p)) \quad (2.29)$$

and their updated versions $\widehat{\eta}_n$, represent respectively, the one-step predictors and the optimal filters; that is, we have that

$$\eta_n = \text{Law}(X_n \mid Y_{[0,n-1]} = (y_0, \ldots, y_{n-1}))$$
$$\widehat{\eta}_n = \text{Law}(X_n \mid Y_{[0,n]} = (y_0, \ldots, y_n))$$

with $Y_{[0,n]} = (Y_n)_{0 \le p \le n}$. Under our assumptions, η_n and $\widehat{\eta}_n$ are Gaussian distributions

$$\eta_n = \mathcal{N}(\widehat{X}_n^-, P_n^-) \quad \text{and} \quad \widehat{\eta}_n = \mathcal{N}(\widehat{X}_n, P_n)$$

The synthesis of the conditional mean and covariance matrices is carried out using the traditional Kalman-Bucy recursive equations. A short and slightly abusive way to derive these recursions is as follows. To find the prediction step we simply observe that

$$\widehat{X}_{n+1}^- = \mathbb{E}(A_{n+1} X_n + a_{n+1} + B_{n+1} W_{n+1} \mid Y_{[0,n]} = (y_0, \ldots, y_n))$$
$$= A_{n+1} \widehat{X}_n + a_{n+1}$$

and

$$P_{n+1}^-$$
$$= \mathbb{E}((A_{n+1}(X_n - \widehat{X}_n) + B_{n+1} W_{n+1})(A_{n+1}(X_n - \widehat{X}_n) + B_{n+1} W_{n+1})')$$
$$= A_{n+1} P_n A'_{n+1} + B_{n+1} R^w_{n+1} B'_{n+1}$$

In summary, we have proved the

Lemma 2.5.1 *For any $(m, P) \in (\mathbb{R}^p \times \mathbb{R}^{p \times p})$ and $n \ge 1$ the linear prediction step is given by the Markov transport equation*

$$\mathcal{N}(m, P) M_n = \mathcal{N}(m_n, P_n)$$

with the mean vector $m_n \in \mathbb{R}^d$ and covariance matrix $P_n \in \mathbb{R}^{d \times d}$

$$m_n = A_n m + a_n$$
$$P_n = A_n P A'_n + B_n R^w_n B'_n$$

The updating step is partly based on the fact that the $\mathcal{Y}$-martingale difference $(\widehat{X}_n - \widehat{X}_n^-)$ has the representation property with respect to the innovation process; that is, we have

$$\widehat{X}_n - \widehat{X}_n^- = \mathbf{G}_n \ (Y_n - \widehat{Y}_n^-)$$

for some gain matrix $\mathbf{G}_n$, and where

$$\widehat{Y}_n^- \ = \ \mathbb{E}(Y_n | Y_{[0, n-1]}) = C_n \widehat{X}_n^- + c_n$$

Since we have $\mathbb{E}((X_n - \widehat{X}_n)(Y_n - \widehat{Y}_n^-)') = 0$, and

$$(Y_n - \widehat{Y}_n^-) = C_n(X_n - \widehat{X}_n^-) + D_n V_n$$

we find that

$$\mathbb{E}((X_n - \widehat{X}_n^-)(Y_n - \widehat{Y}_n^-)') = \mathbf{G}_n \mathbb{E}((Y_n - \widehat{Y}_n^-)(Y_n - \widehat{Y}_n^-)')$$

We conclude that $\mathbf{G}_n = P_n^- C_n' (C_n P_n^- + D_n R_n^v D_n')^{-1}$. Finally, using the decomposition $X_n - \widehat{X}_n = (X_n - \widehat{X}_n^-) + (\widehat{X}_n^- - \widehat{X}_n)$ and by symmetry argument, we conclude that

$$\begin{aligned} P_n &= P_n^- - \mathbb{E}((\widehat{X}_n^- - \widehat{X}_n)(\widehat{X}_n^- - \widehat{X}_n)') \\ &= P_n^- - \mathbf{G}_n \mathbb{E}((Y_n - \widehat{Y}_n^-)(Y_n - \widehat{Y}_n^-)')\mathbf{G}_n' = P_n^- - \mathbf{G}_n C_n P_n^- \end{aligned}$$

In summary, we have proved the

Lemma 2.5.2 *For any $(m, P) \in (\mathbb{R}^p \times \mathbb{R}^{p \times p})$ and $n \geq 0$ the linear prediction step is given by the Markov transport equation*

$$\Psi_n(\mathcal{N}(m, P)) = \mathcal{N}(m_n, P_n)$$

with the mean vector $m_n \in \mathbb{R}^d$ and covariance matrix $P_n \in \mathbb{R}^{d \times d}$

$$\begin{aligned} m_n &= m + \mathbf{G}_n(y_n - (C_n m + c_n)) \\ P_n &= (I - \mathbf{G}_n C_n) P \end{aligned}$$

with the filter gain matrix $\mathbf{G}_n = PC_n'[C_n P C_n' + D_n R_n^v D_n']^{-1}$.

Finally observe that whenever $A_n = 0 = C_n$ are null matrices then the potential functions are constant G_n and the Feynman-Kac flow reduces to

$$\eta_n = \widehat{\eta}_n = \mathrm{Law}(X_n) = \mathcal{N}(a_n, B_n \ R_n^w \ B_n')$$

2.6 Feynman-Kac Models in Random Media

In this section, we discuss a modeling technique to represent Feynman-Kac formulae in random media. This new level of randomness has different interpretations. In physics, the randomness usually appears in the description of a given absorbing medium. This rather traditional point of view consists in considering random potential functions. In some other instances, such as in filtering problems, the randomness rather comes from a realization of an auxiliary process that influences the evolution of a reference signal. In this situation, the potential functions and the Markov motions are both random. One way of treating these two cases is to consider a nonhomogeneous Markov chain with two components

$$X_n = (X_n^1, X_n^2) \in E_n = E_n^{(1)} \times E_n^{(2)}$$

where $(E_n^{(i)}, \mathcal{E}_n^{(i)})$, $i = 1, 2$, is a pair of measurable spaces. The first and second components represent respectively the random variation of the medium and the "reference" particle motion. We further require that its Markov transitions M_n from E_{n-1} into E_n have the form

$$M_n((x_{n-1}, y_{n-1}), d(x_n, y_n)) = M_n^{(1)}(x_{n-1}, dx_n)\, M_{x_n,n}^{(2)}(y_{n-1}, dy_n)$$

where $M_n^{(1)}$ and $M_{x_n,n}^{(2)}$ are Markov transitions from $E_{n-1}^{(i)}$ into $E_n^{(i)}$ with $i = 1, 2$. Finally, we assume that the distribution of X_0 is given by

$$\eta_0(d(x_0, y_0)) = \eta_0^{(1)}(dx_0)\, \eta_{x_0,0}^{(2)}(dy_0)$$

with $\eta_0^{(1)} \in \mathcal{P}(E_n^{(1)})$ and $\eta_{x_0,0}^{(2)} \in \mathcal{P}(E_n^{(2)})$. By construction the medium is changing randomly at each time but is not influenced by the evolution of the particle. In the reverse angle, the particle transitions depend on the current value of the medium.

Before entering into the precise definition of the Feynman-Kac models associated with this pair Markov chain, it is convenient to fix some notation and to describe with some precision the law of the quenched Markov particle motions. By construction, the nth time marginals $\mathbb{P}_{\eta_0,n}$ of the law $\mathbb{P}_{\eta_0}$ associated with the canonical Markov chain X_n are given by

$$\mathbb{P}_{\eta_0,n}(d((x_0, y_0), \ldots, (x_n, y_n)))$$

$$= \eta_0(d(x_0, y_0)) M_n((x_0, y_0), d(x_1, y_1)) \ldots M_n((x_{n-1}, y_{n-1}), d(x_n, y_n))$$

By direct inspection, we see that X^1 is a Markov chain with transitions $M_n^{(1)}$ and initial distribution $\eta_0^{(1)}$. In other words, the distribution $\mathbb{P}_{\eta_0^{(1)},n}^{(1)}$ of path $(X_0^1, \ldots, X_n^1)$ from 0 up to time n of the first component is given by

$$\mathbb{P}_{\eta_0^{(1)},n}^{(1)}(d(x_0, \ldots, x_n)) = \eta_0^{(1)}(dx_0)\, M_1^{(1)}(x_0, dx_1) \ldots M_n^{(1)}(x_{n-1}, dx_n)$$

We conclude that for any realization of the medium

$$x = (x_n)_{n \geq 0} \in \prod_{n \geq 0} E_n^{(1)}$$

we have the synthetic integral formula

$$\mathbb{P}_{\eta_0,n}(d((x_0,y_0),\ldots,(x_n,y_n))) = \mathbb{P}^{(1)}_{\eta_0^{(1)},n}(d(x_0,\ldots,x_n))\, \mathbb{P}^{(2)}_{[x],n}(d(y_0,\ldots,y_n))$$

with

$$\mathbb{P}^{(2)}_{[x],n}(d(y_0,\ldots,y_n)) = \eta^{(2)}_{x_0,0}(dy_0)\, M^{(2)}_{x_1,1}(y_0,dy_1)\ldots M^{(2)}_{x_n,n}(y_{n-1},dy_n)$$

In other words, given $X^1 = x$, the random sequence X_n^2 is a Markov chain with transitions $M^{(2)}_{x_n,n}$ and initial distribution $\eta^{(2)}_{x_0,0}$.

In the further development of this section, we denote by $\mathbb{E}_{\eta_0}(.)$ the expectation with respect to the law $\mathbb{P}_{\eta_0}$ of the Markov chain X_n. We also simplify notation and we write $\mathbb{E}_{[x]}(.)$ instead of $\mathbb{E}^{(2)}_{[x]}(.)$ for the expectation with respect to the conditional distribution $\mathbb{P}^{(2)}_{[x]}$ of the Markov chain X_n^2 with respect to a realization $X^1 = x$ of the medium.

2.6.1 Quenched and Annealed Feynman-Kac Flows

Let $G_n : (E_n^{(1)} \times E_n^{(2)}) \to (0,\infty)$ be a given collection of bounded measurable functions. We notice that the Feynman-Kac path measures (2.7) associated with the pair (G_n, M_n) with initial distribution η_0 are defined by

$$\mathbb{Q}_{\eta_0,n}(d((x_0,y_0),\ldots,(x_n,y_n)))$$

$$= \frac{1}{\mathcal{Z}_n}\left\{\prod_{p=0}^{n-1} G_p(x_p,y_p)\right\}\, \mathbb{P}^{(1)}_{\eta_0^{(1)},n}(d(x_0,\ldots,x_n))\, \mathbb{P}^{(2)}_{[x],n}(d(y_0,\ldots,y_n))$$

with the normalizing constant $\mathcal{Z}_n = \mathbb{E}_{\eta_0}(\prod_{p=0}^{n-1} G_p(X_p^1, X_p^2)) > 0$.

Definition 2.6.1 • *The quenched Feynman-Kac path measures associated with a realization $X^1 = x$ are defined by the formulae*

$$\mathbb{Q}_{[x],n}(d(y_0,\ldots,y_n)) = \frac{1}{\mathcal{Z}_{[x],n}}\left\{\prod_{p=0}^{n-1} G_p(x_p,y_p)\right\}\, \mathbb{P}^{(2)}_{[x],n}(d(y_0,\ldots,y_n))$$

with normalizing constants $\mathcal{Z}_{[x],n} = \mathbb{E}_{[x]}(\prod_{p=0}^{n-1} G_p(x_p, X_p^2)) > 0$.

• *The annealed path measures are defined by the synthetic integral formula*

$$\overline{\mathbb{Q}}_{\eta_0,n}(d(y_0,\ldots,y_n))$$

$$= \frac{1}{\mathcal{Z}_n}\int \left\{\prod_{p=0}^{n-1} G_p(x_p,y_p)\right\}\, \mathbb{P}^{(2)}_{[x],n}(d(y_0,\ldots,y_n))\, \mathbb{P}^{(1)}_{\eta_0^{(1)}}(dx)$$

84   2. Feynman-Kac Formulae

Note that the quenched quantities $(\mathbb{Q}_{[x],n}, \mathbb{P}^{(2)}_{[x],n}, \mathcal{Z}_{[x],n})$ only depend on the path $(x_0, \ldots, x_n)$ from the origin up to time n. To clarify the presentation, we denote by $G_{x_p,p}(.)$ the "random" potential functions defined by

$$G_{x_n,n} : y_n \in E_n^{(2)} \longrightarrow G_{x_n,n}(y_n) = G_n(x_n, y_n) \in (0, \infty)$$

Definition 2.6.2 *Given a realization $X^1 = x$, the quenched Feynman-Kac distributions $\eta^{(2)}_{[x],n}$ on $E_n^{(2)}$ are defined for any $f_n \in \mathcal{B}_b(E_n^{(2)})$ by*

$$\eta^{(2)}_{[x],n}(f_n) = \gamma^{(2)}_{[x],n}(f_n) / \gamma^{(2)}_{[x],n}(1)$$

with

$$\gamma^{(2)}_{[x],n}(f_n) = \mathbb{E}_{[x]}\left(f_n(X_n^2) \prod_{p=0}^{n-1} G_{x_p,p}(X_p^2) \right) \qquad (2.30)$$

The annealed Feynman-Kac distributions $\eta_n^{(1)}$ on $E_n^{(1)}$ are defined for any $f_n \in \mathcal{B}_b(E_n^{(1)})$ by

$$\eta_n^{(1)}(f_n) = \gamma_n^{(1)}(f_n) / \gamma_n^{(1)}(1)$$

with

$$\gamma_n^{(1)}(f_n) = \mathbb{E}_{\eta_0}\left(f_n(X_n^1) \prod_{p=0}^{n-1} G_p(X_p) \right) \qquad (2.31)$$

Next we have collected two important properties related to the normalized and unnormalized quenched measures. These properties are simple consequences of results presented in Section 2.3 and Section 2.5.2. Their complete proofs are left to the reader. Using Proposition 2.3.1, we prove the following result.

Proposition 2.6.1 *For any $n \geq 1$, $f_n \in \mathcal{B}_b(E_n^{(2)})$ and, given $X^1 = x$, we have the representation formula*

$$\gamma^{(2)}_{[x],n}(f_n) = \eta^{(2)}_{[x],n}(f_n) \times \prod_{p=0}^{n-1} \left[\int_{E_n^{(2)}} G_p(x_p, y) \, \eta^{(2)}_{[x],p}(dy) \right]$$

The dynamical structure of the quenched Feynman-Kac model is defined in terms of a pair of random updating/prediction mechanisms. More precisely, using the same line of argument as the one given in Section 2.5.2, we prove the following result.

Proposition 2.6.2 *The quenched distribution flow $\eta^{(2)}_{[x],n}$ satisfies the nonlinear equation*

$$\eta^{(2)}_{[x],n} = \Phi_n^{(2)}((x_{n-1}, x_n), \eta^{(2)}_{[x],n-1}) \qquad (2.32)$$

The one-step mappings

$$\Phi_n^{(2)} : (E_{n-1}^{(1)} \times E_n^{(1)}) \times \mathcal{P}(E_{n-1}^{(2)}) \longrightarrow \mathcal{P}(E_n^{(2)})$$
$$((u,v), \eta) \longrightarrow \Phi_n^{(2)}((u,v), \eta) = \Psi_{n-1,u}(\eta) \, M_{v,n}^{(2)}$$

are defined in terms of the mappings $\Psi_{n,u} : \mathcal{P}(E_n^{(2)}) \to \mathcal{P}(E_n^{(2)})$ given by

$$\Psi_{n,u}(\eta)(dy_n) = \frac{1}{\eta(G_{u,n})} G_{u,n}(y_n) \eta(dy_n)$$

In some instances, the quenched distribution flow $\eta_{[x],n}^{(2)}$ has a nice explicit description. For instance, given the first component X^1 of a pair signal (X^1, X^2), the signal/observation (X^2, Y) may be a traditional linear/Gaussian model. In this case, the nonlinear quenched equations are solved by the traditional Kalman-Bucy recursive formulae (see Section 2.5.4 and Section 12.6.7). We recommend the interested reader to derive the quenched Kalman-Bucy solution of the filtering problem defined as in (2.28), replacing X_n by X_n^2, (a_n, c_n) by $(a_n(X_n^1), c_n(X_n^1))$, and (A_n, B_n, C_n, D_n) by some functions $(A_n(X_n^1), B_n(X_n^1), C_n(X_n^1), D_n(X_n^1))$.

2.6.2 Feynman-Kac Models in Distribution Space

Our next objective is to introduce a Feynman-Kac model in distribution space that connects the quenched and annealed measures. To this end, we introduce the stochastic sequence

$$X_n' = (X_n^1, \eta_{[X^1],n}^{(2)}) \in E_n' = E_n^{(1)} \times \mathcal{P}(E_n^{(2)}) \qquad (2.33)$$

Using the recursive equations (2.32), we prove the following result.

Proposition 2.6.3 *The stochastic sequence X_n' is a $\mathbb{P}_{\eta_0}$-Markov chain with transitions defined for any $f_n' \in \mathcal{B}_b(E_n')$ and $(u, \eta) \in E_n'$ by*

$$M_n'(f_n')(u, \eta) = \int_{E_n^{(1)}} M_n^{(1)}(u, dv) f_n'(v, \Phi_n^{(2)}((u, v), \eta))$$

and initial distribution $\eta_0' \in \mathcal{P}(E_0') = \mathcal{P}(E_0^{(1)} \times \mathcal{P}(E_0^{(2)}))$ defined by

$$\eta_0'(d(x, \nu)) = \eta_0^{(1)}(dx) \, \delta_{\eta_{x,0}^{(2)}}(d\nu)$$

Proof:
For any $f_n' \in \mathcal{B}_b(E_n')$, we have

$$\begin{aligned}
\mathbb{E}_{\eta_0}(f_n'(X_n') \mid \mathcal{F}_{n-1}) &= \mathbb{E}_{\eta_0}(f_n'(X_n^1, \eta_{[X^1],n}^{(2)}) \mid \mathcal{F}_{n-1}) \\
&= \mathbb{E}_{\eta_0}(f_n'(X_n^1, \Phi_n^{(2)}((X_{n-1}^1, X_n^1), \eta_{[X^1],n-1}^{(2)})) \mid \mathcal{F}_{n-1})
\end{aligned}$$

where $\mathcal{F}_n$ stands for the σ-algebra generated by the random variables X_p', $p \leq n$. Recalling that $X_{n-1}' = (X_{n-1}^1, \eta_{[X^1],n-1}^{(2)})$, we conclude that

$$\begin{aligned}
&\mathbb{E}_{\eta_0}(f_n'(X_n') \mid \mathcal{F}_{n-1}) \\
&= \mathbb{E}_{\eta_0}(f_n'(X_n') \mid X_{n-1}') \\
&= \int_{E_n^{(1)}} f_n'(x_n, \Phi_n^{(2)}((X_{n-1}^1, x_n), \eta_{[X^1],n-1}^{(2)})) \, M_n^{(1)}(X_{n-1}^1, dx_n)
\end{aligned}$$

86   2. Feynman-Kac Formulae

This clearly ends the proof of the proposition. ∎

Similar arguments can be used to check that X'_n is a Markov chain with respect to the distribution $\mathbb{P}^{(1)}_{\eta^{(1)}_0}$. We associate with the Markov chain X'_n the distribution flow on E'_n defined for any $f'_n \in \mathcal{B}_b(E'_n)$ by the Feynman-Kac formulae

$$\eta'_n(f'_n) = \gamma'_n(f'_n)/\gamma'_n(1) \quad \text{with} \quad \gamma'_n(f'_n) = \mathbb{E}_{\eta_0}\left(f'_n(X'_n) \prod_{p=0}^{n-1} G'_p(X'_p)\right) \tag{2.34}$$

with the "annealed" potential functions

$$G'_n : (x,\mu) \in E'_n \longrightarrow G'_n(x,\mu) = \int_{E^{(2)}_n} \mu(dy)\, G_n(x,y) \in (0,\infty) \tag{2.35}$$

Arguing as above, we find that the flow η'_n satisfies the recursive equation

$$\eta'_n = \Phi'_n(\eta'_{n-1}) \tag{2.36}$$

with the one-step mappings

$$\Phi'_n : \mathcal{P}(E'_{n-1}) \longrightarrow \mathcal{P}(E'_n)$$
$$\eta \longrightarrow \Phi'_n(\eta) = \Psi'_{n-1}(\eta)\, M'_n$$

The updating transitions $\Psi'_n : \mathcal{P}(E'_n) \to \mathcal{P}(E'_n)$ are now given by

$$\Psi'_n(\eta)(f'_n) = \eta(G'_n\, f'_n)/\eta(G'_n)$$

In the next proposition, we show that η'_n contains all information on the annealed distributions.

Proposition 2.6.4 *For any functions $f'_n \in \mathcal{B}_b(E'_n)$, $f_n \in \mathcal{B}_b(E^{(2)}_n)$, and $h_n \in \mathcal{B}_b(E^{(1)}_n)$, we have*

$$f'_n(x,\eta) = \eta(f_n) \implies \gamma'_n(f'_n) = \mathbb{E}_{\eta_0}(\gamma^{(2)}_{[X^1],n}(f_n))$$
$$f'_n(x,\eta) = h_n(x) \implies \gamma'_n(f'_n) = \gamma^{(1)}_n(h_n) \quad \text{and} \quad \eta'_n(f'_n) = \eta^{(1)}_n(h_n)$$

Proof:
The proof of the first assertion is based on the fact that, for any $f'_n \in \mathcal{B}_b(E'_n)$ of the form $f'_n(x,\eta) = \eta(f_n)$ with $f_n \in \mathcal{B}_b(E^{(2)}_n)$, we have

$$\begin{aligned}
\gamma'_n(f'_n) &= \mathbb{E}_{\eta_0}\left(f'_n(X'_n) \prod_{p=0}^{n-1} G'_p(X'_p)\right) \\
&= \mathbb{E}_{\eta_0}\left(\eta^{(2)}_{[X^1],n}(f_n) \prod_{p=0}^{n-1} G'_p(X'_p)\right) \\
&= \mathbb{E}_{\eta_0}\left(\eta^{(2)}_{[X^1],n}(f_n)\, \mathbb{E}_{[X^1]}\left(\prod_{p=0}^{n-1} G'_p(X'_p)\right)\right)
\end{aligned}$$

Using Proposition 2.6.1, we find that

$$\gamma^{(2)}_{[X^1],n}(1) = \mathbb{E}_{[X^1]}\left(\prod_{p=0}^{n-1} G'_p(X'_p)\right)$$

and

$$\eta^{(2)}_{[X^1],n}(f_n) = \gamma^{(2)}_{[X^1],n}(f_n)/\gamma^{(2)}_{[X^1],n}(1)$$

This ends the proof of the first assertion. The second implication is proved similarly by noting that, for any $f'_n \in \mathcal{B}_b(E'_n)$ of the form $f'_n(x,\eta) = h_n(x)$ with $h_n \in \mathcal{B}_b(E_n^{(1)})$, we have that

$$\begin{aligned}
\gamma'_n(f'_n) &= \mathbb{E}_{\eta_0}\left(h_n(X_n^1)\, \mathbb{E}_{[X^1]}\left(\prod_{p=0}^{n-1} G_p(X_p^1, X^2)\right)\right) \quad \text{(by Prop. 2.6.1)} \\
&= \mathbb{E}_{\eta_0}\left(h_n(X_n^1) \prod_{p=0}^{n-1} G_{X_p^1,p}(X_p^2)\right) = \gamma_n^{(1)}(h_n)
\end{aligned}$$

The end of the proof is now straightforward. ∎

2.7 Feynman-Kac Semigroups

This section focuses on the semigroup structure of discrete time Feynman-Kac models. We describe these structural properties in an abstract nonhomogeneous framework. We first recall some notation relative to the Markov chain and the potential functions with which the models are built.

We let $(E_n, \mathcal{E}_n)$, $n \in \mathbb{N}$, be a collection of measurable spaces. We consider an arbitrary probability measure η_0 on E_0 and a collection of Markov transitions $M_{n+1}(x_n, dx_{n+1})$ from E_n into E_{n+1}. We associate with these objects a nonhomogeneous Markov chain $(\Omega, \mathcal{F}, (X_n)_{n\in\mathbb{N}}, \mathbb{P}_{\eta_0})$ taking values at each time n on E_n with initial distribution η_0 and elementary transitions M_{n+1} from E_n into E_{n+1}.

For any $p \in \mathbb{N}$ and $x_p \in E_p$, we denote by $\mathbb{P}_{p,x_p}$ the probability distribution of the shifted chain $(X_{p+n})_{n\geq 0}$ starting at x_p and we use the notation $\mathbb{E}_{p,x_p}(.)$ for the expectation with respect to this law. In this notation, we have for instance for any bounded measurable function $f_{p,n}$ on $E_{[p+1,p+n]}$

$$\begin{aligned}
&\mathbb{E}_{p,x_p}(f_{p,n}(X_{p+1},\ldots,X_{p+n})) \\
&= \int f_{p,n}(x_{p+1},\ldots,x_{p+n})\, M_{p+1}(x_p, dx_p)\ldots M_{p+n}(x_{p+n-1}, dx_{p+n})
\end{aligned}$$

When $p = 0$, sometimes we slightly abuse the notation and write $\mathbb{P}_{x_0}$ and $\mathbb{E}_{x_0}$ instead of $\mathbb{P}_{0,x_0}$ and $\mathbb{E}_{0,x_0}$. Given a distribution $\mu_p \in \mathcal{P}(E_p)$, we use the notation $\mathbb{E}_{p,\mu_p}(.)$ for the expectation with respect to the measure

$$\mathbb{P}_{p,\mu_p}(.) = \int_{E_p} \mu_p(dx) \, \mathbb{P}_{p,x_p}(.)$$

In the further development of this section, we denote respectively by x_n, f_n, and μ_n a point in E_n, a bounded and measurable function on E_n, and a probability measure on E_n.

Definition 2.7.1 *We denote by $M_{p,n}$, $0 \le p \le n$, the linear semigroup associated with the Markov kernels M_n and defined by*

$$M_{p,n} = M_{p+1} M_{p+2} \ldots M_n$$

We use the convention $M_{n,n} = \mathrm{Id}$ for $p = n$. This semigroup is alternatively defined by

$$M_{p,n}(f_n)(x_p) = \mathbb{E}_{p,x_p}(f_n(X_n))$$

Let $G_n : E_n \to (0, \infty)$, $n \ge 0$, be a collection of bounded potential functions. The Feynman-Kac prediction model $\eta_n \in \mathcal{P}(E_n)$ associated with (G_n, M_n) is defined by the formulae

$$\eta_n(f_n) = \gamma_n(f_n)/\gamma_n(1) \quad \text{with} \quad \gamma_n(f_n) = \mathbb{E}_{\eta_0}\left(f_n(X_n) \prod_{p=0}^{n-1} G_p(X_p)\right)$$

We also recall that the updated models are given by

$$\widehat{\eta}_n(f_n) = \widehat{\gamma}_n(f_n)/\widehat{\gamma}_n(1) \quad \text{with} \quad \widehat{\gamma}_n(f_n) = \gamma_n(f_n G_n)$$

2.7.1 Prediction Semigroups

The study of the dynamical structure of γ_n and η_n was initiated in Section 2.5. We recall that γ_n satisfies the recursive equation

$$\gamma_{n+1} = \gamma_n Q_{n+1} \quad \text{with} \quad Q_{n+1}(f_{n+1}) = G_n \, M_{n+1}(f_{n+1}) \quad \text{and} \quad \gamma_0 = \eta_0$$

Definition 2.7.2 *We denote by $Q_{p,n}$, $0 \le p \le n$, the linear semigroup associated with γ_n and defined by*

$$Q_{p,n} = Q_{p+1} Q_{p+2} \ldots Q_n$$

We use the convention $Q_{n,n} = \mathrm{Id}$ for $p = n$. This semigroup is alternatively defined by the Feynman-Kac formulae

$$Q_{p,n}(f_n)(x_p) = \mathbb{E}_{p,x_p}\left(f_n(X_n) \prod_{q=p}^{n-1} G_q(X_q)\right)$$

2.7 Feynman-Kac Semigroups

By the definition of η_n and $Q_{p,n}$ we readily observe that

$$\eta_n(f_n) = \frac{\gamma_n(f_n)}{\gamma_n(1)} = \frac{\gamma_p(Q_{p,n}(f_n))}{\gamma_p(Q_{p,n}(1))} = \frac{\eta_p(Q_{p,n}(f_n))}{\eta_p(Q_{p,n}(1))}$$

This representation leads to the following definition.

Definition 2.7.3 *We denote by $\Phi_{p,n}$, $0 \le p \le n$, the linear semigroup associated with η_n and defined by*

$$\Phi_{p,n} = \Phi_n \circ \Phi_{n-1} \circ \ldots \circ \Phi_{p+1}$$

We use the convention $\Phi_{n,n} = Id$ for $p = n$. This semigroup is alternatively defined by the Feynman-Kac formulae

$$\Phi_{p,n}(\mu_p)(f_n) = \frac{\mu_p(Q_{p,n}(f_n))}{\mu_p(Q_{p,n}(1))} = \frac{\mathbb{E}_{p,\mu_p}(f_n(X_n) \prod_{q=p}^{n-1} G_q(X_p))}{\mathbb{E}_{p,\mu_p}(\prod_{q=p}^{n-1} G_q(X_p))}$$

The preceding models will be of constant use in the next chapters. In the second part of this section, we enter more deeply into the dynamical structure of these models. The forthcoming analysis will be used in Chapter 4 when we study the stability properties of Feynman-Kac semigroups. To describe the fine structure of $\Phi_{p,n}$, it is convenient to introduce the following objects.

Definition 2.7.4 *For any $0 \le p \le n$, we denote respectively by $G_{p,n} : E_p \to (0, \infty)$ and $P_{p,n}$ the potential functions on E_p and the Markov kernels from E_p into E_n defined by*

$$G_{p,n} = Q_{p,n}(1) \quad \text{and} \quad P_{p,n}(f_n) = Q_{p,n}(f_n)/Q_{p,n}(1)$$

The next proposition expresses the fact that all the mappings $\Phi_{p,n}$ have the same updating/prediction nature.

Proposition 2.7.1 *The mappings $\Phi_{p,n} : \mathcal{P}(E_p) \to \mathcal{P}(E_n)$, $0 \le p \le n$, satisfy the formula*

$$\Phi_{p,n}(\mu_p) = \Psi_{p,n}(\mu_p)P_{p,n} \tag{2.37}$$

with the Boltzmann-Gibbs transformation $\Psi_{p,n}$ from $\mathcal{P}(E_p)$ into itself defined by

$$\Psi_{p,n}(\mu_p)(f_p) = \mu_p(G_{p,n}f_p)/\mu_p(G_{p,n})$$

In addition, the pairs $(G_{p,n}, P_{p,n})_{p<n}$ satisfy the backward recursive equations

$$G_{p,n} = G_p \times M_{p+1}(G_{p+1,n}) \quad \text{and} \quad P_{p,n} = R^{(n)}_{p+1} P_{p+1,n}$$

with the Markov kernels $R^{(n)}_p$ from E_{p-1} into E_p, $1 \le p \ge n$, defined by

$$R^{(n)}_p(f_p) = M_p(G_{p,n}f_p)/M_p(G_{p,n})$$

Proof:
The proof of the first assertion is a simple consequence of the definition of $G_{p,n}$, $P_{p,n}$, and $\Phi_{p,n}$. To prove the inductive formulae, we simply note that

$$G_{p,n} = Q_{p,n}(1) = Q_{p,p+1}[Q_{p+1,n}(1)] = G_p \, M_{p+1}(G_{p+1,n})$$

To prove the second one, we observe that

$$\begin{aligned}
P_{p,n}(f_n) &= \frac{Q_{p,p+1}Q_{p+1,n}(f_n)}{Q_{p,p+1}Q_{p+1,n}(1)} = \frac{M_{p+1}[Q_{p+1,n}(f_n)]}{M_{p+1}[Q_{p+1,n}(1)]} \\
&= M_{p+1}[G_{p+1,n}\, P_{p+1,n}(f_n)]/M_{p+1}[G_{p+1,n}]
\end{aligned}$$

The end of the proof is now straightforward. ∎

One important consequence of the proposition above is that $P_{p,n}$ can be regarded as the transition from E_p into E_n of a nonhomogeneous Markov chain with elementary transitions $R_q^{(n)}$, $p < q \leq n$. More precisely, using Proposition 2.7.1, we find that

$$P_{p,n} = R_{p+1}^{(n)} R_{p+2}^{(n)} \ldots R_n^{(n)}$$

Definition 2.7.5 *For any $n \in \mathbb{N}$, we denote by $(R_{p,q}^{(n)})_{0 \leq p \leq q \leq n}$, the linear semigroup associated with the Markov kernels $(R_p^{(n)})_{0 \leq p \leq n}$, and defined by*

$$R_{p,q}^{(n)} = R_{p+1}^{(n)} R_{p+2}^{(n)} \ldots R_q^{(n)}$$

We use the convention $R_{p,p}^{(n)} = Id$ for $p = q$.

In this notation, we note that $R_n^{(n)} = M_n$ and $R_{p,n}^{(n)} = P_{p,n}$. We also quote the following technical lemma.

Lemma 2.7.1 *For any $0 \leq p \leq q \leq n$, we have*

$$P_{p,n} = R_{p,q}^{(n)} P_{q,n} \quad \text{and} \quad R_{p,q}^{(n)}(f_q) = Q_{p,q}(f_q G_{q,n})/Q_{p,q}(G_{q,n})$$

This semigroup is alternatively defined by the Feynman-Kac formulae

$$R_{p,q}^{(n)}(f_q)(x_p) = \frac{\mathbb{E}_{p,x_p}(f_q(X_q) \prod_{k=p}^{n-1} G_k(X_k))}{\mathbb{E}_{p,x_p}(\prod_{k=p}^{n-1} G_k(X_k))}$$

Proof:
By the definition of $(P_{p,n}, G_{p,n})$, we have

$$P_{p,n}(f_n) = \frac{Q_{p,n}(f_n)}{Q_{p,n}(1)} = \frac{Q_{p,q}Q_{q,n}(f_n)}{Q_{p,q}Q_{q,n}(1)} = \frac{Q_{p,q}(G_{q,n}P_{q,n}(f_n))}{Q_{p,q}(G_{q,n})}$$

We conclude that $P_{p,n} = R_{p,q}^{(n)} P_{q,n}$ with

$$R_{p,q}^{(n)}(f_q) = Q_{p,q}(G_{q,n}f_q)/Q_{p,q}(G_{q,n})$$

2.7 Feynman-Kac Semigroups

To check the Feynman-Kac formulation of the semigroup, we observe that

$$\begin{aligned}
Q_{p,q}(G_{q,n}f_q)(x_p) &= Q_{p,q}(Q_{q,n}(1)\,f_q)(x_p) \\
&= \mathbb{E}_{p,x_p}\left(Q_{q,n}(1)(X_q)\,f_q(X_q)\prod_{k=p}^{q-1} G_k(X_k)\right) \\
&= \mathbb{E}_{p,x_p}\left(\mathbb{E}_{q,X_q}\left(\prod_{l=q}^{n-1} G_l(X_l)\right)f_q(X_q)\prod_{k=p}^{q-1} G_k(X_k)\right)
\end{aligned}$$

Using the Markov property, we conclude that

$$Q_{p,q}(G_{q,n}f_q)(x_p) = \mathbb{E}_{p,x_p}\left(f_q(X_q)\prod_{k=p}^{n-1} G_k(X_k)\right)$$

The proof of the lemma is now completed. ∎

Next we introduce another semigroup that appears to be useful in the analysis of the fluctuations of particle models.

Definition 2.7.6 *We denote by $\mathbf{Q}_{p,n}$, $0 \le p \le n$, the "normalized" integral operator $\mathbf{Q}_{p+1} = \frac{Q_{p+1}}{\eta_p(G_p)}$ defined by*

$$\mathbf{Q}_{p,n} = \mathbf{Q}_{p+1}\mathbf{Q}_{p+2}\ldots\mathbf{Q}_n$$

We use the convention $\mathbf{Q}_{n,n} = Id$ for $p = n$. This semigroup is alternatively defined by the formulae

$$\mathbf{Q}_{p,n}(f_n) = \frac{Q_{p,n}(f_n)}{\eta_p(Q_{p,n}(1))} = \frac{\gamma_p(1)}{\gamma_n(1)}\,Q_{p,n}(f_n) = [\prod_{q=p}^{n-1}\eta_q(G_q)]^{-1}\,Q_{p,n}(f_n)$$

2.7.2 Updated Semigroups

In this final section, we give a brief discussion on the semigroup structure of the updated models $\widehat{\gamma}_n$ and $\widehat{\eta}_n$. Arguing as in Section 2.5, we immediately notice that $\widehat{\gamma}_n$ satisfies the recursive equation

$$\widehat{\gamma}_{n+1} = \widehat{\gamma}_n\widehat{Q}_{n+1} \quad \text{with} \quad \widehat{Q}_{n+1}(f_{n+1}) = \widehat{G}_n\,\widehat{M}_{n+1}(f_{n+1})$$

and $\widehat{\gamma}_0(dx_0) = G_0(x_0)\eta_0(dx_0)$, with the pair potentials/kernels $(\widehat{G}_n, \widehat{M}_n)$ defined in Proposition 2.5.2. From the discussion given in Section 2.4.3, the representations above show that the updated flow models associated with the pair (G_n, M_n) can be regarded as the prediction flows associated with the pair $(\widehat{G}_n, \widehat{M}_n)$. We conclude that the semigroup structure of the updated models $\widehat{\gamma}_n$ and $\widehat{\eta}_n$ is defined as the one given above by replacing the pair (G_n, M_n) by $(\widehat{G}_n, \widehat{M}_n)$. We use the superscript $\widehat{(.)}$ to denote the corresponding objects. In this notation, the semigroups of $\widehat{\gamma}_n$ and $\widehat{\eta}_n$ are respectively defined by

$$\widehat{Q}_{p,n} = \widehat{Q}_{p+1}\widehat{Q}_{p+2}\ldots\widehat{Q}_n \quad \text{and} \quad \widehat{\Phi}_{p,n} = \widehat{\Phi}_n \circ \widehat{\Phi}_{n-1} \circ \ldots \circ \widehat{\Phi}_{p+1}$$

We use the conventions $\widehat{Q}_{n,n} = Id = \widehat{\Phi}_{n,n}$ for $p = n$. We also have

$$\widehat{\Phi}_{p,n}(\mu_p)(f_n) = \frac{\mu_p(\widehat{Q}_{p,n}(f_n))}{\mu_p(\widehat{Q}_{p,n}(1))} = \frac{\mu_p(\widehat{G}_{p,n}\widehat{P}_{p,n}(f_n))}{\mu_p(\widehat{G}_{p,n})}$$

with the backward formulae

$$\widehat{G}_{p,n} = \widehat{G}_p\ \widehat{M}_{p+1}(\widehat{G}_{p+1,n}) \quad \text{and} \quad \widehat{P}_{p,n} = \widehat{R}^{(n)}_{p+1}\widehat{P}_{p+1,n}$$

where $\widehat{R}^{(n)}_p$ is the Markov kernel from E_{p-1} into E_p, $p \geq 1$, defined by

$$\widehat{R}^{(n)}_p(f_p) = \widehat{M}_p(\widehat{G}_{p,n}f_p)/\widehat{M}_p(\widehat{G}_{p,n})$$

These semigroups also appear to be useful in the analysis of Feynman-Kac models associated with not necessarily strictly positive potentials. Let $\widehat{E}_n = G_n^{-1}(0, \infty)$ be the support of the potential function G_n and, we assume that the accessibility condition ($\mathcal{A}$) introduced on page 67 is met. Rephrasing the results given in Section 2.4.3, the main advantage of this condition is that we can restrict the whole semigroup analysis on the state spaces $(\widehat{E}_n, \widehat{\mathcal{E}}_n)$. We recall that in this case $\widehat{M}_{n+1}$ is a well-defined Markov kernel from $\widehat{E}_n$ into $\widehat{E}_{n+1}$ and $\widehat{G}_n$ are now strictly positive potentials on $\widehat{E}_n$. Another important comment in practice is that the potential functions G_n are usually not homogeneous with respect to the time parameter, and the subsets $\widehat{E}_n$ may vary. This shows that we do need to consider nonhomogeneous state-space models.

From the discussions above, it is also clear that the analysis of updated models is reduced to the analysis of prediction ones by a suitable choice of potentials and Markov kernels.

Nevertheless, in some situations it is more judicious to work directly with the particular Feynman-Kac structure of the updated semigroups. We can use alternatively one of the representations

$$\begin{aligned}\widehat{Q}_{p,n}(f_n)(x_p) &= \mathbb{E}_{p,x_p}\left(f_n(X_n)\prod_{k=p+1}^n G_k(X_k)\right) \\ &= \widehat{\mathbb{E}}_{p,x_p}\left(f_n(X_n)\prod_{k=p}^{n-1}\widehat{G}_k(X_k)\right)\end{aligned}$$

where $\mathbb{E}_{p,x_p}(.)$ (resp. $\widehat{\mathbb{E}}_{p,x_p}(.)$) represents the expectation with respect to the law $\mathbb{P}_{p,x_p}(.)$, respectively $\widehat{\mathbb{P}}_{p,x_p}$, of the shifted Markov chain $(X_n)_{n \geq p}$ starting at x_p at time $n = p$ with elementary transitions M_n (resp. $\widehat{M}_n$).

3
Genealogical and Interacting Particle Models

3.1 Introduction

This chapter is devoted to particle interpretations of Feynman-Kac models. From the discussion given in the introduction, these particle models can be sought in many different ways, depending on the application we have in mind. For this reason, we have chosen to describe these models as an abstract stochastic linearization technique for solving nonlinear and measure-valued equations. The second part of this chapter is concerned with genealogical tree-based particle interpretations of Feynman-Kac models in path space.

In the further development of this chapter and unless otherwise stated, we will assume that the potential functions are bounded and strictly positive. To simplify the presentation, we only discuss the particle model associated with a McKean interpretation of the prediction flow. We recall that the Feynman-Kac prediction model $\eta_n \in \mathcal{P}(E_n)$ is defined for any $f_n \in \mathcal{B}_b(E_n)$ by the equation

$$\eta_n(f_n) = \gamma_n(f_n)/\gamma_n(1) \quad \text{with} \quad \gamma_n(f_n) = \mathbb{E}_{\eta_0}\left(f_n(X_n)\prod_{p=0}^{n-1}G_p(X_p)\right)$$

The process X_n is a nonhomogeneous and E_n-valued Markov chain with Markov transitions M_n from E_{n-1} into E_n, and G_n is a given collection of (strictly positive) $\mathcal{E}_n$-measurable potential functions on E_n. The situation where G_n may take null values can be reduced to this situation, under appropriate accessibility conditions, by replacing η_n by the updated model $\widehat{\eta}_n$

(see (2.16), Section 2.5 and Proposition 2.4.3, page 67). For the convenience of the reader, we recall that η_n satisfy a nonlinear recursive equation

$$\eta_{n+1} = \eta_n K_{n+1,\eta_n}$$

where $K_{n,\eta}$ is a nonunique collection of Markov kernels from E_n into E_{n+1} satisfying the compatibility condition

$$\eta K_{n+1,\eta} = \Psi_n(\eta) M_{n+1} \quad \text{with} \quad \Psi_n(\eta)(dx) = \frac{1}{\eta(G_n)} \ G_n(x) \ \eta(dx)$$

For a more detailed description as well as several worked-out examples of kernels $K_{n,\eta}$, we refer the reader to Section 2.5.2 and Section 2.5.3. In this chapter, we design an abstract strategy to associate with a given collection of compatible kernels $K_{n,\eta}$ a sequence of interacting particle approximation models. We will not describe the whole class of models associated with all the McKean interpretations discussed in Section 2.5.3; that would of course be too much digression. We leave the interested reader to derive the corresponding particle interpretations. In this chapter, we have chosen to concentrate the discussion on the somewhat generic situation where the kernels $K_{n,\eta}$ are a combination of a selection and mutation transition

$$K_{n+1,\eta} = S_{n,\eta} M_{n+1} \tag{3.1}$$

The selection transition $S_{n,\eta}$ on E_n is given by

$$S_{n,\eta}(x_n, .) = \epsilon_n G_n(x_n) \ \delta_{x_n} + (1 - \epsilon_n G_n(x_n)) \ \Psi_n(\eta)$$

where $\epsilon_n \geq 0$ stands for any nonnegative number such that $\epsilon_n G_n \leq 1$.

3.2 Interacting Particle Interpretations

In this section, we associate with a given collection of Markov transitions a sequence of N-interacting particle systems. We will examine in more detail the evolution of the particle model associated with the McKean interpretation (3.1) and we will also compare the two cases $\epsilon_n = 0$ and $\epsilon_n > 0$. At the end of this section, we discuss the situation where G_n is not necessarily strictly positive.

Definition 3.2.1 *The interacting particle model associated with a collection of Markov transitions $K_{n,\eta}$, $\eta \in \mathcal{P}(E_n)$, $n \geq 1$, and with an initial distribution $\eta_0 \in \mathcal{P}(E_0)$ is a sequence of nonhomogeneous Markov chains*

$$\left(\Omega^{(N)} = \prod_{n \geq 0} E_n^N, \ \mathcal{F}^N = (\mathcal{F}_n^N)_{n \in \mathbb{N}}, \ \xi^{(N)} = (\xi_n^{(N)})_{n \in \mathbb{N}}, \ \mathbb{P}_{\eta_0}^N \right)$$

taking values at each time $n \in \mathbb{N}$ in the product space E_n^N, that is

$$\xi_n^{(N)} = (\xi_n^{(N,1)}, \ldots, \xi_n^{(N,N)}) \in E_n^N = \underbrace{E_n \times \ldots \times E_n}_{N \text{ times}}$$

The initial configuration $\xi_0^{(N)}$ consists of N independent and identically distributed random variables with common law η_0. Its elementary transitions from $E_{n_1}^N$ into E_n^N are given in a symbolic integral form by

$$\mathbb{P}_{\eta_0}^N\left(\xi_n^{(N)} \in dx_n \mid \xi_{n-1}^{(N)}\right) = \prod_{p=1}^{N} K_{n, \frac{1}{N} \sum_{i=1}^{N} \delta_{\xi_{n-1}^{(N,i)}}}(\xi_{n-1}^{(N,p)}, dx_n^p) \qquad (3.2)$$

where $dx_n = dx_n^1 \times \ldots \times x_n^N$ is an infinitesimal neighborhood of a point $x_n = (x_n^1, \ldots, x_n^N) \in E_n^N$.

As traditionally, when there is no possible confusion, we simplify notation and suppress the index $(.)^{(N)}$ so that we write (ξ_n, ξ_n^i) instead of $(\xi_n^{(N)}, \xi_n^{(N,i)})$. To clarify the presentation we slightly abuse the notation and sometimes write

$$m(x_n) = \frac{1}{N} \sum_{i=1}^{N} \delta_{x_n^i}$$

for each $x_n = (x_n^i)_{1 \leq i \leq N} \in E_n^N$. In this simplified notation, the elementary transition (3.2) reads

$$\mathbb{P}_{\eta_0}^N(\xi_n \in dx_n \mid \xi_{n-1}) = \prod_{p=1}^{N} K_{n, m(\xi_{n-1})}(\xi_{n-1}^p, dx_n^p)$$

The parameter $N \geq 1$ is called the size of the system or the precision parameter of the particle algorithm. The N-tuple ξ_n represents the configuration of the system at time n of N particles ξ_n^i. The N-particle model associated with the Markov transitions $K_{n,\eta}$ given by (3.1) is the Markov chain ξ_n with elementary transitions

$$\mathbb{P}_{\eta_0}^N(\xi_{n+1} \in dx_{n+1} \mid \xi_n) = \prod_{p=1}^{N} (S_{n,m(\xi_n)} M_{n+1})(\xi_n^p, dx_{n+1}^p) \qquad (3.3)$$

By direct inspection, we see that

$$\mathbb{P}_{\eta_0}^N(\xi_{n+1} \in dx_{n+1} \mid \xi_n) = \int_{E_n^N} S_n(\xi_n, dx_n) \; M_{n+1}(x_n, dx_{n+1})$$

with the Boltzmann-Gibbs transition $\mathcal{S}_n$ from E_n^N into itself and the mutation transition $\mathcal{M}_{n+1}$ from E_n^N into E_{n+1}^N defined by

$$\mathcal{S}_n(\xi_n, dx_n) = \prod_{p=1}^{N} S_{n,m(\xi_n)}(\xi_n^p, dx_n^p)$$

$$\mathcal{M}_{n+1}(x_n, dx_{n+1}) = \prod_{p=1}^{N} M_{n+1}(x_n^p, dx_{n+1}^p)$$

Loosely speaking, this integral decomposition shows that this particle model has the same updating/prediction nature as that of the "limiting" Feynman-Kac model. More precisely, the deterministic two-step updating/prediction transitions in distribution spaces

$$\eta_n \in \mathcal{P}(E_n) \xrightarrow{S_{n,\eta_n}} \widehat{\eta}_n = \eta_n S_{n,\eta_n} \in \mathcal{P}(E_n) \xrightarrow{M_{n+1}} \eta_{n+1} = \widehat{\eta}_n M_{n+1}$$

have been replaced by a two-step selection/mutation transition in product spaces

$$\xi_n \in E_n^N \xrightarrow{\text{selection}} \widehat{\xi}_n \in E_n^N \xrightarrow{\text{mutation}} \xi_{n+1} \in E_{n+1}^N$$

Next we describe in more detail the motion of the particles in terms of the pair transitions $(S_{n,\eta}, M_n)$.

- **Selection:** The selection stage only depends on the current potential function G_n and the parameter ϵ_n. More precisely, given the configuration $\xi_n \in E_n^N$ of the system at time n, the selection transition consists in selecting randomly N particles $\widehat{\xi}_n^i$ with respective distribution

$$S_{n,m(\xi_n)}(\xi_n^i, .) = \epsilon_n G_n(\xi_n^i)\, \delta_{\xi_n^i} + (1 - \epsilon_n G_n(\xi_n^i))\, \Psi_n(m(\xi_n))$$

In other words, with a probability $\epsilon_n G_n(\xi_n^i)$, we set $\widehat{\xi}_n^i = \xi_n^i$; otherwise we select randomly a particle $\widetilde{\xi}_n^i$ with distribution

$$\Psi_n(m(\xi_n)) = \sum_{i=1}^{N} \frac{G_n(\xi_n^i)}{\sum_{j=1}^{N} G_n(\xi_n^j)} \delta_{\xi_n^i} \quad \text{and we set} \quad \widehat{\xi}_n^i = \widetilde{\xi}_n^i$$

- **Mutation:** The mutation stage only depends on the Markov kernel M_{n+1}. During this stage, each selected particle $\widehat{\xi}_n^i$ evolves randomly according to the Markov transition M_{n+1}. In other words, given the selected configuration $\widehat{\xi}_n \in E_n^N$, the mutation transition consists in sampling randomly N independent random particles ξ_{n+1}^i with respective distributions $M_{n+1}(\widehat{\xi}_n^i, .)$.

From the discussion above, we see that the case $\epsilon_n = 0$ corresponds to the so-called "simple genetic algorithm" with mutation and proportional selections. Also notice that in this situation the particle model consists of conditionally N-independent particles. More precisely, in this situation the elementary transitions (3.3) can be rewritten as

$$\mathbb{P}^N_{\eta_0}(\xi_{n+1} \in dx_{n+1} \mid \xi_n) = \prod_{p=1}^{N} \Phi_n(m(\xi_n))(dx^p_{n+1}) \qquad (3.4)$$

where $\Phi_{n+1} : \mathcal{P}(E_n) \to \mathcal{P}(E_{n+1})$ is the one-step mapping of the flow η_n defined by

$$\Phi_{n+1}(\eta) = \Psi_n(\eta) M_{n+1}$$

3.3 Particle models with Degenerate Potential

As promised in the introduction, this section discusses the situation where G_n is not strictly positive. To avoid some unnecessary discussions on degenerate situations, we suppose the accessibility condition $(\mathcal{A})$ introduced in (2.16) on page 67 is met so that the limiting Feynman-Kac model is well-defined at any time.

Two strategies can be underlined. In view of the discussion given in Section 2.5.2, the first idea is to consider the N-particle approximation model associated with some McKean interpretation of the updated model. To be more precise, we recall that the updated measures $\widehat{\eta}_n = \Psi_n(\eta_n)$ can be regarded as a sequence of measures on $\widehat{E}_n$ with $\widehat{E}_n = G_n^{-1}(0, \infty)$. Furthermore, $\widehat{\eta}_n$ coincides with the prediction model starting at $\widehat{\eta}_0$ and associated with the pair of potentials/kernels $(\widehat{G}_n, \widehat{M}_n)$ on the state spaces $\widehat{E}_n$ and defined in Proposition 2.5.2, page 70.

One advantage of this interpretation is that the potential $\widehat{G}_n$ is now a strictly positive function on the restricted space $\widehat{E}_n$. By symmetry arguments, we prove that the updated model $\widehat{\eta}_n \in \mathcal{P}(\widehat{E}_n)$ satisfies the recursive equation

$$\widehat{\eta}_{n+1} = \widehat{\eta}_n \widehat{K}_{n+1,\widehat{\eta}_n} \quad \text{with} \quad \widehat{K}_{n+1,\eta} = \widehat{S}_{n,\eta} \widehat{M}_{n+1}$$

The selection transitions $\widehat{S}_{n,\eta}$ are now Markov kernels from $\widehat{E}_n$ into itself and they are defined for any $x_n \in \widehat{E}_n$ by

$$\widehat{S}_{n,\eta}(x_n, dy_n) = \epsilon_n \widehat{G}_n(x_n)\, \delta_{x_n}(dy_n) + (1 - \epsilon_n \widehat{G}_n(x_n))\, \widehat{\Psi}_n(\eta)(dy_n)$$

The Boltzmann-Gibbs transformation $\widehat{\Psi}_n : \mathcal{P}(\widehat{E}_n) \to \mathcal{P}(\widehat{E}_n)$ associated with $\widehat{G}_n$ is given for any $\eta \in \mathcal{P}(\widehat{E}_n)$ by

$$\widehat{\Psi}_n(\eta)(dx_n) = \frac{1}{\eta(\widehat{G}_n)}\, \widehat{G}_n(x_n)\, \eta(dx_n)$$

In this interpretation, the model $\widehat{\eta}_n$ satisfies the deterministic updating/prediction transitions

$$\widehat{\eta}_n \in \mathcal{P}(\widehat{E}_n) \xrightarrow{\text{updating}} \widetilde{\eta}_n = \widehat{\eta}_n \widehat{S}_{n,\widehat{\eta}_n} \in \mathcal{P}(\widehat{E}_n) \xrightarrow{\text{prediction}} \widehat{\eta}_{n+1} = \widetilde{\eta}_n \widehat{M}_{n+1}$$

The N-particle associated with this McKean interpretation is defined along the same lines as before. It is interesting to observe that in this case the mutation transitions of the particles depend on the current potential functions. This aspect of the particle model is particularly useful in filtering problems (see Section 12.6). In this context, each elementary mutation depends on the likelihood function of the particle with respect to the current observation delivered by the sensors. This intuitively indicates that the resulting algorithm has better "tracking properties" than the one with "free mutations". We also mention that for regular potential functions the sampling of transitions according to $\widehat{M}_n$ can be performed by using for instance an acceptance/rejection simulation technique. More precisely, when G_n has bounded relative oscillations, we have that

$$\widehat{M}_n(x_{n-1}, dx_n) \leq \sup_{y_n, z_n \in \widehat{E}_n} \left(\frac{G_n(y_n)}{G_n(z_n)} \right) M_n(x_{n-1}, dx_n)$$

This shows that we can sample elementary $\widehat{M}_n$-transitions with a traditional acceptance/rejection mechanism based on independent M_n-samples.

An alternative way to produce these potential dependent transitions consists in adding another level of randomness. We proceed as follows. Suppose we want to simulate the elementary $\widehat{M}_n$-transition of a particle, say ξ^i_{n-1}. Then we first produce a collection of N' auxiliary and independent transitions $\xi^i_{n-1} \to \xi^{i,i'}_n$, $1 \leq i' \leq N'$, according to $M_n(\xi^i_{n-1}, \cdot)$. The rationale behind this is that for sufficiently large N' we have in some sense

$$M_n(\xi^i_{n-1}, dx_n) \simeq M_n^{N'}(\xi^i_{n-1}, dx_n) =_{\text{def.}} \frac{1}{N'} \sum_{i'=1}^{N'} \delta_{\xi^{i,i'}_n}$$

Replacing M_n by $M_n^{N'}$ in the definitions of $\widehat{M}_n$ and $\widehat{G}_{n-1}$, we obtain a particle approximation of the desired quantities

$$\widehat{M}_n(\xi^i_{n-1}, dx_n) \simeq \widehat{M}_n^{N'}(\xi^i_{n-1}, dx_n) =_{\text{def.}} \sum_{i'=1}^{N'} \frac{G_n(\xi^{i,i'}_n)}{\sum_{j'=1}^{N'} G_n(\xi^{i,j'}_n)} \delta_{\xi^{i,i'}_n}$$

and

$$\widehat{G}_{n-1}(\xi^i_{n-1}) \simeq \widehat{G}_{n-1}^{N'}(\xi^i_{n-1}) =_{\text{def.}} \frac{1}{N'} \sum_{i'=1}^{N'} G_n(\xi^{i,i'}_n)$$

This strategy provides a "local" N'-particle approximation model for the evaluation of the potential functions and for the sampling of $\widehat{M}_n$ transitions.

3.3 Particle models with Degenerate Potential

The second strategy consists in still working with the McKean interpretation of the prediction flow associated with the collection of transitions

$$K_{n+1,\eta} = S_{n,\eta} M_{n+1} \quad \text{with} \quad \eta \in \mathcal{P}_n(E_n), \quad n \in \mathbb{N} \tag{3.5}$$

It is important to note that the collection of transitions $K_{n,\eta}$ and particularly $S_{n,\eta}$ are not defined on the whole set of distributions $\mathcal{P}(E_n)$ but on the subset $\mathcal{P}_n(E_n)$ of measures η such that $\eta(G_n) > 0$. In this case, the particle interpretation given in Definition 3.2.1 is not well-defined since it may happen that the whole configuration ξ_n moves out of the set $\widehat{E}_n$. To describe rigorously the particle model associated with this McKean kernel, we proceed as in Section 2.5.1. We add a cemetery point Δ to the product space E_n^N and we extend the test functions and the mutation/selection transitions $(\mathcal{S}_n, \mathcal{M}_n)$ on E_n^N to $E_n^N \cup \{\Delta\}$ as follows:

- The test functions $\varphi \in \mathcal{B}_b(E_n^N)$ are extended to $E_n^N \cup \{\Delta\}$ by setting $\varphi_n(\delta) = 0$.

- The selection transitions $\mathcal{S}_n$ from E_n^N into itself are extended into transitions on $E_n^N \cup \{\Delta\}$ by setting $\mathcal{S}_n(x,.) = \delta_\Delta$ as soon as $m(x) = \frac{1}{N}\sum_{i=1}^N \delta_{x^i} \notin \mathcal{P}_n(E_n)$.

- The mutation transitions $\mathcal{M}_{n+1}$ from E_n^N into E_{n+1}^N are extended into transitions from $E_n^N \cup \{\Delta\}$ into $E_{n+1}^N \cup \{\Delta\}$ by setting $\mathcal{M}_{n+1}(\Delta,.) = \delta_\Delta$.

The interacting particle model associated with the McKean interpretation (3.5) and with an initial distribution $\eta_0 \in \mathcal{P}(E_0)$ is a sequence of nonhomogeneous Markov chains

$$\left(\Omega^{(N)} = \prod_{n \geq 0}(E_n^N \cup \{\Delta\}), \; \mathcal{F}^N = (\mathcal{F}_n^N)_{n \in \mathbb{N}}, \; \xi = (\xi_n)_{n \in \mathbb{N}}, \; \mathbb{P}_{\eta_0}^N \right)$$

taking values at each time n in the product space $E_n^N \cup \{\Delta\}$. These models are again defined by a two-step selection/mutation transition of the same nature as before,

$$\xi_n \in E_n^N \cup \{\Delta\} \xrightarrow{\text{selection}} \widehat{\xi}_n \in E_n^N \cup \{\Delta\} \xrightarrow{\text{mutation}} \xi_{n+1} \in E_{n+1}^N \cup \{\Delta\}$$

The only difference is that the chain is killed the first time n we have $m(\xi_n) \notin \mathcal{P}_n(E_n)$. At that time, we set $\widehat{\xi}_n = \Delta$ and $\widehat{\xi}_p = \xi_p = \Delta$ for all $p > n$. In this general context, it is important to estimate the distribution's tails on the date at which the chain is killed

$$\tau_N = \inf\{n \in \mathbb{N} \; ; \; \widehat{\xi}_n = \Delta\} = \inf\{n \in \mathbb{N} \; ; \; m(\xi_n)(G_n) = 0\}$$

In this general context, it may happen that the "limiting" Feynman-Kac model itself is stopped. Let

$$\tau = \inf\{n \in \mathbb{N}, \quad \eta_n(G_n) = 0\}$$

be the first time the set $G_n^{-1}(0, \infty)$ and the support of the measure η_n are disjoint. In this case, the distribution η_n cannot be updated anymore and the flow η_n is only defined up to that time horizon. From the discussion given on page 71, the time τ can also be regarded as the first time n we have $\gamma_{n+1}(1) = 0$. If we set $\mathbf{G}_n(x_0, \ldots, x_n) = \prod_{p=0}^{n} G_p(x_p)$, then we find that τ is the first time n we have

$$\mathbb{P}_{\eta_0, n}(\mathbf{G}_n) = \int \mathbf{G}_n(x_0, \ldots, x_n) \eta_0(dx_0) M_1(x_0, dx_1) \ldots M_n(x_{n-1}, dx_n) = 0$$

This shows that at that time the admissible paths of length n must visit at a given date $p \leq n$ the set $G_p^{-1}(0)$. From these observations, it is intuitively clear that the path-particle genealogical model has the same property and we have $\tau^N \leq \tau$. In Chapter 7, Theorem 7.4.1, we will prove that for any $n \leq \tau$ and $N \geq 1$ we have the exponential estimate

$$\mathbb{P}_{\eta_0}^N(\tau^N \leq n) \leq a(n) \exp\left(-N/b(n)\right) \tag{3.6}$$

In particular, this shows that

$$\lim_{N \to \infty} \mathbb{P}_{\eta_0}^N(\tau^N = \tau) = 1$$

This result indicates that the particle model is asymptotically stopped at time τ.

In some instances, such as in finite graph analysis, one is interested in computing the depth τ of a rooted tree E. In this context, we have are given a Markov chain X_n starting at the root and exploring the tree. When X_n visits a leaf, it is stopped. Otherwise, it chooses randomly one of the deepest vertex in its neighborhood. If we consider the indicator potential functions $G_n = 1_A$ of the leaves subset $A \subset E$, then we clearly have

$$\gamma_{n+1}(1) = \mathbb{P}(\forall p \leq n, \ X_n \notin A) = \mathbb{E}\left(\prod_{p=0}^{n} G_p(X_p)\right)$$

and the depth τ coincides with the first time n we have $\gamma_{n+1}(1) = 0$. The particle interpretation of this model is sometimes called the "go with the winner" [4, 263, 264] It consists in evolving N particles from the root. When a particle reaches a leaf, it is killed and instantly a randomly chosen particle in $E - A$ duplicates. The exponential estimate above implies that the particle model is asymptotically stopped at time τ; that is, when the particles have reached the set of deepest leaves. The running time of the algorithm, to acheive a given success probability, depends on the tree geometry and on the way the probability $\gamma_{n+1}(1)$ to reach a depth n tends to 0 (see for instance Theorem 7.4.1).

3.4 Historical and Genealogical Tree Models

In this section, we show that the particle interpretation of the Feynman-Kac model in path space represents the genealogy of the particle model associated with the time marginal model. We first start with a somehow heuristic description. Then we provide a rigorous mathematical proof. Finally, we show that this result is related to more general transport problems. We end this section around this theme.

3.4.1 Introduction

The best pedagogical way to introduce the historical model associated with a genetic type particle approximation model is to analyze the situation where the underlying Markov motion X_n is a path process. Suppose X'_n is an auxiliary nonhomogeneous Markov chain with initial distribution $\eta_0 \in \mathcal{P}(E'_0)$ and elementary transitions M'_{n+1} from some measurable space $(E'_n, \mathcal{E}'_n)$ into another measurable space $(E'_{n+1}, \mathcal{E}'_{n+1})$. Also let X_n be the stochastic sequence defined by

$$X_n = X'_{[0,n]} (= (X'_0, \ldots, X'_n)) \in E_n = E'_{[0,n]} (= (E'_0 \times \ldots \times E'_n))$$

We recall that X_n forms a nonhomogeneous Markov chain with initial distribution $\eta_0 \in \mathcal{P}(E_0)$ and its transitions M_n, from E_{n-1} into E_n, are defined in (2.3) on page 51.

We consider a sequence of bounded potential functions $G_n : E_n \to (0, \infty)$ and we suppose the energy of a path $x_n = (x'_0, \ldots, x'_n) \in E_n$ only depends on its terminal value. That is, we have $G_n(x_n) = G'_n(x'_n)$ for some function G'_n on E'_n. The Feynman-Kac model η_n associated with the pair (G_n, M_n) can be alternatively written for any $f_n \in \mathcal{B}(E_n)$ as $\eta_n(f_n) = \gamma_n(f_n)/\gamma_n(1)$ with the Feynman-Kac formulae

$$\gamma_n(f_n) = \mathbb{E}_{\eta_0}\left(f_n(X_n) \prod_{p=0}^{n-1} G_p(X_p)\right) = \mathbb{E}_{\eta_0}\left(f_n(X'_{[0,n]}) \prod_{p=0}^{n-1} G'_p(X'_p)\right) \tag{3.7}$$

Let π_n be the canonical projection from E_n into E'_n defined by

$$\pi : x_n = (x'_0, \ldots, x'_n) \in E_n \longrightarrow \pi_n(x_n) = x'_n \in E'_n$$

We associate with π_n the image mapping $\pi_n^{-1} : \mathcal{P}(E_n) \to \mathcal{P}(E'_n)$ defined for any $\eta \in \mathcal{P}(E_n)$ and $A'_n \in \mathcal{E}'_n$ by setting

$$\pi_n^{-1}(\eta)(A'_n) = \eta \circ \pi_n^{-1}(A'_n) = \eta(\pi_n^{-1}(A'_n))$$

The image measures $\eta'_n = \eta_n \circ \pi_n^{-1} \in \mathcal{P}(E'_n)$ are clearly the nth time marginal of η_n. In view of (3.7), the sequence of measures η'_n is the Feynman-Kac distribution flow associated with the pair (G'_n, M'_n) and is defined for any $f'_n \in \mathcal{B}(E'_n)$ by the formulae

$$\eta'_n(f'_n) = \gamma'_n(f'_n)/\gamma'_n(1) \quad \text{with} \quad \gamma'_n(f'_n) = \mathbb{E}_{\eta_0}\left(f'_n(X'_n)\prod_{p=0}^{n-1}G'_p(X'_p)\right)$$

The N-path particle model associated with the McKean interpretation (3.1) of the Feynman-Kac flow η_n in path space consists of N path particles. That is, for any $1 \leq i \leq N$ and $n \in \mathbb{N}$, we have

$$\xi^i_n = (\xi'^i_{p,n})_{0\leq p\leq n} \quad \text{and} \quad \widehat{\xi}^i_n = (\widehat{\xi}'^i_{p,n})_{0\leq p\leq n} \in E_n = E'_{[0,n]}.$$

We recall that the motion of the N-path particles is decomposed in two separate selection/mutation transitions:

$$\xi_n \in E^N_n \xrightarrow{S_{n,m(\xi_n)}} \widehat{\xi}_n \in E^N_n \xrightarrow{M_{n+1}} \xi_{n+1} \in E^N_{n+1}$$

During the selection transition, each path particle ξ^i_n selects a new path particle $\widehat{\xi}^i_n$ randomly chosen according to the distribution

$$S_{n,m(\xi_n)}(\xi^i_n, \cdot) = \epsilon_n G'_n(\xi'^i_{n,n})\,\delta_{\xi^i_n} + (1 - \epsilon_n G'_n(\xi'^i_{n,n}))\,\Psi_n(m(\xi_n))$$

with the Boltzmann-Gibbs distribution

$$\Psi_n(m(\xi_n)) = \sum_{i=1}^{N}\frac{G'_n(\xi'^i_{n,n})}{\sum_{j=1}^{N}G'_n(\xi'^j_{n,n})}\,\delta_{\xi^i_n}$$

At this stage, it is convenient to make a couple of remarks. First, we note that the selection probabilities only depend on the terminal values $\pi_n(\xi^i_n) = \xi'^i_{n,n}$ of the path. From this observation, we check that the terminal values $\widehat{\xi}'^i_{n,n}$ of the selected paths are randomly chosen according to the distribution

$$S'_{n,m(\xi'_{n,n})}(\xi'^i_{n,n}, \cdot) = \epsilon_n G'_n(\xi'^i_{n,n})\,\delta_{\xi'^i_{n,n}} + (1 - \epsilon_n G'_n(\xi'^i_{n,n}))\,\Psi'_n(m(\xi'_{n,n}))$$

with $m(\xi'_{n,n}) = \frac{1}{N}\sum_{i=1}^{N}\delta_{\xi'^i_{n,n}} = m(\xi_n)\circ \pi_n^{-1}$ and

$$\Psi'_n(m(\xi'_{n,n})) = \sum_{i=1}^{N}\frac{G'_n(\xi'^i_{n,n})}{\sum_{j=1}^{N}G'_n(\xi'^j_{n,n})}\,\delta_{\xi'^i_{n,n}}$$

Secondly, if the ith particle $\widehat{\xi}'^i_{n,n}$ has selected, say, the jth terminal state $\xi'^j_{n,n}$, we can alternatively define the ith path selection $\widehat{\xi}^i_n$ by setting $\widehat{\xi}^i_n = \xi^j_n$. In other words, we have in a symbolic form

$$\widehat{\xi}'^i_{n,n} = \xi'^j_{n,n} \Longrightarrow \widehat{\xi}^i_n = (\xi'^j_{p,n})_{0\leq p\leq n}$$

3.4 Historical and Genealogical Tree Models   105

If we interpret the path particles as an ancestral line, we see that the path selection tends to select randomly the ancestral line of a current individual with high G'_n-potential.

During mutation, each selected path $\widehat{\xi}^i_n$ evolves randomly according to the transition M_{n+1} of the path process. By definition, this transition simply consists in extending the selected path with an elementary move randomly performed with M'_n. In other words, we have

$$\begin{aligned}\xi^i_{n+1} &= ((\xi'^i_{p,n+1})_{0 \leq p \leq n}, \xi'^i_{n+1,n+1}) \\ &= ((\widehat{\xi}'^i_{p,n})_{0 \leq p \leq n}, \ \xi'^i_{n+1,n+1}) \in E_{n+1} = E_n \times E'_{n+1}\end{aligned}$$

where $\xi'^i_{n+1,n+1}$ is a random variable with law $M'_{n+1}(\widehat{\xi}'^i_{n,n}, .)$. In connection with the interpretation above, this elementary move can be regarded as the mutation of the selected individual $\widehat{\xi}'^i_{n,n}$ with ancestral line $(\widehat{\xi}'^i_{0,n}, \ldots, \widehat{\xi}'^i_{n,n})$.

From the preceding discussion it should be intuitively clear that the marginal particle model is the two-step selection/mutation Markov model

$$\xi'_{n,n} \in (E'_n)^N \xrightarrow{S'_{n,m(\xi'_{n,n})}} \widehat{\xi}'_{n,n} \in (E'_n)^N \xrightarrow{M'_{n+1}} \xi'_{n+1,n+1} \in (E'_{n+1})^N$$

Furthermore, it coincides with the particle model associated with the McKean interpretation of the flow η'_n with Markov kernels $K'_{n+1,\eta} = S'_{n,\eta} M'_{n+1}$.

3.4.2 A Rigorous Approach and Related Transport Problems

Our next objective is to make rigorous the intuitive statement presented in the introductory Section 3.4.1. To this end, it is convenient to introduce the canonical mappings

$$\begin{aligned}\pi^N_n : \quad E^N_n &\longrightarrow (E'_n)^N \\ (x^1_n, \ldots, x^N_n) &\mapsto \pi^N_n(x^1_n, \ldots, x^N_n) = (\pi_n(x^1_n), \ldots, \pi_n(x^N_n))\end{aligned}$$

Proposition 3.4.1 *Let $(\xi_n, \widehat{\xi}_n)$ be the N-path-particle model associated with the Feynman-Kac distribution flow η_n on path space. The stochastic process defined by*

$$\xi'_n = \pi^N_n(\xi_n) \quad \text{and} \quad \widehat{\xi}'_n = \pi^N_n(\widehat{\xi}_n)$$

coincides with the N-particle model associated with the Feynman-Kac flow η'_n.

Proof:
We first prove that for any $u_n = (u^0_n, \ldots, u'_n) \in E_n$, $n \in \mathbb{N}$, and $\eta \in \mathcal{P}(E_n)$ we have

$$S_{n,\eta}(u_n, .) \circ \pi^{-1}_n = S'_{n,\eta \circ \pi^{-1}_n}(\pi_n(u_n), .) \tag{3.8}$$

$$M_{n+1}(u_n, .) \circ \pi^{-1}_{n+1} = M'_{n+1}(\pi_n(u_n), .) \tag{3.9}$$

Let $f'_n \in \mathcal{B}_b(E'_n)$ and let us denote by $dv_n = dv'_0 \times \ldots \times dv'_n$ the infinitesimal neighborhood of a point $v_n = (v'_0, \ldots, v'_n) \in E_n$. A simple calculation shows that

$$\int_{E_n} f'_n(\pi_n(v_n)) \, S_{n,\eta}(u_n, dv_n)$$
$$= \int_{E_n} f'_n(v'_n) \, (\epsilon_n G'_n(u'_n) \delta_{u_n}(dv_n) + (1 - \epsilon_n G'_n(u'_n)) \frac{G'_n(v'_n)}{\eta \circ \pi_n^{-1}(G'_n)} \eta(dv_n))$$
$$= \int_{E'_n} f'_n(v'_n) \, \epsilon_n G'_n(u'_n) \delta_{u'_n}(dv'_n)$$
$$+ \int_{E'_n} f'_n(v'_n) \, (1 - \epsilon_n G'_n(u'_n)) \frac{G'_n(v'_n)}{\eta \circ \pi_n^{-1}(G'_n)} \, \eta \circ \pi_n^{-1}(dv'_n)$$

This implies that

$$\int_{E_n} f'_n(\pi_n(v_n)) \, S_{n,\eta}(u_n, dv_n) = \int_{E'_n} f'_n(v'_n) \, S'_{n, \eta \circ \pi_n^{-1}}(\pi_n(u_n), dv'_n)$$

and the proof of the first assertion is completed. To prove the second one, we argue in the same way. We first note that for any $u_{n-1} \in E_{n-1}$, $n \geq 1$, we have

$$\int_{E_n} f'_n(\pi_n(v_n)) \, M_n(u_{n-1}, dv_n) = \int_{E'_n} f'_n(v'_n) \, M'_n(\pi_{n-1}(u_{n-1}), dv'_n)$$

This ends the proof of these two preliminary results. We now come to the proof of the proposition. For any $f_n \in \mathcal{B}_b(E_n)$ and $n \geq 0$, we find that

$$\mathbb{E}^N_{\eta_0}(f_n(\pi_n^N(\widehat{\xi}_n)) \mid \xi_n) = \int_{E_n^N} f_n(\pi_n^N(x_n)) \prod_{i=1}^N S_{n,m(\xi_n)}(\xi_n^i, dx_n^i)$$
$$= \int_{(E'_n)^N} f_n(x'_n) \prod_{i=1}^N (S_{n,m(\xi_n)}(\xi_n^i, \cdot) \circ \pi_n^{-1})(dx'^i_n)$$

Using (3.8), we obtain

$$\mathbb{E}^N_{\eta_0}(f_n(\pi_n^N(\widehat{\xi}_n)) \mid \xi_n) = \int_{(E'_n)^N} f_n(x'_n) \prod_{i=1}^N S'_{n,m(\xi'_{n,n})}(\xi'^i_{n,n}, dx'^i_n)$$
$$= \mathbb{E}^N_{\eta_0}(f_n(\widehat{\xi}'_{n,n}) \mid \xi'_{n,n})$$

Finally, we observe that

$$\mathbb{E}^N_{\eta_0}(f_{n+1}(\pi^N_{n+1}(\xi_{n+1})) \mid \widehat{\xi}_n)$$
$$= \int_{E^N_{n+1}} f_{n+1}(\pi^N_{n+1}(x_{n+1})) \prod_{i=1}^N M_{n+1}(\widehat{\xi}^i_n, dx^i_{n+1})$$
$$= \int_{(E'_{n+1})^N} f_{n+1}(x'_{n+1}) \prod_{i=1}^N (M_{n+1}(\widehat{\xi}^i_n, \cdot) \circ \pi_n^{-1})(dx'^i_{n+1})$$

3.4 Historical and Genealogical Tree Models

Using (3.9), we conclude that

$$\mathbb{E}_{\eta_0}^N(f_{n+1}(\pi_{n+1}^N(\xi_{n+1})) \mid \widehat{\xi}_n)$$

$$= \int_{(E'_{n+1})^N} f_{n+1}(x'_{n+1}) \prod_{i=1}^N M'_{n+1}(\widehat{\xi}'^i_{n,n}, dx'^i_{n+1})$$

$$= \mathbb{E}_{\eta_0}^N(f_{n+1}(\xi'_{n+1,n+1}) \mid \widehat{\xi}'_{n,n})$$

From the Markov description of the particle models, the end of the proof of the proposition is now straightforward. ∎

On closer inspection, the proof of the proposition above shows that this result is related to a more general transport problem. Suppose η_n is a given sequence of distributions on some measurable space E_n with a McKean interpretation

$$\eta_{n+1} = \eta_n K_{n+1,\eta_n}$$

where $K_{n+1,\eta}$ is a well-defined collection of Markov transitions from E_n into E_{n+1}. We consider an auxiliary collection of measurable spaces $(E'_n, \mathcal{E}'_n)$ and a measurable mapping $\pi_n : E_n \to E'_n$. As before, we introduce the image measures of η_n with respect to π_n and defined by

$$\eta'_n = \eta_n \circ \pi_n^{-1} \in \mathcal{P}(E'_n)$$

We assume there exists a McKean interpretation

$$\eta'_{n+1} = \eta'_n K'_{n+1,\eta'_n}$$

in terms of a collection of Markov kernels $K'_{n+1,\eta'}$, $\eta' \in \mathcal{P}(E'_n)$, from E'_n into E'_{n+1}. Note that the existence of a McKean measure is equivalent to the fact that the sequence η'_n satisfies a recursive equation. Using the same arguments as the ones used in the proof of Proposition 3.4.1, we prove the following result.

Proposition 3.4.2 *Suppose the collections of Markov kernels $K_{n+1,\eta}$ and $K'_{n+1,\eta'}$ satisfy the compatibility condition*

$$K_{n+1,\eta}(x_n, \cdot) \circ \pi_{n+1}^{-1} = K'_{n+1,\eta \circ \pi_n^{-1}}(\pi_n(x_n), \cdot) \qquad (3.10)$$

for any $(x_n, \eta) \in (E_n \times \mathcal{P}(E_n))$ and $n \in \mathbb{N}$. Let $\xi_n \in E_n^N$ be the N-particle model associated with the collection of transitions $K_{n+1,\eta}$ and with a measure $\eta_0 \in \mathcal{P}(E_0)$. Then, the stochastic process defined by

$$\pi_n^N(\xi_n) = (\pi_n(\xi_n^1), \ldots, \pi_n(\xi_n^1)) \in (E'_n)^N$$

coincides with the N-particle model associated with the collection of transitions $K'_{n+1,\eta'}$ and with the measure $\eta'_0 = \eta_0 \circ \pi_0^{-1} \in \mathcal{P}(E'_0)$.

108   3. Genealogical and Interacting Particle Models

We can use this transport property to construct genealogical tree models of various classes of interacting jump particle models. Although we didn't work out the continuous time version of this proposition, we believe that this strategy can be used to analyze the genealogical structure of Nanbu type and colliding particles interpretations of Boltzmann's rarefied gas models.

3.4.3 Complete Genealogical Tree Models

In this section we give a brief discussion on complete genealogical tree models. We suppose E_n are defined by

$$E_n = E'_{[0,n]} (= E'_0 \times \ldots \times E'_n))$$

We let π_n be the canonical projection from E_n into E'_n and we suppose that $K_{n+1,\eta}$ is a given collection of Markov transitions from E_n into E_{n+1}. When condition (3.10) holds for some Markov transitions $K'_{n+1,\eta'}$ from E'_n into E'_{n+1}, we have two distribution flows:

$$\eta_{n+1} = \eta_n K_{n+1,\eta_n} \in \mathcal{P}(E'_{[0,n]})$$
$$\eta'_{n+1} = \eta'_n K'_{n+1,\eta'_n} = \eta_{n+1} \circ \pi_{n+1}^{-1} \in \mathcal{P}(E'_n)$$

Let $\mathbb{K}'_{n,\eta_0} \in \mathcal{P}(E'_{[0,n]})$ be the McKean measure associated with $K'_{n+1,\eta'}$ and $\eta'_0 = \eta_0 \in \mathcal{P}(E_0)$ and given by

$$\mathbb{K}'_{n,\eta_0}(d(x'_0, \ldots, x'_n)) = \eta'_0(dx'_0) \, K'_{1,\eta'_0}(x'_0, dx'_1) \ldots K'_{n,\eta'_{n-1}}(x'_{n-1}, dx'_n)$$

We observe that the sequence of distributions $\mathbb{K}'_{n,\eta_0}$ satisfies the recursive equation

$$\mathbb{K}'_{n+1,\eta_0} = \mathbb{K}'_{n,\eta_0} \times K'_{n+1,\eta'_n} \quad \text{with} \quad \eta'_n = \mathbb{K}'_{n,\eta_0} \circ \pi_n^{-1} \tag{3.11}$$

If we set for each $\mu_n \in \mathcal{P}(E'_{[0,n]})$

$$\mathbf{K}_{n+1,\mu_n}((x'_0, \ldots, x'_n), d(y'_0, \ldots, y'_{n+1}))$$
$$= \delta_{(x'_0,\ldots,x'_n)}(d(y'_0, \ldots, y'_n)) \, K'_{n+1,\mu_n \circ \pi_n^{-1}}(y'_n, dy'_{n+1})$$

then (3.11) can be rewritten in the following form

$$\mathbb{K}'_{n+1,\eta_0} = \mathbb{K}'_{n,\eta_0} \mathbf{K}_{n+1,\mathbb{K}'_{n,\eta_0}} \in \mathcal{P}(E_{n+1})$$

By definition, the collection of transitions $\mathbf{K}_{n+1,\mu_n}$ from E_n into E_{n+1} satisfies condition (3.10); that is, we have for any $\mu_n \in \mathcal{P}(E_n)$ and $x_n \in E_n$

$$\mathbf{K}_{n+1,\mu_n}(x_n, .) \circ \pi_{n+1}^{-1} = K'_{n+1,\mu_n \circ \pi_n^{-1}}(\pi_n(x_n), .)$$

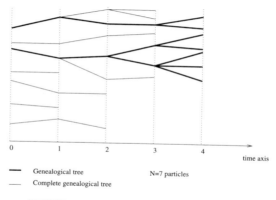

FIGURE 3.1. Complete genealogical tree

To illustrate these abstract constructions, we notice that, in the case of the Feynman-Kac models examined above, the distributions η_n and $\mathbb{K}'_{n,\eta_0}$ represent respectively the Feynman-Kac path measure associated with the pair (G'_n, M'_n) and the McKean measure associated with $K'_{n,\eta'}$. From previous considerations, the N-interacting-particle model

$$\xi_n^i = (\xi'^i_{0,n}, \xi'^i_{1,n} \ldots, \xi'^i_{n,n})$$

associated with $K_{n,\eta}$ gives the genealogies of the N-particle model ξ'^i_n associated with $K'_{n,\eta'}$. Furthermore, the N-path-particle model

$$\zeta_n^i = (\xi'^i_0, \ldots, \xi'^i_n)$$

coincides with the N-path particle model associated with $\mathbf{K}_{n,\mu}$. In contrast with the genealogical tree model ξ_n^i, the system ζ_n^i represents the *complete history* of the particle model ξ'^i_n.

In Figure 3.1, we have represented with a thick line the genealogical tree of the current set of $N = 7$ individuals and by thin lines the complete genealogy of the particle evolution model since its origin.

3.5 Particle Approximation Measures

In Section 3.2 and Section 3.4, we introduced several particle interpretations of Feynman-Kac distribution flows, including path-space and genealogical tree-based models. Chapter 12 is devoted to selected applications in statistics, physics, biology, operations research, and signal processing. In each of these domains particle models have a different interpretation. They can be regarded alternatively as a particle simulation technique or as a microscopic particle interpretation of some physical or biological equations as well as a stochastic and adaptive grid approximation technique. The asymptotic behavior of particle models as the size of the systems and the time parameter tend to infinity will be discussed in the further development of Chapter 7.

The objective of this section is mainly to motivate the abstract constructions developed in earlier sections as well as to illustrate the forthcoming applications and convergence analysis. We first present a nonexhaustive catalog of particle approximation measures. In Section 3.5.1, we provide some convergence estimates. In Section 3.5.2, we discuss the main regularity properties used in this book.

At the risk of repetition, we briefly recall the definition of the main Feynman-Kac models introduced so far. Let $(E'_n, \mathcal{E}'_n)$, $n \in \mathbb{N}$, be a sequence of measurable spaces. We denote by $X_n = X'_{[0,n]}$, $n \in \mathbb{N}$, the historical process associated with a nonhomogeneous Markov chain X'_n with transitions M'_{n+1} from E'_n into E'_{n+1}. By M_n we denote the Markov transitions of the path process X_n. We also consider a collection of bounded and nonnegative potential functions G'_n on E'_n, and we let G_n be the extension of G'_n to the path space $E_n = E'_{[0,n]}$ defined by $G_n(x'_0, \ldots, x'_n) = G'_n(x'_n)$. To clarify the presentation and unless otherwise stated, we assume that the potential functions are strictly positive. The Feynman-Kac prediction flow $\eta'_n \in \mathcal{P}(E'_n)$ associated with the pair (G'_n, M'_n) is defined for any test function $f'_n \in \mathcal{B}_b(E'_n)$ by the formula

$$\eta'_n(f'_n) = \gamma'_n(f'_n)/\gamma'_n(1) \quad \text{with} \quad \gamma'_n(f'_n) = \mathbb{E}_{\eta_0}\left(f'_n(X'_n) \prod_{p=0}^{n-1} G'_p(X'_p)\right)$$

The measures η'_n are the marginal distributions of the Feynman-Kac prediction model $\eta_n \in \mathcal{P}(E_n)$ associated with the pair (G_n, M_n) and defined for any test function $f_n \in \mathcal{B}_b(E_n)$ by

$$\eta_n(f_n) = \gamma_n(f_n)/\gamma_n(1) \quad \text{with} \quad \gamma_n(f_n) = \mathbb{E}_{\eta_0}\left(f_n(X_n) \prod_{p=0}^{n-1} G_p(X_p)\right)$$

The corresponding updated models are defined by

$$\widehat{\eta}'_n(f'_n) = \eta'_n(f'_n G'_n)/\eta'_n(1) = \widehat{\gamma}'_n(f'_n)/\widehat{\gamma}'_n(1) \quad \text{with} \quad \widehat{\gamma}'_n(f'_n) = \gamma'_n(f'_n G'_n)$$

and

$$\widehat{\eta}_n(f_n) = \eta_n(f_n G_n)/\eta_n(1) = \widehat{\gamma}_n(f_n)/\widehat{\gamma}_n(1) \quad \text{with} \quad \widehat{\gamma}_n(f_n) = \gamma_n(f_n G_n)$$

We finally recall (see Proposition 2.3.1) that the unnormalized models $(\gamma_n, \widehat{\gamma}_n)$ can be written in terms of the normalized distribution flows $(\eta_n, \widehat{\eta}_n)$ with the multiplicative formulae

$$\gamma_n(f_n) = \eta_n(f_n) \prod_{p=0}^{n-1} \eta'_p(G'_p) \quad \text{and} \quad \widehat{\gamma}_n(f_n) = \widehat{\eta}_n(f_n) \prod_{p=0}^{n} \eta'_p(G'_p)$$

Suppose we are given a McKean interpretation of the flows

$$\eta'_{n+1} = \eta'_n K'_{n+1,\eta'_n} \quad \text{and} \quad \eta_{n+1} = \eta_n K_{n+1,\eta_n}$$

with a collection of "compatible" Markov kernels $K'_{n+1,\eta'} = S'_{n,\eta'} M'_{n+1}$ and $K_{n+1,\eta} = S_{n,\eta} M_{n+1}$ (see Section 2.5.3 and Section 3.4). The corresponding McKean measures (see Section 2.5.2) are defined by

$$\begin{aligned}
\mathbb{K}_{\eta_0,n}(d(x_0,\ldots,x_n)) &= \eta_0(dx_0)\, \mathcal{K}_{1,\eta_0}(x_0, dx_1)\, \ldots\, \mathcal{K}_{n,\eta_{n-1}}(x_{n-1}, dx_n)\\
\mathbb{K}'_{\eta'_0,n}(d(x'_0,\ldots,x'_n)) &= \eta'_0(dx'_0)\, \mathcal{K}'_{1,\eta'_0}(x'_0, dx'_1)\, \ldots\, \mathcal{K}'_{n,\eta'_{n-1}}(x'_{n-1}, dx'_n)
\end{aligned}$$

The N-particle models associated with $K_{n,\eta}$ consists of N path particles

$$\xi^i_n = (\xi'^i_{p,n})_{0\le p \le n} \quad \text{and} \quad \widehat{\xi}^i_n = (\widehat{\xi}'^i_{p,n})_{0 \le p \le n} \in E_n = E'_{[0,n]}$$

Furthermore, the "marginal systems" $(\xi'^i_{n,n})_{1\le i \le N}$ and $(\widehat{\xi}'^i_{n,n})_{1 \le i \le N}$ on $(E'_n)^N$ coincide with the N-particle model associated with $K'_{n,\eta'}$.

We are now in position to present a list of particle approximation models of the previously defined distributions. We use the superscript ν^N to define the particle approximation measure of a measure ν that is in some sense $\lim_{N \to \infty} \nu^N = \nu$. The particle approximation measures of the McKean distributions $\mathbb{K}'_{\eta'_0,n}$ and $\mathbb{K}_{\eta_0,n}$ are given by

$$\begin{aligned}
\mathbb{K}'^N_{\eta_0,n} &= \frac{1}{N} \sum_{i=1}^{N} \delta_{(\xi'^1_0,\ldots,\xi'^i_n)}\\
\mathbb{K}^N_{\eta_0,n} &= \frac{1}{N} \sum_{i=1}^{N} \delta_{(\xi^1_0,\ldots,\xi^i_n)}\\
&= \frac{1}{N} \sum_{i=1}^{N} \delta_{[(\xi'^i_{0,0}),(\xi'^i_{0,1},\xi'^i_{1,1}),(\xi'^i_{0,2},\xi'^i_{1,2},\xi'^i_{2,2}),\ldots,(\xi'^i_{0,n},\ldots,\xi'^i_{n,n})]}
\end{aligned}$$

The particle approximation measures of the prediction models (γ_n, η_n) are defined by

$$\gamma^N_n(.) = \left[\prod_{p=0}^{n-1} \eta'^N_p(G'_p)\right] \times \eta^N_n(.) \quad \text{with} \quad \eta^N_n = \frac{1}{N} \sum_{i=1}^{N} \delta_{(\xi'^i_{0,n},\ldots,\xi'^i_{n,n})}$$

The resulting particle approximation measures of the "marginal" prediction models (γ'_n, η'_n) are clearly given by

$$\gamma'^N_n(.) = \left[\prod_{p=0}^{n-1} \eta'^N_p(G'_p)\right] \times \eta'^N_n(.) \quad \text{with} \quad \eta'^N_n = \frac{1}{N} \sum_{i=1}^{N} \delta_{\xi'^i_{n,n}}$$

The N-particle approximation of the updated models $(\widehat{\eta}_n, \widehat{\gamma}_n)$ can be defined in two different ways. First we can choose the updated path-particle model

$$\widehat{\gamma}_n^N(.) = \left[\prod_{p=0}^n \eta_p'^N(G_p')\right] \times \widehat{\eta}_n^N(.) \quad \text{with} \quad \widehat{\eta}_n^N = \frac{1}{N}\sum_{i=1}^N \delta_{(\widehat{\xi}_{0,n}'^i, \ldots, \widehat{\xi}_{n,n}'^i)}$$

Alternatively, we can choose the updated versions of the distributions η_n^N and γ_n^N defined respectively in terms of the Boltzmann-Gibbs transformations

$$\widehat{\gamma}_n^N(.) = \left[\prod_{p=0}^n \eta_p^N(G_p)\right] \times \widehat{\eta}_n^N(.) \quad \text{with} \quad \widehat{\eta}_n^N(dx_n) = \frac{G_n(x_n)}{\eta_n^N(G_n)}\eta_n^N(dx_n)$$

When the potential functions are not strictly positive, the particle model is generally stopped the first time τ_N the whole configuration visits the set $G_n^{-1}(0)$ (see the end of Section 3.2). In this situation, we use for instance the particle approximation measures

$$\gamma_n^N(.) = \left[\prod_{p=0}^{n-1} \eta_p'^N(G_p')\right] \times \eta_n^N(.) \quad \text{with} \quad \eta_n^N = 1_{\tau_N \geq n} \cdot \frac{1}{N}\sum_{i=1}^N \delta_{(\xi_{0,n}'^i, \ldots, \xi_{n,n}'^i)}$$

3.5.1 Some Convergence Results

To guide the reader, we provide hereafter a brief and informal discussion on the asymptotic behavior of these particle measures as the size of the system N tends to infinity. This short discussion will help the reader in appreciating the impact and the usefulness of the preceding particle interpretations in the set of applications presented in Chapter 12. We will develop the complete proofs of these results and the precise set of conditions on the pair (G_n, M_n) in the further development of Chapters 7 to 10.

A surprising result at first sight is that γ_n^N is an unbiased estimator; that is, we have for any $f_n \in \mathcal{B}_b(E_n)$

$$\mathbb{E}_{\eta_0}^N(\gamma_n^N(f_n)) = \gamma_n(f_n)$$

In fact, we will see that γ_n^N can be regarded as the terminal value of a martingale. This observation is also the stepping stone of a powerful semigroup approach for studying the fluctuations of these particle models.

In a first stage of analysis, we will develop a collection of $\mathbb{L}_p$-estimates using martingale decompositions and limit theorems for processes. For instance, we will show that

$$\sqrt{N}\mathbb{E}_{\eta_0}^N[|\eta_n^N(f_n) - \eta_n(f_n)|^p]^{1/p} \leq a(p)b(n)\|f\|$$

We will also extend these estimates to the empirical processes $\eta_n^N : f_n \in \mathcal{F}_n \longrightarrow \eta_n^N(f_n)$ associated with a given countable collection of uniformly bounded functions $\mathcal{F}_n \subset \mathcal{B}_b(E_n)$. These results at the process level lead to the version of the Glivenko-Cantelli theorem for a particle model. In particular, we will prove that

$$\sqrt{N} \mathbb{E}_{\eta_0}^N \left[\sup_{f_n \in \mathcal{F}_n} |\eta_n^N(f_n) - \eta_n(f_n)|^p \right]^{1/p} \leq a(p) b(n) \ I(\mathcal{F}_n)$$

for some finite constant $I(\mathcal{F}_n) < \infty$ that only depends on the class $\mathcal{F}_n$. Similar but exponential type estimates will also be covered. For instance, we will check that for any $\epsilon > 0$ and N sufficiently large

$$\mathbb{P}_{\eta_0}^N \left[\sup_{f_n \in \mathcal{F}_n} |\eta_n^N(f_n) - \eta_n(f_n)| > \varepsilon \right] \leq d_n(\epsilon, \mathcal{F}_n) \ e^{-N\epsilon^2/b(n)}$$

with a finite constant $d_n(\epsilon, \mathcal{F}_n)$ depending on ϵ and on the class $\mathcal{F}_n$. From these estimates and using the Borel-Cantelli lemma, we conclude the almost sure convergence result

$$\lim_{N \to \infty} \sup_{f_n \in \mathcal{F}_n} |\eta_n^N(f_n) - \eta_n(f_n)| = 0$$

When the Markov transitions are sufficiently regular, we will obtain a series of uniform convergence results with respect to the time parameter. For instance, we will prove that for sufficiently mixing kernels M_n' and for any parameter $p \geq 1$

$$\sqrt{N} \sup_{n \geq 0} \mathbb{E}_{\eta_0}^N \left[\sup_{f_n' \in \mathcal{F}_n'} |\eta_n'^N(f_n') - \eta_n'(f_n')|^p \right]^{\frac{1}{p}} \leq a(p) \ b \ I(\mathcal{F})$$

for another finite constant $I(\mathcal{F}') < \infty$ whose values depend on the collection of functions $f_n' \in \mathcal{F}_n' \subset \mathcal{B}_b(E_n')$. Nevertheless, without getting into more detail, we mention that the uniform estimates with respect to the time parameter will not apply in the path-valued situation and for genealogical models.

The corresponding fluctuations and large deviations will also be discussed in Chapter 9 and Chapter 10. Roughly speaking, these results will give the exact asymptotic deviations of η_n^N around the limiting distribution η_n.

Another interesting way to measure the performance of these particle models consists in studying the adequacy of laws and the independence between the particles. These propagation-of-chaos results are concerned with the asymptotic behavior of the distributions of the particle paths (see Chapter 8). To describe these results in some detail, it is convenient to introduce a few notation.

For each $n \geq 0$ and $N \geq 1$, the state space $E_0^N \times \ldots \times E_n^N$ represents the set of paths from the origin up to time n of the Markov particle model, while the product space $(E_0 \times \ldots \times E_n)^N$ represents the state space of the N paths of each elementary particle from the origin up to time n. To connect these two spaces, we introduce the change of coordinate mapping

$$\Theta_n^N : (E_0^N \times \ldots \times E_n^N) \longrightarrow (E_0 \times \ldots \times E_n)^N$$

defined by $\Theta_n^N[(x_0^i)_{1\leq i\leq N},\ldots,(x_n^i)_{1\leq i\leq N}] = (x_0^i,\ldots,x_n^i)_{1\leq i\leq N}$. The mapping above connects the paths of the particle Markov chain with the elementary particle paths

$$\Theta_n^N(\xi_0,\ldots,\xi_n) = (\xi_0^i,\ldots,\xi_n^i)_{1\leq i\leq N} =_{\text{def.}} (\xi_{[0,n]}^i)_{1\leq i\leq N} = \xi_{[0,n]}$$

We recall that the particle Markov chain ξ_n is defined on the canonical space

$$\left(\Omega^{(N)} = \prod_{n\geq 0} E_n^N, \ \mathcal{F}^N = (\mathcal{F}_n^N)_{n\in\mathbb{N}}, \ \xi^{(N)} = (\xi_n^{(N)})_{n\in\mathbb{N}}, \ \mathbb{P}_{\eta_0}^N \right)$$

The restriction of $\mathbb{P}_{\eta_0}^N$ on the sequence of paths from the origin up to time n is given by

$$\mathbb{P}_{\eta_0,n}^N = \mathbb{P}_{\eta_0}^N \circ (\xi_0,\ldots,\xi_n)^{-1} \in \mathcal{P}(E_0^N \times \ldots \times E_n^N)$$

The Θ_n^N image of $\mathbb{P}_{\eta_0,n}^N$ on the elementary particle paths $(E_0 \times \ldots \times E_n)^N$ is defined by

$$\mathbb{P}_{\eta_0,n}^{(N)} = \mathbb{P}_{\eta_0,n}^N \circ (\Theta_n^N)^{-1}$$

With some obvious abusive notation, we have

$$\mathbb{P}_{\eta_0,n}^{(N)} = \mathbb{P}_{\eta_0,n}^N \circ \xi_{[0,n]}^{-1} = \text{Law}((\xi_0^i,\ldots,\xi_n^i)_{1\leq i\leq N})$$

We denote by $\mathbb{P}_{\eta_0,n}^{(N,q)}$, $q \leq N$, their marginals on the first q-particles

$$\mathbb{P}_{\eta_0,n}^{(N,q)} = \text{Law}((\xi_0^i,\ldots,\xi_n^i)_{1\leq i\leq q})$$

and we let $\mathbb{P}_{\eta_0,[n]}^{(N,q)} = \text{Law}((\xi_n^i)_{1\leq i\leq q})$ be their n-time marginals. Without any regularity assumptions on M_n, we prove first a rather crude estimate

$$N \ \|\mathbb{P}_{\eta_0,n}^{(N,q)} - \mathbb{K}_{\eta_0,n}^{\otimes q}\|_{tv} \leq b(n) \ q^2$$

When the Markov kernels M_n are regular enough, we will obtain a propagation-of-chaos estimate with respect to the relative entropy criterion

$$N \ \text{Ent}(\mathbb{P}_{\eta_0,n}^{(N,q)} \mid \mathbb{K}_{\eta_0,n}^{\otimes q}) \leq b(n) \ q$$

Under some additional mixing conditions, we will see that

$$N \sup_{n \geq 0} \|\mathbb{P}^{(N,q)}_{\eta_0,[n]} - \eta_n^{\otimes q}\|_{\mathrm{tv}} \leq b(q) \quad \text{and} \quad N \operatorname{Ent}(\mathbb{P}^{(N,q)}_{\eta_0,n} \mid \mathbb{K}^{\otimes q}_{\eta_0,n}) \leq n\,q\,b$$

From these uniform estimates, we obtain increasing propagation of chaos with respect to increasing time horizons and particle block sizes. More precisely, for any increasing sequences $q = q(N) \uparrow \infty$ and $n = n(N) \uparrow \infty$, we have

$$q(N)n(N) = o(N) \implies \lim_{N \to \infty} \operatorname{Ent}(\mathbb{P}^{(N,q(N))}_{\eta_0,n(N)} \mid \mathbb{K}^{\otimes q(N)}_{\eta_0,n(N)}) = 0$$

3.5.2 Regularity Conditions

Besides the modeling and the applications of Feynman-Kac and particle methods, a rather large part of the book is concerned with qualitative and asymptotic properties of the models as the time horizon or the size of the system tends to infinity. We discuss topics covered by traditional books on ordinary Markov chains and Monte Carlo methods: stability of semigroups, existence and uniqueness of invariant measures, annealed and concentration properties, weak and strong laws of large numbers, fluctuation and large-deviation principles, and propagation of chaos.

The study of each of these topics requires a specific mathematical technique and a precise set of regularity conditions on the pair of potentials/kernels (G_n, M_n). Apart from some purely technical assumptions, that we have made to simplify the analysis, most of the book is based on two types of regularity conditions. As a guide to their use, we provide in this section a short discussion on these conditions.

Before entering into more detail, we first recall that the Feynman-Kac distribution flow associated with an abstract and general pair of potentials/kernels (G_n, M_n) is in general well-defined only up to the first (deterministic) time τ we have $\mathbb{E}_{\eta_0}(\prod_{p=0}^{\tau} G_p(X_p)) = 0$. The deterministic horizon τ depends on the triplet (η_0, G_n, M_n). In this situation, we have seen on page 102 that the N-particle model is also stopped the first time $\tau^N (\leq \tau)$ the whole configuration falls outside the support of the current potential function. This general situation, which might appear as a study of Feynman-Kac models with minimal regularity structure, will be discussed in Chapter 7 (see also Remark 9.4.1 on page 301, and Section 12.2.7).

The analysis is technically less involved when the potential functions satisfy the following condition:

(G) There exists a sequence of strictly positive number $\epsilon_n(G) \in (0, 1]$ such that for any $x_n, y_n \in E_n$

$$G_n(x_n) \geq \epsilon_n(G)\, G_n(y_n) > 0$$

3. Genealogical and Interacting Particle Models

Most of the topics developed in this book will be based on this single regularity condition, except the following subjects:

- Stability of Feynman-Kac semigroups.
- Uniform convergence results with respect to the time parameter.
- Relative entropy and $\mathbb{L}_p$ propagation-of-chaos estimates.
- Fluctuations and large-deviation principles on path space.

To study these questions, we will suppose that the collection of distributions $M_{n+1}(x_n, \cdot)$ are absolutely continuous with one another. That is, for each $n \geq 0$ and $x_n, y_n \in E_n$, we have $M_{n+1}(x_n, \cdot) \ll M_{n+1}(y_n, \cdot)$. We will strengthen this condition in three ways:

- $(M)_m$ There exists some integer $m \geq 1$ and some sequence of numbers $\epsilon_p(M) \in (0, 1)$ such that for p and $x_p, y_p \in E_p$ we have

$$M_{p,p+m}(x_p, \cdot) = M_{p+1} M_{p+2} \ldots M_{p+m}(x_p, \cdot) \geq \epsilon_p(M) \; M_{p,p+m}(y_p, \cdot)$$

- $(M)^{(p)}$ For each $n \geq 1$ and $x_{n-1} \in E_{n-1}$, $M_n(x_{n-1}, \cdot)$ is absolutely continuous with respect to η_n and we have

$$M_n(x_{n-1}, dx_n) = k_n(x_{n-1}, x_n) \; \eta_n(dx_n)$$

with $\sup_{x_{n-1} \in E_{n-1}} k_n(x_{n-1}, \cdot) \in \mathbb{L}_p(\eta_n)$.

- $(M)^{\exp}$ For each $n \geq 1$, there exists a reference probability measure $p_n \in \mathcal{P}(E_n)$ and a measurable function a_n on E_n such that for any $x_{n-1} \in E_{n-1}$

$$M_n(x_{n-1}, dx_n) = m_n(x_{n-1}, x_n) \; p_n(dx_n)$$

and $\sup_{x_{n-1} \in E_{n-1}} |\log m_n(x_{n-1}, x_n)| \leq a_n(x_n)$ with $p_n(e^{r a_n}) < \infty$, for any $r \geq 1$.

To guide the reader, we mention the different places in which these conditions are used.

- Under (G), we will study the asymptotic behavior of particle density profiles and genealogical-tree occupation measures:
    - Exponential and $\mathbb{L}_p$ mean error estimates.
    - Weak convergence of empirical processes.
    - Central limit theorems.
    - Propagation-of-chaos estimates for the total variation norm.

- Under (G) and $(M)_m$, we will discuss:
  - Contraction and asymptotic stability properties of Feynman-Kac semigroups.
  - Annealed properties of Feynman-Kac flows.
  - Spectral estimates of Feynman-Kac-Schrödinger semigroups.
  - Uniform estimates for particle density profiles with respect to the time parameter.
  - Long time behavior of genealogical tree-based models.

- Under (G) and $(M)^{(p)}$, we will discuss entropy propagation-of-chaos estimates.

- Under (G) and $(M)^{\exp}$ we will discuss fluctuation and large-deviation principles for particle models in path space.

Note that condition $(M)^{(p)}$ is met as soon as we have

$$M_n(x_{n-1}, dx_n) \leq h_n(x_n) \, M_n(x'_{n-1}, dx_n) \tag{3.12}$$

for some $(0,1]$-valued function $h_n \in \mathbb{L}_p(\eta_n)$ ($\Leftarrow \|M_n(h_n^p)\| < \infty$). To prove this claim we first recall that $\widehat{\eta}_{n-1} M_n = \eta_n$. Integrating (3.12) with respect to $\widehat{\eta}_{n-1}$ we find that

$$k_n(x_{n-1}, x_n) = \frac{dM_n(x_{n-1}, \cdot)}{d\eta_n}(x_n) \in [1/h_n(x_n), h_n(x_n)]$$

In the same way, we conclude that $(M)^{\exp}$ holds true with $(m_n, p_n) = (k_n, \eta_n)$ as soon as $h_n \in \cap_{p \geq 0} \mathbb{L}_p(\eta_n)$.

To illustrate this condition we note that the one dimensional and Gaussian transitions

$$M_n(x_{n-1}, dx_n) = \frac{1}{\sqrt{2\pi}} \exp\left\{-\frac{1}{2}(x_n - a(x_{n-1}))^2\right\} dx_n$$

satisfy (3.12) with $h_n(x_n) = \exp\{\operatorname{osc}(a_n)[|x_n| + \|a_n\|]\}$ as soon as $\|a_n\| < \infty$. We also have the following

Lemma 3.5.1

$$(M)_{m=1} \Longrightarrow (M)^{\exp} \Longrightarrow \left((M)^{(p)} \quad \text{for any } p \geq 1 \right)$$

Proof:
To prove the first implication, we choose a point $x^\star \in E$ and we set

$$p_n(dy) = M_n(x^\star, dy) \quad \text{and} \quad m_n(x, y) = \frac{dM_n(x, \cdot)}{dM_n(x^\star, \cdot)}(y)$$

When the uniform mixing condition $(M)_m$ holds true for $m = 1$, we have for any $x, y \in E$

$$\epsilon_{n-1}(M) \leq m_n(x, y) \leq \frac{1}{\epsilon_{n-1}(M)}$$

We conclude that $(M)^{\exp}$ is met with $a_n(y) = -\log \epsilon_{n-1}(M)$. Let us prove the second assertion. Suppose we have

$$e^{-a_n(y)} \, p_n(dy) \leq M_n(x, dy) = m_n(x, y) \, p_n(dy) \leq e^{a_n(y)} \, p_n(dy)$$

for some pair (m_n, p_n) with, for any $p \geq 1$, $\int \exp(p\, a_n) \, dp_n < \infty$. A simple calculation shows that for any choice of the reference probability measure $p_n \in \mathcal{P}(E)$ and for any $p \geq 1$ we have

$$\int \sup_x \left(\frac{dM_n(x, \cdot)}{d\eta_n} \right)^p d\eta_n = \int \sup_x \left(\frac{dM_n(x, \cdot)}{dp_n} \right)^p \left(\frac{dp_n}{d\eta_n} \right)^{p-1} dp_n \tag{3.13}$$

Since $\frac{d\eta_n}{dp_n}(y) = \frac{\eta_{n-1}(G_{n-1} \, m_n(\cdot, y))}{\eta_{n-1}(G_{n-1})}$, we find that $\left| \log \frac{d\eta_n}{dp_n} \right| \leq a_n$. From (3.13), one concludes that

$$\int \sup_x \left(\frac{dM_n(x, \cdot)}{d\eta_n} \right)^p d\eta_n \leq \int e^{(2p-1)\, a_n} \, dp_n$$

It is now easily checked that $\int e^{(2p-1)a_n(y)} \, p_n(dy) \Rightarrow (M)^{(p)}$ for any $p \geq 1$. The end of the proof is now clear. ∎

We illustrate these conditions with two typical examples. Other situations will be examined on page 148. Note that for time homogeneous models on finite spaces condition $(M)_m$ is met as soon as the Markov chain is aperiodic and irreducible.

Example 3.5.1 *Suppose that $E_n = \mathbb{R}^d$ and M_n is given by*

$$M_n(x, dy) = \frac{1}{((2\pi)^d |Q_n|)^{1/2}} \exp\left(-\frac{1}{2}(y - A_n(x))' \, Q_n^{-1} \, (y - A_n(x)) \right) dy$$

where Q_n is a $d \times d$ symmetric nonnegative matrix and $A_n : \mathbb{R}^d \to \mathbb{R}^d$ is a bounded function. Using previous observations, it is not difficult to check that $(M)^{\exp}$ is satisfied with

$$p_n(dy) = \frac{1}{((2\pi)^d |Q_n|)^{1/2}} \exp\left(-\frac{1}{2} y' \, Q_n^{-1} \, y \right) dy$$

and

$$\log m_n(x, y) = -\frac{1}{2}(y - A_n(x))' \, Q_n^{-1} \, (y - A_n(x)) + \frac{1}{2} y' \, Q_n^{-1} \, y$$
$$= A_n(x) \, Q_n^{-1} \, y - \frac{1}{2} A_n'(x) \, Q_n^{-1} \, A_n(x)$$

To see this claim, it suffices to observe that

$$
\begin{aligned}
|\log m_n(x,y)| &= |A_n(x)' Q_n^{-1} y| + \frac{1}{2} |A_n(x)' Q_n^{-1} A_n(x)| \\
&\leq a_n(y) = \|A_n\| \, \||Q_n^{-1}\||_1 \, \|y\|_1 + \frac{d}{2} \|A_n\|^2 \, \||Q_n^{-1}\||_1
\end{aligned}
$$

where $\|y\|_1 = \sum_{i=1}^d |y^i|$,

$$
\|A_n\| = \sup_{1 \leq i \leq d} \sup_x |A_n^i(x)| \quad \text{and} \quad \||Q_n^{-1}\||_1 = \sup_{1 \leq j \leq d} \sum_{i=1}^d |(Q_n^{-1})_{i,j}|
$$

Example 3.5.2 For $E_n = \mathbb{R}$ and M_n given by

$$
M_n(x, dy) = \frac{c(n)}{2} e^{-c(n)|y - A_n(x)|} \, dy
$$

for some $c(n) > 0$ and $\mathrm{osc}(A_n) < \infty$, condition $(M)_m$ holds true for $m = 1$. Indeed we have

$$
\log \frac{dM_n(x, \cdot)}{dM_n(y, \cdot)}(z) = c(n) \left[\, |z - A_n(y)| - |z - A_n(x)| \,\right]
$$

Recalling that $||z - a| - |z - b|| \leq |b - a|$, we readily find that

$$
\left\| \log \frac{dM_n(x, \cdot)}{dM_n(y, \cdot)} \right\| \leq c(n) \, |A_n(x) - A_n(y)| \leq c(n) \, \mathrm{osc}(A_n)
$$

We conclude that the mixing condition $(M)_m$ holds true for $m = 1$ and

$$
\epsilon_{n-1}(M) = \exp\left(-c(n) \, \mathrm{osc}(A_n)\right)
$$

4
Stability of Feynman-Kac Semigroups

4.1 Introduction

This chapter is devoted to structural and stability properties of Feynman-Kac semigroups. These regularity properties appear to be central in the understanding of various topics discussed in this book. They will be applied in Chapter 5 to analyze the existence and the uniqueness of invariant measures, and in Chapter 12 they are used to study the asymptotic stability of nonlinear filtering equations. In Chapter 6, we will also use these results to analyze the concentration properties of annealed Feynman-Kac semigroups associated with a cooling schedule. Their applications to the study of the long time behavior of particle methods will be the main object of Chapter 7.

To explain and motivate the organization of this chapter, it is convenient to observe that for constant potential functions Feynman-Kac models coincide with traditional Markov semigroups. The study of the stability of Markov models is one of the most active research subjects in probability theory. We refer the reader to traditional textbooks on this theme (see for instance the book [250] and references therein). Among the variety of approaches developed in this field, R.L. Dobrushin introduced in a two-part article [116] in 1956 a powerful measure-theoretic technique for studying the contraction properties of a Markov kernel with respect to the total variation distance. The main feature of this approach is that it applies to Markov transitions on general measurable spaces and it does not use any assumptions on the invariant measures of the chain. These ideas were pursued and extended in [93] with a systematic study of Markov contraction and ergodic constants with respect to a general class of distance-like entropy criteria. The essentials of this study are provided in Section 4.2. Section 4.3 is concerned with the extensions of these results to Feynman-Kac semigroups.

The rest of the chapter has the following structure. In Section 4.3.1, we provide several functional inequalities in terms of the Dobrushin ergodic coefficient of a Feynman-Kac type Markov kernel and in terms of the relative oscillations of a sequence of potential functions. In Section 4.3.2, we propose a semigroup approach to control these two quantities. In Section 4.3.3, we apply these results to derive a collection of strong contraction estimates with respect to a fairly general class of relative entropies. Another important feature of this approach is that it applies to estimates of the "weak" regularity of Feynman-Kac semigroups. These questions are discussed in Section 4.3.4. The kinship between the stability properties of updated and prediction semigroups is studied in Section 4.4. We complete this chapter with an application of these results to the study of asymptotic stability properties of a class of stochastic Feynman-Kac semigroups arising in nonlinear filtering (see Section 4.5).

4.2 Contraction Properties of Markov Kernels

Let $(E, \mathcal{E})$ and $(F, \mathcal{F})$ be a pair of measurable spaces. In this section, we develop general contraction properties of Markov kernels M from $(E, \mathcal{E})$ into $(F, \mathcal{F})$. We provide Lipschitz type estimates with respect to various distance-like criteria. Since the state space does not play a distinguished role, we simplify the presentation and we use the same notation to denote any relative entropy criteria on the set of measures on possibly different state spaces.

4.2.1 h-relative Entropy

Let $h : \mathbb{R}_+^2 \to \mathbb{R} \cup \{\infty\}$ be a convex function satisfying for any $a, x, y \in \mathbb{R}_+$ the following conditions:

$$h(ax, ay) = ah(x, y) \quad \text{and} \quad h(1, 1) = 0$$

We associate with this homogeneous function the h-relative entropy on $\mathcal{M}_+(E)$ defined symbolically as

$$H(\mu, \nu) = \int h\,(d\mu, d\nu)$$

More precisely, by homogeneity arguments, the mapping H is defined in terms of any measure $\lambda \in \mathcal{M}(E)$ dominating μ and ν by the formula

$$H(\mu,\nu) = \int h\left(\frac{d\mu}{d\lambda}, \frac{d\nu}{d\lambda}\right) d\lambda \tag{4.1}$$

To illustrate this abstract definition and motivate the forthcoming analysis, we provide hereafter a collection of classical h-relative entropies arising in the literature. First we come back to the definition of h-entropy. We denote by $h' : \mathbb{R}_+ \to \mathbb{R} \sqcup \{+\infty\}$ the convex function given for any $x \in \mathbb{R}_+$ by $h'(x) = h(x, 1)$.

By homogeneity arguments, we note that h is almost equivalent to h'. More precisely, only the specification of the value $h(1, 0)$ is missing. In most applications, the natural convention is $h(1, 0) = \infty$.

The next lemma connects the h-relative entropy with the h'-divergence in the sense of Csiszar [68].

Lemma 4.2.1 *Assume that $h(1,0) = +\infty$. Then, for any μ and $\nu \in \mathcal{M}_+(E)$, we have*

$$H(\mu,\nu) = \int h'\left(\frac{d\mu}{d\nu}\right) d\nu \tag{4.2}$$

if $\mu \ll \nu$, and $H(\mu,\nu) = \infty$ otherwise.

Proof:
Let $\mu = \mu_1 + \mu_2$ be the Lebesgue decomposition of μ with respect to ν. That is, we have that $\mu_1 \ll \nu$ and $\mu_2 \perp \nu$. Also let $A \in \mathcal{E}$ be such that $\nu(A^c) = 0 = \mu_2(A)$. To compute $H(\mu, \nu)$, we can take $\lambda =_{\text{def.}} \nu + \mu_2$ in (4.1) and we get

$$\begin{aligned}
H(\mu,\nu) &= \int_A h\left(\frac{d\mu_1}{d\nu}, 1\right) d\nu + \int_{A^c} h(1,0)\, d\mu_2 \\
&= H(\mu_1,\nu) + h(1,0)\, \mu_2(E)
\end{aligned}$$

If $\mu_2(E) > 0$, we deduce that $H(\mu, \nu) = +\infty$. Otherwise, we take in (4.1) $\lambda = \nu$ and we get (4.2). This ends the proof of the lemma. ∎

In the reverse angle, suppose $h' : \mathbb{R}_+ \to \mathbb{R} \sqcup \{\infty\}$ is a given convex function. Since $t \in (1, +\infty) \mapsto (h'(t) - h'(1))/(t-1)$ is nondecreasing, the limit $l_0 = \lim_{t \to +\infty} h'(t)/t$ exists and for any $l \in [l_0, +\infty]$ we can prove that any function defined for any $(x, y) \in \mathbb{R}_+^2$ by

$$h(x,y) = \begin{cases} y\, h'(x/y) & \text{, if } y > 0 \\ lx & \text{, if } y = 0 \end{cases} \tag{4.3}$$

is convex.

- If we take $h'(t) = |t-1|^p$, $p \geq 1$, we find the $\mathbb{L}_p$-norm given for any $\mu, \nu \in \mathcal{P}(E)$ by $H(\mu,\nu) = \|1 - d\mu/d\nu\|_{p,\nu}^p$ if $\mu \ll \nu$, and ∞ otherwise.

- The case $h'(t) = t\log(t)$ corresponds to the Boltzmann entropy or Shannon-Kullback information. In this situation, we find for any $\mu, \nu \in \mathcal{P}(E)$

$$H(\mu,\nu) = \text{Ent}(\mu|\nu) = \int \ln\left(\frac{d\mu}{d\nu}\right) d\mu$$

if $\mu \ll \nu$ and ∞ otherwise.

- The Havrda-Charvat entropy of order $p > 1$ corresponds to the choice $h'(t) = \frac{1}{p-1}(t^p - 1)$. In this case, we have for any $\mu \ll \nu$

$$H(\mu,\nu) = \mathcal{C}_p(\mu|\nu) =_{\text{def.}} \frac{1}{p-1}\left[\int \left(\frac{d\mu}{d\nu}\right)^p d\nu - 1\right]$$

Notice that $\mathcal{C}_p(\mu,\nu) \to \text{Ent}(\mu|\nu)$ as p tends to 1_+.

- The Hellinger and Kakutani-Hellinger integrals of order $\alpha \in (0,1)$ correspond to the choice $h'(t) = t - t^\alpha$. For any $\mu, \nu \in \mathcal{P}(E)$, for any dominating measure λ, we have

$$H(\mu,\nu) = \mathcal{H}_\alpha(\mu,\nu) =_{\text{def.}} 1 - \int \left(\frac{d\mu}{d\lambda}\right)^\alpha \left(\frac{d\nu}{d\lambda}\right)^{1-\alpha} d\lambda$$

It is also sometimes written symbolically in the form

$$\mathcal{H}_\alpha(\mu,\nu) =_{\text{def.}} 1 - \int (d\mu)^\alpha (d\nu)^{1-\alpha}$$

Notice that it can be rewritten more simply as $1 - \int \left(\frac{d\mu}{d\nu}\right)^\alpha d\nu$ if $\mu \ll \nu$. In the special case $\alpha = 1/2$, this relative entropy coincides with the Kakutani-Hellinger distance defined by

$$\mathcal{H}_{1/2}(\mu,\nu) =_{\text{def.}} \frac{1}{2}\int \left(\sqrt{\frac{d\mu}{d\lambda}} - \sqrt{\frac{d\nu}{d\lambda}}\right)^2 d\lambda$$

or, symbolically, $\mathcal{H}_{1/2}(\mu,\nu) = \frac{1}{2}\int \left(\sqrt{d\mu} - \sqrt{d\nu}\right)^2$.

- Finally, the case $h'(t) = |t-1|/2$ corresponds to the total variation distance defined for any $\mu, \nu \in \mathcal{M}_+(E)$

$$H(\mu,\nu) = \|\mu - \nu\|_{tv}$$

For later use, we have collected in the next lemma three equivalent representations of the total variation distance.

Lemma 4.2.2 *For any pair of probability measures (m_1, m_2) on E, we have*

$$\begin{aligned}
\|m_1 - m_2\|_{\mathrm{tv}} &= \sup\{|m_1(f) - m_2(f)| \,;\, f \in \mathrm{Osc}_1(E)\} \quad (4.4) \\
&= 1 - \sup_{\nu \leq m_1, m_2} \nu(E) \quad (4.5) \\
&= 1 - \inf \sum_{p=1}^{n}(m_1(A_p) \wedge m_2(A_p)) \quad (4.6)
\end{aligned}$$

where the infimum is taken over all finite resolutions of E into pairs of nonintersecting subsets A_p, $1 \leq p \leq n$, with $n \geq 1$.

Proof:
To prove (4.4), we recall that the total variation distance between two probability measures m_1 and m_2 can alternatively be defined in terms of a Hahn-Jordan orthogonal decomposition

$$m = m_1 - m_2 = m^+ - m^-$$

with $\|m_1 - m_2\|_{\mathrm{tv}} = m^+(E) = m^-(E)$. From this observation, we have for any $f \in \mathrm{Osc}_1(E)$

$$|m_1(f) - m_2(f)|$$

$$= \left|\int f(x)\, m^+(dx) - \int f(y)\, m^-(dy)\right|$$

$$= \|m_1 - m_2\|_{\mathrm{tv}} \left|\int (f(x) - f(y))\, \frac{m^+(dx)}{m^+(E)}\, \frac{m^-(dy)}{m^-(E)}\right|$$

We conclude that

$$|m_1(f) - m_2(f)| \leq \|m_1 - m_2\|_{\mathrm{tv}}$$

By taking the supremum over all $f \in \mathrm{Osc}_1(E)$, we find that

$$\sup\{|m_1(f) - m_2(f)| \,;\, f \in \mathrm{Osc}_1(E)\} \leq \|m_1 - m_2\|_{\mathrm{tv}}$$

The reverse inequality can be checked easily by noting that the indicator functions 1_A, with $A \in \mathcal{E}$, belong to $\mathrm{Osc}_1(E)$. Now we come to the proof of (4.5). By construction, there exist two disjoint subsets E_+ and E_- such that

$$m^+(E_-) = 0 = m^-(E_+)$$

Therefore we have for any $A \in \mathcal{E}$

$$m^+(A) = m(A \cap E_+) \geq 0 \quad \text{and} \quad m^-(A) = -m(A \cap E_-) \geq 0$$

from which we conclude that

$$m_1(A \cap E_+) \geq m_2(A \cap E_+) \quad \text{and} \quad m_2(A \cap E_-) \geq m_1(A \cap E_-) \quad (4.7)$$

Let ν be defined for any $A \in \mathcal{E}$ by

$$\nu(A) = m_1(A \cap E_-) + m_2(A \cap E_+)$$

By construction, we have

$$\nu(A) \leq m_1(A) \wedge m_2(A) \quad \text{and} \quad \nu(E) = m_1(E_-) + m_2(E_+) \quad (4.8)$$

Since

$$\begin{aligned}
\|m_1 - m_2\|_{\text{tv}} &= m^+(E) = m(E_+) \\
&= m_1(E_+) - m_2(E_-) = 1 - (m_1(E_+) + m_2(E_-))
\end{aligned}$$

by (4.8) we obtain

$$1 - \sup_{\mu \leq m_1, m_2} \mu(E) \leq 1 - \nu(E) = \|m_1 - m_2\|_{\text{tv}}$$

The reverse inequality is proved as follows. Let μ be a nonnegative measure such that for any $A \in \mathcal{E}$ we have

$$\mu(A) \leq m_1(A) \wedge m_2(A)$$

If we take $A = E_+$ and then $A = E_-$, we necessarily have

$$\mu(E_+) \leq m_1(E_+) \quad \text{and} \quad \mu(E_-) \leq m_2(E_-)$$

and therefore

$$\mu(E) \leq m_1(E_+) + m_2(E_-) = 1 - \|m_1 - m_2\|_{\text{tv}}$$

We conclude that

$$1 - \mu(E) \geq \|m_1 - m_2\|_{\text{tv}}$$

Taking the infimum over all the distributions $\mu \leq m_1$ and m_2, we find the desired result. To prove (4.6), we use the same ideas and notation as above. First, we note by (4.7) that

$$m_2(E_+) = m_1(E_+) \wedge m_2(E_+) \quad \text{and} \quad m_1(E_-) = m_1(E_-) \wedge m_2(E_-)$$

This implies that

$$\begin{aligned}
\nu(E) &= m_1(E_-) + m_2(E_+) \\
&= (m_1(E_-) \wedge m_2(E_-)) + (m_1(E_+) \wedge m_2(E_+))
\end{aligned}$$

Since E_+, E_- are disjoint, we conclude that

$$\nu(E) \geq \inf \sum_{p=1}^{n}(m_1(A_p) \wedge m_2(A_p))$$

where the infimum is taken over all resolutions of E into pairs of nonintersecting subsets A_p, $1 \leq p \leq n$, $n \geq 1$. To prove the reverse inequality, we come back to the definition of ν. By (4.8), for any finite resolution $A_p \in \mathcal{E}$, $1 \leq p \leq n$, we have

$$\nu(A_p) \leq m_1(A_p) \wedge m_2(A_p)$$

and therefore

$$\nu(E) = \sum_{p=1}^{n} \nu(A_p) \leq \sum_{p=1}^{n}(m_1(A_p) \wedge m_2(A_p))$$

We end the proof of (4.6) by taking the infimum over all resolutions. Since $\nu(E) = 1 - \|m_1 - m_2\|_{\mathrm{tv}}$ the end of the proof of the lemma is now straightforward. ∎

4.2.2 Lipschitz Contractions

In this section, we discuss the regularity properties of a Markov kernel M with respect to the h-relative entropy. We provide a universal Lipschitz inequality in terms of the Dobrushin ergodic coefficient. In Section 4.3, we will use this contraction estimate to study the stability properties of nonlinear Feynman-Kac semigroups. Before getting into the precise description of this inequality, we recall the definition of the Dobrushin coefficient and provide some key properties. We recall that the total variation distance on $\mathcal{M}(E)$ is defined for any $\mu \in \mathcal{M}(E)$ by

$$\|\mu\|_{\mathrm{tv}} = \frac{1}{2} \sup_{(A,B) \in \mathcal{E}^2} (\mu(A) - \mu(B))$$

Definition 4.2.1 *The Dobrushin contraction or ergodic coefficient $\beta(M)$ of Markov kernel M from $(E, \mathcal{E})$ into $(F, \mathcal{F})$ is the quantity defined by*

$$\beta(M) = \sup\{\|M(x, \cdot) - M(y, \cdot)\|_{\mathrm{tv}} \,;\, (x, y) \in E^2\} \in [0, 1]$$

Proposition 4.2.1 *Let M be a Markov kernel from $(E, \mathcal{E})$ into $(F, \mathcal{F})$. For any measure $\mu \in \mathcal{M}(E)$, we have the estimate*

$$\|\mu M\|_{\mathrm{tv}} \leq \beta(M)\,\|\mu\|_{\mathrm{tv}} + (1 - \beta(M))\,|\mu(E)|/2 \qquad (4.9)$$

128   4. Stability of Feynman-Kac Semigroups

In addition, $\beta(M)$ is the operator norm of M on $\mathcal{M}_0(E)$, and we have the equivalent formulations

$$\begin{align}
\beta(M) &= \sup_{\mu \in \mathcal{M}_0(E)} \|\mu M\|_{\mathrm{tv}}/\|\mu\|_{\mathrm{tv}} \tag{4.10}\\
&= \sup\{\mathrm{osc}(M(f)) \,;\, f \in \mathrm{Osc}_1(F)\} \tag{4.11}\\
&= 1 - \inf \sum_{p=1}^{n}(M(x,A_p) \wedge M(y,A_p)) \tag{4.12}
\end{align}$$

where the infimum is taken over all $x, y \in E$ and all finite resolutions of F into pairs of nonintersecting subsets $A_p \in \mathcal{F}$, $1 \le p \le n$, $n \ge 1$.

Proof:
We first prove (4.9) for $\mu \in \mathcal{M}_0(E)$. Arguing as in the proof of Lemma 4.2.2, by the Hahn-Jordan decomposition theorem, we can write any signed measure μ as the difference of two nonnegative and orthogonal measures $\mu = \mu^+ - \mu^-$. Let E_+ and E_- be two disjoint subsets such that

$$\mu^+(E_-) = 0 = \mu^-(E_+)$$

We also recall that in this case

$$\|\mu\|_{\mathrm{tv}} = (\mu^+(E) + \mu^-(E))/2 = (\mu(E_+) - \mu(E_-))/2$$

When μ has a null total mass, we clearly have $\|\mu\|_{\mathrm{tv}} = \mu(E_+) = -\mu(E_-)$. Now, for any $A \in \mathcal{F}$, we observe that

$$\begin{align}
\mu M(A) &= \mu(1_{E_+} M(1_A)) + \mu(1_{E_-} M(1_A)) \\
&\le \mu(1_{E_+} M(1_A)) + \inf_{y \in E}(M(y,A))\, \mu(E_-) \\
&= \int_{E_+} [\sup_{y \in E}(M(x,A) - M(y,A))]\, \mu(dx)
\end{align}$$

Taking the supremum in the r.h.s. integral and then over all $A \in \mathcal{F}$, we conclude that (4.9) holds true for any $\mu \in \mathcal{M}_0(E)$ and

$$\beta(M) \ge \sup_{\mu \in \mathcal{M}_0(E)} \frac{\|\mu M\|_{\mathrm{tv}}}{\|\mu\|_{\mathrm{tv}}}$$

To complete the proof of (4.10), we note that for any $x, y \in E$ we have $(\delta_x - \delta_y) \in \mathcal{M}_0(E)$ and $\|\delta_x - \delta_y\|_{\mathrm{tv}} = 1$. This yields the desired reverse inequality

$$\sup_{\mu \in \mathcal{M}_0(E)} \frac{\|\mu M\|_{\mathrm{tv}}}{\|\mu\|_{\mathrm{tv}}} \ge \beta(M)$$

By homogeneity arguments, we only need to prove (4.9) for any signed measure μ with $\mu(E) \ge 0$. We use the decomposition $\mu = \overline{\mu} + \widetilde{\mu}$ with

$$\overline{\mu} = \frac{\mu(E)}{\mu^+(E)}\, \mu^+ \in \mathcal{M}_+(E) \quad \text{and} \quad \widetilde{\mu} = \mu - \frac{\mu(E)}{\mu^+(E)}\, \mu^+ \in \mathcal{M}_0(E)$$

4.2 Contraction Properties of Markov Kernels

Notice that $\widetilde{\mu}$ has a natural Hahn-Jordan decomposition

$$\widetilde{\mu} = \frac{\mu^-(E)}{\mu^+(E)} \mu^+ - \mu^- \quad (\Rightarrow \|\overline{\mu}\|_{\mathrm{tv}} = \mu(E)/2 \text{ and } \|\widetilde{\mu}\|_{\mathrm{tv}} = \mu^-(E))$$

One advantage of this decomposition is that

$$\|\mu\|_{\mathrm{tv}} = (\mu^+(E) + \mu^-(E))/2 = \|\overline{\mu}\|_{\mathrm{tv}} + \|\widetilde{\mu}\|_{\mathrm{tv}}$$

This implies that

$$\begin{aligned}
\|\mu M\|_{\mathrm{tv}} &\leq \|\overline{\mu} M\|_{\mathrm{tv}} + \|\widetilde{\mu} M\|_{\mathrm{tv}} \\
&\leq \|\overline{\mu}\|_{\mathrm{tv}} + \beta(M)\, \|\widetilde{\mu}\|_{\mathrm{tv}} \\
&= \|\overline{\mu}\|_{\mathrm{tv}} + \beta(M)\, (\|\mu\|_{\mathrm{tv}} - \|\overline{\mu}\|_{\mathrm{tv}})
\end{aligned}$$

and finally we get $\|\mu M\|_{\mathrm{tv}} \leq \beta(M)\, \|\mu\|_{\mathrm{tv}} + (1 - \beta(M))\, \mu(E)/2$. We now come to the proof of (4.11). Using the representation (4.4), we obtain

$$\begin{aligned}
\beta(M) &= \sup_{x,y \in E} \|M(x,\,.\,) - M(y,\,.\,)\|_{\mathrm{tv}} \\
&= \sup_{x,y \in E} \sup \{|M(f)(x) - M(f)(y)| \,;\, f \in \mathrm{Osc}_1(F)\} \\
&= \sup \{\sup_{x,y} |M(f)(x) - M(f)(y)| \,;\, f \in \mathrm{Osc}_1(F)\}
\end{aligned}$$

This ends the proof of (4.11). By the definition of $\beta(M)$, the proof of (4.12) is a simple consequence (4.6) of Lemma 4.2.2. This ends the proof of the proposition. ∎

We are now in a position to state the main result of this section.

Theorem 4.2.1 *For any pair of probability measures μ and $\nu \in \mathcal{P}(E)$ and for any Markov kernel M from E into F, we have the contraction estimate*

$$H(\mu M, \nu M) \leq \beta(M)\, H(\mu, \nu)$$

The proof of this theorem is based on a key technical lemma that provides a strategy to compare integrals of convex functions on $\mathbb{R}_+^2$. Before stating this result, it is convenient to examine the scalar case. This result is essentially Lemma 3.3 in [60] (see also Exercise 249 in [174]) but for noncompactly supported measures. In comparison with these two referenced works, our strategy of proof here is to use monotone convergence arguments instead of uniform convergence.

Lemma 4.2.3 *Let m_1, m_2 be two bounded measures on the Borelian real line $(\mathbb{R}, \mathcal{R})$ admitting a first moment and such that*

- *m_1 and m_2 are acting in the same manner on affine mappings:*

$$m_1(\mathbb{R}) = m_2(\mathbb{R}) \quad \text{and} \quad \int t\, m_1(dt) = \int t\, m_2(dt)$$

- For any $s \in \mathbb{R}$, $\int |t-s|\, m_1(dt) \le \int |t-s|\, m_2(dt)$.

Then for any convex function h', we have $m_1(h') \le m_2(h')$ (the value $+\infty$ is not excluded).

Proof:
Let h' be a given convex function on $\mathbb{R}$. One can find two two-sided sequences $(x_i)_{i \in \mathbb{Z}^*}$ and $(k_i)_{i \in \mathbb{Z}^*}$ of nonnegative reals such that if we denote for $n \ge 1$ and $t \in \mathbb{R}$

$$h'_n(t) = h'(0) + \partial^+ h'(0) t + \sum_{0 < i \le n} k_i (t - x_i)^+ + \sum_{-n \le i < 0} k_i (t + x_i)^-$$

where $\partial^+ h'$ is the right derivative of h', then $(h'_n)_{n \ge 1}$ is an increasing sequence converging towards h'. To see this claim, we note that $t \in \mathbb{R} \mapsto \partial^+ h'(t) - \partial^+ h'(0)$ is nondecreasing and we have

$$h'(t) = h'(0) + \partial^+ h'(0) t + \int_0^t (\partial^+ h'(s) - \partial^+ h'(0))\, ds$$

Then we approximate from below the latter function by nondecreasing step functions (for instance, constant on appropriate dyadic intervals) to conclude at the desired convergence. Coming back to m_1 and m_2, we note that for any $i \ge 1$

$$\begin{aligned}
\int k_i (t - x_i)^+ m_1(dt) &= \frac{1}{2} \int k_i |t - x_i|\, m_1(dt) + \frac{1}{2} \int k_i (t - x_i)\, m_1(dt) \\
&\le \frac{1}{2} \int k_i |t - x_i|\, m_2(dt) + \frac{1}{2} \int k_i (t - x_i)\, m_2(dt) \\
&= \int k_i (t - x_i)^+ m_2(dt)
\end{aligned}$$

In the same way, for nonpositive parts, we find that

$$\int k_{-i}(t + x_{-i})^-\, m_1(dt) \le \int k_{-i}(t + x_{-i})^-\, m_2(dt)$$

This implies that for any $n \ge 1$

$$\int h'_n(t)\, m_1(dt) \le \int h'_n(t)\, m_2(dt)$$

We end the proof by letting n tend to infinity and using the monotone convergence theorem.
∎

In the present form, the previous lemma would only imply Theorem 4.2.1 for probabilities satisfying $\mu \ll \nu$, so let us modify it a little:

4.2 Contraction Properties of Markov Kernels   131

Lemma 4.2.4 *Let m_1, m_2 be two bounded measures on the Borelian quadrant $(\mathbb{R}_+^2, \mathcal{R}_+^{\otimes 2})$ admitting a first moment and such that*

- m_1 *and* m_2 *are acting in the same way on affine mappings with* $m_1(\mathbb{R}_+^2) = m_2(\mathbb{R}_+^2)$,

$$\int s\, m_1(ds, dt) = \int s\, m_2(ds, dt) \text{ and } \int t\, m_1(ds, dt) = \int t\, m_2(ds, dt)$$

- *For any* $a, b \in \mathbb{R}$, $\int |as - bt|\, m_1(ds, dt) \leq \int |as - bt|\, m_2(ds, dt)$

Then for any convex and homogeneous function h on $\mathbb{R}_+^2$ we have the inequality $m_1(h) \leq m_2(h)$ (the value $+\infty$ is again not excluded).

Proof:
We recall that any convex and homogeneous function h on the product space $\mathbb{R}_+^2$ has the form (4.3) for some convex function h' on $\mathbb{R}_+$. Using Lemma 4.2.3, we find that we simply need to check that for all $a, b \in \mathbb{R}$,

$$\int (as - bt)^+\, m_1(ds, dt) \leq \int (as - bt)^+\, m_2(ds, dt)$$

$$\int \mathbf{1}_{\{t=0\}} s\, m_1(ds, dt) \leq \int \mathbf{1}_{\{t=0\}} s\, m_2(ds, dt) \qquad (4.13)$$

The last inequality is needed for the cases where $l > l_0$. Note that this result can be deduced from the first condition by letting $b \to \infty$ with $a = 1$. Finally, under our assumptions, a simple subtraction shows that (4.13) holds true. ∎

Now we come to the proof of the theorem.

Proof of Theorem 4.2.1:
Let $\mu, \nu \in \mathcal{P}(E)$ be given and let $\lambda \in \mathcal{M}(E)$ be a dominating measure. We apply Lemma 4.2.4 to the measures m_1 and m_2 on $(\mathbb{R}_+^2, \mathcal{R}_+^{\otimes 2})$ defined for any $h \in \mathcal{B}_b(\mathbb{R}^2)$ by the formulae

$$m_1(h) = \int h\left(\frac{d\mu M}{d\lambda M}, \frac{d\nu M}{d\lambda M}\right) d\lambda M$$

$$m_2(h) = \beta(M) \int h\left(\frac{d\mu}{d\lambda}, \frac{d\mu}{d\lambda}\right) d\lambda + (1 - \beta(M))\, h(1, 1)$$

The first condition of Lemma 4.2.4 is immediate, and the second one amounts to proving that for all $a, b \in \mathbb{R}$,

$$\int \left| a\frac{d\mu M}{d\lambda M} - b\frac{d\nu M}{d\lambda M} \right| d\lambda M$$
$$\leq \beta(M) \int \left| a\frac{d\mu}{d\lambda} - b\frac{d\nu}{d\lambda} \right| d\nu + (1 - \beta(M))\, |a - b|$$

In other words, in terms of the total variation distance, we need to check that

$$\|(a\mu - b\nu)K\|_{\mathrm{tv}} \leq \beta(M)\,\|a\mu - b\nu\|_{\mathrm{tv}} + (1 - \beta(M))\,|a - b|/2$$

which is clear from Proposition 4.2.1 since $(a\mu - b\nu)(E) = a - b$. ∎

4.3 Contraction Properties of Feynman-Kac Semigroups

In this section, we discuss the contraction properties of the nonlinear semigroup $\Phi_{p,n}$ presented in Section 2.7 with respect to the h-relative entropy criteria introduced in Section 4.2. We recall that $\Phi_{p,n}$ is the nonlinear mapping from $\mathcal{P}(E_p)$ into $\mathcal{P}(E_n)$ defined by

$$\Phi_{p,n}(\mu_p)(f_n) = \frac{\mu_p(Q_{p,n}(f_n))}{\mu_p(Q_{p,n}(1))} = \frac{\mathbb{E}_{p,\mu_p}(f(X_n)\prod_{q=p}^{n-1} G_q(X_p))}{\mathbb{E}_{p,\mu_p}(\prod_{q=p}^{n-1} G_q(X_p))}$$

In the study of regularity properties of $\Phi_{p,n}$, the following notion will play a major role.

Definition 4.3.1 *Let $(E, \mathcal{E})$ and $(F, \mathcal{F})$ be a pair of measurable spaces. We consider an h-relative entropy criterion H on the sets $\mathcal{P}(E)$ and $\mathcal{P}(F)$. The contraction or Lipschitz coefficient $\beta_H(\Phi) \in \mathbb{R}_+ \cup \{\infty\}$ of a mapping $\Phi : \mathcal{P}(E) \to \mathcal{P}(F)$ with respect to H is the best constant such that for any pair of measures $\mu, \nu \in \mathcal{P}(E)$ we have*

$$H(\Phi(\mu), \Phi(\nu)) \leq \beta_H(\Phi)\, H(\mu, \nu)$$

When H represents the total variation distance, we simplify notation and sometimes we write $\beta(\Phi)$ instead of $\beta_H(\Phi)$.

When H is the total variation distance, the parameter $\beta(\Phi)$ coincides with the traditional notion of a Lipschitz constant of a mapping between two metric spaces. In addition, for linear mappings, it coincides with the Dobrushin ergodic coefficient defined in Section 4.2.2.

One of the main objectives of this section will be to estimate the contraction coefficients $\beta_H(\Phi_{p,n})$ of the nonlinear Feynman-Kac transformations $\Phi_{p,n}$. By the semigroup property and the definition of the contraction coefficient, we start by noting that for any $0 \leq p_1 \leq p_2 \leq n$ we have

$$\beta_H(\Phi_{p_1,n}) \leq \beta_H(\Phi_{p_1,p_2})\,\beta_H(\Phi_{p_2,n})$$

4.3 Contraction Properties of Feynman-Kac Semigroups

Such arguments are powerful tools for the study of the asymptotic stability properties of the semigroup $\Phi_{p,n}$. For instance, for any pair of measures with $H(\mu_p, \nu_p) < \infty$, we can check that

$$\exists n \in \mathbb{N} : \quad \beta_H(\Phi_{p,p+n}) < 1 \Longrightarrow \lim_{n \to \infty} H(\Phi_{p+n}(\mu_p), \Phi_{p,p+n}(\nu_p)) = 0$$

Before getting into further details, it is instructive to note that $\Phi_{p,n}$ may have completely different kinds of asymptotic behavior. We examine hereafter two "opposite" situations. When the potential functions G_n are constant functions, then we have $P_{p,n} = M_{p,n}$. In this case, the asymptotic stability properties of $\Phi_{p,n}$ are reduced to that of $M_{p,n}$. On the other hand, if the semigroup $M_n = Id$, then we also have $P_{p,n} = Id$. In this situation $\beta(P_{p,n}) = 1$, and one cannot expect to obtain uniform stability properties. For instance, in the homogeneous case $E_n = E$ with a potential $G_n = e^{-V}$ associated with a nonnegative energy function V, the semigroup $\Phi_{p,n}$ can be rewritten as

$$\Phi_{p,n}(\mu)(f) = \Psi_{p,n}(\mu)(f) = \frac{\mu\left(e^{-(n-p)V} f\right)}{\mu(e^{-(n-p)V})}$$

It is then easily seen that $\Phi_{p,n}(\mu)$ tends as $n \to \infty$ and in a narrow sense to the restriction of μ to the subset

$$V^\star = \{x \in E \; ; \; V(x) = \mu - \operatorname{ess\,inf} V\}$$

Exact calculations as in previous examples are in general not possible, and the question of the regularity properties of general Feynman-Kac semigroups is a difficult nonlinear problem. In the present section, we design a semigroup approach based on the Markov contraction analysis developed in Section 4.2.2 to give some partial answers to this question. The first central idea is to use the alternative description of the semigroup $\Phi_{p,n}$ presented in (2.37), Proposition 2.7.1. More precisely, we recall that $\Phi_{p,n}$ is alternatively defined by the equation

$$\Phi_{p,n}(\mu_p) = \Psi_{p,n}(\mu_p) P_{p,n} \tag{4.14}$$

The Boltzmann-Gibbs transformation $\Psi_{p,n}$ from $\mathcal{P}(E_p)$ into itself is defined by

$$\Psi_{p,n}(\mu_p)(dx_p) = \frac{G_{p,n}(x_p)}{\mu_p(G_{p,n})} \mu_p(dx_p) \quad \text{with} \quad G_{p,n} = Q_{p,n}(1)$$

and the kernel $P_{p,n}$ can be regarded as the transition from E_p into E_n of a nonhomogeneous Markov chain from p to time n with transition semigroup $(R_{p,q}^{(n)})_{0 \leq p \leq q \leq n}$. More precisely, we have that

$$P_{p,n} = R_{p,q}^{(n)} P_{q,n} \quad \text{with} \quad R_{p,q}^{(n)}(f_q) = Q_{p,q}(f_q G_{q,n})/Q_{p,q}(G_{q,n})$$

The next proposition expresses the fact that the Dobrushin ergodic coefficient $\beta(P_{p,n})$ of the Markov kernel $P_{p,n}$ is a measure of the oscillations of the mapping $\Phi_{p,n}$ with respect to the total variation distance.

134   4. Stability of Feynman-Kac Semigroups

Proposition 4.3.1 *For any $0 \leq p \leq n$, we have*

$$\beta(P_{p,n}) = \sup_{\mu_p, \nu_p \in \mathcal{P}(E_p)} \frac{\|\Phi_{p,n}(\mu_p) - \Phi_{p,n}(\nu_p)\|_{\mathrm{tv}}}{\|\Psi_{p,n}(\mu_p) - \Psi_{p,n}(\nu_p)\|_{\mathrm{tv}}}$$
$$= \sup_{\mu_p, \nu_p \in \mathcal{P}(E_p)} \|\Phi_{p,n}(\mu_p) - \Phi_{p,n}(\nu_p)\|_{\mathrm{tv}} \qquad (4.15)$$

In addition, for any h-relative entropy H and for any $\mu_p, \nu_p \in \mathcal{P}(E_p)$, we have

$$H(\Phi_{p,n}(\mu_p), \Phi_{p,n}(\nu_p)) \leq \beta(P_{p,n}) \; H(\Psi_{p,n}(\mu_p), \Psi_{p,n}(\nu_p)) \qquad (4.16)$$

Proof:
To establish the first assertion, we note that for any $x_p, y_p \in E_p$ we have

$$\Phi_{p,n}(\delta_{x_p}) = P_{p,n}(x_p, .) \quad \text{and} \quad \|\delta_{x_p} - \delta_{y_p}\|_{\mathrm{tv}} = 1$$

from which we conclude that the two terms in the r.h.s. of (4.15) are greater than $\beta(P_{p,n})$. The reverse inequality is a simple consequence of (4.14) and Proposition 4.2.1. The final assertion is again a consequence of Theorem 4.2.1 and (4.14). ∎

The inequality (4.16) is of course not sufficient to estimate $\beta_H(\Phi_{p,n})$, but it already shows a natural relation between $\beta_H(\Phi_{p,n})$ and the Dobrushin coefficient $\beta(P_{p,n})$ of $P_{p,n}$. More precisely, from (4.16) we find that

$$\beta_H(\Phi_{p,n}) \leq \beta(P_{p,n}) \; \beta_H(\Psi_{p,n})$$

This inequality is one of the cornerstones of the forthcoming analysis. It underlines the two different roles played by the Markov kernel $P_{p,n}$ and the nonlinear Boltzmann-Gibbs transformation $\Psi_{p,n}$ in the estimation of $\beta_H(\Phi_{p,n})$.

The rest of the section is decomposed into three parts. In the first part, we derive some "local" functional concentration inequalities. We present a class of h'-divergence criteria with respect to which the semigroup $\Phi_{p,n}$ is locally Lipschitz. We illustrate these results with several examples of concentration inequalities with respect to the Boltzmann, the Havrda-Charvat, the Hellinger, and the $\mathbb{L}_2$ relative entropies presented in Section 4.2. The second part of this section is concerned with uniform concentration estimates. We propose a series of sufficient conditions under which the local concentration inequalities can be turned into uniform Lipschitz inequalities. In the third and last part of the section, we use these results to estimate the contraction parameters $\beta_H(\Phi_{p,n})$.

4.3.1 Functional Entropy Inequalities

The next theorem is the main result of this section. It provides a way to estimate the local H-contraction properties of the semigroup $\Phi_{p,n}$ in terms of the Dobrushin coefficient $\beta(P_{p,n})$ and the relative oscillations of the potential functions $G_{p,n}$.

To describe these functional inequalities precisely, it is convenient to introduce some additional notation. When H is the h'-divergence associated with a differentiable $h' \in C^1(\mathbb{R}_+)$, we denote by Δh the function on $\mathbb{R}_+^2$ defined by

$$\Delta h(t,s) = h'(t) - h'(s) - \partial h'(s)\,(t-s) \quad (\geq 0)$$

where $\partial h'(s)$ stands for the derivative of h' at $s \in \mathbb{R}_+$. We will also use the growth condition

$$(\mathcal{H}_a) \quad \forall (r,s,t) \in \mathbb{R}_+^3 \text{ we have } \quad \Delta h(rt,s) \leq a(r)\,\Delta h(t,\theta(r,s)) \quad (4.17)$$

for some nondecreasing function a on $\mathbb{R}_+$ and a mapping θ on $\mathbb{R}_+^2$ such that for any $r \in \mathbb{R}_+$, $\theta(r,\mathbb{R}_+) = \mathbb{R}_+$.

Theorem 4.3.1 *For any $0 \leq p \leq n$ and $\mu_p, \nu_p \in \mathcal{P}(E_p)$, we have*

$$\|\Phi_{p,n}(\mu_p) - \Phi_{p,n}(\nu_p)\|_{\mathrm{tv}} \leq \beta(P_{p,n})\,\frac{\|G_{p,n}\|_{\mathrm{osc}}}{\nu_p(G_{p,n}) \vee \mu_p(G_{p,n})}\,\|\mu_p - \nu_p\|_{\mathrm{tv}} \quad (4.18)$$

In addition, for any h'-divergence H satisfying the growth condition $(\mathcal{H})_a$ for some nondecreasing function a, we have

$$H(\Phi_{p,n}(\mu_p), \Phi_{p,n}(\nu_p)) \leq \beta(P_{p,n})\,\frac{\|G_{p,n}\|}{\nu_p(G_{p,n})}\,a\!\left(\frac{\nu_p(G_{p,n})}{\mu_p(G_{p,n})}\right)\,H(\mu_p,\nu_p) \quad (4.19)$$

The proof of the theorem is a simple consequence of Proposition 4.3.1 and the next technical lemma.

Lemma 4.3.1 *Let G be a strictly positive and measurable function on some measurable space $(E,\mathcal{E})$. We associate with G the Boltzmann-Gibbs transformation Ψ from $\mathcal{P}(E)$ into itself defined by*

$$\Psi(\mu)(dx) = \frac{1}{\mu(G)}\,G(x)\,\mu(dx)$$

For any $\mu, \nu \in \mathcal{P}(E)$, we have

$$\|\Psi(\mu) - \Psi(\nu)\|_{\mathrm{tv}} \leq \frac{\|G\|_{\mathrm{osc}}}{\nu(G) \vee \mu(G)}\,\|\mu - \nu\|_{\mathrm{tv}}$$

In addition, for any h'-divergence H satisfying the growth condition $(\mathcal{H})_a$ for some nondecreasing function a we have

$$H(\Psi(\mu), \Psi(\nu)) \leq \frac{\|G\|}{\nu(G)}\,a\!\left(\frac{\nu(G)}{\mu(G)}\right)\,H(\mu,\nu)$$

Proof:
To prove the first assertion, we use the decomposition

$$\Psi(\mu)(f) - \Psi(\nu)(f) = \frac{1}{\mu(G)} \mu[G\ (f - \Psi(\nu)(f))]$$

for any $\mu, \nu \in \mathcal{P}(E)$ and $f \in \mathcal{B}_b(E)$. Since we have

$$G(x)\ [f(x) - \Psi(\nu)(f)] - G(y)\ [f(y) - \Psi(\nu)(f)]$$
$$= [G(x) - G(y)]\ [f(x) - \Psi(\nu)(f)]\ +\ G(y)\ [f(x) - f(y)]$$

we find that

$$\mathrm{osc}(G[f - \Psi(\nu)(f)]) \le \|G\|_{\mathrm{osc}}\ \mathrm{osc}(f)$$

Suppose next that H is an h'-divergence satisfying the assumptions of the theorem. If $\mu \not\ll \nu$, then we have $\Psi(\mu) \not\ll \Psi(\nu)$, and by Proposition 4.3.1 the result is trivial. Then suppose $\mu \ll \nu$. To prove the result, it is convenient to use the variational representation of H on $\mathcal{P}(E)$

$$H(\mu,\nu) = \inf_{s\in\mathbb{R}_+} \int \Delta h\left(\frac{d\mu}{d\nu}, s\right)\ d\nu$$

It is a simple exercise to check that the infimum is attained at $s=1$ and

$$H(\mu,\nu) = \int \Delta h\left(\frac{d\mu}{d\nu}, 1\right)$$

Using this representation, we notice that

$$H(\Psi(\mu),\Psi(\nu)) = \inf_{s\ge 0} \int \Delta h\left(\frac{\nu(G)}{\mu(G)}\frac{d\mu}{d\nu}, s\right)\frac{G}{\nu(G)}\ d\nu$$

Under our assumptions, we find that

$$H(\Psi(\mu),\Psi(\nu)) \le \frac{\|G\|}{\nu(G)}\ a(r) \inf_{s\ge 0} \int \Delta h\left(\frac{d\mu}{d\nu},\theta(r,s)\right)\ d\nu$$
$$\le \frac{\|G\|}{\nu(G)}\ a(r)\ H(\mu,\nu)$$

with $r = \nu(G)/\mu(G)$. This clearly ends the proof of the lemma. ∎

Corollary 4.3.1 *For any $0 \le p \le n$ and $\mu_p, \nu_p \in \mathcal{P}(E_p)$, we have the following contraction estimates:*

- *Boltzmann relative entropy*

$$\mathrm{Ent}(\Phi_{p,n}(\mu_p)\mid \Phi_{p,n}(\nu_p)) \le \beta(P_{p,n})\ \frac{\|G_{p,n}\|}{\mu_p(G_{p,n})}\ \mathrm{Ent}(\mu_p,\nu_p)$$

4.3 Contraction Properties of Feynman-Kac Semigroups

- *Havrda-Charvat entropy of order* $\alpha > 1$

$$\mathcal{C}_\alpha(\Phi_{p,n}(\mu_p), \Phi_{p,n}(\nu_p)) \leq \beta(P_{p,n}) \frac{\|G_{p,n}\|}{\nu_p(G_{p,n})} \left(\frac{\nu_p(G_{p,n})}{\mu_p(G_{p,n})}\right)^\alpha \mathcal{C}_\alpha(\mu_p, \nu_p)$$

- *Hellinger integrals of order* $\alpha \in (0,1)$

$$\mathcal{H}_\alpha(\Phi_{p,n}(\mu_p), \Phi_{p,n}(\nu_p)) \leq \beta(P_{p,n}) \frac{\|G_{p,n}\|}{\nu_p(G_{p,n})} \left(\frac{\nu_p(G_{p,n})}{\mu_p(G_{p,n})}\right)^\alpha \mathcal{H}_\alpha(\mu_p, \nu_p)$$

- $\mathbb{L}_2$-*relative entropy*

$$\left\|\frac{d\Phi_{p,n}(\mu_p)}{d\Phi_{p,n}(\nu_p)} - 1\right\|^2_{2,\Phi_{p,n}(\nu_p)} \leq \beta(P_{p,n}) \frac{\|G_{p,n}\|}{\nu_p(G_{p,n})} \left(\frac{\nu_p(G_{p,n})}{\mu_p(G_{p,n})}\right)^2 \left\|\frac{d\mu_p}{d\nu_p} - 1\right\|^2_{2,\nu_p}$$

Proof:
The proof of all of these functional inequalities amounts to a check that the respective convex functions h' satisfy the growth condition stated in Theorem 4.3.1. The Boltzmann entropy corresponds to the situation where $h'(t) = t \log t$. In this case, we notice that

$$\begin{aligned} \Delta h(t,s) &= t\log(t) - s\log(s) - (1 + \log(s))\,(t - s) \\ &= t\log(t) - (t-s) - t\log(s) = t\log(t/s) - (t-s) \end{aligned}$$

from which we find that

$$\Delta h(rt, s) = r\,[t\log(t/(s/r)) - (t - (s/r))\,] = r\,\Delta h(t, s/r)$$

We conclude that (4.17) is met with $a(r) = r$ and $\theta(r,s) = s/r$. This clearly ends the proof of the first estimate. For the Havrda-Charvat entropy of order $\alpha > 1$, we have $h'(t) = \frac{1}{\alpha-1}(t^\alpha - 1)$. In this case, we notice that

$$\begin{aligned} (\alpha - 1)\Delta h(rt, s) &= r^\alpha\,(t^\alpha - (s/r)^\alpha - \alpha\,(s/r)^{\alpha-1}\,(t - s/r)) \\ &= r^\alpha\,(\alpha - 1)\Delta h(t, s/r) \end{aligned}$$

from which we conclude that the growth condition (4.17) is now met with $a(r) = r^\alpha$ and $\theta(r,s) = s/r$. This ends the proof of the second estimate. The Hellinger integrals of order $\alpha \in (0,1)$ correspond to the choice $h'(t) = t - t^\alpha$. In this situation, we observe that

$$\begin{aligned} \Delta h(t,s) &= t - t^\alpha - s + s^\alpha - (1 - \alpha s^{\alpha-1})\,(t-s) \\ &= \alpha\,(t-s)\,s^{\alpha-1} + s^\alpha - t^\alpha \end{aligned}$$

from which we conclude that

$$\Delta h(rt, s) = r^\alpha\,\Delta h(t, s/r)$$

Arguing as above, we conclude that the growth condition (4.17) is met with the same parameters. The proof of the third estimate is now completed. The final one corresponds to the case $h'(t) = (t-1)^2$. Since we have

$$\begin{aligned}\Delta h(t,s) &= (t-1)^2 - (s-1)^2 - 2(s-1)\,(t-s) \\ &= (t-s)\,[(t+s-2) - 2(s-1)] = (t-s)^2\end{aligned}$$

we find that (4.17) is met with $a(r) = r^2$ and again $\theta(r,s) = s/r$. This ends the proof of the corollary. ∎

4.3.2 Contraction Coefficients

The local functional inequalities presented in the preceding theorem show the way to estimate the uniform contraction coefficients $\beta_H(\Phi_{p,n})$. To fix the ideas, we recall that

$$\beta_H(\Phi_{p,n}) \leq \beta(P_{p,n})\,\beta_H(\Psi_{p,n}) \tag{4.20}$$

Our immediate objective is to connect more precisely the contraction coefficient $\beta_H(\Psi_{p,n})$ of the Boltzmann-Gibbs transformation $\Psi_{p,n}$ with the relative oscillations of the potential functions $G_{p,n}$.

In the further development of this section, H represents any h'-divergence H satisfying the growth condition $(\mathcal{H})_a$ stated on page 135 or the total variation distance on $\mathcal{P}(E)$. To unify the exposition, it is convenient to introduce the following terminology.

Definition 4.3.2 *For any h'-divergence H satisfying the growth condition $(\mathcal{H})_a$ for some nondecreasing function a on $\mathbb{R}_+$, we denote by $\bar{a}_H$ the function on $\mathbb{R}_+$ defined by $\bar{a}_H(r) = r\,a(r)$. When H is the total variation distance, we denote by $\bar{a}_H$ the function on $\mathbb{R}_+$ defined by $\bar{a}_H(r) = 2r$.*

Proposition 4.3.2 *For any $0 \leq p \leq n$, we have the estimates*

$$\beta_H(\Phi_{p,n}) \leq \bar{a}_H(r_{p,n})\,\beta(P_{p,n}) \quad \text{with} \quad r_{p,n} = \sup_{x_p, y_p}\,(G_{p,n}(x_p)/G_{p,n}(y_p))$$

Proof:
When H is an h'-divergence H (satisfying the growth condition $(\mathcal{H})_a$), the estimate stated in the proposition is a simple consequence of Lemma 4.3.1. We next examine the situation where H is the total variation distance. Applying again Lemma 4.3.1 to $G = G_{p,n}$, we find that

$$\beta_H(\Psi_{p,n}) \leq \frac{\|G_{p,n}\|_{\mathrm{osc}}}{\inf_{E_p} G_{p,n}}$$

Since we have $\|G_{p,p+n}\|_{\mathrm{osc}}/\|G_{p,p+n}\| = (1 + \mathrm{osc}(G_{p,p+n}/\|G_{p,n}\|)) \leq 2$, we conclude that

$$\beta_H(\Psi_{p,n}) \leq 2 r_{p,n}$$

4.3 Contraction Properties of Feynman-Kac Semigroups

In view of (4.20), this implies that $\beta_H(\Phi_{p,n}) \leq 2 r_{p,n}\, \beta(P_{p,n})$, and the proof of the proposition is completed. ∎

After these preliminaries to properly describe the relative entropies and the various parameters that we are using, our next objective is to find a set of sufficient conditions on the semigroups $M_{p,n}$, $Q_{p,n}$, and $R_{p,q}^{(n)}$ and on the potential function G_n under which we have

$$\lim_{n\to\infty} \beta(P_{p,n}) = 0 \quad \text{and} \quad \sup_{x_p, y_p, p \leq n} (G_{p,n}(x_p)/G_{p,n}(y_p)) < \infty$$

We will investigate these questions in terms of three regularity conditions on the semigroups $(M_{p,q}, Q_{p,q}, R_{p,q}^{(n)})$. We say that a given semigroup $I_{p,q}$ satisfies condition $(I)_m$ when we have for some integer parameter $m \geq 1$ and some sequence of numbers $\epsilon_p(I) \in (0,1)$

$$I_{p,p+m}(x_p, .) \geq \epsilon_p(I)\, I_{p,p+m}(y_p, .)$$

for any $(x_p, y_p) \in E_p^2$ and $p \in \mathbb{N}$. In this notation, the semigroups $M_{p,q}$ and $Q_{p,q}$ satisfy respectively conditions $(M)_m$ and $(Q)_m$ when we have for any $p \in \mathbb{N}$ and any pair $(x_p, y_p) \in E_p^2$

$$M_{p,p+m}(x_p, .) \geq \epsilon_p(M)\, M_{p,p+m}(y_p, .)$$
$$\text{and} \quad Q_{p,p+m}(x_p, .) \geq \epsilon_p(Q)\, Q_{p,p+m}(y_p, .)$$

With some obvious abusive notation, the semigroup $R_{p,q}^{(n)}$ satisfies condition $(R^{(n)})_m$ when we have for any $0 \leq p+m \leq n$ and any pair $(x_p, y_p) \in E_p^2$

$$R_{p,p+m}^{(n)}(x_p, .) \geq \epsilon_p(R^{(n)})\, R_{p,p+m}^{(n)}(y_p, .)$$

We will also assume frequently that the potential functions G_n satisfy the condition (G) stated on page 115, for some $\epsilon_n(G) > 0$. For any $0 \leq p \leq n$, we also use the notation

$$\epsilon_{p,n}(G) = \prod_{p \leq k < n} \epsilon_k(G)$$

In this notation, we notice that $\epsilon_{n,n}(G) = 1$ and $\epsilon_{n,n+1}(G) = \epsilon_n(G)$.

In the next proposition, we have collected some key implications between the mixing conditions above as well as some estimations of mixing parameters. We also underline that condition $(Q)_m$ ensures a uniform control on the relative oscillations of the nonhomogeneous potential functions $G_{p,n}$. As mentioned above, these results will be of constant use in the forthcoming analysis of Feynman-Kac semigroup stability and the convergence of particle methods. To emphasize the role of the forthcoming estimates we

140   4. Stability of Feynman-Kac Semigroups

recall that that the Dobrushin ergodic coefficient is an operator norm (see Proposition 4.2.1), that is we have

$$\beta(P_{p,p+q}) \leq \prod_{k=0}^{\lfloor q/m \rfloor - 1} \beta(R^{(n)}_{p+km+1,p+(k+1)m})$$

Proposition 4.3.3 *If condition $(Q)_m$ is satisfied, then $(R^{(n)})_m$ is also met, and for any $0 \leq p + m \leq n$ we have*

$$\epsilon_p(R^{(n)}) \geq \epsilon_p^2(Q) \quad \text{and} \quad \beta(R^{(n)}_{p,p+m}) \leq 1 - \epsilon_p^2(Q)$$

In addition, for any $(x_p, y_p) \in E_p^2$, we have the uniform estimate

$$G_{p,n}(x_p) \geq \epsilon_p(Q) \; G_{p,n}(y_p) \tag{4.21}$$

When (G) and $(M)_m$ are satisfied, then $(Q)_m$ is met and we have

$$\epsilon_p(Q) \geq \epsilon_{p,p+m}(G)\epsilon_p(M) \quad \text{and} \quad \beta(R^{(n)}_{p,p+m}) \leq 1 - \epsilon_{p+1,p+m}(G)\epsilon_p^2(M)$$

Proof:
The proof of the first implication is very simple. Suppose $(Q)_m$ holds true. Then, for any nonnegative function $f_q \in \mathcal{B}_b(E_q)$ and $x_p, y_p \in E_p$, $0 \leq p + m \leq q \leq n$, we have

$$R^{(n)}_{p,q}(f_q)(x_q) = \frac{Q_{p,q}(f_q G_{q,n})(x_p)}{Q_{p,q}(G_{q,n})(x_p)} \geq \epsilon_p^2(Q) \; R^{(n)}_{p,q}(f_q)(y_q)$$

This ends the proof of the first assertion. To prove (4.21), we observe that for any $0 \leq p + m \leq n$

$$\frac{G_{p,n}(x_p)}{G_{p,n}(y_p)} = \frac{Q_{p,n}(1)(x_p)}{Q_{p,n}(1)(y_p)} = \frac{Q_{p,p+m}(G_{p+m,n})(x_p)}{Q_{p,p+m}(G_{p+m,n})(y_p)} \geq \epsilon_p(Q)$$

To prove the second assertion, we observe that

$$\frac{Q_{p,q}(f_q)(x_p)}{Q_{p,q}(f_q)(y_p)} = \frac{G_p(x_p)}{G_p(y_p)} \frac{M_{p+1}(Q_{p+1,q}(f_q))(x_p)}{M_{p+1}(Q_{p+1,q}(f_q))(y_p)}$$

$$\geq \epsilon_p(G) \frac{M_{p+1}(G_{p+1}M_{p+2}Q_{p+2,q}(f_q))(x_p)}{M_{p+1}(G_{p+1}M_{p+2}Q_{p+2,q}(f_q))(y_p)}$$

$$\geq \epsilon_p(G) \, \epsilon_{p+1}(G) \frac{M_{p,p+2}(Q_{p+2,q}(f_q))(x_p)}{M_{p,p+2}(Q_{p+2,q}(f_q))(y_p)}$$

Using a simple induction, we conclude that

$$\frac{Q_{p,q}(f_q)(x_p)}{Q_{p,q}(f_q)(y_p)} \geq \epsilon_{p,p+m}(G) \frac{M_{p,p+m}(Q_{p+m,q}(f_q))(x_p)}{M_{p,p+m}(Q_{p+m,q}(f_q))(y_p)}$$

Finally, we get
$$\frac{Q_{p,q}(f_q)(x_p)}{Q_{p,q}(f_q)(y_p)} \geq \epsilon_{p,p+m}(G) \, \epsilon_p(M)$$
and we conclude that $\epsilon_p(Q) \geq \epsilon_{p,p+m}(G) \, \epsilon_p(M)$. To estimate the Dobrushin coefficient $\beta(R^{(n)}_{p,p+m})$, we observe that, for any $x_p, y_p \in E_p$ and for any nonnegative bounded measurable function f_{p+m} on E_{p+m}, we have

$$R^{(n)}_{p,p+m}(f_{p+m})(x_p) \geq \epsilon_{p+1,p+m}(G) \, \epsilon_p^2(M) \, \frac{M_{p,p+m}(f_{p+m}G_{p+m,n})(y_p)}{M_{p,p+m}(G_{p+m,n})(y_p)}$$

We end the proof using the representation (4.12) in Proposition 4.2.1. ∎

For later use, we have collected in the next two corollaries some simple consequences of the preceding proposition. In the first one, we provide uniform and explicit controls on the contraction coefficients of $\Psi_{p,n}$. The second corollary presents a series of estimations of the Dobrushin coefficient $\beta(P_{p,n})$.

Corollary 4.3.2 *For any $n \geq m \geq 1$ and $p \in \mathbb{N}$, the following conditions are satisfied:*

$$(Q)_m \quad \Longrightarrow \quad \beta_H(\Psi_{p,p+n}) \leq \bar{a}_H(\epsilon_p^{-1}(Q))$$

$$\Uparrow$$

$$(G)_m \text{ and } (M)_m \quad \Longrightarrow \quad \beta_H(\Psi_{p,p+n}) \leq \bar{a}_H(\epsilon_{p,p+m}^{-1}(G)\epsilon_p^{-1}(M))$$

Corollary 4.3.3 *For any $0 \leq p+q \leq n$, we have the following series of estimates and implications:*

$$(R^{(n)})_m \quad \Longrightarrow \quad \beta(P_{p,p+q}) \leq \prod_{k=0}^{\lfloor q/m \rfloor - 1} \left(1 - \epsilon_{p+km}(R^{(n)})\right)$$

$$\Uparrow$$

$$(Q)_m \quad \Longrightarrow \quad \beta(P_{p,p+q}) \leq \prod_{k=0}^{\lfloor q/m \rfloor - 1} \left(1 - \epsilon_{p+km}^2(Q)\right)$$

$$\Uparrow$$

$$(G) \text{ and } (M)_m \quad \Longrightarrow \quad \beta(P_{p,p+q}) \leq \prod_{k=0}^{\lfloor q/m \rfloor - 1} \left(1 - \epsilon_{p+km}^{(m)}(G,M)\right)$$

with $\epsilon_p^{(m)}(G,M) = \epsilon_p^2(M) \, \epsilon_{p+1,p+m}(G)$. In addition, we have that

$$(M)_m \text{ with } m = 1 \quad \Longrightarrow \quad \beta(P_{p,n}) \leq \prod_{k=p}^{n-1} \left(1 - \epsilon_k^2(M)\right) \qquad (4.22)$$

To emphasize the improvements we obtain in strengthening each mixing condition, it is instructive to examine the time-homogeneous situation. We suppose the various mathematical objects are homogeneous with respect to the time parameter, and we suppress the time index in the notation. When condition $(Q)_m$ is met, we have proved that

$$\beta(P_{0,nm}) \leq \left(1 - \epsilon^2(Q)\right)^n$$

Suppose now conditions (G) and $(M)_m$ are met. By (4.21) we have in this case $\epsilon(Q) \geq \epsilon^m(G)\epsilon(M)$, and from a previous estimate we find that

$$\beta(P_{0,nm}) \leq \left(1 - \epsilon^{2m}(G)\epsilon^2(M)\right)^n$$

Nevertheless, under this stronger condition, it is more judicious to estimate $\beta(R_{0,m}^{(n)})$ directly. As stated in corollary 4.3.3, we find that

$$\beta(P_{0,nm}) \leq \left(1 - \epsilon^{(m-1)}(G)\ \epsilon^2(M)\right)^n$$

When the mixing condition $(M)_m$ holds true for $m = 1$, we even get a potential free estimate of the contraction parameter

$$\beta(P_{0,n}) \leq \left(1 - \epsilon^2(M)\right)^n$$

4.3.3 Strong Contraction Estimates

Theorem 4.3.1, Proposition 4.3.3, and Corollaries 4.3.3 and 4.3.2 are powerful weapons to derive several strong contraction estimates. As in Section 4.3.2, in order to unify our statements, we denote by H the total variation distance on $\mathcal{P}(E)$ or any h'-divergence satisfying the growth condition $(\mathcal{H})_a$, stated on page 135. We recall that for these two classes of distance-like criteria, $\bar{a}_H$ is the function on $\mathbb{R}_+$ defined respectively by $\bar{a}_H(r) = 2r$ and $\bar{a}_H(r) = r\, a(r)$. (see Definition 4.3.2).

Proposition 4.3.4 *We suppose condition $(Q)_m$ is met. Then for any $p \in \mathbb{N}$ and $n \geq m$, we have the contraction estimates*

$$\|\Phi_{p,p+n}(\mu_p) - \Phi_{p,p+n}(\nu_p)\|_{\mathrm{tv}} \leq \prod_{k=0}^{\lfloor n/m \rfloor - 1} \left(1 - \epsilon_{p+km}^2(Q)\right) \qquad (4.23)$$

and

$$\beta_H(\Phi_{p,p+n}) \leq \bar{a}_H(\epsilon_p^{-1}(Q)) \prod_{k=0}^{\lfloor n/m \rfloor - 1} \left(1 - \epsilon_{p+km}^2(Q)\right) \qquad (4.24)$$

We can improve the preceding inequalities by strengthening condition $(Q)_m$. Using Theorem 4.3.1, (4.21), and Corollary 4.3.3, we prove the following result.

4.3 Contraction Properties of Feynman-Kac Semigroups

Proposition 4.3.5 *When conditions* (G) *and* $(M)_m$ *are met for some* $m \geq 1$, *then we have for any* $n \geq m$

$$\beta_H(\Phi_{p,p+n}) \leq \overline{a}_H(\epsilon_p^{-1}(M)\ \epsilon_{p,p+m}^{-1}(G)) \prod_{k=0}^{\lfloor n/m \rfloor - 1} \left(1 - \epsilon_{p+km}^{(m)}(G,M)\right) \quad (4.25)$$

with $\epsilon_p^{(m)}(G,M) = \epsilon_p^2(M)\ \epsilon_{p+1,p+m}(G)$.

Proposition 4.3.6 *Suppose the mixing condition* $(M)_m$ *is satisfied with* $m = 1$. *Then we have for any* $n \geq 1$

$$\beta_H(\Phi_{p,p+n}) \leq \overline{a}_H(\epsilon_p^{-1}(G)\epsilon_p^{-1}(M)) \prod_{k=0}^{n-1} \left(1 - \epsilon_{p+k}^2(M)\right) \quad (4.26)$$

and the potential free estimates

$$\|\Phi_{p,p+n}(\mu_p) - \Phi_{p,p+n}(\nu_p)\|_{\mathrm{tv}} \leq \prod_{k=0}^{n-1} \left(1 - \epsilon_{p+k}^2(M)\right) \quad (4.27)$$

Corollary 4.3.4 *Assume that conditions* (G) *and* $(M)_m$ *are met for some* $m \geq 1$. *For any* $(\mu_p, \nu_p) \in \mathcal{P}(E_p)^2$ *with* $H(\mu_p, \nu_p) < \infty$ *and* $p \in \mathbb{N}$, *we have*

$$\sum_{n \geq 0} \epsilon_n(M)\epsilon_{n+1,n+m}(G) = \infty \Rightarrow \lim_{n \to \infty} H(\Phi_{p,p+n}(\mu_p), \Phi_{p,p+n}(\nu_p)) = 0$$

In addition, if we have $\lim_{n \to \infty} \frac{1}{n} \sum_{p=0}^{n-1} \epsilon_p(M)\epsilon_{p+1,p+m}(G) = \overline{\epsilon}$, *then we have the asymptotic exponential decay*

$$\lim_{n \to \infty} \frac{1}{n} \log H(\Phi_{p,p+nm}(\mu_p), \Phi_{p,p+nm}(\nu_p)) < -\overline{\epsilon}$$

Finally, if we assume that $\inf_{n \geq 0} \epsilon_n(G) = \epsilon(G)$ *and* $\inf_{n \geq 0} \epsilon_n(M) = \epsilon(M)$, *then we have*

$$H(\Phi_{p,p+nm}(\mu_p), \Phi_{p,p+nm}(\nu_p)) \leq c\ \exp\left(-n\ \epsilon^2(M)\epsilon^{(m-1)}(G)\right)$$

for some finite constant $c < \infty$ *whose values only depend on the pair* $(\epsilon^m(G), \epsilon(M))$ *and on the relative entropy* $H(\mu_p, \nu_p)$ *between the two measures* μ_p, ν_p.

It is instructive to examine the nature of the various contraction estimates obtained so far for time-homogeneous Feynman-Kac semigroups. We again suppose the various mathematical objects are time-homogeneous and we suppress the time index in the notation. We restrict ourselves to the situation where (G) and $(M)_m$ are satisfied for some $m \geq 1$. In this context, we have proved that

$$\beta_H(\Phi_{0,nm}) \leq \overline{a}_H(\epsilon^{-1}(M)\,\epsilon^{-m}(G))\ \left(1 - \epsilon^2(M)\,\epsilon^{m-1}(G)\right)^n$$

We can alternatively quantify the contraction properties in terms of the first time n at which $\beta_H(\Phi_{0,n}) < 1/e$

$$\Delta_H(\Phi) = \inf\{n \in \mathbb{N}\,;\quad \beta_H(\Phi_{0,n}) \leq 1/e\}$$

This should be thought of as a relaxation time after which the semigroup becomes contractive. Recalling that $\log(1-x) \leq -x$, for any $x \in (0,1)$, we find that

$$\Delta_H(\Phi) \leq m\ \frac{1 + \log \overline{a}_H(\epsilon^{-1}(M)\,\epsilon^{-m}(G))}{\epsilon^2(M)\,\epsilon^{m-1}(G)}$$

4.3.4 Weak Regularity Properties

In previous sections, we studied the regularity properties of the semigroup $\Phi_{p,n}$ with respect to a collection of relative entropies on the set of probability measures. The present section is concerned with regularity properties with respect to the weak topology. More precisely, we want to estimate the measure of contraction of the mappings

$$\Phi_{p,n}(\cdot)(f_n) : \mu_p \in \mathcal{P}(E_p) \longrightarrow \Phi_{p,n}(\mu_p)(f_n) \in \mathbb{R} \qquad (4.28)$$

where $f_n \in \mathcal{B}_b(E_n)$ is a given test function. This problem is intimately related to a natural representation of the oscillations of (4.28). To describe this formula precisely, it is convenient to introduce another key integral operator related to the Boltzmann transformations $\Psi_{p,n}$.

Definition 4.3.3 *For any $0 \leq p \leq n$ and $\mu_p \in \mathcal{P}(E_p)$, we denote by $\tilde{Q}_{p,n}^{\mu_p}$ the integral operator on $\mathcal{M}(E_p)$ and $\mathcal{B}_b(E_p)$ defined by*

$$\tilde{Q}_{p,n}^{\mu_p} = \frac{G_{p,n}}{\mu_p(G_{p,n})}\ (Id - \Psi_{p,n}(\mu_p))$$

In the next technical lemma, we have collected some important properties of $\tilde{Q}_{p,n}^{\mu_p}$.

Lemma 4.3.2 *For any $0 \leq p \leq n$ and $\mu_p \in \mathcal{P}(E_p)$, we have $\mu_p \tilde{Q}_{p,n}^{\mu_p} = 0$, and for any $\eta_p \in \mathcal{P}(E_p)$*

$$\Psi_{p,n}(\eta_p) - \Psi_{p,n}(\mu_p) = \frac{\mu_p(G_{p,n})}{\eta_p(G_{p,n})} \times (\eta_p - \mu_p)\tilde{Q}_{p,n}^{\mu_p} \qquad (4.29)$$

In addition, for any $f_p \in \mathcal{B}_b(E_p)$, we have

$$\mathrm{osc}(\tilde{Q}_{p,n}^{\mu_p}(f_p)) \leq \frac{\|G_{p,n}\|_{\mathrm{osc}}}{\mu_p(G_{p,n})}\ \mathrm{osc}(f_p) \qquad (4.30)$$

4.3 Contraction Properties of Feynman-Kac Semigroups

Proof:
By the definition of $\tilde{Q}_{p,n}^{\mu_p}$, we clearly have $\mu_p \tilde{Q}_{p,n}^{\mu_p} = 0$. Next we observe that

$$[\Psi_{p,n}(\eta_p) - \Psi_{p,n}(\mu_p)](f_p)$$

$$= \Psi_{p,n}(\eta_p)[f_p - \Psi_{p,n}(\mu_p)(f_p)]$$

$$= \frac{\mu_p(G_{p,n})}{\eta_p(G_{p,n})} \eta_p \left[\frac{G_{p,n}}{\mu_p(G_{p,n})} (f_p - \Psi_{p,n}(\mu_p)(f_p)) \right] = \frac{\mu_p(G_{p,n})}{\eta_p(G_{p,n})} \eta_p \tilde{Q}_{p,n}^{\mu_p}(f_p)$$

Since $\mu_p \tilde{Q}_{p,n}^{\mu_p}(f_p) = 0$, we can also write

$$[\Psi_{p,n}(\eta_p) - \Psi_{p,n}(\mu_p)](f_p) = \frac{\mu_p(G_{p,n})}{\eta_p(G_{p,n})} [\eta_p - \mu_p] \tilde{Q}_{p,n}^{\mu_p}(f_p)$$

This ends the proof of (4.29). Recalling the decomposition presented in the proof of Theorem 4.3.1

$$\mu_p(G_{p,n}) \, (\tilde{Q}_{p,n}^{\mu_p}(f_p)(x_p) - \tilde{Q}_{p,n}^{\mu_p}(f_p)(y_p))$$

$$= (G_{p,n}(x_p) - G_{p,n}(y_p)) \, (f_p(x_p) - \Psi_{p,n}(\mu_p)(f_p))$$

$$+ G_{p,n}(y_p) \, (f_p(x_p) - f_p(y_p))$$

and the end of the proof of the lemma is straightforward. ∎

Using the formula (4.29), we find that

$$\Phi_{p,n}(\eta_p) - \Phi_{p,n}(\mu_p) = \frac{\mu_p(G_{p,n})}{\eta_p(G_{p,n})} \times (\eta_p - \mu_p) \tilde{Q}_{p,n}^{\mu_p} P_{p,n}$$

It is sometimes more convenient to write the display above as a second-order development:

$$\Phi_{p,n}(\eta_p) - \Phi_{p,n}(\mu_p) - [\eta_p - \mu_p]\tilde{Q}_{p,n}^{\mu_p} P_{p,n}$$

$$= \frac{1}{\eta_p(G_{p,n})} [\mu_p - \eta_p](G_{p,n}) \times [\eta_p - \mu_p]\tilde{Q}_{p,n}^{\mu_p} P_{p,n}$$

Furthermore, using (4.30) and (4.11), we find that for any $f_n \in \mathcal{B}_b(E_n)$

$$\mathrm{osc}(\tilde{Q}_{p,n}^{\mu_p} P_{p,n}(f_n)) \leq \beta(P_{p,n}) \frac{\|G_{p,n}\|_{\mathrm{osc}}}{\mu_p(G_{p,n})} \, \mathrm{osc}(f_n)$$

and

$$\|\tilde{Q}_{p,n}^{\mu_p} P_{p,n}(f_n)\| \leq \beta(P_{p,n}) \frac{2\|G_{p,n}\|}{\mu_p(G_{p,n})} \, \|f_n\|$$

Summarizing the discussion above, we have proved the following proposition.

Proposition 4.3.7 *For any $0 \leq p \leq n$, $\mu_p \in \mathcal{P}(E_p)$, and $f_n \in \mathcal{B}_b(E_n)$ with $\mathrm{osc}(f_n) \leq 1$, respectively $\|f_n\| \leq 1$, there exists a function $f_{p,n}^{(\mu_p)}$ in $\mathcal{B}_b(E_p)$ with $\mathrm{osc}(f_{p,n}^{(\mu_p)}) \leq 1$, respectively $\|f_{p,n}^{(\mu_p)}\| \leq 1$, such that for any $\eta_p \in \mathcal{P}(E_p)$ we have*

$$|[\Phi_{p,n}(\eta_p) - \Phi_{p,n}(\mu_p)](f_n)| \leq \beta(P_{p,n}) \frac{\|G_{p,n}\|_{\mathrm{osc}}}{\eta_p(G_{p,n})} |(\eta_p - \mu_p)(f_{p,n}^{(\mu_p)})|$$

and respectively

$$|[\Phi_{p,n}(\eta_p) - \Phi_{p,n}(\mu_p)](f_n)| \leq \beta(P_{p,n}) \frac{2\|G_{p,n}\|}{\eta_p(G_{p,n})} |(\eta_p - \mu_p)(f_{p,n}^{(\mu_p)})|$$

4.4 Updated Feynman-Kac Models

The study of regularity properties of the updated Feynman-Kac semigroups

$$\widehat{\Phi}_{p,n} : \mathcal{P}(E_p) \to \mathcal{P}(E_n)$$

can be carried out along the same line of argument as the one used in previous sections. We will of course not rewrite the whole analysis, but we indicate the precise way to transfer these results.

First we notice that $\Phi_{p,n}$ and $\widehat{\Phi}_{p,n}$ have the same structural properties. More precisely, the description of $\widehat{\Phi}_{p,n}$ in terms of the updated linear semigroup $\widehat{Q}_{p,n}$ coincides with that of $\Phi_{p,n}$ by replacing $Q_{p,n}$ by $\widehat{Q}_{p,n}$. To illustrate this assertion, we recall that $\widehat{\Phi}_{p,n}$ can alternatively be written as

$$\widehat{\Phi}_{p,n}(\mu_p)(f_n) = \frac{\mu_p(\widehat{Q}_{p,n}(f_n))}{\mu_p(\widehat{Q}_{p,n}(1))} = \widehat{\Psi}_{p,n}(\mu_p)\widehat{P}_{p,n}$$

with the Markov kernel $\widehat{P}_{p,n}$ from E_p into E_n and Boltzmann-Gibbs transformation $\widehat{\Psi}_{p,n}$ on $\mathcal{P}(E_p)$ associated with the potential $\widehat{G}_{p,n} = \widehat{Q}_{p,n}(1)$ and defined by

$$\widehat{P}_{p,n}(f_n) = \frac{\widehat{Q}_{p,n}(f_n)}{\widehat{Q}_{p,n}(1)} \quad \text{and} \quad \widehat{\Psi}_{p,n}(\mu_p)(dx_p) = \frac{\widehat{G}_{p,n}(x_p)}{\mu_p(\widehat{G}_{p,n})} \mu_p(dx_p)$$

To take the final step, we recall that the updated Feynman-Kac flow associated with the pair (G_n, M_n) can be regarded as the prediction flow model associated with the pair $(\widehat{G}_n, \widehat{M}_n)$ with

$$\widehat{G}_n = M_{n+1}(G_{n+1}) \quad \text{and} \quad \widehat{M}_n(f_n) = M_n(f_n G_n)/M_n(G_n)$$

In addition, using the fact that $\widehat{Q}_n(f_n) = M_n(G_n f_n) = \widehat{G}_{n-1} \widehat{M}_n(f_n)$, we see that the updated semigroups $\widehat{Q}_{p,n}$ are defined as the prediction semigroups $Q_{p,n}$ by replacing the pairs (G_n, M_n) by the pair $(\widehat{G}_n, \widehat{M}_n)$ (with the same labeling indexes). From these two simple observations, we conclude that the whole analysis derived in the previous sections is valid for the updated semigroups by replacing the quantities $(G_{p,n}, P_{p,n}, Q_{p,n}, \Phi_{p,n})$ and the pairs (G_n, M_n) by the corresponding quantities $(\widehat{G}_{p,n}, \widehat{P}_{p,n}, \widehat{Q}_{p,n}, \widehat{\Phi}_{p,n})$ and the pairs $(\widehat{G}_n, \widehat{M}_n)$.

It is instructive at this stage to give an example of the contraction properties that can be transferred using this parallel. We denote by $(\widehat{G})$, $(\widehat{M})_m$, and $(\widehat{Q})_m$ the mixing conditions on the updated objects defined as (G), (M_m), and (Q_m) replacing (G_n, M_n, Q_n) by $(\widehat{G}_n, \widehat{M}_n, \widehat{Q}_n)$. For instance, the regularity condition $(\widehat{G})$ and mixing condition $(\widehat{M})_m$ read:

$(\widehat{G})$: There exists a sequence of numbers $\epsilon_n(\widehat{G}) \in (0,1)$, $n \in \mathbb{N}$, such that for any $(x_n, y_n) \in E_n^2$ we have $\widehat{G}_n(x_p) \geq \epsilon_n(\widehat{G}) \, \widehat{G}_n(y_p) > 0$.

$(\widehat{M})_m$: There exists a sequence of numbers $\epsilon_n(\widehat{M}) \in (0,1)$, $n \in \mathbb{N}$, such that for any $p \in \mathbb{N}$, and $(x_p, y_p) \in E_p^2$, we have

$$\widehat{M}_{p,p+m}(x_p, .) \geq \epsilon_p(\widehat{M}) \, \widehat{M}_{p,p+m}(y_p, .)$$

In the display above, $\widehat{M}_{p,n}$ represents the Markov semigroup associated with the Markov kernels $\widehat{M}_n$. Note that if $(M)_m$ is satisfied for $m=1$, then $(\widehat{G})$ holds true with $\epsilon_n(\widehat{G}) \geq \epsilon_n(M)$.

As usual, in order to unify the presentation, in this section we denote by H the total variation distance on $\mathcal{P}(E)$ or any h'-divergence satisfying the growth condition $(\mathcal{H})_a$, stated on page 135.

Proposition 4.4.1 *When conditions $(\widehat{G})$ and $(\widehat{M})_m$ are met for some $m \geq 1$, then we have for any $n \geq m$*

$$\beta_H(\widehat{\Phi}_{p,p+n}) \leq \overline{a}_H(\epsilon_p^{-1}(\widehat{M}) \, \epsilon_{p,p+m}^{-1}(\widehat{G})) \prod_{k=0}^{\lfloor n/m \rfloor - 1} \left(1 - \epsilon_{p+km}^{(m)}(\widehat{G}, \widehat{M})\right) \quad (4.31)$$

with $\epsilon_p^{(m)}(\widehat{G}, \widehat{M}) = \epsilon_p^2(\widehat{M}) \, \epsilon_{p+1,p+m}(\widehat{G})$.

This way of transferring results from the prediction to the updated models can be extended in a natural way to situations where the potential functions G_n are not strictly positive. In this situation, we write $\widehat{E}_n = G_n^{-1}(0, \infty)$. The main assumption in this context is the accessibility hypothesis $(\mathcal{A})$ introduced in (2.16) on page 67. Rephrasing the discussion given in Section 2.5, the main advantages of this condition are that $\widehat{M}_{n+1}$ are well-defined Markov kernels from $\widehat{E}_n$ into $\widehat{E}_{n+1}$ and the potential functions $\widehat{G}_n$ are strictly positive on $\widehat{E}_n$. If we make this assumption, then $\widehat{\Phi}_{p,n}$ are well-defined semigroups on the sets $\mathcal{P}(\widehat{E}_p)$ and the whole analysis can be conducted as before by replacing E_n by $\widehat{E}_n$. In this context, conditions $(\widehat{G})$

148   4. Stability of Feynman-Kac Semigroups

and $(\widehat{M})_m$ take the form

$(\widehat{G})$: There exists a sequence of strictly positive numbers $\epsilon_n(\widehat{G})$, $n \in \mathbb{N}$, such that

$$\forall (x_n, y_n) \in \widehat{E}_n^2, \quad \widehat{G}_n(x_p) \geq \epsilon_n(\widehat{G})\, \widehat{G}_n(y_p) > 0$$

$(\widehat{M})_m$: There exists a sequence of strictly positive numbers $\epsilon_n(\widehat{M})$, $n \in \mathbb{N}$, such that

$$\forall p \in \mathbb{N}, \quad \forall (x_p, y_p) \in \widehat{E}_p^2, \quad \widehat{M}_{p,p+m}(x_p, .) \geq \epsilon_p(\widehat{M})\, \widehat{M}_{p,p+m}(y_p, .)$$

The next three examples illustrate situations where conditions $(\widehat{G})$ and $(\widehat{M})_m$ are met on the "restricted spaces" $\widehat{E}_n$ but not on E_n.

Example 4.4.1 *Suppose that the state spaces are homogeneous with $E_n = \mathbb{Z}^d$, $d \geq 1$, and $M_n = M$ is the Markov transition on $\mathbb{Z}^d$ defined by*

$$M(x, dy) = \sum_{e \in \mathbb{Z}^d\, :\, |e| \leq 1} p(e)\, \delta_{x+e}(dy)$$

with $p_{\min} = \inf_{|e| \leq 1} p(e) > 0$, and $|x| = \vee_{i=1}^d |x^i|$ for all $x \in \mathbb{Z}^d$. We take the indicator potential function $G = 1_{\widehat{E}}$ associated with the set $\widehat{E} = \{x \in \mathbb{Z}^d\, ;\, |x| \leq q\}$, where q is a given strictly positive integer. Since we have for any $x \in \widehat{E}$ and $|y| > q+1$

$$\widehat{G}(x) = M(x, \widehat{E}) \geq p(0) > 0 \quad \text{and} \quad \widehat{G}(y) = 0$$

we see that condition $(\widehat{G})$ is not satisfied on the whole lattice but it holds true on $\widehat{E}$ with $\epsilon(\widehat{G}) \geq p(0)$. Furthermore, the kernel $\widehat{M}$ defined for any $x \in \widehat{E}$ by

$$\widehat{M}(x, dy) = \frac{1}{M(x, \widehat{E})} \sum_{e \in \mathbb{Z}^d\, :\, |e| \leq 1} p(e)\, 1_{\widehat{E}}(x+e)\, \delta_{x+e}(dy)$$

is a well-defined Markov transition from $\widehat{E}$ into itself. Notice that each coordinate $x^i \in [-q, +q]$ of $x = (x^i)_{1 \leq i \leq d}$ can be joined to any coordinate $z^i \in [-q, +q]$ of $z = (z^i)_{1 \leq i \leq d}$ with an M-admissible path in $\widehat{E}$ of maximal length $2q$. From this observation, we find the rather crude estimate

$$\forall (x, y, z) \in \widehat{E}^3 \quad \widehat{M}^m(x, \{z\}) \geq p_{\min}^m \widehat{M}^m(y, \{z\})$$

with $m = 2dq$. We conclude that the mixing condition $(\widehat{M})_m$ is not met on the whole lattice E but it holds true on $\widehat{E}$ with $\epsilon(\widehat{M}) = p_{\min}^m$.

4.4 Updated Feynman-Kac Models

Example 4.4.2 *Again we assume the state space to be homogeneous, $E_n = \mathbb{R}$, and let $M_n = M$ be the Gaussian transition on the real line defined by*

$$M(x, dy) = \frac{1}{\sqrt{2\pi}} \exp\left(-\frac{1}{2}(y - a(x))^2\right) dy$$

where $a : \mathbb{R} \to \mathbb{R}$ is a given Borel drift function. We let $G = 1_{\widehat{E}}$ be the indicator of a given Borel subset $\widehat{E} \subset \mathbb{R}$ and we suppose

$$|\widehat{E}| = \sup\{|x|, \; x \in \widehat{E}\} < \infty \quad \text{and} \quad \|\widehat{a}\| = \sup\{|a(x)|, \; x \in \widehat{E}\} < \infty$$

In this situation, the Markov kernel $\widehat{M}$ is defined on the whole real line and it is given by

$$\widehat{M}(x, dy) = \frac{1}{\sqrt{2\pi} M(x, \widehat{E})} \exp\left(-\frac{1}{2}(y - a(x))^2\right) 1_{\widehat{E}}(y) \, dy$$

with

$$M(x, \widehat{E}) = \widehat{G}(x) = \int_{\widehat{E}} \frac{1}{\sqrt{2\pi}} \exp\left(-\frac{1}{2}(y - a(x))^2\right) dy$$

After some elementary computations, we find that for any $(x, y, z) \in \widehat{E}$ we have

$$\log \frac{dM(x, \cdot)}{dM(y, \cdot)}(z) = (a(x) - a(y))\left(z - \frac{a(x) + a(y)}{2}\right) \in [-c(a), c(a)]$$

with the crude estimate $c(a) \leq 2\|\widehat{a}\| \, (|\widehat{E}| + \|\widehat{a}\|)$. This clearly implies that

$$\frac{d\widehat{M}(x, \cdot)}{d\widehat{M}(y, \cdot)}(z) = \frac{M(y, \widehat{E})}{M(x, \widehat{E})} \frac{dM(x, \cdot)}{dM(y, \cdot)}(z) \in [e^{-2c(a)}, e^{2c(a)}]$$

We conclude that $(\widehat{G})$ and $(\widehat{M})_m$ are satisfied with $m = 1$ on $\widehat{E}$ with

$$\epsilon(\widehat{G}) \geq e^{-c(a)} \quad \text{and} \quad \epsilon(\widehat{M}) \geq e^{-2c(a)}$$

For an unbounded drift function a and unbounded Borel set $\widehat{E}$, the reader will notice that conditions $(\widehat{G})$ and $(\widehat{M})_m$ are not met.

Example 4.4.3 (One-dimensional neutron model) *The following simplified neutron collision/absorption model is taken from Harris [176]. We assume that $E_n = \mathbb{R}$ and the pair $(G_n, M_n) = (G, M)$ is homogeneous and given by*

$$G(x) = 2 1_{[0, L]}(x) \quad \text{and} \quad M(x, dy) = \frac{c}{2} e^{-c|y - x|} dy$$

150   4. Stability of Feynman-Kac Semigroups

where $L > 0$ and $c > 0$ are given constants. In this situation, we check that $\widehat{E} = [0, L]$, and for any $x \in [0, L]$ we have

$$\widehat{G}(x) = M(G)(x) = 2 - (e^{-cx} + e^{-c(L-x)})$$
$$\widehat{M}(x, dy) = c\, \widehat{G}(x)^{-1}\, e^{-c|y-x|}\, 1_{[0,L]}(y)\, dy$$

We now observe that $\widehat{G}(x) \in [1 - e^{-cL}, 2]$, from which we conclude that conditions $(\widehat{G})$ and $(\widehat{M})_m$ are met on $[0, L]$ with $m = 1$ and

$$\varepsilon(\widehat{G}) = (1 - e^{-cL})/2 \quad \text{and} \quad \varepsilon(\widehat{M}) = e^{-L}\varepsilon(\widehat{G})$$

In general, these two conditions are not easy to check, mainly because the updated kernels $\widehat{M}_n$ are related to the potential function G_n. Our next objective is to give a sufficient condition in terms of the reference pair (G_n, M_n) under which the central mixing hypothesis $(\widehat{Q})_m$ is met. We recall that in the context of updated semigroups we have

$$\widehat{P}_{p,n} = \widehat{R}_{p,q}^{(n)} \widehat{P}_{q,n} \quad \text{with} \quad \widehat{R}_{p,q}^{(n)}(f_q) = \widehat{Q}_{p,q}(f_q \widehat{G}_{q,n})/\widehat{Q}_{p,q}(\widehat{G}_{q,n})$$

Proposition 4.4.2 *Suppose conditions (G) and $(M)_m$ are met for some $m \geq 1$. Then, the mixing condition $(\widehat{Q})_m$ is also met with*

$$\epsilon_p(\widehat{Q}) \geq \epsilon_p(M)\, \epsilon_{p+1, p+m}(G)$$

and we have $0 \leq p + m \leq n$ and $(x_p, y_p) \in E_p^2$ the uniform estimate

$$\widehat{G}_{p,n}(x_p)/\widehat{G}_{p,n}(y_p) \geq \epsilon_p(M)\, \epsilon_{p+1, p+m}(G)$$

In addition, for any $0 \leq p + m \leq q \leq n$, we have

$$\beta(\widehat{R}_{p,q}^{(n)}) \leq 1 - \epsilon_{p+1, p+m}(G)\, \epsilon_p^2(M) \tag{4.32}$$

Proof:
For any nonnegative function $f_n \in \mathcal{B}_b(E_n)$ and $x_p, y_p \in E_p$, $0 \leq p + m \leq n$, we have

$$\frac{\widehat{Q}_{p,n}(f_n)(x_p)}{\widehat{Q}_{p,n}(f_n)(y_p)} = \frac{M_{p+1}(G_{p+1}\widehat{Q}_{p+1,n}(f_n))(x_p)}{M_{p+1}(G_{p+1}\widehat{Q}_{p+1,n}(f_n))(y_p)}$$

$$\geq \epsilon_{p+1}(G) \frac{M_{p,p+2}(G_{p+2}\widehat{Q}_{p+2,n}(f_n))(x_p)}{M_{p,p+2}(G_{p+2}\widehat{Q}_{p+2,n}(f_n))(y_p)}$$

$$\geq \epsilon_{p+1}(G)\, \epsilon_{p+2}(G) \frac{M_{p,p+3}(G_{p+3}\widehat{Q}_{p+2,n}(f_n))(x_p)}{M_{p,p+3}(G_{p+3}\widehat{Q}_{p+2,n}(f_n))(y_p)}$$

Using a clear induction, we find that

$$\frac{\widehat{Q}_{p,n}(f_n)(x_p)}{\widehat{Q}_{p,n}(f_n)(y_p)} \geq \left[\prod_{k=1}^{m-1} \epsilon_{p+k}(G)\right] \frac{M_{p,p+m}(G_{p+m}\widehat{Q}_{p+m,n}(f_n))(x_p)}{M_{p,p+m}(G_{p+m}\widehat{Q}_{p+m,n}(f_n))(y_p)}$$

from which we conclude that

$$\frac{\widehat{Q}_{p,n}(f_n)(x_p)}{\widehat{Q}_{p,n}(f_n)(y_p)} \geq \epsilon_{p+1,p+m}(G)\ \epsilon_p(M)$$

Using similar arguments, for any $0 \leq p+m \leq q \leq n$ we find that

$$\begin{aligned}\widehat{R}_{p,q}^{(n)}(f_q)(x_q) &= \frac{M_{p+1}(G_{p+1}\widehat{Q}_{p+1,q}(f_q\widehat{G}_{q,n}))(x_p)}{M_{p+1}(G_{p+1}\widehat{Q}_{p+1,q}(\widehat{G}_{q,n}))(x_p)} \\ &\geq \epsilon_{p+1}(G)\ \frac{M_{p,p+2}(G_{p+2}\widehat{Q}_{p+2,q}(f_q\widehat{G}_{q,n}))(x_p)}{M_{p,p+2}(G_{p+2}\widehat{Q}_{p+2,q}(\widehat{G}_{q,n}))(x_p)}\end{aligned}$$

and by induction

$$\begin{aligned}\widehat{R}_{p,q}^{(n)}(f_q)(x_q) &\geq \prod_{k=1}^{m-1} \epsilon_{p+k}(G)\ \frac{M_{p,p+m}(G_{p+m}\widehat{Q}_{p+m,q}(f_q\widehat{G}_{q,n}))(x_p)}{M_{p,p+m}(G_{p+m}\widehat{Q}_{p+m,q}(\widehat{G}_{q,n}))(x_p)} \\ &\geq \epsilon_{p+1,p+m}(G)\ \epsilon_p^2(M)\ \frac{M_{p,p+m}(G_{p+m}\widehat{Q}_{p+m,q}(f_q\widehat{G}_{q,n}))(y_p)}{M_{p,p+m}(G_{p+m}\widehat{Q}_{p+m,q}(\widehat{G}_{q,n}))(y_p)}\end{aligned}$$

The proof of (4.32) is now clear. This ends the proof of the proposition. ∎

We end this section with simple contraction estimates that can be deduced from Proposition 4.4.2 using previous considerations (see also Theorem 4.3.4).

Proposition 4.4.3 *Suppose conditions (G) and $(M)_m$ are satisfied. Then, for any $n \geq m$, we have*

$$\beta_H(\widehat{\Phi}_{p,p+n}) \leq \overline{a}_H(\epsilon_p^{-1}(M)\epsilon_{p+1,p+m}^{-1}(G))\prod_{k=0}^{\lfloor n/m \rfloor - 1}\left(1 - \widehat{\epsilon}_{p+km}^{(m)}(G,M)\right) \quad (4.33)$$

with $\widehat{\epsilon}_p^{(m)}(G,M) = \epsilon_p^2(M)\ \epsilon_{p+1,p+m}(G)$. When condition $(M)_m$ is satisfied with $m=1$, we have the potential free estimates

$$\beta_H(\widehat{\Phi}_{p,p+n}) \leq \overline{a}_H(\epsilon_p^{-1}(M))\prod_{k=0}^{n-1}(1 - \epsilon_{p+k}^2(M))$$

Corollary 4.4.1 *Assume that conditions (G) and $(M)_m$ are met for some $m \geq 1$. For any $(\mu_p, \nu_p) \in \mathcal{P}(E_p)^2$ with $H(\mu_p, \nu_p) < \infty$ and $p \in \mathbb{N}$, we have*

$$\sum_{n \geq 0} \epsilon_n(M)\epsilon_{n+1,n+m}(G) = \infty \Rightarrow \lim_{n \to \infty} H(\widehat{\Phi}_{p,p+n}(\mu_p), \widehat{\Phi}_{p,p+n}(\nu_p)) = 0$$

In addition, if we have $\lim_{n\to\infty} \frac{1}{n}\sum_{p=0}^{n-1} \epsilon_p(M)\epsilon_{p+1,p+m}(G) = \bar{\epsilon}$, then we have the asymptotic exponential decay

$$\lim_{n\to\infty} \frac{1}{n} \log H(\widehat{\Phi}_{p,p+nm}(\mu_p), \widehat{\Phi}_{p,p+nm}(\nu_p)) < -\bar{\epsilon}$$

Finally, if we assume that $\inf_{n\geq 0} \epsilon_n(G) = \epsilon(G)$ and $\inf_{n\geq 0} \epsilon_n(M) = \epsilon(M)$, then we have

$$H(\widehat{\Phi}_{p,p+nm}(\mu_p), \widehat{\Phi}_{p,p+nm}(\nu_p)) \leq c \, \exp\left(-n \, \epsilon^2(M)\epsilon^{(m-1)}(G)\right)$$

for some finite constant $c < \infty$ whose values only depend on the pair $(\epsilon^{m-1}(G), \epsilon(M))$ and on the relative entropy $H(\mu_p, \nu_p)$ between the two measures μ_p, ν_p. In particular, when $m = 1$, we have a uniform estimate

$$H(\widehat{\Phi}_{p,p+nm}(\mu_p), \widehat{\Phi}_{p,p+nm}(\nu_p)) \leq c \, \exp\left(-n \, \epsilon^2(M)\right)$$

with a finite constant c that does not depend on $\epsilon(G)$ (nor on $H(\mu_p, \nu_p)$).

We finally examine the impact of these contraction results in the study of time-homogeneous models. It is also instructive to connect the forthcoming discussion with the one given at the end of Section 4.3.3 on the semigroups $\Phi_{p,n}$. As usual, we suppose the various mathematical objects are time-homogeneous and we suppress the time index in the notation. We also restrict ourselves to the situation where (G) and $(M)_m$ are satisfied for some $m \geq 1$. In this context, we have proved the inequalities

$$\beta_H(\widehat{\Phi}_{0,nm}) \leq \bar{a}_H(\epsilon^{-1}(M)\epsilon^{-(m-1)}(G)) \left(1 - \epsilon^2(M) \, \epsilon^{m-1}(G)\right)^n \tag{4.34}$$

We finally introduce the relaxation time

$$\Delta_H(\widehat{\Phi}) = \inf\left\{n \in \mathbb{N} \, ; \quad \beta_H(\widehat{\Phi}_{0,n}) \leq 1/e\right\}$$

Using simple computations, we find that

$$\Delta_H(\widehat{\Phi}) \leq m \, \frac{1 + \log \bar{a}_H(\epsilon^{-1}(M) \, \epsilon^{-(m-1)}(G))}{\epsilon^2(M) \, \epsilon^{m-1}(G)}$$

4.5 A Class of Stochastic Semigroups

In this section, we use the Markov contraction estimates developed in Section 4.2 to study the stability of a class of stochastic Feynman-Kac models arising in nonlinear filtering problems. The sensitivity of filtering equations with respect to initialization errors consists in studying the long time behaviors of an incorrectly initialized filter with the exact optimal filter. In

Chapter 12, Section 12.6, we will see that the semigroup associated with the filter equation is a nonlinear Feynman-Kac semigroup. As a result, we can estimate the asymptotic stability properties of these nonlinear semigroups using the contraction estimates provided in Section 4.3 and Section 4.4.

Under some rather strong regularity and mixing conditions on the pair (G_n, M_n) (see page 139), we have derived several contraction estimates with respect to a class of h-relative entropy criteria. The aim of this section is to replace these conditions by a single hypothesis on the Dobrushin contraction coefficient of M_n. The strategy consists in combining the Markov contraction analysis presented in Section 4.2 with some entropy inequalities recently obtained by Ocone in [260] (see also [59]).

We consider a sequence of measurable spaces $(E_n, \mathcal{E}_n)$, $(F_n, \mathcal{F}_n)$, $n \in \mathbb{Z}$ and a measurable nonnegative potential function $g_n : E_n \times F_n \to (0, \infty)$. We also suppose there exists a nonnegative measure q_n on F_n such that for any $x_n \in E_n$ and $n \in \mathbb{N}$ we have

$$\int_{F_n} g_n(x_n, y_n)\, q_n(dy_n) = 1 \qquad (4.35)$$

let $\eta_0 \in \mathcal{P}(E_0)$, and let M_{n+1} be a Markov kernel from E_n into E_{n+1}. We associate with these objects the Markov chain

$$(\Omega, \mathcal{G}, U_n = (Y_{n-1}, \eta_n)_{n \geq 0}, \mathbb{P})$$

taking values in $(F_{n-1} \times E_n, \mathcal{F}_{n-1} \otimes \mathcal{E}_n)$ with initial distribution $\mu_{-1} \otimes \delta_{\eta_0}$, with an arbitrary $\mu_{-1} \in \mathcal{P}(F_{-1})$, and elementary transitions given for any $F_n \in \mathcal{B}_b(F_n \times \mathcal{P}(E_{n+1}))$ by the formula

$$\mathbb{E}(F_n(Y_n, \eta_{n+1}) \mid (Y_{n-1}, \eta_n))$$

$$= \int_{E_n \times F_n} F_n(y_n, \Psi_{n,y_n}(\eta_n) M_{n+1})\, g_n(y_n, x_n)\, q_n(dy_n)\, \eta_n(dx_n)$$

For each $y_n \in F_n$, $\Psi_{n,y_n} : \mathcal{P}(E_n) \to \mathcal{P}(E_n)$ is the Boltzmann-Gibbs mapping associated with the nonhomogeneous potential function $x_n \in E_n \mapsto g_n(x_n, y_n) \in (0, \infty)$. That is, we have

$$\Psi_{n,y_n}(\eta_n)(dx_n) = \frac{g_n(y_n, x_n)}{\int_{E_n} g_n(x'_n, y_n)\eta_n(dx'_n)}\, \eta_n(dx_n)$$

Notice that the stochastic flow η_n satisfies a nonlinear and random recursive equation starting at η_0 at time $n = 0$

$$\eta_{n+1} = \Psi_{n,Y_n}(\eta_n) M_{n+1}$$

Let η'_n be an auxiliary model defined with the same random equation

$$\eta'_{n+1} = \Psi_{n,Y_n}(\eta'_n) M_{n+1}$$

154   4. Stability of Feynman-Kac Semigroups

but starting at some possibly different $\eta_0' \in \mathcal{P}(E_0)$.

By construction, it is also clear that the triplet $(Y_{n-1}, \eta_n, \eta_n')$, $n \in \mathbb{N}$, forms a Markov chain taking values in $(F_{n-1} \times \mathcal{P}(E_n)^2)$. In nonlinear filtering literature, the distributions η_n and $\Psi_{n,Y_n}(\eta_n) = \widehat{\eta}_n$ are called the one-step predictor and the optimal filter. The flows η_n' and $\Psi_{n,Y_n}(\eta_n') = \widehat{\eta}_n'$ are called the wrong initialized models. The distribution η_0 represents the initial distribution of a Markov signal, and the sequence Y_n represents the noisy and partial observation delivered by the sensors. In practice, η_0 is generally unknown and we traditionally initialized the filter or any kind of approximation scheme with a wrong initial condition. One important problem is clearly to find sufficient conditions insuring that the filtering problem is well posed in the sense that it corrects any wrong initial condition.

Next theorem is a simple application of the Markov contraction estimates developed in Section 4.2.

Theorem 4.5.1 *For any $n \in \mathbb{N}$, we have*

$$\mathbb{E}(\mathrm{Ent}(\widehat{\eta}_n \mid \widehat{\eta}_n')) \leq \mathbb{E}(\mathrm{Ent}(\eta_n \mid \eta_n')) \leq \left[\prod_{p=1}^{n} \beta(M_p)\right] \mathrm{Ent}(\eta_0 \mid \eta_0') \quad (4.36)$$

Proof:
If $\eta_0 \not\ll \eta_0'$, then the r.h.s. in (4.36) is equal to ∞ and the second inequality is trivial. The l.h.s. inequality is also trivial as soon as $\eta_n \not\ll \eta_n'$. Otherwise we notice that

$$\mathrm{Ent}(\widehat{\eta}_n \mid \widehat{\eta}_n') = \int \log \frac{d\Psi_{n,Y_n}(\eta_n)}{d\Psi_{n,Y_n}(\eta_n')} \, d\Psi_{n,Y_n}(\eta_n)$$

Since we have

$$\frac{d\Psi_{n,Y_n}(\eta_n)}{d\Psi_{n,Y_n}(\eta_n')} = \frac{\eta_n'(g_n(.,Y_n))}{\eta_n(g_n(.,Y_n))} \times \frac{d\eta_n}{d\eta_n'}$$

we find

$$\mathrm{Ent}(\widehat{\eta}_n \mid \widehat{\eta}_n') = -\log \frac{\eta_n(g_n(.,Y_n))}{\eta_n'(g_n(.,Y_n))} + \int \log\left(\frac{d\eta_n}{d\eta_n'}\right) \frac{g_n(.,Y_n)}{\eta_n(g_n(.,Y_n))} d\eta_n$$

By construction, we have

$$\mathbb{E}(\mathrm{Ent}(\widehat{\eta}_n \mid \widehat{\eta}_n') \mid (Y_{n-1}, \eta_n))$$

$$= -\int_{F_n} \log\left(\frac{\eta_n(g_n(.,y_n))}{\eta_n'(g_n(.,y_n))}\right) \eta_n(g_n(.,y_n)) \, q_n(dy_n)$$

$$+ \int_{E_n \times F_n} \log\left(\frac{d\eta_n}{d\eta_n'}(x_n)\right) g_n(x_n, y_n) \, q_n(dy_n) \, \eta_n(dx_n)$$

Using (4.35), this implies that

$$\mathbb{E}(\mathrm{Ent}(\widehat{\eta}_n \mid \widehat{\eta}'_n) \mid (Y_{n-1}, \eta_n))$$
$$= \mathrm{Ent}(\eta_n \mid \eta'_n) - \int_{F_n} \log \left(\frac{\eta_n(g_n(.,y_n))}{\eta'_n(g_n(.,y_n))} \right) \eta_n(g_n(.,y_n)) \, q_n(dy_n)$$

On the other hand, applying the Fubini theorem and again (4.35), we also have

$$\int_{F_n} \eta'_n(g_n(.,y_n)) \, q_n(dy_n) = 1$$

Therefore the term

$$\int_{F_n} \log \left(\frac{\eta_n(g_n(.,y_n))}{\eta'_n(g_n(.,y_n))} \right) \eta_n(g_n(.,y_n)) \, q_n(dy_n) \geq 0$$

represents the relative entropy between two equivalent distributions on F_n. Thus, we find the almost sure estimate

$$\mathbb{E}(\mathrm{Ent}(\widehat{\eta}_n \mid \widehat{\eta}'_n) \mid (Y_{n-1}, \eta_n)) \leq \mathrm{Ent}(\eta_n \mid \eta'_n) \qquad (4.37)$$

This ends the proof of the l.h.s. inequality in (4.36). By Theorem 4.2.1, we also have for any $n \geq 1$

$$\begin{aligned} \mathrm{Ent}(\eta_n \mid \eta'_n) &= \mathrm{Ent}(\widehat{\eta}_{n-1} M_n \mid \widehat{\eta}'_{n-1} M_n) \\ &\leq \beta(M_n) \, \mathrm{Ent}(\widehat{\eta}_{n-1} \mid \widehat{\eta}'_{n-1}) \end{aligned} \qquad (4.38)$$

If we combine (4.37) and (4.38), we readily end the proof of the theorem. ∎

5
Invariant Measures and Related Topics

5.1 Introduction

One of the central questions in the theory of time-homogeneous Feynman-Kac semigroups is the existence of invariant measures and the rapidity at which the memory of the initial distribution is lost. This section is centered around this theme. This question is related to different kinds of problems arising in physical, engineering, and applied probability. To guide the reader and give some concrete basis to this section, we have chosen to give a brief introduction on the different ways to interpret and to answer this question.

In physics, the invariant measures may describe the limiting behavior of a nonabsorbed Markov particle evolving in a pocket of obstacles (see Section 2.5.1). In this connection, we also mention that the Lyapunov exponent of Feynman-Kac-Schrödinger semigroups and related spectral quantities is also described in terms of these limiting measures. In biology, Feynman-Kac models can be viewed as distribution flows of infinite population genetic algorithms (see Section 2.5.2). In this interpretation, invariant measures represent the asymptotic concentration of individuals or genes in a genetic population evolution model. In the preceding application areas, the pair of homogeneous potentials/kernels (G, M) is dictated by the problem at hand. The Markov kernel M may represent the physical motion of a particle in some environment $(E; \mathcal{E})$ as well as the mutation of individual or genes in some natural evolution process. In particle trapping problems, the potential G represents the absorption rate and the strength of the obstacles

158   5. Invariant Measures and Related Topics

in the medium. In biology, G is instead interpreted as the selection pressure of the environment.

From a somewhat radically different angle, Feynman-Kac semigroups can be thought of as a natural extension of Markov semigroups. To better understand this point of view, it is useful to recall that Feynman-Kac models also have several nonhomogeneous Markov interpretations (see Section 2.5.2). The essential difference between the corresponding McKean models and traditional homogeneous Markov chains is that their elementary transitions depend on the distribution flow of the random states. In this perspective, one can ask the following question:

Is it possible to build a Feynman-Kac model admitting a given distribution as an invariant measure?

If we restrict this question to the class of Feynman-Kac models with constant potential functions, then the question above is equivalent to that of finding a Markov model having a given invariant measure. This is of course the traditional central question in Monte Carlo Markov chain literature. There have been considerable efforts during the past decades to answer this important question. Several algorithms have been proposed, including the popular Metropolis model and the Gibbs sampler. We refer the reader for instance to the book by D.J. Spielgelhalter, W.R. Gilks and S. Richardson, [291] and to the pioneering articles of W.K. Hastings [178] and N. Metropolis, A.W. Rosenbluth, M.N. Rosenbluth, E. Teller and A.H. Teller [249]. If we extend the question to nonlinear (or nonhomogeneous) Markov models, then one expects to obtain new Monte Carlo simulation methods that work.

Of course, in contrast to traditional Markov chain methods these nonlinear models cannot be sampled perfectly, and another level of approximation is needed. In this connection, the particle methods discussed in this book provide a natural and successful strategy to produce approximate samples according to these McKean models. In the present section, we will not discuss the performance of these particle numerical schemes. Their asymptotic behavior as the size of the systems increases will be discussed in full detail in the further development of Chapters 7 to 10. We also refer the reader to the discussion on particle approximation measures provided in Section 3.5. Here we concentrate our discussion on the modeling of Feynman-Kac semigroups admitting a given distribution as an invariant measure.

This section is organized as follows. In a preliminary section, Section 5.2, we discuss the existence and uniqueness of invariant measures. We connect this question with the contraction analysis developed in Section 4.3. We provide simple sufficient conditions on the pair (G, M) under which the corresponding Feynman-Kac semigroup has a unique invariant measure. We will also transfer the contraction estimates provided in earlier sections to quantify the decays to equilibrium of the corresponding McKean models.

5.1 Introduction

In Section 5.3, we design an original strategy to build Feynman-Kac models admitting a given distribution as an invariant measure. These models are related to a judicious choice of Radon-Nikodym and Metropolis type potential functions. The Feynman-Kac modeling technique presented in this section gives natural powerful tools for developing new particle simulation methods. Because of their importance in practice, we have devoted a separate section, Section 5.5, to these Feynman-Kac-Metropolis models. We already mentioned that these nonlinear models have better decays to equilibrium than traditional Monte Carlo Markov chain algorithms. Furthermore, they are not only useful in drawing samples according to a given target distribution. They also induce new genealogical particle simulation methods for drawing samples according to the law of restricted Markov chains with respect to their terminal values.

In the further development of this chapter, we will use the same terminology as was used in Section 2.7 for abstract Feynman-Kac semigroups. However, in the present homogeneous situation, we have chosen to clarify the presentation, and we adopt a slightly more simple system of notation. Next we discuss these simplifications and take this opportunity to fix some simple but generic properties of invariant measures.

When there is no possible confusion, we suppress the time index and we write (E, G, M) and $(\Phi, \Psi, \widehat{\Phi})$ instead of (E_n, G_n, M_n) and $(\Phi_n, \Psi_n, \widehat{\Phi}_n)$. To avoid repetition, we denote respectively by f and μ a test function in $\mathcal{B}_b(E)$ and a probability measure on E. We also use the letter x to denote a point in E.

We recall that the one-step mappings Φ and $\widehat{\Phi} : \mathcal{P}(E) \to \mathcal{P}(E)$ are defined by the formulae

$$\Phi(\mu) = \Psi(\mu)M \quad \text{and} \quad \widehat{\Phi}(\mu) = \Psi(\mu M) \tag{5.1}$$

where $\Psi : \mathcal{P}(E) \to \mathcal{P}(E)$ is the Boltzmann-Gibbs transformation associated with a bounded nonnegative potential function G on E. That is, we have that

$$\Psi(\mu)(dx) = \frac{1}{\mu(G)} \, G(x) \, \mu(dx) \tag{5.2}$$

Unless otherwise stated, we will assume that G is strictly positive so that Ψ is well-defined on the whole set of distributions on E.

Definition 5.1.1 *Given a mapping $\Theta : \mathcal{P}(E) \to \mathcal{P}(E)$, a measure $\mu \in \mathcal{P}(E)$ is said to be Θ-invariant if $\mu = \Theta(\mu)$. When $\Theta = \Phi$ (or $\Theta = \widehat{\Phi}$), sometimes we say that μ is Φ-invariant/(G, M) (or $\widehat{\Phi}$-invariant/(G, M)) to emphasize for which pair (G, M) the measure μ is Φ-invariant.*

We end this preliminary section with a simple observation.

Proposition 5.1.1 *For any potential function $G : E \to [0, \infty)$ and for any distribution $\mu \in \mathcal{P}(E)$ such that $\mu(G) \wedge \mu M(G) > 0$, the following*

160   5. Invariant Measures and Related Topics

assertions are satisfied.

$$\mu \text{ is } \Phi\text{-invariant}/(G,M) \implies \Psi(\mu) \text{ is } \widehat{\Phi}\text{-invariant}/(G,M)$$
$$\mu \text{ is } \widehat{\Phi}\text{-invariant}/(G,M) \implies \mu M \text{ is } \Phi\text{-invariant}/(G,M)$$

Proof:
Let us assume that μ is Φ-invariant$/(G,M)$. In this case, we have

$$\widehat{\Phi}(\Psi(\mu)) = \Psi(\Psi(\mu)M) = \Psi(\Phi(\mu)) = \Psi(\mu)$$

and the first assertion is proved. To check the second implication, we assume that μ is $\widehat{\Phi}$-invariant$/(G,M)$. In this situation, we observe that

$$\Phi(\mu M) = \Psi(\mu M)M = \widehat{\Phi}(\mu)M = \mu M$$

This clearly ends the proof of the second assertion. ∎

5.2 Existence and Uniqueness

In this short section, we apply the contraction analysis developed in Section 4.3.3 and Section 4.4 to study the existence and the uniqueness of invariant measures. Before presenting the main theorem of this section, we give a brief discussion on some interesting consequences of the existence of these measures.

Suppose $\eta = \Phi(\eta) \in \mathcal{P}(E)$ is a fixed point of Φ and let γ_n and η_n be the unnormalized and the normalized Feynman-Kac model starting at $\gamma_0 = \eta$ and $\eta_0 = \eta$. By construction, we have

$$\eta_n = \Phi^n(\eta) = \eta \quad \text{and} \quad \gamma_n(f) = \mathbb{E}_\eta \left(f(X_n) \prod_{p=0}^{n-1} G(X_p) \right)$$

where $\mathbb{E}_\eta(.)$ is the expectation with respect to the law of a homogeneous Markov chain X_n with transitions M and initial distribution η. In view of Proposition 2.3.1, we find that

$$\mathbb{E}_\eta \left(f(X_n) \prod_{p=0}^{n-1} G(X_p) \right) = \eta(f)\, \eta(G)^n$$

In particular, if we take constant test functions, we have proved that

$$\frac{1}{n} \log \mathbb{E}_\eta \left(\prod_{p=0}^{n-1} G(X_p) \right) = \log \eta(G)$$

The precise physical interpretation of this formula is given in Section 12.4. We have seen that $\log(\eta(G))$ represents the logarithmic Lyapunov exponent of the semigroup $Q(f) = G.M(f)$ on the Banach space $\mathcal{B}_b(E)$. In this connection and whenever G is a $[0,1]$-valued potential, the quantity $\eta(G)^n \in [0,1]$ represents the probability that an absorbed particle motion is still alive at time n (see Section 2.5.1). The following theorem, which is a direct consequence of Proposition 4.3.5 and Proposition 4.4.3, is often useful in applications.

Theorem 5.2.1 *Suppose conditions (G) and $(M)_m$ are met for some integer parameter $m \geq 1$ and some numbers $\epsilon(G) > 0$ and $\epsilon(M) > 0$. Then there exists a unique invariant measure $\eta = \Phi(\eta) \in \mathcal{P}(E)$ and for any $n \in \mathbb{N}$ we have*

$$\mathbb{E}_\eta \left(f(X_n) \prod_{p=0}^{n-1} G(X_p) \right) = \eta(f) \ \eta(G)^n$$

where $\mathbb{E}_\eta(.)$ is the expectation with respect to the law of a homogeneous Markov chain X_n with transitions M and initial distribution η. Furthermore, if we denote by H the total variation distance on the set $\mathcal{P}(E)$ or any h'-divergence H satisfying the growth condition $(\mathcal{H})_a$, then for any $n \geq m$ we have the estimate

$$H(\Phi^n(\mu), \eta) \leq c_1 \ \left(1 - \epsilon^2(M) \ \epsilon^{m-1}(G)\right)^{\lfloor n/m \rfloor} \ H(\mu, \eta)$$

for some finite constant $c_1 < \infty$ whose values only depend on $\epsilon(M)$ and $\epsilon^m(G)$. In addition, $\widehat{\eta} = \Psi(\eta) \in \mathcal{P}(E)$ is the unique $\widehat{\Phi}$-invariant measure and for any $n \geq m$ we have the estimate

$$H(\widehat{\Phi}^n(\mu), \widehat{\eta}) \leq c_2 \ \left(1 - \epsilon^2(M) \ \epsilon^{m-1}(G)\right)^{\lfloor n/m \rfloor} \ H(\mu, \widehat{\eta})$$

for some finite constant $c_2 < \infty$ whose values only depend on $\epsilon(M)$ and $\epsilon^{m-1}(G)$. In particular, when $(M)_m$ is satisfied for $m = 1$, we have a uniform estimate

$$H(\widehat{\Phi}^n(\mu), \widehat{\eta}) \leq c \ \left(1 - \epsilon^2(M)\right)^{\lfloor n/m \rfloor} \ H(\mu, \widehat{\eta})$$

with a finite constant $c < \infty$ that only depends on $\epsilon(M)$.

5.3 Invariant Measures and Feynman-Kac Modeling

In this section, we design a natural strategy to construct a Feynman-Kac model admitting a given distribution as an invariant measure. To describe

5. Invariant Measures and Related Topics

this method precisely, it is convenient to introduce another round of notation. One key idea is to enlarge the state space. We suppose $E = (S \times S)$ is the twofold product of a given measurable space $(S, \mathcal{S})$ and we denote by $K(y, dz)$ a given Markov kernel on S. We associate with K the Markov kernel M^K on E defined in the synthetic integral form

$$M^K((y, z), d(y', z')) = \delta_z(dy') \ K(y', dz')$$

In other words, if $(Y_n)_{n \geq 0}$ is the S-valued Markov chain with elementary transition K, then M^K is the Markov transition of the Markov chain defined by

$$X_n = (Y_n, Y_{n+1}) \in E = S \times S$$

Finally, we associate with any $\pi \in \mathcal{P}(S)$ and with any Markov kernel K on S the distributions $(\pi \times K)_1$ and $(\pi \times K)_2$ on E defined by

$$(\pi \times K)_1(d(y, y')) = \pi(dy) \ K(y, dy')$$
$$(\pi \times K)_2(d(y, y')) = \pi(dy') \ K(y', dy)$$

Sometimes we simplify notation and we write $\pi \times K$ instead of $(\pi \times K)_1$. The following proposition is pivotal.

Proposition 5.3.1 *Let M be a Markov kernel on E. For any pair of distributions $\mu, \eta \in \mathcal{P}(E)$, we have*

$$\mu \ll \eta \ \text{and} \ \mu M = \eta \Longrightarrow \begin{cases} \eta & \text{is} \ \Phi\text{-invariant}/ \left(\frac{d\mu}{d\eta}, M\right) \\ \mu & \text{is} \ \widehat{\Phi}\text{-invariant}/ \left(\frac{d\mu}{d\eta}, M\right) \end{cases}$$

In particular, for any $\pi \in \mathcal{P}(S)$ and any pair of Markov kernels (K, L) on S, we have

$$(\pi \times L)_2 \ll (\pi \times K)_1 \Rightarrow \begin{cases} (\pi \times K)_1 & \text{is} \ \Phi\text{-invariant}/ \left(\frac{d(\pi \times L)_2}{d(\pi \times K)_1}, M^K\right) \\ (\pi \times L)_2 & \text{is} \ \widehat{\Phi}\text{-invariant}/ \left(\frac{d(\pi \times L)_2}{d(\pi \times K)_1}, M^K\right) \end{cases}$$

Proof:
Suppose $\mu \ll \eta$ and $\mu M = \eta$ for some Markov transition M on E. Let Ψ be the Boltzmann-Gibbs transformation associated with the Radon-Nikodym potential $G = \frac{d\mu}{d\eta}$. By construction, we have

$$\Psi(\eta) = \mu \quad \text{and} \quad \Phi(\eta) = \Psi(\eta)M = \mu M = \eta$$

and in the same way $\widehat{\Phi}(\mu) = \Psi(\mu M) = \Psi(\eta) = \mu$. This shows that μ and η satisfy the desired invariance property, and the proof of the first assertion is completed. The second result is a direct consequence of the preceding one. From the latter, it suffices to check that $(\pi \times L)_2 M^K = (\pi \times K)_1$. A simple calculation shows that

$$\begin{aligned}(\pi \times L)_2 M^K(d(y,y')) &= \int \pi(dz') \times L(z',dz)\; \delta_{z'}(dy)\; K(y,dy') \\ &= \int \pi(dy)\; K(y,dy') = (\pi \times K)_1(d(y,dy'))\end{aligned}$$

This completes the proof of the proposition. ∎

Such arguments are powerful tools to construct Feynman-Kac models having a prescribed invariant measure. Let us give a couple of examples to illustrate this assertion. Let π be a given distribution on S and let (K, L) be a pair of Markov kernels such that $L(y,.) \ll \pi \ll K(y,.)$, for any $y \in S$. From the preceding proposition the measures $(\pi \times K)_1$ and $(\pi \times L)_2$ are respectively Φ-and $\widehat{\Phi}$-invariant with respect to the pair potential/kernel

$$(G, M) = \left(\frac{d(\pi \times L)_2}{d(\pi \times K)_1},\; M^K \right)$$

One important target distribution arising in practice is the Boltzmann-Gibbs measure associated with a given nonnegative energy function V on S. These distributions have the form

$$\pi(dy) = \frac{1}{\nu(e^{-V})}\; e^{-V(y)}\; \nu(dy) \tag{5.3}$$

where $\nu \in \mathcal{M}_+(S)$ is a reference measure such that $\nu(e^{-V}) > 0$. For any pair of Markov kernels (K, L) such that $(\nu \times L)_2 \ll (\nu \times K)_1$, we have $(\pi \times L)_2 \ll (\pi \times K)_1$. Using Proposition 5.3.1, we conclude that $(\pi \times K)_1$ is Φ-invariant with respect to the pair

$$(G, M) = \left(\frac{d(\pi \times L)_2}{d(\pi \times K)_1},\; M^K \right)$$

In this situation, we notice that

$$G(y, y') = e^{-(V(y')-V(y))}\; \frac{d(\nu \times L)_2}{d(\nu \times K)_1}(y, y')$$

Suppose the measure ν is reversible with respect to K. If we choose $K = L$, the corresponding potential function takes the form

$$G(y, y') = e^{-(V(y')-V(y))}$$

164   5. Invariant Measures and Related Topics

5.4 Feynman-Kac and Metropolis-Hastings Models

The pair potential/kernel $(G, M) = \left(\frac{d(\pi \times L)_2}{d(\pi \times K)_1}, M^K\right)$ also arises in the construction of the Metropolis-Hastings Markov chain. This model is very popular in Monte Carlo Markov chain literature mainly because it provides a kind of universal strategy to construct a homogeneous Markov chain with target limiting distribution π. To underline the similarities and the differences between these two models, we end this section around this theme.

The Feynman-Kac and Metropolis-Hastings models associated with the pair (G, M^K) correspond to two different ways to associate with the Radon-Nikodym ratio G a Markov kernel on the transition phase $E = S^2$:

1. One of the central objects in the construction of the Feynman-Kac model is the Boltzmann-Gibbs transformation Ψ associated with G. We recall that for any $\eta \in \mathcal{P}(E)$ the measure $\Psi(\eta)$ is given by

$$\Psi(\eta)(d(y, y')) = \frac{1}{\eta(G)} G(y, y')\ \eta(d(y, y'))$$

   There exist many different ways to connect the distribution $\Psi(\eta)$ with a Markov kernel. These different choices correspond to different McKean interpretations of the Feynman-Kac model. We refer the reader to Section 2.5.2 for a precise definition of a McKean model (see also Section 2.5.3 for a collection of examples). We can choose for instance the decomposition

$$\Psi(\eta) = \eta S_\eta$$

   where S_η, $\eta \in \mathcal{P}(E)$, is the collection of Markov transitions on E defined by

$$S_\eta((y, y'), .) = \epsilon\ G(y, y')\ \delta_{(y, y')}(.) + (1 - \epsilon\ G(y, y'))\ \Psi(\eta)(.)$$

   The parameter $\epsilon \geq 0$ is chosen so that $\epsilon G \leq 1$. Under a suitably defined McKean measure, the nonlinear recursive equation

$$\eta_{n+1} = \eta_n S_{\eta_n} M^K \qquad (5.4)$$

   can be interpreted as the evolution of the laws of a nonhomogeneous Markov chain on the transition phase $E = (S \times S)$ with elementary transitions $K_{n+1, \eta_n} = S_{\eta_n} M^K$. Notice that the random evolution of the chain is decomposed into two separate selection/mutation mechanisms. The selection transition S_{η_n} is intended to favor phase regions with high Metropolis ratio G, while the mutation transition consists in exploring the transition phase according to M^K.

2. The Metropolis-Hastings model associated with the pair (G, M^K) is again based on two separate transitions. The first one consists

in exploring the phase space of all transitions $E = S^2$ according to the Markov kernel M^K. The second transition **S** is an acceptance/rejection mechanism on E. It is defined in terms of G by the expression

$$\mathbf{S}((y,y'), .) = (1 \wedge G(y,y'))\, \delta_{(y,y')}(.) + (1 - (1 \wedge G(y,y')))\, \delta_{(y,y)}(.)$$

It is instructive to examine the "advantages and drawbacks" of these two models.

One advantage of the Metropolis-Hastings model is that its semigroup structure is homogeneous and linear so that the algorithm can be sampled perfectly. One drawback is that the acceptance/rejection transition **S** described above tends to slow down the convergence to equilibrium of the chain. For instance, for the Boltzmann-Gibbs limiting distribution (5.3) and in the reversible situation, the transition **S** has the form

$$\mathbf{S}((y,y'), .) = e^{-(V(y')-V(y))^+}\, \delta_{(y,y')}(.) + \left(1 - e^{-(V(y')-V(y))^+}\right) \delta_{(y,y)}(.)$$

In this situation, the rejection probability is close to one in transition phase regions where the difference $(V(y') - V(y))$ is high. On the other hand, we recall that each pair (y, y') has to be interpreted as an elementary transition $(y \to y')$ with distribution $K(y, dy')$. In practice, $K(y, dy')$ is often a distribution on some local neighborhood of y. These two observations indicate that the algorithm may be trapped for a long period of time in the neighborhood of some local minimum of the energy function V.

At first sight, one drawback of the Feynman-Kac model is that its semigroup structure is nonlinear and nonhomogeneous so that it cannot be sampled directly. Indeed, to sample random transitions of this chain, we need to compute the solution to the nonlinear equation (5.4). Nevertheless, one advantage of the nonlinear selection transition S_{η_n} is that the resulting Markov model is not slowed down by a rejection mechanism. Furthermore, using any particle interpretation of the Feynman-Kac model (5.4), this nonlinearity is turned into an interaction mechanism between the particle transitions. In this way, the drawbacks discussed above are turned into a natural and advantageous way to define a sequence of interacting Metropolis type models. In contrast to the classical Metropolis model, a rejected transition is here instantly replaced by a better-fitted one randomly chosen in the current particle transition configuration. From this discussion, it is intuitively clear that the Feynman-Kac model is not slowed down by a rejection stage and it should have better asymptotic properties. More interestingly, the law of the states of this nonhomogeneous Markov model has a nice explicit description in terms of the Feynman-Kac formulae

$$\eta_n(f) = \gamma_n(f)/\gamma_n(1) \quad \text{with} \quad \gamma_n(f) = \mathbb{E}_{\eta_0}^K\left(f(X_n) \prod_{p=0}^{n-1} G(X_p)\right)$$

where $\mathbb{E}_{\eta_0}^K(.)$ is the expectation with respect to the law $\mathbb{P}_{\eta_0}^K$ of a homogeneous Markov chain X_n with transitions M^K and initial distribution η_0. In addition, η_n is the nth time marginal of the Feynman-Kac path measure

$$\mathbb{Q}_{\eta_0,n}(d(x_0,\ldots,x_n)) = \frac{1}{\gamma_n(1)} \left\{ \prod_{p=0}^{n-1} G(x_p) \right\} \mathbb{P}_{\eta_0,n}^K(d(x_0,\ldots,x_n))$$

where $\mathbb{P}_{\eta_0,n}^K$ is the distribution of the path $(X_0,\ldots,X_n)$

$$\mathbb{P}_{\eta_0,n}^K(d(x_0,\ldots,x_n)) = \eta_0(d(x_0))\, M^K(x_0,dx_1)\, \ldots\, M^K(x_{n-1},dx_n)$$

We recall that the measures η_n and $\mathbb{Q}_{\eta_0,n}$ are the limiting measures of the "marginal" and "genealogical" particle approximation models associated with the flow (5.4) (see Chapter 3, Section 11.2, Section 11.4, and the end of Section 5.5.1). On the other hand if we start with the initial distribution $\eta_0 = (\delta_y \times K)_1 \in \mathcal{P}(E)$ for some $y \in S$, we prove that

$$\mathbb{Q}_{(\delta_y \times K)_1,n} = \mathbb{P}_{(\pi \times K)_2}^L(((Y_{n+1},Y_n),\ldots,(Y_1,Y_0)) \in . \mid Y_{n+1} = y) \quad (5.5)$$

where $\mathbb{P}_{(\pi \times K)_2}^L$ is the law of a homogeneous Markov chain $X_n = (Y_n, Y_{n+1})$ with transitions M^L and initial distribution $(\pi \times K)_2$. We also notice that

$$\mathbb{P}_{(\pi \times K)_2}^L((Y_{n+1},\ldots,Y_1) \in . \mid Y_{n+1} = y)$$

$$= \mathbb{P}_{(\pi \times L)_1}^L((Y_n,\ldots,Y_0) \in . \mid Y_n = y) = \mathbf{P}_\pi^L((Y_n,\ldots,Y_0) \in . \mid Y_n = y)$$

where $\mathbf{P}_\pi$ stands for the law of an S-valued Markov chain Y_n with initial distribution π and transitions L. Finally, we observe that for sufficiently regular transitions L we expect that in some sense

$$\eta_n = \mathbb{P}_{(\pi \times K)_2}^L((Y_1, Y_0) \in . \mid Y_{n+1} = y) \xrightarrow[n \to \infty]{} (\pi \times K)_1$$

These results show that the particle model is not only designed to sample according to π, but its genealogical tree also allows us to produce approximate samples according to the law of a restricted Markov chain with respect to its terminal values. The precise analysis of the preceding assertions is out of the scope of this section. Because of its importance in practice, we devote a separate section, Section 5.5, to the full analysis of these models.

5.5 Feynman-Kac-Metropolis Models

5.5.1 Introduction

In this section, we use the same notation and the same terminology as in Section 5.3. We recall that $K(y, dy')$ and $L(y, dy')$ are two given Markov kernels on a measurable space $(S, \mathcal{S})$ and M^K is the Markov kernel on the transition phase $E = S^2$ defined by

$$M^K((y,z), d(y', z')) = \delta_z(dy')\ K(y', dz')$$

We further assume for simplicity that $\pi \in \mathcal{P}(S)$ and the pair kernels (K, L) are chosen so that

$$(\pi \times L)_2(d(y, y')) = \pi(dy')\ L(y', dy) \ll (\pi \times K)_1(d(y, y')) = \pi(dy)\ K(y, dy')$$

and the Radon-Nikodym potential

$$(y, y') \in E = (S \times S) \mapsto G(y, y') = \frac{d(\pi \times L)_2}{d(\pi \times K)_1}(y, y')$$

is a strictly positive and bounded function on the product space $E = S^2$.

In reference to the discussion given in Section 5.4, we adopt the following terminology.

Definition 5.5.1 *The Feynman-Kac model associated with the pair potential/kernel*

$$(G, M) = \left(\frac{d(\pi \times L)_2}{d(\pi \times K)_1}, M^K \right)$$

is called the Feynman-Kac-Metropolis model (associated with (G, M)).

To compare the ways the Radon-Nikodym potential enters into the Feynman-Kac-Metropolis particle approximation models or the Metropolis-Hastings Markov model, we provide hereafter a brief description of these two algorithms.

The Metropolis-Hastings model is a homogeneous Markov chain

$$Z_n = (Y_n, Y_n') \in E = S^2$$

with a two-step selection/mutation transition $\mathbf{S}M^K$. By construction, the first component Y_n is an S-valued and homogeneous Markov chain. Its elementary transitions are given by the familiar expression

$$\mathbf{K}(y, dy') = (1 \wedge G(y, y'))\ K(y, dy') + \left(1 - \int_S (1 \wedge G(y, z)) K(y, dz)\right) \delta_y(dy') \tag{5.6}$$

To prove this assertion, we first note that

$$\mathbb{E}(f(Y_{n+1}) \mid (Y_n, Y_n'))$$
$$= \mathbf{S}(1 \otimes f)(Y_n, Y_n')$$
$$= (1 \wedge G(Y_n, Y_n'))\ f(Y_n') + (1 - (1 \wedge G(Y_n, Y_n')))\ f(Y_n)$$

Since we have $\mathbb{E}(f(Y'_n) \mid Y_n) = K(f)(Y_n)$ for any $n \geq 1$, we conclude that

$\mathbb{E}(f(Y_{n+1}) \mid Y_n = y)$

$= \int_S (1 \wedge G(y,y')) \; f(y') \; K(y,dy') + \left(1 - \int_S (1 \wedge G(y,z)) K(y,dz)\right) \; f(y)$

This ends the proof of the desired result. It is also well-known that if we can choose $K = L$, then $\pi = \pi \mathbf{M}$ is the invariant measure of this chain. By construction, the random evolution of Y_n is decomposed into two separate mechanisms

$$Y_n \xrightarrow{\text{exploration}} Y'_n \xrightarrow{\text{selection}} Y_{n+1}$$

During the exploration stage, the particle Y_n makes an elementary move according to the Markov transition K. In other words, the single particle Y_n randomly chooses a new location Y'_n with distribution $K(Y_n, .)$. During the selection stage we accept the transition $(Y_n \to Y'_n)$ with a probability $(1 \wedge G(Y_n, Y'_n))$ and we set $Y_{n+1} = Y'_n$. Otherwise we reject the transition and we stay in the same location; that is, we set $Y_{n+1} = Y_n$.

The Feynman-Kac-Metropolis model is a nonhomogeneous Markov chain on the transition phase

$$Z_n = (Y_n, Y'_n) \in E = S^2$$

with a two-step selection/mutation transition $K_{n,\eta_{n-1}} = S_{\eta_{n-1}} M^K$. The sequence of distributions $\eta_n \in \mathcal{P}(E)$ is the solution of the nonlinear recursive equation

$$\eta_n = \eta_{n-1} S_{\eta_{n-1}} M^K \; (= \Psi(\eta_{n-1}) M^K) \tag{5.7}$$

where

- S_η, $\eta \in \mathcal{P}(E)$, is the collection of Markov transitions

$$S_\eta((y,y'), .) = \epsilon \; G(y,y') \; \delta_{(y,y')}(.) + (1 - \epsilon \; G(y,y')) \; \Psi(\eta)(.)$$

- $\Psi : \mathcal{P}(E) \to \mathcal{P}(E)$ is the Boltzmann-Gibbs transformation associated with Metropolis potential function G, and it is defined by

$$\Psi(\eta)(d(y,y')) = \frac{1}{\eta(G)} \; G(y,y') \; \eta(d(y,y'))$$

We recall that the distribution flow η_n is alternatively defined in terms of the Feynman-Kac formulae

$$\eta_n(f) = \gamma_n(f)/\gamma_n(1) \quad \text{with} \quad \gamma_n(f) = \mathbb{E}^K_{\eta_0}\left(f(X_n) \prod_{p=0}^{n-1} G(X_p)\right)$$

where $\mathbb{E}_{\eta_0}^K(.)$ is the expectation with respect to the law $\mathbb{P}_{\eta_0}^K$ of a homogeneous Markov chain X_n with transitions M^K and initial distribution η_0. Under a suitably chosen McKean measure, the distribution η_n represents the law of the random state of the chain $Z_n = (Y_n, Y_n')$ at each time n (see Section 2.5.2 for a precise construction of these McKean measures). Arguing as before, we see that the random evolution is again decomposed into two separate mechanisms

$$Y_n \xrightarrow{\text{exploration}} Y_n' \xrightarrow{\text{selection}} Y_{n+1}$$

The exploration stage coincides with that of the Metropolis model but the selection stage is different. Here we accept the transition $(Y_n \to Y_n')$ with a probability $\epsilon G(Y_n, Y_n')$, and we set $Y_{n+1} = Y_n'$. Otherwise, we select randomly a new one $(\widetilde{Y}_n, \widetilde{Y}_n')$ according to $\Psi(\eta_{n-1})$, and we set $Y_{n+1} = \widetilde{Y}_n'$. As we mentioned above, this nonhomogeneous Markov model cannot be sampled perfectly mainly because the distributions η_{n-1} and a fortiori $\Psi(\eta_{n-1})$ are generally unknown.

The interacting Metropolis model is the N-particle approximation model

$$\xi_n^i = (Y_n^i, Y_n'^i) \in E = (S \times S)$$

associated with the McKean interpretation (5.7) of the Feynman-Kac distribution flow. We refer the reader to Section 2.5.2 for a precise description of these particle interpretations. For the convenience of the reader and to better connect this particle simulation method with the preceding Metropolis-Hastings model, we provide hereafter a brief presentation of this model. Suppose the initial distribution is given by

$$\eta_0 = \delta_y \times K$$

with an arbitrary point $y \in S$. In this situation, the initial system is given by $\xi_0^i = (Y_0^i, Y_0'^i) = (y, Y_0'^i)$, where $Y_0'^i$ are independent and identically distributed random variables with common law $K(y, .)$. The N-particle model associated with the McKean kernels $S_\eta M^K$ is a mutation/selection algorithm

$$(Y_n^i)_{1 \leq i \leq N} \xrightarrow{\text{mutation}} (Y_n'^i)_{1 \leq i \leq N} \xrightarrow{\text{selection}} (Y_{n+1}^i)_{1 \leq i \leq N}$$

During the mutation stage, each particle Y_n^i evolves randomly and independently according to the Markov kernel K to some new locations $Y_n'^i$. These new locations $Y_n'^i$ are accepted or rejected according to a mechanism that depends on the set of sampled transitions $(Y_n^j, Y_n'^j)$, $1 \leq j \leq N$. With a probability $\epsilon G(Y_n^i, Y_n'^i)$, we accept the ith state $Y_n'^i$ and we set $Y_{n+1}^i = Y_n'^i$. Otherwise, we select randomly a state $\widetilde{Y}_n'^i$ with distribution

$$\sum_{j=1}^N \frac{G(Y_n^j, Y_n'^j)}{\sum_{k=1}^N G(Y_n^k, Y_n'^k)} \delta_{Y_n'^j}$$

and we set $Y_{n+1}^i = \widetilde{Y}_n^{\prime i}$. Loosely speaking the selection transition is intended to improve the quality of the configuration by allocating more reproductive opportunities to pair particles $(Y_n^j, Y_n^{\prime j})$ with higher Metropolis ratio. From the preceding description, this N-particle model can clearly be interpreted as a sequence of N interacting Metropolis algorithms.

The choice of the McKean selection transition is not unique (see Sectionmckean). In practice, it is desirable to choose a McKean interpretation which the highest acceptance probability. In this connection, if one chooses the selection model (2.27) then the acceptance probability $\epsilon G(Y_n^i, Y_n^{\prime i})$ is replaced by $G(Y_n^i, Y_n^{\prime i}) / \vee_j G(Y_n^j, Y_n^{\prime j})$.

We finally note that each equivalent formulation of a given Feynman-Kac measure induces a different interacting Metropolis type model. For instance, we have for any $x_0 \in E$ and any $f_n \in \mathcal{B}_b(E^{n+1})$

$$\mathbb{E}_{x_0}^K \left(f_n(X_0, \ldots, X_n) \prod_{p=1}^{n} G(X_p) \right) = \mathbb{E}_{x_0}^{\widehat{K}} \left(f_n(X_0, \ldots, X_n) \prod_{p=0}^{n-1} \widehat{G}(X_p) \right)$$

where $\widehat{G} = M^K(G)$ and $\mathbb{E}_{\eta_0}^{\widehat{K}}(.)$ is the expectation with respect to the law of a homogeneous Markov chain X_n with transitions

$$M^{\widehat{K}}(x, dx') = \frac{M^K(x, dx') G(x')}{M^K(G)(x)}$$

For instance, for the Boltzmann-Gibbs distribution (5.3), and in the reversible situation, the pair potential/kernel $(\widehat{G}, M^{\widehat{K}})$ takes the form

$$\widehat{G}(y, y') = e^{V(y')} K(e^{-V})(y')$$

and $M^{\widehat{K}}((y, y'), d(z, z')) = \delta_{y'}(dz) \widehat{K}(z, dz')$ with

$$\widehat{K}(z, dz') = \frac{K(z, dz') e^{-V(z')}}{K(e^{-V})(z)}$$

The corresponding N-particle model is again a two-step Markov chain

$$(\widehat{Y}_n^i)_{1 \le i \le N} \xrightarrow{\text{mutation}} (\widehat{Y}_n^{\prime i})_{1 \le i \le N} \xrightarrow{\text{selection}} (\widehat{Y}_{n+1}^i)_{1 \le i \le N}$$

with a selection/mutation procedure defined as above by replacing the pair (G, M^K) by the $(\widehat{G}, M^{\widehat{K}})$.

5.5.2 The Genealogical Metropolis Particle Model

Another important feature of the evolutionary particle scheme described in the introductory section concerns its birth and death interpretation: The individual $Y_{n-1}^{\prime j}$ selected by the ith individual Y_n^i can be seen as the parent

5.5 Feynman-Kac-Metropolis Models

of Y_n^i. Recalling that $Y_{n-1}^{\prime j}$ has itself been sampled according to $K(Y_{n-1}^j, \cdot)$, we can interpret Y_{n-1}^j as the ancestor $Y_{n-1,n}^i$ of Y_n^i at level $(n-1)$. Running this construction back in time, we can trace back the complete ancestral line of each current individual $Y_{n,n}^i = Y_n^i$

$$Y_{0,n}^i \longleftarrow Y_{1,n}^i \longleftarrow \ldots \longleftarrow Y_{n-1,n}^i \longleftarrow Y_{n,n}^i$$

The study of the genealogical structure of the interacting Metropolis model (starting at $y \in S$) is not of pure mathematical interest. This object is in fact a powerful particle simulation method for drawing path samples of restricted Markov models with respect to their terminal value. More precisely, we have

$$\lim_{N \to \infty} \frac{1}{N} \sum_{i=1}^N \delta_{(Y_{0,n}^i, Y_{1,n}^i, \ldots, Y_{n,n}^i)}(d(y_n, \ldots, y_0))$$

$$= \mathbf{P}_\pi^L((Y_n, Y_{n-1}, \ldots, Y_0) \in d(y_n, \ldots, y_0) \mid Y_n = y) \tag{5.8}$$

where $\mathbf{P}_\pi^L$ represents the distribution of a time-homogeneous Markov chain Y_n with initial distribution π and elementary transition L. For a precise meaning of this convergence result, we refer the reader to Chapter 7. The preceding convergence result indicates that the Feynman-Kac-Metropolis model is related to a time reversal of a Markov chain. To prove that the limiting distribution is precisely the one given above, we need to analyze the Feynman-Kac model on path space. More precisely, let $\xi_n^i = (Y_n^i, Y_n^{\prime i})$ be the N-particle model associated with the McKean interpretation of the equation (5.7). Using the same line of reasoning as above, we can interpret this particle model as a birth and death process and we can trace back the complete ancestral line

$$\xi_{0,n}^i \longleftarrow \xi_{1,n}^i \longleftarrow \ldots \longleftarrow \xi_{n-1,n}^i \longleftarrow \xi_{n,n}^i$$

of each pair individual $\xi_n^i = (Y_n^i, Y_n^{\prime i})$. From the results presented in Section 3.4, we know that this path-particle model can also be regarded as the N-particle model associated with the Feynman-Kac distribution flow on path space. More precisely, the occupation measures of the N-particle genealogical model

$$\mathbb{Q}_{\eta_0, n}^N = \frac{1}{N} \sum_{i=1}^N \delta_{(\xi_{0,n}^i, \ldots, \xi_{n,n}^i)}$$

converge as the size of the systems N tends to ∞ to the path measures $\mathbb{Q}_{\eta_0, n} \in \mathcal{P}(E^{n+1})$ defined by

$$\mathbb{Q}_{\eta_0, n}(d(x_0, \ldots, x_n)) = \frac{1}{\gamma_n(1)} \left\{ \prod_{p=0}^{n-1} G(x_p) \right\} \mathbb{P}_{\eta_0, n}^K(d(x_0, \ldots, x_n))$$

where $\mathbb{P}^K_{\eta_0,n}$ is the distribution of the path from zero to time n of an E-valued Markov chain with initial distribution η_0 and Markov transition M^K. That is, we have that

$$\mathbb{P}^K_{\eta_0,n}(d(x_0,\ldots,x_n)) = \eta_0(dx_0)\, M^K(x_0,dx_1)\, \ldots\, M^K(x_{n-1},dx_n)$$

If we take a test function $f_n \in \mathcal{B}_b((S \times S)^{n+1})$ of the form

$$f_n((y_0,y'_0),\ldots,(y_n,y'_n)) = \varphi_n(y_0,\ldots,y_n)$$

for some $\varphi_n \in \mathcal{B}(S^{n+1})$, then we find that

$$\mathbb{Q}^N_{\eta_0,n}(f_n) = \frac{1}{N}\sum_{i=1}^N \varphi_n(Y^i_{0,n},Y^i_{1,n},\ldots,Y^i_{n,n}) \xrightarrow[N \to \infty]{} \mathbb{Q}_{\eta_0,n}(f_n) \quad (5.9)$$

with

$$\mathbb{Q}_{\eta_0,n}(f_n) = \frac{\mathbf{E}^K_y\left(\varphi_n(Y_0,Y_1,\ldots,Y_n)\prod_{p=0}^{n-1}G(Y_p,Y_{p+1})\right)}{\mathbf{E}^K_y\left(\prod_{p=0}^{n-1}G(Y_p,Y_{p+1})\right)}$$

In the formula displayed above, $\mathbf{E}^K_y(.)$ represents the expectation with respect to the law $\mathbf{P}^K_y$ of a time-homogeneous Markov chain Y_n starting at $Y_0 = y$ with elementary transition K.

To prove that the Feynman-Kac path measure in the previous display coincides with the conditional distribution (5.8), we need to analyze more deeply the structure of these path measures. The complete proof of this result will be provided in the next section.

Using the same line of argument, the conditional distribution (5.8) can be approximated using the ancestral lines $(\widehat{Y}^i_{p,n})_{p\leq n}$ of the particle model associated to the pair $(\widehat{G},M^{\widehat{K}})$ defined in the end of section 5.5.1, and starting at $(\widehat{Y}_0,\widehat{Y}'_0) = (y_0,y_1)$. That is, we have

$$\lim_{N\to\infty} \frac{1}{N}\sum_{i=1}^N \delta_{(\widehat{Y}^i_{1,n},\ldots,\widehat{Y}^i_{n,n})}$$

$$= \mathbf{P}^L_\pi((Y_n,Y_{n-1},\ldots,Y_0) \in . \mid Y_n = y_1)$$

5.5.3 Path Space Models and Restricted Markov Chains

In this section, we connect the Feynman-Kac path measures associated with the pair potential/kernel

$$(G,M) = \left(\frac{d(\pi \times L)_2}{d(\pi \times K)_1},\, M^K\right)$$

with the distributions of a restricted Markov chain with respect to its terminal values. Without further mention, we suppose in this section that the triplet (π, K, L) is chosen such that

- The Radon-Nikodym potential G is a strictly positive and bounded function.

- The measures π and πL are mutually absolutely continuous, and for any $n \in \mathbb{N}$ the Radon-Nikodym derivative $\frac{d\pi L^n}{d\pi}$ is a bounded and strictly positive function on S.

Our first task is to describe more rigorously the Feynman-Kac path measures on a canonical space. Let $\mu \in \mathcal{P}(S^2)$ and let K be a given Markov kernel on S. We denote by

$$(E^{\mathbb{N}}, \mathcal{F} = (\mathcal{F}_n)_{n \geq 0}, (X_n)_{n \geq 0}, \mathbb{P}_\mu^K)$$

the canonical Markov chain

$$X_n = (Y_n, Y_{n+1}) \in E = (S \times S) \qquad (5.10)$$

with initial distribution μ and transition M^K. More rigorously, we should have set $X_n = (Y_n, Y_n')$ instead of (5.10), but by definition of the Markov transition M^K we have $Y_n' = Y_{n+1}$ and therefore $X_n = (Y_n, Y_{n+1})$ as soon as $n \geq 1$. To simplify the presentation, it seems more appropriate to use the notation (5.10). In this obvious abusive notation, we have for instance that

$$\mathbb{P}_\mu^K((Y_0, \ldots, Y_n) \in d(y_0, \ldots, y_n))$$
$$= \mu(d(y_0, dy_1)) \, K(y_1, dy_2) \, \ldots \, K(y_{n-1}, dy_n)$$

As usual we denote by $\mathbb{E}_\mu^K(.)$ the expectation with respect to the law $\mathbb{P}_\mu^K$. When $\mu = \delta_{(x,y)}$ is concentrated at a single point, we simplify notation and we write $(\mathbb{P}_{(x,y)}^K, \mathbb{E}_{(x,y)}^K)$ instead of $(\mathbb{P}_{\delta_{(x,y)}}^K, \mathbb{E}_{\delta_{(x,y)}}^K)$. We finally recall that the Feynman-Kac path measure $\mathbb{Q}_{\eta_0, n} \in \mathcal{P}(E^{n+1})$ associated with the pair (G, M^K) and its terminal time marginal distributions $\eta_n \in \mathcal{P}(E)$ are defined by

$$\mathbb{Q}_{\eta_0, n}(d(x_0, \ldots, x_n)) = \frac{1}{\gamma_n(1)} \left\{ \prod_{p=0}^{n-1} G(x_p) \right\} \mathbb{P}_{\eta_0, n}^K(d(x_0, \ldots, x_n))$$

and

$$\eta_n(f) = \gamma_n(f)/\gamma_n(1) \quad \text{with} \quad \gamma_n(f) = \mathbb{E}_{\eta_0}^K \left(f(X_n) \prod_{p=0}^{n-1} G(X_p) \right)$$

By the definition of G, we readily obtain the time reversal formula

$$\pi(dy_0) K(y_0, dy_1) \ldots K(y_n, dy_{n+1}) \prod_{p=0}^{n} G(y_p, y_{p+1})$$

$$= \pi(dy_{n+1}) L(y_{n+1}, dy_1) \ldots L(y_1, dy_0)$$

In other words we have

$$\frac{d\mathbb{P}^L_{\pi \times L}((Y_{n+1}, Y_n, \ldots, Y_0) \in \cdot)}{d\mathbb{P}^K_{\pi \times K}((Y_0, Y_1, \ldots, Y_{n+1}) \in \cdot)}(y_0, \ldots, y_{n+1}) = \prod_{p=0}^{n} G(y_p, y_{p+1})$$

and we find the following pivotal lemma.

Lemma 5.5.1 (time reversal formula) *For any $n \geq 0$ and any $\varphi_n \in \mathcal{B}_b(S^{n+2})$ we have*

$$\mathbb{E}^L_{\pi \times L}(\varphi_n(Y_{n+1}, Y_n, \ldots, Y_0))$$

$$= \mathbb{E}^K_{\pi \times K}\left(\varphi_n(Y_0, Y_1, \ldots, Y_{n+1}) \prod_{p=0}^{n} G(Y_p, Y_{p+1})\right)$$

In the next theorem, we provide another couple of time reversal formulae. They allow us to interpret the multiplicative structure of the Feynman-Kac model as a combination of a change of probability transitions with a time reversal of a Markov chain.

Theorem 5.5.1 *For any $\varphi_n \in \mathcal{B}_b(S^{n+2})$ and $n \in \mathbb{N}$, we have the following Feynman-Kac formulae:*

$$\mathbb{E}^K_{(y_0, y_1)}\left(\varphi_n(Y_0, \ldots, Y_{n+1}) \prod_{p=1}^{n} G(Y_p, Y_{p+1})\right)$$

$$= \frac{d\pi L^n}{d\pi}(y_1) \, \mathbb{E}^L_{(\pi \times L)_1}\left(\varphi_n(Y_{n+1}, \ldots, Y_0) \mid (Y_{n+1}, Y_n) = (y_0, y_1)\right) \quad (5.11)$$

and

$$\mathbb{E}^K_{(y_0, y_1)}\left(\varphi_n(Y_0, \ldots, Y_{n+1}) \prod_{p=1}^{n-1} G(Y_p, Y_{p+1})\right)$$

$$= \frac{d\pi L^{n-1}}{d\pi}(y_1) \, \mathbb{E}^L_{(\pi \times K)_2}\left(\varphi_n(Y_{n+1}, \ldots, Y_0) \mid (Y_{n+1}, Y_n) = (y_0, y_1)\right) \quad (5.12)$$

Proof:
First we use the time reversal formula presented in Lemma 5.5.1 to check

that for any $\varphi \in \mathcal{B}_b(E)$

$$\mathbb{E}^K_{\pi \times K}(\varphi(Y_0, Y_1) \prod_{p=1}^n G(Y_p, Y_{p+1}))$$

$$= \mathbb{E}^K_{\pi \times K}(\varphi(Y_0, Y_1) \, G(Y_0, Y_1)^{-1} \prod_{p=0}^n G(Y_p, Y_{p+1}))$$

$$= \mathbb{E}^L_{\pi \times L}(\varphi(Y_{n+1}, Y_n) \, G(Y_{n+1}, Y_n)^{-1})$$

$$= \int \frac{d\pi L^n}{d\pi}(y_n) \pi(dy_{n+1}) K(y_{n+1}, dy_n) \varphi(y_{n+1}, y_n)$$

This yields that

$$\mathbb{E}^K_{\pi \times K}\left(\varphi(Y_0, Y_1) \prod_{p=1}^n G(Y_p, Y_{p+1})\right) = \mathbb{E}^K_{\pi \times K}\left(\varphi(Y_0, Y_1) \frac{d\pi L^n}{d\pi}(Y_1)\right)$$

Since this formula holds true for any $\varphi \in \mathcal{B}_b(E)$, we conclude that for any $(y_0, y_1) \in E$

$$\mathbb{E}^K_{(y_0, y_1)}\left(\prod_{p=1}^n G(Y_p, Y_{p+1})\right) = \frac{d\pi L^n}{d\pi}(y_1) \tag{5.13}$$

From Lemma 5.5.1, we find that for any $\varphi' \in \mathcal{B}_b(E)$

$$\mathbb{E}^L_{\pi \times L}(\varphi'(Y_{n+1}, Y_n) \, \varphi_n(Y_{n+1}, \ldots, Y_0))$$

$$= \mathbb{E}^K_{\pi \times K}(\varphi'(Y_0, Y_1) \, \varphi_n(Y_0, \ldots, Y_{n+1}) \prod_{p=0}^n G(Y_p, Y_{p+1}))$$

$$= \mathbb{E}^K_{\pi \times K}(\varphi'(Y_0, Y_1) \, \mathbb{E}^K_{(Y_0, Y_1)}[\varphi_n(Y_0, \ldots, Y_{n+1}) \prod_{p=0}^n G(Y_p, Y_{p+1})] \,)$$

Using (5.13), we also prove that

$$\mathbb{E}^L_{\pi \times L}(\varphi'(Y_{n+1}, Y_n) \, \varphi_n(Y_{n+1}, \ldots, Y_0))$$

$$= \mathbb{E}^K_{\pi \times K}\left(\varphi'(Y_0, Y_1) \, \mathbb{E}^K_{(Y_0, Y_1)}\left[\prod_{p=0}^n G(Y_p, Y_{p+1})\right]\right.$$

$$\left. \times \frac{d\pi}{d\pi L^n}(Y_1) \, \mathbb{E}^K_{(Y_0, Y_1)}\left[\varphi_n(Y_0, \ldots, Y_{n+1}) \prod_{p=1}^n G(Y_p, Y_{p+1})\right] \,\right)$$

$$= \mathbb{E}^K_{\pi \times K}\left(\varphi'(Y_0, Y_1) \prod_{p=0}^n G(Y_p, Y_{p+1})\right.$$

$$\left. \times \frac{d\pi}{d\pi L^n}(Y_1) \, \mathbb{E}^K_{(Y_0, Y_1)}\left[\varphi_n(Y_0, \ldots, Y_{n+1}) \prod_{p=1}^n G(Y_p, Y_{p+1})\right] \,\right)$$

176   5. Invariant Measures and Related Topics

Then, by Lemma 5.5.1, we get

$$\mathbb{E}^L_{\pi \times L}(\varphi'(Y_{n+1}, Y_n)\, \varphi_n(Y_{n+1}, \ldots, Y_0))$$

$$= \mathbb{E}^L_{\pi \times L}\left(\varphi'(Y_{n+1}, Y_n)\, \frac{d\pi}{d\pi L^n}(Y_n)\right.$$

$$\left. \times\; \mathbb{E}^K_{(Y_{n+1}, Y_n)}[\varphi_n(Y_0, \ldots, Y_{n+1}) \prod_{p=1}^n G(Y_p, Y_{p+1})]\right)$$

Since this formula is valid for any φ', we conclude that for any $(y_0, y_1) \in E$

$$\mathbb{E}^K_{(y_0, y_1)}(\varphi_n(Y_0, \ldots, Y_{n+1}) \prod_{p=1}^n G(Y_p, Y_{p+1}))$$

$$= \frac{d\pi L^n}{d\pi}(y_1)\, \mathbb{E}^L_{\pi \times L}(\varphi_n(Y_{n+1}, \ldots, Y_0) \mid (Y_{n+1}, Y_n) = (y_0, y_1))$$

This ends the proof of the first assertion of the proposition. To prove the second formula, we first observe that

$$\mathbb{E}^L_{(\pi \times K)_2}(\varphi_n(Y_{n+1}, \ldots, Y_0))$$

$$= \mathbb{E}^L_{(\pi \times L)_1}\left(\varphi_n(Y_{n+1}, \ldots, Y_0)\, \frac{d(\pi \times K)_2}{d(\pi \times L)_1}(Y_0, Y_1)\right)$$

By Lemma 5.5.1, we find that

$$\mathbb{E}^L_{(\pi \times K)_2}(\varphi_n(Y_{n+1}, \ldots, Y_0))$$

$$= \mathbb{E}^K_{(\pi \times K)_1}\left(\varphi_n(Y_0, \ldots, Y_{n+1})\, \frac{d(\pi \times K)_2}{d(\pi \times L)_1}(Y_{n+1}, Y_n) \prod_{p=0}^n G(Y_p, Y_{p+1})\right)$$

$$= \mathbb{E}^K_{(\pi \times K)_1}\left(\varphi_n(Y_0, \ldots, Y_{n+1})\, G(Y_n, Y_{n+1})^{-1} \prod_{p=0}^n G(Y_p, Y_{p+1})\right)$$

and therefore

$$\mathbb{E}^L_{(\pi \times K)_2}(\varphi_n(Y_{n+1}, \ldots, Y_0))$$

$$= \mathbb{E}^K_{(\pi \times K)_1}\left(\varphi_n(Y_0, \ldots, Y_{n+1}) \prod_{p=0}^{n-1} G(Y_p, Y_{p+1})\right) \qquad (5.14)$$

From this formula, we prove that for any $\varphi' \in \mathcal{B}_b(E)$

$$\mathbb{E}^L_{(\pi \times K)_2}(\varphi'(Y_{n+1}, Y_n)\, \varphi_n(Y_{n+1}, \ldots, Y_0))$$

$$= \mathbb{E}^K_{(\pi \times K)_1}\left(\varphi'(Y_0, Y_1)\, \varphi_n(Y_0, \ldots, Y_{n+1}) \prod_{p=0}^{n-1} G(Y_p, Y_{p+1})\right)$$

$$= \mathbb{E}^K_{(\pi \times K)_1}\left(\varphi'(Y_0, Y_1)\, \mathbb{E}^K_{(Y_0, Y_1)}\left(\varphi_n(Y_0, \ldots, Y_{n+1}) \prod_{p=0}^{n-1} G(Y_p, Y_{p+1})\right)\right)$$

5.5 Feynman-Kac-Metropolis Models

On the other hand, by (5.13) we have for any $(y_0, y_1) \in E$ and $n \geq 1$

$$\mathbb{E}^K_{(y_0,y_1)}\left(\prod_{p=1}^{n-1} G(Y_p, Y_{p+1})\right) = \frac{d\pi L^{n-1}}{d\pi}(y_1)$$

This yields that

$$\mathbb{E}^L_{(\pi \times K)_2}\left(\varphi'(Y_{n+1}, Y_n)\, \varphi_n(Y_{n+1}, \ldots, Y_0)\right)$$

$$= \mathbb{E}^K_{(\pi \times K)_1}\left(\varphi'(Y_0, Y_1)\, \mathbb{E}^K_{(Y_0,Y_1)}\left[\prod_{p=0}^{n-1} G(Y_p, Y_{p+1})\right]\right.$$

$$\left.\times \frac{d\pi}{d\pi L^{n-1}}(Y_1)\, \mathbb{E}^K_{(Y_0,Y_1)}\left[\varphi_n(Y_0, \ldots, Y_{n+1})\prod_{p=1}^{n-1} G(Y_p, Y_{p+1})\right]\right)$$

$$= \mathbb{E}^K_{(\pi \times K)_1}\left(\varphi'(Y_0, Y_1)\prod_{p=0}^{n-1} G(Y_p, Y_{p+1})\right.$$

$$\left.\times \frac{d\pi}{d\pi L^{n-1}}(Y_1)\, \mathbb{E}^K_{(Y_0,Y_1)}\left[\varphi_n(Y_0, \ldots, Y_{n+1})\prod_{p=1}^{n-1} G(Y_p, Y_{p+1})\right]\right)$$

Finally, using (5.14), we arrive at

$$\mathbb{E}^L_{(\pi \times K)_2}\left(\varphi'(Y_{n+1}, Y_n)\, \varphi_n(Y_{n+1}, \ldots, Y_0)\right)$$

$$= \mathbb{E}^L_{(\pi \times K)_2}\left(\varphi'(Y_{n+1}, Y_n)\frac{d\pi}{d\pi L^{n-1}}(Y_n)\right.$$

$$\left.\times \mathbb{E}^K_{(Y_{n+1},Y_n)}[\varphi_n(Y_0, \ldots, Y_{n+1})\, \textstyle\prod_{p=1}^{n-1} G(Y_p, Y_{p+1})]\right)$$

Since this formula holds true for any φ', we conclude that for any pair $(y_0, y_1) \in E$ we have

$$\mathbb{E}^K_{(y_0,y_1)}(\varphi_n(Y_0, \ldots, Y_{n+1})\, \textstyle\prod_{p=1}^{n-1} G(Y_p, Y_{p+1}))$$

$$= \frac{d\pi L^{n-1}}{d\pi}(y_1)\, \mathbb{E}^L_{(\pi \times K)_2}\left(\varphi_n(Y_{n+1}, \ldots, Y_0) \mid (Y_{n+1}, Y_n) = (y_0, y_1)\right)$$

This ends the proof of the theorem. ∎

178   5. Invariant Measures and Related Topics

Corollary 5.5.1 *For any $\varphi_n \in \mathcal{B}_b(S^{n+2})$, $n \in \mathbb{N}$, and $y \in S$, we have the Feynman-Kac formulae*

$$\mathbb{E}^K_{(\delta_y \times K)_1}\left(\varphi_n(Y_0,\ldots,Y_{n+1})\prod_{p=0}^{n}G(Y_p,Y_{p+1})\right)$$

$$= \frac{d\pi L^{n+1}}{d\pi}(y)\,\mathbb{E}^L_{(\pi \times L)_1}(\varphi_n(Y_{n+1},\ldots,Y_0)\mid Y_{n+1}=y) \qquad (5.15)$$

and

$$\mathbb{E}^K_{(\delta_y \times K)_1}\left(\varphi_n(Y_0,\ldots,Y_{n+1})\prod_{p=0}^{n-1}G(Y_p,Y_{p+1})\right)$$

$$= \frac{d\pi L^n}{d\pi}(y)\,\mathbb{E}^L_{(\pi \times K)_2}(\varphi_n(Y_{n+1},\ldots,Y_0)\mid Y_{n+1}=y) \qquad (5.16)$$

Proof:
The proof of (5.15) is based on (5.11). Indeed, in view of (5.11), we have for any $(y_0,y_1) \in S^2$

$$\mathbb{E}^K_{(\delta_y \times K)_1}(\varphi_n(Y_0,\ldots,Y_{n+1})\prod_{p=0}^{n}G(Y_p,Y_{p+1}))$$

$$= \mathbb{E}^K_{(y_0,y)}(\varphi_n(Y_1,\ldots,Y_{n+2})\prod_{p=1}^{n+1}G(Y_p,Y_{p+1}))$$

$$= \frac{d\pi L^{n+1}}{d\pi}(y)\,\mathbb{E}^L_{(\pi \times L)_1}(\varphi_n(Y_{n+1},\ldots,Y_0)\mid Y_{n+1}=y_1)$$

This ends the proof of the first assertion. In much the same way, we prove (5.16) using (5.12). We simply note that for any $(y_0,y_1) \in S^2$

$$\mathbb{E}^K_{(\delta_y \times K)_1}(\varphi_n(Y_0,\ldots,Y_{n+1})\prod_{p=0}^{n-1}G(Y_p,Y_{p+1}))$$

$$= \mathbb{E}^K_{(y_0,y)}(\varphi_n(Y_1,\ldots,Y_{n+2})\prod_{p=1}^{n}G(Y_p,Y_{p+1}))$$

$$= \frac{d\pi L^n}{d\pi}(y)\,\mathbb{E}^L_{(\pi \times K)_2}(\varphi_n(Y_{n+1},\ldots,Y_0)\mid Y_{n+1}=y_1)$$

This ends the proof of the corollary. ∎

This corollary shows that the limiting distribution of the genealogical particle model presented in (5.5), (5.8), and (5.9) coincides with the desired conditional distribution. More precisely, using (5.15), we conclude that for any $\varphi_n \in \mathcal{B}_b(S^{n+1})$

$$\frac{\mathbb{E}^K_{(\delta_y \times K)_1}\left(\varphi_n(Y_0, Y_1, \ldots, Y_n) \prod_{p=0}^{n-1} G(Y_p, Y_{p+1})\right)}{\mathbb{E}^K_{(\delta_y \times K)_1}\left(\prod_{p=0}^{n-1} G(Y_p, Y_{p+1})\right)}$$

$$= \mathbb{E}^L_{(\pi \times L)_1}(\varphi_n(Y_n, Y_{n-1}, \ldots, Y_0) \mid Y_n = y)$$

Furthermore, by (5.16), we also find that

$$\mathbb{Q}_{(\delta_y \times K)_1, n} = \mathbb{P}^L_{(\pi \times K)_2}(((Y_{n+1}, Y_n), \ldots, (Y_1, Y_0)) \in . \mid Y_{n+1} = y)$$

5.5.4 Stability Properties

In this section, we analyze in some detail the stability properties of Feynman-Kac-Metropolis models and we improve the contraction inequalities presented in Theorem 5.2.1. We present a natural mixing condition under which the decays to equilibrium do not depend on the nature of the target invariant distribution. Without further mention, we assume that the triplet (π, K, L) is chosen such that the Metropolis potential function G satisfies the following condition:

(G) : There exists an $\epsilon(G) > 0$ such that for any pair $(x, x') \in E^2$ we have

$$G(x) \geq \epsilon(G) \, G(x') \, (> 0)$$

It is instructive to observe that this condition can alternatively be written in terms of the pairs (π, K) and (π, L) as follows: for any $x, x' \in E = S^2$

$$\epsilon(G) \leq G(x)/G(x') = \frac{d(\pi \times L)_2}{d(\pi \times K)_1}(x) \, \frac{d(\pi \times K)_1}{d(\pi \times L)_2}(x') \leq 1/\epsilon(G)$$

Also notice that the latter is equivalent to the following inequalities

$$\sqrt{\epsilon(G)} \, (\pi \times K)_1 \leq (\pi \times L)_2 \leq \frac{1}{\sqrt{\epsilon(G)}} (\pi \times K)_1$$

Using a clear induction on the time parameter $n \in \mathbb{N}$, we also find that, for any $n \in \mathbb{N}$, we have $\pi L^n \ll \pi$, and for any $y \in S$

$$\frac{d\pi L^n}{d\pi}(y) \in [\epsilon(G)^{n/2}, \epsilon(G)^{-n/2}]$$

In this section, we use the same terminology as that used in Section 2.7, but to clarify the presentation, we simplify the presentation and we use the superscript $(.)^{n-p}$ instead of the subscript $(.)_{n,p}$ to represent the $(n-p)$ iterates of a semigroup (from time p to time n). For instance, we write

$$(Q^{n-p}, \widehat{Q}^{n-p}, P^{n-p}, \widehat{P}^{n-p}, \Phi^{n-p}, \widehat{\Phi}^{n-p})$$

instead of

$$(Q_{p,n}, \widehat{Q}_{p,n}, P_{p,n}, \widehat{P}_{p,n}, \Phi_{p,n}, \widehat{\Phi}_{p,n})$$

We recall that the Feynman-Kac semigroups Φ^n and $\widehat{\Phi}^n$ associated with the pair (G, M) can alternatively be defined by the expressions

$$\Phi^n(\mu)(f) = \frac{\mathbb{E}^K_\mu(f(X_n) \prod_{p=0}^{n-1} G(X_p))}{\mathbb{E}^K_\mu(\prod_{p=0}^{n-1} G(X_p))} = \frac{\mu Q^n(f)}{\mu Q^n(1)}$$

and

$$\widehat{\Phi}^n(\mu)(f) = \frac{\mathbb{E}^K_\mu(f(X_n) \prod_{p=1}^{n} G(X_p))}{\mathbb{E}^K_\mu(\prod_{p=1}^{n} G(X_p))} = \frac{\mu \widehat{Q}^n(f)}{\mu \widehat{Q}^n(1)}$$

Let G^n and $\widehat{G}^n$ denote the potential functions on E defined by

$$G^n(x) = Q^n(1)(x) \quad \text{and} \quad \widehat{G}^n(x) = \widehat{Q}^n(1)(x)$$

From the previous displayed formulae, we see that Φ^n and $\widehat{\Phi}^n$ can be rewritten as

$$\Phi^n(\mu) = \Psi^n(\mu) P^n \quad \text{and} \quad \widehat{\Phi}^n(\mu) = \widehat{\Psi}^n(\mu) \widehat{P}^n \qquad (5.17)$$

with the Boltzmann-Gibbs transformations Ψ^n and $\widehat{\Psi}^n$ associated with G^n and $\widehat{G}^n$ and defined by

$$\Psi^n(\mu)(dx) = \frac{G^n(x)}{\mu(G^n)} \mu(dx) \quad \text{and} \quad \widehat{\Psi}^n(\mu)(dx) = \frac{\widehat{G}^n(x)}{\mu(\widehat{G}^n)} \mu(dx)$$

The notations $(G^n, P^n, \widehat{G}^n, \widehat{P}^n)$ may be somewhat confusing. To prevent any kind of confusion, we emphasize that even for time-homogeneous models this pair sequence of transformations as well as the Markov kernels P^n and $\widehat{P}^n$ are not the n iterates of Ψ, $\widehat{\Psi}$, P^1, and $\widehat{P}^1$. We also mention that the decompositions (5.17) coincide with the ones presented in Section 2.7 when replacing $(G_{p,n}, \Psi_{p,n}, P_{p,n})$ and $(\widehat{G}_{p,n}, \widehat{\Psi}_{p,n}, \widehat{P}_{p,n})$ by $(G^{n-p}, \Psi^{n-p}, P^{n-p})$ and $(\widehat{G}^{n-p}, \widehat{\Psi}^{n-p}, \widehat{P}^{n-p})$.

Our immediate objective is to provide an "explicit" description of the Feynman-Kac semigroups Q^n and $\widehat{Q}^n$ in terms of the triplet (π, K, L).

Using (5.12), we prove that

$$\begin{aligned}
Q^n(f)(y_0, y_1) &= \mathbb{E}^K_{(y_0,y_1)}\left(f(X_n) \prod_{p=0}^{n-1} G(X_p)\right) \\
&= \mathbb{E}_{(y_0,y_1)}\left(f(Y_n, Y_{n+1}) \prod_{p=0}^{n-1} G(Y_p, Y_{p+1})\right) \\
&= G(y_0, y_1) \frac{d\pi L^{n-1}}{d\pi}(y_1) \, \mathbb{E}^L_{(\pi \times K)_2}(f(Y_1, Y_0) \mid Y_n = y_1)
\end{aligned}$$

Now, by (5.11), we find that

$$\begin{aligned}
\widehat{Q}^n(f)(y_0, y_1) &= \mathbb{E}^K_{(y_0,y_1)}\left(f(X_n) \prod_{p=1}^{n} G(X_p)\right) \\
&= \mathbb{E}_{(y_0,y_1)}\left(f(Y_n, Y_{n+1}) \prod_{p=1}^{n} G(Y_p, Y_{p+1})\right) \\
&= \frac{d\pi L^n}{d\pi}(y_1) \, \mathbb{E}^L_{(\pi \times L)_1}(f(Y_1, Y_0) \mid Y_n = y_1)
\end{aligned}$$

From these two formulae, we conclude that

$$\begin{aligned}
P^n(f)(y_0, y_1) &= \mathbb{E}^L_{(\pi \times K)_2}(f(Y_1, Y_0) \mid Y_n = y_1) \\
\widehat{P}^n(f)(y_0, y_1) &= \mathbb{E}^L_{(\pi \times L)_1}(f(Y_1, Y_0) \mid Y_n = y_1)
\end{aligned}$$

For a sufficiently regular kernel L and for n large enough, we also find the following "explicit" descriptions of the nonhomogeneous potential functions

$$G^n(y_0, y_1) = G(y_0, y_1) \frac{d\pi L^{n-1}}{d\pi}(y_1) \quad \text{and} \quad \widehat{G}^n(y_0, y_1) = \frac{d\pi L^n}{d\pi}(y_1) \tag{5.18}$$

From previous considerations, we see that the semigroups Φ^n and $\widehat{\Phi}^n$ have a "nice" explicit formulation in terms of the triplet (π, K, L). Thus, one expects to improve the stability results discussed in Section 4.3.3 in the context of abstract nonhomogeneous models. We will examine this question through three different angles related respectively to the oscillations of the semigroups and to the weak and the strong contraction properties. As usual, in order to unify the statements of this section, we denote by H the total variation distance on $\mathcal{P}(E)$ or any h'-divergence satisfying the growth condition $(\mathcal{H})_a$, stated on page 135. We recall that, for these two classes of distance-like criteria, $\bar{a}_H$ is the function on $\mathbb{R}_+$ defined respectively by $\bar{a}_H(r) = 2r$ and $\bar{a}_H(r) = ra(r)$ (see Definition 4.3.2).

1. First we observe from Proposition 4.3.1 that $\beta(P^n)$ is a measure of the oscillations of the mapping Φ^n with respect to the total variation

distance. That is, we have that

$$\beta(P^n) = \sup_{\mu,\nu \in \mathcal{P}(E)} \|\Phi^n(\mu) - \Phi^n(\nu)\|_{\text{tv}}$$

2. The Dobrushin ergodic coefficient also appears useful in the analysis of the weak regularity properties of the semigroups. Using Proposition 4.3.7, we prove that for any $(n, f, \mu) \in (\mathbb{N} \times \text{Osc}_1(E) \times \mathcal{P}(E))$ there exists a function $f_n^\mu \in \text{Osc}_1(E)$ such that for any $\eta \in \mathcal{P}(E)$

$$|[\Phi^n(\eta) - \Phi^n(\mu)](f)|$$

$$\leq 2\beta(P^n) \sup_{x,x'}(G^n(x)/G^n(x')) \;|(\eta - \mu)(f_n^{(\mu)})|$$

3. The preceding weak regularity properties can be turned into strong ones. From Theorem 4.3.1, we have

$$H(\Phi^n(\mu), \Phi^n(\nu)) \leq \beta(P^n) \; \bar{a}_H\left(\sup_{x,x'}[G^n(x)/G^n(x')]\right) \; H(\mu, \nu)$$

From the arguments given in Section 4.4, we emphasize that these three types of regularity properties remain valid if we replace in the preceding inequalities the triplet (G^n, P^n, Φ^n) by the corresponding updated objects $(\widehat{G}^n, \widehat{P}^n, \widehat{\Phi}^n)$.

In these three routes, the stability properties of the semigroups Φ^n and $\widehat{\Phi}^n$ are expressed in terms of the regularity properties of Markov kernels $(P^n, \widehat{P}^n)$ and the relative oscillations of the potential functions $(G^n, \widehat{G}^n)$. To estimate these properties, we introduce the following condition:

$(L)_m$: There exist an integer parameter $m \geq 1$ and some $\epsilon(L) > 0$ such that for any pair $(y, y') \in S^2$ we have

$$L^m(y, .) \geq \epsilon(L) \; L^m(y', .)$$

Under the latter, it follows from (4.12) that $\beta(L^m) \leq (1-\epsilon(L))$. This shows in particular that L has a unique invariant measure $\nu_L = \nu_L L \in \mathcal{P}(S)$, and using (4.10) we prove that

$$\|\mu L^{mn} - \nu_L\|_{\text{tv}} \leq (1 - \epsilon(L))^n \; \|\mu - \nu_L\|_{\text{tv}}$$

Furthermore, recalling that $L(y, .) \ll \pi$, we also find that $\nu_L \ll \pi$. Another simplification due to this condition is that the conditional Markov kernels P^{m+n+1} and $\widehat{P}^{m+n+1}$ described have the integral representations

$$P^{m+n+1}(f)(y_0, y_1) = \int \pi(dy_1') K(y_1', dy_0') \frac{dL^{m+n}(y_1', .)}{d\pi L^{m+n}}(y_1) \; f(y_1', y_0')$$

$$\widehat{P}^{m+n+1}(f)(y_0, y_1) = \int \pi(dy_0') L(y_0', dy_1') \frac{dL^{m+n}(y_1', .)}{d(\pi L) L^{m+n}}(y_1) \; f(y_1', y_0')$$

(5.19)

Finally, under $(L)_m$ we also find that for any $y \in S$ and $n \geq m$ we have $L^n(y,\cdot) \ll \pi L^n$, and for any $y' \in S$

$$\frac{dL^n(y,\cdot)}{d\pi L^n}(y') \in [1/l_m, l_m]$$

The following proposition is pivotal.

Proposition 5.5.1 *When the Markov kernel L satisfies condition $(L)_m$, then we have for any $n \in \mathbb{N}$*

$$\beta(P^{n+m+1}) \vee \beta(\widehat{P}^{n+m+1}) \leq 2\epsilon^{-1}(L)\ \beta(L^n)$$

and for any $x, x' \in E$

$$\widehat{G}^{n+m}(x) \geq \epsilon^2(L)\ h(\nu_L, \pi)\ \widehat{G}^{n+m}(x')$$

In addition, suppose condition (G) holds true and the pair of measures (π, ν_L) on S are such that

$$h(\nu_L, \pi) = \inf_{y, y' \in S} \left(\frac{d\nu_L}{d\pi}(y) \Big/ \frac{d\nu_L}{d\pi}(y')\right) > 0 \qquad (5.20)$$

Then we have the uniform estimate

$$G^{n+m+1}(x) \geq \epsilon(G)\ \epsilon^2(L)\ h(\nu_L, \pi)\ G^{n+m+1}(x')$$

The proof of Proposition 5.5.1 is based on the following technical lemma.

Lemma 5.5.2 *Let $\mu \in \mathcal{P}(S)$, and let (M_1, M_2) be a pair of Markov kernels on $(S, \mathcal{S})$ such that for any $y \in S$ we have*

$$M_2(y, \cdot) \ll \mu M_1 M_2 \ll M_1(y, \cdot)$$

Then we have for any $(y, y') \in S^2$

$$\left| \frac{dM_1 M_2(y, \cdot)}{d\mu M_1 M_2}(y') - 1 \right| \leq 2\beta(M_1) \sup_{z, z' \in S} \frac{dM_2(z, \cdot)}{dM_2(z', \cdot)}(y')$$

Proof:
We first use the decomposition

$$\frac{dM_1 M_2(y, \cdot)}{d\mu M_1 M_2}(z) - 1 = \int \mu(dy')\ [M_1(y, dy'') - M(y', dy'')]\ \frac{dM_2(y'', \cdot)}{d\mu M_1 M_2}(z)$$

to check that

$$\left| \frac{dM_1 M_2(y, \cdot)}{d\mu M_1 M_2}(z) - 1 \right| \leq 2\beta(M_1) \sup_{y'' \in S} \frac{dM_2(y'', \cdot)}{d\mu M_1 M_2}(z)$$

Since we also have

$$\frac{d\mu M_1 M_2}{dM_2(y'',\,.)}(z) = \int \mu M_1(dy) \frac{dM_2(y,\,.)}{dM_2(y'',\,.)}(z) \geq \inf_{u,u' \in S} \frac{dM_2(u,\,.)}{dM_2(u',\,.)}(z)$$

the end of the proof is straightforward. ∎

Proof of Proposition 5.5.1: The proof of the first assertion is a direct consequence of Lemma 5.5.2. By (5.19), we have for any $f \in \mathcal{B}_b(E)$ with $\|f\| \leq 1$ and $n \in \mathbb{N}$

$$|P^{m+n+1}(f)(y_0, y_1) - (\pi \times K)_1(f)| \leq \int \pi(dy_1') \left| \frac{dL^{m+n}(y_1',\,.)}{d\pi L^{m+n}}(y_1) - 1 \right|$$

$$|\widehat{P}^{m+n+1}(f)(y_0, y_1) - (\pi \times L)_2(f)| \leq \int \pi L(y_1') \left| \frac{dL^{m+n}(y_1',\,.)}{d(\pi L)L^{m+n}}(y_1) - 1 \right|$$

Under $(L)_m$, we have for any $y \in S$

$$L^m(y,\,.) \ll \pi L^{m+n} \ll L^n(y,\,.)$$

Applying Lemma 5.5.2 to the pair $(M_1, M_2) = (L^n, L^m)$ and to the measures $\mu \in \{\pi, \pi L\}$, the proof of the first assertion is clear. To check the last two inequalities, we use the descriptions of the nonhomogeneous potential functions given in (5.18). We have for any $n \in \mathbb{N}$, $x = (y_0, y_1) \in E$, and $x' = (y_0', y_1') \in E$

$$\frac{G^{m+n+1}(x)}{G^{m+n+1}(x')} = \frac{G(x)}{G(x')} \frac{d\pi L^{m+n}}{d\pi}(y_1) \frac{d\pi}{d\pi L^{m+n}}(y_1')$$

$$\frac{\widehat{G}^{m+n}(x)}{\widehat{G}^{m+n}(x')} = \frac{d\pi L^{m+n}}{d\pi}(y_1) \frac{d\pi}{d\pi L^{m+n}}(y_1')$$

On the other hand, under $(L)_m$ we have for any $(y, y') \in E$

$$L^m(y,\,.) \ll \nu_L \quad \text{and} \quad \frac{dL^m(y,\,.)}{d\nu_L}(y') \in [\epsilon(L), \epsilon^{-1}(L)]$$

This implies that

$$\epsilon(L) \frac{d\nu_L}{d\pi}(y') \leq \frac{d\pi L^{n+m}}{d\pi}(y') = \int \pi L^n(dy) \frac{dL^m(y,\,.)}{d\pi}(y') \leq \epsilon^{-1}(L) \frac{d\nu_L}{d\pi}(y')$$

from which we conclude that

$$\epsilon^2(L) \, h(\nu_L, \pi) \leq \frac{d\pi L^{n+m}}{d\pi}(y) \frac{d\pi}{d\pi L^{n+m}}(y') \leq \epsilon^{-2}(L) \, h^{-1}(\nu_L, \pi)$$

The end of the proof of the proposition is now clear. ∎

It is useful at this stage to make some remarks concerning Proposition 5.5.1. First we mention that the quantity $h(\nu_L, \pi)$ represents the inverse of the Hilbert projective distance between ν_L and π. In this interpretation, condition (5.20) means that the Hilbert distance between the target distribution and the L-invariant measure ν_L is finite. We also note that condition (G) is not a new condition. It is only the time-homogeneous version of the hypothesis (G) introduced in Section 4.3.2. Suppose that the target distribution π is a Boltzmann-Gibbs measure associated with a given nonnegative energy function V on S

$$\pi(dy) = \frac{1}{\nu(e^{-V})} e^{-V(y)} \nu(dy)$$

where ν is a reference nonnegative measure such that $\nu(e^{-V}) > 0$. Further assume that the pair of Markov kernels (K, L) is chosen such that $(\nu \times L)_2 \ll (\nu \times K)_1$. In this situation, we notice that

$$G(y, y') = e^{-(V(y') - V(y))} \frac{d(\nu \times L)_2}{d(\nu \times K)_1}(y, y')$$

If V has finite oscillations $\mathrm{osc}(V) < \infty$, then condition (G) is met with

$$\epsilon(G) \geq e^{-\mathrm{osc}(V)} \ h((\nu \times L)_2, (\nu \times K)_1)$$

In addition, we find that

$$h(\nu_L, \pi) \geq e^{-\mathrm{osc}(V)} \ h(\nu_L, \nu)$$

When the measure ν is L-reversible and $L = K$, then we have $\nu = \nu_L$ and

$$h(\nu_L, \pi) \wedge \epsilon(G) \geq \exp\left(-\mathrm{osc}(V)\right)$$

More interestingly, the mixing condition $(L)_m$, which allows control of the Dobrushin ergodic coefficients $(\beta(P^n), \beta(\widehat{P}^n))$, is not directly related to the traditional mixing condition $(M)_m$ usually made on the Markov kernel $M = M^K$ associated with the Feynman-Kac model. The former condition is dictated by the particular nature of the Metropolis potential functions. Finally, we note that the estimates on the Dobrushin ergodic coefficients $(\beta(P^n), \beta(\widehat{P}^n))$ as well as the uniform control of the relative oscillations of $\widehat{G}^n$ do not depend on condition (G).

If we combine the preceding proposition with the three routes described earlier, we get three types of stability theorems.

Theorem 5.5.2 (asymptotic stability) *Suppose condition $(L)_m$ is satisfied for some $m \geq 1$. Then, for any $n \in \mathbb{N}$ and $(\mu, \eta) \in \mathcal{P}(E)$, we have*

$$\|\Phi^{m+n+1}(\mu) - \Phi^{m+n+1}(\eta)\|_{\mathrm{tv}} \leq 2\epsilon^{-1}(L) \ \beta(L^n)$$

and

$$\|\widehat{\Phi}^{m+n+1}(\mu) - \widehat{\Phi}^{m+n+1}(\eta)\|_{\mathrm{tv}} \leq 2\epsilon^{-1}(L) \ \beta(L^n)$$

186   5. Invariant Measures and Related Topics

Theorem 5.5.3 (weak contraction) *Suppose condition (G) is satisfied. Also assume that $(L)_m$ is met for some $m \geq 1$ and the pair distribution (ν_L, π) is such that*

$$h(\nu_L, \pi) = \inf_{y,y' \in S} \left(\frac{d\nu_L}{d\pi}(y) / \frac{d\nu_L}{d\pi}(y') \right) > 0$$

Then, for any $(n, f, \mu) \in (\mathbb{N} \times \mathrm{Osc}_1(E) \times \mathcal{P}(E))$ there exists a pair of functions $(f_n^\mu, \widehat{f}_n^\mu) \in \mathrm{Osc}_1(E)$ such that for any $\eta \in \mathcal{P}(E)$

$$|[\Phi^{m+n+1}(\eta) - \Phi^{m+n+1}(\mu)](f)| \leq c(\pi, G, L)\ \beta(L^n)\ |(\eta - \mu)(f_n^{(\mu)})|$$

$$|[\widehat{\Phi}^{m+n+1}(\eta) - \widehat{\Phi}^{m+n+1}(\mu)](f)| \leq \widehat{c}(\pi, L)\ \beta(L^n)\ |(\eta - \mu)(\widehat{f}_n^{(\mu)})|$$

for some finite constants $c(\pi, G, L)$ and $\widehat{c}(\pi, L)$ such that

$$c(\pi, G, L) \leq 4/[\epsilon(G)\epsilon^3(L)h(\nu_L, \pi)] \quad \text{and} \quad \widehat{c}(\pi, L) \leq 4/[\epsilon^3(L)h(\nu_L, \pi)]$$

Theorem 5.5.4 (strong contraction) *Under the assumptions of Theorem 5.5.3, the distributions $(\pi \times K)_1$ and $(\pi \times L)_2$ are the unique invariant measures of Φ and $\widehat{\Phi}$. Furthermore, we have for any $n \in \mathbb{N}$ and any pair $(\mu, \eta) \in \mathcal{P}(E)$*

$$H(\Phi^{m+n+1}(\mu), \Phi^{m+n+1}(\eta)) \leq c_H(\pi, G, L)\ \beta(L^n)\ H(\mu, \eta)$$

$$H(\widehat{\Phi}^{m+n+1}(\mu), \widehat{\Phi}^{m+n+1}(\eta)) \leq \widehat{c}_H(\pi, L)\ \beta(L^n)\ H(\mu, \eta)$$

for some finite constants $c_H(\pi, G, L)$ and $\widehat{c}_H(\pi, L)$ such that

$$c_H(\pi, G, L) \leq 2\epsilon^{-1}(L)\ \bar{a}_H\left((\epsilon(G)\epsilon^2(L)h(\nu_L, \pi))^{-1}\right)$$
$$\widehat{c}_H(\pi, L) \leq 2\epsilon^{-1}(L)\ \bar{a}_H\left((\epsilon^2(L)h(\nu_L, \pi))^{-1}\right)$$

6
Annealing Properties

6.1 Introduction

This chapter is concerned with the long time behavior of a Feynman-Kac model associated with a potential function and a cooling schedule. In contrast to the annealed and quenched Feynman-Kac models discussed in Section 2.6, the word "annealed" is not related to an integration with respect to the law of a given random medium but on a freezing medium. We have chosen to examine this question for updated rather than for prediction Feynman-Kac models. There are two main reasons that explain this choice.

First, the stability properties of Feynman-Kac semigroups presented in Section 4.3 and in Section 4.4 seem to indicate that the updated semigroups have "better" contraction properties than the prediction ones. For instance, when the underlying Markov kernel is sufficiently mixing, the contraction coefficients of the updated semigroup do not depend on the potential function (compare for instance Proposition 4.3.6 and Proposition 4.4.3). On the other hand, the prediction models are connected to the updated ones by a simple Markov integral operation. We can use this observation to transfer annealed properties of updated models to prediction models. In the reverse angle, the probability mass repartition of updated distributions is better related to the energy function. We will examine two types of Feynman-Kac models.

The first one is the Feynman-Kac-Metropolis model introduced in Section 5.5. We recall that this model can be interpreted as an interacting Metropolis Markov chain admitting a given distribution as an invariant measure. The interaction comes from the fact that the elementary transitions of this nonhomogeneous Markov chain depend on the laws of the random states. This model is defined in terms of a Metropolis potential function and a Markov kernel on the space of all possible transitions. Suppose the invariant measure is the Gibbs distribution associated with an energy function. In this situation, the Metropolis potential is a measure of the difference of energy in a given elementary transition. In this context, the cooling schedule corresponds to the selection pressure in the random exploration search of the space of transitions. In this connection, the resulting Markov chain can be interpreted as a nonlinear simulated annealing model. In this context, it is important to find sufficient conditions on the exploration transitions and on the cooling schedule that ensure that the model concentrates on the global minima of the energy function. One advantage of this model is that the limiting measure is a Gibbs measure. As a result, the concentration properties of the annealed model are reduced to the contraction properties of its semigroup.

The second model studied in this chapter is related to the trapping problem described in Section 2.5.1. In this context, the Feynman-Kac model represents the law evolution of a particle motion in an absorbing environment. In this situation, the obstacles are related to an energy function and the cooling schedule is interpreted as the strength of the obstacles. The more the temperature decreases, the more stringent become the obstacles. In this situation, the exploration transition kernel of the particle and the energy function are dictated by the physical problem at hand. This trapping problem was formulated and studied in a recent article [101]. The asymptotic concentration regions correspond to the limiting energy levels visited by the particle. In contrast to the first model, the main difficulty here comes from the fact that the invariant measure of the Feynman-Kac model is generally unknown and the concentration analysis is much more involved. We already mention that in some situations a particle with off-diagonal transitions may be asymptotically attracted by some trapping regions.

This chapter is organized as follows. Section 6.2 focuses on the annealed properties of Feynman-Kac-Metropolis models. As mentioned earlier, the corresponding annealed model can be regarded as a nonlinear simulated annealing random search. For traditional linear models, it is known that for judicious logarithmic cooling schedules the random search concentrates in probability to the global minima of the energy function. We show that the nonlinear model has the same concentration properties for any "sublinear" temperature schedules. In Section 6.3, we discuss the annealed properties of a Feynman-Kac model arising in trapping analysis. In Section 6.3.2, we exhibit two different types of cooling schedules for which the annealed model has the same concentration property as the invariant distribution flow. In Section 6.3.4, we characterize the limiting concentration regions in terms of a variational problem in distribution space. We show that the concentration level is related to a competition between the exploration transition and the selection potential. In Section 6.3.5, we show that annealed models with diagonal exploration transitions do concentrate on the global minima of the energy function. We also provide an off-diagonal model in finite space that concentrates on an obstacle.

6.2 Feynman-Kac-Metropolis Models

6.2.1 Description of the Model

Let V be a nonnegative and bounded potential function on some measurable space $(S, \mathcal{S})$ and let $\nu \in \mathcal{P}(S)$. We consider the collection of Gibbs measures π_α, $\alpha \in \mathbb{R}_+$, on S defined by

$$\pi_\alpha(dy) = \frac{1}{\nu(e^{-\alpha V})} \, e^{-\alpha V(y)} \, \nu(dy)$$

The parameter α is interpreted as an inverse temperature parameter. We associate with a given pair of Markov kernels (K, L) on S the distributions $(\nu \times K)_1$ and $(\nu \times L)_2$ on the product space $E = S \times S$. We recall that these distributions are defined by

$$\begin{aligned}(\nu \times K)_1(d(y, y')) &= \nu(dy) \, K(y, dy') \\ (\nu \times L)_2(d(y, y')) &= \nu(dy') \, L(y', dy)\end{aligned}$$

We further require that the triplet (ν, K, L) be chosen such that for any $x, x' \in E$ we have

$$(\nu \times L)_2 \ll (\nu \times K)_1 \quad \text{and} \quad \frac{d(\nu \times L)_2}{d(\nu \times K)_1}(x) \leq g \, \frac{d(\nu \times L)_2}{d(\nu \times K)_1}(x')$$

for some finite $g < \infty$. We denote by G_α the Radon-Nikodym derivative of $(\pi_\alpha \times L)_2$ with respect to $(\pi_\alpha \times K)_1$

$$G_\alpha = \frac{d(\pi_\alpha \times L)_2}{d(\pi_\alpha \times K)_1}$$

Notice that G_α can be alternatively rewritten as

$$G_\alpha(y, y') = e^{-\alpha(V(y') - V(y))} G_0(y, y') \quad \text{with} \quad G_0 = \frac{d(\nu \times L)_2}{d(\nu \times K)_1}$$

Under our assumptions on (ν, K, L, V), we observe that for any $x, x' \in E$

$$G_\alpha(x) \leq g \, e^{2\alpha \, \mathrm{osc}(V)} \, G_\alpha(x')$$

This shows that the collection of potential functions G_α satisfies the condition (G) introduced on page 115 with

$$\epsilon(G_\alpha) \geq g^{-1} \, e^{-2\alpha \, \mathrm{osc}(V)}$$

6. Annealing Properties

To go one step further, we suppose that the Markov kernel L satisfies the regularity condition $(L)_m$ presented on page 182. That is, we have, for some $m \geq 1$, $\epsilon(L) > 0$ and for any $y, y' \in S$

$$L^m(y, .) \geq \epsilon(L) \, L^m(y', .)$$

We recall that under $(L)_m$ the Markov kernel L has a unique invariant measure $\nu_L = \nu_L L \in \mathcal{P}(S)$ and for any $n \in \mathbb{N}$ we have that

$$\|\mu L^{mn} - \nu_L\|_{\text{tv}} \leq (1 - \epsilon(L))^n \, \|\mu - \nu_L\|_{\text{tv}}$$

Finally, we further require that the pair of measures (ν, ν_L) be chosen so that ν and ν_L are mutually absolutely continuous and

$$h(\nu_L, \nu) = \inf_{y, y' \in S} \left(\frac{d\nu_L}{d\nu}(y) \Big/ \frac{d\nu_L}{d\nu}(y') \right) > 0$$

Under these conditions, we know from Theorem 5.5.4 that the distributions $(\pi_\alpha \times L)_2$, $\alpha \in \mathbb{R}_+$, are the unique fixed points of the mappings

$$\widehat{\Phi}_\alpha : \mathcal{P}(E) \to \mathcal{P}(E)$$

defined for any $(\mu, f) \in (\mathcal{P}(E) \times \mathcal{B}_b(E))$ by

$$\widehat{\Phi}_\alpha(\mu)(f) = \frac{\mu M^K(G_\alpha \, f)}{\mu M^K(G_\alpha)}$$

where M^K is the Markov kernel from E into itself defined by

$$M^K((y, y'), d(z, z')) = \delta_{y'}(dz) \, K(z, dz')$$

To simplify the presentation, we slightly abuse the notation and denote by $\alpha : \mathbb{N} \to \mathbb{R}_+$ a given nondecreasing function. The annealed Feynman-Kac-Metropolis model associated with the pair (G_α, M^K) is the distribution flow $\widehat{\eta}_n \in \mathcal{P}(E)$ defined by the recursive equation

$$\widehat{\eta}_n = \widehat{\Phi}_{\alpha(n)}(\widehat{\eta}_{n-1})$$

with a given initial distribution $\widehat{\eta}_0$ on E. We emphasize that $\widehat{\eta}_n$ is alternatively defined by the Feynman-Kac formulae

$$\widehat{\eta}_n(f) = \widehat{\gamma}_n(f) / \widehat{\gamma}_n(1) \quad \text{and} \quad \widehat{\gamma}_n(f) = \mathbb{E}^K_{\widehat{\eta}_0}\left(f(X_n) \prod_{p=1}^{n} G_{\alpha(p)}(X_p) \right)$$

where $\mathbb{E}^K_{\widehat{\eta}_0}(.)$ represents the expectation with respect to the law $\mathbb{P}^K_{\widehat{\eta}_0}$ of a Markov chain X_n with initial distribution $\widehat{\eta}_0$ and Markov transitions M^K.

In this context, M^K represents an exploration Markov kernel on the transition space $E = S^2$ and $G_{\alpha(n)}$ represents the energy landscape at temperature $1/\alpha(n)$. The precise description of the nonlinear simulated annealing algorithm associated with this flow can be found in Section 5.5 on page 168.

In this section, we consider the problem of finding a nondecreasing cooling schedule α such that

$$\lim_{n\to\infty} \|\widehat{\eta}_n - (\pi_{\alpha(n)} \times L)_2\|_{\text{tv}} = 0 \tag{6.1}$$

Since the distributions π_α concentrate as $\alpha \to \infty$ to the set of ν-essential infima of the energy function V, the preceding convergence result implies that the annealed Feynman-Kac-Metropolis model has the same concentration properties. More precisely, any McKean interpretation of this annealed model converges in probability to the set of $(\nu \times L)_2$-essential infima of the energy function $(1 \otimes V)$ on E.

We decompose this problem into two parts. First we estimate the relaxation times of the semigroups

$$\widehat{\Phi}_\alpha^{n+1} = \widehat{\Phi}_\alpha^n \circ \widehat{\Phi}_\alpha$$

associated with the one-step mappings $\widehat{\Phi}_\alpha$, $\alpha \in \mathbb{R}_+$. The second step consists in estimating the oscillations of the mappings

$$\alpha \in \mathbb{R}_+ \longrightarrow (\pi_\alpha \times L)_2 \in \mathcal{P}(E)$$

with respect to the total variation distance. These two regularity properties are studied in Section 6.2.2. In Section 6.2.3, we prove that we have the desired convergence result (6.1) for any "sublinear" increasing cooling schedule α.

6.2.2 Regularity Properties

To clarify the presentation, it is convenient to introduce the nonnegative and finite constants

$$b_m(L) = -\frac{1}{\log \beta(L^m)}$$
$$c_L(\nu) = 1 - \log\left(\epsilon^3(L) h(\nu_L, \nu)/4\right)$$

with the convention $b_m(L) = 0$ when $\beta(L^m) = 0$ and for any $\alpha \in \mathbb{R}_+$

$$\Delta(\alpha) = (m+1)\ (2 + \lfloor b_m(L)\ (c_L(\nu) + \alpha\ \text{osc}(V)) \rfloor)$$

Lemma 6.2.1 *For any positive measure λ on a measurable space $(E, \mathcal{E})$ and for any measurable function $U : E \to \mathbb{R}_+$ such that $\lambda(e^{-U}) > 0$, we set*

$$\mu_U(dx) = Z_U^{-1} \, e^{-U(x)} \, \lambda(dx) \quad \text{with} \quad Z_U = \lambda(e^{-U})$$

Then for any pair of nonnegative measurable functions (U_1, U_2) such that $\lambda(e^{-(U_1+U_2)/2}) > 0$, we have that

$$2 \, \|\mu_{U_1} - \mu_{U_2}\|_{\mathrm{tv}} \leq \mathrm{osc}(U_1 - U_2)$$

Proof:
We use the decomposition

$$\mu_{U_1} = Z_{U_1}^{-1} \, Z_{\frac{U_1+U_2}{2}} \, e^{\frac{U_2-U_1}{2}} \, \mu_{\frac{U_1+U_2}{2}}$$

and the fact that

$$Z_{U_1}^{-1} \, Z_{\frac{U_1+U_2}{2}} = \mu_{U_1}\left(\exp\left[-\frac{U_2-U_1}{2}\right]\right) \geq \exp\left[-\frac{1}{2}\sup_E(U_2-U_1)\right]$$

to check that for any $A \in \mathcal{E}$ we have

$$\mu_{U_1}(A) \geq \mu_{\frac{U_1+U_2}{2}}(A) \, \exp\left[-\frac{1}{2}\mathrm{osc}(U_2-U_1)\right]$$

Since we have $\mathrm{osc}(U_2 - U_1) = \mathrm{osc}(U_1 - U_2)$, by symmetry arguments we prove that for any $A \in \mathcal{E}$

$$\mu_{U_2}(A) \geq \mu_{\frac{U_1+U_2}{2}}(A) \, \exp\left[-\frac{1}{2}\mathrm{osc}(U_2-U_1)\right]$$

Using (4.5) in Lemma 4.2.2, we conclude that

$$\|\mu_{U_1} - \mu_{U_2}\|_{\mathrm{tv}} \leq 1 - \exp\left[-\frac{1}{2}\mathrm{osc}(U_2-U_1)\right] \leq \frac{1}{2}\,\mathrm{osc}(U_1-U_2)$$

∎

Proposition 6.2.1 *For any $\alpha_1 \leq \alpha_2$, we have*

$$\|(\pi_{\alpha_1} \times L)_2 - (\pi_{\alpha_2} \times L)_2\|_{\mathrm{tv}} \leq \frac{1}{2}\,(\alpha_2 - \alpha_1)\,\mathrm{osc}(V) \qquad (6.2)$$

For any $\alpha \in \mathbb{R}_+$ and for any pair of distributions (μ, η) on E, we have

$$\|\widehat{\Phi}_\alpha^{\Delta(\alpha)}(\mu) - \widehat{\Phi}_\alpha^{\Delta(\alpha)}(\eta)\|_{\mathrm{tv}} \leq \frac{1}{e}\,\|\mu - \eta\|_{\mathrm{tv}} \qquad (6.3)$$

Proof:
To prove the first assertion, we observe that for each $\alpha \in \mathbb{R}_+$ the measure $(\pi_\alpha \times L)_2$ on $E = S^2$ can be rewritten in the Gibbs form

$$(\pi_\alpha \times L)_2(dx) = \frac{1}{\lambda(e^{-U})} e^{-U(x)} \lambda(dx)$$

with the function $U : E \to \mathbb{R}_+$ and the measure λ on E defined by

$$U(y, y') = V(y') \quad \text{and} \quad \lambda = (\nu \times L)_2$$

Thus, the proof of the first assertion is a clear consequence of Lemma 6.2.1. To prove the contraction estimate, we use Theorem 5.5.4 to first check that for any $\alpha \in \mathbb{R}_+$

$$\|\widehat{\Phi}_\alpha^{(k+1)(m+1)}(\mu) - \widehat{\Phi}_\alpha^{(k+1)(m+1)}(\eta)\|_{tv}$$

$$\leq 4\epsilon^{-3}(L) h(\nu_L, \pi_\alpha)^{-1} \ \beta(L^{k(m+1)}) \ \|\mu - \eta\|_{tv}$$

Since we also have that

$$h(\nu_L, \pi_\alpha) \geq e^{-\alpha \ \text{osc}(V)} \ h(\nu_L, \nu) \quad \text{and} \quad \beta(L^{k(m+1)}) \leq \beta(L^{km}) \leq \beta(L^m)^k$$

we conclude that

$$\|\widehat{\Phi}_\alpha^{(k+1)(m+1)}(\mu) - \widehat{\Phi}^{(k+1)\alpha(m+1)}(\eta)\|_{tv}$$

$$\leq 4\epsilon^{-3}(L) h(\nu_L, \nu)^{-1} \ e^{\alpha \ \text{osc}(V)} \ \beta(L^m)^k \ \|\mu - \eta\|_{tv}$$

The case $\beta(L^m) = 0$ corresponds to the situation where $\widehat{\Phi}_\alpha^n(\mu) = (\pi_\alpha \times L)_2$ for any $n \geq m$ and $\mu \in \mathcal{P}(E)$. In this situation, we have $\Delta(\alpha) = 2(m+1)$ and the proof of (6.3) is trivial. If $\beta(L^m) > 0$, we observe that

$$[4\epsilon^{-3}(L)h(\nu_L, \nu)^{-1}] \ e^{\alpha \ \text{osc}(V)} \ \beta(L^m)^k \leq e^{-1}$$

$$\iff 1 + \alpha \ \text{osc}(V) \leq k b_m^{-1}(L) + 1 - c_L(\nu)$$

$$\iff k \geq b_m(L) \ (c_L(\nu) + \alpha \ \text{osc}(V))$$

The end of the proof of the proposition is now clear. ∎

6.2.3 Asymptotic Behavior

We use the same notations as in Section 6.2.2. Let $\alpha' : \mathbb{N} \to \mathbb{R}_+$ be a nondecreasing function. We associate with α' the time mesh

$$t(n+1) = t(n) + \Delta'(n) \quad \text{with} \quad t(0) = 0, \quad \Delta'(n) = \Delta(\alpha'(n))$$

and the piecewise constant cooling schedule $\alpha : \mathbb{N} - \{0\} \to \mathbb{R}_+$ given for any $n \geq 1$ by

$$\alpha(p) = \alpha'(n) \quad \text{for} \quad t(n) < p \leq t(n+1)$$

By construction, the annealed Feynman-Kac flow $\widehat{\eta}_p$ associated with the preceding cooling schedule is defined for any $n \in \mathbb{N}$ by

$$\widehat{\eta}_p = \widehat{\Phi}_{\alpha'(n)}(\widehat{\eta}_{p-1}) \quad \text{for each} \quad t(n) < p \leq t(n+1)$$

In other words, $\widehat{\eta}_p$ is the Feynman-Kac model with a constant inverse temperature parameter $\alpha'(n)$ between the dates $t(n)$ and $t(n+1)$. That is, we have for each $0 \leq p \leq t_m(n+1) - t(n)$

$$\widehat{\eta}_{t(n)+p}(f) = \frac{\mathbb{E}^K_{\widehat{\eta}_{t(n)}}\left(f(X_p) \prod_{q=1}^p G_{\alpha'(n)}(X_q)\right)}{\mathbb{E}^K_{\widehat{\eta}_{t(n)}}\left(\prod_{q=1}^p G_{\alpha'(n)}(X_q)\right)}$$

Theorem 6.2.1 *For any nondecreasing cooling schedule α' we have*

$$\lim_{n \to \infty} (\alpha'(n+1) - \alpha'(n)) = 0 \Longrightarrow \lim_{n \to \infty} \|\widehat{\eta}_{t_n} - (\pi_{\alpha'(n)} \times L)_2\|_{\mathrm{tv}} = 0$$

In particular, if we choose for some $a \in (0,1)$ $\alpha'(n) = (n+1)^a$, then we have $t(n) = O(n^{1+a})$ and for any $n \in \mathbb{N}$

$$(n+1)^{1-a} \|\widehat{\eta}_{t_n} - (\pi_{\alpha'(n)} \times L)_2\|_{\mathrm{tv}} \leq \frac{1}{e} + \frac{7a}{2} \mathrm{osc}(V)$$

For the proof of this theorem, we need the following lemma.

Lemma 6.2.2 (Toeplitz-Kronecker) *For any sequence of strictly positive numbers a_n and for any converging sequence of numbers x_n, we have*

$$\sum_{n \geq 1} a_n = \infty \quad \text{and} \quad \lim_{n \to \infty} x_n = x \Longrightarrow \lim_{n \to \infty} \frac{\sum_{p=1}^n a_p x_p}{\sum_{p=1}^n a_p} = x$$

Whenever a_n is strictly increasing, we have

$$\lim_{n \to \infty} a_n = \infty \quad \text{and} \quad \sum_{n \geq 1} x_n < \infty \Longrightarrow \lim_{n \to \infty} \frac{1}{a_n} \sum_{p=1}^n a_p x_p = 0$$

Proof:
For any $\epsilon > 0$, we first choose an integer $n(\epsilon) \geq 1$ such that $|x_n - x| \leq \epsilon$ for any $n \geq n(\epsilon)$. Since $\sum_{p=1}^n a_p$ converges to infinity as $n \to \infty$, we can also find an integer $n'(\epsilon) \geq 1$ such that $\sum_{p=1}^{n(\epsilon)} a_p |x_p - x| \leq \epsilon \sum_{p=1}^{n'(\epsilon)} a_p$. Now we use the estimate

$$\left| \frac{\sum_{p=1}^n a_p x_p}{\sum_{p=1}^n a_p} - x \right| \leq \frac{1}{\sum_{p=1}^n a_p} \sum_{p=1}^n a_p |x_p - x|$$

For any $n \geq n(\epsilon) \vee n'(\epsilon)$, we have that

$$\sum_{p=1}^n a_p \, |x_p - x| = \sum_{p=1}^{n(\epsilon)} a_p \, |x_p - x| + \sum_{p=n(\epsilon)+1}^n a_p \, |x_p - x|$$

$$\leq \epsilon \left[\sum_{p=1}^{n'(\epsilon)} a_p + \sum_{p=n(\epsilon)+1}^n a_p \right] \leq 2\epsilon \sum_{p=1}^n a_p$$

This yields that for any $n \geq n(\epsilon) \vee n'(\epsilon)$

$$\left| \frac{\sum_{p=1}^n a_p \, x_p}{\sum_{p=1}^n a_p} - x \right| \leq 2\epsilon$$

and the proof of the first assertion is completed. To prove the second one, we put

$$\overline{x}_n = \sum_{p=1}^n x_p \quad \text{and} \quad \overline{a}_n = \sum_{p=1}^n a_p$$

and, for any sequence of numbers u_n, $\Delta(u)_n = u_n - u_{n-1}$. From previous calculations, we get

$$\begin{aligned} a_n \, x_n &= a_n \, (\overline{x}_n - \overline{x}_{n-1}) \\ &= (a_n \, \overline{x}_n - a_{n-1} \, \overline{x}_{n-1}) - \overline{x}_{n-1} \, (a_n - a_{n-1}) \\ &= \Delta(a\overline{x})_n - \overline{x}_{n-1} \, \Delta a_n \end{aligned}$$

Therefore we have that $\sum_{p=1}^n a_p \, x_p = a_n \, \overline{x}_n - \sum_{p=1}^n \overline{x}_{p-1} \, \Delta a_p$, from which we conclude that

$$\frac{1}{a_n} \sum_{p=1}^n a_p \, x_p = \overline{x}_n - \frac{1}{a_n} \sum_{p=1}^n \overline{x}_{p-1} \, \Delta a_p$$

Since a_n is strictly positive, we have $a_n = \sum_{p=1}^n \Delta a_p > 0$. We deduce from the first assertion that

$$\lim_{n \to \infty} \frac{1}{a_n} \sum_{p=1}^n \overline{x}_{p-1} \, \Delta a_p = \lim_{n \to \infty} \overline{x}_n$$

Using the preceding decomposition, this implies that $\lim_{n \to \infty} \frac{1}{a_n} \sum_{p=1}^n a_p \, x_p = 0$. This ends the proof of the lemma. ∎

Proof of Theorem 6.2.1:

The proof of the first assertion results from the contraction and oscillation properties presented in Proposition 6.2.1. We use the decomposition

$$\widehat{\eta}_{t(n+1)} - (\pi_{\alpha'(n+1)} \times L)_2$$

$$= [\widehat{\Phi}_{\alpha'(n)}^{\Delta'(n)}(\widehat{\eta}_{t(n)}) - (\pi_{\alpha'(n)} \times L)_2] + [(\pi_{\alpha'(n)} \times L)_2 - (\pi_{\alpha'(n+1)} \times L)_2]$$

Since we have

$$(\pi_{\alpha'(n)} \times L)_2 = \widehat{\Phi}_{\alpha'(n)}^{\Delta'(n)}((\pi_{\alpha'(n)} \times L)_2)$$

then by (6.3) we prove that

$$\|\widehat{\Phi}_{\alpha'(n)}^{\Delta'(n)}(\widehat{\eta}_{t(n)}) - (\pi_{\alpha'(n)} \times L)_2\|_{\mathrm{tv}} \leq \frac{1}{e} \|\widehat{\eta}_{t(n)} - (\pi_{\alpha'(n)} \times L)_2\|_{\mathrm{tv}}$$

On the other hand, using (6.2) we find that

$$\|(\pi_{\alpha'(n)} \times L)_2 - (\pi_{\alpha'(n+1)} \times L)_2\|_{\mathrm{tv}} \leq \frac{1}{2} (\alpha'(n+1) - \alpha'(n)) \operatorname{osc}(V)$$

If we put

$$I_n = \|\widehat{\eta}_{t_n} - (\pi_{\alpha'(n)} \times L)_2\|_{\mathrm{tv}}$$

from the preceding estimates, we find that

$$I_{n+1} \leq \frac{1}{e} I_n + \frac{1}{2} (\alpha'(n+1) - \alpha'(n)) \operatorname{osc}(V)$$

By simple calculations, we conclude that

$$e^{n+1} I_{n+1} \leq 1 + \frac{e}{2} \operatorname{osc}(V) \sum_{p=0}^{n} e^p (\alpha'(p+1) - \alpha'(p))$$

By the Toeplitz lemma, we have

$$\lim_{n \to \infty} (\alpha'(n+1) - \alpha'(n)) = 0 \implies \lim_{n \to \infty} \sum_{p=0}^{n} \frac{e^p}{\sum_{q=0}^{n} e^q} (\alpha'(p+1) - \alpha'(p)) = 0$$

Since $e^{-(n+1)} \sum_{q=0}^{n} e^q = (1 - e^{-(n+1)})/(e-1)$, we readily get

$$\lim_{n \to \infty} (\alpha'(n+1) - \alpha'(n)) = 0 \implies \lim_{n \to \infty} I_n = 0$$

To prove the second assertion, we recall that $x^a - y^a \leq a y^{a-1} (x - y)$ for any $x, y \geq 0$, and from this inequality we get

$$\sum_{p=1}^{n} e^p ((p+1)^a - p^a) \leq a \sum_{p=1}^{n} \frac{e^p}{p^{1-a}} \leq ae \left(1 + \sum_{p=2}^{n} \frac{e^{p-1}}{p^{1-a}}\right)$$

Next, we observe that for any $p \geq 2$

$$\frac{e^{p-2}}{(p-1)^{1-a}} = \frac{1}{e} \frac{e^{p-1}}{p^{1-a}} \frac{p^{1-a}}{(p-1)^{1-a}} \leq \frac{2^{1-a}}{e} \frac{e^{p-1}}{p^{1-a}} \leq \frac{2}{e} \frac{e^{p-1}}{p^{1-a}}$$

Therefore we have

$$\sum_{p=2}^{n} \frac{e^{p-1}}{p^{1-a}} \leq (1-2/e)^{-1} \sum_{p=2}^{n} \left(\frac{e^{p-1}}{p^{1-a}} - \frac{e^{p-2}}{(p-1)^{1-a}} \right) \leq \frac{2e^n}{n^{1-a}}$$

This yields that

$$e^{(n+1)} I_{n+1} \leq 1 + \frac{1}{2} \operatorname{osc}(V) \sum_{p=1}^{n+1} e^p ((p+1)^a - p^a)$$

$$\leq 1 + \frac{ae}{2} \operatorname{osc}(V) + ae \operatorname{osc}(V) \frac{e^{n+1}}{(n+1)^{1-a}}$$

Recalling that $ne^{-n} \leq 1$, for any $n \geq 1$, we conclude that

$$(n+1)^{1-a} I_{n+1} \leq 1/e + a (e+1/2) \operatorname{osc}(V)$$

To prove that $t(n) = O(n^{1+a})$, we use the fact that

$$\frac{\Delta'(p)}{(m+1)} \leq 3 + b_m(L) (c_L(\nu) + (p+1)^a \operatorname{osc}(V))$$

This implies that

$$\begin{aligned} t(n) &= \sum_{p=0}^{n-1} \Delta'(p) \\ &\leq (m+1) \left([3 + b_m(L) \, c_L(\nu)] \, n + b_m(L) \operatorname{osc}(V) \, n^{1+a} \right) \\ &\leq (m+1) \, n^{1+a} \, (3 + b_m(L) \, (c_L(\nu) + \operatorname{osc}(V))) \end{aligned}$$

This establishes the theorem. ∎

6.3 Feynman-Kac Trapping Models

6.3.1 Description of the Model

Let M be a Markov kernel on some measurable space $(E, \mathcal{E})$. Also, let V be a nonnegative and measurable function on E with bounded oscillations

198   6. Annealing Properties

osc$(V) < \infty$ and let $\alpha : \mathbb{N} \to \mathbb{R}_+$ be a nondecreasing function. We associate with the triplet (α, V, M) the annealed Feynman-Kac updated model

$$\widehat{\eta}_n = \widehat{\Phi}_{\alpha(n)}(\widehat{\eta}_{n-1})$$

where $\widehat{\eta}_0$ is an arbitrary distribution on E and $\widehat{\Phi}_{\alpha'}$, $\alpha' \in \mathbb{R}_+$, is the collection of mappings

$$\widehat{\Phi}_{\alpha'} : \mathcal{P}(E) \to \mathcal{P}(E)$$

defined for any $(\mu, f) \in (\mathcal{P}(E) \times \mathcal{B}_b(E))$ by

$$\widehat{\Phi}_{\alpha'}(\mu)(f) = \frac{\mu M(e^{-\alpha' V} f)}{\mu M(e^{-\alpha' V})}$$

Notice that $\widehat{\eta}_n$ is alternatively defined by the Feynman-Kac formulae

$$\widehat{\eta}_n(f) = \widehat{\gamma}_n(f)/\widehat{\gamma}_n(1) \quad \text{and} \quad \widehat{\gamma}_n(f) = \mathbb{E}_{\widehat{\eta}_0}(f(X_n) \, e^{-\sum_{p=1}^n \alpha(p) \, V(X_p)})$$

where $\mathbb{E}_{\widehat{\eta}_0}(.)$ represents the expectation with respect to the law $\mathbb{P}_{\widehat{\eta}_0}$ of a Markov chain X_n with initial distribution $\widehat{\eta}_0$ and Markov transitions M.

This model arises in a natural way in trapping analysis. We refer the reader to Section 2.5.1 for a detailed discussion on this subject. In this context, the Markov kernel M represents the transitions of a particle evolving in a medium E. The potential function V represents the energy landscape and the strength of the obstacles. The exponential term $e^{-\alpha(n)V(x)}$ represents the probability at which the particle at site x is not absorbed. In this interpretation, the cooling schedule represents the temperature of the medium. The more the temperature decreases, the more stringent become the obstacles.

6.3.2 Regularity Properties

This section is concerned with the regularity properties of the semigroups $\widehat{\Phi}_\alpha^n$, $\alpha \in \mathbb{R}_+$. This question is clearly connected with the study of the contraction properties of updated Feynman-Kac semigroups presented in Section 4.4. In our further development, we assume that the Markov kernel M satisfies condition $(M)_m$ for some integer parameter $m \geq 1$ and some $\epsilon(M) > 0$. That is, we have for any pair $(x, x') \in E^2$

$$M^m(x, .) \geq \epsilon(M) \, M^m(x', .)$$

To clarify the presentation, we introduce the nonnegative constants

$$\begin{aligned} c(M) &= 1 - \log(\epsilon(M)/2) \\ \delta(m) &= (m-1) \operatorname{osc}(V) \end{aligned}$$

and for any $\alpha \in \mathbb{R}_+$

$$\Delta(\alpha) = m \, (1 + \lfloor e^{\alpha \delta(m)}(\alpha \delta(m) + c(M))/\epsilon^2(M) \rfloor)$$

Notice that for $m = 1$ we have $\delta(m) = 0$ and

$$\Delta(\alpha) = \Delta(0) = (1 + \lfloor c(M)/\epsilon^2(M) \rfloor)$$

We observe that for any fixed $\alpha \in \mathbb{R}_+$ the semigroup defined by the inductive formulae

$$\widehat{\Phi}_\alpha^n = \widehat{\Phi}_\alpha^{n-1} \circ \widehat{\Phi}_\alpha$$

is the updated Feynman-Kac semigroup associated with the pair of potential/kernel $(e^{-\alpha V}, M)$. Since V has finite oscillations we see that the time-homogeneous potential function $G_\alpha = e^{-\alpha V}$ satisfies condition (G) with $\epsilon(G) = e^{-\alpha \operatorname{osc}(V)}$. That is, we have for any $(x, x') \in E^2$

$$G_\alpha(x) \geq e^{-\alpha \operatorname{osc}(V)} G_\alpha(x')$$

Using Proposition 4.4.3, we get the contraction inequality

$$\|\widehat{\Phi}_\alpha^{nm}(\mu) - \widehat{\Phi}_\alpha^{nm}(\eta)\|_{\operatorname{tv}} \leq 2\epsilon^{-1}(M)\, e^{\alpha\delta(m)} \left(1 - \epsilon^2(M)\, e^{-\alpha\delta(m)}\right)^n \|\mu - \eta\|_{\operatorname{tv}}$$

for any pair $(\mu, \eta) \in \mathcal{P}(E)^2$ and for any $n \in \mathbb{N}$. By the Banach fixed point theorem we conclude that each mapping $\widehat{\Phi}_\alpha$ has a unique fixed point

$$\mu_\alpha = \widehat{\Phi}_\alpha(\mu_\alpha) \in \mathcal{P}(E)$$

Proposition 6.3.1 *For any $\alpha \in \mathbb{R}_+$ and for any pair $(\mu, \eta) \in \mathcal{P}(E)^2$, we have*

$$\|\widehat{\Phi}_\alpha^{\Delta(\alpha)}(\mu) - \widehat{\Phi}_\alpha^{\Delta(\alpha)}(\eta)\|_{\operatorname{tv}} \leq \frac{1}{e} \|\mu - \eta\|_{\operatorname{tv}} \tag{6.4}$$

In addition, for any $\alpha_1 \leq \alpha_2$, we have the oscillation estimate

$$\|\mu_{\alpha_1} - \mu_{\alpha_2}\|_{\operatorname{tv}} \leq \operatorname{osc}(V)\, \Delta(\alpha_1)\, (\alpha_2 - \alpha_1)$$

Proof:
In view of the preceding contraction estimate, we have

$$\|\widehat{\Phi}_\alpha^{\Delta(\alpha)}(\mu) - \widehat{\Phi}_\alpha^{\Delta(\alpha)}(\eta)\|_{\operatorname{tv}}$$

$$\leq 2\epsilon^{-1}(M)\, \exp\left(\alpha\delta(m) - \epsilon^2(M)\frac{\Delta(\alpha)}{m}\, e^{-\alpha\delta(m)}\right)\|\mu - \eta\|_{\operatorname{tv}}$$

Since

$$\epsilon^2(M)\, \Delta(\alpha) \geq m\, e^{\alpha\delta(m)}(\alpha\delta(m) + c(M))$$

we get

$$\alpha\delta(m) - \epsilon^2(M)\frac{\Delta(\alpha)}{m}\, e^{-\alpha\delta(m)} \leq -c(M) = \log\left(\epsilon(M)/2\right) - 1$$

from which the proof of (6.4) is clear. To prove the second assertion, we use the decomposition

$$\mu_{\alpha_1} - \mu_{\alpha_2} = \widehat{\Phi}_{\alpha_1}^{\Delta(\alpha_1)}(\mu_{\alpha_1}) - \widehat{\Phi}_{\alpha_1}^{\Delta(\alpha_1)}(\mu_{\alpha_2}) + \widehat{\Phi}_{\alpha_1}^{\Delta(\alpha_1)}(\mu_{\alpha_2}) - \widehat{\Phi}_{\alpha_2}^{\Delta(\alpha_1)}(\mu_{\alpha_2})$$

By (6.4), we find that

$$\|\mu_{\alpha_1} - \mu_{\alpha_2}\|_{\text{tv}} \leq \frac{1}{e}\|\mu_{\alpha_1} - \mu_{\alpha_2}\|_{\text{tv}} + \|\widehat{\Phi}_{\alpha_1}^{\Delta(\alpha_1)}(\mu_{\alpha_2}) - \widehat{\Phi}_{\alpha_2}^{\Delta(\alpha_1)}(\mu_{\alpha_2})\|_{\text{tv}}$$

from which we conclude that

$$\|\mu_{\alpha_1} - \mu_{\alpha_2}\|_{\text{tv}} \leq \frac{e}{e-1}\,\|\widehat{\Phi}_{\alpha_1}^{\Delta(\alpha_1)}(\mu_{\alpha_2}) - \widehat{\Phi}_{\alpha_2}^{\Delta(\alpha_1)}(\mu_{\alpha_2})\|_{\text{tv}}$$

It is convenient to recall at this stage that for any fixed parameter $\alpha_2 \in \mathbb{R}_+$ and for any $\alpha \in \mathbb{R}_+$, $n \in \mathbb{N}$, $f \in \mathcal{B}_b(E)$, we have

$$\widehat{\Phi}_\alpha^n(\mu_{\alpha_2}) = \frac{\mathbb{E}_{\mu_{\alpha_2}}(f(X_n)\,e^{-\alpha \sum_{p=1}^n V(X_p)})}{\mathbb{E}_{\mu_{\alpha_2}}(e^{-\alpha \sum_{p=1}^n V(X_p)})}$$

We also see that each distribution $\widehat{\Phi}_\alpha^n(\mu_{\alpha_2})$ is the n-time marginal of the Gibbs-Boltzmann measure on E^n defined by

$$\mu_\alpha^{(n)}(dx) = \frac{e^{-\alpha V_n(x)}}{\lambda^{(n)}(e^{-\alpha V_n})}\,\lambda^{(n)}(dx)$$

with the reference distribution $\lambda_\alpha^{(n)}$ and the potential V_n from E^n into $\mathbb{R}_+$ defined for any $x = (x_1, \ldots, x_n)$ by

$$V_n(x_1, \ldots, x_n) = \sum_{p=1}^n V(x_p)$$
$$\lambda^{(n)}(d(x_1, \ldots, x_n)) = (\mu_{\alpha_2} M)(dx_1)\,M(x_1, dx_2) \ldots M(x_{n-1}, dx_n)$$

By Lemma 6.2.1, we have for any $\alpha_1 \leq \alpha_2$

$$\|\mu_{\alpha_1}^{(n)} - \mu_{\alpha_2}^{(n)}\|_{\text{tv}} \leq \frac{1}{2}\,(\alpha_2 - \alpha_1)\,\text{osc}(V_n) \leq \frac{n}{2}\,(\alpha_2 - \alpha_1)\,\text{osc}(V)$$

We conclude that

$$\|\widehat{\Phi}_{\alpha_1}^{\Delta(\alpha_1)}(\mu_{\alpha_2}) - \widehat{\Phi}_{\alpha_2}^{\Delta(\alpha_1)}(\mu_{\alpha_2})\|_{\text{tv}} \leq \frac{\Delta(\alpha_1)}{2}(\alpha_2 - \alpha_1)\,\text{osc}(V)$$

We end the proof of the proposition using the bound $e \leq 2(e-1)$. ∎

6.3.3 Asymptotic Behavior

We use the same notations as in Section 6.3.2. To define the annealed model, we associate with a given nondecreasing function $\alpha' : \mathbb{N} \to \mathbb{R}_+$ the time mesh

$$t(n+1) = t(n) + \Delta(\alpha'(n)) \quad \text{with} \quad t(0) = 0, \quad \Delta'(n) = \Delta(\alpha'(n))$$

and the piecewise constant cooling schedule $\alpha : \mathbb{N} - \{0\} \to \mathbb{R}_+$ given for any $n \geq 1$ by

$$\alpha(p) = \alpha'(n) \quad \text{for} \quad t(n) < p \leq t(n+1)$$

The annealed Feynman-Kac flow $\widehat{\eta}_p$ associated with this cooling schedule is defined for any $n \in \mathbb{N}$ and $t(n) < p \leq t(n+1)$ by

$$\widehat{\eta}_p = \widehat{\Phi}_{\alpha'(n)}(\widehat{\eta}_{p-1})$$

We emphasize that $\widehat{\eta}_p$ is the Feynman-Kac model with a constant inverse temperature parameter $\alpha'(n)$ between the dates $t(n)$ and $t(n+1)$. For each $0 \leq p \leq \Delta'(n)$, we have that

$$\widehat{\eta}_{t(n)+p}(f) = \frac{\mathbb{E}_{\widehat{\eta}_{t(n)}}\left(f(X_p)\, e^{-\alpha'(n) \sum_{q=1}^{p} V(X_q)}\right)}{\mathbb{E}_{\widehat{\eta}_{t(n)}}\left(e^{-\alpha'(n) \sum_{q=1}^{p} V(X_q)}\right)}$$

To connect the uniform contraction estimates with the oscillations of the fixed point measures presented in Proposition 6.3.1, we introduce the decomposition

$$\widehat{\eta}_{t(n+1)} - \mu_{\alpha'(n+1)} = (\widehat{\Phi}_{\alpha'(n)}^{\Delta'(n)}(\widehat{\eta}_{t(n)}) - \widehat{\Phi}_{\alpha'(n)}^{\Delta'(n)}(\mu_{\alpha'(n)})) + (\mu_{\alpha'(n)} - \mu_{\alpha'(n+1)})$$

Using Proposition 6.3.1, we find that

$$\|\widehat{\eta}_{t(n+1)} - \mu_{\alpha'(n+1)}\|_{\text{tv}}$$
$$\leq \frac{1}{e}\, \|\widehat{\eta}_{t(n)} - \mu_{\alpha'(n)}\|_{\text{tv}} + \text{osc}(V)\ \Delta'(n)\ (\alpha'(n+1) - \alpha'(n))$$

and thus it appears that

$$e^{n+1}\, \|\widehat{\eta}_{t(n+1)} - \mu_{\alpha'(n+1)}\|_{\text{tv}}$$
$$\leq 1 + \text{osc}(V)\ \sum_{p=0}^{n} e^{p+1}\, \Delta'(p)\, (\alpha'(p+1) - \alpha'(p)) \qquad (6.5)$$

We are now in a position to state the main result of this section.

6. Annealing Properties

Theorem 6.3.1 *Suppose condition* $(M)_m$ *is satisfied for some* $m \geq 1$. *Then we have*
$$\lim_{n \to \infty} \|\widehat{\eta}_{t(n)} - \mu_{\alpha'(n)}\|_{\mathrm{tv}} = 0$$
for any increasing cooling schedule α' *such that*
$$\lim_{n \to \infty} e^{\alpha'(n)\delta(m)}(1 + \alpha'(n)\delta(m)) \, [\alpha'(n+1) - \alpha'(n)] = 0$$
We have the two distinguished cases

- *If* $m = 1$, *then we have* $t(n) = O(n)$ *and we can choose for any* $a \in (0,1)$
$$\alpha'(n) = (n+1)^a$$
In this case, we have for some $c(a) < \infty$ *and any* $n \geq 1$
$$\|\widehat{\eta}_{t(n)} - \mu_{\alpha'(n)}\|_{\mathrm{tv}} \leq c(a)/n^{1-a} \qquad (6.6)$$

- *If* $m > 1$, *then we can choose*
$$\alpha'(n) = \alpha'(0) \log(n+e), \quad \text{with} \quad b = \delta(m)\alpha'(0) < 1$$
In this case we have $t(n) = O(n^{b+1} \log n)$ *and for some* $c(b) < \infty$ *and any* $n \geq 1$
$$\|\widehat{\eta}_{t(n)} - \mu_{\alpha'(n)}\|_{\mathrm{tv}} \leq c(b) \, \log n \, / n^{1-b}$$

Proof:
If we put $I_n = \|\widehat{\eta}_{t(n)} - \mu_{\alpha'(n)}\|_{\mathrm{tv}}$, then by (6.5) we have
$$e^{n+1} I_{n+1} \leq 1 + \operatorname{osc}(V) \sum_{p=0}^{n} e^{p+1} \Delta'(p) \, (\alpha'(p+1) - \alpha'(p))$$

By the definition of $\Delta'(p)$, we have
$$\begin{aligned} \Delta'(p) &\leq m \, (2 + e^{\delta(m)\alpha'(p)}[\alpha'(p)\delta(m) + c(M)]/\epsilon^2(M)) \\ &\leq m \, (2 + c'(M) e^{\delta(m)\alpha'(p)}[1 + \alpha'(p)\delta(m)]) \end{aligned}$$

with $c'(M) = (1 \vee c(M))/\epsilon^2(M)$. This readily yields that
$$\Delta'(p) \leq 3m \, c'(M) \, e^{\delta(m)\alpha'(p)}(1 + \alpha'(p)\delta(m))$$

and therefore

$e^{n+1} I_{n+1}$
$$\leq 1 + c_m(V, M) \sum_{p=0}^{n} e^{p+1} \, e^{\delta(m)\alpha'(p)}(1 + \alpha'(p)\delta(m)) \, (\alpha'(p+1) - \alpha'(p))$$

with $c_m(V, M) = 3m\, c'(M)\, \mathrm{osc}(V)$. We use the Toeplitz lemma as in the proof of Theorem 6.2.1 to prove the first assertion of the present theorem. Next we examine the two cases $m = 1$ and $m > 1$. When $m = 1$, we have $\delta(m) = 0$ and we find that

$$e^{n+1}\, I_{n+1} \leq 1 + c_1(V, M) \sum_{p=0}^{n} e^{p+1}\, (\alpha'(p+1) - \alpha'(p))$$

If we take $\alpha'(n) = (n+1)^a$ for some $a \in (0, 1)$, then we argue as in the proof of Theorem 6.2.1 and we prove that

$$e^{n+1}\, I_{n+1} \leq 1 + c_1(V, M)\, a\, e\, \left(1 + 2 \frac{e^{n+1}}{(n+1)^{1-a}}\right)$$

from which we conclude that

$$(n+1)^{1-a} I_{n+1} \leq 1 + 3ae\, c_1(V, M)$$

Also notice that in this situation we have $\Delta'(p) \leq 3c'(M)$ and therefore

$$t(n) = \sum_{p=0}^{n-1} \Delta'(p) \leq 3c'(M) n$$

This completes the proof of the second assertion. Next we examine the situation where $m > 1$. We use the rather crude estimate

$$e^{n+1}\, I_{n+1}$$

$$\leq 1 + c_m(V, M)\, [1 + \alpha'(n)\delta(m)] \sum_{p=0}^{n} e^{p+1}\, e^{\delta(m)\alpha'(p)}\, (\alpha'(p+1) - \alpha'(p))$$

If we choose $\alpha'(n) = \alpha'(0) \log(n + e)$ with $b = \delta(m)\alpha'(0) < 1$, then we find that

$$e^{n+1}\, I_{n+1}$$

$$\leq 1 + (1+b)c_m(V, M)\, \log(n+e) \sum_{p=0}^{n} e^{p+1}\, (p+e)^b \log\left(1 + \tfrac{1}{p+e}\right)$$

Recalling that $\log(1 + x) \leq x$, for all $x \in (0, \infty)$ we prove that

$$e^{n+1}\, I_{n+1} \leq 1 + (1+b)c_m(V, M)\, \log(n+e) \sum_{p=0}^{n} \frac{e^{p+1}}{(p+1)^{1-b}}$$

On the other hand, from the estimates given in the proof of Theorem 6.2.1, we have

$$e^{-n} \sum_{p=1}^{n} \frac{e^p}{p^{1-b}} \leq e^{-n} e \left(1 + 2 \frac{e^n}{n^{1-b}}\right) \leq \frac{9}{n^{1-b}}$$

Thus we find that
$$e^{n+1} I_{n+1} \leq 1 + (1+b)c_m(V,M) \log(n+e) \frac{9e^{n+1}}{(n+1)^{1-b}}$$

We conclude easily that
$$\frac{(n+1)^{1-b}}{\log(n+e)} I_{n+1} \leq 1 + 9(1+b)c_m(V,M)$$

We end the proof of noting that
$$t(n) = \sum_{p=0}^{n-1} \Delta'(p) \leq 3m\, c'(M)\, (1+b)\, \log(n+e) \sum_{p=0}^{n-1} (p+e)^b$$
$$\leq 3m\, c'(M)\, (1+b)\, \log(n+e)\, (n+e)^{1+b}$$

∎

6.3.4 Large-Deviation Analysis

This section is concerned with the concentration properties of the fixed point distributions μ_α as α tends to infinity. We use large deviation arguments, and it is convenient to reduce the analysis to Polish state-space models. More precisely, we further assume that E is a separate topological space whose topology is generated by a metric that is supposed to be complete. We also assume that V is a continuous and bounded potential function on E.

The interplay between μ_α and the quantities (α, M, V) is described by the fixed point formula
$$\mu_\alpha(f) = \mu_\alpha(\widehat{Q}_\alpha(f))/\mu_\alpha(\widehat{Q}_\alpha(1)) \quad \text{with} \quad \widehat{Q}_\alpha(f) = M(e^{-\alpha V} f)$$

Under the uniform mixing condition $(M)_m$, we recall that the Markov kernel M has a unique invariant measure $\nu = \nu M \in \mathcal{P}(E)$, and the sequence of occupation measures $L_n = \frac{1}{n} \sum_{p=1}^n \delta_{X_p}$ of the chain X_n under $\mathbb{P}_{\widehat{\eta}_0}$ satisfies as $n \to \infty$ a large deviation principle with good rate function

$$I(\mu) = \inf \left\{ \int_E \mu(dx) \, \text{Ent}(K(x,.) \mid M(x,.)) \right\} \tag{6.7}$$

where the infimum is taken over all Markov kernels K with invariant measure μ.

In the most naive view, we could think that the Feynman-Kac simulated annealing model converges in probability to the ν-essential infimum V_ν of the potential V defined by

$$V_\nu = \sup \{v \in \mathbb{R}_+ \,;\, V \geq v \quad \nu \text{ a.e.}\}$$

This intuitive idea appears to be true for regular Markov transitions M with a diagonal term $M(x,x) > 0$, but it is false in more general situations.

To better introduce our strategy to study the concentration properties of μ_α, we need a more physical interpretation of the Feynman-Kac models. If we interpret the potential V as the absorption rate for a Markov particle with transition M evolving in a medium with obstacles, the normalizing constant

$$\mathbb{E}_{\widehat{\eta}_0}(e^{-\alpha \sum_{p=1}^n V(X_p)})$$

represents the probability that a Markov particle starting with distribution $\widehat{\eta}_0$ performs a long crossing of length n without being absorbed. For a more precise description of this interpretation, we refer the reader to Section 2.5.1. The cost attached to performing long crossings is measured in terms of the logarithmic Lyapunov exponents $\Lambda(-\alpha V)$ of the semigroup $\widehat{Q}_\alpha$ on the Banach space $\mathcal{B}_b(E)$ defined by the formulae

$$\Lambda(-\alpha V) = \lim_{n\to\infty} \frac{1}{n} \log \|\widehat{Q}_\alpha^n(1)\| = \lim_{n\to\infty} \frac{1}{n} \log \sup_x \mathbb{E}_x(e^{-\alpha \sum_{p=1}^n V(X_p)})$$

The next lemma shows that these Lyapunov exponents coincide with the exponential moments of the fixed point measures μ_α. It also enters the large-deviation rate I in the concentration properties of μ_α. Informally it shows that

$$\mu_\alpha(e^{\alpha V}) \simeq e^{\alpha V_I}$$

where V_I is the value of the variational problem

$$V_I = \inf \{\mu(V) \,;\, \mu \in \mathcal{P}(E) \text{ s.t. } I(\mu) < \infty\} \qquad (6.8)$$

Loosely speaking, the concentration properties of the limiting measures μ_α as α tends to infinity are related to a competition in $\mathcal{P}(E)$ between the mean potential $\mu(V)$ and the I-entropy $I(\mu)$. Recall that $I(\mu) < \infty$ iff we can find a kernel K such that $\mu = \mu K$ and $K(x,.) \ll M(x,.)$.

The next lemma also shows that the concentration of μ_α is related to a variational problem in which the competition with the entropy I becomes less and less severe as α tends to infinity.

Lemma 6.3.1 *For any $\alpha \in \mathbb{R}_+$, we have the formulae*

$$-\frac{\Lambda(-\alpha V)}{\alpha} = \frac{1}{\alpha} \log \mu_\alpha(e^{\alpha V}) = \inf_{\eta \in \mathcal{P}(E)} \left(\eta(V) + \frac{1}{\alpha} I(\eta)\right) \xrightarrow[\beta\to\infty]{} V_I \geq V_\nu$$

Proof:
If we take $f = \widehat{Q}_\alpha^n(1)$ in the fixed point formula, we readily find the recursive equation

$$\mu_\alpha(\widehat{Q}_\alpha^{n+1}(1)) = \mu_\alpha(\widehat{Q}_\alpha^n(1)) \, \mu_\alpha(\widehat{Q}_\alpha(1))$$

6. Annealing Properties

Thus we have for each $n \geq 0$

$$\mu_\alpha(\widehat{Q}_\alpha^n(1)) = (\mu_\alpha(\widehat{Q}_\alpha(1)))^n = \mathbb{E}_{\mu_\alpha}(e^{-\alpha \sum_{p=1}^n V(X_p)}) \qquad (6.9)$$

Now if we take $f = e^{\alpha V}$ in the fixed point equation, we get

$$\mu_\alpha(e^{\alpha V}) \, \mu_\alpha(\widehat{Q}_\alpha(1)) = 1 \qquad (6.10)$$

Recalling that under condition $(M)_m$ the Laplace transformation

$$\Lambda(-\alpha V) = \lim_{n \to \infty} \frac{1}{n} \log \mathbb{E}_\mu(e^{nL_n(-\alpha V)})$$

doesn't depend on the choice of the initial distribution μ, we deduce that

$$-\Lambda(-\alpha V) = -\log \mu_\alpha(Q(1)) = \log \mu_\alpha(e^{\alpha V})$$

Since $\Lambda(-\alpha V)$ is also given as the Fenchel transformation of I

$$\Lambda(-\alpha V) = \sup_{\eta \in \mathcal{P}(E)} (\eta(-\alpha V) - I(\eta)) \qquad (6.11)$$

the end of the proof of the first assertion is clear. To end the proof, we note that

$$V_I \leq \inf_{\eta \in \mathcal{P}(E)} \left(\eta(V) + \frac{1}{\alpha} I(\eta) \right) \leq \eta(V) + \frac{1}{\alpha} I(\eta)$$

for each distribution η such that $I(\eta) < \infty$. Letting $\alpha \to \infty$, we find that

$$V_I \leq \limsup_{\alpha \to \infty} \inf_{\eta \in \mathcal{P}(E)} \left(\eta(V) + \frac{1}{\alpha} I(\eta) \right) \leq \eta(V)$$

Taking the infimum over all distributions η such that $I(\eta) < \infty$, we obtain

$$\lim_{\alpha \to \infty} \frac{1}{\alpha} \log \mu_\alpha(e^{\alpha V}) = V_I$$

To see that $V_I \geq V_\nu$, it is clearly sufficient to show that for any probability μ, $I(\mu) < +\infty$ implies that $\mu \ll \nu$. One easy way to obtain this assertion in our context is to note that if $I(\mu) < +\infty$, then there exists a kernel K verifying $\mu = \mu K$ and $K(x, \cdot) \ll M(x, \cdot)$ for μ-a.s. all $x \in E$. But since for all $x \in E$, $M^m(x, \cdot)$ is equivalent to ν, due to the condition $(M)_m$, we get that $\mu = \mu K^m \ll \mu M^m \sim \nu$. This ends the proof of the lemma. ∎

Using the exponential version of Markov's inequality, Lemma 6.3.1 provides a concentration property of μ_α in the level sets $(V < V_I + \delta)$, $\delta > 0$. More precisely, we have for any $\delta > 0$

$$\mu_\alpha(V \geq V_I + \delta) = \mu_\alpha(e^{\alpha(V - V_I)} \geq e^{\alpha \delta}) \leq e^{-\alpha \delta} \mu_\alpha(e^{\alpha(V - V_I)})$$

One concludes that

$$\lim_{\alpha \to \infty} \frac{1}{\alpha} \log \mu_\alpha(V \geq V_I + \delta) \leq -\delta$$

Combining this concentration property with Theorem 6.3.1, we prove the following asymptotic convergence result.

Proposition 6.3.2 *Suppose condition $(M)_m$ holds true for some $m \geq 1$, and let $t(n)$ and $\alpha'(n)$ be respectively the time mesh sequence and the cooling schedule described in Theorem 6.3.1. Then the corresponding annealed Feynman-Kac distribution flow $\widehat{\eta}_{t(n)}$ concentrates as $n \to \infty$ to regions with potential less than V_I; that is, for each $\delta > 0$, we have that*

$$\lim_{n \to \infty} \widehat{\eta}_{t(n)}(V \geq V_I + \delta) = 0$$

The topological hypotheses that E is Polish and that V is continuous are only necessary to obtain (6.11); see for instance [112]. So except for the definition (6.8), the concentration analysis developed in this section is true under the assumptions that $(E, \mathcal{E})$ is a measurable space and V is a non-negative bounded and measurable potential. In particular, under this extended setting, we can consider

$$V_* =_{\text{def.}} -\lim_{\alpha \to +\infty} \frac{1}{\alpha} \lim_{n \to \infty} \frac{1}{n} \log \mathbb{E}_x\left[\exp\left(-\alpha \sum_{p=1}^n V(X_p)\right)\right]$$

which always exists and does depend on the initial condition $x \in E$. Indeed, if we denote for all $n \in \mathbb{N}$ and $\alpha \in \mathbb{R}_+$

$$\lambda_n(\alpha) = \inf_{x \in E} \log \mathbb{E}_x\left[\exp\left(-\alpha \sum_{p=1}^n V(X_p)\right)\right]$$

then it is quite clear via the Markov property that $(\lambda_n(\alpha))_{n \in \mathbb{N}}$ is super-additive so that the following limit exists:

$$\lambda(\alpha) =_{\text{def.}} \lim_{n \to \infty} \frac{1}{n} \lambda_n(\alpha)$$

(this is just a rewriting of the traditional existence of the Lyapunov exponent of the underlying unnormalized Feynman-Kac operator). Now taking into account condition $(M)_m$, it appears that for any $n \geq m$ and $x, x' \in E$,

$$\mathbb{E}_x[\exp(-\alpha \sum_{p=1}^n V(X_p))]$$
$$\geq [\epsilon^2 \exp(-(m-1)\alpha \operatorname{osc}(V))] \; \mathbb{E}_{x'}[\exp(-\alpha \sum_{p=1}^n V(X_p))]$$

thus we see that

$$\lim_{n \to \infty} \frac{1}{n} \log\left(\frac{\mathbb{E}_x[\exp(-\alpha \sum_{p=1}^n V(X_p))]}{\mathbb{E}_{x'}[\exp(-\alpha \sum_{p=1}^n V(X_p))]}\right) = 0$$

208   6. Annealing Properties

In particular, for any initial distribution $\widehat{\eta}_0$, we have

$$\lambda(\alpha) = \lim_{n\to\infty} \frac{1}{n} \log \left(\mathbb{E}_{\widehat{\eta}_0} \left[\exp\left(-\alpha \sum_{p=1}^{n} V(X_p)\right) \right] \right)$$

As a limit of convex functions, the l.h.s. term in the preceding display is a convex function in α. Thus, we are ensured of the existence of

$$-\lim_{\alpha\to+\infty} \frac{\lambda(\alpha)}{\alpha} = -\lim_{\alpha\to+\infty} \frac{\lambda(\alpha)-\lambda(0)}{\alpha} = -\sup_{\alpha>0} \frac{\lambda(\alpha)-\lambda(0)}{\alpha}$$

a priori in $\mathbb{R} \sqcup \{-\infty\}$, but as V is nonnegative and bounded, we conclude that $V_* \in \mathbb{R}_+$. In this context, Lemma 6.3.1 can be rewritten as saying that under the topological hypotheses that E is Polish and that V is continuous, we have $V_* = V_I \geq V_\nu$.

6.3.5 Concentration Levels

In this section, we discuss the concentration regions of μ_α as α tends to infinity. In a first subsection, we examine Feynman-Kac models where the Markov kernel M satisfies condition $(M)_m$ with $m = 1$ or has a regular diagonal term. We show that in this case the concentration level V_I coincides with the essential infimum of the potential with respect to the invariant measure of M. The second subsection focuses on Feynman-Kac models on finite state spaces. We relate the exponential concentration of μ_α with a collection of Bellman's fixed point equations. We propose an alternative characterization of the concentration level V_I. We show that V_I can be seen as the minimal mean potential value over all closed cycles on E. Thanks to this representation, we prove that $V_I = V_\nu$ iff there exists a closed cycle on $V^{-1}(V_\nu)$. For more general off-diagonal mutation transitions, we have $V_I > V_\nu$. We illustrate this assertion with a simple three-point example, showing furthermore that μ_α does not concentrate on "neighborhoods" of $V^{-1}(V_\nu)$.

Diagonal Mutations

The easiest way to ensure that $V_I = V_\nu$ is to impose loops on every point of E for M. This assertion is based on the following simple upper bound.

Proposition 6.3.3

$$V_I \leq V_M = \inf \{V(x), \quad M(x,x) > 0\}$$

Proof:
Let us prove that for any $x \in E$ with $M(x,x) > 0$ we have $V_I \leq V(x)$. By

the definition of the Markov chain X, we find that for any $\alpha \in \mathbb{R}$ and any $n \in \mathbb{N}$,

$$\mathbb{E}_x[\exp(-\alpha \sum_{p=1}^n V(X_p))]$$
$$\geq \mathbb{E}_x[1_{X_1=x,\, X_2=x,\, \cdots,\, X_n=x} \exp(-\alpha \sum_{p=1}^n V(X_p))]$$
$$= (M(x,x))^n \exp(-n\alpha V(x))$$

This yields that

$$\Lambda(-\alpha V) \geq \lim_{n \to \infty} \frac{1}{n} \log \mathbb{E}_x \left[\exp\left(-\alpha \sum_{p=1}^n V(X_p)\right) \right]$$
$$\geq \log M(x,x) - \alpha V(x)$$

from which we conclude that

$$V_I = -\lim_{\alpha \to +\infty} \frac{\Lambda(-\alpha V)}{\alpha} \leq V(x)$$

This ends the proof of the proposition. ∎

As a simple corollary, we have $V_I = V_\nu$ as soon as we can find a sequence $(x_n)_{n \in \mathbb{N}}$ such that $\lim_{n \to \infty} V(x_n) = V_\nu$ with $M(x_n, x_n) > 0$ for all $n \in \mathbb{N}$. This clearly holds true when M is chosen so that $M(x,x) > 0$ for any $x \in E$.

Also notice that $V_I = V_\nu$ as soon as $(M)_m$ is satisfied for $m = 1$. To see this claim, we use the fixed point equation to check that, for any $\alpha \in \mathbb{R}_+$ and any nonnegative measurable function f on E, we have

$$\epsilon^2(M) \frac{\nu(e^{-\alpha V} f)}{\nu(e^{-\alpha V})} \leq \mu_\alpha(f) \leq \frac{1}{\epsilon^2(M)} \frac{\nu(e^{-\alpha V} f)}{\nu(e^{-\alpha V})}$$

Finite State Space

We further suppose that M is an irreducible Markov kernel on a finite state space E. In this case, M has a unique invariant measure ν and, for any $x \in E$, we have $\nu(x) > 0$. As an aside, we note that in this situation condition $(M)_m$ is met if and only if M is aperiodic. Our immediate objective is to give an explicit representation of V_I. For any $\mu \in \mathcal{P}(E)$, we have that

$$V_I = -\lim_{\alpha \to +\infty} \frac{1}{\alpha} \lim_{n \to \infty} \frac{1}{n} \log \mathbb{E}_\mu \left[\exp\left(-\alpha \sum_{p=1}^n V(X_p)\right) \right] \quad (6.12)$$

210   6. Annealing Properties

Definition 6.3.1 *A finite collection $P = (y_1, ..., y_n)$ of elements of E is called an M-path of length $l(P) = n \in \mathbb{N}$ if for any $1 \leq i < n$ we have $M(y_i, y_{i+1}) > 0$. The mean potential of an M-path $P = (y_1, ..., y_n)$ is defined by*

$$V(P) =_{\text{def.}} \frac{1}{n} \sum_{i=1}^{n} V(x_i)$$

An M-cycle of length $n \in \mathbb{N} - \{0\}$ is an M-path $(x_1, \cdots, x_n) \in E^n$ such that $x_i \neq x_{i+1}$ for any $1 \leq i < n$ and $M(x_n, x_1) > 0$.

Proposition 6.3.4

$$V_I = V_\mathcal{C} =_{\text{def.}} \min_{C \in \mathcal{C}} V(C)$$

In particular, we have $V_I = V_\nu$ if and only if there exists an M-cycle inside $V^{-1}(V_\nu)$.

Proof:
We first prove that $V_I \geq V_\mathcal{C}$. Let $P = (y_1, ..., y_n)$ be an M-path of length $n \in \mathbb{N}$. We can find k M-cycles $C_1, ..., C_k$ and a subpath R of P (not necessarily of the form $(y_r, y_{r+1}, ..., y_{r+l(R)}))$ of length $l(R)$ less than card(E) such that

$$l(P)V(P) = \sum_{1 \leq i \leq k} l(C_i)V(P_i) + l(R)V(R) \qquad (6.13)$$

To be convinced of the existence of such a decomposition, we look for the first return of the path P on itself: let $s = \min\{t \geq 2 : y_t \in \{y_1, ..., y_{t-1}\}\}$ and $1 \leq r < s$ be such that $y_s = y_r$. Then we define

$$C_1 =_{\text{def.}} (y_r, y_{r+1}, ..., y_{s-1})$$

and we consider the new path $P' =_{\text{def.}} (y_1, ..., y_{r-1}, y_s, y_{s+1}, ..., y_n)$ (one would have noted that $M(y_{r-1}, y_s) > 0$). Next, recursively applying the previous procedure, we construct/remove the M-cycles $C_2,...,C_k$ and we end up with a path R whose elements are all different. From formula (6.13), we deduce that

$$l(P)V(P) \geq \sum_{1 \leq i \leq k} l(C_i)V_\mathcal{C} - \text{card}(E) \|V\|_\infty$$
$$\geq l(P)V_\mathcal{C} - 2\text{card}(E) \|V\|_\infty$$

Thus, for any $x \in E$ and $n \in \mathbb{N}^*$, we have

$$\mathbb{E}_x \left[\exp\left(-\alpha \sum_{p=1}^{n} V(X_p)\right) \right] \leq \exp(n\alpha V_\mathcal{C} - 2\text{card}(E)\alpha \|V\|_\infty)$$

and the announced bound follows at once. To prove the reverse inequality, let us consider $C \in \mathcal{C}$ such that $V(C) = V_\mathcal{C}$. If an initial point x and a large enough length n are given, we construct a path P_n by first going from x to a point of C by a self-avoiding path (whose existence is ensured by irreducibility) and next always following C (in the direction included in its definition and jumping from its last element to the first one). Then it is quite clear that $\lim_{n\to\infty} V(P_n) = V(C)$, thus denoting $q = \min_{x,y\in E\,:\,M(x,y)>0} M(x,y)$ and taking into account the bound

$$\mathbb{E}_x\left[\exp\left(-\alpha\sum_{p=1}^n V(X_p)\right)\right] \geq q^n \exp(n\alpha V(P_n))$$

We conclude by an argument similar to the one given in the proof of Proposition 6.3.4. ∎

In fact, the equality of the preceding proposition remains valid if M admits a unique recurrence class (but in this situation ν does not necessarily charge all points of E). In the most general case, the initial point x in (6.12) plays a role: $V_I(x)$ is the minimal mean potential of M-cycles included in the recurrence classes that can be reached from x.

Remark 6.3.1 *Let $\mathcal{A}_{\mathcal{C}}$ be the set of positive functions f defined on E that are of the form $f = \sum_{C\in\mathcal{C}} a_C\, 1_C$ with $(a_C)_{C\in\mathcal{C}} \in \mathbb{R}_+^{\mathbb{N}}$. In view of the preceding result, we note that*

$$V_I = \inf\{\nu(fV)/\nu(f)\,;\, f \in \mathcal{A}_{\mathcal{C}}\}$$

This expression should be compared with the general formula for V_ν:

$$V_\nu = \inf\{\nu(fV)/\nu(f)\,;\, f \in \mathcal{A}_+\}$$

where $\mathcal{A}_+$ denotes the set of positive bounded measurable functions defined on $(E,\mathcal{E})$.

To understand precisely the concentration phenomenon for μ_α we would like to obtain a large-deviation principle; that is, to find a function $U : E \to \mathbb{R}_+$ such that for any $x \in E$

$$U(x) = -\lim_{\alpha\to+\infty}\frac{1}{\alpha}\log(\mu_\alpha(x))$$

(necessarily $\min_E U = 0$, in analogy with the generalized simulated annealing, we would say that U is the virtual energy). Unfortunately we have not been able to prove such a convergence, even under the condition $(M)_m$, but we are still trying to get this result. Nevertheless, we note that under $(M)_m$ the family of mappings $(\log(\mu_\alpha(\cdot))/\alpha)_{\alpha\geq 1}$ is compact. Indeed, we have for any $\alpha > 0$ and $x \in E$,

$$\mu_\alpha(x) = \frac{\mu_\alpha(\widehat{Q}_\alpha^m(1_{\{x\}}))}{\mu_\alpha(\widehat{Q}_\alpha^m(1))} \geq \epsilon^2(M)\, e^{-(m-1)\alpha\,\mathrm{osc}(V)}\frac{\nu(e^{-\alpha V}\,1_{\{x\}})}{\nu(e^{-\alpha V})}$$

$$\geq \epsilon^2(M)\, e^{-m\alpha\,\mathrm{osc}(V)}\nu(x)$$

and therefore

$$0 \leq -\frac{1}{\alpha}\log \mu_\alpha(x) \leq m \operatorname{osc}(V) - \frac{1}{\alpha}\log\left(\epsilon^2(M)\min_{x\in E}\nu(x)\right)$$

We can consider the accumulation functions U of $-\log(\mu_\alpha(x))/\alpha$ for large α. In order to derive the corresponding Bellman's equations, we introduce for $n \in \mathbb{N}^*$ and $x,y \in E$ the n-communication cost function

$$V^{(n)}(x,y) =_{\mathrm{def.}} \min_{P \in \mathcal{P}_{x,y}^{(n)}} V(P)$$

where $\mathcal{P}_{x,y}^{(n)}$ is the set of M-paths of length n from x to y. In particular, for any $x,y \in E$, $V^{(1)}(x,y) = V(y)$. As in the proof of Proposition 6.3.3, we prove that for any $x,y \in E$, $\liminf_{n\to\infty} V^{(n)}(x,y) = V_\mathcal{C}$ (and this is a true limit if M is aperiodic, the difference of the two terms being at most of order $1/n$).

For a subset $A \subset E$, we also define the M-boundary of A as the subset of all possible sites that are accessible from A; that is,

$$\partial_M(A) = \{y \in E - A\ ;\ \exists x \in A\ \ M(x,y) > 0\}$$

Now we can state the following proposition

Proposition 6.3.5 *Let $U \in \mathbb{R}_+^E$ be any accumulation point as above, then it satisfies the Bellman's fixed point equations*

$$U(y) = \inf_{x\in E}(U(x) + nV^{(n)}(x,y)) - nV_I \qquad (6.14)$$

for any $n \in \mathbb{N}^*$ and $nV_I = \inf_{x,y\in E}(U(x) + nV(x,y))$. Furthermore, we have the inclusions

$$U^{-1}(0) \subset (V \leq V_I) \quad \text{and} \quad \partial_M U^{-1}(0) \subset (V > V_I) \qquad (6.15)$$

Before getting into the proof of this proposition, let us pause for a while and give some comments on the consequence of these results. The inclusions (6.15) show that a point $x \in \{V \leq V_I\}$ with energy $U(x) > 0$ cannot be reached from $U^{-1}(0)$ (the reverse being in general true). This shows that when all pairs of points $x,y \in \{V \leq V_I\}$ can be joined by a path in this level set, then $U^{-1}(0) = \{V \leq V_I\}$.

Proof of Proposition 6.3.5:
Bellman's equations are immediate consequences of the fixed point equation (see the proof of Lemma 6.3.1). We have for any $n \in \mathbb{N} - \{0\}$, $x \in E$, and $\alpha > 0$

$$\mu_\alpha(x) = (\mu_\alpha[\exp(\alpha V)])^n \sum_{y\in E}\mu_\alpha(y)\mathbb{E}_y\left[1_{\{x\}}(X_n)\exp\left(-\alpha\sum_{p=1}^n V(X_p)\right)\right]$$

6.3 Feynman-Kac Trapping Models

Taking the logarithm, dividing by α, and letting α tend to infinity, we get the desired formulae. To prove the inclusions (6.15), we suppose on the contrary that we can find a pair $(x,y) \in E^2$ such that

$$U(x) = 0, \quad M(x,y) > 0, \quad U(y) > 0, \quad \text{and} \quad V(y) \leq V_I$$

From Bellman's equation, this will give that

$$\begin{aligned} U(y) &= \inf\{U(z) + V(y) - V_I \; ; \; z \in E, \; M(z,y) > 0\} \\ &\leq \inf\{U(z) \; ; \; M(z,y) > 0\} \leq U(x) = 0 \end{aligned}$$

and we obtain a contradiction with the fact that $U(y) > 0$. ∎

We end this section with a simple three-point example in which $V_I > V_\nu$ and $V^{-1}(V_\nu) \not\subset U^{-1}(0)$. So we take for state space $E = \{0,1,2\}$ and we consider the Markov kernel defined by

$$M = \begin{pmatrix} p & 1-p & 0 \\ 0 & 0 & 1 \\ 1 & 0 & 0 \end{pmatrix} \quad \text{with} \quad p \in (0,1)$$

It is clear that M is irreducible and aperiodic, and we check that its unique invariant probability ν is given by

$$\nu(0) = \frac{1}{3-2p} \quad \text{and} \quad \nu(1) = \nu(2) = \frac{1-p}{3-2p}$$

Let $V : E \to \mathbb{R}_+$ be a potential function such that

$$V(0) > \frac{V(0) + V(1) + V(2)}{3} > V(2) > V(1) = 0 \tag{6.16}$$

So the ν-essential infimum V_ν is given by $V_\nu = 0 = V(1)$, and by Proposition 6.3.4 we have

$$V_I = (V(0) + V(1) + V(2))/3$$

This could also be deduced from the fact that here the rate function I satisfies

$$I(\mu) < \infty \iff \exists r \in [0,1] : \mu = r(\delta_0 + \delta_1 + \delta_2)/3 + (1-r)\delta_0$$

a property that reflects that trajectories of X are concatenations of the words [0] and [1,2,0] (except for a possible start with [2]). Our next objective is to solve explicitly Bellman's fixed point equation (6.14) for $n = 1$:

$$\begin{cases} U(0) &= \min\{U(0), U(2)\} + V(0) - V_I \\ U(1) &= U(0) + V(1) - V_I \\ U(2) &= U(1) + V(2) - V_I \end{cases}$$

214   6. Annealing Properties

By (6.16), we see that in the first equality the minimum cannot be $U(0)$ (otherwise $V(0) = V_I$), so $U(0) = U(2) + V(0) - V_I$ and this shows that $U(2) < U(0)$. The last equation also implies that $U(2) < U(1)$ and necessarily $U(2) = 0$, from which we obtain that U is unique and that it is given by

$$\begin{cases} U(0) &= V(0) - V_I \\ U(1) &= V_I - V(2) \\ U(2) &= 0 \end{cases}$$

One concludes that $\lim_{\alpha \to \infty} \mu_\alpha(2) = 1$ and that this convergence is exponentially fast. In particular, μ_α does not concentrate for large α on the unique point 1, where the "essential" infimum is achieved (this latter assertion could also be deduced directly from the observation (6.15)).

7
Asymptotic Behavior

7.1 Introduction

This chapter provides an introduction to the asymptotic behavior of particle approximation models as the size of the systems and/or the time horizon tends to infinity. In the following picture, we have illustrated the random evolution of the simple N-genetic approximation model described in (3.4). This picture gives a sound basis to the main questions related to the asymptotic analysis of the particle approximation scheme.

$$
\begin{array}{ccccccccc}
\eta_0 & \to & \eta_1 = \Phi_1(\eta_0) & \to & \eta_2 = \Phi_{0,2}(\eta_0) & \to & \cdots & \to & \Phi_{0,n}(\eta_0) \\
\Downarrow & & & & & & & & \\
\eta_0^N & \to & \Phi_1(\eta_0^N) & \to & \Phi_{0,2}(\eta_0^N) & \to & \cdots & \to & \Phi_{0,n}(\eta_0^N) \\
& & \Downarrow & & & & & & \\
& & \eta_1^N & \to & \Phi_2(\eta_1^N) & \to & \cdots & \to & \Phi_{1,n}(\eta_1^N) \\
& & & & \Downarrow & & & & \\
& & & & \eta_2^N & \to & \cdots & \to & \Phi_{2,n}(\eta_2^N) \\
& & & & & & \Downarrow & & \vdots \\
& & & & & & \eta_{n-1}^N & \to & \Phi_n(\eta_{n-1}^N) \\
& & & & & & & & \Downarrow \\
& & & & & & & & \eta_n^N
\end{array}
$$

Intuitively, we first observe that the sampling error (represented by the implication sign "$\Downarrow$") does not propagate but stabilizes as soon as the semigroup $\Phi_{p,n}$ is sufficiently stable. This intuitive idea is made clear by the pivotal formula

$$\eta_n^N - \eta_n = \sum_{q=0}^n [\Phi_{q,n}(\eta_q^N) - \Phi_{q,n}(\Phi_q(\eta_{q-1}^N))]$$

with the convention $\Phi_0(\eta_{-1}^N) = \eta_0$ for $p = 0$. Note that each term on the r.h.s. represents the propagation of the pth sampling local error $\Phi_p(\eta_{p-1}^N) \Rightarrow \eta_p^N$. This observation indicates that the numerical analysis of the particle algorithm or any numerical approximation model (based on local approximations) is intimately related to the stability property of the nonlinear semigroup of the limiting model. The picture also suggests that the fluctuations of the flow of local errors (properly renormalized) behave asymptotically as a sequence of independent and identically distributed Gaussian random variables. These questions will be made clear in the further development of this chapter (see also [86, 87, 100, 220, 219]).

The chapter is organized as follows. In Section 7.2, we provide a short discussion on Feynman-Kac models and their particle interpretations. We also take the opportunity to fix some of the notation and some regularity conditions we shall be using in the further development of this book. Section 7.3 focuses on independent sequences of random variables (which we shall abbreviate iid). In the first Section 7.3.1, we discuss some general inequalities such as a refined version of the inequalities of Khinchine/Bürkholder/Marcinkiewicz-Zygmund. We already mention that these original inequalities provide a natural and simple way to estimate the moment-generating functions of the empirical measures associated with independent random variables. In Section 7.3.2, we derive some more or less well-known $\mathbb{L}_p$ and exponential inequalities for empirical processes. These estimates extend the corresponding statements for sums of iid to the convergence of empirical processes with respect to some Zolotarev type seminorm. The inequalities presented in this section will be of use in the further development of this chapter. Special attention is also paid to deriving as soon as possible precise and sharp inequalities. This choice is not only for mathematical elegance but in some instances it is essential to start with a precise estimate to work out another analytical result with exact rates of decay. For instance, the complement of the $\mathbb{L}_p$-inequalities of Bürkholder presented in this section provide precise and sharp constants. We will use this estimate in the proof of strong propagation-of-chaos estimates. In this particular situation, we propose a strategy of analysis in which the exact decay rates in the propagation-of-chaos are related to the precision of these $\mathbb{L}_p$-inequalities.

The strong law of large numbers for interacting particle systems is discussed in Section 7.4. In Section 7.4.1 and Section 7.4.2, we study a fairly general class of interacting processes, including the situation where the potential functions may take null values and the algorithm may be stopped when the system dies. In this connection, we derive in Section 7.4.1 several types of exponential bounds to estimate the probability of extinction. We also mention that Section 7.4.2 contains some key martingale type decompositions that are essential on our way to proving central limit theorems. The final Section 7.4.3 focuses on time-uniform estimates with respect to the time parameter. We examine this question from different angles related to a graduate set of regularity conditions. These estimates are probably one of the most important results in practice. They allow us to quantify the size of the particle approximation models that ensures a given precision.

7.2 Some Preliminaries

For the convenience of the reader, we have collected hereafter some essential results on Feynman-Kac semigroups and their interacting particle interpretations. Let $(E_n, \mathcal{E}_n)_{n\geq 0}$ be a collection of measurable spaces. For any $p \leq n$, we recall that $E_{[p,n]} = (E_p \times \ldots \times E_n)$ and $\mathcal{E}_{(p,n]} = E_{[p+1,n]}$. Also let $\eta_0 \in \mathcal{P}(E_0)$ and $M_n(x_{n-1}, dx_n)$ be a sequence of Markov transitions from E_{n-1} into E_n, $n \geq 1$. We denote by $G_n : E_n \to (0, \infty)$ a collection of nonnegative and bounded $\mathcal{E}_n$-measurable functions, and we associate with the triplet (η_0, G_n, M_n) the Feynman-Kac measures $\eta_n \in \mathcal{P}(E_n)$ defined for any $f_n \in \mathcal{B}_b(E_n)$ and $n \in \mathbb{N}$ by the formulae

$$\eta_n(f_n) = \gamma_n(f_n)/\gamma_n(1) \quad \text{with} \quad \gamma_n(f_n) = \mathbb{E}_{\eta_0}\left(f_n(X_n) \prod_{p=0}^{n-1} G_n(X_p)\right) \tag{7.1}$$

where $\mathbb{E}_{\eta_0}$ stands for the expectation with respect to the distribution of an E_n-valued Markov chain X_n with transitions M_n. Without further mention, we will suppose that G_n satisfy condition (G) for some $\epsilon_n(G) > 0$ (see page 115). By the definition of η_n, no generality is lost and much convenience is gained by supposing, as will be done in this chapter and unless otherwise stated, that the potential functions G_n take values in $(0, 1]$ (see also Section 2.5 and page 77 Section 2.5.3). We recall that the distribution flow η_n satisfies the nonlinear equation $\eta_{n+1} = \Phi_{n+1}(\eta_n)$, where the mapping $\Phi_{n+1} : \mathcal{P}(E_n) \to \mathcal{P}(E_{n+1})$ is defined for any $\eta \in \mathcal{P}(E_n)$ by

$$\Phi_{n+1}(\eta) = \Psi_n(\eta) M_{n+1} \quad \text{with} \quad \Psi_n(\eta)(dx) = \frac{1}{\eta(G_n)} G_n(x)\,\eta(dx) \tag{7.2}$$

We let $Q_{p,n}$ and $\Phi_{p,n}$, $p \leq n$, be the semigroups associated respectively with the Feynman-Kac distribution flows γ_n and η_n defined in (7.1),

$$Q_{p,n} = Q_{p+1} \ldots Q_{n-1} Q_n \quad \text{and} \quad \Phi_{p,n} = \Phi_n \circ \Phi_{n-1} \circ \ldots \circ \Phi_{p+1}$$

with $Q_n(x_{n-1}, dx_n) = G_{n-1}(x_{n-1}) M_n(x_{n-1}, dx_n)$. We use the convention $Q_{n,n} = Id$ and $\Phi_{n,n} = Id$ for $p = n$. We recall that $\Phi_{p,n}$ is a nonlinear integral operator from $\mathcal{P}(E_p)$ into $\mathcal{P}(E_n)$. For any $(\mu_p, f_n) \in (\mathcal{P}(E_p) \times \mathcal{B}_b(E_n))$, it can be written in terms of a Boltzmann-Gibbs transformation

$$\Phi_{p,n}(\mu_p)(f_n) = \mu_p(G_{p,n}\, P_{p,n}(f_n))/\mu_p(G_{p,n})$$

with the pair potential/transition $(G_{p,n}, P_{p,n})$ defined by

$$G_{p,n} = Q_{p,n}(1) \quad \text{and} \quad P_{p,n}(f_n) = Q_{p,n}(f_n)/Q_{p,n}(1)$$

The next two parameters

$$r_{p,n} = \sup_{x_p, y_p \in E_p} (G_{p,n}(x_p)/G_{p,n}(y_p))$$

and

$$\beta(P_{p,n}) = \sup_{x_p, y_p \in E_p} \|P_{p,n}(x_p, \cdot) - P_{p,n}(y_p, \cdot)\|_{\mathrm{tv}} \qquad (7.3)$$

measure respectively the relative oscillations of the potential functions $G_{p,n}$ and the contraction properties of the Markov transition $P_{p,n}$. Various asymptotic estimates on particle models derived in the forthcoming sections will be expressed in terms of these parameters.

7.2.1 McKean Interpretations

The flow η_n can alternatively be described by a nonlinear equation of the form $\eta_{n+1} = \eta_n K_{n+1,\eta_n}$, where $K_{n+1,\eta}$, $\eta \in \mathcal{P}(E_n)$, is a (nonunique) collection of Markov transitions satisfying the compatibility condition

$$\eta K_{n+1,\eta} = \Psi_n(\eta) M_{n+1} = \Phi_{n+1}(\eta)$$

We associate with a given pair $(\eta_0, K_{n,\eta})$ the McKean measure $\mathbb{K}_{\eta_0}$ on the canonical space $(\Omega = \prod_{n \geq 0} E_n, \mathcal{F} = (\mathcal{F}_n)_{n \geq 0})$ with marginals

$$\mathbb{K}_{\eta_0, n}(d(x_0, \ldots, x_n))$$
$$= \eta_0(dx_0)\, K_{1,\eta_0}(x_0, dx_1)\, \ldots\, K_{n-1,\eta_n}(x_{n-1}, dx_n) \in \mathcal{P}(E_{[0,n]})$$

Given a McKean measure, the Feynman-Kac flow η_n can be interpreted as the (marginal) distributions of a nonhomogeneous Markov chain with transitions $K_{n,\eta}$ and initial distribution η_0. The corresponding N-particle model is defined as a sequence of nonhomogeneous and E_n^N-valued Markov chains

$$\left(\Omega^{(N)} = \prod_{n \in \mathbb{N}} E_n^N,\ \mathcal{F}^N = (\mathcal{F}_n^N)_{n \in \mathbb{N}},\ (\xi_n)_{n \in \mathbb{N}},\ \mathbb{P}_{\eta_0}^N \right)$$

The initial configuration ξ_0 consists of N independent and identically distributed random variables with common law η_0, and its elementary transitions from E_{n-1}^N into E_n^N are given in a symbolic integral form by

$$\mathbb{P}_{\eta_0}^N\left(\xi_n \in dx_n \mid \xi_{n-1}\right) = \prod_{p=1}^N K_{n,m(\xi_{n-1})}(x_{n-1}^p, dx_n^p) \qquad (7.4)$$

where $m(\xi_{n-1}) = \frac{1}{N}\sum_{i=1}^N \delta_{\xi_{n-1}^i}$ and $dx_n = dx_n^1 \times \ldots \times dx_n^N$ is an infinitesimal neighborhood of a point $x_n = (x_n^1, \ldots, x_n^N) \in E_n^N$.

Several examples of McKean models are described in Section 2.5.3. Two generic situations arising in practice can be underlined:

- **Case 1**: $K_{n+1,\eta}(x, .) = \Phi_{n+1}(\eta)$
- **Case 2**: $K_{n+1,\eta}(x, .) = G_n(x)\, M_{n+1}(x, .) + (1 - G_n(x))\, \Phi_{n+1}(\eta)$

We recall that these two situations belong to the same class of McKean transitions defined by

$$K_{n+1,\eta}(x, .) = \varepsilon_n(\eta)G_n(x)\, M_{n+1}(x, .) + (1 - \varepsilon_n(\eta)G_n(x))\, \Phi_{n+1}(\eta)$$

for some constant $\varepsilon_n(\eta)$ that may depend on the current pair of parameters (n, η) and such that $\varepsilon_n(\eta)G_n \le 1$. The two cases above correspond to the situation where, respectively, $\varepsilon_n(\eta) = 0$ and $\varepsilon_n(\eta) = 1$.

Except of few situations, such as the fluctutaions on path-space and propagation-of-chaos analysis with respect to the total variation distance, the asymptotic theory developed in this book applies to any kind of McKean interpretation model. To give a practical sound basis to the forthcoming analysis, sometimes we illustrate our results on the two cases described above. We shall distinguish the corresponding particle and McKean models with mentioning that they are related to the first and second cases (McKean interpretation).

7.2.2 Vanishing Potentials

Let us now take up the problem where the potential functions may vanish on some regions of the state spaces. In this situation, the Feynman-Kac model represents the distributions of a single Markov particle model evolving in an absorbing medium with hard obstacles (see Section 2.4.3, Section 2.5, Section 3.3 and Section 4.4). Let $\widehat{E}_n = E_n - G_n^{-1}(0)$. We recall (see Section 2.5, page 71) that the limiting flow η_n is well-defined only up to the first time τ we have $\eta_\tau(\widehat{E}_\tau) = 0 (= \eta_\tau(G_\tau 1_{\widehat{E}_\tau}))$; that is, up to the deterministic time horizon

$$\begin{aligned}\tau &= \inf\{n \in \mathbb{N} : \gamma_{n+1}(1) = 0\} \\ &= \inf\{n \in \mathbb{N} : \eta_n(G_n) = 0\} \in [0, \infty]\end{aligned}$$

220   7. Asymptotic Behavior

Note that
$$\tau = \infty \iff \forall n \in \mathbb{N} \;\; \gamma_n(1) > 0$$
As an aside, since G_n are assumed to be $[0,1]$-valued potential functions, we find that
$$0 < \gamma_{n+1}(1) \leq \gamma_n(1) \leq 1$$
Consequently, we have $\gamma_n(1) > 0$ if and only if we have $\gamma_p(1) > 0$ for any $0 \leq p \leq n$. Next we present two sufficient conditions under which the normalizing constant $\gamma_n(1)$ is well-defined for any $n \in \mathbb{N}$.

($\mathcal{A}$) For any $n \geq 1$ and $x_n \in \widehat{E}_n$, $\eta_0(\widehat{E}_0) > 0$ and $M_{n+1}(x_n, \widehat{E}_{n+1}) > 0$.

($\mathcal{B}$) There exists a sequence of positive numbers $\alpha(n)$ such that for any $n \geq 0$ and $x_n \in \widehat{E}_n$ we have
$$\eta_0(\widehat{E}_0) > 1 - e^{-\alpha(0)} \quad \text{and} \quad M_{n+1}(x_n, \widehat{E}_{n+1}) \geq 1 - e^{-\alpha(n+1)}$$
It is easily checked that $(\mathcal{B}) \Longrightarrow (\mathcal{A}) \Longrightarrow \tau = \infty$.

When $\widehat{E}_n = E_n$, condition ($\mathcal{B}$) holds true for any choice of $\alpha(n)$. Also recall that if ($\mathcal{A}$) is met, the updated distribution flow model $\widehat{\eta}_n = \Psi_n(\eta_n)$ can be regarded as the prediction flow model with initial distribution $\widehat{\eta}_0$ and associated with the pair potential/transition $(\widehat{G}_n, \widehat{M}_n)$ defined in Proposition 2.5.2 on page 70. In other words under ($\mathcal{A}$) the analysis of the updated flow $\widehat{\eta}_n$ reduces to that of a prediction flow with strictly positive potentials.

In accordance with previous comments, the N-interacting particle systems
$$\xi_n \longrightarrow \widehat{\xi}_n \longrightarrow \xi_{n+1}$$
associated with a general class of Feynman-Kac models with $[0,1]$-valued potential functions are only defined up to the time $\tau^N = n$ the whole configuration $\xi_n \in E_n^N$ first hits the hard obstacle set $(E_n - \widehat{E}_n)^N$:
$$\tau^N = \inf\{n \in \mathbb{N} : m(\xi_n)(G_n) = 0\} \in [0, \infty]$$
These stopped algorithms are defined as Markov chains taking values in $E_n^N \cup \{\Delta\}$ where $\Delta = (\partial, \ldots, \partial)$ (N times) represents a cemetery point and $\widehat{\xi}_p = \xi_{p+1} = \Delta$, for any $p \geq \tau^N$. Notice that we have $\xi_n = \Delta$ if and only if $\xi_n^i = \partial$ for all $1 \leq i \leq N$. We refer the reader to Section 3.3 for a precise construction of the probability space associated with these models.

It follows from the definition of τ^N that
$$\tau^N = n \iff \widehat{\xi}_0 \in \widehat{E}_0, \ldots, \widehat{\xi}_{n-1} \in \widehat{E}_{n-1} \text{ and } \widehat{\xi}_n = \Delta$$
$$\iff \xi_0 \in \widehat{E}_0, \ldots, \xi_{n-1} \in \widehat{E}_{n-1} \text{ and } \xi_n \notin \widehat{E}_n$$
and $\tau^N \geq n \iff \xi_0 \in \widehat{E}_0, \ldots, \xi_{n-1} \in \widehat{E}_{n-1}$. This indicates that τ^N is a predictable Markov time with respect to the natural filtration $\mathcal{F}_n^N$ associated with the Markov chain ξ_n in the sense that $\{\tau^N = n\} \in \mathcal{F}_n^N$ and

$\{\tau^N \geq n\} \in \mathcal{F}_{n-1}^N$. In this context, the N-particle density profiles associated with the Feynman-Kac flows (γ_n, η_n) are given by

$$\eta_n^N = \frac{1}{N} \sum_{i=1}^N \delta_{\xi_n^i} \in \mathcal{P}(E_n \cup \{\partial\})$$

$$\gamma_n^N(.) = \eta_n^N(.) \times \prod_{p=0}^{n-1} \eta_p^N(G_p) \in \mathcal{M}_+(E_n \cup \{\partial\})$$

Since test functions $f_n \in \mathcal{B}_b(E_n)$ are extended to $E_n \cup \{\partial\}$ by setting $f_n(\partial) = 0$, we shall identify the null measure on $E_n \cup \{\partial\}$ with the Dirac measures δ_∂. In these conventions, we have

$$1_{\tau^N < n} \times \eta_n^N = \delta_\partial = 1_{\tau^N < n} \times \gamma_n^N$$

More interestingly, in the event that $\tau^N \geq n$, the N-particle model $\xi_n \in E_n^N$ at time n has not been killed and we have

$$1_{\tau^N \geq n} \times \eta_n^N \in \mathcal{M}_+(E_n) \quad \text{and} \quad 1_{\tau^N \geq n} \times \gamma_n^N \in \mathcal{M}_+(E_n)$$

When no confusion can be made, we shall clarify notations suppressing the superscript $(.)^N$ in the expectation operators. For instance, we shall often denote by $\mathbb{E}(.)$ instead of $\mathbb{E}_{\eta_0}^N(.)$ the expectation operator associated with the distribution $\mathbb{P}_{\eta_0}^N$ of an N-particle type model.

7.3 Inequalities for Independent Random Variables

Let $(\mu_i)_{i \geq 1}$ be a sequence of probability measures on a given measurable state space $(E, \mathcal{E})$. We also consider a sequence of $\mathcal{E}$-measurable functions $(h_i)_{i \geq 1}$ such that $\mu_i(h_i) = 0$ for all $i \geq 1$. During the further development of this section, we fix an integer $N \geq 1$. To clarify the presentation, we slightly abuse the notation and we denote respectively by

$$m(X) = \frac{1}{N} \sum_{i=1}^N \delta_{X^i} \quad \text{and} \quad \mu = \frac{1}{N} \sum_{i=1}^N \mu_i$$

the N-empirical measure associated with a collection of independent random variables $X = (X^i)_{i \geq 1}$, with respective distributions $(\mu_i)_{i \geq 1}$, and the N-averaged measure associated with the sequence of measures $(\mu_i)_{i \geq 1}$. To clarify the presentation, when we are given N-sequences of points $x = (x^i)_{1 \leq i \leq N} \in E^N$ and functions $(h_i)_{1 \leq i \leq N} \in \mathcal{B}_b(E)^N$, we shall often use the abusive notations

$$m(x)(h) = \frac{1}{N} \sum_{i=1}^N h_i(x^i) \quad \text{and} \quad \sigma^2(h) = \frac{1}{N} \sum_{i=1}^N \operatorname{osc}^2(h_i)$$

For any pair of integers (p, n) with $1 \leq p \leq n$, we denote by

$$(n)_p = n!/(n-p)!$$

the number of one-to-one mappings from a set of p elements into another set of n elements.

7.3.1 $\mathbb{L}_p$ and Exponential Inequalities

We start with two elementary and well-known results.

Lemma 7.3.1 *Let X be a real-valued random variable X with $a \leq X \leq b$ and $\mathbb{E}(X) = 0$. Then for any $t \geq 0$*

$$\mathbb{E}(e^{tX}) \leq e^{t^2(b-a)^2/8}$$

Proof:
To prove this assertion, we use the convexity property of the exponential function to check that for any $x \in [a, b]$

$$\frac{e^{tx} - e^{ta}}{x - a} \leq \frac{e^{tb} - e^{ta}}{b - a}$$

or equivalently $e^{tx} \leq \frac{x-a}{b-a} e^{tb} + \frac{b-x}{b-a} e^{ta}$. This convexity inequality readily bounds the moment-generating function of a random variable in terms of its mean and support

$$\begin{aligned}
\mathbb{E}(e^{tX}) &\leq \frac{\mathbb{E}(X) - a}{b - a} e^{tb} + \frac{b - \mathbb{E}(X)}{b - a} e^{ta} \\
&= e^{ta} \left[1 + \frac{a}{b - a}(1 - e^{t(b-a)})\right] = e^{\varphi(t(b-a))} \qquad (7.5)
\end{aligned}$$

with

$$\varphi(s) = d\,s + \log\left[1 + d\,(1 - e^s)\right] \quad \text{and} \quad d = \frac{a}{(b - a)}$$

By straightforward calculations, we find the derivatives

$$\varphi'(s) = d + \frac{d}{d - (1 + d)e^{-s}} \quad \text{and} \quad \varphi''(s) = \frac{d(1 + d)e^{-s}}{(d - (1 + d)e^{-s})^2}$$

Since $d(1 + d) = ab/(b - a)^2 \leq 1/4$, using Taylor's formula we find that $\varphi(s) \leq \frac{s^2}{8}$, and by (7.5) we end the proof of the lemma. ∎

By Markov's inequality, for any random variable U and for any pair $s, t \geq 0$, we have

$$\mathbb{P}(U \geq s) = \mathbb{P}(e^{tU} \geq e^{st}) \leq e^{-st}\,\mathbb{E}(e^{tU})$$

7.3 Inequalities for Independent Random Variables

This inequality is also known as Bernstein's exponential bound. Chernov's method consists in finding the parameter $t \geq 0$ that minimizes the upper bound.

In the context of the sum of N independent random variables X^i with respective distributions μ_i, the Bernstein-Chernov inequality yields that

$$\mathbb{P}(N\ m(X)(h) \geq s) \leq e^{-s\,t} \prod_{i=1}^{N} \mu_i(e^{th_i}) \leq e^{-s\,t+t^2 N\ \sigma^2(h)/8}$$

(recall that $\mu_i(h_i) = 0$ and $\sigma^2(h) = \frac{1}{N}\sum_{i=1}^{N} \operatorname{osc}^2(h_i)$). By choosing $t = 4s/(N\sigma^2(h))$ and $s = N\epsilon$, we conclude that for any $\epsilon > 0$

$$\mathbb{P}(m(X)(h) \geq \epsilon) \leq e^{-2N\ \epsilon^2/\sigma^2(h)}$$

In the same way, we prove that $\mathbb{P}(-m(X)(h) \geq \epsilon) \leq e^{-2N\epsilon^2/\sigma^2(h)}$. This readily implies the following lemma.

Lemma 7.3.2 (Chernov-Hoeffding)

$$\mathbb{P}(|m(X)(h)| \geq \epsilon) \leq 2\,e^{-2N\epsilon^2/\sigma^2(h)}$$

These exponential bounds were originally proved by Chernov in 1952 for binomial distributions and extended to general bounded random variables in 1963 by W. Hoeffding.

The next lemma is a complement of the inequalities of Khinchine, Bürkholder, Davis and Marcinkiewicz-Zygmund.

Lemma 7.3.3 *The following assertions are satisfied for any sequence of $\mathcal{E}$-measurable functions $(h_i)_{i\geq 1}$ such that $\mu_i(h_i) = 0$ for all $i \geq 1$.*

- *If the functions h_i have finite oscillations, then for any $p \geq 1$ we have*

$$\sqrt{N}\ \mathbb{E}(|m(X)(h)^p|)^{\frac{1}{p}} \leq d(p)^{\frac{1}{p}}\ \sigma(h) \qquad (7.6)$$

with the sequence of finite constants $(d(n))_{n\geq 0}$ defined for any $n \geq 1$ by the formulae

$$d(2n) = (2n)_n\ 2^{-n} \quad \text{and} \quad d(2n-1) = \frac{(2n-1)_n}{\sqrt{n-1/2}}\ 2^{-(n-1/2)} \qquad (7.7)$$

- *If we have $\mu(h^{2n}) < \infty$ for some $n \geq 1$, then*

$$N^n\ \mathbb{E}(m(X)(h)^{2n}) \leq d(2n)\ \mu((2h)^{2n})$$
$$N^{n-1/2}\ \mathbb{E}(|m(X)(h)|^{2n-1}) \leq d(2n-1)\ \mu((2h)^{2n})^{1-\frac{1}{2n}}$$

As we mentioned in the introduction, this technical lemma will be of use in this chapter, including in $\mathbb{L}_p$-mean errors, in increasing strong propagation-of-chaos analysis and in the derivation of a Berry-Esseen inequality for particle models. In this context, the use of Burkholder type estimates will lead to different conclusions and very coarse properties.

There are a number of significant and related estimates in the literature on martingales that apply to our context. For instance, using Burkholder's inequality (see for instance [287]), we would find that

$$N^n \, \mathbb{E}(m(X)(h)^{2n}) \leq (18 B_{2n})^{2n} \, \sigma(h)^{2n}$$

with $(2n) \leq B_{2n} = 2n\sqrt{n/(n-1/2)} \leq \sqrt{2} \, (2n)$. This would lead to the estimate

$$N^n \, \mathbb{E}(m(X)(h)^{2n}) \leq 2^n \, 18^{2n} \, (2n)^{2n} \, \sigma(h)^{2n}$$

The next inequality gives a quick and simple way to measure the improvements obtained in Lemma 7.3.3:

$$\frac{2^{-n} \, (2n)_n}{2^n \, 18^{2n} \, (2n)^{2n}} = \frac{1}{64^n \, (2n)^n} \prod_{p=1}^{n-1} \left(1 - \frac{p}{2n}\right) \leq \frac{1}{64^n \, (2n)^n}$$

On the other hand, for homogeneous pairs $(h_i, \mu_i) = (h, \mu)$, the central limit theorem applies and we have the asymptotic result

$$\left(\sqrt{N} \, m(X)[h/\|h\|_{2,\mu}]\right)^{2n} \xrightarrow{d} W^{2n}$$

where W is a centered and Gaussian random variable with $\mathbb{E}(W^2) = 1$ and the superscript $\xrightarrow{d}$ stands for the convergence in distribution as N tends to infinity. In this connection, if we have $\mu(h^{2n}) < \infty$ for some integer $n \geq 1$, then it is well-known that

$$\lim_{N \to \infty} N^n \, \mathbb{E}\left(m(X)[h/\|h\|_{2,\mu}]\right)^{2n} = \mathbb{E}(W^{2n}) = (2n)_n \, 2^{-n}$$

This asymptotic result already indicates that in this sense the estimates presented in Lemma 7.3.3 are sharp. As a final illustration of the impact of these inequalities, we provide hereafter an estimation of the moment-generating function of the empirical measures $m(X)$.

Theorem 7.3.1 *For any sequence of $\mathcal{E}$-measurable functions $(h_i)_{i \geq 1}$ such that $\mu_i(h_i) = 0$, for all $i \geq 1$ we have for any $\varepsilon > 0$*

$$\sigma(h) < \infty \implies \mathbb{E}(e^{\varepsilon \sqrt{N} |m(X)(h)|}) \leq \left(1 + \frac{\varepsilon}{\sqrt{2}} \, \sigma(h)\right) e^{\frac{\varepsilon^2}{2} \sigma^2(h)}$$

Proof:
The $\mathbb{L}_n$-inequalities stated in Lemma 7.3.3 clearly imply that for any $\varepsilon > 0$

$$\mathbb{E}(e^{\varepsilon|m(X)(h)|}) = \sum_{n\geq 0} \frac{\varepsilon^{2n}}{(2n)!} \mathbb{E}(m(X)(h)^{2n})$$

$$+ \sum_{n\geq 0} \frac{\varepsilon^{2n+1}}{(2n+1)!} \mathbb{E}(|m(X)(h)|^{2n+1})$$

$$\leq \sum_{n\geq 0} \frac{1}{n!} \left(\frac{\varepsilon^2 \sigma(h)^2}{2N}\right)^n + \sum_{n\geq 0} \frac{1}{n!} \left(\frac{\varepsilon^2 \sigma(h)^2}{2N}\right)^{n+1/2}$$

from which we conclude that

$$\mathbb{E}(e^{\varepsilon|m(X)(h)|}) = \left(1 + \frac{\varepsilon \sigma(h)}{\sqrt{2N}}\right) \sum_{n\geq 0} \frac{1}{n!} \left(\frac{\varepsilon^2 \sigma(h)^2}{2N}\right)^n$$

$$= \left(1 + \frac{\varepsilon}{\sqrt{2N}} \sigma(h)\right) e^{\frac{\varepsilon^2}{2N}\sigma^2(h)}$$

We end the proof of the theorem by replacing ε by $\varepsilon\sqrt{N}$. ∎

Proof of Lemma 7.3.3: We first use a symmetrization technique. We consider a collection of independent copies $X' = (X'^i)_{i\geq 1}$ of the random variables $X = (X^i)_{i\geq 1}$. We also assume that (X, X') are independent. As usual, we slightly abuse the notation and we denote by $m(X') = \frac{1}{N}\sum_{i=1}^{N} \delta_{X'^i}$ the N-empirical distribution associated with X'. We observe that

$$m(X)(h) = \mathbb{E}(m(X)(h) - m(X')(h) \mid X)$$

This clearly implies that, for any $p \geq 1$, we have that

$$\mathbb{E}(|m(X)(h)|^p) \leq \mathbb{E}(|m(X)(h) - m(X')(h)|^p)$$

We first examine the case $p = 2n$ with $n \geq 0$. In this situation, we have

$$N^{2n}\mathbb{E}(|m(X)(h) - m(X')(h)|^{2n})$$

$$= \sum_{k=1}^{2n} \sum_{p_1+\ldots+p_k=2n} \frac{(2n)!}{p_1!\ldots p_k!} \sum_{\alpha \in \langle k, N \rangle} \prod_{i=1}^{k} \mathbb{E}((h_{\alpha(i)}(X^{\alpha(i)}) - h_{\alpha(i)}(X'^{\alpha(i)}))^{p_i})$$

where $\sum_{p_1+\ldots+p_k=2n}$ indicates summation over all ordered sets of strictly positive integers $p_i \geq 1$ such that $p_1 + \ldots + p_k = 2n$, and $\langle k, N \rangle$ is the set of all one-to-one mappings from $\langle k \rangle =_{\text{def.}} \{1,\ldots,k\}$ into $\langle N \rangle$. Since we have

$$\mathbb{E}([h_j(X^j) - h_j(X'^j)]^p) = -\mathbb{E}([h_j(X^j) - h_j(X'^j)]^p) = 0$$

226   7. Asymptotic Behavior

for any $1 \leq j \leq N$ and any odd integer p, we check easily that

$$N^{2n}\mathbb{E}(|m(X)(h) - m(X')(h)|^{2n})$$

$$= \sum_{k=1}^{n} \sum_{p_1+\ldots+p_k=n} \frac{(2n)!}{(2p_1)!\ldots(2p_k)!}$$

$$\times \sum_{\alpha \in \langle k,N \rangle} \mathbb{E}\left(\prod_{i=1}^{k}[h_{\alpha(i)}(X^{\alpha(i)}) - h_{\alpha(i)}(X'^{\alpha(i)})]^{2p_i}\right)$$

$$\leq (2n)_n \left(\sup_{1 \leq k \leq n} \sup_{p_1+\ldots+p_k=n} \prod_{i=1}^{k}(2p_i)_{p_i}^{-1}\right) \mathbb{E}((\sum_{i=1}^{N}[h_i(X^i) - h_i(X'^i)]^2)^n)$$

Using the fact that for any $p \geq 1$ we have

$$(2p)_p = (2p)!/p! = 2p\,(2p-1)\ldots(2p-(p-1))$$
$$= \prod_{k=1}^{p}(p+k) \geq 2^p$$

we conclude that

$$N^n \mathbb{E}(|m(X)(h) - m(X')(h)|^{2n})$$

$$\leq (2n)_n\,2^{-n}\,\mathbb{E}\left(\left(\tfrac{1}{N}\sum_{i=1}^{N}[h_i(X^i) - h_i(X'^i)]^2\right)^n\right)$$

and therefore

$$N^n \mathbb{E}(|m(X)(h)|^{2n}) \leq (2n)_n\,2^{-n}\,\mathbb{E}\left(\left(\frac{1}{N}\sum_{i=1}^{N}[h_i(X^i) - h_i(X'^i)]^2\right)^n\right)$$

This implies that

$$N^n \mathbb{E}(|m(X)(h)|^{2n}) \leq (2n)_n\,2^{-n}\,\sigma(h)^{2n}$$

as soon as $\sigma(h) < \infty$. In the same way, if we have $\mu(h^{2n}) < \infty$, then

$$N^n \mathbb{E}(|m(X)(h)|^{2n}) \leq (2n)_n\,\mathbb{E}((m(X)(h^2) + m(X')(h^2))^n)$$
$$\leq (2n)_n\,2^n\,\mathbb{E}(m(X)(h^2)^n) \leq (2n)_n\,2^n\,\mu(h^{2n})$$

For odd integers $p = 2n+1$, we use the Cauchy-Schwartz inequality to check that

$$\mathbb{E}(|m(X)(h)|^{2n+1})^2 \leq \mathbb{E}(|m(X)(h)|^{2n})\,\mathbb{E}(|m(X)(h)|^{2(n+1)})$$

From previous estimates, we find that

$$N^{2n+1}\,\mathbb{E}(|m(X)(h)|^{2n+1})^2 \leq (2n)_n\,(2(n+1))_{n+1}\,2^{-(2n+1)}\,\sigma(h)^{2(2n+1)}$$

as soon as $\sigma(h) < \infty$. Since

$$(2(n+1))_{n+1} = \frac{(2(n+1))!}{(n+1)!} = 2\,\frac{2n+1!}{n!} = 2\,(2n+1)_{n+1}$$

$$(2n)_n = \frac{2n!}{n!} = \frac{1}{2n+1}\,\frac{2n+1!}{n!} = \frac{(2n+1)_{n+1}}{(2n+1)}$$

we get

$$N^{n+1/2}\,\mathbb{E}(|m(X)(h)|^{2n+1}) \leq \frac{(2n+1)_{n+1}}{\sqrt{n+1/2}}\,2^{-(n+1/2)}\,\sigma(h)^{(2n+1)}$$

In the same way, for any h such that $\mu(h^{2(n+1)}) < \infty$, we have

$$N^{2n+1}\,\mathbb{E}(|m(X)(h)|^{2n+1})^2 \leq \frac{(2n+1)_{n+1}^2}{n+1/2}\,2^{2n+1}\,\mu(h^{2n})\mu(h^{2(n+1)})$$

Since

$$\mu(h^{2n})\mu(h^{2(n+1)}) \leq \mu(h^{2(n+1)})^{2-\frac{1}{n+1}}$$

we conclude that

$$N^{n+1/2}\mathbb{E}(|m(X)(h)|^{2n+1}) \leq \frac{(2n+1)_{n+1}}{\sqrt{n+1/2}}\,2^{n+1/2}\,\mu(h^{2(n+1)})^{1-\frac{1}{2(n+1)}}$$

and the proof of the lemma is now completed. ∎

7.3.2 Empirical Processes

Let $\mathcal{F}$ be a given collection of measurable functions $f : E \to \mathbb{R}$ such that $\|f\| \leq 1$. We associate with $\mathcal{F}$ the Zolotarev seminorm on $\mathcal{P}(E)$ defined by

$$\|\mu - \nu\|_{\mathcal{F}} = \sup\{|\mu(f) - \nu(f)|;\ f \in \mathcal{F}\},$$

(see for instance [276]). No generality is lost and much convenience is gained by supposing that the unit constant function $f = 1 \in \mathcal{F}$. Furthermore, to avoid some unnecessary technical measurability questions, we shall also suppose that $\mathcal{F}$ is separable in the sense that it contains a countable and dense subset.

To measure the size of a given class $\mathcal{F}$, one considers the covering numbers $N(\varepsilon, \mathcal{F}, L_p(\mu))$ defined as the minimal number of $L_p(\mu)$-balls of radius $\varepsilon > 0$

needed to cover $\mathcal{F}$. By $\mathcal{N}(\varepsilon, \mathcal{F})$, $\varepsilon > 0$, and by $I(\mathcal{F})$ we denote the uniform covering numbers and entropy integral given by

$$\mathcal{N}(\varepsilon, \mathcal{F}) = \sup\{\mathcal{N}(\varepsilon, \mathcal{F}, L_2(\eta)); \eta \in \mathcal{P}(E)\}$$
$$I(\mathcal{F}) = \int_0^1 \sqrt{\log(1 + \mathcal{N}(\varepsilon, \mathcal{F}))} \, d\varepsilon$$

Various examples of classes of functions with finite covering and entropy integral are given in the book of Van der Vaart and Wellner [311] (see for instance p. 86, p. 135, and exercise 4 on p.150). The estimation of the quantities introduced above depends on several deep results on combinatorics that are not discussed here. To illustrate these covering numbers, we content ourselves with mentioning that, for the set of indicator functions $\mathcal{F} = \{1_{\prod_{i=1}^d (0, x_i]} \, ; \, (x_i)_{1 \leq i \leq d} \in \mathbb{R}^d\}$ of cells in $E = \mathbb{R}^d$, we have

$$\mathcal{N}(\varepsilon, \mathcal{F}) \leq c(d+1)(4e)^{d+1} \epsilon^{-2d}$$

Since $\int_0^1 \log(1/\epsilon) d\epsilon < \infty$, we readily check that $I(\mathcal{F}) < \infty$. The exponential estimates and the $\mathbb{L}_p$-mean errors discussed hereafter will depend respectively on $\mathcal{N}(\varepsilon, \mathcal{F})$ and $I(\mathcal{F})$. Although it is usually claimed in the Monte Carlo literature that the convergence of Monte Carlo methods is dimension-free, previous considerations clearly indicate that this assertion is far from being true for empirical approximation processes.

Let $(E_n, \mathcal{E}_n)_{n=0,1}$ be a pair of measurable spaces and let $\mathcal{F} \subset \mathcal{B}_b(E_1)$. Also let M be a Markov kernel from $(E_0, \mathcal{E}_0)$ into $(E_1, \mathcal{E}_1)$ and $G : E_0 \to \mathbb{R}$ an $\mathcal{E}_0$-measurable function with $\|G\| \leq 1$. We associate with the triplet $(\mathcal{F}, G, M)$ the collection of $\mathcal{E}_0$-measurable functions

$$G \cdot M\mathcal{F} = \{G \cdot M(f); f \in \mathcal{F}\} \subset \mathcal{B}_b(E_0)$$

Lemma 7.3.4 *For any $p \geq 1$, $\varepsilon > 0$, and $\nu \in \mathcal{P}(E_0)$, we have*

$$\mathcal{N}(\varepsilon, G \cdot M\mathcal{F}, L_p(\nu)) \leq \mathcal{N}(\varepsilon, \mathcal{F}, L_p(\nu M))$$

Therefore we find that

$$\mathcal{N}(\varepsilon, G \cdot M\mathcal{F}) \leq \mathcal{N}(\varepsilon, \mathcal{F}) \quad \text{and} \quad I(G \cdot M\mathcal{F}) \leq I(\mathcal{F})$$

Proof:
Lemma 7.3.4 follows from the fact that

$$\mathcal{N}(\varepsilon, \, G \cdot \mathcal{F}, \, L_p(\nu)) \leq \mathcal{N}(\varepsilon, \, \mathcal{F}, \, L_p(\nu))$$
$$\mathcal{N}(\varepsilon, \, M\mathcal{F}, \, L_p(\nu)) \leq \mathcal{N}(\varepsilon, \, \mathcal{F}, \, L_p(\nu M))$$

The first assertion is obvious. To establish the second inequality, simply note that, for every function f, $|M(f)|^p \leq M(|f|^p)$ and go back to the definition of the covering numbers. This ends the proof of the lemma. ∎

7.3 Inequalities for Independent Random Variables

Lemma 7.3.5 *For any $p \geq 1$, we have*
$$\sqrt{N}\, \mathbb{E}\left(\|m(X)-\mu\|_{\mathcal{F}}^p\right)^{\frac{1}{p}} \leq c\, [p/2]!\, I(\mathcal{F})$$

Proof:
We consider a collection of independent copies $X' = (X'^i)_{i\geq 1}$ of the random variables $X = (X^i)_{i\geq 1}$. Let $\varepsilon = (\varepsilon_i)_{i\geq 1}$ constitute a sequence that is independent and identically distributed with $P(\varepsilon_1 = +1) = P(\varepsilon_1 = -1) = 1/2$. We also assume that (ϵ, X, X') are independent. We associate with the pairs (ϵ, X) and (ϵ, X') the random measures $m_\epsilon(X) = \frac{1}{N}\sum_{i=1}^N \epsilon_i\, \delta_{X^i}$ and $m_\epsilon(X') = \frac{1}{N}\sum_{i=1}^N \epsilon_i\, \delta_{X'^i}$. Notice that

$$\|m(X)-\mu\|_{\mathcal{F}}^p = \sup_{f\in\mathcal{F}} |m(X)(f) - \mathbb{E}(m(X')(f))|^p \leq \mathbb{E}(\|m(X)-m(X')\|_{\mathcal{F}}^p \mid X)$$

and in view of the symmetry of the random variables $(f(X^i) - f(X'^i))_{i\geq 1}$ we have

$$\mathbb{E}(\|m(X)-m(X')\|_{\mathcal{F}}^p) = \mathbb{E}(\|m_\epsilon(X)-m_\epsilon(X')\|_{\mathcal{F}}^p)$$

This implies that

$$E\left(\|m(X)-\mu\|_{\mathcal{F}}^p\right) \leq 2^p\, E\left(\|m_\epsilon(X)\|_{\mathcal{F}}^p\right)$$

By using the Chernov-Hoeffding inequality for any $x^1,\ldots,x^N \in E$, the empirical process

$$f \longrightarrow \sqrt{N}\, m_\varepsilon(x)(f)$$

is sub-Gaussian for the norm $\|f\|_{L_2(m(x))} = m(x)(f^2)^{1/2}$. Namely, for any $f, g \in \mathcal{F}$ and $\delta > 0$

$$P\left(\sqrt{N}\, |m_\varepsilon(x)(f) - m_\varepsilon(x)(g)| \geq \delta\right) \leq 2\, e^{-\frac{1}{2}\delta^2/\|f-g\|_{L_2(m(x))}^2}$$

Using the maximal inequality for sub-Gaussian processes and the fact that $0 \in \mathcal{F}$, we arrive at

$$\sqrt{N}\, \pi_\psi(\|m_\varepsilon(x)\|_{\mathcal{F}}) \leq c \int_0^\infty \sqrt{\log\left(1 + \mathcal{D}(\varepsilon, \mathcal{F}, \|\cdot\|_{L_2(m(x))})\right)}\, d\varepsilon$$

where

- $\pi_\psi[Y]$ is the Orlicz norm of a random variable Y associated with the increasing convex function $\psi(u) = e^{u^2} - 1$ (and $\psi^{-1}(u) = \sqrt{\log(1+u)}$) and defined by

$$\pi_\psi(Y) = \inf\{a \in (0,\infty) : \mathbb{E}(\psi(|Y|/c)) \leq 1\}$$

- $\mathcal{D}(\varepsilon, \mathcal{F}, \|\cdot\|_{L_2(\mu)})$ is the maximum number of ε-separated points in the metric space $(\mathcal{F}, \|\cdot\|_{L_2(\mu)})$ and c is a universal constant (see for instance Corollary 2.2.8 in [311]).

230   7. Asymptotic Behavior

On the other hand, by a simple calculation, we see that

$$\mathcal{D}(2\,\varepsilon, \mathcal{F}, \|\cdot\|_{L_2(m(x))}) \leq \mathcal{N}\left(\varepsilon, \mathcal{F}, \|\cdot\|_{L_2(m(x))}\right) \leq \sup_{y \in E^N} \mathcal{N}\left(\varepsilon, \mathcal{F}, \|\cdot\|_{L_2(m(y))}\right)$$

Recalling that $\mathbb{E}(|Y|^p)^{1/p} \leq [p/2]!\,\pi_\psi(Y)$, for $p \geq 1$, we arrive at

$$\sqrt{N}\,\mathbb{E}\left(\|m_\varepsilon(X)\|_\mathcal{F}^p\right)^{1/p}$$
$$\leq\ c\,[p/2]! \int_0^\infty \sup_{y \in E^N} \sqrt{\log\left(1 + \mathcal{N}(\tfrac{\varepsilon}{2}, \mathcal{F}, \|\cdot\|_{L_2(m(y))})\right)}\,d\varepsilon$$

and therefore

$$\mathbb{E}\left(\|m(X) - \mu\|_\mathcal{F}^p\right)^{1/p}$$
$$\leq \tfrac{c}{\sqrt{N}}\,[p/2]! \int_0^\infty \sup_{y \in E^N} \sqrt{\log\left(1 + \mathcal{N}(\varepsilon, \mathcal{F}, \|\cdot\|_{L_2(m(y))})\right)}\,d\varepsilon$$

Under our assumptions, if ε is larger than 1, then $\mathcal{F}$ fits in a single $L_2(\mu)$-ball of radius ε around the origin, for any $\mu \in \mathcal{P}(E)$. The end of the proof is now straightforward. ∎

Lemma 7.3.6 *For any $\varepsilon > 0$ and $\sqrt{N} \geq 4\varepsilon^{-1}$, we have that*

$$\mathbb{P}\left(\|m(X) - \mu\|_\mathcal{F} > 8\,\varepsilon\right) \leq 8\,\mathcal{N}(\varepsilon, \mathcal{F})\,e^{-N\varepsilon^2/2}$$

Proof:
Using classical symmetrization inequalities (see Lemma 2.3.7 in [311] or pp. 14-15 in [271]), for any $\varepsilon > 0$ and $\sqrt{N} \geq 4\varepsilon^{-1}$,

$$\mathbb{P}\left(\|m(X) - \mu\|_\mathcal{F} > \varepsilon\right) \leq 4\mathbb{P}\left(\|m_\varepsilon(X)\|_\mathcal{F} > \frac{\varepsilon}{4}\right) \tag{7.8}$$

where $m_\varepsilon(X)$ denotes the signed measure $m_\varepsilon(X) = \frac{1}{N}\sum_{i=1}^N \varepsilon_i \delta_{X^i}$ and $\{\varepsilon_1, \ldots, \varepsilon_N\}$ are symmetric Bernoulli random variables, independent of the X^i's. Conditionally on the X^i's, and by the definition of the covering numbers, we easily get by a standard argument that

$$\mathbb{P}(\,\|m_\varepsilon(X)\|_\mathcal{F} > \delta\,|\,X)$$
$$\leq \mathcal{N}(\delta/2, \mathcal{F}, L^1(m(X)))\,\sup_{f \in \mathcal{F}} P\left(|m_\varepsilon(X)(f)| > \delta/2\,\big|\,X\right) \tag{7.9}$$

Indeed, let $\{f_p;\ 1 \leq p \leq \mathcal{N}(\delta/2, \mathcal{F}, L^1(m(X)))\}$ be a $(\delta/2)$-coverage of $\mathcal{F}$ for the $L^1(m(X))$-norm. Then

$$\mathbb{P}\Big(\|m_\varepsilon(X)\|_\mathcal{F} > \delta\,\big|\,X\Big) \leq P\Big(\sup_p |m_\varepsilon(X)(f_p)| > \delta/2\ \big|\,X\Big)$$

Therefore,

$$\mathbb{P}\Big(\|m_\varepsilon(X)\|_{\mathcal{F}} > \delta \,|\, X\Big)$$

$$\leq \mathcal{N}\big(\delta/2, \mathcal{F}, L^1(m^N(X))\big) \sup_p \mathbb{P}\Big(|m_\varepsilon(X)(f_p)| > \delta/2 \,|\, X\Big)$$

By the Chernov-Hoeffding inequality, for any $f \in \mathcal{F}$ and $\delta > 0$,

$$\mathbb{P}\Big(|m_\varepsilon(X)(f)| > \delta/2 \,|\, X\Big) \leq 2\, e^{-N\delta^2/8}$$

As a consequence, we see that $\mathbb{P}\Big(\|m_\varepsilon(X)\|_{\mathcal{F}} > \frac{\varepsilon}{4} \,|\, X\Big)$ is bounded above by

$$2\mathcal{N}\big(\varepsilon/8, \mathcal{F}, L_1(m(X))\big)\, e^{-N\varepsilon^2/128} \leq 2\,\mathcal{N}(\varepsilon/8, \mathcal{F})\, e^{-N\varepsilon^2/128}$$

From (7.8), it follows that

$$\mathbb{P}\left(\|m(X) - \mu\|_{\mathcal{F}} > \varepsilon\right) \leq 8\,\mathcal{N}(\varepsilon/8, \mathcal{F})\, e^{-N\varepsilon^2/128}$$

as soon as $\sqrt{N} \geq 4\varepsilon^{-1}$, which is the result. This ends the proof of the desired estimate. ∎

7.4 Strong Law of Large Numbers

7.4.1 Extinction Probabilities

The objective of this short section is to estimate the probability of extinction of a class of particle models associated with potential functions that may take null values. The forthcoming developement is valid for any McKean interpretation model of the form

$$K_{n+1,\eta} = S_{n,\eta} M_{n+1}$$

where $S_{n,\eta}$ is a selection transition satisfying the compatibility condition $\eta S_{n,\eta} = \Psi_n(\eta)$, for any distribution η such that $\eta(G_n) > 0$. We also require that $S_{n,\eta}\left(x_n, \widehat{E}_n\right) = 1$, as soon as $\eta(\widehat{E}_n) > 0$, with $\widehat{E}_n =_{\text{def.}} G_n^{-1}(0, \infty)$. These technical requirements are clearly met in the two cases examined on page 219.

The analysis of the extinction probability arises for instance in physics when the particle model evolves in an environment with hard obstacles. For discrete time models, it may happen that at a given date all particles enter into a hard obstacle. At that time, the system dies and the algorithm stops.

232   7. Asymptotic Behavior

The analysis of these stopping times is far from complete, and many questions remain to be answered. In this section, we content ourselves with proving the following rather crude but reassuring result.

Theorem 7.4.1 *Suppose we have $\gamma_n(1) > 0$ for any $n \geq 0$. Then, for any $N \geq 1$ and $n \geq 0$, we have the estimate*

$$\mathbb{P}(\tau^N \leq n) \leq a(n) \, e^{-N/b(n)}$$

In addition, when assumption $(\mathcal{B})$ is satisfied for some collection of positive numbers $(\alpha(n))_{n\geq 0}$, then for any $n \geq 0$ we have

$$\mathbb{P}\left(\tau^N \leq n\right) \leq \sum_{p=0}^{n} e^{-N\alpha(p)}$$

Proof:
Let $\Omega_N(n+1)$ be the set of events defined by

$$\Omega_N(n+1) = \{\forall 0 \leq p < q \leq n+1, \quad |\eta_p^N(Q_{p,q}1) - \eta_p(Q_{p,q}1)| \leq \gamma_q(1)/2\}$$

By the definition of $Q_{p,q}$, we have $\eta_p(Q_{p,q}1) = \gamma_p(Q_{p,q}1)/\gamma_p(1) = \gamma_q(1)/\gamma_p(1)$. Since $\gamma_p(1) \leq 1$, we find that $\eta_p(Q_{p,q}1) \geq \gamma_q(1)$. For the set of events $\Omega_N(n+1)$, the following inequalities hold true for any $0 \leq p < q \leq (n+1)$:

$$0 < \frac{\gamma_q(1)}{2} \leq \eta_p(Q_{p,q}1) - \frac{\gamma_q(1)}{2} \leq \eta_p^N(Q_{p,q}1) \leq \eta_p(Q_{p,q}1) + \frac{\gamma_q(1)}{2} \leq 2$$

Consequently, for $\Omega_N(n+1)$ we have $\eta_p^N(g_p) \geq \frac{\gamma_{n+1}(1)}{2} > 0$ for any $0 \leq p \leq n$. This yields the inclusion $\Omega_N(n+1) \subset \{\tau^N > n\}$. On the other hand, we notice that $\Omega_N(n+1) = \Omega_N(n) \cup \Omega'_N(n)$ with

$$\Omega'_N(n) = \{\forall 0 \leq p \leq n, \quad |\eta_p^N(Q_{p,n+1}1) - \eta_p(Q_{p,n+1}1)| \leq \gamma_{n+1}(1)/2\}$$

For $n=0$, we also find that $\Omega_N(1) = \{|\eta_0^N(G_0) - \eta_0(G_0)| \leq \gamma_1(1)/2\}$. By the definition of η_0^N (and since $\mathrm{osc}(G_0) \leq 1$), using the Chernov-Hoeffding inequality (see Lemma 7.3.2), we prove that $\mathbb{P}(\Omega_N(1)) \geq 1 - 2\exp\left(-N\gamma_1^2(1)/2\right)$. To go a step further, we use the decomposition

$$\begin{aligned}\mathbb{P}(\Omega_N(n+1)) &= \mathbb{P}(\Omega_N(n) \cap \Omega'_N(n)) \\ &= \mathbb{P}(\Omega_N(n)) - \mathbb{P}(\Omega_N(n) \cap \Omega''_N(n))\end{aligned}$$

with

$$\Omega''_N(n) = \{\exists 0 \leq p \leq n, \quad |\eta_p^N(Q_{p,n+1}1) - \eta_p(Q_{p,n+1}1)| > \gamma_{n+1}(1)/2\}$$

It follows that $\mathbb{P}(\Omega_N(n+1)) \geq \mathbb{P}(\Omega_N(n)) - \sum_{p=0}^n \mathbb{P}(\Omega_N(n) \cap \Omega''_N(p,n))$ with

$$\Omega''_N(p,n) = \{|\eta_p^N(Q_{p,n+1}1) - \eta_p(Q_{p,n+1}1)| > \gamma_{n+1}(1)/2\}$$

For the set of events $\Omega_N(n)$, we have for each $0 \leq k < p \leq n$

$$\eta_k^N(Q_{k,p}1) \geq \gamma_p(1)/2 > 0$$

Thus, for the set $\Omega_N(n)$, we have for each $0 < k < p \leq n$

$$\Phi_k(\eta_{k-1}^N)(Q_{k,p}1) = \frac{\eta_{k-1}^N(Q_{k-1,p}(1))}{\eta_{k-1}^N(G_{k-1})} \geq \eta_{k-1}^N(Q_{k-1,p}(1)) \geq \frac{\gamma_p(1)}{2}$$

and for $k = p$ and $k = 0$ we have respectively

$$Q_{p,p}(1) = 1 \implies \Phi_p(\eta_{p-1}^N)(Q_{p,p}1) = 1 \geq \frac{\gamma_p(1)}{2}$$

and $\Phi_0(\eta_{-1}^N)(Q_{0,p}1) = \eta_0(Q_{0,p}1) = \gamma_p(1) \geq \frac{\gamma_p(1)}{2}$. Observe that for the set $\Omega_N(n)$, we have the decomposition

$$\eta_p^N - \eta_p = \sum_{k=0}^{p} [\Phi_{k,p}(\eta_k^N) - \Phi_{k,p}(\Phi_k(\eta_{k-1}^N))]$$

with the conventions $\Phi_{p,p} = Id$ and $\Phi_{-1}(\eta_{-1}^N) = \eta_0$ for $k = p$ and $k = 0$. In addition, we have for each $0 \leq k \leq p$, $f \in \mathcal{B}_b(E_p)$, and $\eta_1, \eta_2 \in \mathcal{P}(E_k)$, such that $\eta_1(Q_{k,p}1) > 0$ and $\eta_2(Q_{k,p}1) > 0$

$$\Phi_{k,p}(\eta_1)(f) - \Phi_{k,p}(\eta_2)(f) = \frac{1}{\eta_2(Q_{k,p}1)} [(\eta_1(Q_{k,p}f) - \eta_2(Q_{k,p}f)) + \Phi_{k,p}(\eta_1)(f) (\eta_2(Q_{k,p}1) - \eta_1(Q_{k,p}1))]$$

We conclude that, for the set of events $\Omega_N(n)$, we have for any $0 \leq p \leq n$

$$\eta_p^N(Q_{p,n+1}1) - \eta_p(Q_{p,n+1}1)$$

$$= \sum_{k=0}^{p} \frac{1}{\Phi_k(\eta_{k-1}^N)(Q_{k,p}1)} [(\eta_k^N(Q_{k,n+1}1) - \Phi_k(\eta_{k-1}^N)(Q_{k,n+1}1))$$

$$+ \Phi_{k,p}(\eta_k^N)(Q_{p,n+1}1) \, (\Phi_k(\eta_{k-1}^N)(Q_{k,p}1) - \eta_k^N(Q_{k,p}1))]$$

This yields that (for the set $\Omega_N(n)$) for any $0 \leq p \leq n$

$$|\eta_p^N(Q_{p,n+1}1) - \eta_p(Q_{p,n+1}1)|$$

$$\leq \frac{2}{\gamma_p(1)} \sum_{k=0}^{p} [|\eta_k^N(Q_{k,n+1}1) - \Phi_k(\eta_{k-1}^N)(Q_{k,n+1}1)|$$

$$+ |\eta_k^N(Q_{k,p}1) - \Phi_k(\eta_{k-1}^N)(Q_{k,p}1)|]$$

It also follows that for the set $\Omega_N(n) \cap \Omega_N''(p,n)$ there exists some index k, $0 \le k \le p \le n$, such that

$$|\eta_k^N(Q_{k,n+1}1) - \Phi_k(\eta_{k-1}^N)(Q_{k,n+1}1)| \ge \frac{\gamma_{n+1}(1)\gamma_p(1)}{8(p+1)} \ge \frac{\gamma_{n+1}^2(1)}{8(n+1)}$$

and $|\Phi_k(\eta_{k-1}^N)(Q_{k,p}1) - \eta_k^N(Q_{k,p}1)| \ge \frac{\gamma_{n+1}^2(1)}{8(n+1)}$. If it will not be the case, we will have a contradiction. We conclude that for any $0 \le p \le n$

$$\mathbb{P}(\Omega_N(n) \cap \Omega_N''(p,n))$$

$$\le \sum_{k=0}^{p} \left[\mathbb{P}\left(\Omega_N(n) \cap \left\{|\eta_k^N(Q_{k,n+1}1) - \Phi_k(\eta_{k-1}^N)(Q_{k,n+1}1)| \ge \frac{\gamma_{n+1}^2(1)}{8(n+1)}\right\}\right) \right.$$

$$\left. + \mathbb{P}\left(\Omega_N(n) \cap \left\{|\eta_k^N(Q_{k,p}1) - \Phi_k(\eta_{k-1}^N)(Q_{k,p}1)| \ge \frac{\gamma_{n+1}^2(1)}{8(n+1)}\right\}\right) \right]$$

To end the proof, we recall that $\Omega_N(n) \subset \{\tau^N \ge n\} \subset \{\tau^N \ge k\}$ for any $0 \le k \le n$ so that

$$\mathbb{P}\left(\Omega_N(n) \cap \left\{|\eta_k^N(Q_{k,n+1}1) - \Phi_k(\eta_{k-1}^N)(Q_{k,n+1}1)| \ge \frac{\gamma_{n+1}^2(1)}{8(n+1)}\right\}\right)$$

$$\le \mathbb{P}\left(\tau^N \ge k \text{ and } |\eta_k^N(Q_{k,n+1}1) - \Phi_k(\eta_{k-1}^N)(Q_{k,n+1}1)| \ge \frac{\gamma_{n+1}^2(1)}{8(n+1)}\right)$$

$$\le \mathbb{E}\left(\mathbb{P}\left(\left\{|\eta_k^N(Q_{k,n+1}1) - \Phi_k(\eta_{k-1}^N)(Q_{k,n+1}1)| \ge \frac{\gamma_{n+1}^2(1)}{8(n+1)}\right\} \mid \eta_{k-1}^N\right) 1_{\tau^N \ge k}\right)$$

$$\le 2\exp\left(-2N \left(\gamma_{n+1}^2(1)/[8(n+1)\mathrm{osc}(Q_{k,n+1}1)]\right)^2\right) \mathbb{P}(\tau^N \ge k)$$

and since $\mathrm{osc}(Q_{k,n+1}1) \le 1$

$$\mathbb{P}\left(\Omega_N(n) \cap \left\{|\eta_k^N(Q_{k,n+1}1) - \Phi_k(\eta_{k-1}^N)(Q_{k,n+1}1)| \ge \frac{\gamma_{n+1}^2(1)}{8(n+1)}\right\}\right)$$

$$\le 2\exp\left(-\tfrac{N}{32}\left(\gamma_{n+1}^2(1)/(n+1)\right)^2\right) \mathbb{P}(\tau^N \ge k)$$

The last displayed estimates are proved by using the Chernov-Hoeffding inequality. This readily implies that for each $0 \le k \le n$

$$\mathbb{P}\left(\Omega_N(n) \cap \left\{|\eta_k^N(Q_{k,n+1}1) - \Phi_k(\eta_{k-1}^N)(Q_{k,n+1}1)| \ge \frac{\gamma_{n+1}^2(1)}{8(n+1)}\right\}\right)$$

$$\le 2\exp\left(-\tfrac{N}{32}\left(\gamma_{n+1}^2(1)/(n+1)\right)^2\right)$$

and similarly for each $0 \leq k \leq p \leq n$

$$\mathbb{P}\left(\Omega_N(n) \cap \left\{|\eta_k^N(Q_{k,p}1) - \Phi_k(\eta_{k-1}^N)(Q_{k,p}1)| \geq \frac{\gamma_{n+1}^2(1)}{8(n+1)}\right\}\right)$$

$$\leq 2\exp\left(-\frac{N}{32}\left(\gamma_{n+1}^2(1)/(n+1)\right)^2\right)$$

Using these two upper bounds, we find that for any $0 \leq p \leq n$

$$\mathbb{P}(\Omega_N(n) \cap \Omega_N''(p,n)) \leq 4(n+1)\exp\left(-\frac{N}{32}\left(\gamma_{n+1}^2(1)/(n+1)\right)^2\right)$$

and finally

$$\mathbb{P}(\Omega_N(n+1)) \geq \mathbb{P}(\Omega_N(n)) - 4(n+1)^2 \exp\left(-\frac{N}{32}\left(\gamma_{n+1}^2(1)/(n+1)\right)^2\right)$$

$$\geq \mathbb{P}(\Omega_N(1)) - 4(n+1)^3 \exp\left(-\frac{N}{32}\left(\gamma_{n+1}^2(1)/(n+1)\right)^2\right)$$

$$\geq 1 - 8(n+1)^3 \exp\left(-\frac{N}{32}\left(\gamma_{n+1}^2(1)/(n+1)\right)^2\right)$$

This ends the proof of the first assertion. We prove the second one by a simple induction argument. First we observe that

$$\mathbb{P}\left(\tau^N > n\right) = \mathbb{P}\left(\tau^N > n-1 \text{ and } \eta_n^N(G_n) > 0\right)$$
$$= \mathbb{E}\left(\mathbb{P}\left(\exists\, 1 \leq i \leq N \text{ s.t. } \xi_n^i \in \widehat{E}_n \,\middle|\, \mathcal{F}_{n-1}^N\right) 1_{\tau^N > n-1}\right)$$
$$= \mathbb{P}\left(\tau^N > n-1\right)$$
$$- \mathbb{E}\left(\mathbb{P}\left(\forall\, 1 \leq i \leq N \quad \xi_n^i \notin \widehat{E}_n \,\middle|\, \mathcal{F}_{n-1}^N\right) 1_{\tau^N > n-1}\right)$$

By the definition of the particle models, we have

$$\mathbb{P}\left(\forall\, 1 \leq i \leq N \quad \xi_n^i \notin \widehat{E}_n \,\middle|\, \mathcal{F}_{n-1}^N\right) 1_{\tau^N > n-1}$$
$$= \prod_{i=1}^{N} (S_{n-1,m(\xi_{n-1})} M_n)\left(\xi_{n-1}^i, E_n - \widehat{E}_n\right) 1_{\tau^N > n-1}$$

Since, for any $\eta \in \mathcal{P}(E_n)$ with $\eta(\widehat{E}_n) > 0$, we have $S_{n,\eta}\left(x_n, \widehat{E}_n\right) = 1$, for any $x_n \in E_n$, we readily get the estimate $S_{n-1,\eta} M_n\left(x_{n-1}, E_n - \widehat{E}_n\right) \leq e^{-\alpha(n)}$, for any $x_{n-1} \in E_{n-1}$, from which we conclude that

$$\mathbb{P}\left(\tau^N > n\right) \geq \mathbb{P}\left(\tau^N > n-1\right) - \mathbb{P}\left(\tau^N > n-1\right) e^{-N\alpha(n)}$$
$$\geq \mathbb{P}\left(\tau^N > n-1\right) - e^{-N\alpha(n)} \geq 1 - \sum_{p=0}^{n} e^{-N\alpha(p)}$$

and the end of the proof of the theorem is completed. ∎

7.4.2 Convergence of Empirical Processes

This section contains several martingale decompositions for a general class of particle approximation Feynman-Kac models. These key martingales will provide precise estimates on the convergence of particle density profiles when the size of the system tends to infinity. They also introduce martingale techniques and stochastic calculus tools into the numerical analysis of these algorithms. The asymptotic analysis presented in this section is valid for any McKean interpretation model statisfying the technical requirements stated in the beginning of Section 7.4.1.

We start with the analysis of the unnormalized particle models and we show that this approximation particle model has no bias. The central idea consists in expressing the difference between the particle measures and the limiting Feynman-Kac ones as end values of martingale sequences. These natural martingales are built using the semigroup structure of the unnormalized Feynman-Kac flow. We also examine the consequences of this result in the estimation of the extinction probabilities and in the analysis of the normalized particle model.

Proposition 7.4.1 *For each $n \geq 0$ and $f_n \in B_b(E_n)$, we let $\Gamma^N_{\bullet,n}(f_n)$ be the $\mathbb{R}$-valued process defined by*

$$\Gamma^N_{\bullet,n}(f_n) : p \in \{0, \ldots, n\} \to \Gamma^N_{p,n}(f_n) = \gamma^N_p (Q_{p,n} f_n) \; 1_{\tau^N \geq p} - \gamma_p (Q_{p,n} f_n)$$

For any $p \leq n$, $\Gamma^N_{\bullet,n}(f)$ has the $\mathcal{F}^N$-martingale decomposition

$$\Gamma^N_{p,n}(f_n) = \sum_{q=0}^{p} \gamma^N_q(1) \, 1_{\tau^N \geq q} \left[\eta^N_q (Q_{q,n} f_n) - \eta^N_{q-1} K_{q, \eta^N_{q-1}} (Q_{q,n} f_n) \right]$$
(7.10)

and its angle bracket is given by

$$\left\langle \Gamma^N_{\bullet,n}(f_n) \right\rangle_p$$

$$= \frac{1}{N} \sum_{q=0}^{p} \gamma^N_q(1)^2 \; 1_{\tau^N \geq q} \; \eta^N_{q-1}(K_{q, \eta^N_{q-1}}[Q_{q,n} f_n - K_{q, \eta^N_{q-1}} Q_{q,n} f_n]^2)$$
(7.11)

with the convention for $q = 0$, $\eta^N_{-1} = \Phi\left(\eta^N_{-1}\right) = K_{\eta^N_{-1}} = \eta_0$.

Before getting into the proof of this proposition, it is interesting to make some remarks:

- We first observe that these martingale decompositions provide sharp estimates of $\mathbb{L}^2$ mean error between the particle approximation measures γ^N_n and the unnormalized Feynman-Kac measures γ_n. Indeed by (7.11) we have that for any $n \in \mathbb{N}$ and $f_n \in \mathcal{B}_b(E_n)$

$$\sup_{0 \leq q \leq n} N \, \mathbb{E}([\gamma_q^N(Q_{q,n}f_n) \, 1_{\tau^N \geq q} - \gamma_q(Q_{q,n}f_n)]^2)$$

$$= \sum_{q=0}^{n} \mathbb{E}(\gamma_q^N(1)^2 \, 1_{\tau^N \geq q} \, \eta_{q-1}^N(K_{q,\eta_{q-1}^N}[Q_{q,n}f_n - K_{q,\eta_{q-1}^N}Q_{q,n}f_n]^2))$$
(7.12)

- The estimates presented in Proposition 7.4.1 also allow us to initiate a comparison between the particle approximation models associated with the two cases introduced on page 219. In the first situation, we observe that for any $q \geq 1$, $\eta \in \mathcal{P}(E_{q-1})$, and $\varphi \in \mathcal{B}_b(E_q)$, we have

$$K_{q,\eta}[\varphi - K_{q,\eta}(\varphi)]^2 = \Phi_q(\eta)[\varphi - \Phi_q(\eta)(\varphi)]^2$$

while in the second one

$$\eta K_{q,\eta}[\varphi - K_{q,\eta}\varphi]^2$$

$$= \eta K_{q,\eta}[\varphi - \Phi_q(\eta)(\varphi)]^2 - \eta[\Phi_q(\eta)(\varphi) - K_{q,\eta}(\varphi)]^2$$

$$= \Phi_q(\eta)[\varphi - \Phi_q(\eta)(\varphi)]^2 - \eta(G^2 \, [\Phi_q(\eta)(\varphi) - M_q(\varphi)]^2)$$

$$\leq \Phi_q(\eta)[\varphi - \Phi_q(\eta)(\varphi)]^2 \tag{7.13}$$

This simple observation indicates that the particle model in the second case is more accurate than the other one. For instance, suppose that the potential functions reduce to $G_n = 1$ and the mutation transitions are "trivial" in the sense that $(E_n, M_n) = (E, Id)$. In this rather degenerate situation, we have in the first case

$$N \, \mathbb{E}([\eta_n^N(f) - \eta_n(f)]^2)$$

$$= \eta_0([f - \eta_0(f)]^2) + \sum_{q=0}^{n-1} \mathbb{E}(\eta_q^N[f - \eta_q^N(f)]^2)$$

Using the fact that

$$\mathbb{E}(\eta_n^N(f)) = \eta_0(f)$$
$$\mathbb{E}(\eta_n^N(f)^2) = \left(1 - \frac{1}{N}\right) \mathbb{E}(\eta_{n-1}^N(f)^2) + \frac{1}{N} \eta_0(f^2)$$

after some elementary manipulations, we find that

$$N \, \mathbb{E}([\eta_n^N(f) - \eta_n(f)]^2) = \eta_0([f - \eta_0(f)]^2) \left(1 + \sum_{p=0}^{n-1}(1 - \frac{1}{N})^p\right)$$

In the second case, we have $K_{n,\eta} = Id$, and therefore

$$N \, \mathbb{E}([\eta_n^N(f) - \eta_n(f)]^2) = \eta_0([f - \eta_0(f)]^2)$$

- The third, more classical observation is that we can reverse the expectation and the supremum operators in (7.12) by a simple application of Doob's maximal inequality. More precisely, for any $p > 1$, we have that

$$\mathbb{E}\left(\sup_{0 \leq q \leq n} |\Gamma_{q,n}^N(f_n)|^p\right)^{\frac{1}{p}} \leq \frac{p}{p-1} \mathbb{E}(|\Gamma_{n,n}^N(f_n)|^p)^{\frac{1}{p}}$$

$$= \frac{p}{p-1} \mathbb{E}(|\gamma_n^N(f_n) 1_{\tau^N \geq n} - \gamma_n(f_n)|^p)^{\frac{1}{p}}$$

(7.14)

To prove this traditional martingale inequality, we use the fact that for any nonnegative random variable U and for any $p > 0$ we have

$$\mathbb{E}(U^p) = p \int_0^\infty t^{p-1} \mathbb{P}(U \geq t) \, dt$$

If we set $U_n^\star = \sup_{0 \leq q \leq n} U_{q,n}$ and $U_{q,n} = |\Gamma_{q,n}^N(f_n)|$ then we readily check that $(U_{q,n})_{q \leq n}$ is an $(\mathcal{F}_q^N)_{q \leq n}$-submartingale and by Doob's maximal inequality we have for any $t > 0$

$$t \, \mathbb{P}(U_n^\star \geq t) \leq \mathbb{E}(U_{n,n} 1_{U_n^\star \geq t})$$

This yields that

$$\mathbb{E}((U_n^\star)^p) = p \int_0^\infty t^{p-1} \mathbb{P}(U_n^\star \geq t) \, dt$$

$$\leq p \, \mathbb{E}\left[U_{n,n} \int_0^\infty t^{p-2} 1_{U_n^\star \geq t} \, dt\right]$$

$$= p \, \mathbb{E}\left[U_{n,n} \int_0^{U_n^\star} t^{p-2} \, dt\right] \leq \frac{p}{p-1} \mathbb{E}(U_{n,n}(U_n^\star)^{p-1})$$

Now, by Holder's inequality, we conclude that

$$\mathbb{E}((U_n^\star)^p) \leq \frac{p}{p-1} \mathbb{E}(U_{n,n}^p)^{\frac{1}{p}} \mathbb{E}((U_n^\star)^p)^{1-\frac{1}{p}}$$

which ends the proof of (7.14). Using Lemma 7.3.3, we immediately obtain the crude estimate

$$\sqrt{N} \, \mathbb{E}(\sup_{0 \leq q \leq n} |\Gamma_{q,n}^N(f_n)|^p)^{\frac{1}{p}} \leq \frac{p \, d^{\frac{1}{p}}(p)}{p-1} \sum_{q=0}^n \mathrm{osc}(Q_{q,n}(f_n))$$

Proof of Proposition 7.4.1: We use the decomposition for each $\varphi \in \mathcal{B}_b(E_p)$

$$\gamma_p^N(\varphi) 1_{\tau^N \geq p} - \gamma_p(\varphi) = \sum_{q=0}^p \left[\gamma_q^N(Q_{q,p}\varphi) 1_{\tau^N \geq q} - \gamma_{q-1}^N(Q_{q-1,p}\varphi) 1_{\tau^N \geq q-1}\right]$$

(7.15)

with the convention for $q = 0$, $\gamma_{-1}^N (Q_{-1,p}\varphi) \, 1_{\tau^N \geq -1} = \gamma_p(\varphi)$. Observe that

$$\gamma_q^N (Q_{q,p}\varphi) \, 1_{\tau^N \geq q} = \gamma_q^N (1) \, 1_{\tau^N \geq q} \times \eta_q^N (Q_{q,p}\varphi) \qquad (7.16)$$

and

$$\gamma_{q-1}^N (Q_{q-1,p}\varphi) \, 1_{\tau^N \geq q-1} = \gamma_{q-1}^N (G_{q-1} M_q (Q_{q,p}\varphi)) \, 1_{\tau^N \geq q-1}$$
$$= \gamma_{q-1}^N (1) \, 1_{\tau^N \geq q-1} \times \eta_{q-1}^N (G_{q-1} M_q (Q_{q,p}\varphi))$$

Since $1 = 1_{\eta_{q-1}^N(G)=0} + 1_{\eta_{q-1}^N(G)>0}$, $1_{\eta_{q-1}^N(G)>0} \times 1_{\tau^N \geq q-1} = 1_{\tau^N \geq q}$, and

$$\eta_{q-1}^N (G_{q-1} M_q (Q_{q,p}\varphi)) \, 1_{\eta_{q-1}^N(G)=0} = 0$$

we conclude that

$$\gamma_{q-1}^N (Q_{q-1,p}\varphi) \, 1_{\tau^N \geq q-1} = \gamma_{q-1}^N (1) \, 1_{\tau^N \geq q} \, \eta_{q-1}^N (Q_q (Q_{q,p}\varphi))$$
$$= \gamma_q^N (1) \, 1_{\tau^N \geq q} \, \Phi_q (\eta_{q-1}^N) (Q_{q,p}\varphi)$$

If we set $\varphi = Q_{p,n}(f)$, for some $f \in \mathcal{B}_b(E_n)$, we find that

$$\gamma_p^N (Q_{p,n}f) \, 1_{\tau^N \geq p} - \gamma_p (Q_{p,n}f)$$
$$= \sum_{q=0}^{p} \gamma_q^N (1) 1_{\tau^N \geq q} \left[\eta_q^N (Q_{q,n}f) - \eta_{q-1}^N K_{q,\eta_{q-1}^N} (Q_{q,n}f) \right]$$

The end of the proof is now clear. ∎

Theorem 7.4.2 *For each $p \geq 1$, $n \in \mathbb{N}$, and for any (separable) collection $\mathcal{F}_n$ of measurable functions $f : E_n \to \mathbb{R}$ such that $\|f\| \leq 1$ (and $1 \in \mathcal{F}_n$), we have for any $f \in \mathcal{F}_n$*

$$\mathbb{E}\left(\gamma_n^N (f) 1_{\tau^N \geq n}\right) = \gamma_n (f) \qquad (7.17)$$

and for any $r \leq n$

$$\sqrt{N} \, \mathbb{E}(\|1_{\tau^N \geq r} \gamma_r^N Q_{r,n} - \gamma_r Q_{r,n}\|_{\mathcal{F}_n}^p)^{1/p} \leq c \, (n+1) \, [p/2]! \, I(\mathcal{F}_n) \qquad (7.18)$$

In addition, for any $\epsilon \geq 4/\sqrt{N}$, we have the exponential estimate

$$\mathbb{P}\left(\|1_{\tau^N \geq r} \gamma_r^N Q_{r,n} - \gamma_r Q_{r,n}\|_{\mathcal{F}_n} > \varepsilon\right) \leq 8(n+1) \, \mathcal{N}(\varepsilon_n, \mathcal{F}_n) \, e^{-N\varepsilon_n^2/2} \qquad (7.19)$$

with $\varepsilon_n = \varepsilon/(n+1)$.

Proof:
The first assertion is a simple consequence of Proposition 7.4.1. Using the martingale decomposition (7.10), we find that

$$\|1_{\tau^N \geq r} \gamma_r^N Q_{r,n} - \gamma_r Q_{r,n}\|_{\mathcal{F}_n}$$

$$\leq \sum_{q=0}^{r} \gamma_q^N(1) \, 1_{\tau^N \geq q} \, \|\eta_q^N - \eta_{q-1}^N K_{q,\eta_{q-1}^N}\|_{\mathcal{F}_{q,n}}$$

$$\leq \sum_{q=0}^{r} 1_{\tau^N \geq q} \, \|\eta_q^N - \eta_{q-1}^N K_{q,\eta_{q-1}^N}\|_{\mathcal{F}_{q,n}}$$

with $\mathcal{F}_{q,n} = \{Q_{q,n}(f) : f \in \mathcal{F}_n\}$. This implies that

$$\mathbb{E}(\|1_{\tau^N \geq r}\gamma_r^N Q_{r,n} - \gamma_r Q_{r,n}\|_{\mathcal{F}_n}^p)^{1/p}$$

$$\leq \sum_{q=0}^{r} \mathbb{E}(1_{\tau^N \geq q} \|\eta_q^N - \eta_{q-1}^N K_{q,\eta_{q-1}^N}\|_{\mathcal{F}_{q,n}}^p)^{1/p}$$

By Lemma 7.3.5, we have for any $r \leq n$

$$\sqrt{N} \; \mathbb{E}(\|\eta_r^N - \eta_{r-1}^N K_{r,\eta_{r-1}^N}\|_{\mathcal{F}_{r,n}}^p \mid F_{r-1}^N)^{1/p} \, 1_{\tau^N \geq r} \leq c \, [p/2]! \, I(\mathcal{F}_{r,n})$$

Now, by Lemma 7.3.4, we find that $I(\mathcal{F}_{r,n}) \leq I(\mathcal{F}_n)$ and we conclude that

$$\sqrt{N} \; \mathbb{E}(\|1_{\tau^N \geq r}\gamma_r^N Q_{r,n} - \gamma_r Q_{r,n}\|_{\mathcal{F}_n}^p)^{1/p} \leq c \, (n+1) \, [p/2]! \, I(\mathcal{F}_n)$$

Using the inequality (7.4.2), we prove that for every $\varepsilon > 0$

$$\mathbb{P}(\|1_{\tau^N \geq r}\gamma_r^N Q_{r,n} - \gamma_r Q_{r,n}\|_{\mathcal{F}_n} > (n+1)\varepsilon)$$

$$\leq (n+1) \sup_{0 \leq q \leq n} \mathbb{P}(1_{\tau^N \geq q} \|\eta_q^N - \eta_{q-1}^N K_{q,\eta_{q-1}^N}\|_{\mathcal{F}_{q,n}} > \varepsilon)$$

By Lemma 7.3.6 and Lemma 7.3.4, we have for any $q \leq n$ the exponential estimate

$$\mathbb{P}(1_{\tau^N \geq q} \|\eta_q^N - \eta_{q-1}^N K_{q,\eta_{q-1}^N}\|_{\mathcal{F}_{q,n}} > \varepsilon \mid \mathcal{F}_{q-1}^N) \leq 8 \, \mathcal{N}(\varepsilon, \mathcal{F}_n) \, e^{-N\varepsilon^2/2}$$

as soon as $\sqrt{N} \geq 4 \, \varepsilon^{-1}$. This clearly yields that

$$\mathbb{P}(\|1_{\tau^N \geq r}\gamma_r^N Q_{r,n} - \gamma_r Q_{r,n}\|_{\mathcal{F}_n} > (n+1)\varepsilon) \leq 8(n+1) \, \mathcal{N}(\varepsilon, \mathcal{F}_n) \, e^{-N\varepsilon^2/2}$$

and the proof of the theorem is completed. ■

Corollary 7.4.1 *For any $p \geq 1$, we have*

$$\mathbb{P}\left(\gamma_n^N(1) \, 1_{\tau^N \geq n} \geq \gamma_n(1)/2\right) \geq 1 - a(p) \, \frac{b^p(n)}{N^{p/2}}$$

for some finite constant $b(n) \leq (n+1)/\gamma_n(1)$. In addition, for any pair (n, N) such that $\sqrt{N} \geq 8/\gamma_n(1)$, we have the exponential estimate

$$\mathbb{P}\left(1_{\tau^N \geq n}\ \gamma_n^N(1) \geq \gamma_n(1)/2\right) \geq 1 - 8(n+1)\ \mathcal{N}(\varepsilon_n, \mathcal{F}_n)\ e^{-N\varepsilon_n^2/2}$$

with $\varepsilon_n = \gamma_n(1)/(2(n+1))$.

Theorem 7.4.3 *Suppose assumption $(\mathcal{B})$ (see p. 220) is satisfied for some constants $\alpha(n) > 0$. For each $n \in \mathbb{N}$ and for any (separable) collection $\mathcal{F}_n$ of measurable functions $f : E_n \to \mathbb{R}$ such that $\|f\| \leq 1$ and $1 \in \mathcal{F}_n$, we have*

$$\sup_{f \in \mathcal{F}_n} \left|\mathbb{E}\left(\eta_n^N(f) 1_{\tau^N \geq n}\right) - \eta_n(f)\right| \leq \frac{b^2(n)}{N} + \sum_{q=0}^{n-1} e^{-N\alpha(q)}$$

and

$$\mathbb{E}(\|1_{\tau^N \geq n}\ \eta_n^N - \eta_n\|_{\mathcal{F}_n}^2)^{1/2} \leq \frac{b(n)}{\sqrt{N}}\ (1 + I(\mathcal{F}_n)) + \sum_{q=0}^{n-1} e^{-N\alpha(q)} \quad (7.20)$$

for some finite constant $b(n) \leq c \cdot (n+1)/\gamma_n(1)$. In addition, for any $\varepsilon \in (0,1)$ and $\sqrt{N} \geq 48/(\varepsilon \gamma_n(1))$, we have the exponential estimate

$$\mathbb{P}\left(\|1_{\tau^N \geq n}\ \eta_n^N - \eta_n\|_{\mathcal{F}_n} > \varepsilon\right) \leq 16(n+1)\ \mathcal{N}(\varepsilon_n, \mathcal{F}_n)\ e^{-N\varepsilon_n^2/2} + \sum_{q=0}^{n-1} e^{-\alpha(q)N}$$
(7.21)

with $\varepsilon_n = \varepsilon\ \gamma_n(1)/(12(n+1))$.

Before getting into the proof of the theorem, it is convenient to note that for strictly positive potential functions we have $\widehat{E}_n = E_n$ and condition $(\mathcal{B})$ is met for any $\alpha(n)$. In this particular situation, we have $\tau^N = \infty$ and the estimates (7.20) and (7.21) are valid for any $\alpha(n)$. Letting $\alpha(n) \to \infty$, we find that these estimates hold true without the very r.h.s. term. Another simple consequence of Theorem 7.4.3 is the following extension of the Glivenko-Cantelli theorem to particle models.

Corollary 7.4.2 *Assume that condition $(\mathcal{B})$ is satisfied, and let $\mathcal{F}_n$ be a countable collection of functions f such that $\|f_n\| \leq 1$ and $\mathcal{N}(\varepsilon_n, \mathcal{F}_n)$ for any $\varepsilon > 0$. Then, for any time $n \geq 0$, $\|1_{\tau^N \geq n} \eta_n^N - \eta_n\|_{\mathcal{F}}$ converges almost surely to 0 as $N \to \infty$.*

Proof of Theorem 7.4.3: We use the decomposition

$$\left(\eta_n^N(f) - \eta_n(f)\right) 1_{\tau^N \geq n} = \left(\frac{\gamma_n^N(f)}{\gamma_n^N(1)} - \frac{\gamma_n(f)}{\gamma_n(1)}\right) 1_{\tau^N \geq n}$$

$$= \frac{\gamma_n(1)}{\gamma_n^N(1)}\ \gamma_n^N\left(\frac{1}{\gamma_n(1)}\ (f - \eta_n(f))\right) 1_{\tau^N \geq n}$$
(7.22)

242   7. Asymptotic Behavior

If we set $f_n = \frac{1}{\gamma_n(1)}(f - \eta_n(f))$, then, since $\gamma_n(f_n) = 0$, (7.22) also reads

$$\left(\eta_n^N(f) - \eta_n(f)\right) 1_{\tau^N \geq n} = \frac{\gamma_n(1)}{\gamma_n^N(1)} \left(\gamma_n^N(f_n) 1_{\tau^N \geq n} - \gamma_n(f_n)\right) 1_{\tau^N \geq n}$$

By Proposition 7.4.1, we have $\mathbb{E}\left(\gamma_n^N(f_n) 1_{\tau^N \geq n}\right) = \gamma_n(f_n)$. This implies that

$$\mathbb{E}\left(\left(\eta_n^N(f) - \eta_n(f)\right) 1_{\tau^N \geq n}\right)$$

$$= \mathbb{E}\left(\left(\frac{\gamma_n(1)}{\gamma_n^N(1)} - 1\right) \left(\gamma_n^N(f_n) 1_{\tau^N \geq n} - \gamma_n(f_n)\right) 1_{\tau^N \geq n}\right)$$

$$= \mathbb{E}\left(\frac{\gamma_n(1)}{\gamma_n^N(1)} \left(1 - \frac{\gamma_n^N(1)}{\gamma_n(1)}\right) \left(\gamma_n^N(f_n) 1_{\tau^N \geq n} - \gamma_n(f_n)\right) 1_{\tau^N \geq n}\right)$$

$$= \mathbb{E}\left(\frac{\gamma_n(1)}{\gamma_n^N(1)} \left(1 - \frac{\gamma_n^N(1)}{\gamma_n(1)}\right) 1_{\tau^N \geq n} \left(\gamma_n^N(f_n) 1_{\tau^N \geq n} - \gamma_n(f_n)\right)\right)$$

If we set $h_n = \frac{1}{\gamma_n(1)} - 1$, we get the formula

$$\mathbb{E}\left(\left(\eta_n^N(f) - \eta_n(f)\right) 1_{\tau^N \geq n}\right) = -\mathbb{E}\left(\frac{\gamma_n(1)}{\gamma_n^N(1)} \left(\gamma_n^N(h_n) 1_{\tau^N \geq n} - \gamma_n(h_n)\right) \right.$$
$$\left. \times \left(\gamma_n^N(f_n) 1_{\tau^N \geq n} - \gamma_n(f_n)\right) 1_{\tau^N \geq n}\right)$$
(7.23)

Let Ω_n^N be the set of events

$$\Omega_n^N = \left\{\gamma_n^N(1) 1_{\tau^N \geq n} \geq \gamma_n(1)/2\right\}$$
$$= \left\{\gamma_n^N(1) \geq \gamma_n(1)/2 \text{ and } \tau^N \geq n\right\}$$
$$= \left\{\frac{\gamma_n(1)}{\gamma_n^N(1)} \leq 2 \text{ and } \tau^N \geq n\right\} \subset \left\{\tau^N \geq n\right\}$$

We recall by Corollary 7.4.1 that

$$\mathbb{P}\left(\Omega_n^N\right) \geq 1 - \frac{b(n)^2}{N}$$

with $b(n) \leq c.((n+1)/\gamma_n(1))^2$. If we combine this estimate with (7.23), we find that for any $f \in \mathcal{B}_b(E_n)$, with $\|f\| \leq 1$,

$$\left|\mathbb{E}\left(\left(\eta_n^N(f) - \eta_n(f)\right) 1_{\tau^N \geq n}\right)\right|$$

$$\leq \left|\mathbb{E}\left(\left(\eta_n^N(f) - \eta_n(f)\right) 1_{\Omega_n^N}\right)\right| + 2\mathbb{P}\left((\Omega_n^N)^c\right)$$

$$\leq 2\mathbb{E}\left(\left|\gamma_n^N(h_n) 1_{\tau^N \geq n} - \gamma_n(h_n)\right| \left|\gamma_n^N(f_n) 1_{\tau^N \geq n} - \gamma_n(f_n)\right|\right) + \frac{b(n)^2}{N}$$

7.4 Strong Law of Large Numbers

By Theorem 7.4.2 and the Cauchy-Schwartz inequality, this implies that

$$\left|\mathbb{E}\left((\eta_n^N(f) - \eta_n(f))\,1_{\tau^N \geq n}\right)\right| \leq \frac{b(n)^2}{N}$$

Finally, by Theorem 7.4.1, we conclude that

$$\left|\mathbb{E}\left(\eta_n^N(f)\,1_{\tau^N \geq n} - \eta_n(f)\right)\right| \leq \frac{b(n)^2}{N} + \sum_{q=0}^{n-1} e^{-N\alpha(q)}$$

To prove the second assertion, we first observe that

$$(2 \geq) \;\; \|(\eta_n^N - \eta_n)\,1_{\tau^N \geq n}\|_{\mathcal{F}_n} = 1_{\tau^N \geq n}\,\frac{1}{\gamma_n^N(1)}\,\|1_{\tau^N \geq n}\,\gamma_n^N - \gamma_n\|_{\mathcal{F}_n'}$$

with $\mathcal{F}_n' = \{(f - \eta_n(f)) \,:\, f \in \mathcal{F}_n\}$. Arguing as above, we also prove that

$$\|(\eta_n^N - \eta_n)\,1_{\tau^N \geq n}\|_{\mathcal{F}_n} \leq 1_{\Omega_n^N}\,\frac{1}{\gamma_n^N(1)}\,\|1_{\tau^N \geq n}\,\gamma_n^N - \gamma_n\|_{\mathcal{F}_n'} + 2\,1_{(\Omega_n^N)^c}$$

$$\leq \frac{2}{\gamma_n(1)}\,\|1_{\tau^N \geq n}\,\gamma_n^N - \gamma_n\|_{\mathcal{F}_n'} + 2\,1_{(\Omega_n^N)^c}$$

Since by Lemma 7.3.4 we have $I(\mathcal{F}_n') \leq 1 + I(\mathcal{F}_n)$, a simple application of Theorem 7.4.2 now yields that

$$\mathbb{E}(\|(\eta_n^N - \eta_n)\,1_{\tau^N \geq n}\|_{\mathcal{F}_n}^2)^{\frac{1}{2}}$$

$$\leq \frac{2}{\gamma_n(1)}\,\mathbb{E}(\|1_{\tau^N \geq n}\,\gamma_n^N - \gamma_n\|_{\mathcal{F}_n'}^2)^{\frac{1}{2}} + 2\mathbb{P}((\Omega_n^N)^c)^{\frac{1}{2}} \leq \frac{b(n)}{\sqrt{N}}\,(1 + I(\mathcal{F}_n))$$

with $b(n) \leq c.(n+1)/\gamma_n(1)$. A simple manipulation now gives that

$$\mathbb{E}(\|1_{\tau^N \geq n}\,\eta_n^N - \eta_n\|_{\mathcal{F}_n}^2)^{\frac{1}{2}} \leq \frac{b(n)}{\sqrt{N}}\,(1 + I(\mathcal{F}_n)) + \sum_{q=0}^{n-1} e^{-N\alpha(q)}$$

We prove the final assertion of the theorem using the inequality

$$\|1_{\tau^N \geq n}\,\eta_n^N - \eta_n\|_{\mathcal{F}_n} \leq \frac{2}{\gamma_n(1)}\,\|1_{\tau^N \geq n}\,\gamma_n^N - \gamma_n\|_{\mathcal{F}_n'} + 2\,1_{(\Omega_n^N)^c} + 1_{\tau^N < n}$$

Arguing as in the proof of Theorem 7.4.2, for every $\varepsilon \in (0,1)$ we have

$$\mathbb{P}\left(\|1_{\tau^N \geq n}\,\eta_n^N - \eta_n\|_{\mathcal{F}_n} > 3\varepsilon\right)$$

$$\leq \mathbb{P}(\tfrac{2}{\gamma_n(1)}\,\|1_{\tau^N \geq n}\,\gamma_n^N - \gamma_n\|_{\mathcal{F}_n'} > \varepsilon) + \mathbb{P}(2\,1_{(\Omega_n^N)^c} > \varepsilon) + \mathbb{P}(1_{\tau^N < n} > \varepsilon)$$

$$= \mathbb{P}(\|1_{\tau^N \geq n}\,\gamma_n^N - \gamma_n\|_{\mathcal{F}_n'} > \gamma_n(1)\varepsilon/2) + \mathbb{P}((\Omega_n^N)^c) + \mathbb{P}(\tau^N < n)$$

From previous estimations and Theorem 7.4.2, we find that

$$\mathbb{P}\left(\|1_{\tau^N \geq n}\, \eta_n^N - \eta_n\|_{\mathcal{F}_n} > 3\varepsilon\right) \leq 16(n+1)\,\mathcal{N}(\varepsilon_n, \mathcal{F}_n)\, e^{-N\varepsilon_n^2/2} + \sum_{q=0}^{n-1} e^{-\alpha(q)N}$$

as soon as $\sqrt{N} \geq 16/(\varepsilon \gamma_n(1))$ with $\varepsilon_n = \gamma_n(1)\varepsilon/(4(n+1))$. This ends the proof of the theorem. ∎

7.4.3 Time-Uniform Estimates

This section is concerned with the long time behavior of N-particle approximation models. The uniform estimates presented in this section are valid for any McKean interpretation model.

Our strategy will be to connect this problem with the stability properties discussed in Section 4.3. Unless otherwise stated, we shall suppose that the pair (G_n, M_n) satisfies the regularity conditions (G) and $(M)_m$ stated on page 116, for some parameters $\epsilon_n(G)$ and $\epsilon_n(M) > 0$. When these conditions are met the nonlinear Feynman-Kac semigroup $\Phi_{p,n}$ has several regularity and asymptotic stability properties. These properties are often expressed in terms of the regularity parameters $(r_{p,n}, \beta(P_{p,n}))$ introduced in (7.3) on page 218. We also refer the interested reader to Chater 4 for a systematic study of these quantities. For instance, we have seen that, for any fixed $p \geq 0$ and for any $n \geq p + m$, we have

$$\beta(P_{p,n}) \leq \prod_{k=0}^{[(n-p)/m]-1} (1 - \epsilon_{p+km}^{(m)}(G, M))$$

$$r_{p,n} \leq \epsilon_{p,n}^{-1}(G) \wedge [\epsilon_p(M)\, \epsilon_{p,p+m}(G)]^{-1}$$

with $\epsilon_p^{(m)}(G, M) = \epsilon_p^2(M)\, \epsilon_{p+1,p+m}(G)$ and $\epsilon_{p,n}(G) = \prod_{p \leq q < n} \epsilon_q(G)$.

We recall that the central formula that allows us to connect the stability properties of $\Phi_{p,n}$ with the long time behavior of the particle density profiles is the following decomposition

$$\eta_n^N - \eta_n = \sum_{q=0}^{n} [\Phi_{q,n}(\eta_q^N) - \Phi_{q,n}(\Phi_q(\eta_{q-1}^N))] \tag{7.24}$$

The rationale behind this decomposition is as follows. When the semigroup $\Phi_{p,n}$ is asymptotically stable, it naturally forgets any erroneous initial conditions. This property ensures that in some sense for each elementary term

$$[\Phi_{q,n}(\eta_q^N) - \Phi_{q,n}(\Phi_q(\eta_{q-1}^N))] \longrightarrow 0 \quad \text{as} \quad (n-q) \to \infty$$

Consequently, we expect to prove a uniform estimate (w.r.t. the time parameter) of the sum of the "small errors" induced by replacing at each step

$\Phi_p(\eta_p^N)$ by the N-approximation density profiles η_p^N. This strategy is not restricted to Feynman-Kac and particle models and applies to any kind of approximation schemes conducted by local approximations of the one-step mappings Φ_p. We also mention that this technique is well-known in the literature on numerical analysis of dynamical systems. In this context, and in contrast to chaotic systems the stability of the limiting model ensures that the local errors induced for instance by numerical roundoffs do not propagate. Because of its importance in practice, we have not chosen to house this strategy in the proof of some theorem.

It is first convenient to introduce the random potential functions

$$G_{q,n}^N : x_q \in E_q \longrightarrow G_{q,n}^N(x_q) = \frac{G_{q,n}}{\Phi_q(\eta_{q-1}^N)(G_{q,n})} \in (0, \infty)$$

and the random bounded operators $P_{q,n}^N$ from $\mathcal{B}_b(E_n)$ into $\mathcal{B}_b(E_q)$ defined for any $(f_n, x_q) \in (\mathcal{B}_b(E_n) \times E_q)$ by

$$\begin{aligned} P_{q,n}^N(f_n)(x_q) &= P_{q,n}\left(f_n - \Phi_{q,n}\left(\Phi_q(\eta_{q-1}^N)\right)(f_n)\right)(x_q) \\ &= \int (P_{q,n}f(x_q) - P_{q,n}f(y_q)) \; G_{q,n}^N(y_q) \; \Phi_q(\eta_{q-1}^N)(dy_q) \end{aligned}$$

We associate with the pair $(G_{q,n}^N, P_{q,n}^N)$, the random bounded and integral operator $Q_{q,n}^N$ from $\mathcal{B}_b(E_n)$ into $\mathcal{B}_b(E_q)$ defined for any $(f_n, x_q) \in (\mathcal{B}_b(E_n) \times E_q)$ by

$$Q_{q,n}^N(f_n)(x_q) = G_{q,n}^N(x_q) \times P_{q,n}^N(f_n)(x_q) \qquad (7.25)$$

Each "local" term in (7.24) can be expressed in terms of $Q_{q,n}^N$ as follows. For any $q \le n$ and $f_n \in \mathcal{B}_b(E_n)$ with $\mathrm{osc}(f_n) \le 1$, we have

$$\Phi_{q,n}(\eta_q^N)([f_n - \Phi_{q,n}(\Phi_q(\eta_{q-1}^N))(f_n)])$$

$$= \frac{1}{\eta_q^N(G_{q,n})} \; \eta_q^N \left(G_{q,n} \; P_{q,n}[f_n - \Phi_{q,n}(\Phi_q(\eta_{q-1}^N))(f_n)]\right) = \frac{\eta_q^N Q_{q,n}^N(f_n)}{\eta_q^N Q_{q,n}^N(1)}$$

By construction, we also observe that

$$\Phi_q(\eta_{q-1}^N)\left(G_{q,n}^N\right) = 1 \quad \text{and} \quad \Phi_q(\eta_{q-1}^N)\left(Q_{q,n}^N(f_n)\right) = 0$$

From the considerations above, we have the decomposition

$$\Phi_{q,n}(\eta_q^N) - \Phi_{q,n}(\Phi_q(\eta_{q-1}^N)) = \frac{1}{\eta_q^N(G_{q,n}^N)} \; [\eta_q^N - \Phi_q(\eta_{q-1}^N)]Q_{q,n}^N$$

Using the properties of Dobrushin's contraction coefficient, we have

$$\begin{aligned} \|P_{q,n}^N(f_n)\| &\le \mathrm{osc}(P_{q,n}f) \le \beta(P_{q,n}) \\ \frac{\|Q_{q,n}^N(f_n)\|}{\eta_q^N(G_{q,n}^N)} &\le \frac{\|G_{q,n}^N\|}{\eta_q^N(G_{q,n}^N)} \; \|P_{q,n}^N(f_n)\| \le r_{q,n} \; \beta(P_{q,n}) \end{aligned}$$

From these estimates, we readily prove the inequality

$$|[\Phi_{q,n}(\eta_q^N) - \Phi_{q,n}(\Phi_q(\eta_{q-1}^N))](f_n)| \leq r_{q,n}\,\beta(P_{q,n})|[\eta_q^N - \Phi_q(\eta_{q-1}^N)]\overline{Q}_{q,n}^N(f_n)|$$

with $\overline{Q}_{q,n}^N(f_n) = Q_{q,n}^N(f_n)/\|Q_{q,n}^N(f_n)\|$. Now, using Lemma 7.3.3, we check that for any $p \geq 1$ we have

$$\sqrt{N}\,\mathbb{E}(|[\eta_q^N - \Phi_q(\eta_{q-1}^N)]Q_{q,n}^N(f_n)|^p \mid F_{q-1}^N)^{1/p} \leq 2\,d(p)^{1/p}\,r_{q,n}\,\beta(P_{q,n})$$

with the sequence of finite constants $d(p)$ introduced in (7.7). Using the decomposition (7.24), these "local" estimations readily yield the following theorem.

Theorem 7.4.4 *For any $n \geq 0$, $p \geq 1$, and $f_n \in \mathrm{Osc}_1(E_n)$, we have*

$$\sqrt{N}\,\mathbb{E}\left(|[\eta_n^N - \eta_n](f_n)|^p\right)^{\frac{1}{p}} \leq 2\,d(p)^{1/p}\,\sum_{q=0}^{n} r_{q,n}\,\beta(P_{q,n})$$

with the sequence of finite constants $d(p)$ given for any $p \geq 1$ by

$$d(2p) = (2p)_p\,2^{-p} \quad \text{and} \quad d(2p-1) = \frac{(2p-1)_p}{\sqrt{p-1/2}}\,2^{-(p-1/2)}$$

In addition, suppose that conditions (G) and $(M)_m$ hold true for some integer $m \geq 1$ and some pair parameters $(\epsilon_n(G), \varepsilon_n(M))$ such that $\epsilon(G) = \wedge_n \epsilon_n(G)$ and $\varepsilon(M) = \wedge_n \varepsilon_n(M) > 0$. Then we have the uniform estimate

$$\sup_{n \geq 0}\,\sup_{f_n \in \mathrm{Osc}_1(E_n)} \sqrt{N}\,\mathbb{E}\left(|[\eta_n^N - \eta_n](f_n)|^p\right)^{\frac{1}{p}} \leq \frac{2\,d(p)^{1/p}\,m}{\epsilon^3(M)\,\varepsilon(G)^{2m-1}} \qquad (7.26)$$

Proof:
Note that for any $p \leq n$ we have the estimates

$$\beta(P_{p,n}) \leq \left(1 - \epsilon^2(M)\epsilon(G)^{m-1}\right)^{[(n-p)/m]}$$

$$r_{p,n} \leq \left(\varepsilon^{-(n-p)}(G) \wedge (\varepsilon^{-1}(M)\varepsilon^{-m}(G))\right) \leq \varepsilon^{-1}(M)\varepsilon^{-m}(G)$$

Since

$$\sum_{q=0}^{n}\left(1 - \epsilon^2(M)\,\epsilon(G)^{m-1}\right)^{[q/m]} \leq m\sum_{k=0}^{[n/m]}\left(1 - \varepsilon^2(M)\,\varepsilon(G)^{m-1}\right)^k$$

$$\leq \frac{m}{\epsilon^2(M)\,\varepsilon(G)^{m-1}}$$

the end of the proof is clear. ∎

Theorem 7.4.4 can be regarded as the extension of the first part of Lemma

7.3.3 to interacting particle models. Arguing as in the proof of Corollary 7.3.1, from the $\mathbb{L}_p$-inequalities stated in Theorem 7.4.4 we easily estimate the moment-generating function of the particle density profiles. The proof of the exponential estimate results from a simple application of Markov inequality.

Corollary 7.4.3 *For any $n \geq 0$, $f_n \in \mathrm{Osc}_1(E_n)$, and any $\varepsilon > 0$, we have*

$$\mathbb{E}(e^{\varepsilon\sqrt{N}|\eta_n^N(f_n)-\eta_n(f_n)|}) \leq (1+\varepsilon\, b(n)/\sqrt{2})\, e^{(\varepsilon b(n))^2/2} \quad (7.27)$$

$$\mathbb{P}(|\eta_n^N(f_n) - \eta_n(f_n)| > \varepsilon) \leq (1+\varepsilon\sqrt{N/2})\, e^{-N\frac{\varepsilon^2}{2b(n)^2}} \quad (7.28)$$

for some finite constant $b(n)$ such that $b(n) \leq 2\sum_{q=0}^{n} r_{q,n}\,\beta(P_{q,n})$. In addition, under the regularity conditions of Theorem 7.4.4, we have

$$\sup_{n\geq 0}\mathbb{E}(e^{\varepsilon\sqrt{N}|\eta_n^N(f_n)-\eta_n(f_n)|}) \leq (1+\varepsilon\, b/\sqrt{2})\, e^{(\varepsilon b)^2/2}$$

$$\sup_{n\geq 0}\mathbb{P}(|\eta_n^N(f_n) - \eta_n(f_n)| > \varepsilon) \leq (1+\varepsilon\sqrt{N/2})\, e^{-N\frac{\varepsilon^2}{2b^2}}$$

for some finite constant $b \leq 2m/(\epsilon^3(M)\,\varepsilon(G)^{2m-1})$.

Proof (sketched):
We deduce the exponential probability estimate (7.28) from the first inequality (7.27). Note that there is no loss of generality in assuming that $b(n) \geq 1$. On the other hand, by Markov's inequality, we have for any ε and $t > 0$

$$\mathbb{P}(|\eta_n^N(f_n) - \eta_n(f_n)| \geq \varepsilon) = \mathbb{P}(e^{t|\eta_n^N(f_n)-\eta_n(f_n)|} \geq e^{t\varepsilon})$$
$$\leq e^{-t\varepsilon}\,\mathbb{E}(e^{t|\eta_n^N(f_n)-\eta_n(f_n)|})$$

Using the first estimate stated in the theorem, we find that

$$\mathbb{P}(|\eta_n^N(f_n) - \eta_n(f_n)| \geq \varepsilon) \leq (1 + b(n)t/\sqrt{2N})\, e^{\frac{b(n)^2 t^2}{2N} - t\varepsilon}$$

Choosing $t = (N\varepsilon)/b^2(n)$, we arrive at

$$\mathbb{P}(|\eta_n^N(f_n) - \eta_n(f_n)| \geq \varepsilon) \leq (1+\varepsilon\sqrt{N}/(b(n)\sqrt{2}))\, e^{-\frac{N\varepsilon^2}{2b^2(n)}}$$
$$\leq (1+\varepsilon\sqrt{N/2})\, e^{-\frac{N\varepsilon^2}{2b^2(n)}}$$

∎

Corollary 7.4.4 *Let $\mathcal{F}_n$ be a countable collection of functions f_n with $\|f_n\| \leq 1$ and finite entropy $I(\mathcal{F}_n) < \infty$. Suppose that the Markov transitions M_n have the form $M_n(u,dv) = m_n(u,v)\,p_n(dv)$ for some measurable* function m_n on $(E_{n-1} \times E_n)$ and some $p_n \in \mathcal{P}(E_n)$. Also assume that we have $\sup_{u \in E_{n-1}} |\log m_n(u,v)| \le \theta_n(v)$ with $p_n(e^{3\theta_n}) < \infty$ and for some collection of mappings θ_n on E_n. Then for any $n \ge 0$ and $p \ge 1$ we have

$$\mathbb{E}\left(\|\eta_n^N - \eta_n\|_{\mathcal{F}_n}^p\right)^{\frac{1}{p}} \le a(p)\,[I(\mathcal{F}_n) + b(n)]/\sqrt{N} \tag{7.29}$$

with $b(0) = 0$ and $b(n+1) \le r_n\, p_{n+1}(e^{3\theta_{n+1}}) \sum_{q=0}^{n} r_{q,n}\, \beta(P_{q,n})$ and $a(p) \le c.[p/2]!$.

Proof:
We use the decomposition

$$\eta_n^N - \eta_n = [\eta_n^N - \Phi_n(\eta_{n-1}^N)] + [\Phi_n(\eta_{n-1}^N) - \Phi_n(\eta_{n-1})]$$

to check that

$$\|\eta_n^N - \eta_n\|_{\mathcal{F}_n} = \|\eta_n^N - \Phi_n(\eta_{n-1}^N)\|_{\mathcal{F}_n} + \int \left|\frac{d\Phi_n(\eta_{n-1}^N)}{d\Phi_n(\eta_{n-1})} - 1\right| d\eta_n$$

By Lemma 7.3.5, we have

$$\sqrt{N}\,\mathbb{E}\left(\|\eta_n^N - \Phi_n(\eta_{n-1}^N)\|_{\mathcal{F}_n}^p\right)^{\frac{1}{p}} \le a(p)\,I(\mathcal{F}_n) \tag{7.30}$$

To estimate the second term, we observe that for any $\mu \in \mathcal{P}(E_n)$ we have

$$\frac{d\Phi_n(\mu)}{dp_n}(v) = \frac{\mu(G_{n-1}\, m_n(.,v))}{\mu(G_{n-1})}$$

From this observation, we find the following decomposition. For any pair $(\mu,\eta) \in \mathcal{P}(E_n)$ and for any $v \in E_{n-1}$, we have

$$\frac{d\Phi_n(\mu)}{d\Phi_n(\eta)}(v) - 1$$

$$= \frac{\eta(G_{n-1})}{\mu(G_{n-1})}\frac{\mu(G_{n-1}\, m_n(.,v))}{\eta(G_{n-1}\, m_n(.,v))} - 1$$

$$= \left[\frac{\eta(G_{n-1})}{\mu(G_{n-1})} - 1\right]\left[\frac{\mu(G_{n-1}\, m_n(.,v))}{\eta(G_{n-1}\, m_n(.,v))}\right] + \left[\frac{\mu(G_{n-1}\, m_n(.,v))}{\eta(G_{n-1}\, m_n(.,v))} - 1\right] \tag{7.31}$$

Under our assumptions, we have the estimates

$$\left|\frac{d\Phi_n(\mu)}{d\Phi_n(\eta)}(v) - 1\right| \le e^{2\theta_n(v)}\left|\frac{\mu(G_{n-1})}{\eta(G_{n-1})} - 1\right| + \left|\frac{\mu(G_{n-1}\, m_n(.,v))}{\eta(G_{n-1}\, m_n(.,v))} - 1\right|$$

Consequently, we find that

$$\left|\frac{d\Phi_n(\mu)}{d\Phi_n(\eta_{n-1})}(v) - 1\right| \le \left|\mu(f_{n,v}^{(1)}) - \eta_{n-1}(f_{n,v}^{(1)})\right| + \left|\mu(f_{n,v}^{(2)}) - \eta_{n-1}(f_{n,v}^{(2)})\right| \tag{7.32}$$

with

$$f_{n,v}^{(1)}(u) = e^{2\theta_n(v)} \frac{G_{n-1}(u)}{\eta_{n-1}(G_{n-1})} \quad \text{and} \quad f_{n,v}^{(2)}(u) = \frac{G_{n-1}(u)\, m_n(u,v)}{\eta_{n-1}(G_{n-1}\, m_n(.,v))}$$

By Theorem 7.4.4, we get for $i = 1, 2$ and any $p \geq 1$ the estimates

$$\sqrt{N}\, \mathbb{E}(|\eta_{n-1}^N(f_{n,v}^{(i)}) - \eta_{n-1}(f_{n,v}^{(i)})|^p)^{1/p}$$

$$\leq a(p)\, r_{n-1}\, e^{2\theta_n(v)} \sum_{q=0}^{n-1} r_{q,n-1}\, \beta(P_{q,n-1})$$

On the other hand, we have

$$\frac{d\eta_n}{dp_n}(v) = \frac{d\Phi_n(\eta_{n-1})}{dp_n}(v) = \frac{\eta_{n-1}(G_{n-1}\, m_n(.,v))}{\eta_{n-1}(G_{n-1})} \leq e^{\theta_n(v)}$$

Consequently, from (7.32) we find that

$$\mathbb{E}\left(\left[\int \left|\frac{d\Phi_n(\eta_{n-1}^N)}{d\Phi_n(\eta_{n-1})} - 1\right|\, d\eta_n\right]^p\right)^{1/p} \leq a(p)\, b(n) \qquad (7.33)$$

with $b(n) \leq r_{n-1}\, p_n(e^{3\theta_n}) \sum_{q=0}^{n-1} r_{q,n-1}\, \beta(P_{q,n-1})$. Thus, if we combine (7.30) with (7.33), we readily prove (7.29). ∎

Using the same line of argument as in the proof of Theorem 7.4.4, we prove the following uniform estimate.

Corollary 7.4.5 *Assume that the regularity conditions stated in Corollary 7.4.4 are met with*

$$p(e^{3\theta}) =_{\text{def.}} \sup_{n \geq 1} p_n(e^{(3a_n)}) < \infty \quad \text{and} \quad I(\mathcal{F}) =_{\text{def.}} \sup_{n \geq 0} I(\mathcal{F}_n) < \infty$$

In addition, suppose that (G) and $(M)_m$ hold true for some $m \geq 1$ and some pair parameters $(\epsilon_n(G), \varepsilon_n(M))$ with $\epsilon(G) = \wedge_n \epsilon_n(G)$ and $\varepsilon(M) = \wedge_n \varepsilon_n(M) > 0$. Then for any $p \geq 1$ we have

$$\sqrt{N} \sup_{n \geq 0} \mathbb{E}\left(\|\eta_n^N - \eta_n\|_{\mathcal{F}_n}^p\right)^{\frac{1}{p}} \leq a(p)\left[I(\mathcal{F}) + \frac{m\, p(e^{3\theta})}{\varepsilon^3(M)\epsilon^{2m}(G)}\right]$$

Theorem 7.4.5 *For any $n \in \mathbb{N}$, we let $\mathcal{F}_n$ be a countable collection of functions f_n such that $\|f_n\| \leq 1$ and satisfying the uniform entropy condition $I(\mathcal{F}) = \sup_{n \geq 0} I(\mathcal{F}_n) < \infty$. Assume moreover that the semigroup $\Phi_{p,n}$ is asymptotically stable with respect to the sequence $(\mathcal{F}_n)_{n \geq 0}$ in the sense that*

$$\lim_{n \to \infty} \sup_{\mu_q, \nu_q \in \mathcal{P}(E_q)} \sup_{q \geq 0} \|\Phi_{q,q+n}(\mu_q) - \Phi_{q,q+n}(\nu_q)\|_{\mathcal{F}_{q+n}} = 0$$

When condition (G) holds true with $\inf_{n\geq 1} \varepsilon_n(G) \stackrel{\text{def.}}{=} \varepsilon(G) > 0$, then we have the following uniform convergence result with respect to time:

$$\lim_{N\to\infty} \sup_{n\geq 0} E\left(\left\|\eta_n^N - \eta_n\right\|_{\mathcal{F}_n}\right) = 0 \qquad (7.34)$$

In addition, let us assume that the semigroup $\Phi_{p,n}$ is exponentially stable in the sense that there exist some positive constant $\lambda > 0$ and $n_0 \geq 0$ such that for any $n \geq n_0$

$$\sup_{\mu_q, \nu_q \in \mathcal{P}(E_q)} \sup_{q \geq 0} \left\| \Phi_{q,q+n}(\mu_q) - \Phi_{q,q+n}(\nu_q) \right\|_{\mathcal{F}_{q+n}} \leq e^{-\lambda n}$$

Then for any $p \geq 1$ we have the uniform estimate

$$\sup_{n\geq 0} N^{\alpha/2} \, \mathbb{E}\left(\left\|\eta_n^N - \eta_n\right\|_{\mathcal{F}_n}^p\right)^{\frac{1}{p}} \leq a(p)\,(1 + e^{\lambda'} I(\mathcal{F})) \qquad (7.35)$$

as soon as $N \geq \exp(2n_0\,(\lambda + \lambda'))$ with

$$\alpha = \frac{\lambda}{\lambda + \lambda'} \quad \text{and} \quad \lambda' = 1 + \log(1/\varepsilon(G))$$

Proof:
By Lemma 7.3.5 and arguing as in the beginning of the proof of Theorem 7.4.2, one proves that for any $0 \leq q \leq n$ and $p \geq 1$

$$\sqrt{N}\,\mathbb{E}\left(\left\|\Phi_{q,n}(\eta_q^N) - \Phi_{q,n}\left(\Phi_q(\eta_{q-1}^N)\right)\right\|_{\mathcal{F}_n}^p\right)^{\frac{1}{p}} \leq a(p)\, I(\mathcal{F})/\varepsilon^{(n-q)}(G)$$

By the decomposition (7.24), one concludes that for any $0 \leq n \leq T$

$$\sqrt{N}\,\mathbb{E}\left(\left\|\eta_n^N - \eta_n\right\|_{\mathcal{F}_n}^p\right)^{\frac{1}{p}} \leq a(p)\, I(\mathcal{F})\,(T+1)/\epsilon^T(G) \qquad (7.36)$$

On the other hand, for any $q \geq 0$ we have

$$\left\|\eta_{q+T}^N - \eta_{q+T}\right\|_{\mathcal{F}_n} \leq \sum_{r=q+1}^{q+T} \left\|\Phi_{r,q+T}(\eta_r^N) - \Phi_{r,q+T}\left(\Phi_r(\eta_{r-1}^N)\right)\right\|_{\mathcal{F}_n}$$
$$+ \left\|\Phi_{q,q+T}(\eta_q^N) - \Phi_{q,q+T}(\eta_q)\right\|_{\mathcal{F}_n}$$

Under our assumptions, we find that

$$\left\|\eta_{q+T}^N - \eta_{q+T}\right\|_{\mathcal{F}_n} \leq \sum_{r=q+1}^{q+T} \left\|\Phi_{r,q+T}(\eta_r^N) - \Phi_{r,q+T}\left(\Phi_r(\eta_{r-1}^N)\right)\right\|_{\mathcal{F}_n} + e^{-\lambda T}$$

and arguing as above one gets that for any $T \geq n_0$

$$\sup_{q\geq 0} \mathbb{E}\left(\left\|\eta_{q+T}^N - \eta_{q+T}\right\|_{\mathcal{F}}^p\right)^{\frac{1}{p}} \leq e^{-\lambda T} + a(p)\,(T+1)\,\frac{\varepsilon^T(G)}{\sqrt{N}}\, I(\mathcal{F}) \qquad (7.37)$$

Combining (7.36) and (7.37), we readily prove the uniform estimate

$$\sup_{n\geq 0} \mathbb{E}\left(\|\eta_n^N - \eta_n\|_{\mathcal{F}_n}^p\right)^{\frac{1}{p}} \leq e^{-\lambda T} + a(p)\,\frac{e^{\lambda' T}}{\sqrt{N}}\,I(\mathcal{F})$$

for any $T \geq n_0$ and where $\lambda' = 1 - \log \varepsilon(G)$. Obviously, if we choose $N \geq 1$ and

$$T = \left[\frac{1}{2}\frac{\log N}{\lambda + \lambda'}\right] + 1 \geq n_0$$

where $[r]$ denotes the integer part of $r \in \mathbb{R}$, we get (7.35). The end of the proof of the theorem is now clear. ∎

We end this section with some brief comments on the long time behavior of interacting processes. For a more thorough study we refer the reader to section 12.4. For time homogeneous Feynman-Kac models and in the context of the statement of Corollary 7.4.5, the measure-valued process η_n admits a unique invariant measure η_∞ (see for instance Chapter 4 or Chapter 5). In addition, we have for any $p \geq 1$ and $n \geq 0$

$$\mathbb{E}\left(\|\eta_n^N - \eta_\infty\|_{\mathcal{F}}^p\right)^{\frac{1}{p}} \leq b\left(\frac{a(p)}{\sqrt{N}} + (1-\rho)^{\lfloor n/m \rfloor}\right)$$

for some constants $\rho \in (0,1)$ and $b < \infty$ that depend on the pair (G, M). On the other hand, under the regularity conditions of Corollary 7.4.5, the Markov chain ξ_n has a unique invariant measure on the product space E^N. The estimate above provide an asymptotic estimate of the limiting empirical measures in terms of η_∞; that is, we have

$$\lim_{N\to\infty}\lim_{n\to\infty}\mathbb{E}\left(\|\eta_n^N - \eta_\infty\|_{\mathcal{F}}\right) = 0 = \lim_{n\to\infty}\lim_{N\to\infty}\mathbb{E}\left(\|\eta_n^N - \eta_\infty\|_{\mathcal{F}}\right)$$

8
Propagation of Chaos

8.1 Introduction

This chapter is concerned with propagation-of-chaos properties of particle models. These properties measure the adequacy of the laws of the particles with the desired limiting distribution. They also allows us to quantify the independence between particles. Loosely speaking, the initial configuration of an N-particle model consists of N independent particles in a "complete chaos." Then they evolve and interact with one another. The nature of the interactions depends on the McKean interpretation of the limiting process (see Section 2.5.3). For any fixed time horizon n, when the size of the system N, tends to infinity, any finite block of $q(\leq N)$ particles asymptotically behaves as a collection of independent particles. In other words, the law of any q particle paths of length n converges as $N \to \infty$ towards the q tensor product of the n-path McKean measure.

The interpretations of propagation of chaos differ from the different application particle model areas we consider.

From the physical point of view, particle algorithms are often related to some microscopic particle interpretation of some physical evolution equation. In this context, the limiting distribution flow model is regarded as an infinite particle model. Here propagation-of-chaos estimates give precise information on the degree of interaction between the particles. They justify in some sense the well-founded microscopic particle interpretations.

From a statistical point of view, particle methods are rather regarded as particle simulation techniques of complex path distributions. In this context, propagation-of-chaos properties offer precise information on the numerical quality of these simulation techniques. First of all, they make it possible to quantify independence between the simulated variables. Moreover, they guarantee the adequacy of their laws with the desired target distribution. For instance, in engineering applications such as in nonlinear filtering or global optimization problem, propagation of chaos ensures the adaptation of the stochastic grid with the signal conditional distributions or the Boltzmann-Gibbs concentration laws.

From the biological perspective, the propagation-of-chaos of genetic models gives precise information on their genealogical structure. More precisely, they quantify the degree of interaction between the ancestral lines of evolution of a group of individuals. They provide information not only on current populations but also on the complete genealogies of ancestral lines that have disappeared.

We design three strategies with different precision levels. In the first one, we examine the propagation of chaos of the particle model associated to the McKean interpretation model

$$K_{n+1,\eta}(x,.) = \varepsilon_n G_n(x) \, M_{n+1}(x,.) + (1 - \varepsilon_n G_n(x)) \, \Phi_{n+1}(\eta) \qquad (8.1)$$

where ε_n are nonnegative constants such that $\varepsilon_n G_n \leq 1$. Note that the pair of examples provided on page 219 fit into this model, and the case $\varepsilon_n = 0$ corresponds to the traditional mutation/selection genetic algorithm. We present a general and basic strategy that probably works for other McKean interpretations but does not give any information on the rate of propagation of chaos. Another drawback of this technique is that it is restricted to locally compact and separable metric state spaces.

One important question arising in practice is to estimate the rate of propagation of chaos with respect to the pair parameters (q, n). This led for instance to propagation-of-chaos properties with respect to increasing particle block sizes and/or time horizons. We first derive strong propagation-of-chaos estimates with respect to the relative entropy criterion. This strategy is based on an inequality of Csiszar on exchangeable measures. It allows us to restrict the analysis to profile measures. The only drawback of this elegant entropy technique is that it requires some regularity on the mutation transitions. As a result, it doesn't apply to path-space and genealogical tree models.

The third strategy is not based on any kind of regularity property of the Feynman-Kac model but it is restricted, as presented, to the simple mutation/selection genetic model. We use as a tool a natural tensor product Feynman-Kac semigroup approach with respect to time horizons and particle block sizes. We derive several propagation-of-chaos estimates for Boltzmann-Gibbs measures from a precise moment analysis of empirical measures and from an original transport equation relating q-tensor product and symmetric statistic type empirical measures. This analysis applies to the study of the asymptotic behavior of genetic historical processes and their complete genealogical tree evolution. In contrast to traditional studies on q-symmetric statistics, here the particles are nonindependent but interact with one another according to precise mutation and selection genetic rules. In this sense, these results can also be considered as an extension of the traditional asymptotic theory of q-symmetric statistics to interacting random sequences.

8.2 Some Preliminaries

To get an overview on Feynman-Kac and McKean particle interpretations, we recommend the reader to start his/her study of the propagation-of-chaos properties with the introductory section, Section 7.2.

Definition 8.2.1 *We say that the distribution $\mathbb{P}_{\eta_0}^N$ is weakly $\mathbb{K}_{\eta_0}$-chaotic if we have for any $n \in \mathbb{N}$, $q \geq 1$, and $(F_n^i)_{i \geq 1} \in \mathcal{C}_b(E_{[0,n]})^{\mathbb{N}}$*

$$\lim_{N \to \infty} \mathbb{E}_{\eta_0}^N \left(\prod_{i=1}^q F_n^i(\xi_{[0,n]}^i) \right) = \prod_{i=1}^q \mathbb{K}_{\eta_0,n}(F_n^i)$$

This property can be restated in terms of the law of the first q path particles. To present this alternative description, let (q, N) be a pair of integers with $1 \leq q \leq N$ and for each $0 \leq p \leq n$ we set

$$E_{[p,n]}^{(q)} = E_p^q \times E_{p+1}^q \cdots \times E_n^q \quad \text{and} \quad E_{(p,n]}^{(q)} = E_{[p+1,n]}^{(q)}$$

These sets represent the path space of a block of q particles from time p to the current time n. They are connected to the product spaces $E_{[p,n]}^p = (E_{[p,n]})^q$ by the mapping $\Theta_{p,n}^q : E_{[p,n]}^{(q)} \longrightarrow E_{[p,n]}^q$ defined by

$$\Theta_{p,n}^q[(x_p^i)_{1 \leq i \leq q}, \ldots, (x_n^i)_{1 \leq i \leq q}] = (x_p^i, \ldots, x_n^i)_{1 \leq i \leq q} \quad (8.2)$$

For $p = 0$, we slightly abuse the notation and write Θ_n^q instead of $\Theta_{0,n}^q$. By $\mathbb{P}_{\eta_0,n}^{(N,q)}$ we denote the distribution of the first q-path particles

$$\mathbb{P}_{\eta_0,n}^{(N,q)} = \text{Law}((\xi_{[0,n]}^i)_{1 \leq i \leq q}) \in \mathcal{P}(E_{[0,n]}^q)$$

with $\xi_{[0,n]}^i = (\xi_0^i, \ldots, \xi_n^i) \in E_{[0,n]}$ and we let $\mathbb{P}_{\eta_0,[n]}^{(N,q)} = \text{Law}((\xi_n^i)_{1 \leq i \leq q}) \in \mathcal{P}(E_n^q)$ be their nth time marginals. For $q = N$, we simplify the notation and we write $\mathbb{P}_{\eta_0,n}^{(N)}$ instead of $\mathbb{P}_{\eta_0,n}^{(N,N)}$. In this notation, we see that $\mathbb{P}_{\eta_0}^N$ is weakly $\mathbb{K}_{\eta_0}$-chaotic if and only if we have that

$$\lim_{N \to \infty} \mathbb{P}_{\eta_0,n}^{(N,q)}(F) = \mathbb{K}_{\eta_0,n}^{\otimes q}(F)$$

for any q-tensor product function $F = \otimes_{i=1}^{q} F_n^i \in \mathcal{C}_b(E_{[0,n]}^q)$. It is often more simple to derive this type of weak law of large numbers in terms of the path empirical measure

$$\mathbb{K}_n^N = \frac{1}{N} \sum_{i=1}^{N} \delta_{\xi_{[0,n]}^i} \in \mathcal{P}(E_{[0,n]})$$

To be more precise, let $\langle N \rangle^{\langle q \rangle}$ be the set of all mappings from $\langle q \rangle = \{1, \ldots, q\}$ into $\langle N \rangle = \{1, \ldots, N\}$ and $\langle q, N \rangle \subset \langle N \rangle^{\langle q \rangle}$ the subset of all $(N)_q = N!/(N-q)!$ one-to-one mappings. We associate with $\mathbb{K}_n^N$ a pair of q-tensor and symmetric statistic type empirical distributions

$$(\mathbb{K}_n^N)^{\otimes q} = \frac{1}{N^q} \sum_{\alpha \in \langle N \rangle^{\langle q \rangle}} \delta_{(\xi_{[0,n]}^{\alpha(1)}, \ldots, \xi_{[0,n]}^{\alpha(q)})}$$

$$(\mathbb{K}_n^N)^{\odot q} = \frac{1}{(N)_q} \sum_{\alpha \in \langle q, N \rangle} \delta_{(\xi_{[0,n]}^{\alpha(1)}, \ldots, \xi_{[0,n]}^{\alpha(q)})} \quad (8.3)$$

In contrast to traditional q-symmetric statistics, the N-random paths $\xi_{[0,n]}^i$ are non independent but they interact with each other according to some precise mutation and genetic selection rules. By symmetry arguments, we observe that for any $F \in \mathcal{B}_b(E_{[0,n]}^q)$ we have

$$\mathbb{P}_{\eta_0, n}^{(N,q)}(F) = \mathbb{E}_{\eta_0}^N(F((\xi_{[0,n]}^i)_{1 \le i \le q})) = \mathbb{E}_{\eta_0}^N((\mathbb{K}_n^N)^{\odot q}(F))$$

The next central observation is that the empirical measures (8.3) are connected by a Markov transport equation of the form

$$(\mathbb{K}_n^N)^{\otimes q} = (\mathbb{K}_n^N)^{\odot q} R_N^{(q)} \quad \text{where} \quad R_N^{(q)} = \frac{(N)_q}{N^q} Id + (1 - \frac{(N)_q}{N^q}) \tilde{R}_N^{(q)}$$

and $\tilde{R}_N^{(q)}$ a Markov transition on $E_{[0,n]}^q$. We will give the proof of this result with a precise and explicit description of $\tilde{R}_N^{(q)}$ in the Section 8.6. One easy consequence of this formula is that

$$\|(\mathbb{K}_n^N)^{\otimes q} - (\mathbb{K}_n^N)^{\odot q}\|_{\text{tv}} \le (1 - (N)_q/N^q) \le (q-1)^2/N \quad (8.4)$$

Using this property, we have the following lemma.

Lemma 8.2.1 *The sequence of distributions $\mathbb{P}_{\eta_0}^N$ is weakly $\mathbb{K}_{\eta_0}$-chaotic if and only if for any $n \in \mathbb{N}$ the random distributions $\mathbb{K}_n^N$ converge in law to the deterministic measure $\mathbb{K}_{\eta_0, n}$.*

Proof:
We first suppose that $\mathbb{P}_{\eta_0}^N$ is weakly $\mathbb{K}_{\eta_0}$-chaotic. In this situation, we have for any $f \in \mathcal{C}_b(E_{[0,n]})$

$$\mathbb{E}_{\eta_0}^N((\mathbb{K}_n^N(f) - \mathbb{K}_{\eta_0,n}(f))^2) = \tfrac{1}{N}[\mathbb{E}_{\eta_0}^N(f^2(\xi_{[0,n]}^1)) - \mathbb{E}_{\eta_0}^N(f(\xi_{[0,n]}^1)f(\xi_{[0,n]}^2))]$$

$$+ \mathbb{E}_{\eta_0}^N(f(\xi_{[0,n]}^1)f(\xi_{[0,n]}^2)) - 2\mathbb{K}_{\eta_0,n}(f)\mathbb{E}_{\eta_0}^N(f(\xi_{[0,n]}^1)) + \mathbb{K}_{\eta_0,n}(f)^2$$

Since $\mathbb{P}_{\eta_0}^N$ is weakly $\mathbb{K}_{\eta_0}$-chaotic, we easily find that

$$\lim_{N\to\infty} \mathbb{E}_{\eta_0}^N((\mathbb{K}_n^N(f) - \mathbb{K}_{\eta_0,n}(f))^2) = 0$$

and we conclude that $\mathbb{K}_n^N$ converge in law to $\mathbb{K}_{\eta_0,n}$. In the reverse angle, if $\mathbb{K}_n^N$ converge in law to $\mathbb{K}_{\eta_0,n}$, we have for any $F = \otimes F_n^i \in \mathcal{C}_b(E_{[0,n]}^q)$ with $\|F\| \leq 1$

$$|\mathbb{E}_{\eta_0}^N(\textstyle\prod_{i=1}^q F_n^i(\xi_{[0,n]}^i)) - \prod_{i=1}^q \mathbb{K}_{\eta_0,n}(F_n^i)|$$

$$\leq |\mathbb{E}_{\eta_0}^N((\mathbb{K}_n^N)^{\odot q}(F)) - \mathbb{E}_{\eta_0}^N((\mathbb{K}_n^N)^{\otimes q}(F))| + |\mathbb{E}_{\eta_0}^N((\mathbb{K}_n^N)^{\otimes q}(F)) - \mathbb{K}_{\eta_0,n}^{\otimes q}(F)|$$

$$= |\mathbb{E}_{\eta_0}^N([(\mathbb{K}_n^N)^{\odot q} - (\mathbb{K}_n^N)^{\otimes q}](F))| + |\mathbb{E}_{\eta_0}^N(H(\mathbb{K}_n^N)) - H(\mathbb{K}_{\eta_0,n})|$$

with the bounded continuous function H on $\mathcal{P}(E_{[0,n]})$ defined for any $\mu \in \mathcal{P}(E_{[0,n]})$ by $H(\mu) = \prod_{i=1}^q \mu(F_n^i)$. By (8.4) we get

$$\left|\mathbb{E}_{\eta_0}^N\left(\prod_{i=1}^q F_n^i(\xi_{[0,n]}^i)\right) - \prod_{i=1}^q \mathbb{K}_{\eta_0,n}(F_n^i)\right| \leq \frac{(q-1)^2}{N}$$
$$+ |\mathbb{E}_{\eta_0}^N(H(\mathbb{K}_n^N)) - H(\mathbb{K}_{\eta_0,n})|$$

Since $\mathbb{K}_n^N$ converge in law to $\mathbb{K}_{\eta_0,n}$, the proof of the lemma is easily completed. ∎

A stronger version of the propagation-of-chaos property is presented in the next definition.

Definition 8.2.2 *We say that the distribution $\mathbb{P}_{\eta_0}^N$ is strongly $\mathbb{K}_{\eta_0}$-chaotic if we have for any $n \in \mathbb{N}$ and $q \geq 1$*

$$\lim_{N\to\infty} \|\mathbb{P}_{\eta_0,n}^{(N,q)} - \mathbb{K}_{\eta_0,n}^{\otimes q}\|_{\mathrm{tv}} = 0$$

By symmetry arguments, we see that $\mathbb{P}_{\eta_0}^N$ is strongly $\mathbb{K}_{\eta_0}$-chaotic if and only if we have for any $n \geq 0$

$$\lim_{N\to\infty} \sup |\mathbb{E}_{\eta_0}^N((\mathbb{K}_n^N)^{\odot q}(F)) - \mathbb{K}_{\eta_0,n}^{\otimes q}(F)| = 0$$

where the supremum in the display above is taken over all functions $F \in \mathcal{B}_b(E_{[0,n]}^q)$ such that $\|F\| \leq 1$.

8.3 Outline of Results

As mentionned in the introductory section, Section 8.4 is only concerned with weak propagation-of-chaos properties. We examine the particle model associated with the McKean transitions (8.1) introduced on page 254. When the state spaces $(E_n, \mathcal{E}_n)$ are locally compact and separable metric spaces, we prove that the sequence of distributions $\mathbb{P}_{\eta_0}^N$ is weakly $\mathbb{K}_{\eta_0}$-chaotic. To describe with some precision our main result, we let $\Pi_n \subset \mathcal{C}_b(E_{[0,n]})$ be the subset of all tensor product functions of the form

$$F_n = f_0 \otimes \ldots \otimes f_n, \quad \text{where} \quad f_0 \in \mathcal{C}_b(E_0), \ldots, f_n \in \mathcal{C}_b(E_n)$$

with $\vee_{p=0}^n \|f_p\| \leq 1$. We also denote by $\Pi_n^q \subset \mathcal{C}_b(E_{[0,n]}^q)$, with $q \geq 1$, the subset of q-tensor product functions of the following form

$$H_n = F_n^1 \otimes \ldots \otimes F_n^q, \quad \text{where} \quad F_n^i \in \Pi_n$$

Theorem 8.3.1 *For any $n \in \mathbb{N}$ and $p \geq 1$, we have*

$$\sup_{F_n \in \Pi_n} \mathbb{E}_{\eta_0}^N(|\mathbb{K}_n^N(F_n) - \mathbb{K}_{\eta_0,n}(F_n)|^p)^{\frac{1}{p}} \leq a(p)\ b(n)/\sqrt{N}$$

and

$$\sup_{H_n \in \Pi_n^p} |\mathbb{E}_{\eta_0}^N((\mathbb{K}_n^N)^{\otimes p}(H_n)) - \mathbb{K}_{\eta_0,n}^{\otimes p}(H_n)| \leq a(p)\ b(n)/N \quad (8.5)$$

By the Stone-Weierstrass theorem, the set of all finite linear combinations of functions in Π_n is a dense subset of $\mathcal{C}_b(E_{[0,n]})$ as soon as the "marginal state spaces" E_n are locally compact and separable metric spaces. Using this density argument, we conclude that $\mathbb{P}_{\eta_0}^N$ is weakly $\mathbb{K}_{\eta_0}$-chaotic.

The next sections, Section 8.5 to Section 8.9, cover general measurable state-space models and discuss strong propagation-of-chaos estimates. To describe these results precisely, we let $Q_{p,n}$, respectively $Q_{p,n}^{(q)}$, be the linear semigroup associated with the unnormalized Feynman-Kac distributions γ_n and respectively $\gamma_n^{\otimes q}$. Notice that $Q_{p,n}(f_n) = G_{p,n}\ P_{p,n}(f_n)$ with the potential function $G_{p,n}$ and the Markov transition $P_{p,n}$

$$G_{p,n} = Q_{p,n}(1) \quad \text{and} \quad P_{p,n}(f_n) = Q_{p,n}(f_n)/Q_{p,n}(1)$$

(see for instance Section 7.2 for a brief overview on Feynman-Kac semigroups). Let $(G_{p,n}^{(q)}, P_{p,n}^{(q)})$ be the corresponding pair potential and Markov transition associated with the semigroup $Q_{p,n}^{(q)}$.

As usual, the asymptotic estimates developed in this chapter are expressed in terms of the parameters $(r_{p,n}, \beta(P_{p,n}))$ introduced on page 218. To simplify the notation, sometimes we write r_n instead of $r_{n,n+1}$.

In Section 8.5, we discuss increasing propagation-of-chaos estimates with respect to the relative entropy criterion. The first main result is the following theorem.

8.3 Outline of Results   259

Theorem 8.3.2 *Suppose the Markov transitions M_n satisfy the regularity condition $(M)^{(p)}$ stated on page 116, for $p = 2$ and some functions k_n. Then for any $q \leq N$ we have that*

$$N \operatorname{Ent}(\mathbb{P}^{(N,q)}_{\eta_0,n} \mid \mathbb{K}^{\otimes q}_{\eta_0,n}) \leq b(n)\, q$$

for some finite constant

$$b(n) \leq c \sum_{p=0}^{n-1} r_p^2 \, (1 + \eta_{p+1}(|k_{p+1}|^2))\, [\sum_{q=0}^{p} r_{q,p}\, \beta(P_{q,p})]^2$$

with $|k_n| = \sup_{x_{n-1} \in E_{n-1}} k_n(x_{n-1}, \cdot) \in \mathbb{L}_2(\eta_n)$.

To illustrate another impact of this result in practice, we present hereafter an easily derived consequence of Theorem 8.3.2. For simplicity, we further assume that the Feynman-Kac model (7.1) is time-homogeneous $(E_n, G_n, M_n) = (E, G, M)$ and the following regularity condition is met for any $x, y \in E$ and for some $m \geq 1$ and $\epsilon(G), \epsilon(M) \in (0, 1]$:

$$(G, M): \quad G(x) \geq \epsilon(G)\, G(y) \quad \text{and} \quad M^m(x, \cdot) \geq \epsilon(M)\, M^m(y, \cdot) \tag{8.6}$$

Combining Theorem 8.3.2 with some well-known results on the stability of Feynman-Kac semigroup, we will prove the following increasing propagation-of-chaos properties.

Let $n(N)$ and $q(N)$, $N \geq 1$, be respectively a nondecreasing sequence of time horizons and particle block sizes such that $\lim_{N \to \infty} n(N)q(N)/N = 0$. In this situation, we have

$$\overline{\lim}_{N \to \infty} \frac{N}{q(N)n(N)} \operatorname{Ent}(\mathbb{P}^{(N,q)}_{\eta_0,n} \mid \mathbb{K}^{\otimes q}_{\eta_0,n}) \leq c\, \frac{m^2\, \eta(|k|^2)}{\epsilon^6(M)\, \varepsilon(G)^{4m}}$$

as soon as $\eta(|k|^2) =_{\text{def.}} \sup_{n \geq 1} \eta_n(|k_n|^2) < \infty$.

As mentioned in the introduction, to analyze precisely the limiting behavior of the path-space distributions $(\mathbb{K}_n^N)^{\odot q}$, we develop in Section 8.6 to Section 8.9 an original approach based on q-tensor product and path-space Feynman-Kac semigroups. This strategy enters in a natural way the dynamical structure of interactions in the study of the propagation-of-chaos properties. It allows us to use the stability properties of the limiting system to derive precise and uniform estimates with respect to the time parameter. In Section 8.8, we express precise strong propagation-of-chaos estimates in terms of Dobrushin's ergodic coefficient associated with a Markovian and Feynman-Kac type transition on a product space. This approach to strong propagation-of-chaos is restricted to the McKean interpretation model (8.1) with $\varepsilon_n = 0$, and the corresponding simple genetic mutation/selection model. Our first main result is the following theorem.

8. Propagation of Chaos

Theorem 8.3.3 *For any $N \geq q \geq 1$, we have*

$$N \, \|\mathbb{P}^{(N,q)}_{\eta_0,[n]} - \eta_n^{\otimes q}\|_{\text{tv}} \leq c \, q^2 \, \left(1 + \sum_{p=0}^{n} \beta(P^{(q)}_{p,n}) \, [1 + e_{p,n}\,(2q^2/N)]\right) \quad (8.7)$$

where $\beta(P^{(q)}_{p,n}) \in [0,1]$, represents the Dobrushin ergodic coefficient associated with the Markov transition $P^{(q)}_{p,n}$ and $e_{p,n} : (0,\infty) \to (0,\infty)$ is the collection of mappings defined by

$$e_{p,n}(u) \;=\; (r_{p,n}-1)^2 (1 + (r_{p,n}-1)\sqrt{u}) \, \exp\left((r_{p,n}-1)^2 \, u\right) \tag{8.8}$$

The estimate (8.7) holds true for a fairly general and abstract class of Feynman-Kac models. It can be used to analyze the strong propagation-of-chaos properties of genetic particle systems as well as those of the corresponding genealogical tree models.

We further suppose the regularity condition (G,M) is satisfied for some $m \geq 1$ and $\epsilon(G), \epsilon(M) \in (0,1]$. In this case we will deduce from Theorem 8.3.3 the following increasing propagation-of-chaos properties.

Let $n(N)$ and $q(N)$, $N \geq 1$, be respectively a nondecreasing sequence of time horizons and particle block sizes such that $\lim_{N \to \infty} n(N) q^2(N)/N = 0$. In this situation, we have

$$\overline{\lim}_{N \to \infty} \frac{N}{q^2(N) n(N)} \|\mathbb{P}^{(N,q(N))}_{\eta_0,[n(N)]} - \eta_{n(N)}^{\otimes q(N)}\|_{\text{tv}} \leq c \, /(\epsilon^m(G)\epsilon(M))^2$$

Note that Theorem 8.3.3 does not apply to the complete N-genealogical particle model $\xi_{[0,n]}$. Our second main result is the following theorem.

Theorem 8.3.4 *For any $n, q, N \geq 1$ such that $(n+1)q \leq N$ we have*

$$\|\mathbb{P}^{(N,q)}_{\eta_0,n} - (\eta_0 \otimes \ldots \otimes \eta_n)^q\|_{\text{tv}} \leq c \, \frac{q^2}{N}(n+1)^3 \left[1 + e_n(\frac{2(q(n+1))^2}{N})\right] \quad (8.9)$$

with the mapping $e_n(u)$ defined as in (8.8) by replacing the constants $r_{p,n}$ by $\overline{r}_n = \sup_{p \leq n} r_{p,n}$.

This second estimate readily implies the following increasing propagation-of-chaos property: If we have $\lim_{N \to \infty} q^2(N)/N = 0$, then for any $n \in \mathbb{N}$

$$\overline{\lim}_{N \to \infty} \frac{N}{q^2(N)} \left\|\mathbb{P}^{(N,q(N))}_{\eta_0,n} - (\eta_0 \otimes \ldots \otimes \eta_n)^{\otimes q(N)}\right\|_{\text{tv}} \leq b(n)$$

with $b(n) \leq c \, (n+1)^3 (1 + (\overline{r}_n - 1)^2)$. In the case of time-homogeneous models satisfying condition (G,M) for some $m \geq 1$ and $\epsilon(G), \epsilon(M) \in (0,1)$, we shall also prove that

$$b(n) \leq c \, (n+1)^3 /(\epsilon^m(G)\epsilon(M))^2$$

In Section 8.7, we measure the propagation-of-chaos properties of Boltzmann-Gibbs transformations. The complete proofs of Theorem 8.3.3 and Theorem 8.3.4 are housed in Section 8.9.

8.4 Weak Propagation of Chaos

This section is concerned with the following proof.

Proof of Theorem 8.3.1: For any function $F_n = \otimes_{p=0}^n f_p \in \Pi_n$, we have the decomposition

$$\mathbb{K}_n^N(F_n) - \mathbb{K}_{\eta_0,n}(F_n) = \mathbb{K}_{n-1}^N(F_{n-1,n}) - \mathbb{K}_{\eta_0,n-1}(F_{n-1,n}) + I_1 + I_2 \quad (8.10)$$

with $F_{n-1,n} = f_0 \otimes \ldots \otimes [f_{n-1} K_{n,\eta_{n-1}}(f_n)] \in \mathcal{C}_b(E_{[0,n-1]})$ and

$$I_1 = \frac{1}{N} \sum_{i=1}^{N} \left[\prod_{p=0}^{n-1} f_p(\xi_p^i) \right] [f_n(\xi_n^i) - K_{n,\eta_{n-1}^N}(f_n)(\xi_{n-1}^i)]$$

$$I_2 = \frac{1}{N} \sum_{i=1}^{N} \left[\prod_{p=0}^{n-1} f_p(\xi_p^i) \right] [K_{n,\eta_{n-1}^N}(f_n)(\xi_{n-1}^i) - K_{n,\eta_{n-1}}(f_n)(\xi_{n-1}^i)]$$

Conditionally on $\mathcal{F}_{n-1}^N$, the N random variables

$$U_n^i = \left[\prod_{p=0}^{n-1} f_p(\xi_p^i) \right] [f_n(\xi_n^i) - K_{n,\eta_{n-1}^N}(f_n)(\xi_{n-1}^i)]$$

are independent and $\mathbb{E}_{\eta_0}^N(U_n^i | \mathcal{F}_{n-1}^N) = 0$ for any $i \leq N$. Thus, by Lemma 7.3.3, we have $\sqrt{N} \, \mathbb{E}(|I_1|^p)^{\frac{1}{p}} \leq a(p)$, for some finite universal constant $a(p)$. To estimate I_2, we observe that

$$[K_{n,\eta_{n-1}^N} - K_{n,\eta_{n-1}}](f_n) = (1 - \varepsilon_{n-1} G_{n-1}) \left[\Phi_n(\eta_{n-1}^N) - \Phi_n(\eta_{n-1}) \right](f_n)$$

Using the $\mathbb{L}_p$ mean error estimates presented in Section 7.4, we find that

$$\sqrt{N} \, \mathbb{E}_{\eta_0}^N(\|K_{n,\eta_{n-1}^N}(f_n) - K_{n,\eta_{n-1}}(f_n)\|^p)^{\frac{1}{p}} \leq a(p) \, b(n)$$

This yields that $\sqrt{N} \, \mathbb{E}_{\eta_0}^N(|I_2|^p)^{\frac{1}{p}} \leq a(p) \, b(n)$. Let J_n be defined

$$J_n = \sqrt{N} \sup_{F_n \in \Pi_n} \mathbb{E}_{\eta_0}^N(|\mathbb{K}_n^N(F_n) - \mathbb{K}_{\eta_0,n}(F_n)|^p)^{\frac{1}{p}}$$

From previous calculations, we find that $J_n \leq a(p) \, C(n) + J_{n-1}$, and the end of the proof of the first assertion is now straightforward. To prove the mean value estimate, we note that

$$I_2 = \mathbb{K}_{n-1}^N(F'_{n-1}) \left[\Phi_n(\eta_{n-1}^N) - \Phi_n(\eta_{n-1}) \right](f_n)$$

with $F'_{n-1} = f_0 \otimes \ldots \otimes [f_{n-1}(1 - \varepsilon_{n-1} G_{n-1})]$. We use the decomposition

$$\mathbb{E}_{\eta_0}^N(I_2) = \mathbb{E}_{\eta_0}^N([\mathbb{K}_{n-1}^N - \mathbb{K}_{n-1,\eta_0}](F'_{n-1}) \left[\Phi_n(\eta_{n-1}^N) - \Phi_n(\eta_{n-1}) \right](f_n))$$

$$+ \mathbb{K}_{n-1,\eta_0}(F'_{n-1}) \, \mathbb{E}_{\eta_0}^N([\Phi_n(\eta_{n-1}^N) - \Phi_n(\eta_{n-1})](f_n))$$

By the Cauchy-Schwartz inequality, we find that $|\mathbb{E}_{\eta_0}^N(I_2)| \leq b(n)/N$. Consequently, if we set

$$\overline{J}_n = N \sup_{F_n \in \Pi_n} |\mathbb{E}_{\eta_0}^N(\mathbb{K}_n^N(F_n)) - \mathbb{K}_{\eta_0,n}(F_n)|$$

by (8.10) we find that $\overline{J}_n \leq \overline{J}_{n-1} + c(n)$. This clearly ends the proof (8.5) for $q = 1$. Suppose (8.5) holds true at rank $p = (q-1)$. We use the decomposition

$$\prod_{i=1}^q u_i - \prod_{i=1}^q v_i = (u_1 - v_1) \prod_{i=2}^q v_i + (u_1 - v_1)\left(\prod_{i=2}^q u_i - \prod_{i=2}^q v_i\right)$$
$$+ v_1 \left(\prod_{i=2}^q u_i - \prod_{i=2}^q v_i\right)$$

the induction hypothesis, and the Cauchy-Schwartz inequality to prove that for any $(F_n^i)_{1 \leq i \leq q} \in \Pi_n^q$

$$|\mathbb{E}_{\eta_0}^N(\prod_{i=1}^q \mathbb{K}_n^N(F_n^i)) - \prod_{i=1}^q \mathbb{K}_{n,\eta_0}(F_n^i)|$$
$$\leq |\mathbb{E}_{\eta_0}^N(\mathbb{K}_n^N(F_n^1)) - \mathbb{K}_{n,\eta_0}(F_n^1)| + \frac{b(n)}{N}(1 + a(q-1))$$

The end of the proof is now clear. ∎

8.5 Relative Entropy Estimates

In this section, we provide strong propagation estimates with respect to the relative entropy criterion for the interacting particle system associated with the McKean interpretation model defined in (8.1). Without further mention, we assume the Markov transitions M_n satisfy the regularity condition $(M)^{(p)}$ stated on page 116, for $p = 2$ and some functions k_n.

The main simplification due to this condition is that the law of the N-particle model is absolutely continuous with respect to the laws of N independent copies of the limiting distribution model. In this context, a natural tool for the analysis of a strong version of the propagation of chaos for mean field interacting particle systems is the following inequality due to Csiszar [69].

Lemma 8.5.1 (Csiszar) *Let $(E, \mathcal{E})$ be a measurable space and let $\mu^{(N)}$ be an exchangeable measure on the product space E^N such that $\mu^{(N)} << \eta^{\otimes N}$ for some $\eta \in \mathcal{P}(E)$. If $\mu^{(N,q)}$, $1 \leq q \leq N$, are the marginals of $\mu^{(N)}$ on the first q-coordinates, then we have*

$$\mathrm{Ent}(\mu^{(N,q)} \mid \eta^{\otimes q}) \leq \frac{q}{N}\left(1 + \frac{\{N/q\}}{[N/q]}\right) \mathrm{Ent}(\mu^{(N)} \mid \eta^{\otimes N}) \qquad (8.11)$$

where $[a]$ is the integer part of $a \in \mathbb{R}$ and $\{a\} = a - [a]$.

Proof:
The proof of (8.11) is quite simple. From the variational definition of the relative entropy

$$\mathrm{Ent}(\mu|\eta) = \sup_{f \in \mathcal{C}_b(E)} \{\mu(f) - \log \eta(\exp(f))\}$$

we have already $\mathrm{Ent}(\mu^{(N)} \mid \eta^{\otimes N}) \geq (\mu^{(N)}(f^{(q)}) - \log \eta^{\otimes N}(\exp f^{(q)}))$ with

$$f^{(q)}(x_1, \ldots, x_N) = \sum_{p=1}^{[N/q]} \varphi(x_{(p-1)q+1}, \ldots, x_{(p-1)q+q}), \qquad \varphi \in \mathcal{C}_b(E^q)$$

Since $\mu^{(N)}(f^{(q)}) = [N/q] \, \mu^{(N,q)}(\varphi)$ and $\eta^{\otimes N}(\exp f^{(q)}) = (\eta^{\otimes q}(\varphi))^{[N/q]}$, taking the supremum over $\varphi \in \mathcal{C}_b(E^q)$, one concludes that

$$\mathrm{Ent}(\mu^{(N)} \mid \eta^{\otimes N}) \geq [N/q] \, \mathrm{Ent}(\mu^{(N,q)} \mid \eta^{\otimes q})$$

We end the proof by noting that

$$\forall a \in [1, \infty], \quad \frac{1}{[a]} = \frac{1}{a} \frac{[a] + \{a\}}{[a]} = \frac{1}{a}\left(1 + \frac{\{a\}}{[a]}\right) \left(\leq \frac{2}{a}\right)$$

∎

Lemma 8.5.2 *If μ is absolutely continuous with respect to η and $\frac{d\mu}{d\eta} \in \mathbb{L}_2(\eta)$, then we have $\mathrm{Ent}(\mu|\eta) \leq \|1 - d\mu/d\eta\|_{2,\eta}^2$.*

Proof:
Using the standard inequality, $\log u \leq u - 1$, which is valid for any $u \geq 0$, we clearly have

$$\mathrm{Ent}(\mu|\eta) = \int \log \frac{d\mu}{d\eta} \, d\mu \leq \int \left(\frac{d\mu}{d\eta} - 1\right) d\mu = \int \left(\frac{d\mu}{d\eta} - 1\right) \frac{d\mu}{d\eta} \, d\eta$$

from which one concludes that $\mathrm{Ent}(\mu|\eta) \leq \int \left(\frac{d\mu}{d\eta} - 1\right)^2 d\eta = \left\|\frac{d\mu}{d\eta} - 1\right\|_{2,\eta}^2$.
∎

As we mentioned above, our strategy to obtain relative entropy estimates is based on the observation that

$$\mathbb{P}_{\eta_0, n}^{(N)} = \mathrm{Law}((\xi_{[0,n]}^i)_{1 \leq i \leq N}) \in \mathcal{P}(E_{[0,n]}^N)$$

is a "regular" mean field Gibbs measure. To describe the potential function, we use the the change of coordinate mapping

$$\Theta_n^N(\xi_0, \ldots, \xi_n)) = (\xi_{[0,n]}^i)_{1 \leq i \leq N}$$

264   8. Propagation of Chaos

defined in (8.2) to check that for any $F \in \mathcal{B}_b(E_{[0,n]}^N)$ we have

$$\mathbb{E}\left(F((\xi_{[0,n]}^i)_{1\leq i\leq N})\right)$$

$$= \mathbb{E}(F(\Theta_n^N(\xi_0,\ldots,\xi_n))) = \int_{E_{[0,n]}^{(N)}} F(\Theta_n^N(x_0,\ldots,x_n))$$

$$\times \left[\prod_{p=0}^{n-1}\prod_{i=1}^{N} \frac{dK_{p+1,m(x_p)}(x_p^i,\cdot)}{dK_{p+1,\eta_p}(x_p^i,\cdot)}(x_{p+1}^i)\right] \mathbb{K}_{n,\eta_0}^{\otimes N} \circ (\Theta_n^N)^{-1}(d(x_0,\ldots,x_n))$$

with $x_p = (x_p^i)_{1\leq i\leq N} \in E_p^N$ and $m(x_p) = \frac{1}{N}\sum_{i=1}^{N}\delta_{x_p^i}$. Therefore we find that

$$\mathbb{E}(F((\xi_{[0,n]}^i)_{1\leq i\leq N}))$$

$$= \mathbb{E}(F(\Theta_n^N(\xi_0,\ldots,\xi_n))) = \int_{E_{[0,n]}^{(N)}} F(\Theta_n^N(x_0,\ldots,x_n))$$

$$\times \exp\{H_n^{(N)}(\theta_n^N(x_0,\ldots,x_n))\} \mathbb{K}_{n,\eta_0}^{\otimes N} \circ (\Theta_n^N)^{-1}(d(x_0,\ldots,x_n))$$

with the interaction Hamiltonian function $H_n^{(N)} : E_{[0,n]}^N \to [0,\infty)$ given for any $x_{[0,n]} = (x_{[0,n]}^i)_{1\leq i\leq N}$ with $x_{[0,n]}^i = (x_0^i,\ldots,x_n^i)$ by

$$H_n^{(N)}(x_{[0,n]})$$

$$= N\sum_{p=0}^{n-1}\int_{E_p\times E_{p+1}} m(x_p,x_{p+1})(d(u,v))\ \log\frac{dK_{p+1,m(x_p)}(u,\cdot)}{dK_{p+1,\eta_p}(u,\cdot)}(v)$$

One concludes that the distribution $\mathbb{P}_{\eta_0,n}^{(N)} \in \mathcal{P}(E_{[0,n]}^N)$ is absolutely continuous with respect to the tensor product measure $\mathbb{K}_{\eta_0,n}^{\otimes N} \in \mathcal{P}(E_{[0,n]}^N)$, and we have

$$\frac{d\mathbb{P}_{\eta_0,n}^{(N)}}{d\mathbb{K}_{\eta_0,n}^{\otimes N}} = \exp H_n^{(N)} \qquad \mathbb{K}_{\eta_0,n}^{\otimes N}\text{- a.e.}$$

To get one step further in our discussion we consider the function

$$\mathbb{M}_n : \mu \in \mathcal{P}(E_{[0,n]}) \longrightarrow \mathbb{M}_n(\mu) \in \mathcal{P}(E_{[0,n]})$$

that associates to a given path measure μ on $E_{[0,n]}$ the McKean distribution associated with the flow μ_k of kth time marginal and given by

$$\mathbb{M}_n(\mu)(d(u_0,\ldots,u_n)) = \eta_0(du_0)\times K_{1,\mu_0}(u_0,du_1)\times\ldots\times K_{n,\mu_{n-1}}(u_{n-1},du_n)$$

Note that the limiting McKean measure $\mathbb{K}_{n,\eta_0} = \mathbb{M}_n(\mathbb{K}_{n,\eta_0})$ is a fixed point of $\mathbb{M}_n$. It is now easy to check that $H_n^{(N)}$ can be rewritten as

$$H_n^{(N)}(x_{[0,n]}) = N\ \mathcal{H}_n(m(x_{[0,n]}))$$

8.5 Relative Entropy Estimates

with the potential function $\mathcal{H}_n$ on $\mathcal{P}(E_{[0,n]})$ defined for any $\mu \in \mathcal{P}(E_{[0,n]})$ by

$$\mathcal{H}_n(\mu) = \int \log \frac{d\mathbb{M}_n(\mu)}{d\mathbb{K}_{n,\eta_0}} \, d\mu$$

$$= \sum_{p=0}^{n-1} \int_{E_p \times E_{p+1}} \mu_{p,p+1}(d(u,v)) \, \log \frac{dK_{p+1,\mu_p}(u,\cdot)}{dK_{p+1,\eta_p}(u,\cdot)}(v)$$

where $\mu_{p,p+1}$ stands for the $(p, p+1)$-th pair-time marginal of μ. This readily shows that $\mathbb{P}_{\eta_0,n}^{(N)}$ is an exchangeable distribution on the product and path space $E_{[0,n]}^N$.

Lemma 8.5.1 highlights the relations between the relative entropy and the mean value of the potential function $H_n^{(N)}$. More precisely, according to Lemma 8.5.1, we have that

$$\operatorname{Ent}\left(\mathbb{P}_{\eta_0,n}^{(N,q)} \,\big|\, \mathbb{K}_{\eta_0,n}^{\otimes q}\right) \leq 2\frac{q}{N} \operatorname{Ent}\left(\mathbb{P}_{\eta_0,n}^{(N)} \,\big|\, \mathbb{K}_{\eta_0,n}^{\otimes N}\right)$$

$$= 2\frac{q}{N} \, \mathbb{E}(H_n^{(N)}(\Theta_n^N(\xi_0,\ldots,\xi_n))) \quad (8.12)$$

We end this section with some regularity properties of these potential functions. First note that $\mathcal{H}_n(\mathbb{K}_{n,\eta_0}) = 0$ and for any $\mu \in \mathcal{P}(E_{[0,n]})$ we have

$$\mathbb{E}(\mathcal{H}_n(m(\xi_{[0,n]})))$$

$$= \sum_{p=0}^{n-1} \mathbb{E}\left(\int_{E_p} K_{p+1,m(\xi_p)}(\xi_p^1, dv) \, \log \frac{dK_{p+1,m(\xi_p)}(\xi_p^1,\cdot)}{dK_{p+1,\eta_p}(\xi_p^1,\cdot)}(v)\right)$$

$$= \sum_{p=0}^{n-1} \mathbb{E}\left[\operatorname{Ent}\left(K_{p+1,m(\xi_p)}(\xi_p^1,\cdot) \,\big|\, K_{p+1,\eta_p}(\xi_p^1,\cdot)\right)\right]$$

By lemma 8.5.2, we find that

$$|\mathbb{E}(\mathcal{H}_n(m(\xi_{[0,n]})))|$$

$$\leq \sum_{p=0}^{n-1} \mathbb{E}\left(\int_{E_{p+1}} K_{p+1,\eta_p}(\xi_p^1, dv) \, \left|\frac{dK_{p+1,m(\xi_p)}(\xi_p^1,\cdot)}{dK_{p+1,\eta_p}(\xi_p^1,\cdot)}(v) - 1\right|^2\right)$$

$$\leq \sum_{p=0}^{n-1} \mathbb{E}\left(\int_{E_{p+1}} d\eta_{p+1} \frac{d\eta_{p+1}}{dK_{p+1,\eta_p}(\xi_p^1,\cdot)}\right.$$

$$\left. \times \left|\frac{dK_{p+1,m(\xi_p)}(\xi_p^1,\cdot)}{d\eta_{p+1}} - \frac{dK_{p+1,\eta_p}(\xi_p^1,\cdot)}{d\eta_{p+1}}\right|^2\right) \quad (8.13)$$

On the other hand, we have

$$\frac{dK_{n+1,\mu_n}(u,\cdot)}{d\eta_{n+1}}(v)$$

$$= \varepsilon_n G_n(u)\, k_{n+1}(u,v) + (1 - \varepsilon_n G_n(u))\, \frac{\mu_n(G_n k_{n+1}(\cdot,v))}{\mu_n(G_n)}$$

Recalling that

$$\frac{\eta_n(G_n k_{n+1}(\cdot,v))}{\eta_n(G_n)} = \frac{d\Phi_{n+1}(\eta_n)}{d\eta_{n+1}}(v) = 1$$

we find that

$$\frac{dK_{n+1,\eta_n}(u,\cdot)}{d\eta_{n+1}}(v) = \varepsilon_n G_n(u)\, k_{n+1}(u,v) + (1 - \varepsilon_n G_n(u))\ (\geq (1 - \varepsilon_n G_n(u)))$$

This implies that

$$\frac{dK_{n+1,\mu_n}(u,\cdot)}{d\eta_{n+1}}(v) - \frac{dK_{n+1,\eta_n}(u,\cdot)}{d\eta_{n+1}}(v)$$

$$= (1 - \varepsilon_n G_n(u))\, [\frac{\mu_n(G_n k_{n+1}(\cdot,v))}{\mu_n(G_n)} - \frac{\eta_n(G_n k_{n+1}(\cdot,v))}{\eta_n(G_n)}]$$

and for any transition (x_n, x_{n+1})

$$(1 - \varepsilon_n G_n(u))\, \frac{d\eta_{n+1}}{dK_{n+1,\eta_n}(x_n,\cdot)}(x_{n+1}) \leq 1$$

By (8.13), we readily obtain the estimate

$$|\mathbb{E}(\mathcal{H}_n(m(\xi_{[0,n]})))| \leq \sum_{p=0}^{n-1} \int_{E_{p+1}} \eta_{p+1}(dv)\, \mathbb{E}\left(\left|\frac{m(\xi_p)(G_p k_{p+1}(\cdot,v))}{m(\xi_p)(G_p)} - 1\right|^2\right)$$

Since we have

$$|\frac{m(\xi_p)(G_p k_{p+1}(\cdot,v))}{m(\xi_p)(G_p)} - 1| = |\frac{1}{m(\xi_p)(\tilde{G}_p)}|$$

$$\times |m(\xi_p)(\tilde{G}_p[k_{p+1}(\cdot,v) - 1])|$$

$$\leq r_p^2 (1 + |k_{p+1}|(v))\, |m(\xi_p)(f_p^v) - \eta_p(f_p^v)|$$

with

$$f_p^v = \frac{1}{r_p(1 + |k_{p+1}|(v))}\, \tilde{G}_p\, [k_{p+1}(\cdot,v) - 1] \quad \text{and} \quad \tilde{G}_p = G_p/\eta_p(G_p)$$

we finally arrive at

$$|\mathbb{E}(\mathcal{H}_n(m(\xi_{[0,n]})))|$$

$$\leq c \sum_{p=0}^{n-1} r_p^2 \int_{E_{p+1}} \eta_{p+1}(dv)(1 + |k_{p+1}|^2(v))\, \mathbb{E}(|m(\xi_p)(f_p^v) - \eta_p(f_p^v)|^2)$$

This simple estimate allows to apply most of the convergence results presented in Section 7.4.2 and Section 7.4.3. For instance, by Theorem 7.4.4, we conclude that

$$N\,|\mathbb{E}(\mathcal{H}_n(m(\xi_{[0,n]})))| \le c \sum_{p=0}^{n-1} r_p^2 \left(1 + \eta_{p+1}(|k_{p+1}|^2)\right) \sum_{q=0}^{p} r_{q,p} \beta(P_{q,p})$$

This ends the proof of the theorem. ∎

Arguing as in the proof of (7.26), we prove the following corollary.

Corollary 8.5.1 *Suppose that* $\eta(|k|^2) =_{\text{def.}} (1 + \sup_{n \ge 1} \eta_n(|k_n|^2)) < \infty$ *and conditions* (G) *and* $(M)_m$ *hold true for some* $m \ge 1$ *with* $\epsilon(G) = \wedge_n \epsilon_n(G) > 0$ *and* $\epsilon(M) = \wedge_n \epsilon_n(M) > 0$. *Then, we have*

$$\mathrm{Ent}(\mathbb{P}^{(N,q)}_{\eta_0,n} \mid \mathbb{K}^{\otimes q}_{\eta_0,n}) \le c\,\frac{q\,n}{N}\,\frac{m^2\,\eta(|k|^2)}{\epsilon^6(M)\,\varepsilon(G)^{4m}}$$

8.6 A Combinatorial Transport Equation

Throughout this section, $(E, \mathcal{E})$ denotes an arbitrary measurable space. In the further development of this section, we fix the integer $N \ge 1$ and for any $x = (x^1, \ldots, x^N) \in E^N$ we slightly abuse the notation and we denote by $m(x) = \frac{1}{N} \sum_{i=1}^{N} \delta_{x^i} \in \mathcal{P}(E)$ the empirical measure associated with the N-tuple x. For any $1 \le q \le N$, we introduce the empirical measures on E^q defined by

$$m(x)^{\otimes q} = \frac{1}{N^q} \sum_{\alpha \in \langle N \rangle^{\langle q \rangle}} \delta_{(x^{\alpha(1)}, \ldots, x^{\alpha(q)})}$$

$$m(x)^{\odot q} = \frac{1}{(N)_q} \sum_{\alpha \in \langle q, N \rangle} \delta_{(x^{\alpha(1)}, \ldots, x^{\alpha(q)})}$$

with the sets $\langle q, N \rangle$, $\langle q \rangle$, and $\langle N \rangle$ defined on page 256. Note that each mapping $\alpha \in \langle N \rangle^{\langle q \rangle}$ induces a unique equivalence relation $\sim_\alpha$ on $\langle q \rangle$ defined for any $i, j \in \langle q \rangle$ by

$$i \sim_\alpha j \iff \alpha(i) = \alpha(j)$$

The corresponding set of equivalence classes $\langle q \rangle_\alpha$ can alternatively be regarded as a partition π_α of the set $\langle q \rangle$. More precisely, if $b(\pi_\alpha)$ stands for the cardinality of the set $\alpha(\langle q \rangle)$, then we have

$$\pi_\alpha = \{\pi_\alpha(1), \ldots, \pi_\alpha(b(\pi_\alpha))\} \quad \text{with} \quad \pi_\alpha(i) \ne \pi_\alpha(j) \quad \text{for any} \quad i \ne j$$

and

$$\langle q \rangle = \cup_{i=1}^{b(\pi_\alpha)} \pi_\alpha(i) \quad \text{with} \quad \pi_\alpha(i) = \{j \in \langle q \rangle\,:\, \alpha(j) = \alpha(i)\}$$

Inversely to each partition π of the set $\langle q \rangle$ with $b(\pi)$ blocks, we can associate in a unique way $(N)_q$ different mappings $\alpha \in \langle N \rangle^{\langle q \rangle}$. To be more precise, let $\leq$ be the order relation on the subsets of $\langle q \rangle$ defined for any $A, B \subset \langle q \rangle$ by

$$A \leq B \iff \inf\{i \,:\, i \in A\} \leq \inf\{i \,:\, i \in B\}$$

Notice that the $b(\pi_\alpha)$ blocks of partition π of $\langle q \rangle$ can be written in the increasing order

$$\pi_1 \leq \pi_2 \leq \ldots \leq \pi_{b(\pi_\alpha)}$$

We associate with π and with each one-to-one mapping $\beta \in \langle b(\pi), N \rangle$ the mapping $\alpha_\beta^\pi \in \langle N \rangle^{\langle q \rangle}$ defined by

$$\alpha_\beta^\pi = \sum_{i=1}^{b(\pi)} \beta(i) 1_{\pi_i}$$

From these one-to-one associations, we find the decomposition

$$\langle N \rangle^{\langle q \rangle} = \cup_{p=1}^{q} \cup_{\pi : b(\pi) = p} \{\alpha_\beta^\pi \,:\, \beta \in \langle p, N \rangle\}$$

In this notation, for any $x \in E^N$ and any numerical function f on E^q, we have that

$$\begin{aligned}
m(x)^{\otimes q}(f) &= \frac{1}{N^q} \sum_{p=1}^{q} \sum_{\pi : b(\pi) = p} \sum_{\beta \in \langle p, N \rangle} f(x^{\alpha_\beta^\pi(1)}, \ldots, x^{\alpha_\beta^\pi(q)}) \\
&= \frac{1}{N^q} \sum_{p=1}^{q} \sum_{\pi : b(\pi) = p} \sum_{\beta \in \langle p, N \rangle} C_\pi^{p,q}(f)(x^{\beta(1)}, \ldots, x^{\beta(p)})
\end{aligned}$$

with the Markov kernel $C_\pi^{p,q}$ from E^p into E^q defined by

$$C_\pi^{p,q}(f)(x^1, \ldots, x^p) = f\left(\sum_{i=1}^{p} x^i 1_{\pi_i}(1), \ldots, \sum_{i=1}^{p} x^i 1_{\pi_i}(q)\right)$$

It is now convenient to observe that for any $p \leq q$ we have

$$\frac{1}{(N)_p} \sum_{\beta \in \langle p, N \rangle} C_\pi^{p,q}(f)(x^{\beta(1)}, \ldots, x^{\beta(p)})$$

$$= \frac{1}{(N)_q} \sum_{\beta \in \langle q, N \rangle} C_{q,\pi}^{p}(f)(x^{\beta(1)}, \ldots, x^{\beta(q)})$$

with the extended Markov kernel $C_{q,\pi}^p$ from E^q into E^p defined by

$$C_{q,\pi}^p(f)(x^1, \ldots, x^q) = C_\pi^{p,q}(f)(x^1, \ldots, x^p)$$

From previous considerations, we arrive at

$$m(x)^{\otimes q} = \frac{1}{N^q} \sum_{p=1}^{q} (N)_p \; S(p,q) \; m(x)^{\odot q} C_q^p$$

with the Markov transitions C_q^p, $p \leq q$, on E^q defined by the formula

$$C_q^p = \frac{1}{(N)_p} \sum_{\beta \in \langle q, N \rangle} \frac{1}{S(p,q)} \sum_{\pi : b(\pi) = p} C_{q,\pi}^p$$

In the formulae displayed above, $S(p,q)$ stands for the Stirling number of the second kind corresponding to the number of partitions of q elements in p blocks. Using the fact that $S(q,q) = 1$ and $C_{q,\pi}^q = Id$, we prove easily the following result.

Proposition 8.6.1 *For any $x \in E^N$ and $1 \leq q \leq N$, we have*

$$m(x)^{\otimes q} = m(x)^{\odot q} R_N^{(q)} \quad \text{with} \quad R_N^{(q)} = \frac{(N)_q}{N^q} Id + \left(1 - \frac{(N)_q}{N^q}\right) \tilde{R}_N^{(q)}$$

and the Markov kernel $\tilde{R}_N^{(q)}$ on E^q defined by

$$\tilde{R}_N^{(q)} = \frac{1}{N^q - (N)_q} \sum_{p=1}^{q-1} (N)_p \, S(p,q) \, C_q^p$$

One easy consequence of this formula is that

$$\begin{aligned}
\|m(x)^{\otimes q} - m(x)^{\odot q}\|_{\text{tv}} &= \left(1 - \frac{(N)_q}{N^q}\right) \|m(x)^{\odot q}(\tilde{R}_N^{(q)} - Id)\|_{\text{tv}} \\
&\leq \left(1 - \frac{(N)_q}{N^q}\right) \\
&\leq 1 - \left(1 - \frac{q-1}{N}\right)^{q-1} \leq \frac{(q-1)^2}{N} \quad (8.14)
\end{aligned}$$

We end this discussion with a more probabilistic connection between $m(x)^{\otimes q}$ and $m(x)^{\odot q}$. We first observe that for any $q \geq 1$ and any f on E^{q+1}

$$\begin{aligned}
&(m(x) \otimes m(x)^{\odot q})(f) \\
&= \frac{1}{N(N)_q} \sum_{i=1}^{N} \sum_{\alpha \in \langle q, N \rangle} f(x^i, x^{\alpha(1)}, x^{\alpha(2)}, \ldots, x^{\alpha(q)}) \\
&= \frac{1}{N(N)_q} \sum_{\alpha \in \langle q+1, N \rangle} f(x^{\alpha(1)}, x^{\alpha(2)}, \ldots, x^{\alpha(q+1)}) \\
&\quad + \frac{1}{N(N)_q} \sum_{\alpha \in \langle q, N \rangle} \sum_{i=1}^{q} f(x^{\alpha(i)}, x^{\alpha(1)}, x^{\alpha(2)}, \ldots, x^{\alpha(q)}) \\
&= \left(1 - \frac{q}{N}\right) m(x)^{\odot(q+1)}(f) + \frac{q}{N} \, m(x)^{\odot(q+1)}(\tilde{r}^{(q+1)}(f))
\end{aligned}$$

with the Markov transition $\tilde{r}_{q+1}$ on E^{q+1} defined by

$$\tilde{r}^{(q+1)}(f)(x^0, x^1, \ldots, x^q) = \frac{1}{q} \sum_{i=1}^{q} f(x^i, x^1, \ldots, x^q)$$

This readily yields that

$$m(x) \otimes m(x)^{\odot q} = m(x)^{\odot(q+1)} r_N^{(q+1)} \quad \text{with} \quad r_N^{(q+1)} = \left(1 - \frac{q}{N}\right) Id + \frac{q}{N} \tilde{r}^{(q+1)}$$

The probabilistic interpretation of $r_N^{(q+1)}$ is quite elementary. Starting from a given configuration $(x^0, x^1, \ldots, x^q) \in E^{q+1}$, the Markov transition consists of keeping this $(q+1)$-tuple with a probability $\left(1 - \frac{q}{N}\right)$ and otherwise replacing the first component x^0 by choosing randomly and uniformly one of the other components $x^1, \ldots, x^q$. To develop an inductive construction, we associate with a given transition r on some product space E^q a transition $\text{Ext}(r)$ on some product space E^{q+1} by setting

$$\text{Ext}(r)((x^0, x^1, \ldots, x^q), d(y^0, x^1, \ldots, y^q))$$

$$= \delta_{x^0}(dy^0) \; r((x^1, \ldots, x^q), d(x^1, \ldots, y^q))$$

In this somewhat abusive notation, we have for instance

$$\begin{aligned} m(x)^{\otimes 2} &= m(x)^{\odot 2} r_N^{(2)} \\ m(x)^{\otimes 3} &= m(x) \otimes m(x)^{\otimes 2} \\ &= m(x) \otimes (m(x)^{\odot 2} r_N^{(2)}) \\ &= (m(x) \otimes m(x)^{\odot 2}) \text{Ext}(r_N^{(2)}) = m(x)^{\odot 3} r_N^{(3)} \text{Ext}(r_N^{(2)}) \end{aligned}$$

More generally, if we define using backward induction

$$\mathcal{R}_N^{(q+1)} = r_N^{(q+1)} \text{Ext}(\mathcal{R}_N^{(q)}) \quad \text{with} \quad \mathcal{R}_N^{(2)} = r_N^{(2)}$$

then we conclude that $m(x)^{\otimes q} = m(x)^{\odot q} \mathcal{R}_N^{(q)}$. To describe more precisely the Markov transition $\mathcal{R}_N^{(q)}$, we introduce a sequence $\epsilon^{(q)} = (\epsilon_1^{(q)}, \ldots, \epsilon_q^{(q)})$ of q independent and $\{0, 1\}$-valued random variables with respective distributions

$$\mathbb{P}(\epsilon_i^{(q)} = 0) = 1 - \mathbb{P}(\epsilon_i^{(q)} = 1) = \frac{q-i}{N}$$

Notice that for $i = q$ we have $\epsilon_q^{(q)} = 1$. We also associate with a given configuration $(x^1, \ldots, x^q) \in E^q$ a collection of independent (and independent of $\epsilon^{(q)}$) random variables $(\tilde{x}^{(q,1)}, \ldots, \tilde{x}^{(q,q)})$ with respective distributions $\mathbb{P}(\tilde{x}^{(q,i)} \in dy) = \frac{1}{q-i} \sum_{j=i+1}^{q} \delta_{x^j}$ with the convention $\tilde{x}^{(q,q)} = x^q$. From the inductive construction of $\mathcal{R}_N^{(q)}$, we observe that the E^q-valued random variable

$$\widehat{x}^{(q)} = (\widehat{x}^{(q,1)}, \ldots, \widehat{x}^{(q,q)}) \quad \text{with} \quad \widehat{x}^{(q,i)} = \epsilon_i^{(q)} x^i + (1 - \epsilon_i^{(q)}) \tilde{x}^{(q,i)}$$

is distributed according to $\mathcal{R}_N^{(q)}((x^1,\ldots,x^q),.)$. ∎

8.7 Asymptotic Properties of Boltzmann-Gibbs Distributions

Let μ be a probability measure on a given measurable state space $(E, \mathcal{E})$. During the further development of this section, we fix an integer $N \geq 1$ and we denote by $m(X) = \frac{1}{N} \sum_{i=1}^{N} \delta_{X^i}$ the N-empirical measure associated with a collection of independent and identically distributed random variables $X = (X^i)_{i \geq 1}$, with common law μ. We denote by $m(X)^{\otimes q}$ and $m(X)^{\odot q}$, $q \leq N$, the random distributions on E^q defined by

$$m(X)^{\otimes q} = \frac{1}{N^q} \sum_{\alpha \in \langle N \rangle^{\langle q \rangle}} \delta_{(X^{\alpha(1)},\ldots,X^{\alpha(q)})}$$

$$m(X)^{\odot q} = \frac{1}{(N)_q} \sum_{\alpha \in \langle q, N \rangle} \delta_{(X^{\alpha(1)},\ldots,X^{\alpha(q)})}$$

(with the sets $\langle q, N \rangle$, $\langle q \rangle$, and $\langle N \rangle$ defined on page 256). Let $g = (g_i)_{i \geq 1}$ be a collection of $\mathcal{E}$-measurable and nonnegative functions on E such that $\mu(g_i) \in (0, \infty)$ for each $i \geq 1$. For any fixed integer $q \geq 1$, we denote by $g^{(q)}$ the q-tensor product function on E^q defined by

$$g^{(q)} = g_1 \otimes \ldots \otimes g_q : (x^1,\ldots,x^q) \in E^q \longrightarrow g_1(x_1)\ldots g_q(x^q) \in (0, \infty)$$

We note that

$$m(X)^{\otimes q}(g^{(q)}) = \prod_{i=1}^{q} m(X)(g_i) \quad \text{and} \quad \mu^{\otimes q}(g^{(q)}) = \prod_{i=1}^{q} m(X)(g_i)$$

It is also convenient to introduce the mapping $e_{\mu,g}$ from $(0, \infty)$ into itself and defined by

$$e_{\mu,g}(u) = \operatorname{osc}_\mu^2(g)(1 + \operatorname{osc}_\mu(g) \sqrt{u}) \exp\left(\operatorname{osc}_\mu^2(g) u\right)$$

with $\operatorname{osc}_\mu(g) = \sup_{i \geq 1} \operatorname{osc}(g_i/\mu(g_i))$. When the potential functions g are chosen such that $\mu(g_i) = 1$ for any $i \geq 1$, we simplify the notation and we write e_g instead of $e_{\mu,g}$ to emphasize that the function does not depend on μ.

We associate with the pair (g, q) the Boltzmann-Gibbs transformation $\Psi^{(q)} : \mathcal{P}(E^q) \to \mathcal{P}(E^q)$ defined for any $(\eta, f) \in \mathcal{P}(E^q) \times \mathcal{B}_b(E^q)$ by the formula

$$\Psi^{(q)}(\eta)(f) = \eta(g^{(q)} f)/\eta(g^{(q)})$$

272   8. Propagation of Chaos

The main object of this section is to analyze the asymptotic properties of the random distributions $\Psi^{(q)}(m(X)^{\otimes q})$ as the pair parameter (q, N) tends to infinity. Our main result is the following theorem.

Theorem 8.7.1 *Let $(g_i)_{i \geq 1}$ be a collection of measurable functions g_i with uniformly bounded oscillations $\mathrm{osc}(g) = \sup_{i \geq 1} \mathrm{osc}(g_i) < \infty$. For any $N \geq q \geq 1$ and $f \in \mathcal{B}_b(E^q)$ with $\mathrm{osc}(f) \leq 1$, we have*

$$|\mathbb{E}(\Psi^{(q)}(m(X)^{\otimes q})(f)) - \Psi^{(q)}(\mu^{\otimes q})(f)| \leq c \frac{q^2}{N} \left[1 + e_{\mu,g}\left(\frac{2q^2}{N}\right) \right] \quad (8.15)$$

and for any $n \geq 1$

$$\mathbb{E}([\Psi^{(q)}(m(X)^{\otimes q})(f) - \Psi^{(q)}(\mu^{\otimes q})(f)]^{2n})$$

$$\leq c\, 2^{4n}\, \frac{(nq)^2}{N} \left[1 + e_{\mu,g}\left(\frac{2(nq)^2}{N}\right) \right] \quad (8.16)$$

Theorem 8.7.1 will be proved at the end of the section. In order to prepare for its proof, we first present a technical lemma of separate interest.

Lemma 8.7.1 *Let $(g_i)_{i \geq 1}$ be a collection of measurable functions g_i with uniformly bounded oscillations $\mathrm{osc}(g) = \sup_{i \geq 1} \mathrm{osc}(g_i) < \infty$ and such that $\mu(g_i) = 1$ for any $i \geq 1$. Then, for any $n \geq 1$ we have*

$$| \mathbb{E}([m(X)^{\otimes q}(g^{(q)}) - 1]^n) | \leq 2^{n-1} \frac{(nq)^2}{N} e_g\left(\frac{(nq)^2}{2N}\right) \quad (8.17)$$

Proof:
We first prove (8.17) for $n = 1$. Using the decomposition

$$\prod_{i=1}^{q}(1 + a_i) = 1 + \sum_{1 \leq p \leq q} \sum_{1 \leq i_1 < \cdots < i_p \leq q} \prod_{j=1}^{p} a_{i_j}$$

which is valid for any $q \geq 0$ and any collection of real numbers $(a_i)_{i \geq 1}$, we find that

$$\mathbb{E}\left(\prod_{i=1}^{q} m(X)(g_i)\right) - 1 = \sum_{2 \leq p \leq q} \sum_{1 \leq i_1 < \cdots < i_p \leq q} \mathbb{E}\left(\prod_{j=1}^{p}[m(X)(g_j) - 1]\right)$$

Using Holder's inequality, we find that

$$\left| \mathbb{E}\left(\prod_{i=1}^{q} m(X)(g_i)\right) - 1 \right| = \sum_{2 \leq p \leq q} C_q^p\, \mathbb{E}(|m(X)(g) - 1|^p)$$

with

$$\mathbb{E}(|m(X)(g) - 1|^p) = \sup_{i \geq 1}\, \mathbb{E}(|m(X)(g_j) - 1|^p)$$

8.7 Asymptotic Properties of Boltzmann-Gibbs Distributions

Suppose $q = 2q'$ is an even integer. In this case, using the first part of Lemma 7.3.3, we find that

$$|\mathbb{E}(\prod_{i=1}^{2q'} m(X)(g_i)) - 1| \leq \sum_{p=1}^{q'} C_{2q'}^{2p} (2p)_p \left(\frac{\operatorname{osc}^2(g)}{2N}\right)^p$$

$$+ \sum_{p=1}^{q'-1} C_{2q'}^{2p+1} \frac{(2p+1)_{p+1}}{\sqrt{p+1/2}} \left(\frac{\operatorname{osc}^2(g)}{2N}\right)^{p+1/2}$$

In the display above, we have used the notation $\operatorname{osc}(g) = \sup_{i \geq 1} \operatorname{osc}(g_i)$. Since we have the estimates

$$C_{2q'}^{2p} (2p)_p = \frac{1}{p!} \frac{(2q')!}{(2q'-2p)!} = \frac{(2q')_{2p}}{p!} \leq \frac{(2q')^{2p}}{p!} = \frac{q^{2p}}{p!}$$

$$C_{2q'}^{2p+1} (2p+1)_{p+1} = \frac{1}{p!} \frac{(2q')!}{(2q'-(2p+1))!} = \frac{(2q')_{2p+1}}{p!} \leq \frac{q^{2p+1}}{p!}$$

this also yields that

$$|\mathbb{E}(\prod_{i=1}^{2q'} m(X)(g_i)) - 1|$$

$$\leq \sum_{p=1}^{q'} \frac{1}{p!} \left(\frac{q^2}{2N}\operatorname{osc}^2(g)\right)^p + \sum_{p=1}^{q'-1} \frac{1}{p!} \left(\frac{q^2}{2N}\operatorname{osc}^2(g)\right)^{p+1/2}$$

$$\leq (1 + \operatorname{osc}(g) \, q/\sqrt{2N}) \sum_{p=1}^{q/2} \frac{1}{p!} \left(\frac{q^2}{2N}\operatorname{osc}^2(g)\right)^p$$

Recalling that $\sum_{p=1}^{n} \frac{\epsilon^p}{p!} \leq \epsilon \sum_{p=0}^{n-1} \frac{\epsilon^p}{p!} \leq \epsilon \, e^\epsilon$ for any $n \geq 0$ and $\epsilon \geq 0$, we arrive at

$$\left|\mathbb{E}\left(\prod_{i=1}^{q} m(X)(g_i)\right) - 1\right| \leq \frac{q^2}{2N} e_g\left(\frac{q^2}{2N}\right)$$

with $e_g(u) = \operatorname{osc}^2(g)(1 + \operatorname{osc}(g) \sqrt{u}) \exp{(\operatorname{osc}^2(g) \, u)}$. The proof for odd integers $q = 2q'+1$ is derived in a completely analogous fashion. This ends the proof of (8.17) when $n = 1$. Next we prove (8.17) for even integers $n = 2n'$, $n' \in \mathbb{N}$. We use the decomposition

$$\mathbb{E}([m(X)^{\otimes q}(g^{(q)}) - 1]^{2n'}) = \sum_{p=0}^{2n'} C_{2n'}^p \, (-1)^p \, \mathbb{E}([m(X)^{\otimes q}(g^{(q)})]^p)$$

$$= I_1 + I_2 + I_3$$

8. Propagation of Chaos

with

$$I_1 = \sum_{p=0}^{n'} C_{2n'}^{2p} \left[\mathbb{E}([m(X)^{\otimes q}(g^{(q)})]^{2p}) - 1\right]$$

$$I_2 = -\sum_{p=0}^{n'-1} C_{2n'}^{2p+1} \left[\mathbb{E}([m(X)^{\otimes q}(g^{(q)})]^{2p+1}) - 1\right]$$

$$I_3 = \sum_{p=0}^{n'} C_{2n'}^{2p} - \sum_{p=0}^{n'-1} C_{2n'}^{2p+1} = 0$$

Next we observe that for any $n \geq 1$ we have

$$\mathbb{E}([m(X)^{\otimes q}(g^{(q)})]^n) = \mathbb{E}([m(X)^{\otimes(q,n)}(g^{(q,n)})])$$

with

$$m(X)^{\otimes(q,n)} = \underbrace{m(X)^{\otimes q} \otimes \ldots \otimes m(X)^{\otimes q}}_{n \text{ times}} \quad \text{and} \quad g^{(q,n)} = \underbrace{g^{(q)} \otimes \ldots \otimes g^{(q)}}_{n \text{ times}}$$

From previous considerations, we find that

$$|I_1| = \sum_{p=1}^{n'} C_{2n'}^{2p} \left|\mathbb{E}([m(X)^{\otimes(q,2p)}(g^{(q,2p)})]) - 1\right|$$

$$\leq \sum_{p=1}^{n'} C_{2n'}^{2p} \frac{(2pq)^2}{2N} e_g\left(\frac{(2qp)^2}{2N}\right) \leq \frac{(nq)^2}{2N} e_g\left(\frac{(nq)^2}{2N}\right) \sum_{p=1}^{n'} C_{2n'}^{2p}$$

Using similar arguments, we find that

$$|I_2| \leq \frac{(nq)^2}{2N} e_g\left(\frac{(nq)^2}{2N}\right) \sum_{p=0}^{n'-1} C_{2n'}^{2p+1}$$

from which we conclude that

$$|\mathbb{E}([m(X)^{\otimes q}(g^{(q)}) - 1]^n)| \leq 2^{n-1} \frac{(nq)^2}{N} e_g\left(\frac{(nq)^2}{2N}\right)$$

The proof of this estimate for odd integers $n = 2n' + 1$ follows the same arguments. This completes the proof of the lemma. ∎

The proof of Theorem 8.7.1 will be easily established as follows.

8.7 Asymptotic Properties of Boltzmann-Gibbs Distributions 275

Proposition 8.7.1 Let $(g_i)_{i\geq 1}$ be a collection of measurable functions g_i with uniformly bounded oscillations $\mathrm{osc}(g) = \sup_{i\geq 1} \mathrm{osc}(g_i) < \infty$ and such that $\mu(g_i) = 1$ for any $i \geq 1$. For any $n \geq 1$, $N \geq q \geq 1$, and $f \in \mathcal{B}_b(E^q)$ with $\|f\| \leq 1$ and $\mathrm{osc}(f) \leq 1$, we have

$$|\mathbb{E}([m(X)^{\otimes q}(g^{(q)}f) - \mu^{\otimes q}(g^{(q)}f)]^n)| \leq 2^{n+1} \frac{(nq)^2}{N} \left[1 + e_g\left(\frac{(nq)^2}{2N}\right)\right] \tag{8.18}$$

Proof:
From Proposition 8.6.1, we have the Markovian transport equation

$$m(X)^{\otimes q} = m(X)^{\odot q} R_N^{(q)} \quad \text{with} \quad R_N^{(q)} = \frac{(N)_q}{N^q} \, Id + \left(1 - \frac{(N)_q}{N^q}\right) \tilde{R}_N^{(q)}$$

for some Markov kernel $\tilde{R}_N^{(q)}$ on E^q and for any $q \leq N$. Since

$$\left(R_N^{(q)} - Id\right) = (1 - (N)_q/N^q) \left(\tilde{R}_N^{(q)} - Id\right)$$

and recalling that $\mathbb{E}(m(X)^{\odot q}(g^{(q)}f)) = \mu^{\otimes q}(g^{(q)}f)$, we readily prove that

$$\mathbb{E}(m(X)^{\otimes q}(g^{(q)}f)) - \mu^{\otimes q}(g^{(q)}f)$$

$$= \mathbb{E}\left(m(X)^{\odot q}[R_N^{(q)} - Id](g^{(q)}f)\right)$$

$$= (1 - (N)_q/N^q) \, \mu^{\otimes q}[\tilde{R}_N^{(q)} - Id](g^{(q)}f)$$

To estimate the r.h.s. term in the display above, we use the decomposition

$$\mu^{\otimes q}[\tilde{R}_N^{(q)} - Id](g^{(q)}f) = I_1 + I_2$$

with

$$I_1 = \mu^{\otimes q}\left(\tilde{R}_N^{(q)}(g^{(q)}) \left[\frac{\tilde{R}_N^{(q)}(g^{(q)}f)}{\tilde{R}_N^{(q)}(g^{(q)})} - \mu^{\otimes q}(g^{(q)}f)\right]\right)$$

$$I_2 = \mu^{\otimes q}(g^{(q)}f) \, [\mu^{\otimes q}\tilde{R}_N^{(q)}(g^{(q)}) - 1]$$

We observe that

$$|I_2| \leq |\mu^{\otimes q}\tilde{R}_N^{(q)}(g^{(q)}) - 1| = |\mu^{\otimes q}[\tilde{R}_N^{(q)} - Id](g^{(q)})|$$
$$|I_1| \leq \mu^{\otimes q}\tilde{R}_N^{(q)}(g^{(q)}) \leq 1 + |\mu^{\otimes q}[\tilde{R}_N^{(q)} - Id](g^{(q)})|$$

From these estimates, we find that

$$|\mathbb{E}(m(X)^{\otimes q}(g^{(q)}f) - \mu^{\otimes q}(g^{(q)}f)|$$

$$\leq (1 - (N)_q/N^q) \, [1 + 2|\mu^{\otimes q}[\tilde{R}_N^{(q)} - Id](g^{(q)})|]$$

$$= (1 - (N)_q/N^q) + 2 \, |\mu^{\otimes q}[R_N^{(q)} - Id](g^{(q)})|$$

Consequently we have

$$|\mathbb{E}(m(X)^{\otimes q}(g^{(q)}f)) - \mu^{\otimes q}(g^{(q)}f)| \leq (1-(N)_q/N^q) + 2|\mathbb{E}(m(X)^{\otimes q}(g^{(q)}))-1|$$

and by Lemma 8.7.1, this implies that

$$\begin{aligned}|\mathbb{E}(m(X)^{\otimes q}(g^{(q)}f)) - \mu^{\otimes q}(g^{(q)}f)| &\leq (1-(N)_q/N^q) + \frac{2q^2}{N} e_g\left(\frac{q^2}{2N}\right) \\ &\leq \frac{2q^2}{N}\left[1+e_g\left(\frac{q^2}{2N}\right)\right]\end{aligned}$$

Using the same line of reasoning as at the end of the proof of Lemma 8.7.1, we also prove that for any $n \geq 1$

$$|\mathbb{E}([m(X)^{\otimes q}(g^{(q)}f) - \mu^{\otimes q}(g^{(q)}f)]^n)| \leq 2^{n+1}\frac{(nq)^2}{N}\left[1+e_g\left(\frac{(nq)^2}{2N}\right)\right]$$

This ends the proof of the proposition. ∎

Proof of Theorem 8.7.1:

By the definition of $\Psi^{(q)}$, no generality is lost and much convenience is gained by supposing (as will be done) that we have $\mu(g_i) = 1$ for each $i \geq 1$. To prove (8.16), we use the decomposition

$$\begin{aligned}\Psi^{(q)}(m(X)^{\otimes q})(f) - \Psi^{(q)}(\mu^{\otimes q})(f) &= \Psi^{(q)}(m(X)^{\otimes q})(f - \mu^{\otimes q}(g^{(q)}f)) \\ &= I_1 + I_2 \end{aligned} \quad (8.19)$$

with $I_1 = m(X)^{\otimes q}(g^{(q)}(f - \mu^{\otimes q}(g^{(q)}f)))$ and

$$I_2 = \Psi^{(q)}(m(X)^{\otimes q})(f - \mu^{\otimes q}(g^{(q)}f))\left(1 - m(X)^{\otimes q}(g^{(q)})\right)$$

It is now convenient to observe that

$$\begin{aligned}\mu^{\otimes q}(g^{(q)}(f - \mu^{\otimes q}(g^{(q)}f))) &= 0 \\ \|f - \mu^{\otimes q}(g^{(q)}f)\| &\leq \operatorname{osc}(f) = \operatorname{osc}(f - \mu^{\otimes q}(g^{(q)}f)) \leq 1\end{aligned}$$

and for any $n \geq 1$ we have

$$\mathbb{E}([\Psi^{(q)}(m(X)^{\otimes q})(f) - \Psi^{(q)}(\mu^{\otimes q})(f)]^{2n}) \leq 2^{2n-1}\left(\mathbb{E}(I_1^{2n}) + \mathbb{E}(I_2^{2n})\right)$$

Therefore, using Proposition 8.7.1 and Lemma 8.7.1, we check that

$$\mathbb{E}([\Psi^{(q)}(m(X)^{\otimes q})(f) - \Psi^{(q)}(\mu^{\otimes q})(f)]^{2n}) \leq c\, 2^{4n}\frac{(nq)^2}{N}\left[1+e_g\left(\frac{2(nq)^2}{N}\right)\right]$$

This ends the proof of (8.16). To prove (8.15), we use again the decomposition (8.19). By (8.18), we find that

$$|\mathbb{E}(I_1)| \leq c\,\frac{q^2}{N}\,\left[1 + e_g\left(\frac{q^2}{2N}\right)\right]$$

To estimate the mean value of I_2, we first use the Cauchy-Schwartz inequality to check that

$$|\mathbb{E}(I_2)|^2 \leq \mathbb{E}([\Psi^{(q)}(m(X)^{\otimes q})(f - \mu^{\otimes q}(g^{(q)}f))]^2)\,\mathbb{E}([1 - m(X)^{\otimes q}(g^{(q)})]^2)$$

Via (8.17) and (8.16), this implies that

$$|\mathbb{E}(I_2)| \leq c\,\frac{q^2}{N}\,\left[1 + e_g(\frac{2q^2}{N})\right]$$

from which we conclude that

$$|\mathbb{E}(\Psi^{(q)}(m(X)^{\otimes q})(f)) - \Psi^{(q)}(\mu^{\otimes q})(f)| \leq c\,\frac{q^2}{N}\,\left[1 + e_g\left(\frac{2q^2}{N}\right)\right]$$

This ends the proof of the theorem. ∎

8.8 Feynman-Kac Semigroups

To analyze propagation-of-chaos properties in path space, it is convenient to consider the Feynman-Kac tensor product distributions on path space

$$\mathbb{K}_n = (\eta_0 \otimes \ldots \otimes \eta_n) \in \mathcal{P}(E_{[0,n]})$$

By the definition of $\Phi_{p,n}$, we have $\mathbb{K}_n = \Omega_{p,n}(\mathbb{K}_p)$, for any $p \leq n$, with the (nonlinear) semigroup $\Omega_{p,n} : \mathcal{P}(E_{[0,p]}) \longrightarrow \mathcal{P}(E_{[0,n]})$ defined for any $\mu \in \mathcal{P}(E_{[0,p]})$ by

$$\Omega_{p,n}(\mu) = \mu \otimes \Phi_{p,p+1}(\mu_p) \otimes \Phi_{p,p+2}(\mu_p) \otimes \ldots \otimes \Phi_{p,n}(\mu_p) \qquad (8.20)$$

In the display above, $\mu_p \in \mathcal{P}(E_p)$ stands for the pth time marginal of μ defined for any $\varphi_p \in \mathcal{B}_b(E_p)$ by

$$\mu_p(\varphi_p) = \mu(\underbrace{1 \otimes \ldots \otimes 1}_{p\ \text{times}} \otimes \varphi_p)$$

Again we use the convention $\Omega_{n,n} = Id$ for $p = n$. To check that $\Omega_{p,n}$ is a well-defined semigroup, we observe that for any $\mu \in \mathcal{P}(E_{[0,p]})$ we have

$$\Omega_{p,p+1}(\mu) = \mu \otimes \Phi_{p,p+1}(\mu_p)$$

278   8. Propagation of Chaos

It follows that

$$\Omega_{p+1,n}(\Omega_{p,p+1}(\mu))$$
$$= \Omega_{p+1,n}(\mu \otimes \Phi_{p,p+1}(\mu_p))$$
$$= \mu \otimes \Phi_{p,p+1}(\mu_p) \otimes \Phi_{p+1,p+2}(\Phi_{p,p+1}(\mu_p)) \otimes \ldots \otimes \Phi_{p+1,n}(\Phi_{p,p+1}(\mu_p))$$
$$= \mu \otimes \Phi_{p,p+1}(\mu_p) \otimes \Phi_{p,p+2}(\mu_p) \otimes \ldots \otimes \Phi_{p,n}(\mu_p) = \Omega_{p,n}(\mu)$$

In the forthcoming development of this section, we fix a positive integer $q \geq 1$ and we denote by $\mathbb{K}_n^{(q)}$ the q-tensor product Feynman-Kac measures defined by

$$\mathbb{K}_n^{(q)} = \eta_0^{\otimes q} \otimes \ldots \otimes \eta_n^{\otimes q} \in \mathcal{P}(E_{[0,n]}^{(q)}) \quad \text{with} \quad E_{[0,n]}^{(q)} = E_0^q \times \ldots \times E_n^q$$

Notice that $\mathbb{K}_n^{(q)} = \mathbb{K}_n^{\otimes q} \circ (\Theta_n^q)^{-1}$ with the change of coordinate mapping $\Theta_n^q : E_{[0,n]}^{(q)} \longrightarrow E_{[0,n]}^q$ defined in (8.2). The next two subsections are devoted respectively to the study of the dynamical structure of the tensor product distributions $\eta_n^{\otimes q}$ and $\mathbb{K}_n^{(q)}$.

8.8.1   Marginal Models

We observe that $\eta_n^{\otimes q}$ can alternatively be defined for any $f \in \mathcal{B}_b(E_n)$ by the Feynman-Kac formulae

$$\eta_n^{\otimes q}(f) = \gamma_n^{\otimes q}(f)/\gamma_n^{\otimes q}(1) \quad \text{with} \quad \gamma_n^{\otimes q}(f) = \mathbb{E}_{\eta_0^{\otimes q}}^{(q)}\left(f(X_n^{(q)}) \prod_{p=0}^{n-1} G_p^{(q)}(X_p^{(q)})\right)$$

where

- $\mathbb{E}_{\eta_0^{\otimes q}}^{(q)}(.)$ represents the integration with respect to the law $\mathbb{P}_{\eta_0^{\otimes q}}^{(q)}$ of q independent copies $X_n^{(q)} = (X_n^{(q,1)}, X_n^{(q,2)}, \ldots, X_n^{(q,q)}) \in E_n^q$ of a Markov chain with initial distribution $\eta_0 \in \mathcal{P}(E_0)$ and Markov transitions M_n. In other words, $X_n^{(q)}$ is a nonhomogeneous and E_n^q-valued Markov chain with transitions

$$M_n^{(q)}((x_{n-1}^1, \ldots, x_{n-1}^q), d(x_n^1, \ldots, x_n^q)) = \prod_{i=1}^{q} M_n(x_{n-1}^i, dx_n^i)$$

- $G_n^{(q)} : E_n^q \to (0, \infty)$, $n \geq 0$, is the sequence of q-tensor product potential functions defined for any $(x_n^1, \ldots, x_n^q) \in E_n^q$ by

$$G_n^{(q)}(x_n^1, \ldots, x_n^q) = \prod_{i=1}^{q} G_n(x_n^i)$$

8.8 Feynman-Kac Semigroups

This rather simple representation indicates that the sequence of distribution flows $\eta_n^{\otimes q}$ and $\gamma_n^{\otimes q}$, $q \geq 1$, has exactly the same semigroup structure.

Let $Q_{n+1}^{(q)}$ and respectively $\Phi_{n+1}^{(q)}$ be the bounded integral operator from E_n^q into E_{n+1}^q and the mapping from $\mathcal{P}(E_n^q)$ into $\mathcal{P}(E_{n+1}^q)$ defined for any $(\eta, f) \in \mathcal{P}(E_n^q) \times \mathcal{B}_b(E_{n+1}^q)$ by

$$Q_{n+1}^{(q)}(f) = G_n^{(q)} \, M_{n+1}^{(q)}(f) \quad \text{and} \quad \Phi_{n+1}^{(q)}(\eta) = \Psi_n^{(q)}(\eta) M_{n+1}^{(q)}$$

with the Boltzmann-Gibbs transformations $\Psi_n^{(q)}$ on $\mathcal{P}(E_n^q)$ given by

$$\Psi_n^{(q)}(\eta)(dx_n) = \frac{1}{\eta(G_n^{(q)})} \, G_n^{(q)}(x_n) \, \eta(dx_n)$$

By the Markov property and the multiplicative form of the Feynman-Kac models, we prove that the distribution flows $\gamma_n^{\otimes q}$ and $\eta_n^{\otimes q}$ satisfy the recursions

$$\gamma_{n+1}^{\otimes q} = \gamma_n^{\otimes q} Q_{n+1}^{(q)} \quad \text{and} \quad \eta_{n+1}^{\otimes q} = \Phi_{n+1}^{(q)}(\eta_n^{\otimes q})$$

We let $Q_{p,n}^{(q)}$ and $\Phi_{p,n}^{(q)}$, $p \leq n$, be the semigroups associated respectively with $\gamma_n^{\otimes q}$ and $\eta_n^{\otimes q}$. That is, we have that

$$Q_{p,n}^{(q)} = Q_{p+1}^{(q)} \ldots Q_{n-1}^{(q)} Q_n^{(q)} \quad \text{and} \quad \Phi_{p,n}^{(q)} = \Phi_n^{(q)} \circ \Phi_{n-1}^{(q)} \circ \ldots \circ \Phi_{p+1}^{(q)}$$

As usual, we use the convention $Q_{n,n}^{(q)} = Id$ and $\Phi_{n,n}^{(q)} = Id$ for $p = n$.

Our final objective is to provide a Boltzmann-Gibbs representation of the semigroup $\Phi_{p,n}^{(q)}$. To this end, we let $G_{p,n}^{(q)} : E_p^q \to (0, \infty)$ and $P_{p,n}^{(q)}$ be respectively the potential function and the Markov transition from E_p^q into E_n^q defined for any $f \in \mathcal{B}_b(E_n^q)$ by the formulae

$$G_{p,n}^{(q)} = Q_{p,n}^{(q)}(1) \quad \text{and} \quad P_{p,n}^{(q)}(f_n) = Q_{p,n}^{(q)}(f_n)/Q_{p,n}^{(q)}(1)$$

If we set $G_{p,n} = Q_{p,n}(1)$, then we find that for any $(x_p^1, \ldots, x_p^q) \in E_p^q$

$$\begin{aligned} G_{p,n}^{(q)}(x_p^1, \ldots, x_p^q) &= Q_{p,n}(1)(x_p^1) \ldots Q_{p,n}(1)(x_p^q) \\ &= G_{p,n}(x_p^1) \ldots G_{p,n}(x_p^q) \end{aligned}$$

From previous considerations, we readily see that for any $\mu \in \mathcal{P}(E_p^q)$ we have

$$\Phi_{p,n}^{(q)}(\mu) = \Psi_{p,n}^{(q)}(\mu) P_{p,n}^{(q)} \tag{8.21}$$

with the Boltzmann-Gibbs transformations $\Psi_{p,n}^{(q)}$ on $\mathcal{P}(E_p^q)$ associated with the potential function $G_{p,n}^{(q)}$ and defined for any $(\mu, f) \in \mathcal{P}(E_p^q) \times \mathcal{B}_b(E_p^q)$ by $\Psi_{p,n}^{(q)}(\mu)(f) = \mu(G_{p,n}^{(q)} \, f)/\mu(G_{p,n}^{(q)})$.

8.8.2 Path-Space Models

To describe the dynamical structure of the semigroups $\Omega_{p,n}$ introduced in (8.20), we first observe that for $\eta \in \mathcal{P}(E_p)$ and $F \in \mathcal{B}_b(E_{(p,n]})$ we have

$$(\Phi_{p,p+1}(\eta) \otimes \Phi_{p,p+2}(\eta) \otimes \ldots \otimes \Phi_{p,n}(\eta))(F)$$

$$= \left[\prod_{k=1}^{n} \frac{1}{\eta Q_{p,p+k}(1)}\right] (\eta Q_{p,p+1} \otimes \ldots \otimes \eta Q_{p,n})(F) = \frac{\eta^{\otimes(n-p)}(T_{p,n}(F))}{\eta^{\otimes(n-p)}(T_{p,n}(1))}$$

with the bounded integral operator $T_{p,n}$ from $E_p^{(n-p)}$ into $E_{(p,n]}$ defined for any $(x_p^1, \ldots, x_p^{(n-p)}) \in E_p^{(n-p)}$ by

$$T_{p,n}(F)(x_p^1, \ldots, x_p^{(n-p)}) = \int_{E_{(p,n]}} \prod_{k=1}^{n-p} Q_{p,p+k}(x_p^k, dx_{p+k}) \ F(x_{p+1}, \ldots, x_n)$$

Also observe that the mapping $T_{p,n}(1)$ coincides with the $(n-q)$-tensor product potential function

$$T_{p,n}(1)(x_p^1, \ldots, x_p^{(n-p)}) = \prod_{k=1}^{n-p} Q_{p,p+k}(1)(x_p^k)$$

In other words, in terms of the potential functions $G_{p,n} = Q_{p,n}(1)$, we have that
$$T_{p,n}(1) = G_{p,p+1} \otimes G_{p,p+2} \otimes \ldots \otimes G_{p,n} \tag{8.22}$$

In this notation, (8.20) can be rewritten for any $\mu \in \mathcal{P}(E_{[0,p]})$ as follows

$$\Omega_{p,n}(\mu)(.) = \mu \otimes \frac{\mu_p^{\otimes(n-p)} T_{p,n}(.)}{\mu_p^{\otimes(n-p)} T_{p,n}(1)} = \mu \otimes (B_{p,n}[\mu_p^{\otimes(n-p)}]U_{p,n})$$

with

- the pth time marginal distribution $\mu_p \in \mathcal{P}(E_p)$ of $\mu \in \mathcal{P}(E_{[0,p]})$

- the Boltzmann-Gibbs transformation $B_{p,n}$ on $\mathcal{P}(E_p^{(n-p)})$ and the Markov transition $U_{p,n}$ from $E_p^{(n-p)}$ into $E_{(p,n]}$ defined for any pair $(\nu, f) \in (\mathcal{P}(E_p^{(n-p)}) \times \mathcal{B}_b(E_p^{(n-p)}))$ and $F \in \mathcal{B}_b(E_{(p,n]})$ by

$$B_{p,n}(\nu)(f) = \frac{\nu(T_{p,n}(1) \ f)}{\nu(T_{p,n}(1))} \quad \text{and} \quad U_{p,n}(F) = \frac{T_{p,n}(F)}{T_{p,n}(1)}$$

This updating-prediction type representation of the semigroup $\Omega_{p,n}$ provides a precise description of the dependence of $\Omega_{p,n}(\nu)$ with respect to the measure ν. Next we present a formula that emphasizes the role of the one-step mappings Φ_p in the dynamical structure of these transformations.

8.8 Feynman-Kac Semigroups

Lemma 8.8.1 *For any $p \geq 1$ and $\eta \in \mathcal{P}(E_{p-1})$, we have*
$$B_{p-1,n}[\eta^{\otimes(n-p+1)}]U_{p-1,n} = \Phi_p(\eta) \otimes (B_{p,n}[\Phi_p(\eta)^{\otimes(n-p)}]U_{p,n})$$

Proof:
By the definition of the operator $T_{p,n}$, we have
$$\begin{aligned}
\eta^{\otimes(n-p+1)}T_{p-1,n} &= [\eta Q_{p-1,p}] \otimes [\eta Q_{p-1,p+1}] \otimes \ldots \otimes [\eta Q_{p-1,n}] \\
&= (\eta Q_p) \otimes [(\eta Q_p)Q_{p,p+1}] \otimes \ldots \otimes [(\eta Q_p)Q_{p,n}] \\
&= (\eta Q_p) \otimes [(\eta Q_p)^{\otimes(n-p)}T_{p,n}]
\end{aligned}$$

This implies that
$$\eta^{\otimes(n-p+1)}T_{p-1,n}(1) = \eta Q_p(1) \times [(\eta Q_p)^{\otimes(n-p)}T_{p,n}(1)]$$

On the other hand, for any $\varphi_1 \in \mathcal{B}_b(E_p)$ and $\varphi_2 \in \mathcal{B}_b(E(p,n])$, we have $\Phi_p(\eta)(\varphi_1) = \eta Q_p(\varphi_1)/\eta Q_p(1)$ and
$$\begin{aligned}
\frac{(\eta Q_p)^{\otimes(n-p)}T_{p,n}(\varphi_2)}{(\eta Q_p)^{\otimes(n-p)}T_{p,n}(1)} &= \frac{\Phi_p(\eta)^{\otimes(n-p)}T_{p,n}(\varphi_2)}{\Phi_p(\eta)^{\otimes(n-p)}T_{p,n}(1)} \\
&= B_{p,n}(\Phi_p(\eta)^{\otimes(n-p)})U_{p,n}(\varphi_2)
\end{aligned}$$

From these observations, we find that for any $f \in \mathcal{B}_b(E_p \times E_{(p,n]})$
$$\begin{aligned}
B_{p-1,n}(\eta^{\otimes(n-p+1)})U_{p-1,n} &= \frac{\eta^{\otimes(n-p+1)}T_{p-1,n}(f)}{\eta^{\otimes(n-p+1)}T_{p-1,n}(1)} \\
&= \Phi_p(\eta) \otimes [B_{p,n}(\Phi_p(\eta)^{\otimes(n-p)})U_{p,n}]
\end{aligned}$$

This ends the proof of the lemma. ∎

From the Feynman-Kac representation of q-tensor marginal distributions given in Section 8.8.1, we see that the semigroup structure of the q-tensor product measures on path space
$$\mathbb{K}_n^{(q)} = \eta_0^{\otimes q} \otimes \ldots \otimes \eta_n^{\otimes q} \in \mathcal{P}(E_{[0,n]}^{(q)})$$
can be studied using the same line of argument as above by replacing the pair semigroups $(Q_{p,n}, \Phi_{p,n})$ by the q-tensor product semigroups $(Q_{p,n}^{(q)}, \Phi_{p,n}^{(q)})$.

We will use the superscript $(.)^{(q)}$ to define the corresponding mathematical quantities. To be more precise, let $\Omega_{p,n}^{(q)}$, $0 \leq p \leq n$, be the (nonlinear) semigroup associated with the distribution flow $\mathbb{K}_n^{(q)}$ and given by $\mathbb{K}_n^{(q)} = \Omega_{p,n}^{(q)}(\mathbb{K}_p^{(q)})$. From the preceding construction we check that $\Omega_{p,n}^{(q)}$ can be described for any $\mu \in \mathcal{P}(E_0^q \times \ldots E_p^q)$ by the formula
$$\Omega_{p,n}^{(q)}(\mu) = \mu \otimes B_{p,n}^{(q)}(\mu_p^{\otimes(n-p)})U_{p,n}^{(q)}$$
where

- $\mu_p^{\otimes(n-p)} \in \mathcal{P}(E_p^{q(n-p)})$ is the $(n-p)$-tensor product distribution of the pth time marginal $\mu_p \in \mathcal{P}(E_p^q)$ of μ

- $U_{p,n}^{(q)}$ is the Markov transition from $E_p^{q(n-p)}$ into $E_{(p,n]}^q (= E_{p+1}^q \times \ldots \times E_n^q)$ and defined by $U_{p,n}^{(q)}(F) = T_{p,n}^{(q)}(F)/T_{p,n}^{(q)}(1)$ for any $F \in \mathcal{B}_b(E_{(p,n]}^q)$ with

$$T_{p,n}^{(q)}(F)(x_p^1, \ldots, x_p^{n-p})$$

$$= \int_{E_{(p,n]}^{(q)}} \prod_{k=1}^{n-p} Q_{p,p+k}^{(q)}(x_p^k, dx_{p+k}) \, F(x_{p+1}, \ldots, x_n)$$

- $B_{p,n}^{(q)}$ is the Boltzmann-Gibbs transformation on $\mathcal{P}(E_p^{q(n-p)})$ defined for any pair $(\nu, f) \in (\mathcal{P}(E_p^{q(n-p)}) \times \mathcal{B}_b(E_p^{q(n-p)}))$ by

$$B_{p,n}^{(q)}(\nu)(f) = \nu(T_{p,n}^{(q)}(1) \, f)/\nu(T_{p,n}^{(q)}(1))$$

As in (8.22), we note that the nonhomogeneous potential functions $T_{p,n}^{(q)}(1)$ are given by

$$T_{p,n}^{(q)}(1) = G_{p,p+1}^{(q)} \otimes G_{p,p+2}^{(q)} \otimes \ldots \otimes G_{p,n}^{(q)}$$

In other words, in terms of the potential functions $G_{p,n}$ for any

$$(x_p^1, \ldots, x_p^{(n-p)}) \in E_p^{q(n-p)} \quad \text{with} \quad x_p^k = (x^{k,1}, \ldots, x^{k,q}) \in E_p^q$$

for each $1 \le k \le (n-p)$, we have

$$T_{p,n}^{(q)}(1)(x_p^1, \ldots, x_p^{(n-p)}) = \prod_{k=1}^{(n-p)} \prod_{i=1}^q G_{p,p+k}(x_p^{k,i})$$

We end this section with the version of Lemma 8.8.1 in the context of q-tensor product semigroups.

Lemma 8.8.2 *For any $q, p \ge 1$ and $\eta \in \mathcal{P}(E_{p-1}^q)$ we have*

$$B_{p-1,n}^{(q)}[\eta^{\otimes(n-p+1)}]U_{p-1,n}^{(q)} = \Phi_p^{(q)}(\eta) \otimes (B_{p,n}^{(q)}[\Phi_p^{(q)}(\eta)^{\otimes(n-p)}]U_{p,n}^{(q)})$$

8.9 Total Variation Estimates

This section is mainly concerned with the proofs of Theorem 8.3.3 and Theorem 8.3.4. We recall that the forthcoming analysis is restricted to the simple mutation/selection genetic model.

8.9 Total Variation Estimates

Proof of Theorem 8.3.3: We use the decomposition

$$(\eta_n^N)^{\otimes q} - \eta_n^{\otimes q} = \sum_{p=0}^{n} \left[\Phi_{p,n}^{(q)}((\eta_p^N)^{\otimes q}) - \Phi_{p,n}^{(q)}(\Phi_{p-1,p}^{(q)}((\eta_{p-1}^N)^{\otimes q})) \right] \quad (8.23)$$

with the convention $\Phi_{-1,0}^{(q)}((\eta_{-1}^N)^{\otimes q}) = \eta_0^{\otimes q}$ for $p = 0$. Our next objective is to estimate the differences of measures

$$I_{p,n}^{(q)} =_{\text{def.}} \left[\Phi_{p,n}^{(q)}((\eta_p^N)^{\otimes q}) - \Phi_{p,n}^{(q)}(\Phi_{p-1,p}^{(q)}((\eta_{p-1}^N)^{\otimes q})) \right]$$

Using (8.21), we first observe that $\Phi_{p-1,p}^{(q)}((\eta_{p-1}^N)^{\otimes q}) = \Phi_p(\eta_{p-1}^N)^{\otimes q}$ and for any $f \in \mathcal{B}_b(E_n^q)$

$$I_{p,n}^{(q)}(f) = \left[\Psi_{p,n}^{(q)}((\eta_p^N)^{\otimes q}) - \Psi_{p,n}^{(q)}(\Phi_p(\eta_{p-1}^N)^{\otimes q}) \right] P_{p,n}^{(q)}(f)$$

The conclusion now follows from Theorem 8.7.1. First we notice that for any $\mu \in \mathcal{P}(E_p)$ we have

$$\text{osc}_\mu(G_{p,n}) = \frac{\text{osc}(G_{p,n})}{\mu(G_{p,n})} \leq (r_{p,n} - 1) \quad \text{with} \quad r_{p,n} = \sup_{x_p, y_p \in E_p} \frac{G_{p,n}(x_p)}{G_{p,n}(y_p)} \quad (8.24)$$

Therefore, recalling that η_p^N is the empirical measure associated with a collection of N conditionally independent and identically distributed random variables with common law $\Phi_p(\eta_{p-1}^N)$, we find from (8.15) the $\mathbb{P}_{\eta_0}^N$-a.s. estimate

$$|\mathbb{E}_{\eta_0}^N(I_{p,n}^{(q)}(f)|\mathcal{F}_{p-1}^N)| \leq c\, \frac{q^2}{N} \left[1 + e_{p,n}\left(\frac{2q^2}{N}\right)\right] \text{osc}(P_{p,n}^{(q)}(f))$$

We recall that, for any Markov transition M from a measurable space $(E, \mathcal{E})$ into a (possibly different) measurable space $(E', \mathcal{E}')$ and for any $f \in \mathcal{B}_b(E')$, we have the inequality $\text{osc}(M(f)) \leq \beta(M)\, \text{osc}(f)$ (see (4.11) on page 128). From this property, we conclude that, for any $f \in \mathcal{B}_b(E_n^q)$ with $\text{osc}(f) \leq 1$, we have

$$|\mathbb{E}_{\eta_0}^N(I_{p,n}^{(q)}(f)|\mathcal{F}_{p-1}^N)| \leq c\, \frac{q^2}{N} \beta(P_{p,n}^{(q)}) \left[1 + e_{p,n}\left(\frac{2q^2}{N}\right)\right] \quad \mathbb{P}_{\eta_0}^N\text{-a.s.}$$

By (8.23), it follows that, for any $f \in \mathcal{B}_b(E_n^q)$ with $\text{osc}(f) \leq 1$, we have

$$|\mathbb{E}_{\eta_0}^N((\eta_n^N)^{\otimes q}(f)) - \eta_n^{\otimes q}(f)| = c\, \frac{q^2}{N} \sum_{p=0}^{n} \beta(P_{p,n}^{(q)}) \left[1 + e_{p,n}\left(\frac{2q^2}{N}\right)\right]$$

Taking into account that

$$\mathbb{P}_{\eta_0,[n]}^{(N,q)}(f) = \mathbb{E}_{\eta_0}^N(f(\xi_n^1, \ldots, \xi_n^q)) = \mathbb{E}_{\eta_0}^N(m(\xi_n)^{\odot q})$$

Lemma 8.6.1 ensures that, for any $f \in \mathcal{B}_b(E_n^q)$ with $\mathrm{osc}(f) \le 1$, we have

$$|\mathbb{P}_{\eta_0,[n]}^{(N,q)}(f) - \eta_n^{\otimes q}(f)|$$

$$\le \frac{(q-1)^2}{N} + c\frac{q^2}{N}\sum_{p=0}^n \beta(P_{p,n}^{(q)})\left[1 + e_{p,n}\left(\frac{2q^2}{N}\right)\right]$$

$$\le c\frac{q^2}{N}\left(1 + \sum_{p=0}^n \beta(P_{p,n}^{(q)})\left[1 + e_{p,n}\left(\frac{2q^2}{N}\right)\right]\right)$$

This ends the proof of Theorem 8.3.3. ∎

To illustrate the impact of Theorem 8.3.3, we present hereafter some easily derived strong and uniform propagation-of-chaos estimates.

Corollary 8.9.1 *Let us suppose that the triplet* (E_n, G_n, M_n) *is time-homogeneous and the regularity condition* (G, M) *introduced in (8.6) on page 259 is met for some* $\epsilon(G), \epsilon(M) > 0$, $m \ge 1$. *Then we have*

$$\|\mathbb{P}_{\eta_0,[n]}^{(N,q)} - \eta_n^{\otimes q}\|_{\mathrm{tv}} \le c\frac{q^2}{N}\left(1 + d_m^{(\epsilon)}(q,n)\left[1 + e_m^{(\epsilon)}(2q^2/N)\right]\right)$$

where

$$d_m^{(\epsilon)}(q,n) = \sum_{p=0}^n (1 - \epsilon_m^q(G,M))^{\lfloor p/m \rfloor} \le (n+1) \wedge \left(m\,\epsilon_m^{-q}(G,M)\right)$$

with $\epsilon_m(G,M) = \epsilon^{(m-1)}(G)\,\epsilon(M)$ *and* $e_m^{(\epsilon)}(u)$ *is the mapping defined as* $e_{p,n}(u)$ *(see (8.8)) by replacing the constant* $r_{p,n}$ *by* $r_m^{(\epsilon)} = \epsilon^{-m}(G)\epsilon^{-1}(M)$.

Proof:
When the regularity conditions (8.6) are met, we recall that for any $0 \le p+m \le n$ we have the uniform estimate $r_{p,n} \le \epsilon^{-m}(G)\epsilon^{-1}(M)$ and for any $x_p, y_p \in E_p^q$ and any nonnegative function φ on E_{p+m}^q

$$\frac{Q_{p,p+m}^{(q)}(\varphi)(x_p)}{Q_{p,p+m}^{(q)}(1)(x_p)} \ge \epsilon_m^q(G,M)\,\frac{M_{p,p+m}^{(q)}(\varphi)(y_p)}{M_{p,p+m}^{(q)}(1)(y_p)}$$

(see for instance Proposition 4.3.3). By the definition of Dobrushin's ergodic coefficient, this yields that

$$\beta(P_{p,n}^{(q)}) \le (1 - \epsilon_m^q(G,M))^{\lfloor (n-p)/m \rfloor}$$

Recalling that $r_{p,n} \le \epsilon^{-(n-p)}(G)$, we observe that for any $p \le n$

$$r_{p,n} \le \epsilon^{-m}(G)(\epsilon^{-1}(M) \vee 1) = \epsilon^{-m}(G)\epsilon^{-1}(M)$$

and consequently $\sup_{p\leq n} e_{p,n}(u) \leq e_m^{(\epsilon)}(u)$. From previous calculations, we easily find that

$$\|\mathbb{P}_{\eta_0,[n]}^{(N,q)} - \eta_n^{\otimes q}\|_{\text{tv}} \leq c\frac{q^2}{N}\left(1 + [1 + e_m^{(\epsilon)}(2q^2/N)]\, d_m^{(\epsilon)}(q,n)\right)$$

This ends the proof of the corollary. ∎

Corollary 8.9.2 *Assume that the regularity assumptions stated in Corollary 8.9.1 are met for some $\epsilon(G), \epsilon(M) > 0$, and $m \geq 1$. Then, using the same notation as there, we have the uniform propagation-of-chaos estimate*

$$\sup_{n\geq 0}\left\|\mathbb{P}_{\eta_0,[n]}^{(N,q)} - \eta_n^{\otimes q}\right\|_{\text{tv}} \leq c\frac{q^2}{N}\left(1 + m\,\epsilon_m^{-q}(G,M)\,[1+e_m^{(\epsilon)}(2q^2/N)]\right)$$

In addition, for any nondecreasing sequence of time horizons $n(N)$ and particle block sizes $q(N)$ such that $\lim_{N\to\infty} n(N)q^2(N)/N = 0$, we have the increasing propagation-of-chaos property

$$\limsup_{N\to\infty} \frac{N}{q^2(N)n(N)}\|\mathbb{P}_{\eta_0,[n(N)]}^{(N,q(N))} - \eta_{n(N)}^{\otimes q(N)}\|_{\text{tv}} \leq c/(\epsilon^m(G)\epsilon(M))^2$$

The end of this section is concerned with the proof of Theorem 8.3.4. Our first task is to better connect the distributions

$$\mathbb{P}_{\eta_0,n}^{(N,q)} = \text{Law}((\xi_{[0,n]}^i)_{1\leq i\leq q}) \in \mathcal{P}(E_{[0,n]}^q)$$

with the Markovian structure of the interacting particle model defined in (7.4). We recall that for each $0 \leq p \leq n$ and $1 \leq q \leq N$ the state space

$$E_{[p,n]}^{(q)} = E_p^q \times \ldots \times E_n^q$$

represents the set of the first q paths from time p to time n of the Markov particle model, while the product space $E_{[p,n]}^q = (E_p \times \ldots \times E_n)^q$ represents the state space of the paths of each of the first q elementary particles from time p up to time n. We shall also use the notation $E_{(p,n]}^{(q)} = E_{[p+1,n]}^{(q)}$. We recall that $E_{[0,n]}^{(q)}$ and $E_{[0,n]}^q$ are connected by the mapping Θ_n^q defined in (8.2). For instance, for $q = N$ we have $\Theta_n^N(\xi_0,\ldots,\xi_n) = \xi_{[0,n]}$ and $\mathbb{P}_{\eta_0,n}^{(N)} = \mathbb{P}_{\eta_0,n}^N \circ (\Theta_n^N)^{-1}$, where $\mathbb{P}_{\eta_0,n}^N = \mathbb{P}_{\eta_0}^N \circ (\xi_0,\ldots,\xi_n)^{-1} \in \mathcal{P}(E_{[0,n]}^{(N)})$. In this connection, it is also convenient to associate with the pair of path measures $((\mathbb{K}_n^N)^{\otimes q}, (\mathbb{K}_n^N)^{\odot q})$ defined in (8.3) the distributions

$$\mathbb{K}_n^{(N,q)} = (\mathbb{K}_n^N)^{\otimes q} \circ (\Theta_n^q)^{-1} \quad \text{and} \quad \mathbb{K}_n^{\langle N,q\rangle} = (\mathbb{K}_n^N)^{\odot q} \circ (\Theta_n^q)^{-1} \in \mathcal{P}(E_{[0,n]}^{(q)})$$

In other words, we have with some obvious abusive notation

$$\mathbb{K}_n^{(N,q)} = \frac{1}{N^q} \sum_{\alpha \in \langle N \rangle^{\langle q \rangle}} \delta_{((\xi_0^{\alpha(i)})_{1\leq i \leq q},\ldots,(\xi_n^{\alpha(i)})_{1\leq i \leq q})}$$

$$\mathbb{K}_n^{\langle N,q \rangle} = \frac{1}{(N)_q} \sum_{\alpha \in \langle q, N \rangle} \delta_{((\xi_0^{\alpha(i)})_{1\leq i \leq q},\ldots,(\xi_n^{\alpha(i)})_{1\leq i \leq q})}$$

Lemma 8.9.1 *For any pair of integers $1 \leq q \leq N$ and any test function $F \in \mathcal{B}_b(E_{[0,n]}^{(q)})$ with $\|F\| \leq 1$, we have*

$$|\mathbb{E}_{\eta_0}^N(\mathbb{K}_n^{(N,q)}(F)) - \mathbb{E}_{\eta_0}^N(F((\xi_0^i)_{1\leq i \leq q},\ldots,(\xi_n^i)_{1\leq i \leq q}))| \leq (q-1)^2/N$$

Proof:
By Lemma 8.6.1, we observe that

$$\|(\mathbb{K}_n^N)^{\odot q} - (\mathbb{K}_n^N)^{\otimes q}\|_{\text{tv}} \leq (q-1)^2/N \qquad (8.25)$$

By the exchangeability property of the particle model, we also have that for any $\alpha \in \langle q, N \rangle$ and $F \in \mathcal{B}_b(E_{[0,n]}^{(q)})$

$$\mathbb{E}_{\eta_0}^N(F((\xi_0^{\alpha(i)})_{1\leq i \leq q},\ldots,(\xi_n^{\alpha(i)})_{1\leq i \leq q})) = \mathbb{E}_{\eta_0}^N(F((\xi_0^i)_{1\leq i \leq q},\ldots,(\xi_n^i)_{1\leq i \leq q}))$$

This implies that

$$\mathbb{E}_{\eta_0}^N(\mathbb{K}_n^{\langle N,q \rangle}(F)) = \mathbb{E}_{\eta_0}^N(F((\xi_0^i)_{1\leq i \leq q},\ldots,(\xi_n^i)_{1\leq i \leq q}))$$

However, (8.25) also ensures that

$$\|\mathbb{K}_n^{(N,q)} - \mathbb{K}_n^{\langle N,q \rangle}\|_{\text{tv}} \leq (q-1)^2/N \qquad (8.26)$$

from which the end of the proof of the lemma is easily completed. ∎

We are now in a position to prove the theorem.

Proof of Theorem 8.3.4: In a similar fashion as in the proof of Theorem 8.3.3, we use the decomposition

$$\mathbb{K}_n^{(N,q)} - \mathbb{K}_n^{(q)} = \sum_{p=0}^{n} \left[\Omega_{p,n}^{(q)}(\mathbb{K}_p^{(N,q)}) - \Omega_{p,n}^{(q)}(\Omega_{p-1,p}^{(q)}(\mathbb{K}_{p-1}^{(N,q)})) \right] \qquad (8.27)$$

As usual, we take the convention for $p = 0$, $\Omega_{-1,0}^{(q)}(\mathbb{K}_{-1}^{(N,q)}) = \eta_0^{\otimes q}$. To describe more precisely each term in the summand above, we first observe that, for the q-tensor product measure, $(\eta_p^N)^{\otimes q} \in \mathcal{P}(E_p^q)$ is the pth time marginal of $\mathbb{K}_n^{(N,q)}$. On the other hand, by the definition of the semigroup $\Omega_{p,n}^{(q)}$, we have

$$\Omega_{p,n}^{(q)}(\mathbb{K}_p^{(N,q)}) = \mathbb{K}_p^{(N,q)} \otimes B_{p,n}^{(q)}((\eta_p^N)^{\otimes q(n-p)})U_{p,n}^{(q)}$$

and

$$\Omega^{(q)}_{p-1,p}(\mathbb{K}^{(N,q)}_{p-1}) = \mathbb{K}^{(N,q)}_{p-1} \otimes \Phi^{(q)}_{p-1,p}((\eta^N_{p-1})^{\otimes q}) = \mathbb{K}^{(N,q)}_{p-1} \otimes \Phi_p(\eta^N_{p-1})^{\otimes q}$$

This implies that for any $1 \leq p \leq n$ we have

$$\begin{aligned}\Omega^{(q)}_{p-1,n}(\mathbb{K}^{(N,q)}_{p-1}) &= \Omega^{(q)}_{p,n}(\Omega^{(q)}_{p-1,p}(\mathbb{K}^{(N,q)}_{p-1}))\\ &= \mathbb{K}^{(N,q)}_{p-1} \otimes \Phi_p(\eta^N_{p-1})^{\otimes q} \otimes [B^{(q)}_{p,n}(\Phi_p(\eta^N_{p-1})^{\otimes q(n-p)})U^{(q)}_{p,n}]\end{aligned}$$

Let $\Omega^{\langle q,N \rangle}_{p,n}$ be the random measures defined by

$$\Omega^{\langle q,N \rangle}_{p,n} = \mathbb{K}^{\langle N,q \rangle}_p \otimes B^{(q)}_{p,n}((\eta^N_p)^{\otimes q(n-p)})U^{(q)}_{p,n}$$

Using Lemma 8.6.1, we find that for any $p \leq n$

$$\|\Omega^{\langle q,N \rangle}_{p,n} - \Omega^{(q)}_{p,n}(\mathbb{K}^{(N,q)}_p)\|_{\mathrm{tv}} \leq (q-1)^2/N \tag{8.28}$$

As an aside, using Lemma 8.8.2, we already notice that

$$\begin{aligned}\Omega^{\langle q,N \rangle}_{p-1,n} &= \mathbb{K}^{\langle N,q \rangle}_{p-1} \otimes B^{(q)}_{p-1,n}((\eta^N_{p-1})^{\otimes q(n-p+1)})U^{(q)}_{p-1,n}\\ &= \mathbb{K}^{\langle q,N \rangle}_{p-1} \otimes \Phi_p(\eta^N_{p-1})^{\otimes q} \otimes B^{(q)}_{p,n}(\Phi_p(\eta^N_{p-1})^{\otimes q(n-p)})U^{(q)}_{p,n}\end{aligned} \tag{8.29}$$

Now, by (8.27), the estimates (8.28) imply that

$$\left\|\mathbb{K}^{(N,q)}_n - \mathbb{K}^{(q)}_n - \sum_{p=0}^n (\Omega^{\langle q,N \rangle}_{p,n} - \Omega^{\langle q,N \rangle}_{p-1,n})\right\|_{\mathrm{tv}} \leq 2(n+1)(q-1)^2/N \tag{8.30}$$

with the convention, for $p = 0$, $\Omega^{\langle q,N \rangle}_{-1,n} = \mathbb{K}^{(q)}_n = \eta_0^{\otimes q} \otimes B^{(q)}_{0,n}(\eta_0^{\otimes qn})U^{(q)}_{0,n}$. By symmetry arguments, it is now convenient to observe that, for any $p \leq n$ and any test function $\varphi \in \mathcal{B}_b(E^{(q)}_{[p,n]})$, the sequence of random variables

$$\mathbb{E}^N_{\eta_0}\left(\int [B^{(q)}_{p,n}((\eta^N_p)^{\otimes q(n-p)})U^{(q)}_{p,n}](dy)\ \varphi((\xi_p^{\alpha(i)})_{1 \leq i \leq q},y)|F^N_{p-1}\right)$$

$$= \mathbb{E}^N_{\eta_0}\left(\int [B^{(q)}_{p,n}((\eta^N_p)^{\otimes q(n-p)})U^{(q)}_{p,n}](dy)\ \varphi((\xi_p^i)_{1 \leq i \leq q},y)|F^N_{p-1}\right)$$

$$= \mathbb{E}^N_{\eta_0}\left(\int ((\eta^N_n)^{\odot q} \otimes [B^{(q)}_{p,n}((\eta^N_p)^{\otimes q(n-p)})U^{(q)}_{p,n}])(d(x,y))\ \varphi(x,y)|F^N_{p-1}\right)$$

does not depend on the choice of $\alpha \in \langle q,N \rangle$, where the integral is taken over the product space $E^{(q)}_{(p,n]}$. Using this property, we prove that for any

$f \in \mathcal{B}_b(E_{[0,n]}^{(q)})$

$$\mathbb{E}_{\eta_0}^N(\Omega_{p,n}^{\langle q,N\rangle}(f) \mid F_{p-1}^N)$$

$$= \mathbb{E}_{\eta_0}^N([\mathbb{K}_p^{\langle N,q\rangle} \otimes B_{p,n}^{(q)}((\eta_p^N)^{\otimes q(n-p)})U_{p,n}^{(q)}](f) \mid F_{p-1}^N)$$

$$= \mathbb{E}_{\eta_0}^N([\mathbb{K}_{p-1}^{\langle N,q\rangle} \otimes (\eta_p^N)^{\odot q} \otimes B_{p,n}^{(q)}((\eta_p^N)^{\otimes q(n-p)})U_{p,n}^{(q)}](f) \mid F_{p-1}^N)$$

On the other hand, using the fact that $\|(\eta_p^N)^{\odot q} - (\eta_p^N)^{\otimes q}\|_{tv} \leq (q-1)^2/N$, we find the $\mathbb{P}_{\eta_0}^N$-a.s. estimate

$$|\mathbb{E}_{\eta_0}^N([\Omega_{p,n}^{\langle q,N\rangle} - \widetilde{\Omega}_{p,n}^{\langle q,N\rangle}](f) \mid F_{p-1}^N)| \leq \|f\|\,(q-1)^2/N \qquad (8.31)$$

with

$$\widetilde{\Omega}_{p,n}^{\langle q,N\rangle} =_{\text{def.}} \mathbb{K}_{p-1}^{\langle N,q\rangle} \otimes (\eta_p^N)^{\otimes q} \otimes B_{p,n}^{(q)}((\eta_p^N)^{\otimes q(n-p)})U_{p,n}^{(q)}$$

In the display above and for $p=0$, we have used the convention

$$\widetilde{\Omega}_{0,n}^{\langle q,N\rangle} = (\eta_0^N)^{\otimes q} \otimes B_{0,n}^{(q)}((\eta_0^N)^{\otimes qn})U_{0,n}^{(q)}$$

In this notation, we have by (8.29) the formula

$$\widetilde{\Omega}_{p,n}^{\langle q,N\rangle} - \Omega_{p-1,n}^{\langle q,N\rangle} = \mathbb{K}_{p-1}^{\langle q,N\rangle} \otimes [(\eta_p^N)^{\otimes q} \otimes B_{p,n}^{(q)}((\eta_p^N)^{\otimes q(n-p)})$$

$$- \Phi_p(\eta_{p-1}^N)^{\otimes q} \otimes B_{p,n}^{(q)}(\Phi_p(\eta_{p-1}^N)^{\otimes q(n-p)})]U_{p,n}^{(q)}$$

Let $\widetilde{B}_{p,n}^{(q)}$ be the extended Boltzmann-Gibbs transformation on $\mathcal{P}(E_p^{q(n-p+1)})$ defined for any pair $(\nu,\varphi) \in (\mathcal{P}(E_p^{q(n-p+1)}) \times \mathcal{B}_b(E_p^{q(n-p+1)}))$ by

$$\widetilde{B}_{p,n}^{(q)}(\nu)(\varphi) = \nu(\widetilde{T}_{p,n}^{(q)}(1)\,\varphi)/\nu(\widetilde{T}_{p,n}^{(q)}(1))$$

with the potential functions $\widetilde{T}_{p,n}^{(q)}(1)$ on $E_p^{q(n-p+1)}$ given by

$$\widetilde{T}_{p,n}^{(q)}(1) = G_{p,p}^{(q)} \otimes G_{p,p+1}^{(q)} \otimes \ldots \otimes G_{p,n}^{(q)}$$

In this notation, recalling that $G_{p,p}=1$, we find that, for any $\mu \in \mathcal{P}(E_p^q)$ and $\nu \in \mathcal{P}(E_p^{q(n-p)})$, we have $\mu \otimes B_{p,n}^{(q)}(\nu) = \widetilde{B}_{p,n}^{(q)}(\mu \otimes \nu)$. This readily yields that

$$\widetilde{\Omega}_{p,n}^{\langle q,N\rangle} - \Omega_{p-1,n}^{\langle q,N\rangle}$$

$$= \mathbb{K}_{p-1}^{\langle q,N\rangle} \otimes [\widetilde{B}_{p,n}^{(q)}((\eta_p^N)^{\otimes q(n-p+1)}) - \widetilde{B}_{p,n}^{(q)}(\Phi_p(\eta_{p-1}^N)^{\otimes q(n-p+1)})]U_{p,n}^{(q)}$$

Using Theorem 8.7.1 and arguing as in (8.24), we obtain the $\mathbb{P}_{\eta_0}^N$-a.s. estimate

$$|\mathbb{E}_{\eta_0}^N([\widetilde{\Omega}_{p,n}^{\langle q,N\rangle} - \Omega_{p-1,n}^{\langle q,N\rangle}](f)|F_{p-1}^N)|$$

$$\leq c\,\frac{(q(n-p+1))^2}{N}\left[1 + e_{p,n}\left(\frac{2(q(n-p+1))^2}{N}\right)\right]$$

as soon as $(n+1)q \leq N$ and $\|f\| \leq 1$. This readily implies the rather crude and almost sure upper bound

$$|\mathbb{E}_{\eta_0}^N([\widetilde{\Omega}_{p,n}^{\langle q,N\rangle} - \Omega_{p-1,n}^{\langle q,N\rangle}](f)|F_{p-1}^N)| \leq c \, \frac{(q(n+1))^2}{N} \left[1 + e_n\left(\frac{2(q(n+1))^2}{N}\right)\right]$$
(8.32)

with the mapping $e_n(u)$ defined as in (8.8) by replacing the constants $r_{p,n}$ by $\bar{r}_n = \sup_{p \leq n} r_{p,n}$. Combining (8.30), (8.31), and (8.32), we conclude that for any $f \in \mathcal{B}_b(E_{[0,n]}^{(q)})$ with $\|f\| \leq 1$

$$\left|\mathbb{E}_{\eta_0}^N([\mathbb{K}_n^{(N,q)} - \mathbb{K}_n^{(q)}](f))\right| \leq c \, (n+1) \, \frac{(q(n+1))^2}{N} \, [1 + e_n(\frac{2(q(n+1))^2}{N})]$$

By the definition of $\mathbb{P}_{\eta_0,n}^{(N,q)}$ and $\mathbb{K}_n^{(N,q)}$, the total variation estimate (8.9) is now a simple application of Lemma 8.9.1. This completes the proof of Theorem 8.3.4. ∎

9
Central Limit Theorems

9.1 Introduction

The central limit theorem (abbreviated CLT) is one of the most startling results in probability theory. Loosely speaking, it expresses the fact that the sums of local and small independent disturbances (with finite variances) behave asymptotically, at least as Gaussian variables. The first CLT was stated and proved for symmetric and Bernoulli independent disturbances by A. De Moivre in the 18th century (Miscellanea analytica supplementum, 1730). This result was extended by P.S. Laplace in 1812 to general Bernoulli trials in his celebrated treatise *Théorie analytique des probabilités*.

These two pioneering studies have been further developed by several mathematicians, such as, in alphabetical order, Donsker, Dynkin, Feller, Jacod, Lindeberg, Lyapunov, Mandelbaum, and Shiryaev. These developments were followed in various directions going from multidimensional models, symmetric statistics sequences, and empirical processes to fairly general classes of nonidentically distributed and dependent random variables such as properly scaled random triangular arrays or martingale sequences. There also exist several approximation tools, such as the so-called δ-method or Slutsky's technique, to deduce various weak limit results from a given CLT. It is of course out of the scope of this book to review in detail all of these developments, and rather we refer the interested reader to any classical textbook on probability and limit theorems for stochastic processes.

Most of these CLTs are derived using extended versions of the celebrated Levy's convergence theorem, which basically says that the weak conver-

9. Central Limit Theorems

gence of distributions corresponds exactly to the pointwise convergence of their characteristic functions. In this connection, we also mention that the inequality of Berry and Esseen provides an estimation of the maximum difference between distribution functions by means of an average difference between their characteristic functions. This inequality allows us to quantify the speed of convergence of the CLT.

The aim of this chapter is to extend these results to interacting particle approximation models. These fluctuation results provide precise asymptotic information on the various $\mathbb{L}_p$ mean error bounds and the propagation of chaos derived respectively in Chapter 7 and Chapter 8. The reader will find a deep study of topics such as the multidimensional CLTs for normalized and unnormalized particle approximation measures, extended versions of the theorems of Berry and Esseen and of Donsker to interacting particle models, and a fluctuation result for particle McKean measures on path space. While our approach to CLTs is well suited to analyze any sufficiently regular McKean interpretation model, to simplify the presentation, we restrict our study to McKean models of the form

$$K_{n+1,\eta}(x,.) = \varepsilon_n G_n(x)\, M_{n+1}(x,.) + (1 - \varepsilon_n G_n(x))\, \Phi_{n+1}(\eta) \qquad (9.1)$$

where ε_n are nonnegative constants such that $\varepsilon_n G_n \leq 1$. As in Chapter 8, we note that the pair of examples provided on page 219 can be cast in the form above, and the case $\varepsilon_n = 0$ corresponds to the traditional mutation/selection genetic algorithm. As usual, we shall suppose that the potential functions G_n satisfy the traditional condition (G) introduced in page 115 for some sequence of parameters $\epsilon_n(G) > 0$.

The chapter is organized as follows. Section 9.3 and Section 9.4 focus on multidimensional CLTs for particle density profiles. We examine the particle approximation models associated with unnormalized and normalized Feynman-Kac measures. We design an elegant strategy based on martingales and semigroups techniques which allows us to reduce the fluctuation analysis to local sampling errors. In Section 9.5 we study the rate of convergence of these CLTs and provide a Berry-Esseen type theorem for abstract martingale sequences and interacting processes. We provide simple regularity conditions on the increasing processes underwhich we can derive precise estimations of the characteristic functions. In Section 9.6, we discuss the fluctuations of particle random fields and prove an extended version of Donsker's theorem for interacting particle models. We develop an empirical process technique based on sub-Gaussian maximal inequalities to prove the asymptotic tightness of random field errors. All of these fluctuation results for interacting processes are restricted to uniformly bounded classes of test functions. Some extensions to unbounded functions can be found in [89]. We emphasize that the strategy for CLTs presented in Section 9.4 to Section 9.6 is not related to any kind of regularity conditions on the mutation transition M_n. Thus, it applies to path space and genealogical particle models. In Section 9.7, we analyze the fluctuations of particle

McKean measures under the regularity condition $(M)^{\text{exp}}$ introduced on page 116. The approach is essentially based on a theorem of Dynkin and Mandelbaum and a formula of Shiga and Tanaka. In the final section, Section 9.8, we provide an explicit description of the inverse of the $\mathbb{L}_2$ integral operator which characterizes these fluctuations on path-space.

9.2 Some Preliminaries

In the context of interacting particle models, the local errors induced by the approximation sampling transitions of each particle can be interpreted as small disturbances in the time evolution of the particle density profiles. To be more precise, let us consider the particle approximation model $\xi_n = (\xi_n^i)_{1 \leq i \leq N} \in E_n^N$, $n \in \mathbb{N}$, associated with a nonlinear measure-valued equation of the form

$$\eta_n = \eta_{n-1} K_{n,\eta_{n-1}} \in \mathcal{P}(E_n) \tag{9.2}$$

where $K_{n,\eta}$ is a collection of Markov transitions from a measurable space E_{n-1} into another E_n. The nth sampling error is the $M(E_n)$-valued random variable V_n^N defined by the formula

$$\eta_n^N = \eta_{n-1}^N K_{n,\eta_{n-1}^N} + V_n^N/\sqrt{N} \tag{9.3}$$

Notice that V_n^N is itself the sum of the local errors induced by the random elementary transitions $\xi_{n-1}^i \rightsquigarrow \xi_n^i$ of the N particles; that is, we have

$$V_n^N = \sqrt{N}[\eta_n^N - \eta_{n-1}^N K_{n,\eta_{n-1}^N}] = \sum_{i=1}^N \Delta_i V_n^N$$

with the "local" terms given for any $\varphi_n \in \mathcal{B}_b(E_n)$ by

$$\Delta_i V_n^N(\varphi_n) = \frac{1}{\sqrt{N}}[\varphi_n(\xi_n^i) - K_{n,\eta_{n-1}^N}(\varphi_n)(\xi_{n-1}^i)] \tag{9.4}$$

By the definition of the particle model η_n^N is the empirical measure associated with a collection of conditionally independent random variables ξ_n^i with distributions $K_{n,\eta_{n-1}^N}(\xi_{n-1}^i, \cdot)$. From this simple observation, we readily find that $\mathbb{E}(V_n^N(\varphi_n)) = 0$ and

$$\mathbb{E}(V_n^N(\varphi_n)^2) = \mathbb{E}(\eta_{n-1}^N(K_{n,\eta_{n-1}^N}[\varphi_n - K_{n,\eta_{n-1}^N}(\varphi_n)]^2))$$

In addition, for sufficiently regular McKean interpretation models, the asymptotic results developed in Section 7.4 apply, and we find that

$$\lim_{N \to \infty} \mathbb{E}(V_n^N(\varphi_n)^2) = \eta_{n-1}(K_{n,\eta_{n-1}}[\varphi_n - K_{n,\eta_{n-1}}(\varphi_n)]^2)$$

More precise regularity conditions are provided in the preliminary section, Section 9.3. Roughly speaking, the results above indicate that the random particle sampling error is globally unbiased, it has a finite variance, and the asymptotic limiting variance is known. Furthermore, the formula (9.3) also shows that the particle density profiles η_n^N satisfy "almost" the same equation (9.2) as the limiting measures η_n. To get one step further in our discussion, we introduce the following double triangular array:

$$\Delta_1 V_0^N(\varphi_0), \quad \Delta_2 V_0^N(\varphi_0), \quad \ldots, \quad \Delta_N V_0^N(\varphi_0)$$

$$\Delta_1 V_1^N(\varphi_1), \quad \Delta_2 V_1^N(\varphi_1), \quad \ldots, \quad \Delta_N V_1^N(\varphi_1)$$

$$\ldots\ldots\ldots \quad \ldots\ldots\ldots \quad \ldots \quad \ldots\ldots\ldots$$

$$\Delta_1 V_n^N(\varphi_n), \quad \Delta_2 V_n^N(\varphi_n), \quad \ldots, \quad \Delta_N V_n^N(\varphi_n)$$

The variables $(\Delta_i V_n^N(\varphi_n))_{1 \le i \le N}$ on each nth row are independent conditionally with respect to the rows with levels $p < n$. In addition, and in view of (9.4), each term $\Delta_i V_n^N(\varphi_p)$ is "negligible" in comparison with the sum $V_n^N(\varphi_n)$. Applying the classical CLT for triangular arrays (see for instance Theorem 4 on p. 543 in [287]), we find that the sum of the terms $V_n^N(\varphi_n)$ of each row converges in law as $N \to \infty$ to a Gaussian random variable $V_n(\varphi_n)$ such that

$$\mathbb{E}(V_n(\varphi_n)) = 0 \quad \text{and} \quad \mathbb{E}(V_n(\varphi_n)^2) = \eta_{n-1}(K_{n,\eta_{n-1}}[\varphi_n - K_{n,\eta_{n-1}}(\varphi_n)]^2) \tag{9.5}$$

These elementary observations are the first steps to our study of CLT for particle models. In Section 9.3, we will make precise this fluctuation result and will prove that the complete random sequence $(V_p^N(\varphi_p))_{p=0,\ldots,n}$ formed by the sums of each row converges in law to a sequence of independent and Gaussian random variables $(V_p(\varphi_p))_{p=0,\ldots,n}$ with means and variances prescribed by the formulae (9.5). In this preliminary section, we also present some elementary tools to transfer CLTs such as the Slutsky technique and the so-called δ-method.

Although these elementary fluctuations give some insight on the asymptotic normal behavior of the local errors accumulated by the sampling scheme, they do not give directly a CLT result for the difference between the particle measures η_n^N or γ_n^N and the corresponding limiting measures η_n and γ_n. Nevertheless, as the reader may have certainly noticed, the martingale decompositions exhibited in Proposition 7.4.1 in Section 7.4.2, are expressed in terms of the sequence of local errors V_n^N. In Section 9.4, we shall combine this important observation together with the approximation tools provided in Section 9.3 to analyze the fluctuations of η_n^N or γ_n^N.

9.3 Some Local Fluctuation Results

Let $\mathcal{F}^N = \{\mathcal{F}_n^N \ ; \ n \geq 0\}$ be the natural filtration associated with the N-particle system ξ_n. The first class of martingales that arises naturally in our context is the $\mathbb{R}^d$-valued and $\mathcal{F}^N$-martingale $M_n^N(f)$ defined by

$$M_n^N(f) = \sum_{p=0}^{n} \left[\eta_p^N(f_p) - \Phi_p(\eta_{p-1}^N)(f_p)\right] \quad (9.6)$$

where $f_p : x_p \in E_p \mapsto f_p(x_p) = (f_p^u(x_p))_{u=1,\ldots,d} \in \mathbb{R}^d$ is a d-dimensional and bounded measurable function. By direct inspection, we see that the vth component of the martingale $M_n^N(f) = (M_n^N(f^u))_{u=1,\ldots,d}$ is the d-dimensional and F^N-martingale defined for any $u = 1, \ldots, d$ by the formula

$$\begin{aligned}M_n^N(f^u) &= \sum_{p=0}^{n} \left[\eta_p^N(f_p^u) - \Phi_p(\eta_{p-1}^N)(f_p^u)\right] \\ &= \sum_{p=0}^{n} \left[\eta_p^N(f_p^u) - \eta_{p-1}^N K_{p,\eta_{p-1}^N}(f_p^u)\right]\end{aligned}$$

with the usual convention $K_{0,\eta_{-1}^N} = \eta_0 = \Phi_0(\eta_{-1}^N)$ for $p = 0$.

Most of the results presented here are based on the following CLT for the martingale $M_n^N(f)$.

Theorem 9.3.1 *For any $d \geq 1$ and for any sequence of bounded measurable functions $f_p = (f_p^u)_{u=1,\ldots,d} \in \mathbb{R}^d$ and $p \geq 0$, the $\mathbb{R}^d$-valued and $\mathcal{F}^N$-martingale $\sqrt{N}\, M_n^N(f)$ converges in law to an $\mathbb{R}^d$-valued and Gaussian martingale $M_n(f) = (M_n(f^u))_{u=1,\ldots,d}$ such that for any $n \geq 0$ and $1 \leq u, v \leq d$*

$$\langle M(f^u), M(f^v) \rangle_n$$

$$= \sum_{p=0}^{n} \eta_{p-1}[K_{p,\eta_{p-1}}\left(\left(f_p^u - K_{p,\eta_{p-1}} f_p^u\right)\left(f_p^v - K_{p,\eta_{p-1}} f_p^v\right)\right)]$$

with the convention $K_{0,\eta_{-1}} = \eta_0$ for $p = 0$.

Proof:
The idea of the proof due to Jean Jacod consists in using the CLT for triangular arrays of $\mathbb{R}^d$-valued random variables (Theorem 3.33, p. 437 in [189]). We first rewrite the martingale $\sqrt{N}\, M_n^N(f)$ in the following form:

$$\sqrt{N}\, M_n^N(f) = \sum_{i=1}^{N} \sum_{p=0}^{n} \frac{1}{\sqrt{N}} \left(f_p(\xi_p^i) - K_{p,\eta_{p-1}^N}(f_p)(\xi_{p-1}^i) \right)$$

This readily yields $\sqrt{N}\, M_n^N(f) = \sum_{k=1}^{(n+1)N} U_k^N(f)$ where for any $1 \leq k \leq (n+1)N$ with $k = pN + i$ for some $i = 1, \ldots, N$ and $p = 0, \ldots, n$

$$U_k^N(f) = \frac{1}{\sqrt{N}} \left(f_p(\xi_p^i) - K_{p,\eta_{p-1}^N}(f_p)(\xi_{p-1}^i) \right)$$

We further denote by $\mathcal{G}_k^N$ the σ-algebra generated by the random variables ξ_p^j for any pair index (j,p) such that $pN+j \leq k$. It is readily checked that, for any $1 \leq u < v \leq d$ and for any $1 \leq k \leq (n+1)N$ with $k = pN+i$ for some $i = 1, \ldots, N$ and $p = 0, \ldots, n$, we have $\mathbb{E}(U_k^N(f^u) \mid \mathcal{G}_{k-1}^N) = 0$ and

$$\mathbb{E}(U_k^N(f^u)U_k^N(f^v) \mid \mathcal{G}_{k-1}^N)$$
$$= \tfrac{1}{N} K_{p,\eta_{p-1}^N}[(f_p^u - K_{p,\eta_{p-1}^N} f_p^u)(f_p^v - K_{p,\eta_{p-1}^N} f_p^v)](\xi_{p-1}^i)$$

This also yields that

$$\sum_{k=pN+1}^{pN+N} \mathbb{E}(U_k^N(f^u)U_k^N(f^v) \mid \mathcal{F}_{k-1}^N)$$
$$= \eta_{p-1}^N[K_{p,\eta_{p-1}^N}[(f_p^u - K_{p,\eta_{p-1}^N} f_p^u)(f_p^v - K_{p,\eta_{p-1}^N} f_p^v)]]$$

Our aim is now to describe the limiting behavior of the martingale $\sqrt{N} M_n^N(f)$ in terms of the process $X_t^N(f) \stackrel{\text{def.}}{=} \sum_{k=1}^{[Nt]+N} U_k^N(f)$. By the definition of the particle model associated with a given mapping Φ_n and using the fact that $\left[\frac{[Nt]}{N}\right] = [t]$, one gets that for any $1 \leq u, v \leq d$

$$\sum_{k=1}^{[Nt]+N} E\left(U_k^N(f^u)U_k^N(f^v) \mid \mathcal{F}_{k-1}^N\right)$$
$$= C_{[t]}^N(f^u, f^v) + \tfrac{[Nt]-N[t]}{N} \left(C_{[t]+1}^N(f^u, f^v) - C_{[t]}^N(f^u, f^v)\right)$$

where, for any $n \geq 0$ and $1 \leq u, v \leq d$,

$$C_n^N(f^u, f^v)$$
$$= \sum_{p=0}^n \eta_{p-1}^N \left[K_{p,\eta_{p-1}^N}\left(\left(f_p^u - K_{p,\eta_{p-1}^N} f_p^u\right)\left(f_p^v - K_{p,\eta_{p-1}^N} f_p^v\right)\right)\right]$$

This implies that for any $1 \leq i, j \leq d$,

$$\sum_{k=1}^{[Nt]+N} E\left(U_k^N(f^u)U_k^N(f^v) \mid \mathcal{F}_{k-1}^N\right) \xrightarrow[N \to \infty]{P} C_t(f^u, f^v)$$

with

$$C_n(f^u, f^v)$$
$$= \sum_{p=0}^n \eta_{p-1}[K_{p,\eta_{p-1}}\left(\left(f_p^u - K_{p,\eta_{p-1}} f_p^u\right)\left(f_p^v - K_{p,\eta_{p-1}} f_p^v\right)\right)]$$

and for any $t \in \mathbb{R}_+$

$$C_t(f^u, f^v) = C_{[t]}(f^u, f^v) + \{t\}\left(C_{[t]+1}(f^u, f^v) - C_{[t]}(f^u, f^v)\right)$$

Since $\|U_k^N(f)\| \leq \frac{2}{\sqrt{N}} (\vee_{p\leq n}\|f_p\|)$, for any $1 \leq k \leq [Nt] + N$, the conditional Lindeberg condition is clearly satisfied and therefore one concludes that the $\mathbb{R}^d$-valued martingale $\{X_t^N(f)\ ;\ t \in \mathbb{R}_+\}$ converges in law to a continuous Gaussian martingale $\{X_t(f)\ ;\ t \in \mathbb{R}_+\}$ such that, for any $1 \leq u, v \leq d$ and $t \in \mathbb{R}_+$, $\langle X(f^u), X(f^v) \rangle_t = C_t(f^u, f^v)$. Recalling that $X_{[t]}^N(f) = \sqrt{N}\, M_{[t]}^N(f)$, the proof of the theorem is completed. ∎

A first consequence of Theorem 9.3.1 is a CLT for the random vector

$$\mathcal{V}_n^N(\varphi) = (V_0^N(\varphi_0), \ldots, V_n^N(\varphi_n)) \in \mathbb{R}^{d_0} \times \ldots \times \mathbb{R}^{d_n}$$

defined for any $0 \leq p \leq n$ by

$$V_p^N(\varphi_p) = \sqrt{N}\, [\eta_p^N(\varphi_p) - \eta_{p-1}^N K_{p,\eta_{p-1}^N}(\varphi_p)]$$

with $\varphi = (\varphi_n)_{n\geq 0}$ and $\varphi_n \in \mathcal{B}_b(E_n)^{d_n}$, $d_n \geq 1$, for all $n \in \mathbb{N}$. Loosely speaking, and as we already mentioned in the introductory section the next corollary expresses the fact that the local errors associated with the particle approximation sampling steps behave asymptotically as a sequence of independent and centered Gaussian random variables.

Corollary 9.3.1 *The sequence of random fields* $\mathcal{V}_n^N = (V_p^N)_{0\leq p\leq n}$ *converges in law, as $N \to \infty$, to a sequence $\mathcal{V}_n = (V_p)_{0\leq p\leq n}$ of $(n+1)$ independent and Gaussian random fields V_p with, for any $\varphi_p^1, \varphi_p^2 \in \mathcal{B}_b(E_p)$, $\mathbb{E}(V_p(\varphi_p^1)) = 0$ and*

$$\mathbb{E}(V_p(\varphi_p^1) V_p(\varphi_p^2))$$
$$= \eta_{p-1}(K_{p,\eta_{p-1}}[\varphi_p^1 - K_{p,\eta_{p-1}}(\varphi_p^1)]K_{p,\eta_{p-1}}[\varphi_p^2 - K_{p,\eta_{p-1}}(\varphi_p^2)])$$

Proof:
Let φ_n be a sequence of bounded measurable functions in $\mathcal{B}_b(E_n)^{d_n}$. We associate with $\varphi = (\varphi_n)_{n\geq 0}$ the sequence of functions

$$f_p = (f_p^1, \ldots, f_p^{n+1}) \in \mathcal{B}_b(E_p)^{d_1} \times \ldots \times \mathcal{B}_b(E_p)^{d_{n+1}}$$

$0 \leq p \leq n$ on E_p defined for any $0 \leq p, u \leq n$ by

$$f_p^{u+1} = 1_p(u)\, \varphi_p \in \mathcal{B}_b(E_p)^{d_p}$$

By construction, we have for any $0 \leq p \leq n$

$$\sqrt{N}\, M_n^N(f^{p+1}) = \sqrt{N}\, [\eta_p^N(\varphi_p) - \eta_{p-1}^N K_{p,\eta_{p-1}^N}(\varphi_p)] = V_p^N(\varphi_p)$$

and therefore

$$\sqrt{N}\, M_n^N(f) = (\sqrt{N}\, M_n^N(f^{p+1}))_{0\leq p\leq n} = (V_p^N(\varphi_p))_{0\leq p\leq n} = \mathcal{V}_n^N(\varphi)$$

By Theorem 9.3.1, we conclude that $\mathcal{V}_n^N(\varphi)$ converges in law to an $(n+1)$-dimensional and centered Gaussian random field $\mathcal{V}_n(\varphi) = (V_p(\varphi_p))_{0 \leq p \leq n}$ with, for any $0 \leq p, q \leq n$,

$$\mathbb{E}(V_p(\varphi_p^1) V_q(\varphi_q^2))$$

$$= \langle M(f^{p+1,1}), M(f^{q+1,2}) \rangle_n$$

$$= 1_p(q) \ \eta_{p-1}[K_{p,\eta_{p-1}}\left(\varphi_p^1 - K_{p,\eta_{p-1}}\varphi_p^1\right) K_{p,\eta_{p-1}}\left(\varphi_p^2 - K_{p,\eta_{p-1}}\varphi_p^2\right)]$$

This ends the proof of the corollary. ∎

The next two lemmas provide some important approximation tools that can be used in conjunction with Corollary 9.3.1.

Theorem 9.3.2 (Slutsky) *Suppose that $(V_1^N)_{N \geq 1}$ and $(V_2^N)_{N \geq 1}$ are random variables taking values in some separable metric space (E, d). If V_1^N converges in law, as $N \to \infty$, to some random variable V_1, and $d(V_1^N, V_2^N)$ converges to 0 in probability, then V_2^N converges in law, as $N \to \infty$, to V_1.*

Proof:
For any $\delta > 0$, we let $A^\delta = \{x \in E \ : \ d(x, A) = \inf_{y \in A} d(x, y) \leq \delta\}$ be the closed blowup of a closed subset $A \subset E$. Observe that

$$\mathbb{P}(V_2^N \in A)$$

$$= \mathbb{P}(V_2^N \in A, \ d(V_1^N, V_2^N) \leq \delta) + \mathbb{P}(V_2^N \in A, \ d(V_1^N, V_2^N) > \delta)$$

$$\leq \mathbb{P}(V_1^N \in A^\delta) + \mathbb{P}(d(V_1^N, V_2^N) > \delta)$$

Under our assumptions, we deduce that

$$\limsup_{N \to \infty} \mathbb{P}(V_1^N \in A) = \limsup_{N \to \infty} \mathbb{P}(V_2^N \in A)$$
$$\leq \limsup_{N \to \infty} \mathbb{P}(V_1^N \in A^\delta) \leq \mathbb{P}(V_1 \in A^\delta)$$

Letting $\delta \to 0$, we conclude that $\limsup_{N \to \infty} \mathbb{P}(V_2^N \in A) \leq \mathbb{P}(V_1 \in A)$ for any closed subset A. This clearly ends the proof of the theorem. ∎

Recalling that the convergence in law to some constant implies the convergence in probability, we have the following corollary.

Corollary 9.3.2 *Suppose that V_1^N converges in law to some finite constant v_1 and V_2^N converges in law to some random variable V_2. Then (V_1^N, V_2^N) converges in law to (v_1, V_2). In addition, if V_3^N converges in law to some finite constant v_3, then $V_1^N V_2^N + V_3^N$ converges in law to $v_1 V_2 + v_3$.*

We also quote without proof the following traditional theorem.

9.3 Some Local Fluctuation Results

Theorem 9.3.3 (continuous mapping theorem) *Let V^N be a sequence of random variables taking values in some metric space (E_1, d_1). Also let F be a mapping from E_1 into an auxiliary metric space (E_2, d_2). Assume that V^N converges in law to some random variable V. If F is almost continuous with respect to the distribution of V (i.e., $\mathbb{P}(V \in \mathcal{C}(F)) = 1$, where $\mathcal{C}(F)$ is the continuity set of F), then $F(V^N)$ converges in law to $F(V)$.*

Theorem 9.3.4 (Skorohod) *Let P and $(P_N)_{N \geq 1}$ be a collection of probability measures on some separable metric space E. Suppose that P_N weakly converges to P as N tends to infinity. Then there exists a probability space $(\Omega, \mathcal{F}, \mathbb{P})$ and a collection of random variables X and $(X_N)_{N \geq 1}$ defined on it such that $(\mathbb{P} \circ X^{-1}, \mathbb{P} \circ X_N^{-1}) = (P, P_N)$ and $\lim_{N \to \infty} X_N = X$, $\mathbb{P}$ almost surely.*

The following useful lemma is also known as the δ-method.

Lemma 9.3.1 (δ-method) *Let $(U_0^N, \ldots, U_n^N)_{N \geq 1}$ be a sequence of $\mathbb{R}^{n+1}$-valued random variables defined on some probability space and $(u_p)_{0 \leq p \leq n}$ be a given point in $\mathbb{R}^{n+1}$. Suppose that*

$$\sqrt{N}(U_0^N - u_0, \ldots, U_n^N - u_n) \qquad (9.7)$$

converges in law, as $N \to \infty$, to some random vector $(U_0, \ldots, U_n)$. Then, for any differentiable function $F_n : \mathbb{R}^{n+1} \to \mathbb{R}$ at the point $(u_p)_{0 \leq p \leq n}$ the sequence

$$\sqrt{N}[F_n(U_0^N(\omega), \ldots, U_n^N(\omega)) - F_n(u_0, \ldots, u_n)]$$

converges in law as $N \to \infty$ to the random variable $\sum_{p=0}^{n} \frac{\partial F_n}{\partial u_i}(u_0, \ldots, u_n) U_p$.

Proof:
By Skorohod's theorem, in $\mathbb{R}^{n+1}$ it is legitimate to suppose that all random variables are defined on a common probability space $(\Omega, \mathcal{F}, \mathbb{P})$ and (9.7) is the ordinary convergence in $\mathbb{R}^{n+1}$ for each sample $\omega \in \Omega$. Since F_n is assumed to be differentiable and $((U_p^N - u_p))_{0 \leq p \leq n}$ goes to zero, we have for each ω

$$\sqrt{N}[F_n(U_0^N(\omega), \ldots, U_n^N(\omega)) - F_n(u_0, \ldots, u_n)]$$

$$= \sum_{p=0}^{n} \frac{\partial F}{\partial u_i}(u_0, \ldots, u_n) \sqrt{N}(U_p^N(\omega) - u_p)$$

$$+ \varepsilon_n^N(\omega) \left[\sum_{p=0}^{n}(\sqrt{N}(U_p^N(\omega) - u_p))^2\right]^{1/2}$$

with $\lim_{N \to \infty} \varepsilon_n^N(\omega) = 0$. Under our assumptions, we have for each ω

$$\lim_{N \to \infty} \sqrt{N}[F_n(U_0^N(\omega), \ldots, U_n^N(\omega)) - F_n(u_0, \ldots, u_n)]$$

$$= \sum_{p=0}^{n} \frac{\partial F}{\partial u_i}(u_0, \ldots, u_n) U_p(\omega)$$

This ends the proof of the lemma. ∎

9.4 Particle Density Profiles

This section is concerned with the fluctuations of the particle approximation measures γ_n^N and η_n^N. For a precise description of the Feynman-Kac semigroups involved in this study, we refer the reader to Section 7.2 or Section 2.7.

9.4.1 Unnormalized Measures

We recall that the "unnormalized" approximation measures γ_n^N are defined for any $\varphi_n \in \mathcal{B}_b(E_n)$ by

$$\gamma_n^N(\varphi_n) = \gamma_n^N(1)\, \eta_n^N(\varphi_n) \quad \text{with} \quad \gamma_n^N(1) = \prod_{p=0}^{n-1} \eta_p^N(G_p)$$

By Proposition 7.4.1, the real-valued process defined by

$$\Gamma_{\cdot,n}^N(\varphi_n) \,:\, p \in \{0,\ldots,n\} \to \Gamma_{p,n}^N(\varphi_n) = \gamma_p^N(Q_{p,n}\varphi_n) - \gamma_p(Q_{p,n}\varphi_n)$$

has the $\mathcal{F}^N$-martingale decomposition

$$\Gamma_{p,n}^N(\varphi_n) = \sum_{q=0}^{p} \gamma_q^N(1)\left[\eta_q^N(Q_{q,n}\varphi_n) - \eta_{q-1}^N K_{q,\eta_{q-1}^N}(Q_{q,n}\varphi_n)\right]$$

and its angle bracket is given by

$$\langle \Gamma_{\cdot,n}^N(\varphi_n)\rangle_p = \frac{1}{N}\sum_{q=0}^{p} \gamma_q^N(1)^2\, \eta_{q-1}^N(K_{q,\eta_{q-1}^N}[Q_{q,n}\varphi_n - K_{q,\eta_{q-1}^N}Q_{q,n}\varphi_n]^2)$$

with the convention for $q = 0$, $\eta_{-1}^N = \Phi\left(\eta_{-1}^N\right) = K_{\eta_{-1}^N} = \eta_0$. Let $\overline{\Gamma}_{p,n}^N(\varphi_n)$, $0 \le p \le n$, be the random sequence defined as in (7.10) by replacing in the summation the terms $\gamma_q^N(1)$ by their limiting values $\gamma_q(1)$. In order to combine the CLT stated in Corollary 9.3.1 with the δ-method stated in Lemma 9.3.1, we rewrite the resulting random sequence as

$$\begin{aligned}
\sqrt{N}\,\overline{\Gamma}_{n,n}^N(\varphi_n) &= \sqrt{N}\,(\gamma_n^N(\varphi_n) - \gamma_n(\varphi_n)) \\
&= \sqrt{N}\sum_{q=0}^{n}\gamma_q(1)\left[\eta_q^N - \eta_{q-1}^N K_{q,\eta_{q-1}^N}\right](Q_{q,n}\varphi_n) \\
&= \sqrt{N}\,F_n(U_{0,n}^N,\ldots,U_{n,n}^N)
\end{aligned}$$

with the random sequence $U_{p,n}^N$, $0 \le p \le n$, and the function F_n given by

$$U_{p,n}^N = V_p^N(Q_{p,n}\varphi_n)/\sqrt{N} \quad \text{and} \quad F_n(v_0,\ldots,v_n) \stackrel{\text{def.}}{=} \sum_{q=0}^{n}\gamma_q(1)\,v_q$$

Since for any $n \geq 0$ we have $\lim_{N \to \infty} \gamma_n^N(1) = \gamma_n(1)$ in probability, we easily deduce from corollary 9.3.1 and Lemmas 9.3.2 and 9.3.1 that the real-valued random variable $\sqrt{N}\left(\gamma_n^N(\varphi_n) - \gamma_n(\varphi_n)\right)$ converges in law to the Gaussian random variable $\sum_{q=0}^{n} \gamma_q(1) V_p(Q_{p,n}\varphi_n)$. The extension of this result to multidimensional sequences is proved as before using Slutsky's theorem or the same techniques as used in the proof of Theorem 9.3.1.

Proposition 9.4.1 *For any $n \geq 0$, $d \geq 1$, and $f \in \mathcal{B}_b(E_n)^d$, the sequence of d-dimensional and $\mathcal{F}^N$-martingales $\sqrt{N}\Gamma_{p,n}^N(f)$, $0 \leq p \leq n$, converges in law to an $\mathbb{R}^d$-valued and Gaussian martingale $\Gamma_{p,n}(f)$, $0 \leq p \leq n$, such that for any $1 \leq u, v \leq d$, and $0 \leq p \leq n$*

$$\langle \Gamma_{\cdot,n}(f^u), \Gamma_{\cdot,n}(f^v) \rangle_p = \sum_{q=0}^{p} (\gamma_q(1))^2$$

$$\times \eta_{q-1}\left(K_{q,\eta_{q-1}}\left[Q_{q,n}f^u - K_{q,\eta_{q-1}}Q_{q,n}f^u\right]\left[Q_{q,n}f^v - K_{q,\eta_{q-1}}Q_{q,n}f^v\right]\right)$$

One immediate consequence of this multidimensional CLT is that the sequence of random fields

$$W_n^{\gamma,N}(\varphi_n) =_{\text{def.}} \sqrt{N}\left(\gamma_n^N(\varphi_n) - \gamma_n(\varphi_n)\right), \quad \varphi_n \in \mathcal{B}_b(E_n)$$

converges in law as $N \to \infty$, in the sense of convergence of finite-dimensional distributions, to a centered Gaussian field W_n^γ satisfying for any $f^1, f^2 \in \mathcal{B}_b(E_n)$

$$\mathbb{E}\left(W_n^\gamma(f^1)W_n^\gamma(f^2)\right) = \langle \Gamma_{\cdot,n}(f^1), \Gamma_{\cdot,n}(f^2) \rangle_n$$

Remark 9.4.1 *The martingale approach to CLTs we have developed can be extended to Feynman-Kac models with general nonnegative potential functions G_n as soon as the normalized constants $\gamma_n(1)$ are strictly positive. More precisely, using Theorem 7.4.1, we can easily prove that the CLT stated in Proposition 9.4.1 holds true for the martingale $\sqrt{N}\Gamma_{p,n}^N(f)$ introduced in (7.10) in Proposition 7.4.1.*

9.4.2 Normalized Measures

Observing that for any $n \in \mathbb{N}$ and $f \in \mathcal{B}_b(E_n)$ we have

$$\eta_n^N(f) - \eta_n(f) = \frac{\gamma_n(1)}{\gamma_n^N(1)} \gamma_n^N\left(\frac{1}{\gamma_n(1)}(f - \eta_n(f))\right)$$

and $\lim_{N \to \infty} \gamma_n(1)/\gamma_n^N(1) = 1$, in probability. We easily prove, using Slutsky's Lemma (Lemma 9.3.2), that the sequence of real-valued random variables

$$W_n^{\eta,N}(f) = \sqrt{N}\left(\eta_n^N(f) - \eta_n(f)\right)$$

converges to the Gaussian random variable W_n^η given by

$$W_n^\eta(f) = W_n^\gamma\left(\frac{1}{\gamma_n(1)}(f - \eta_n(f))\right)$$

To extend this CLT to multidimensional sequences, we follow the same arguments as those used in the proof of Proposition 9.4.1 by replacing $Q_{p,n}$ by the semigroup $\overline{Q}_{p,n}$ defined by

$$\overline{Q}_{p,n}(\varphi_n) = \frac{Q_{p,n}(\varphi_n)}{\eta_p Q_{p,n}(1)} = \frac{\gamma_p(1)}{\gamma_n(1)} Q_{p,n}(\varphi_n)$$

To clarify the presentation, it is convenient to introduce some additional notation. For any $0 \leq p \leq n$ and $f = (f^1, \ldots, f^d) \in \mathcal{B}_b(E_n)^d$, we write

$$f_{p,n} = \overline{Q}_{p,n}(f - \eta_n f) \tag{9.8}$$

Proposition 9.4.2 *For any $n \geq 0$ and $f = (f^1, \ldots, f^d) \in \mathcal{B}_b(E_n)^d$, the $\mathbb{R}^d$-valued process $W_p^N(f_{p,n})$, $0 \leq p \leq n$, given by*

$$W_p^{\eta,N}(f_{p,n}) \stackrel{\text{def.}}{=} \sqrt{N}\, \eta_p^N(f_{p,n})$$

converges in law to an $\mathbb{R}^d$-valued Gaussian martingale $W_p^\eta(f_{p,n})$, $0 \leq p \leq n$, such that for any $1 \leq u, v \leq d$, and $0 \leq p \leq n$

$$\langle W^\eta(f^u_{\bullet,n}), W^\eta(f^v_{\bullet,n}) \rangle_p$$

$$= \sum_{q=0}^{p} \eta_{q-1}\left(K_{q,\eta_{q-1}}[f^u_{q,n} - K_{q,\eta_{q-1}} f^u_{q,n}][f^v_{q,n} - K_{q,\eta_{q-1}} f^v_{q,n}]\right)$$

Proof:
For any $\varphi = (\varphi^1, \ldots, \varphi^d) \in \mathcal{B}_b(E_n)^d$, we have the decomposition

$$\eta_p^N(\overline{Q}_{p,n}\varphi) = \eta_0^N(\overline{Q}_{0,n}\varphi) + \sum_{q=1}^{p} \left[\eta_q^N(\overline{Q}_{q,n}\varphi) - \eta_{q-1}^N(\overline{Q}_{q-1,n}\varphi)\right]$$

If we choose $\varphi = (f - \eta_n f)$ with $f = (f^1, \ldots, f^d) \in \mathcal{B}_b(E_n)^d$, this yields

$$\eta_p^N(f_{p,n}) = B_p^N(f_{\bullet,n}) + M_p^N(f_{\bullet,n}) \tag{9.9}$$

where

$$B_p^N(f_{\bullet,n}) = \sum_{q=1}^{p} \left[1 - \eta_{q-1}^N(\overline{Q}_{q-1,q}1)\right] \Phi_q(\eta_{q-1}^N)(f_{q,n}) \tag{9.10}$$

$$M_p^N(f_{\bullet,n}) = \sum_{q=0}^{p} \left[\eta_q^N(f_{q,n}) - \Phi_q(\eta_{q-1}^N)f_{q,n}\right] \tag{9.11}$$

with the usual convention $\Phi_0(\eta_{-1}^N) = \eta_0$. Since for any $0 \leq q \leq p$

$$\eta_q(f_{q,n}) = \eta_n(f - \eta_n f) = 0 \quad \text{and} \quad \eta_{q-1}\left(\overline{Q}_{q-1,q}1\right) = \eta_q(1) = 1$$

then (9.10) can also be written in the form

$$B_p^N(f.,n)$$
$$= \sum_{q=1}^{p} \left[\eta_{q-1}(\overline{Q}_{q-1,q}1) - \eta_{q-1}^N(\overline{Q}_{q-1,q}1)\right] \left[\Phi_q(\eta_{q-1}^N)f_{q,n} - \Phi_q(\eta_{q-1})f_{q,n}\right]$$

Using the error estimates presented in previous sections, one gets after some tedious but easy calculations

$$N \, \mathbb{E}\left[\sup_{0 \leq p \leq n} \|B_p^N(f.,n)\|\right] \leq b(n) \, \|f\| \qquad (9.12)$$

for some finite constant $b(n) < \infty$, which only depends on the parameter n (we recall that $\|f\| = \sum_{i=1}^{d} \|f^i\|$ for any $f = (f^1, \ldots, f^d)$). By Theorem 9.3.1, we know that the F^N-martingale $\sqrt{N} \, M_p^N(f.,n)$ converges in law to the desired Gaussian martingale $W_p^\eta(f.,n)$. Arguing as in the proof of Theorem 9.3.1 and using (9.12), we easily complete the proof of the proposition. ∎

9.4.3 Killing Interpretations and Related Comparisons

One of the best ways to interpret the fluctuation variances developed in Section 9.4.1 and Section 9.4.2 is to use the Feynman-Kac killing interpretations provided in Section 2.5.1. In this context X_n is regarded as a Markov particle evolving in an absorbing medium with obstacles related to $[0,1]$-valued potentials. Using the same notation and terminology as was used in Section 2.5.1, the Feynman-Kac semigroups $(Q_{p,n}, P_{p,n})$ have the following interpretation

$$Q_{p,n}(x_p, dx_n) = \int \left\{\prod_{q=p}^{n-1} G_q(x_q)\right\} M_{p+1}(x_p, dx_{p+1}) \ldots M_n(x_{n-1}, dx_n)$$
$$= \mathbb{P}^c_{p,x_p}(X_n \in dx_n, \; T \geq n)$$

and, for any $x_p \in \widehat{E}_p = G_p^{-1}((0,1])$

$$P_{p,n}(x_p, dx_n) = \mathbb{P}^c_{p,x_p}(X_n \in dx_n \mid T \geq n)$$

where $\mathbb{P}^c_{p,x_p}$ represents the distribution of the absorbed particle evolution model starting at $X_p = x_p$ at time p. In this context, the variance of the fluctuation variable $W_n^\gamma(1)$ associated with the McKean interpretation model $K_{n,\eta}(x_{n-1}, \cdot) = \Phi_n(\eta)$ is given by

$$\mathbb{E}(W_n^\gamma(1)^2)$$

$$= \gamma_n(1)^2 \sum_{p=0}^n \eta_p \left([1 - Q_{p,n}(1)/\eta_p Q_{p,n}(1)]^2 \right)$$

$$= \mathbb{P}^c(T \geq n)^2 \sum_{p=0}^n \int_{E_p} \mathbb{P}^c(X_p \in dx_p \mid T \geq p) \left[\frac{\mathbb{P}^c_{p,x_p}(T \geq n)}{\mathbb{P}^c(T \geq n \mid T \geq p)} - 1 \right]^2$$

We further assume that for every $n \geq p$ and η_p-a.e. $x_p, y_p \in \widehat{E}_p$, we have

$$\mathbb{P}^c_{p,x_p}(T \geq n) \geq \delta\, \mathbb{P}^c_{p,y_p}(T \geq n)$$

for some $\delta > 0$. By Proposition 4.3.3, this condition is met as soon as the conditions (G) and $(M)_m$ introduced on page 115 hold true for some integer $m \geq 1$ and some pair parameters $(\epsilon_n(G), \varepsilon_n(M))$ such that $\epsilon(G) = \wedge_n \epsilon_n(G)$ and $\varepsilon(M) = \wedge_n \varepsilon_n(M) > 0$. In this case, we have

$$\mathbb{E}(W_n^\gamma(1)^2) \leq b(\delta)\, (n+1)\, \mathbb{P}^c(T \geq n)^2$$

for some finite constant $b(\delta) < \infty$. In general, the non absorption probabilities $\gamma_n(1) = \mathbb{P}^c(T \geq n)$ are very small and rather difficult to estimate (see also Section 12.2.5). The killing interpretation also suggests another approximation model based on N iid copies X^i of the absorbed particle evolution model.

$$X_n^i \xrightarrow{\text{killing}} \widehat{X}_n^i \xrightarrow{\text{exploration}} X_{n+1}^i$$

The Monte Carlo approximation is now given by $\frac{1}{N} \sum_{i=1}^N 1_{T^i \geq n}$, where T^i represents the absorption time of the ith particle. It is well-known that the fluctuation variance $\sigma_n^{\mathrm{MC}}(1)^2$ of this scheme is given by

$$\sigma_n^{\mathrm{MC}}(1)^2 = \mathbb{P}^c(T \geq n)\, (1 - \mathbb{P}^c(T \geq n))$$

From previous considerations we find that

$$\frac{\sigma_n^{\mathrm{MC}}(1)^2}{\mathbb{E}(W_n^\gamma(1)^2)} \geq \frac{1}{b(\delta)(n+1)} \frac{1 - \mathbb{P}^c(T \geq n)}{\mathbb{P}^c(T \geq n)} \longrightarrow \infty$$

as soon as $\mathbb{P}^c(T \geq n) = o(1/n)$.

If we choose the McKean transitions $K_{n,\eta}(x_{n-1}, \cdot) = \Phi_n(\eta)$, the variance of the random field W_n^η can also be described for any f^1 and $f^2 \in \mathcal{B}_b(E_n)$ as

$$\mathbb{E}\left(W_n^\eta(f^1) W_n^\eta(f^2) \right) = \sum_{p=0}^n \eta_p(f_{p,n}^1 f_{p,n}^2)$$

$$= \sum_{p=0}^n \eta_p \left([\overline{Q}_{p,n}(f^1 - \eta_n f^1)]\, [\overline{Q}_{p,n}(f^1 - \eta_n f^2)] \right)$$

(9.13)

This covariance function can be formulated using the potential functions $G_{p,n}$ and the Markov transitions $P_{p,n}$. More precisely, since for any $f \in \mathcal{B}_b(E_n)$ we have

$$\overline{Q}_{p,n}(f - \eta_n f) = \overline{Q}_{p,n}(1) \left(\frac{\overline{Q}_{p,n}(f)}{\overline{Q}_{p,n}(1)} - \eta_n f \right) = \frac{G_{p,n}}{\eta_p(G_{p,n})} (P_{p,n} f - \eta_n f)$$

we conclude that

$$\mathbb{E}\left(W_n^\eta(f^1) W_n^\eta(f^2)\right)$$

$$= \sum_{p=0}^n \int \left(\frac{G_{p,n}}{\eta_p(G_{p,n})}\right)^2 (P_{p,n} f^1 - \eta_n f^1)(P_{p,n} f^2 - \eta_n f^2) \, d\eta_p \qquad (9.14)$$

If the Markov kernels M_n are trivial in the sense that $M_n(x_{n-1}, \cdot) = \mu_n$ for some $\mu_n \in \mathcal{P}(E_n)$, then one can readily check that $\eta_n = \mu_n$ and $P_{p,n}(x_p, dx_n) = \mu_n(dx_n)$. In this particular situation, W_n^η is the classical μ_n-Brownian bridge. Namely, W_n^η is the centered Gaussian process with covariance

$$\mathbb{E}\left(W_n^\eta(f^1) W_n^\eta(f^2)\right) = \mu_n\left((f^1 - \mu_n f^1)(f^2 - \mu_n f^2)\right)$$

If we choose the McKean transitions

$$K_{n,\eta}(x, dy) = G_{n-1}(x) \, M_n(x, dy) + (1 - G_{n-1}(x)) \, \Phi_n(\eta)(dx)$$

then, using the fact that $\eta_q(f_{q,n}^1) = 0$, we first observe that

$$K_{q,\eta_{q-1}}(f_{q,n}^1) = G_{q-1} \, M_q(f_{q,n}^1)$$

In this situation, we conclude that the variance of the random field W_n^η is defined for any f^1 and $f^2 \in \mathcal{B}_b(E_n)$ by the formula

$$\mathbb{E}\left(W_n^\eta(f^1) W_n^\eta(f^2)\right)$$

$$= \sum_{p=0}^n \eta_p\left(f_{p,n}^1 f_{p,n}^2\right) - \sum_{p=1}^n \eta_{p-1}(G_{p-1}^2 \, M_p(f_{p,n}^1) M_p(f_{p,n}^2))$$

If we take $f^1 = f^2$, it is readily seen that the variance of the corresponding CLT is strictly smaller than the one associated with the McKean interpretation $K_{n,\eta}(x_{n-1}, \cdot) = \Phi_n(\eta)$.

Since we have

$$f_{p,n}^1(x_p) = \frac{G_{p,n}(x_p)}{\eta_p(G_{p,n})} \int_{E_p} (P_{p,n}(f^1)(x_p) - P_{p,n}(f^1)(y_p)) \frac{G_{p,n}(y_p)}{\eta_p(G_{p,n})} \eta_p(dy_p)$$

we readily observe that $\|f_{p,n}^1\| \leq r_{p,n} \, \beta(P_{p,n})$ with the pair of parameters $(r_{p,n}, \beta(P_{p,n}))$ introduced on page 218. Arguing as in the proof of Theorem 7.4.4 we obtain the uniform estimate

$$\sup_{n \geq 0} \mathbb{E}\left(W_n^\eta(f^1)^2\right) \leq m/(\epsilon^3(M) \, \varepsilon(G)^{2m-1})$$

306   9. Central Limit Theorems

as soon as conditions (G) and $(M)_m$ hold true for some integer $m \geq 1$ and some pair parameters $(\epsilon_n(G), \varepsilon_n(M))$ such that $\epsilon(G) = \wedge_n \epsilon_n(G)$ and $\varepsilon(M) = \wedge_n \varepsilon_n(M) > 0$.

9.5 A Berry-Esseen Type Theorem

In earlier sections, we have established various CLTs for the particle approximation measures γ_n^N and η_n^N. These results lead inevitably to the question of the rate of convergence of these fluctuations. In the present section, we analyze the CLT and the rate of convergence of the corresponding fluctuations for martingale sequences. The following theorem, which we give without proof, is due to Berry and Esseen (for a detailed proof, we refer the reader to any textbook on applied probability, such as [55]). It provides a simple way to estimate how close two distribution functions are in terms of their characteristic functions.

Theorem 9.5.1 (Berry-Esseen) *Let (F_1, F_2) be a pair of distribution functions with characteristic functions (f_1, f_2). Also assume that F_2 has a derivative with $\|\frac{\partial F_2}{\partial x}\| < \infty$. Then, for any $a > 0$, we have*

$$\|F_1 - F_2\| \leq \frac{2}{\pi} \int_0^a \frac{|f_1(x) - f_2(x)|}{x}\, dx + \frac{24}{a\pi} \left\|\frac{\partial F_2}{\partial x}\right\|$$

In a CLT, we have a given sequence of distribution functions F_1^N that converge as $N \to \infty$ to a centered Gaussian distribution function F_2 with a prescribed variance, say σ^2. As an aside, we note that in this case we have that $\|\frac{\partial F_2}{\partial x}\| = 1/\sqrt{2\pi\sigma^2}$. Theorem 9.5.1 reduces the study of the rapidity of convergence in the CLT to the estimation of an average difference between the corresponding characteristic functions. In the case of iid random variables $(\zeta^i)_{i \geq 1}$ with mean zero, unit variance, and finite third moment, the Berry-Esseen inequality implies that

$$\sup_{x \in \mathbb{R}} \left| \mathbb{P}\left(\frac{1}{\sqrt{N}} \sum_{i=1}^N \zeta^i \leq x\right) - \int_{-\infty}^x \frac{e^{-z^2/2}}{\sqrt{2\pi}}\, dz \right| \leq \frac{c}{\sqrt{N}}$$

In the display above and in the further development of this section, the letter c represents any finite constant whose values may vary from line to line. The method for obtaining the estimate above is based on Taylor's expansions of characteristic functions. Several extensions to not necessarily identically distributed random sequences, such as martingale sequences ζ^i, can be found in the literature; see for instance [55], p. 228 and [54], Section 9.3, p. 318. The extension to interacting particle models is a little more involved. Next we present a rather general strategy based on martingale techniques and on the following technical lemma, whose proof is

9.5 A Berry-Esseen Type Theorem

based on Stein's approach for CLTs, which is not discussed here; thus its proof will be omitted (see for instance Lemma 1.3 in the textbook by G.R. Shorack [288]).

Lemma 9.5.1 *Let F_Z be the distribution function associated with a real-valued random variable Z, and let W be a centered Gaussian random variable with unit variance. For any pair of random variables (X, Y), we have*

$$\|F_{X+Y} - F_W\| \leq \|F_X - F_W\| + 4\mathbb{E}(|XY|) + 4\mathbb{E}(|Y|) \qquad (9.15)$$

Loosely speaking, the lemma above is well-suited to deduce Berry-Esseen estimates for stochastic sequences of the form $X^N + Y^N$ when the corresponding statement has been proved for X^N and the random "perturbation" Y^N tends to zero "sufficiently fast".

We let $(\Omega^N, \mathcal{F}^N = (\mathcal{F}_n^N)_{n \geq 0}, \mathbb{P}^N)$, $N \geq 1$, be a collection of probability spaces with a distinguished nondecreasing collection of σ-algebras $\mathcal{F}_n^N \subset \mathcal{F}_{n+1}^N$, $n \geq 0$. For a given sequence of $\mathcal{F}^N$-adapted random variables X_n^N, we write for each $n \geq 0$

$$\Delta X_n^N = X_n^N - X_{n-1}^N$$

with the convention $\Delta X_0^N = X_0^N$ for $n = 0$. We recall that a square integrable and $\mathcal{F}^N$-martingale $M^N = (M_n^N)_{n \geq 0}$ is an $\mathcal{F}^N$-adapted sequence such that $\mathbb{E}((M_n^N)^2) < \infty$ for all $n \geq 0$ and

$$\mathbb{E}(M_{n+1}^N \mid \mathcal{F}_n^N) = M_n^N \qquad (\mathbb{P}^N\text{-a.s.})$$

The sequence of random variables $(\Delta M_n^N)_{n \geq 0}$ is also called an F^N-martingale difference and the predictable quadratic characteristic of M^N is the sequence of random variables $<M^N> = (<M^N>_n)_{n \geq 0}$ defined by

$$<M^N>_n = \sum_{p=0}^{n} \mathbb{E}((\Delta M_n^N)^2 \mid \mathcal{F}_{n-1}^N)$$

with the convention $\mathbb{E}((\Delta M_0^N)^2 \mid \mathcal{F}_{-1}^N) = \mathbb{E}((M_0^N)^2)$ for $n = 0$. The stochastic sequence $<M^N>$ is also called the angle bracket of M^N and is the unique predictable increasing process such that the sequence $((M_n^N)^2 - <M^N>_n)_{n \geq 0}$ is an $\mathcal{F}^N$-martingale. For each $n \geq 0$ and $N \geq 1$, we write

$$C_n^N = N <M^N>_n$$

the angle bracket of the $\mathcal{F}^N$-martingale sequence $(\sqrt{N} M_n^N)_{n \geq 0}$. The typical situation we have in mind is the one-dimensional $\mathcal{F}^N$-martingale

$$M_n^N(f) = \sum_{p=0}^{n} [\eta_p^N(f_p) - \Phi_p(\eta_{p-1}^N)(f_p)] \qquad (9.16)$$

introduced in (9.6), where $f_p : x_p \in E_p \mapsto f_p(x_p) \in \mathbb{R}$, $p \geq 0$, stands for some collection of measurable and bounded functions. No generality is lost and much convenience is gained by supposing (as will be done in this section) that the test functions f_p are such that $\|f_p\| \leq 1$ for all $p \geq 0$. Note that in this situation $\mathcal{F}_n^N = \sigma(\xi_0, \ldots, \xi_n)$ is the filtration generated by the particle model and we have that

$$N \langle M^N(f) \rangle_n = C_n^N(f) = \sum_{p=0}^n \eta_{p-1}^N [K_{p,\eta_{p-1}^N}((f_p - K_{p,\eta_{p-1}^N} f_p)^2)]$$

Also note that, in the first case McKean interpretation defined on page 219 we have

$$\begin{aligned} \Delta C_n^N(f) &= \eta_{n-1}^N [K_{n,\eta_{n-1}^N}((f_n - K_{n,\eta_{n-1}^N} f_n)^2)] \\ &= \Phi_n(\eta_{n-1}^N)(f_n^2) - \Phi_n(\eta_{n-1}^N)(f_n)^2 \end{aligned} \quad (9.17)$$

while in the second case

$$\begin{aligned} \Delta C_n^N(f) &= [\Phi_n(\eta_{n-1}^N)(f_n^2) - \Phi_n(\eta_{n-1}^N)(f_n)^2] \\ &\quad - \eta_{n-1}^N [G_{n-1}^2 (M_n(f_n) - \Phi_n(\eta_{n-1}^N)(f_n))^2] \end{aligned}$$

Our approach to the Berry-Esseen theorem for $\mathcal{F}^N$-martingale sequences M^N is based on the following conditions:

- ($H1$) For any $n \geq 0$, there exists some constants $a_1(n) < \infty$ and $0 < c_1(n) \leq 1$ such that for any $n \geq 0$ and $\lambda^3 \leq c_1(n) \sqrt{N}$ we have

$$|\mathbb{E}(e^{i\lambda\sqrt{N}\Delta M_n^N + \frac{\lambda^2}{2}\Delta C_n^N} \mid \mathcal{F}_{n-1}^N) - 1| \leq a_1(n)\lambda^3/\sqrt{N} \qquad (\mathbb{P}^N\text{-a.s})$$

- ($H2$) For any $n \geq 0$, there exists some finite constant $a_2(n) < \infty$ such that for any $N \geq 1$, $\lambda > 0$, and $n \geq 0$

$$|\mathbb{E}(e^{i\lambda\sqrt{N}M_n^N})| \leq \mathbb{E}(e^{-\frac{\lambda^2}{2}\Delta C_n^N}) \; e^{\lambda^3 a_2(n)/\sqrt{N}}$$

- ($H3$) There exists a nonnegative and strictly increasing deterministic process $C = (C_n)_{n \geq 0}$ as well as some finite constants $0 < a_3(n) < \infty$ such that for any $\varepsilon > 0$ we have

$$\mathbb{E}(e^{\varepsilon\sqrt{N}|\Delta C_n^N - \Delta C_n|}) \leq (1 + \varepsilon a_3(n)) \; e^{\varepsilon^2 a_3^2(n)}$$

We readily observe that $|\mathbb{E}(e^{i\lambda\sqrt{N}M_n^N})| \leq \mathbb{E}(|\mathbb{E}(e^{i\lambda\sqrt{N}\Delta M_n^N} \mid \mathcal{F}_{n-1}^N)|)$ from which we find that condition ($H2$) is met as soon as we have the following almost sure estimates

$$|\mathbb{E}(e^{i\lambda\sqrt{N}\Delta M_n^N + \frac{\lambda^2}{2}\Delta C_n^N} \mid \mathcal{F}_{n-1}^N)| \leq e^{\lambda^3 a_2(n)/\sqrt{N}}$$

9.5 A Berry-Esseen Type Theorem

Conditions $(H1)$ and $(H2)$ are rather classical. They are usually checked using simple asymptotic expansions of characteristic functions. The regularity condition $(H3)$ is more tricky to check in practice. It can be regarded as an exponential continuity condition on the increasing process C_n^N. The next two lemmas illustrate these three regularity conditions and their consequences. Their proofs are rather technical and are housed at the end of the section.

Lemma 9.5.2 *The $\mathcal{F}^N$-martingale $M_n^N(f)$ defined in (9.16) satisfies conditions (Hj), $j = 1, 2, 3$, for some universal constants*

$$(a_1(n), a_2(n)) = (a_1, a_2)$$

and with the nonnegative increasing process $C(f) = (C_n(f))_{n \geq 0}$ defined by

$$C_n(f) = \sum_{p=0}^{n} \eta_{p-1}[K_{p,\eta_{p-1}}((f_p - K_{p,\eta_{p-1}}f_p)^2)]$$

as soon as the mapping $n \to C_n(f)$ is strictly increasing. In addition, the constant $a_3(n)$ in $(H3)$ can be chosen such that for any $n \geq 0$

$$0 < a_3(n) \leq c \sum_{p=0}^{n} r_{p,n} \, \beta(P_{p,n})$$

with the collection of parameters $(r_{p,n}, \beta(P_{p,n}))$ introduced in (7.3).

Lemma 9.5.3 *Suppose we are given a sequence of $\mathcal{F}_n^N$-martingales $M^N = (M_n^N)_{n \geq 0}$ satisfying conditions (Hj) with $j = 1, 2, 3$. Then, for any $n \geq 0$, there exists a finite constant $a(n) < \infty$, a positive constant $b(n)$, and some $N(n) \geq 1$ such that for any $N \geq N(n)$ and $0 < \lambda \leq b(n)\sqrt{N}$*

$$|\mathbb{E}(e^{i\lambda\sqrt{N}M_n^N}) - e^{-\frac{\lambda^2}{2} C_n}| \leq a(n) \, e^{-\frac{\lambda^2}{4} \Delta C_n} \, \lambda^2(1+\lambda)/\sqrt{N}$$

Lemma 9.5.3 shows that whenever the regularity conditions (Hj), $j = 1, 2, 3$, are met, for any fixed time parameter $n \geq 0$ the sequence of random variables $\sqrt{N}M_n^N$ converges in law as $N \to \infty$ to a Gaussian random variable M_n with

$$\mathbb{E}(M_n) = 0 \quad \text{and} \quad \mathbb{E}(M_n^2) = C_n$$

In the context of particle models, we readily deduce from Lemma 9.5.2 that the sequence of random variables $\sqrt{N}M_n^N(f)$ converges in law as $N \to \infty$ to a centered Gaussian random variable $M_n(f)$ with

$$\mathbb{E}(M_n(f)^2) = \sum_{p=0}^{n} \eta_{p-1}[K_{p,\eta_{p-1}}((f_p - K_{p,\eta_{p-1}}f_p)^2)]$$

as soon as $C_n(f) > 0$ for any $n \geq 0$. This result is clearly much weaker than the multidimensional CLT presented in Theorem 9.3.1. Nevertheless, by a simple application of Theorem 9.5.1, we find the following fluctuation decays.

Theorem 9.5.2 *Let $M^N = (M_n^N)_{n \geq 0}$ be a sequence of $\mathcal{F}^N$-martingales satisfying conditions (Hj) with $j = 1, 2, 3$ for some nonnegative and increasing process C_n. Also let F_n^N be the distribution function of the random variable $\sqrt{N} M_n^N$ and let F_n be the distribution function of a centered Gaussian random variable with variance C_n. Then, for any $n \geq 0$ there exists some $N(n) \geq 1$ and some finite constant $c(n) < \infty$ such that for any $N \geq N(n)$ we have*

$$\|F_n^N - F_n\| \leq c(n)/\sqrt{N}$$

Proof:
By Theorem 9.5.1 and Lemma 9.5.3, we have for any $N \geq N(n)$

$$\sqrt{N}\|F_n^N - F_n\| \leq \frac{2a(n)}{\pi} \int_0^{b(n)\sqrt{N}} e^{-\frac{\lambda^2}{4} \Delta C_n} \lambda(1+\lambda)\, d\lambda$$
$$+ \frac{24}{b(n)\sqrt{2e\pi^3 C_n}}$$
$$\leq \frac{2a(n)}{\pi} \int_0^{\infty} e^{-\frac{\lambda^2}{4} \Delta C_n} \lambda(1+\lambda)\, d\lambda + \frac{4}{b(n)\sqrt{C_n}}$$

for some $N(n) \geq 1$ and some finite positive constant $0 < b(n) < \infty$. This ends the proof of the theorem. ■

In the context of particle models, we conclude that for each $n \geq 0$ the distribution function $\overline{F}_n^N$ of the normalized random variables

$$\sqrt{N} M_n^N(f)/\sqrt{C_n(f)}$$

weakly converges to the distribution function $\overline{F}_n$ of the centered and normalized Gaussian random variable $M_n(f)/\sqrt{C_n(f)}$ and for any $N \geq N(n)$ and some $N(n) \geq 1$

$$\sqrt{N}\|\overline{F}_n^N - \overline{F}_n\| \leq b(n) \qquad (9.18)$$

One strategy to deduce a Berry-Esseen estimate for the fluctuations of the particle density profiles η_n^N is to use the semimartingale decomposition presented in the proof of Proposition 9.4.2. More precisely, we fix a time horizon $n \geq 0$ and we associate with the test function $f \in \mathcal{B}_b(E_n)$ the sequence of functions $(f_{p,n})_{p \leq n}$ defined by

$$f_{p,n} = \overline{Q}_{p,n}(f - \eta_n f) \in \mathcal{B}_b(E_p)$$

with the normalized Feynman-Kac semigroup $\overline{Q}_{p,n}$ given for any $\varphi_n \in \mathcal{B}_b(E_n)$ and $x_p \in E_p$ by the equations

$$\overline{Q}_{p,n}(\varphi_n) = \frac{Q_{p,n}(\varphi_n)}{\eta_p Q_{p,n}(1)} \quad \text{and} \quad Q_{p,n}(\varphi_n)(x_p) = \mathbb{E}_{p,x_p}\left(\varphi_n(X_n)\prod_{q=p}^{n-1} G_q(X_q)\right)$$

Now we recall from (9.9) that

$$W_p^{\eta,N}(f_{p,n}) \stackrel{\text{def.}}{=} \sqrt{N}\,[\eta_p^N(f_{p,n}) - \eta_p(f_{p,n})]$$
$$= \sqrt{N} B_p^N(f_{\cdot,n}) + \sqrt{N} M_p^N(f_{\cdot,n})$$

with the $\mathcal{F}^N$-martingale sequence $M_p^N(f_{\cdot,n})$ defined by

$$M_p^N(f_{\cdot,n}) = \sum_{q=0}^{p}\,[\eta_q^N(f_{q,n}) - \Phi_q(\eta_{q-1}^N)(f_{q,n})]$$

and the $\mathcal{F}^N$-predictable sequence $B_p^N(f_{\cdot,n})$ defined for any $p \leq n$ by the formula

$$B_p^N(f_{\cdot,n})$$
$$= \sum_{q=1}^{p}\,[\eta_{q-1}(\overline{Q}_{q-1,q}1) - \eta_{q-1}^N(\overline{Q}_{q-1,q}1)]\,[\Phi_q(\eta_{q-1}^N)f_{q,n} - \Phi_q(\eta_{q-1})f_{q,n}]$$

Note that for $p = n$ we have $f_{n,n} = (f - \eta_n f)$ and

$$W_n^{\eta,N}(f_{n,n}) = \sqrt{N}\,[\eta_n^N(f) - \eta_n(f)]$$

Theorem 9.5.3 *For any $n \geq 0$, there exists some $N(n) \geq 1$ and some finite constant $b(n) < \infty$ such that for any $N \geq N(n)$*

$$\sup_{u \in \mathbb{R}}\left|\mathbb{P}(W_n^{\eta,N}(f_{n,n}) \leq u\sqrt{C_n(f)}) - \frac{1}{\sqrt{2\pi}}\int_{-\infty}^{u} e^{-v^2/2}\,dv\right| \leq \frac{b(n)}{\sqrt{N}}$$

Proof:
By arguments that should be now familiar to the reader, we find that

$$N\,\mathbb{E}(|B_n^N(f_{\cdot,n})|^2)^{1/2} \leq b(n) \quad \text{and} \quad N\,\mathbb{E}(|B_n^N(f_{\cdot,n})|) \leq b(n)$$

By the definition of the martingale term, we also easily check that

$$\sqrt{N}\,\mathbb{E}(|M_n^N(f_{\cdot,n})|^2)^{1/2} \leq b(n)$$

To apply Lemma 9.5.1, we set

$$X = \sqrt{N} M_n^N(f_{\cdot,n})/\sqrt{C_n(f)} \quad \text{and} \quad Y = \sqrt{N} B_n^N(f_{\cdot,n})/\sqrt{C_n(f)}$$

From previous estimates, we deduce that

$$\mathbb{E}(|XY|) \leq \frac{N}{C_n(f)} \mathbb{E}(|M_n^N(f_{.,n})|^2)^{1/2} \mathbb{E}(|B_n^N(f_{.,n})|^2)^{1/2} \leq b(n)/\sqrt{N}$$

and $\mathbb{E}(|Y|) \leq b(n)/\sqrt{N}$. The end of the proof is now a simple consequence of (9.15) and (9.18). ∎

We now come to the proofs of Lemma 9.5.3 and Lemma 9.5.2.

Proof of Lemma 9.5.3:

Let I_n^N be the function defined for any $\lambda \geq 0$ by

$$I_n^N(\lambda) = \mathbb{E}(e^{i\lambda\sqrt{N}M_n^N + \frac{\lambda^2}{2}C_n}) - 1$$

We have the easily verified recursive equations

$$I_n^N(\lambda) - I_{n-1}^N(\lambda)$$
$$= \mathbb{E}(e^{i\lambda\sqrt{N}M_{n-1}^N + \frac{\lambda^2}{2}C_{n-1}}[\mathbb{E}(e^{i\lambda\sqrt{N}\Delta M_n^N + \frac{\lambda^2}{2}\Delta C_n}|\mathcal{F}_{n-1}^N) - 1])$$
$$= \mathbb{E}(e^{i\lambda\sqrt{N}M_{n-1}^N + \frac{\lambda^2}{2}C_{n-1}}$$
$$\times [\mathbb{E}(e^{i\lambda\sqrt{N}\Delta M_n^N + \frac{\lambda^2}{2}\Delta C_n^N}|\mathcal{F}_{n-1}^N) - 1][e^{\frac{\lambda^2}{2}(\Delta C_n - \Delta C_n^N)}])$$
$$+ \mathbb{E}(e^{i\lambda\sqrt{N}M_{n-1}^N + \frac{\lambda^2}{2}C_{n-1}}[e^{\frac{\lambda^2}{2}(\Delta C_n - \Delta C_n^N)} - 1])$$

Using this, we obtain

$$I_n^N(\lambda) - I_{n-1}^N(\lambda)$$
$$\leq e^{\frac{\lambda^2}{2}C_{n-1}} \left[\mathbb{E}(|\mathbb{E}(e^{i\lambda\sqrt{N}\Delta M_n^N + \frac{\lambda^2}{2}\Delta C_n^N}|\mathcal{F}_{n-1}^N) - 1| \, e^{\frac{\lambda^2}{2}(\Delta C_n - \Delta C_n^N)}) \right.$$
$$\left. + \mathbb{E}(e^{\frac{\lambda^2}{2}|\Delta C_n - \Delta C_n^N|} - 1) \right]$$

9.5 A Berry-Esseen Type Theorem

Under conditions $(H1)$ and $(H3)$, we find that

$$|I_n^N(\lambda) - I_{n-1}^N(\lambda)|$$

$$\leq e^{\frac{\lambda^2}{2} C_{n-1}} \left[\frac{a_1(n)\lambda^3}{\sqrt{N}} \left(1 + \frac{\lambda^2 a_3(n)}{2\sqrt{N}}\right) e^{\frac{\lambda^4}{4N} a_3^2(n)} \right.$$

$$\left. + \left(1 + \frac{\lambda^2 a_3(n)}{2\sqrt{N}}\right) e^{\frac{\lambda^4}{4N} a_3^2(n)} - 1 \right]$$

$$\leq e^{\frac{\lambda^2}{2} C_{n-1}} \left[\frac{a_1(n)\lambda^3}{\sqrt{N}} \left(1 + \frac{\lambda^2 a_3(n)}{2\sqrt{N}}\right) e^{\frac{\lambda^4}{4N} a_3^2(n)} \right.$$

$$\left. + (e^{\frac{\lambda^4}{4N} a_3^2(n)} - 1) + \frac{\lambda^2 a_3(n)}{2\sqrt{N}} e^{\frac{\lambda^4}{4N} a_3^2(n)} \right]$$

for any $0 < \lambda^3 \leq c_1(n) \sqrt{N}$. Since for these pairs of parameters (λ, N) we have $\lambda^2 \leq \sqrt{N}$ (and therefore $\lambda^4 \leq N$), we find that

$$|I_n^N(\lambda) - I_{n-1}^N(\lambda)| \leq d(n) \, e^{\frac{\lambda^2}{2} C_{n-1}} \lambda^2 (1+\lambda)/\sqrt{N}$$

for some finite constant $d(n) < \infty$ whose values only depend on $a_i(n)$, $i = 1, 3$, and such that

$$d(n) = c \, e^{\frac{a_3^2(n)}{4}} (1 \vee a_1(n) \vee a_3(n))^2$$

If we set

$$c_\star(n) = \wedge_{p=0}^n c_1(p) \, (\leq 1) \quad \text{and} \quad d^\star(n) = \vee_{p=0}^n d(p)$$

then for any $0 \leq p \leq n$ and any $0 < \lambda^3 \leq c_\star(n) \sqrt{N}$ we have that

$$|I_p^N(\lambda) - I_{p-1}^N(\lambda)| \leq d^\star(n) \, e^{\frac{\lambda^2}{2} C_{n-1}} \lambda^2 (1+\lambda)/\sqrt{N}$$

It is now easily verified from these estimates that

$$|I_n^N(\lambda)| \leq (n+1) d^\star(n) e^{\frac{\lambda^2}{2} C_{n-1}} \lambda^2 (1+\lambda)/\sqrt{N}$$

from which we conclude that for any $0 < \lambda^3 \leq c_\star(n) \sqrt{N}$

$$|\mathbb{E}(e^{i\lambda\sqrt{N} M_n^N}) - e^{-\frac{\lambda^2}{2} C_n}| \leq (n+1) d^\star(n) \frac{\lambda^2}{\sqrt{N}}(1+\lambda) \, e^{-\frac{\lambda^2}{2} \Delta C_n} \quad (9.19)$$

On the other hand, we have for any pair (λ, N)

$$|\mathbb{E}(e^{i\lambda\sqrt{N} M_n^N}) - e^{-\frac{\lambda^2}{2} C_n}| \leq |\mathbb{E}(e^{i\lambda\sqrt{N} M_n^N})| + e^{-\frac{\lambda^2}{2} C_n} \quad (9.20)$$

and under condition $(H2)$

$$|\mathbb{E}(e^{i\lambda\sqrt{N} M_n^N})| \leq \mathbb{E}(e^{-\frac{\lambda^2}{2} \Delta C_n^N}) \, e^{\lambda^3 a_2(n)/\sqrt{N}}$$

Again using $(H3)$, we also find that

$$\begin{aligned}|\mathbb{E}(e^{i\lambda\sqrt{N}M_n^N})| &\leq e^{-\frac{\lambda^2}{2}\Delta C_n}\left(1+\frac{\lambda^2 a_3(n)}{2\sqrt{N}}\right)e^{\frac{\lambda^4 a_3^2(n)}{4N}}e^{\frac{\lambda^3 a_2(n)}{\sqrt{N}}} \\ &= e^{-\frac{\lambda^2}{2}[\Delta C_n - \frac{\lambda}{\sqrt{N}}(2a_2(n)+a_3^2(n)\frac{\lambda}{2\sqrt{N}})]}\left(1+\frac{\lambda^2 a_3(n)}{2\sqrt{N}}\right)\end{aligned}$$

Observe that for any pair (λ, N) such that

$$\lambda \leq c^\star(n)\sqrt{N} \quad \text{with} \quad c^\star(n) = [2a_3^{-2}(n) \wedge (2^{-1}\Delta C_n(1+2a_2(n)))^{-1}]$$

we have

$$\frac{\lambda}{\sqrt{N}}\left[2a_2(n)+a_3^2(n)\frac{\lambda}{2\sqrt{N}}\right] \leq \frac{\lambda}{\sqrt{N}}(2a_2(n)+1) \leq \frac{\Delta C_n}{2}$$

This yields that

$$|\mathbb{E}(e^{i\lambda\sqrt{N}M_n^N})| \leq \left[1 \vee \frac{a_3(n)}{2}\right]\left(1+\frac{\lambda^2}{\sqrt{N}}\right)e^{-\lambda^2\frac{\Delta C_n}{4}} \quad (9.21)$$

and hence, by (9.20) and for any $\lambda \leq c^\star(n)\sqrt{N}$, we find that

$$|\mathbb{E}(e^{i\lambda\sqrt{N}M_n^N}) - e^{-\frac{\lambda^2}{2}C_n}| \leq e^{-\frac{\lambda^2}{4}\Delta C_n}[2 \vee a_3(n)]\left(1+\frac{\lambda^2}{\sqrt{N}}\right) \quad (9.22)$$

To take the final step, we observe that for any

$$N \geq c_\star(n)/c^\star(n)^3 \quad \text{and} \quad c_\star^{1/3}(n)N^{1/6} \leq \lambda \leq c^\star(n)\sqrt{N}$$

we have $1 = c_\star(n)/c_\star(n) \leq c_\star^{-1}(n)\lambda^3/\sqrt{N}$ and by (9.22)

$$|\mathbb{E}(e^{i\lambda\sqrt{N}M_n^N}) - e^{-\frac{\lambda^2}{2}C_n}| \leq c_\star^{-1}(n)[2 \vee a_3(n)]\frac{\lambda^2}{\sqrt{N}}(1+\lambda)e^{-\lambda^2\frac{\Delta C_n}{4}} \quad (9.23)$$

In conjunction with (9.19), we conclude that for any

$$N \geq N(n) = c_\star(n)/c^\star(n)^3$$

and any $\lambda \leq c^\star(n)\sqrt{N}$

$$|\mathbb{E}(e^{i\lambda\sqrt{N}M_n^N}) - e^{-\frac{\lambda^2}{4}C_n}| \leq a(n)\frac{\lambda^2}{\sqrt{N}}(1+\lambda)e^{-\frac{\lambda^2}{4}\Delta C_n}$$

with $a(n) = [(n+1)d^\star(n)] \vee [c_\star^{-1}(n)(2 \vee a_3(n))]$. This ends the proof of the lemma. ∎

Proof of Lemma 9.5.2:

We first check that the regularity condition $(H3)$ is satisfied. For the McKean interpretation model $K_{n,\eta}(x,.) = \Phi_n(\eta)$, we have

$$\begin{aligned} \Delta C_n(f) &= \eta_{n-1}[K_{n,\eta_{n-1}}((f_n - K_{n,\eta_{n-1}}f_n)^2)] \\ &= \Phi_n(\eta_{n-1})(f_n^2) - \Phi_n(\eta_{n-1})(f_n)^2 = \eta_n(f_n^2) - \eta_n(f_n)^2 \end{aligned}$$

and by (9.17) we easily prove that

$$\begin{aligned} |\Delta C_n^N(f) - \Delta C_n(f)| &\leq |\Phi_n(\eta_{n-1}^N)(f_n^2) - \Phi_n(\eta_{n-1})(f_n^2)| \\ &\quad + 2|\Phi_n(\eta_{n-1}^N)(f_n) - \Phi_n(\eta_{n-1})(f_n)| \end{aligned}$$

By symmetry arguments, we have

$$\begin{aligned} |\Delta C_n^N(f) - \Delta C_n(f)| &\leq \mathbb{E}(|\eta_n^N(f_n^2) - \eta_n(f_n^2)| \mid \mathcal{F}_{n-1}^N) \\ &\quad + 2\mathbb{E}(|\eta_n^N(f_n) - \eta_n(f_n)| \mid \mathcal{F}_{n-1}^N) \end{aligned}$$

Applying Jensen's inequality, we find that for any $\varepsilon > 0$

$$\mathbb{E}(e^{\varepsilon\sqrt{N}|\Delta C_n^N(f) - \Delta C_n(f)|})$$

$$\leq \mathbb{E}(e^{\varepsilon\sqrt{N}\{\mathbb{E}(|\eta_n^N(f_n^2) - \eta_n(f_n^2)| \mid \mathcal{F}_{n-1}^N) + 2\mathbb{E}(|\eta_n^N(f_n) - \eta_n(f_n)| \mid \mathcal{F}_{n-1}^N)\}})$$

$$\leq \mathbb{E}(\mathbb{E}(e^{\varepsilon\sqrt{N}\{|\eta_n^N(f_n^2) - \eta_n(f_n^2)| + 2|\eta_n^N(f_n) - \eta_n(f_n)|\}} \mid \mathcal{F}_{n-1}^N))$$

$$\leq \mathbb{E}(e^{\varepsilon\sqrt{N}\{|\eta_n^N(f_n^2) - \eta_n(f_n^2)| + 2|\eta_n^N(f_n) - \eta_n(f_n)|\}})$$

Now applying the Cauchy-Schwartz inequality, we obtain

$$\mathbb{E}(e^{\varepsilon\sqrt{N}|\Delta C_n^N(f) - \Delta C_n(f)|})$$

$$\leq \mathbb{E}(e^{8\varepsilon\sqrt{N}|\eta_n^N(f_n^2/4) - \eta_n(f_n^2/4)|})^{1/2} \, \mathbb{E}(e^{8\varepsilon\sqrt{N}|\eta_n^N(f_n/2) - \eta_n(f_n/2)|})^{1/2}$$

Using Corollary 7.4.3 (and recalling that $f_n^2/4$ and $f_n/2 \in \mathrm{Osc}_1(E_n)$ for any $\|f_n\| \leq 1$), we conclude that

$$\mathbb{E}(e^{\varepsilon|\Delta C_n^N(f) - \Delta C_n(f)|}) \leq \left(1 + \varepsilon\, a_3(n)/\sqrt{N}\right) e^{\varepsilon^2 a_3^2(n)/N}$$

for some finite constant $a_3(n)$ such that $a_3(n) \leq 9\sum_{q=0}^n r_{q,n}\, \beta(P_{q,n})$. Using the same line of argument we prove that $(H3)$ also holds true for any McKean interpretation of the form (9.1). To prove that $(H2)$ is met, we first recall that

$$|\mathbb{E}(e^{i\lambda\sqrt{N}M_n^N})| \leq \mathbb{E}(|\mathbb{E}(e^{i\lambda\sqrt{N}\Delta M_n^N} \mid \mathcal{F}_{n-1}^N)|) \qquad (9.24)$$

Then we use a standard symmetrization technique. Given the particle model ξ_p up to time $p \leq (n-1)$, we let $\overline{\eta}_n^N$ be an auxiliary independent copy of η_n^N. In other words, $\overline{\eta}_n^N$ is the empirical measure associated with an independent copy $\overline{\xi}_n$ of the configuration of the system ξ_n at time n. With some obvious abusive notation, we readily check that

$$|\mathbb{E}(e^{i\lambda\sqrt{N}\Delta M_n^N} \mid \mathcal{F}_{n-1}^N)|^2 = \mathbb{E}(e^{i\lambda\sqrt{N}[\Delta M_n^N - \Delta \overline{M}_n^N]}) \mid \mathcal{F}_{n-1}^N)$$

where $\Delta \overline{M}_n^N = [\overline{\eta}_n^N(f_n) - \Phi_n(\eta_{n-1}^N)(f_n)]$. We deduce from this that

$$|\mathbb{E}(e^{i\lambda\sqrt{N}\Delta M_n^N} \mid \mathcal{F}_{n-1}^N)|^2 = \prod_{j=1}^{N} \mathbb{E}(e^{i\frac{\lambda}{\sqrt{N}}[f_n(\xi_n^j)-f_n(\overline{\xi}_n^j)]} \mid \mathcal{F}_{n-1}^N)$$

Since the random variables $[f_n(\xi_n^j) - f_n(\overline{\xi}_n^j)]$ and $-[f_n(\xi_n^j) - f_n(\overline{\xi}_n^j)]$ have the same law, their characteristic functions are real and we have

$$\mathbb{E}(e^{i\frac{\lambda}{\sqrt{N}}[f_n(\xi_n^j)-f_n(\overline{\xi}_n^j)]} \mid \mathcal{F}_{n-1}^N) = \mathbb{E}\left(\cos\left(\frac{\lambda}{\sqrt{N}}[f_n(\xi_n^j)-f_n(\overline{\xi}_n^j)]\right) \mid \mathcal{F}_{n-1}^N\right)$$

Using the elementary inequalities

$$\cos u \leq 1 - u^2/2 + |u|^3/3!, \quad 1 + u \leq e^u, \quad \text{and} \quad |u-v|^3 \leq 4(|u|^3 + |v|^3)$$

we prove that

$$\mathbb{E}(e^{i\frac{\lambda}{\sqrt{N}}[f_n(\xi_n^j)-f_n(\overline{\xi}_n^j)]} \mid \mathcal{F}_{n-1}^N)$$

$$\leq 1 - \frac{\lambda^2}{N} K_{n,\eta_{n-1}^N}(f_n - K_{n,\eta_{n-1}^N}(f_n))^2(\xi_n^j) + \frac{c\lambda^3}{N\sqrt{N}}$$

$$\leq e^{-\frac{\lambda^2}{N} K_{n,\eta_{n-1}^N}(f_n - K_{n,\eta_{n-1}^N}(f_n))^2(\xi_n^j) + \frac{c\lambda^3}{N\sqrt{N}}}$$

Multiplying over j, we obtain

$$|\mathbb{E}(e^{i\lambda\sqrt{N}\Delta M_n^N} \mid \mathcal{F}_{n-1}^N)|^2 \leq e^{-\lambda^2 \Delta C_n^N(f) + \frac{c\lambda^3}{\sqrt{N}}}$$

and by (9.24) we conclude that condition $(H2)$ is met with $a_2(n) = c/2$. We now come to the proof of $(H1)$. By the definition of the particle model associated with a given collection of transitions $K_{n,\eta}$, we have

$$\mathbb{E}(\,e^{i\lambda\sqrt{N}\Delta M_n^N(f) + \frac{\lambda^2}{2}\Delta C_n^N(f)} \mid \mathcal{F}_{n-1}^N)$$

$$= \prod_{j=1}^{N} K_{n,\eta_{n-1}^N}(e^{i\frac{\lambda}{\sqrt{N}}\tilde{f}_n^j + \frac{\lambda^2}{2N}\Delta C_n^N(f)})(\xi_{n-1}^j)$$

with the random function $\tilde{f}_n^j = (f_n - K_{n,\eta_{n-1}^N}(f_n))(\xi_{n-1}^j)$. Using the elementary inequality

$$|e^z - (1 + z + z^2/2)| \leq e^{|z|}\,|z|^3/3!$$

9.5 A Berry-Esseen Type Theorem

after some computations we see that for any $\lambda \leq \sqrt{N}$ we have that

$$e^{i\frac{\lambda}{\sqrt{N}}\tilde{f}_n^j + \frac{\lambda^2}{2N}\Delta C_n^N(f)} = 1 + i\frac{\lambda}{\sqrt{N}}\tilde{f}_n^j + \frac{\lambda^2}{2N}[\Delta C_n^N(f) - (\tilde{f}_n^j)^2] + r_{n,1}^N(f)$$

with $|r_{n,1}^N(f)| \leq c\,\lambda^3/(N\sqrt{N})$. This clearly implies that for any $\lambda \leq \sqrt{N}$

$$K_{n,\eta_{n-1}^N}(e^{i\frac{\lambda}{\sqrt{N}}\tilde{f}_n^j + \frac{\lambda^2}{2N}\Delta C_n^N(f)})(\xi_{n-1}^j)$$

$$= 1 + \frac{\lambda^2}{2N}[\Delta C_n^N(f) - K_{n,\eta_{n-1}^N}(\tilde{f}_n^j)^2(\xi_{n-1}^j)] + r_{n,2}^N(f)$$

with $|r_{n,2}^N(f)| \leq c\,\lambda^3/(N\sqrt{N})$. It is now convenient to note that for any $\lambda \leq \sqrt{N}$

$$\left|\frac{\lambda^2}{2N}[\Delta C_n^N(f) - K_{n,\eta_{n-1}^N}(\tilde{f}_n^j)^2(\xi_{n-1}^j)] + r_{n,2}^N(f)\right| \leq c\,\lambda/\sqrt{N}$$

On the other hand, for any $|z| \leq 1/2$ and with the principal value of the logarithm, we recall that

$$\log(1+z) = z - \int_0^z \frac{u}{1+u}\,du = z - z^2 \int_0^1 \frac{t}{1+tz}\,dt$$

Since for any $|z| \leq 1/2$ and $t \in [0,1]$ we have $|1+tz| \geq 1/2$, we find that for any $|z| \leq 1/2$ we have $|\log(1+z) - z| \leq |z^2|$. From previous computation, we conclude that there exists some universal constant $c_0 \in (0,1)$ such that for any $\lambda \leq c_0\sqrt{N}$ we have

$$\log K_{n,\eta_{n-1}^N}(e^{i\frac{\lambda}{\sqrt{N}}\tilde{f}_n^j + \frac{\lambda^2}{2N}\Delta C_n^N(f)})(\xi_{n-1}^j)$$

$$= \frac{\lambda^2}{2N}[\Delta C_n^N(f) - K_{n,\eta_{n-1}^N}(\tilde{f}_n^j)^2(\xi_{n-1}^j)] + r_{n,3}^N(f)$$

with $|r_{n,3}^N(f)| \leq c\,\lambda^3/(N\sqrt{N})$. Summing over j, we see that for any $\lambda \leq c_0\sqrt{N}$

$$\left|\sum_{j=1}^N \log K_{n,\eta_{n-1}^N}(e^{i\frac{\lambda}{\sqrt{N}}\tilde{f}_n^j + \frac{\lambda^2}{2N}\Delta C_n^N(f)})(\xi_{n-1}^j)\right| \leq c\,\lambda^3/\sqrt{N}$$

Finally, using the elementary inequality $|e^z - 1| \leq |z|e^{|z|}$, we conclude that for any $\lambda \leq c_0\sqrt{N}$

$$\left|\mathbb{E}(e^{i\lambda\sqrt{N}\Delta M_n^N(f) + \frac{\lambda^2}{2}\Delta C_n^N(f)} \mid F_{n-1}^N) - 1\right| \leq c\,\frac{\lambda^3}{\sqrt{N}}\,e^{c\,\frac{\lambda^3}{\sqrt{N}}}$$

This readily implies that there exists some universal positive constant c_1 such that for any $\lambda^3 \leq c_1 \sqrt{N}$ we have

$$|\mathbb{E}(e^{i\lambda\sqrt{N}\Delta M_n^N(f) + \frac{\lambda^2}{2}\Delta C_n^N(f)} \mid F_{n-1}^N) - 1| \leq c\,\frac{\lambda^3}{\sqrt{N}}$$

This proves that condition $(H1)$ is met with $a_1(n) = c$ and $c_1(n) = 1$. ∎

9.6 A Donsker Type Theorem

The random fields $W_n^{\eta,N} = \sqrt{N}\,(\eta_n^N - \eta_n)$ introduced in Section 9.4 can alternatively be regarded as an empirical process indexed by the collection of bounded measurable functions. In this interpretation, the fluctuation results presented earlier simply say that the marginals of the $\mathcal{B}_b(E_n)$-indexed empirical process weakly converge to the marginals of a centered Gaussian process W_n^η. To simplify the notation, we suppress the superscript $(.)^\eta$ and we write W_n^N and W_n instead of $W_n^{\eta,N}$ and W_n^η.

In this empirical process interpretation, one natural question we may ask is whether there exists a functional convergence result for an $\mathcal{F}_n$-indexed empirical process $\{W_n^N(f); f \in \mathcal{F}_n\}$ where $\mathcal{F}_n \subset \mathcal{B}_b(E_n)$. We recall that weak convergence in $l^\infty(\mathcal{F}_n)$ can be characterized as the convergence of the marginals together with the asymptotic tightness of the process $f \in \mathcal{F}_n \to W_n^N(f) \in \mathbb{R}$. The asymptotic tightness is related to the entropy condition $I(\mathcal{F}_n) < \infty$.

Lemma 9.6.1 *If $\mathcal{F}_n$ is a countable collection of functions f such that $\|f\| \leq 1$ and $I(\mathcal{F}_n) < \infty$, then the $\mathcal{F}_n$-indexed process $W_n^N(f)$, $f \in \mathcal{F}_n$, is asymptotically tight.*

From previous comments, one concludes the following.

Theorem 9.6.1 (Donsker) *When condition (G) holds true and $I(\mathcal{F}_n) < \infty$, the empirical process*

$$W_n^N : f \in \mathcal{F}_n \longrightarrow W_n^N(f) = \sqrt{N}\,(\eta_n^N(f) - \eta_n(f))$$

converges in law in $l^\infty(\mathcal{F}_n)$ to a centered Gaussian process $W_n(f)$, $f \in \mathcal{F}_n$.

Proof of lemma 9.6.1: Using the Cauchy-Schwartz inequality, we have

$$E\Big(|W_n(f) - W_n(h)|^2\Big) \leq c(n)\|f - h\|_{L_2(\eta_n)}^2$$

for some constant $c(n) < \infty$ and all $f, h \in L_2(\eta_n)$. In particular, to prove asymptotic tightness it is enough to establish the asymptotic equicontinuity

in probability of $W_n^N(f)$, $f \in \mathcal{F}_n$ with respect to the semi-norm on $\mathcal{F}_n$ given by
$$f \in \mathcal{F}_n \to \eta_n\left((f - \eta_n(f))^2\right)^{1/2}$$
(see for instance Chap. 1.5 and Example 1.5.10 on p. 40 in [311]). Since $I(\mathcal{F}_n) < \infty$, the class $\mathcal{F}_n$ is totally bounded in $L_2(\eta_n)$ for any $n \geq 0$. According to the preceding comment, it will be enough to show that

$$\lim_{N \to \infty} E\left(\|W_n^N\|_{\mathcal{F}^{(n)}(\delta_N)}\right) = 0 \qquad (9.25)$$

for all sequences $\delta_N \downarrow 0$ where, for any $\delta > 0$ (possibly infinite),

$$\mathcal{F}^{(n)}(\delta) \stackrel{\text{def}}{=} \{f - h;\ f, h \in \mathcal{F}_n : \|f - h\|_{L_2(\eta_n)} < \delta\}$$

For this task, we use the traditional decomposition

$$\eta_n^N(f) - \eta_n(f) = \sum_{p=0}^{n} \left[\Phi_{p,n}(\eta_p^N)(f) - \Phi_{p,n}(\Phi_p(\eta_{p-1}^N))(f)\right]$$

with

$$\Phi_{p,n}(\mu)(f) - \Phi_{p,n}(\eta)(f) = \frac{1}{\eta(G_{p,n})}\Big[(\mu(G_{p,n}P_{p,n}(f)) - \eta(G_{p,n}P_{p,n}(f)))$$
$$+ \Phi_{p,n}(\mu)(f)\left(\eta(G_{p,n}) - \mu(G_{p,n})\right)\Big]$$

to prove that

$$\|\eta_n^N - \eta_n\|_{\mathcal{F}^{(n)}(\delta)} \leq \sum_{p=0}^{n} \frac{1}{r_{p,n}} \Big[\|\eta_p^N - \Phi_p(\eta_{p-1}^N)\|_{\mathcal{F}_{p,n}(\delta)}$$
$$+ \|\eta_p^N\|_{\mathcal{F}_{p,n}(\delta)} |\eta_p^N(\overline{G}_{p,n}) - \Phi_p(\eta_{p-1}^N)(\overline{G}_{p,n})| \Big]$$

where

$$\mathcal{F}_{p,n}(\delta) = \{\overline{G}_{p,n} P_{p,n}(f);\ f \in \mathcal{F}^{(n)}(\delta)\},$$

$\overline{G}_{p,n} = G_{p,n}/\|G_{p,n}\|$, and $r_{p,n} = \prod_{q=p}^{n-1} \epsilon_p^{-1}(G)$. Since for any $0 \leq p \leq n$

$$E\left(\left[\eta_p^N(\overline{G}_{p,n}) - \Phi_p(\eta_{p-1}^N)(\overline{G}_{p,n})\right]^2\right) \leq \frac{1}{N}$$

to prove (9.25) it suffices to check that for any $0 \leq p \leq n$ and $\delta_N \downarrow 0$

1) $\quad \lim_{N \to \infty} E\left(\sqrt{N} \|\eta_p^N - \Phi_p(\eta_{p-1}^N)\|_{\mathcal{F}_{p,n}(\delta_N)}\right) = 0$

2) $\quad \lim_{N \to \infty} E\left(\|\eta_p^N\|_{\mathcal{F}_{p,n}^2(\delta_N)}\right) = 0$

where $\mathcal{F}_{p,n}^2(\delta) = \{f^2; f \in \mathcal{F}_{p,n}(\delta)\}$. Let us prove 1). Let $\varepsilon = (\varepsilon_i)_{i \geq 1}$ constitute a sequence of independent and identically distributed with $P(\varepsilon_1 = +1) = P(\varepsilon_1 = -1) = 1/2$. We also assume that ϵ and the particle model $(\xi_n)_{n \geq 0}$ are independent. By the symmetrization inequalities, for any N,

$$E\left(\|\eta_p^N - \Phi_p(\eta_{p-1}^N)\|_{\mathcal{F}_{p,n}(\delta_N)}\right) \leq 2E\left(\|m_\varepsilon^N(\xi_p)\|_{\mathcal{F}_{p,n}(\delta_N)}\right)$$

where $m_\varepsilon^N(\xi_p) = \frac{1}{N}\sum_{i=1}^N \varepsilon_i \delta_{\xi_p^i}$. Fix $\xi_p = (\xi_p^1, \ldots, \xi_p^N)$. By the Chernov-Hoeffding inequality (see Lemma 7.3.2), the process $f \to \sqrt{N} m_\varepsilon^N(\xi_p)(f)$ is sub-Gaussian with respect to the norm $\|\cdot\|_{L_2(\eta_p^N)}$. Namely, for any $f, h \in \mathcal{F}_{p,n}(\delta_N)$ and $\gamma > 0$,

$$P\left(\left|\sqrt{N}(m_\varepsilon^N(\xi_p)(f) - m_\varepsilon^N(\xi_p)(h))\right| > \gamma \,\Big|\, \xi_p\right) \leq 2\, e^{-\frac{1}{2}\gamma^2/\|f-h\|_{L_2(\eta_p^N)}^2}$$

Using the maximal inequality for sub-Gaussian processes (see for instance [311, 217]), we get the quenched inequality

$$E\left(\|m_\varepsilon^N(\xi_p)\|_{\mathcal{F}_{p,n}(\delta_N)} \,\Big|\, \xi_p\right)$$
$$\leq \frac{c}{\sqrt{N}} \int_0^{\theta_{p,n}(N)} \sqrt{\log\left(1 + N(\delta, \mathcal{F}_{p,n}(\delta_N), L_2(\eta_p^N))\right)}\, d\delta \tag{9.26}$$

where $\theta_{p,n}(N) = \|\eta_p^N\|_{\mathcal{F}_{p,n}^2(\delta_N)}$. On the other hand, we clearly have that, for every $\delta > 0$,

$$N(\delta, \mathcal{F}_{p,n}(\delta), L_2(\eta_p^N)) \leq N(\delta, \mathcal{F}_{p,n}(\infty), L_2(\eta_p^N)) \leq \mathcal{N}^2(\delta/2, \mathcal{F}_{p,n}, L_2(\eta_p^N))$$

where we recall that $\mathcal{F}_{p,n} = \overline{G}_{p,n} \cdot P_{p,n} \mathcal{F}_n$. Under our assumptions, it thus follows from Lemma 7.3.4 in Section 7.3 that

$$N(\delta, \mathcal{F}_{p,n}(\delta), L_2(\eta_p^N)) \leq \mathcal{N}^2(\delta/2, \mathcal{F}_n)$$

Using (9.26), one concludes that, for every $N \geq 1$,

$$E\left(\|m_\varepsilon^N(\xi_p)\|_{\mathcal{F}_{p,n}(\delta_N)}\right) \leq \frac{c}{\sqrt{N}} E\left(\int_0^{\theta_{p,n}(N)} \sqrt{\log[1 + \mathcal{N}(\delta/2, \mathcal{F})]}\, d\delta\right)$$

and therefore

$$E\left(\|\eta_p^N - \Phi_p(\eta_{p-1}^N)\|_{\mathcal{F}_{p,n}(\delta_N)}\right)$$
$$\leq \frac{c}{\sqrt{N}} E\left(\int_0^{\theta_{n|p}(N)} \sqrt{\log[1 + \mathcal{N}(\delta/2, \mathcal{F})]}\, d\delta\right)$$

By the dominated convergence theorem, to prove 1) it suffices to check that

$$\lim_{N\to\infty} \theta_{p,n}(N) = \lim_{N\to\infty} \|\eta_p^N\|_{\mathcal{F}_{p,n}^2(\delta_N)} = 0 \quad \mathbb{P}\text{-a.s.} \tag{9.27}$$

We establish this property by proving that

a) $\|\eta_p\|_{\mathcal{F}_{p,n}^2(\delta_N)} \leq \delta_N^2$

b) $\lim_{N\to\infty} \|\eta_p^N - \eta_p\|_{\mathcal{F}_{p,n}^2(\delta_N)} = 0 \quad \mathbb{P}\text{-a.s.}$

Let $f, h \in \mathcal{F}_n$ be chosen so that $\eta_n((f-h)^2) < \delta^2$ (i.e., $f - h \in \mathcal{F}^{(n)}(\delta)$). Use the Cauchy-Schwartz inequality to see that

$$\eta_p\left(\left[\overline{G}_{p,n} P_{p,n}(f-h)\right]^2\right) \leq \eta_p\left(\overline{G}_{p,n}^2 P_{p,n}((f-h)^2)\right) \tag{9.28}$$

Since $0 \leq \overline{G}_{p,n} \leq 1$, the right-hand side of (9.28) is bounded above by

$$\frac{1}{\|G_{p,n}\|} \eta_p\left(G_{p,n} P_{p,n}((f-h)^2)\right) = \frac{\eta_p(G_{p,n})}{\|G_{p,n}\|} \eta_n((f-h)^2)$$

which is less that δ^2. This ends the proof of a). To prove b), first note that

$$\|\eta_p^N - \eta_p\|_{\mathcal{F}_{p,n}^2(\delta_N)} \leq \|\eta_p^N - \eta_p\|_{\mathcal{F}_{p,n}^2(\infty)}$$

Now, to prove b), it certainly suffices to show that

$$\sup_\mu N\big(\delta, \mathcal{F}_{p,n}^2(\infty), L_2(\mu)\big) < \infty$$

for every $\delta > 0$. Since all functions in $\mathcal{F}_n$ have norm less than or equal to 1, we have $\|f^2 - h^2\|_{L_2(\mu)} \leq 4 \|f - h\|_{L_2(\mu)}$, for any f, h in $\mathcal{F}_{p,n}(\infty)$ and any $\mu \in \mathcal{P}(E_p)$. It follows that, for every $\delta > 0$,

$$N\big(\delta, \mathcal{F}_{p,n}^2(\infty), L_2(\mu)\big) \leq N\big(\delta/4, \mathcal{F}_{p,n}(\infty), L_2(\mu)\big)$$

Since $N\big(\delta, \mathcal{F}_{p,n}(\infty), L_2(\mu)\big) \leq N^2\big(\delta/2, \mathcal{F}_{p,n}, L_2(\mu)\big)$, one concludes, using Lemma 7.3.4, that

$$\sup_\mu N\big(\delta, \mathcal{F}_{p,n}^2(\infty), L_2(\mu)\big) \leq \sup_\mu N^2\big(\delta/8, \mathcal{F}_n, L_2(\mu)\big)$$

This ends the proof of b) and 1). In the same way, by dominated convergence, the proof of 2) is an immediate consequence of (9.27). This completes the proof of the lemma. ∎

9.7 Path-Space Models

This section discusses the fluctuations of the path-particle McKean measures $\mathbb{K}_n^N = \frac{1}{N}\sum_{i=1}^N \delta_{(\xi_0^i,\ldots,x_n^i)}$. We restrict our study to the simple mutation/selection genetic model associated with the first McKean interpretation model $K_{n,\eta}(x,\,.) = \Phi_n(\eta)$ and the limiting McKean measures reduce to the tensor product measures $\mathbb{K}_n = (\eta_0 \otimes \ldots \otimes \eta_n)$. To simplify the presentation, we further assume that the state spaces are homogeneous, $E_n = E$, and the Markov transitions M_n satisfy the regularity condition $(M)^{\exp}$ introduced on page 116, for some kernels k_n and some reference measures p_n. Under these conditions we recall (see page 264) that the law of the N-path particles

$$\mathbb{P}_n^{(N)} = \mathrm{Law}((\xi_0^i,\ldots,\xi_n^i)_{1\le i \le N}) \in \mathcal{P}(\Omega_n^N), \qquad \Omega_n^N = (E^{n+1})^N$$

is absolutely continuous with respect to the tensor product measure $\mathbb{K}_n^{\otimes N}$ and

$$\frac{d\mathbb{P}_n^{(N)}}{d\mathbb{K}_n^{\otimes N}} = \exp H_n^{(N)} \qquad \mathbb{K}_n^{\otimes N}\text{-a.e.}$$

The interaction potential function $H^{(N)}$ is defined by

$$H_n^{(N)}((x_0^i,\ldots,x_n^i)_{1\le i \le N}) = N \sum_{p=1}^n \int \log \frac{d\Phi_p(m(x_{p-1}))}{d\Phi_p(\eta_{p-1})}\, dm(x_p)$$

with $m(x_p) = \frac{1}{N}\sum_{i=1}^N \delta_{x_p^i}$.

To clarify the presentation, we shall simplify the notation suppressing the time parameter n so that we write $(\mathbb{K}, \mathbb{P}^{(N)}, H^{(N)}, \Omega^N)$ instead of $(\mathbb{K}_n, \mathbb{P}_n^{(N)}, H_n^{(N)}, \Omega_n^N)$. We also write $\Omega = E^{n+1}$, $\xi^i = (\xi_0^i,\ldots,\xi_n^i)$, $\xi = (\xi^i)_{1\le i \le N}$, and $\overline{\mathbb{E}}_N(.)$ (resp. $\mathbb{E}_N(.)$) denotes the expectation with respect to the measure $\mathbb{K}^{\otimes N}$ (resp. $\mathbb{P}^{(N)}$) on Ω^N.

To get the fluctuations of the particle McKean measures, it is enough to study the limit of their characteristic functions

$$\varphi \in \mathbb{L}_2(\mathbb{K}) \to \mathbb{E}_N(\exp(iW^N(\varphi))) \qquad \text{where} \quad W^N = \sqrt{N}\,[\mathbb{K}^N - \mathbb{K}]$$

Writing

$$\mathbb{E}_N(\exp(iW^N(\varphi))) = \overline{\mathbb{E}}_N(\exp(iW^N(\varphi) + H^{(N)}(\xi)))$$

one finds that the convergence of $\mathbb{E}_N(\exp(iW^N(\varphi)))$ follows from the convergence in law and the uniform integrability of the sequence

$$\exp(iW^N(\varphi) + H^{(N)}(\xi))$$

under the product law $\mathbb{K}^{\otimes N}$. The last point is clearly equivalent to the uniform integrability of $\exp H^{(N)}(\xi)$ under $\mathbb{K}^{\otimes N}$. The proof of the uniform integrability of $\exp H^{(N)}(\xi)$ then relies on a classical result that says that if a sequence of nonnegative random variables X_N converges almost surely towards some random variable X as $N \to \infty$, then we have

$$\lim_{N \to \infty} \mathbb{E}(X_N) = \mathbb{E}(X) < \infty \iff \{X_N \, ; \, N \geq 1\} \text{ is uniformly integrable}$$

The equivalence still holds if X_N only converges in distribution by Skorohod's theorem. Since $\overline{\mathbb{E}}_N(\exp H^{(N)}(\xi)) = 1$, it is clear that the uniform integrability of $\exp H^{(N)}(\xi)$ follows from the convergence in distribution under $(\mathbb{K})^{\otimes N}$ of $H^{(N)}(\xi)$ towards a random variable H such that $\overline{\mathbb{E}}(\exp H) = 1$. Thus, it suffices to study the convergence in distribution of $iW^N(\varphi) + H^{(N)}(\xi)$ for $\mathbb{L}_2(\mathbb{K})$ functions φ to conclude. To state such a result, it is first convenient to introduce another round of notation. Under our assumptions, for any $x = (x_0, \ldots, x_n)$ and $z = (z_0, \ldots, z_n) \in \Omega$ we set

$$q(x,z) = \sum_{p=1}^{n} q_p(x,z) \quad \text{and} \quad a(x,z) = q(x,z) - \int_\Omega q(x',z)\, \mathbb{K}(dx')$$

with

$$q_p(x,z) = G_{p-1}(z_{p-1})\, k_p(z_{p-1}, x_p)/\eta_{p-1}(G_{p-1}k_p(\cdot, x_p)) \qquad (9.29)$$

One consequence of $(M)^{\exp}$ is that the integral operator A given for any $\varphi \in \mathbb{L}_2(\mathbb{K})$ by

$$A(\varphi)(x) = \int_\Omega a(z,x)\, \varphi(z)\, \mathbb{K}(dz)$$

is a Hilbert-Schmidt operator on $\mathbb{L}_2(\mathbb{K})$.

We are now in a position to state the main result of this section.

Theorem 9.7.1 *Assume that condition $(M)^{\exp}$ is satisfied. The integral operator $I - A$ is invertible and the random field $W^N : \varphi \in \mathbb{L}_2(\mathbb{K}) \to W^N(\varphi)$ converges as $N \to \infty$ to a centered Gaussian field $W : \varphi \in \mathbb{L}_2(\mathbb{K}) \to W(\varphi)$ satisfying*

$$\mathbb{E}\left(W(\varphi_1)W(\varphi_2)\right)$$
$$= \left((I-A)^{-1}(\varphi_1 - \mathbb{K}(\varphi_1)), (I-A)^{-1}(\varphi_2 - \mathbb{K}(\varphi_2))\right)_{\mathbb{L}_2(\mathbb{K})}$$

for any $\varphi_1, \varphi_2 \in \mathbb{L}_2(\mathbb{K})$ in the sense of convergence of finite-dimensional distributions.

The basic tools for studying the convergence in law of $H^{(N)}(\xi)$ are the Dynkin-Mandelbaum theorem on symmetric statistics and Shiga and Tanaka's formula of Lemma 1.3 in [286]. The detailed proof of Theorem 9.7.1 is given in [84]. Here we merely content ourselves with describing the main line of this approach. Let us first recall how one can see that $I - A$ is invertible. This is in fact classical now (see [286] for instance). First one notices that, under our assumptions, A^n, $n \geq 2$, and $A A^*$ are trace class operators with

$$\mathrm{tr} A^n = \int \cdots \int_{\Omega^N} a(x^1, x^2) \ldots a(x^n, x^1) \, \mathbb{K}(dx^1) \ldots \mathbb{K}(dx^n)$$

$$\mathrm{tr} A A^* = \int_\Omega a(x, z)^2 \, \mathbb{K}(dx) \, \mathbb{K}(dz) = \|a\|^2_{\mathbb{L}_2(\mathbb{K} \otimes \mathbb{K})}$$

Furthermore, by the definition of a and the fact that $\mathbb{K}$ is a product measure, it is easily checked that $\mathrm{tr} A^n = 0$ for any $n \geq 2$. Standard spectral theory then shows that $\det_2(I - A)$ is equal to one and therefore that $I - A$ is invertible.

The identification of the $\mathbb{K}^{\otimes N}$-weak limit of $H^{(N)}(\xi)$ relies on $\mathbb{L}_2$-techniques and more precisely Dynkin-Mandelbaum construction of multiple Wiener integrals as a limit of symmetric statistics. To state such a result, we first introduce Wiener integrals. Let $\{I_1(\varphi) \; ; \; \varphi \in \mathbb{L}_2(\mathbb{K})\}$ be a centered Gaussian field satisfying $\mathbb{E}\left(I_1(\varphi_1) I_1(\varphi_2)\right) = (\varphi_1, \varphi_2)_{\mathbb{L}_2(\mathbb{K})}$. If we set, for each $\varphi \in \mathbb{L}_2(\mathbb{K})$ and $m \geq 1$,

$$h_0^\varphi = 1 \qquad h_m^\varphi(z_1, \ldots, z_m) = \varphi(z_1) \ldots \varphi(z_m)$$

the multiple Wiener integrals $\{I_m(h_m^\varphi) \; ; \; \varphi \in \mathbb{L}_2(\mathbb{K})\}$ with $m \geq 1$ are defined by the relation

$$\sum_{m \geq 0} \frac{t^m}{m!} I_m(h_m^\varphi) = \exp\left(t I_1(\varphi) - \frac{t^2}{2} \|\varphi\|^2_{\mathbb{L}_2(\mathbb{K})}\right)$$

The multiple Wiener integral $I_m(\phi)$ for $\phi \in \mathbb{L}_{2,\mathrm{sym}}(\mathbb{K}^{\otimes m})$ is then defined by a completion argument. Theorem 9.7.1 is therefore a consequence of the following lemma.

Lemma 9.7.1 *Let $\xi = (\xi^i)_{i \geq 1}$ be a collection of independent and identically distributed random variables with common law $\mathbb{K}$. We have*

$$\lim_{N \to \infty} H^{(N)}(\xi) \stackrel{\mathrm{law}}{=} \frac{1}{2} I_2(f) - \frac{1}{2} \mathrm{tr} A A^* \tag{9.30}$$

where f is given by

$$f(y, z) = a(y, z) + a(z, y) - \int_\Omega a(u, y) \, a(u, z) \, \mathbb{K}(du) \tag{9.31}$$

In addition, for any $\varphi \in \mathbb{L}_2(\mathbb{K})$,

$$\lim_{N \to \infty} \left(H^{(N)}(x) + i W^N(\varphi)\right) \stackrel{\mathrm{law}}{=} \frac{1}{2} I_2(f) + i I_1(\varphi) - \frac{1}{2} \mathrm{tr} A A^*$$

Following the observations above, we get for any $\varphi \in \mathbb{L}_2(\mathbb{K})$,

$$\lim_{N\to\infty} \mathbb{E}_N \left(\exp iW^N(\varphi) \right) = \lim_{N\to\infty} \overline{\mathbb{E}}_N \left(\exp \left(iW^N(\varphi) + H^{(N)}(\xi) \right) \right)$$
$$= \mathbb{E} \left(\exp \left(iI_1(\varphi) + \frac{1}{2}I_2(f) - \frac{1}{2}\mathrm{tr}\, AA^* \right) \right)$$

Moreover, Shiga and Tanaka's formula of Lemma 1.3 in [286] shows that for any $\varphi \in \mathbb{L}_{2,\mathrm{sym}}(\mathbb{K})$,

$$\mathbb{E}\left(\exp \left(iI_1(\varphi) + \frac{1}{2}I_2(f) - \frac{1}{2}\mathrm{tr}\, AA^* \right) \right) = \exp\left(-\frac{1}{2} \|(I-A)^{-1}\varphi\|_{\mathbb{L}_2(\mathbb{K})}^2 \right) \tag{9.32}$$

The proof of Theorem 9.7.1 is thus complete. The proof of Lemma 9.7.1 relies entirely on a construction of multiple Wiener integrals as a limit of symmetric statistics. For completeness and to guide the reader, we present this result.

Let $\{\zeta^i\,;\, i \geq 1\}$ be a sequence of independent and identically distributed random variables with values in an arbitrary measurable space $(\mathcal{X}, \mathcal{B})$. To every symmetric function $h(z_1, \ldots, z_m)$, there corresponds a statistic

$$\sigma_m^N(h) = \sum_{1 \leq i_1 < \ldots < i_m \leq N} h(\zeta^{i_1}, \ldots, \zeta^{i_m})$$

with the convention $\sigma_m^N = 0$ for $m > N$. Every integrable symmetric statistic $S(\zeta^1, \ldots, \zeta^N)$ has a unique representation of the form

$$S(\zeta^1, \ldots, \zeta^N) = \sum_{m \geq 0} \sigma_m^N(h_m) \tag{9.33}$$

where $h_m(z_1, \ldots, z_m)$ are symmetric functions subject to the condition

$$\int h_m(z_1, \ldots, z_{m-1}, u)\, \mu(du) = 0 \tag{9.34}$$

where μ is the probability distribution of ζ^1. We call such functions h_m, $m \geq 0$ canonical. Finally, we denote by $\mathcal{H}$ the set of all sequences

$$h = (h_0, h_1(z_1), \ldots, h_m(z_1, \ldots, z_m), \ldots)$$

where h_m are canonical and $\sum_{m \geq 0} \frac{1}{m!} \mathbb{E}(h_m^2(\zeta^1, \ldots, \zeta^m)) < \infty$. As in [286], the proof of Lemma 9.7.1 is essentially based on the following theorem.

Theorem 9.7.2 (Dynkin-Mandelbaum [129]) *For $h \in \mathcal{H}$, the sequence of random variables $Z_N(h) = \sum_{m \geq 0} N^{-m/2} \sigma_m^N(h_m)$ converges in law and as $N \to \infty$, to $W(h) = \sum_{m \geq 0} I_m(\bar{h}_m)/m!$*

9. Central Limit Theorems

Proof of Lemma 9.7.1: (*Sketched*)
It is first useful to observe that for any $\mu \in \mathcal{P}(E)$ and $p \geq 1$ we have that

$$\frac{d\Phi_p(\mu)}{d\eta_p}(x) = \frac{d\Phi_p(\mu)}{d\Phi_p(\eta_{p-1})}(x) = \frac{\mu(G_{p-1}(.)k_p(.,x))}{\eta_{p-1}(G_{p-1}(.)k_p(.,x))} \Big/ \frac{\mu(G_{p-1})}{\eta_{p-1}(G_{p-1})}$$

By the definition of $\mathbb{K}$ and q_p (see (9.29)), we note that

$$\int_\Omega q_p(x,z)\, \mathbb{K}(dx) = \int_\Omega \frac{G_{p-1}(z_{p-1})\, k_p(z_{p-1}, x_p)}{\eta_{p-1}(G_{p-1}\, k_p(.,x_p))}\, \eta_p(dx_p)$$

$$= \int_E \frac{G_{p-1}(z_{p-1})\, k_p(z_{p-1}, x_p)}{\eta_{p-1}(G_{p-1}\, k_p(.,x_p))}$$

$$\times \frac{\eta_{p-1}(G_{p-1}\, k_p(.,x_p))}{\eta_{p-1}(G_{p-1})}\, p_p(dx_p)$$

$$= \frac{G_{p-1}(z_{p-1})}{\eta_{p-1}(G_{p-1})}$$

Therefore the symmetric statistics $x \in \Omega^N \longrightarrow H^{(N)}(x)$ can be rewritten as

$$H^{(N)}(x) = \sum_{p=1}^n \sum_{i=1}^N \left[\log\left(\frac{1}{N}\sum_{j=1}^N q_p(x^i, x^j)\right) - \log\left(\frac{1}{N}\sum_{j=1}^N \bar{q}_p(x^j)\right) \right]$$

where

$$\bar{q}_p(x^j) = G_{p-1}(x^j_{p-1})/\eta_{p-1}(G_{p-1}) = \int_\Omega q_p(y, x^j)\, \mathbb{K}(dy)$$

By the representation

$$\log u = (u-1) - \frac{(u-1)^2}{2} + \frac{(u-1)^3}{3(\varepsilon u + (1-\varepsilon))^3}$$

which is valid for all $u > 0$ with $\varepsilon = \varepsilon(u)$ such that $\varepsilon(u) \in [0,1]$, we obtain the decomposition

$$H^{(N)}(x) = \frac{1}{N}\sum_{i=1}^N \sum_{j=1}^N a(x^i, x^j) - \frac{1}{2}\sum_{p=1}^n \sum_{i=1}^N \left(\frac{1}{N}\sum_{j=1}^N q_p(x^i, x^j) - 1\right)^2$$

$$+ \frac{N}{2}\sum_{p=1}^n \left(\frac{1}{N}\sum_{j=1}^N \bar{q}_p(x^j) - 1\right)^2 + R^{(N)} \quad (9.35)$$

where the remainder term $R^{(N)}$ cancels as N tends to ∞. The technical trick is to decompose each term as in (9.33) in order to identify the limit by applying Theorem 9.7.2. For instance, the first term can readily be written as

$$\frac{1}{N} \sum_{i=1}^{N} a(x^i, x^i) + \frac{1}{N} \sum_{i<j} (a + a^*)(x^i, x^j)$$

with

$$\int_\Omega a(z,z)\,\mathbb{K}(dz) = \int_\Omega a(x,z)\,\mathbb{K}(dz) = \int_\Omega a(z,x)\,\mathbb{K}(dz) = 0$$

for any $x \in \Omega$, and therefore a clear application of Theorem 9.7.2 yields that it converges in law as $N \to \infty$ to $\frac{1}{2} I_2(a + a^*)$. ∎

9.8 Covariance Functions

We use the same notation and the same regularity conditions as the ones used in Section 9.7 We have proved that the random field

$$\varphi \in \mathbb{L}^2(\mathbb{K}) \to W^N(\varphi) = \sqrt{N}\left(\frac{1}{N}\sum_{i=1}^{N} \varphi(\xi_0^i, \dots, \xi_n^i) - \mathbb{K}(\varphi)\right)$$

converges as $N \to \infty$, in the sense of convergence of finit- dimensional distributions, to a centered Gaussian field $\{W(\varphi);\, \varphi \in \mathbb{L}^2(\mathbb{K})\}$ with covariance

$$\mathbb{E}\left(W(\varphi_1)W(\varphi_2)\right)$$
$$= \Big((I - A)^{-1}(\varphi_1 - \mathbb{K}(\varphi_1)), (I - A)^{-1}(\varphi_2 - \mathbb{K}(\varphi_2))\Big)_{\mathbb{L}^2(\mathbb{K})}$$

for $\varphi_1, \varphi_2 \in \mathbb{L}^2(\mathbb{K})$. From the observations above, it follows that the process

$$f \in \mathbb{L}_2(\eta_n) \to W_n^{\eta,N}(f) = \sqrt{N}\left(\eta_n^N(f) - \eta_n(f)\right)$$

converges in the sense of convergence of finite-dimensional distributions and as $N \to \infty$ to a centered Gaussian field $\{W_n^\eta(f);\, f \in \mathbb{L}^2(\eta_n)\}$ satisfying

$$\mathbb{E}\big(W_n^\eta(f) W_n^\eta(h)\big)$$
$$= \Big((I - A)^{-1}(f - \eta_n(f))^{\otimes 1}, (I - A)^{-1}(h - \eta_n(h))^{\otimes 1}\Big)_{\mathbb{L}^2(\eta_n)}$$

328   9. Central Limit Theorems

for any $f, h \in L^2(\eta_n)$, where $f^{\otimes 1} \stackrel{\text{def}}{=} \underbrace{1 \otimes \ldots \otimes 1}_{(n-1) \text{ times}} \otimes f$, for all $f \in L^2(\eta_n)$.

From (9.14) we also know that the covariance function is given by

$$\mathbb{E}\big(W_n^\eta(f) W_n^\eta(h)\big)$$

$$= \sum_{p=0}^n \int \left(\frac{G_{p,n}}{\eta_p(G_{p,n})}\right)^2 \big(P_{p,n}(f) - \eta_n(f)\big)\big(P_{p,n}(h) - \eta_n(h)\big) d\eta_p$$

In the next proposition, we invert the integral operator and check that these two expressions do coincide. This reassuring lemma gives some precise information on the decompositions to use to obtain fluctuations of particle density profiles. As an aside, we mention that it was in fact at the origin of our study of CLTs for particle density profiles.

Proposition 9.8.1 *For every $f \in L^2(\eta_n)$, and every $z_0, \ldots, z_n \in E$,*

$$(I - A)^{-1}\big(f - \eta_n(f)\big)^{\otimes 1}(z_0, \ldots, z_n) = \sum_{p=0}^n \tilde{f}_{p,n}(z_p)$$

where the functions $\tilde{f}_{p,n}$, $0 \leq p \leq n$, are given for any $p \leq n$ by

$$\tilde{f}_{p,n} = \frac{G_{p,n}}{\eta_p(G_{p,n})}\big(P_{p,n}(f) - \eta_n(f)\big)$$

with the convention $G_{n,n} = 1$ and $M_{n,n} = \text{Id}$.

Proof: We first note that

$$\tilde{f}_{p-1,n} = \frac{G_{p-1} M_p(G_{p,n})}{\eta_{p-1}(G_{p-1,n})} \left(\frac{M_p(G_{p,n} P_{p,n}(f))}{M_p(G_{p,n})} - \eta_n(f)\right)$$

$$= \frac{G_{p-1}}{\eta_{p-1}(G_{p-1,n})} \big(M_p(G_{p,n} P_{p,n}(f)) - M_p G_{p,n} \eta_n(f)\big)$$

and therefore

$$\tilde{f}_{p-1,n} = \frac{G_{p-1}\, \eta_p(G_{p,n})}{\eta_{p-1}(G_{p-1,n})} M_p\left(\frac{G_{p,n}}{\eta_p(G_{p,n})}\big(P_{p,n}(f) - \eta_n(f)\big)\right)$$

$$= G_{p-1} \frac{\eta_p(G_{p,n})}{\eta_{p-1}(G_{p-1,n})} M_p(\tilde{f}_{p,n})$$

Then, using the fact that

$$\eta_p(G_{p,n}) = \frac{\eta_{p-1}(G_{p-1} M_p(G_{p,n}))}{\eta_{p-1}(G_{p-1})} = \frac{\eta_{p-1}(G_{p-1,n})}{\eta_{p-1}(G_{p-1})}$$

one easily gets the backward recursion equations, for each $1 \leq p \leq n$,

$$\tilde{f}_{p-1,n} = \frac{G_{p-1}}{\eta_{p-1}(G_{p-1})} M_p(\tilde{f}_{p,n}) \qquad (9.36)$$

By the definition of A, for any $\varphi \in L^2(\mathbb{K})$, we have that

$$A\varphi(z_0,\ldots,z_n) = \sum_{m=1}^n \int \varphi(x_0,\ldots,x_n) a_m(x_m, z_{m-1}) \mathbb{K}(dx_0,\ldots,dx_n)$$

where, for every $1 \le m \le n$,

$$a_m(x_m, z_{m-1}) = \frac{G_{m-1}(z_{m-1}) k_m(z_{m-1}, x_m)}{\eta_{m-1}(G_{m-1} k_m(\cdot, x_m))} - \frac{G_{m-1}(z_{m-1})}{\eta_{m-1}(G_{m-1})}$$

On the other hand, we observe that since

$$\eta_m(dx_m) = \frac{\eta_{m-1}(G_{m-1} k_m(\cdot, x_m))}{\eta_{m-1}(G_{m-1})} p_m(dx_m)$$

and $K_m(z_{m-1}, dx_m) = k_m(z_{m-1}, x_m) p_m(dx_m)$, we have that

$$\frac{G_{m-1}(z_{m-1}) k_m(z_{m-1}, x_m)}{\eta_{m-1}(G_{m-1} k_m(\cdot, x_m))} \eta_m(dx_m) = \frac{G_{m-1}(z_{m-1})}{\eta_{m-1}(G_{m-1})} M_m(z_{m-1}, dx_m)$$

Therefore,

$$(I - A)\varphi(z_0,\ldots,z_n) = \varphi(z_0,\ldots,z_n)$$

$$+ \sum_{m=1}^n \frac{G_{m-1}(z_{m-1})}{\eta_{m-1}(G_{m-1})} \int \varphi(x_0,\ldots,x_n) \tilde{\eta}^m_{[0,n]}(z_{m-1}; dx_0,\ldots,dx_n) \quad (9.37)$$

where

$$\tilde{\eta}^m_{[0,n]}(z_{m-1}; dx_0,\ldots,dx_n) = (\eta_0 \otimes \ldots \eta_{m-1})(dx_0,\ldots,dx_{m-1})$$
$$\times (\eta_m - M_m(z_{m-1}, \cdot))(dx_m) \times (\eta_{m+1} \otimes \ldots \otimes \eta_n)(dx_{m+1},\ldots,dx_n)$$

Now choose φ given for any $z_0,\ldots,z_n \in E$ by

$$\varphi(z_0,\ldots,z_n) = \sum_{p=0}^n \tilde{f}_{p,n}(z_p)$$

Then, we get

$$(I - A)\varphi(z_0,\ldots,z_n) = \varphi(z_0,\ldots,z_n) - \sum_{m=1}^n \frac{G_{m-1}(z_{m-1})}{\eta_{m-1}(G_{m-1})} M_m(\tilde{f}_{m,n})(z_{m-1})$$

Finally, using the backward equation (9.36), one concludes that

$$(I - A)\varphi(z_0,\ldots,z_n) = \varphi(z_0,\ldots,z_n) - \sum_{m=1}^n \tilde{f}_{m-1,n}(z_{m-1})$$
$$= \sum_{m=0}^n \tilde{f}_{m,n}(z_m) - \sum_{m=0}^{n-1} \tilde{f}_{m,n}(z_m)$$

so that the result follows from

$$(I - A)\varphi(z_0, \ldots, z_n) = f_{n,n}(z_n) = f(z_n) - \eta_n(f)$$

This ends the proof of the proposition. ∎

10
Large-Deviation Principles

This chapter focuses on large-deviation principles for interacting particle models. The main object of the theory of large deviations is to provide sharp exponential estimates of the deviant behavior of random rare events. For instance, in the context of interacting particle models, these events represent the deviations of particle approximation measures around the "nondeviant" limiting McKean distribution or around the solution of the limiting measure-valued equation. We have already analyzed some rather crude exponential decays of these deviation probabilities in Section 7.4. Although these estimates were not asymptotic, they were far from being sharp. These exponential decays are sometimes called strong large-deviation estimates by some authors. In this context, a large-deviation analysis will provide sharp and precise estimates. Before entering into more details on this subject, it is useful to give some comments on the origins of the theory of large deviations and its connections with other scientific disciplines.

The theory of large deviations probably started around the 1930s with the works of A.I. Khintchin [201], N. Smirnoff [290], and H. Crámer [62]. These pioneering studies in this subject were motivated by refining the estimates provided by the central limit theorem. Since that period, the range of applications of large-deviations analysis has constantly increased, going from particle physics [300], dynamical systems [15, 301], stochastic search algorithms [15, 44], statistics [19, 20, 70, 211], statistical mechanics [61, 133] and pure probability [1, 2, 3, 118, 119, 120, 293, 300]. The foundations of the modern theory of large deviations in abstract topological function spaces were laid in celebrated articles by S.R.S. Varadhan [300] and by A.A. Borovkov [36]. There are actually several interesting and complementary textbooks devoted to this subject. We refer the reader to [112, 114, 127, 133] for a detailed account on this topic and a complete list of references.

More recently in the late 1980s, V.P. Maslov presented in [240] a theory of integration of functions taking values in an idempotent semiring. The development of this idempotent analysis was motivated by the study of some Hamilton-Jacobi equations arising in particle physics. This idempotent integration theory is built in the same fashion as the traditional Lebesgue integration. In this context, lower semicontinuous functions can be regarded as idempotent measures, and good rate functions governing large-deviation principles correspond to idempotent probability measures. These interpretations shed some new light on the connections between probability theory and stochastic processes on the one hand and deterministic optimization theory and controlled dynamical systems on the other hand. They also allow probabilistic intuition to enter into deterministic optimization and decision processes. For more details on idempotent probability measures, we refer the reader to [77, 78, 79, 205, 272]. From these points of views, the theory of large deviations can be interpreted as an analytical bridge between idempotent functional analysis and deterministic decision dynamical systems and the traditional theory of probability and stochastic processes.

The chapter has the following structure. In Section 10.1, we present our main large-deviation results on discrete generation interacting processes, namely a large-deviation principle for particle McKean measures on metric spaces and the extension of the Sanov theorem in the strong topology to interacting processes in general Hausdorff topological spaces. In this preliminary section, we also introduce the general definition of what is meant by a large-deviation principle. We interpret these principles in the context of particle approximation models and connect these sharp estimates with the exponential inequalities presented in earlier sections. We shall see that these asymptotic estimates are strongly related to the choice of the topological structure of the state spaces. We illustrate these topological questions with a brief discussion on the weak and the strong topologies in distribution spaces. We already mention that the strong τ-topology on the set of probability measures over some measurable space is in general not metrizable and not even first countable. This simple observation indicates that the large deviation analysis of these models has to be conducted in abstract and general topological spaces.

In Section 10.2, we have collected some essential topological properties of lower semicontinuous functions with an emphasis on their idempotent measure interpretation. These elementary results, such as the well-known contraction principle, will be of current use in various places of this chapter.

In Sections 10.3, 10.4, and 10.5, we underline three generic strategies to obtain a large-deviation principle, namely Crámer's method, Laplace-Varadhan integral techniques, and Dawson-Gärtner projective limit techniques. The first and third techniques will be used in Section 10.8 to analyze the deviations of flows of particle density profiles in the τ-topology, extending a strong version of Sanov's theorem due to Groeneboom, Oosterhoff, and Ruyggaart [170]. The second strategy will be used in Section 10.7 to develop a large-deviation principle for particle McKean measures simplifying and extending a joint work of the author with A. Guionnet [83]. In each of these three sections, we propose a comprehensive and self-contained treatment on these methods (except for the large-deviation lower bound in the generalized Crámer's method,, see Theorem 10.3.1). In addition, we complement these traditional techniques with some recent results. Section 10.4 contains a complement of the classical integral lemma of Varadhan recently presented in a joint work of the author with T. Zajic [109, 110]. In Section 10.5, we revisit Dawson-Gärtner techniques. We also propose a new, simple proof of the strong version of Sanov's theorem based on these projective ideas and multinomial expansions in the spirit of Groeneboom, Oosterhoff, and Ruyggaart [170].

10.1 Introduction

We start with some familiar notation on discrete generation measure-valued processes and interacting particle models. We let E_n, $n \in \mathbb{N}$, be a collection of Polish spaces[1] with Borel σ-fields $\mathcal{E}_n$. We also consider a probability measure $\eta_0 \in \mathcal{P}(E_0)$ and a collection of measurable mappings Φ_{n+1} from $\mathcal{P}(E_n)$ into $\mathcal{P}(E_{n+1})$. We associate with the latter sequence of mappings the nonlinear measure-valued equation

$$\eta_n = \Phi_n(\eta_{n-1}) \qquad (10.1)$$

At this stage, it is worthwhile to recall that the (nonlinear) distribution flow $\eta_n \in \mathcal{P}(E_n)$ can be interpreted as the laws of the random states of a non-homogeneous Markov chain X_n. Each of these Markovian interpretations corresponds to a possibly different sequence of Markov transitions $K_{n+1,\eta}$ from E_{n-1} into E_n and satisfying the compatibility condition

$$\eta K_{n,\eta} = \Phi_n(\eta) \qquad (10.2)$$

for any $\eta \in \mathcal{P}(E_n)$ and $n \geq 1$. We refer the reader to Section 2.5.3 for several examples of compatible transitions in the context of Feynman-Kac semigroups. We associate with a given compatible sequence of transitions $K_{n,\eta}$ a filtered and canonical probability space

$$(\Omega_n = E_{[0,n]}, (\mathcal{F}_p)_{0 \leq p \leq n}, (X_p)_{0 \leq p \leq n}, \mathbb{K}_{\eta_0,n})$$

with the McKean probability measure $\mathbb{K}_{\eta_0,n} \in \mathcal{P}(E_{[0,n]})$ defined by

$$\mathbb{K}_{\eta_0,n}(d(u_0,\ldots,u_n)) = \eta_0(du_0)\, K_{1,\eta_0}(u_0,du_1)\, \ldots\, K_{n,\eta_{n-1}}(u_{n-1},du_n)$$

---
[1] i.e., complete and separable metric space.

Under $\mathbb{K}_{\eta_0,n}$, the canonical sequence $(X_p)_{0 \le p \le n}$ forms a Markov chain with transitions K_{p+1,η_p} and $\eta_p = \mathrm{Law}(X_p)$, $0 \le p \le n$. The N-particle associated with the McKean interpretation above is a Markov chain $\xi_p = (\xi_p^i)_{1 \le i \le N} \in E_p^N$ with initial distribution $\eta_0^{\otimes N}$ and elementary transitions

$$\mathrm{Proba}(\xi_p \in d(x_p^1, \ldots, x_p^N) \mid \xi_{p-1}) = \prod_{i=1}^N K_{p,m(\xi_{p-1})}(\xi_{p-1}^i, dx_p^i)$$

The N-particle approximation McKean measures $\mathbb{K}_{n,\eta_0}^N \in \mathcal{P}(E_{[0,n]})$ and the corresponding density profiles $\eta_n^N \in \mathcal{P}(E_n)$ are defined by

$$\mathbb{K}_{n,\eta_0}^N = \frac{1}{N} \sum_{i=1}^N \delta_{(\xi_0^i, \ldots, \xi_n^i)} \quad \text{and} \quad \eta_n^N = \frac{1}{N} \sum_{i=1}^N \delta_{\xi_n^i}$$

In earlier sections, we developed a series of asymptotic results on these random distributions as the size of the systems N tends to infinity, including weak and strong laws of large numbers, $\mathbb{L}_p$ mean error analysis, exponential decays, and central limit theorems. The latter fluctuation analysis provides sharp asymptotic estimates of the $\mathbb{L}_p$ mean errors

$$\sqrt{N} \, [\mathbb{K}_{n,\eta_0}^N - \mathbb{K}_{n,\eta_0}] \quad \text{and} \quad \sqrt{N} \, [\eta_n^N - \eta_n]$$

In Section 7.4.2, we also derived a collection of exponential estimates (see for instance (7.19) in Theorem 7.4.2 or (7.21) in Theorem 7.4.3). This concentration analysis deals with events where η_n^N differs from η_n by an amount of order N, well beyond the fluctuation order $\sqrt{N}$ that is described by the central limit theorem. One of the challenging questions we address here is to obtain tight and asymptotically sharp exponential bounds for the asymptotic decay rate as $N \to \infty$ of

$$\mathbb{P}(|[\mathbb{K}_{n,\eta_0}^N - \mathbb{K}_{n,\eta_0}](F_n)| \ge \epsilon) \qquad (10.3)$$

or

$$\mathbb{P}(|[\eta_0^N - \eta_0](f_0)| \ge \epsilon_0, \ldots, |[\eta_n^N - \eta_n](f_n)| \ge \epsilon_n) \qquad (10.4)$$

where $\varepsilon, \varepsilon_0, \ldots, \varepsilon_n > 0$, $F_n \in \mathcal{B}_b(E_{[0,n]})$, and $f_p \in \mathcal{B}_b(E_p)$, $p \le n$, is an arbitrary sequence of test functions.

More generally, we can consider the particle McKean measures $\mathbb{K}_{n,\eta_0}^N$ and the approximation flow $(\eta_p^N)_{0 \le p \le n}$ as sequences of random variables taking values in $\mathcal{P}(E_{[0,n]})$ and in the Cartesian product $\prod_{p=0}^n \mathcal{P}(E_p)$. In this interpretation, it is first convenient to equip these distribution spaces with an appropriate topology so that open and closed subsets are well-defined. Furthermore, these topologies must also be "compatible" with the previous analysis in the sense that the deviant events in (10.4) have to be the complementary subset of some open neighborhoods $\mathcal{V}(\eta_n) \subset \mathcal{P}(E_n)$ of η_n

and we have that $\lim_{N\to\infty} \mathbb{P}(\eta_n^N \not\in \mathcal{V}(\eta_n)) = 0$. In this topological context, large deviations describe precisely the increasing "deviant" behavior of the event $\{\eta_n^N \not\in \mathcal{V}(\eta_n)\}$.

At this stage, the reader may wonder what exactly is meant by a large-deviation principle. For later purposes, it is convenient to start here with an abstract and general definition.

Definition 10.1.1 *Let M be a topological space equipped with a σ-field $\sigma(M)$. We say that a sequence of probability measures $(P^N)_{N\geq 1}$ on the measurable space $(M, \sigma(M))$ satisfies the large-deviation principle (abbreviated LDP) with rate function H or, equivalently, that the rate function H governs the LDP of P^N if the following two assertions are met:*

- *H is a lower semicontinuous mapping from M into $[0, \infty]$ (the value ∞ is not excluded)*

- *For any $A \in \sigma(M)$*

$$-H(\mathring{A}) \leq \liminf_{N\to\infty} \log \frac{1}{N} P^N(A) \leq \limsup_{N\to\infty} \log \frac{1}{N} P^N(A) \leq -H(\overline{A}) \tag{10.5}$$

where $\mathring{A}$ and $\overline{A}$ denote respectively the interior and the closure of A and for any $A \subset M$ we have used the notation $H(A) = \inf_{m \in A} H(m)$.

The function H is said to be a good rate function when its level sets $H^{-1}([0, a])$, $a \in [0, \infty)$, are compact subsets in M.

Note that this definition strongly depends on the topology of the state space as well as on the choice of the σ-algebra. Before attempting to get into more details, we begin by presenting one traditional example of topology currently used in the context of interacting particle models. Let $\mathcal{M}(E)$, the space of all finite and signed measures on a Polish space $(E, \mathcal{E})$ equipped with the weak topology generated by $\mathcal{C}_b(E)$. For the convenience of a non-initiated reader, we recall that each $f \in \mathcal{C}_b(E)$, can be identified as a point p_f of the algebraic dual $\mathcal{M}(E)'$ of $\mathcal{M}(E)$ via the map $p_f : \mu \to \mu(f)$. Since the collection of these functionals $\mu \to \mu(f) \in \mathbb{R}$, $f \in \mathcal{C}_b(E)$, is a separating vector space in $\mathcal{M}(E)$, the $\mathcal{C}_b(E)$-topology makes $\mathcal{M}(E)$ into a locally convex topological space and the set $\mathcal{C}_b(E) \simeq \mathcal{M}(E)^\star$ can be identified with the topological dual of $\mathcal{M}(E)$ with the duality relation

$$(f, \mu) \in (\mathcal{C}_b(E) \times \mathcal{M}(E)) \longrightarrow \int_E f \, d\mu \in \mathbb{R}$$

In addition, the set $\mathcal{P}(E)$ of all probability measures on $(E, \mathcal{E})$ as a convex subset of the locally convex topological space $\mathcal{M}(E)$ is a closed subset in $\mathcal{M}(E)$, and it is again a Polish space when endowed with the relative weak topology. This weak topology (also called the vague topology) is generated by the sets

$$\mathcal{V}_{f,\varepsilon}(\mu) = \{\eta \in \mathcal{P}(E) \; : \; |\eta(f) - \mu(f)| < \varepsilon\}$$

336   10. Large-Deviation Principles

where $f \in \mathcal{C}_b(E)$, $\mu \in \mathcal{P}(E)$, and $\varepsilon \in (0, \infty)$. As the reader has certainly noticed the terminology "weak convergence" is not a correct mathematical notion. In a pure mathematical sense, since $\mathcal{C}_b(E) \simeq \mathcal{M}(E)^\star$, the $\mathcal{M}(E)^\star$-topology on $\mathcal{M}(E)$ coincides with the traditional and official weak-$\star$ topology. Since all pure and applied probabilists seem to have adopted this convention, we shall follow this abusive mathematical terminology.

If we take $E = \Omega_n = E_{[0,n]}$ and $f = F_n \in \mathcal{C}_b(E_{[0,n]})$, then the deviant event presented in (10.3) is equivalently expressed in terms of a basis neighborhood of the McKean measure; that is,

$$\mathbb{P}(|[\mathbb{K}_{n,\eta_0}^N - \mathbb{K}_{n,\eta_0}](F_n)| \geq \epsilon) = \mathbb{P}(\mathbb{K}_{n,\eta_0}^N \not\in \mathcal{V}_{F_n,\varepsilon}(\mathbb{K}_{n,\eta_0}))$$

One apparently innocent observation is that the function F_n has to be continuous on the path space $E_{[0,n]}$ so that the deviant event is a closed set in this weak topology. This shows that the LDP analysis on $\mathcal{P}(E_{[0,n]})$ furnished with the weak topology fails to describe the desired deviant behavior of the particle approximation measures on all bounded measurable functions. Therefore it fails to describe the desired deviant behavior on indicator functions of measurable sets! We will return to this "topological" trouble in the further development of this section.

With these preliminaries out of the way, we are now in a position to describe with some precision one of the main results developed in this section. Let Σ_n be the mapping defined by

$$\Sigma_n : \mu \in \mathcal{P}(E_{[0,n]}) \longrightarrow \Sigma_n(\mu) = (\mu_p)_{0 \leq p \leq n} \in \mathcal{P}^n(E) = \prod_{p=0}^{n} \mathcal{P}(E_p) \quad (10.6)$$

where μ_p stands for the pth time marginal of μ with $0 \leq p \leq n$. We associate with a given sequence of distributions $\nu = (\nu_p)_{0 \leq p \leq n} \in \mathcal{P}^n(E)$ the measure $\mathbb{Q}_\nu \in \mathcal{P}(E_{[0,n]})$ defined by

$$\mathbb{Q}_\nu(d(u_0, \ldots, u_n)) = \eta_0(du_0) \, K_{1,\nu_0}(u_0, du_1) \ldots K_{n,\nu_{n-1}}(u_{n-1}, du_n)$$

It is important to note that the McKean distribution defined by $\mathbb{K}_{\eta_0,n}$ is a fixed point $\mathbb{M}(\mathbb{K}_{\eta_0,n}) = \mathbb{K}_{\eta_0,n}$ of the mapping

$$\mathbb{M} : \mu \in \mathcal{P}(E_{[0,n]}) \to \mathbb{M}(\mu) = \mathbb{Q}_{\Sigma_n(\mu)} \in \mathcal{P}(E_{[0,n]})$$

We are now in a position to state the first result of this section.

Theorem 10.1.1 *Suppose that, for each pair of measures for $0 \leq p \leq n$ and $u_{p-1} \in E_{p-1}$, the measures $(K_{p,\mu}(u_{p-1}, \cdot))_{\mu \in \mathcal{P}(E_{p-1})}$ are mutually absolutely continuous. Also suppose that for any $\alpha \in \mathbb{R}$ the mappings*

$$\mu \in \mathcal{P}(E_{p-1}) \to \int_{E_{p-1}} \mu(du_{p-1}) \, \log Z_p^{(\alpha)}(\mu, \eta)(u_{p-1}) \quad (10.7)$$

with

$$Z_p^{(\alpha)}(\mu,\eta)(u_{p-1}) = \int_{E_p} \left[\frac{dK_{p,\mu}(u_{p-1},\cdot)}{dK_{p,\eta}(u_{p-1},\cdot)}(u_p)\right]^\alpha K_{p,\eta}(u_{p-1},du_p) \in (0,\infty)$$

are bounded and continuous at each $\mu = \eta$ *for the weak topology. Then the law of the N-particle measures* $\mathbb{K}_{n,\eta_0}^N$ *satisfies the LDP in* $\mathcal{P}(E_{[0,n]})$ *equipped with the weak topology and with the good rate function* I_n *given by*

$$I_n : \mu \in \mathcal{P}(E_{[0,n]}) \longrightarrow I(\mu) = H_\mu(\mu) = \mathrm{Ent}(\mu \mid \mathbb{M}(\mu)) \in [0,\infty]$$

In addition, we have $I_n(\mu) = 0$ *if and only if* $\mu = \mathbb{K}_{\eta_0,n}$.

In the context of Feynman-Kac models, we provide in Section 10.7 a set of simple sufficient conditions on the pair (G_n, M_n) under which the regularity conditions stated in the theorem above hold true.

To relax the regularity conditions needed in Theorem 10.1.1, we will simplify the analysis and we will work with the flow of the particle density profiles $(\eta_p^N)_{0 \leq p \leq n}$ associated with the first McKean interpretation $K_{n,\eta}(u_{n-1},\cdot) = \Phi_n(\eta)$. In Section 10.2 (see Theorem 10.2.1), we shall see that the LDP is preserved under continuous mappings. Since Σ_n is a continuous mapping, we readily deduce from Theorem 10.1.1 that

$$\Sigma_n(\mathbb{K}_{n,\eta_0}^N) = (\eta_p^N)_{0 \leq p \leq n}$$

satisfies the LDP on $\mathcal{P}(\mathcal{P}^n(E))$ (equipped with the weak topology) with the good rate function J_n defined by

$$J_n((\mu_p)_{0\leq p\leq n}) = \sum_{p=0}^n \mathrm{Ent}(\mu_p \mid \Phi_p(\mu_{p-1}))$$

with the convention $\Phi_0(\mu_{-1}) = \eta_0$ for $p = 0$. Note that $J_n((\mu_p)_{p\leq n}) = 0$ if and only if $(\mu_p)_{p\leq n}$ satisfies the nonlinear equation (10.1) starting at $\mu_0 = \eta_0$. Of course, this result does not imply the LDP for the particle McKean measures but, as we mentioned earlier, we will push forward the LDP analysis to another natural and much stronger topology than the previous one. More precisely, we shall assume that the state spaces E_n are Hausdorff topological spaces furnished with the Borel σ-field $\mathcal{E}_n$ and the set of probability measures $\mathcal{P}(E_n)$ are endowed with the (strong) τ-topology.

We recall that the τ-topology on the set of probability measures $\mathcal{P}(E)$ on an Hausdorff topological space $(E, \mathcal{E})$ equipped with the Borel σ-field $\mathcal{E}$ is the topology generated by the sets

$$\mathcal{U}_{f,\varepsilon}(\mu) = \{\eta \in \mathcal{P}(E) \ : \ |\eta(f) - \mu(f)| < \varepsilon\}$$

where $f \in \mathcal{B}_b(E)$, $\mu \in \mathcal{P}(E)$, and $\varepsilon \in (0,\infty)$. Since the functions f appearing in this definition are measurable and bounded, the τ-topology is finer (or stronger) than the weak topology. This natural topology is in general strictly finer than a topology associated with a given Zolotarev type seminorm (see Section 7.3). For instance, when $E = \mathbb{R}^d$ and $\mathcal{F} = \{1_{(-\infty,x]} \; ; \; x \in \mathbb{R}^d\}$, the topology induced by the supremum distance

$$\|\mu - \eta\|_{\mathcal{F}} = \sup_{x \in \mathbb{R}^d} |\mu((-\infty, x]) - \eta((-\infty, x])|$$

is strictly coarser than the τ-topology (see [170]). Consequently, we note that the sets $\mathcal{U}_\varepsilon(\eta) = \{\mu \in \mathcal{P}(E) \; : \; \|\mu - \eta\|_{\mathcal{F}} < \varepsilon\}$, where $\eta \in \mathcal{P}(E)$, $\varepsilon > 0$, and $\mathcal{F} \subset \mathcal{B}_b(E)$ runs through all countable collections of functions, form a basis and not merely a subbasis of the τ-topology on $\mathcal{P}(E)$.

From the previous discussion, the large-deviation analysis in the τ-topology will provide precise information on the deviant behavior of the particle approximation measures on any countable collection of bounded measurable functions. For instance, the extended version of the deviant events (10.4) with respect to Zolotarev type neighborhoods now reads

$$\mathbb{P}(\|\eta_0^N - \eta_0\|_{\mathcal{F}_0} \geq \epsilon_p, \ldots, \|\eta_n^N - \eta_n\|_{\mathcal{F}_n} \geq \epsilon_n)$$

$$= \mathbb{P}\left((\eta_p^N)_{0 \leq p \leq n} \in \prod_{p=0}^n \mathcal{U}_{p,\varepsilon_p}(\eta_p)^c\right)$$

To avoid some unnecessary discussions on measurability questions, we will suppose that a given set of probability measures $\mathcal{P}(E)$ on a Hausdorff topological space $(E, \mathcal{E})$ and equipped with the τ-topology is always endowed with the smallest σ-field that makes measurable all functionals $p_f : \mu \in \mathcal{P}(E) \to \mu(f) \in \mathbb{R}$ with $f \in \mathcal{B}_b(E)$.

Theorem 10.1.2 *Let η_n^N be the particle density profiles associated with a collection of τ-continuous mappings $\Phi_n : \mathcal{P}(E_{n-1}) \to \mathcal{P}(E_n)$. We also assume that the mappings Φ_n satisfy the following regularity condition*

(H) For any $n \geq 1$, there exists some reference measure $\lambda_n \in \mathcal{P}(E_n)$ and some parameters $\rho_n > 0$ such that for any $\mu_{n-1} \in (\mathcal{P}(E_{n-1})$ we have

$$\rho_n \, \lambda_n \leq \Phi_n(\mu_{n-1}) \ll \lambda_n$$

Then, the law of $(\eta_p^N)_{0 \leq p \leq n}$ satisfies the LDP in $\mathcal{P}^n(E)$ with the product τ-topology (and hence for the weak topology) with rate function J_n.

In the context of Feynman-Kac models, the one-step mappings Φ_n are clearly τ-continuous as soon as the potential functions are strictly positive. Also note that condition (H) is met as soon as for any $x_{n-1} \in E_{n-1}$ we have

$$\rho_n \, \lambda_n \leq M_n(x_{n-1}, .) \ll \lambda_n$$

For instance let us suppose that $E_n = \mathbb{R}$ and

$$M_n(x, dy) = \frac{1}{\sqrt{2\pi}} \, e^{-\frac{1}{2}(y - a_n(x))^2} \, dy$$

where a_n is a bounded measurable drift function on $\mathbb{R}$. In this case condition (H) is met with the reference measure

$$\lambda_n(dy) = p_n(dy)/p_n(\mathbb{R}) \quad \text{with} \quad p_n(dy) =_{\text{def.}} \frac{1}{\sqrt{2\pi}} e^{-\frac{y^2}{2} - |y| \, \|a_n\|} dy$$

and with the parameters $\rho_n = e^{-\|a_n\|^2/2} \, p_n(\mathbb{R})$.

On page 351, we will check that the laws of the particle density profiles $(\eta_p^N)_{0 \leq p \leq n}$ are exponentially tight for the product weak topology on $\mathcal{P}^n(E)$ as soon as E_n are Polish. As we will note on page 350, the LDP lower bounds for the weak topology combined with the exponential tightness also imply that J_n is a good rate function for the weak topology.

10.2 Some Preliminary Results

10.2.1 Topological Properties

Let M be a Hausdorff and regular topological space. We recall that a topology on M is a family of open subsets; that is, a family of subsets that is stable by *finite* intersections and unions. Since the union (resp. intersection) of an empty family of sets in M is $\emptyset$ (resp. M), the sets $\emptyset$ and M are open. A subset A is said to be closed if $M - A$ is open. This implies that the sets $\emptyset$ and M are also closed. The Hausdorff property ensures that single points are closed and every two distinct points have disjoint neighborhoods. The regularity property refers to the fact that any closed set and any point outside this set have disjoint neighborhoods. In this situation, for any neighborhood A of $u \in M$, we can find a subneighborhood B of u such that $\overline{B} \subset A$.

For a detailed account on different classes of topological spaces, we refer the reader to the comprehensive introductory book on topology by Dugundji [126] (see for instance p. 311 for a detailed picture on different topological spaces).

In general, a σ-field $\sigma(M)$ on M is too small and the open and closed sets $\overset{\circ}{A}$ and $\overline{A}$ are not necessarily measurable. Nevertheless, there always exists a unique smallest σ-field $B(M)$ containing the topology of M. This σ-field is generated by the set of all open (or all closed) subsets of M. $B(M)$ is traditionally called the Borel σ-field. When $B(M) \subset \sigma(M)$, it is readily checked that the bounds (10.5) are equivalent to the following:

1. (Upper bound) For any closed subset $A \subset M$

$$\limsup_{N \to \infty} \frac{1}{N} \log P^N(A) \leq -H(A) \qquad (10.8)$$

2. (Lower bound) For any open subset $A \subset M$

$$\liminf_{N \to \infty} \frac{1}{N} \log P^N(A) \geq -H(A) \qquad (10.9)$$

Definition 10.2.1 *When a sequence of distributions P^N only satisfies the LDP upper bounds (10.8) for a compact set, we say the P^N satisfies a weak LDP.*

Definition 10.2.2 *A function $V : M \to \mathbb{R} \cup \{-\infty\}$ is lower semicontinuous (which we abbreviate l.s.c.) if, for each $a \in \mathbb{R} \cup \{-\infty\}$ and $u \in M$ such that $V(u) > a$, there corresponds a neighborhood A of u such that $V(v) > a$ for any $v \in A$. The upper semicontinuity (which we abbreviate u.s.c.) is defined in the same way by reversing the sense of the inequality in both cases. When the level sets of a nonnegative l.s.c. function H are compact, the function H is called a good rate function.*

It is evident that V is l.s.c. iff its level sets $V^{-1}([-\infty, a])$, $a \in \mathbb{R}$, are closed. We also clearly have that V is l.s.c. iff $(-V)$ is u.s.c. One important property is that l.s.c. functions always achieve their infimum over compact sets. Note that for good rate functions this property is also met over all closed sets. Many examples of l.s.c. can be provided using the simple fact that the pointwise supremum of a family of l.s.c. (and hence continuous) maps is an l.s.c. function. In particular, if $V_n \uparrow V$, with each V_n l.s.c., then V is l.s.c.

10.2.2 Idempotent Analysis

Nonnegative l.s.c. functions arise in a variety of research areas, including convex analysis, game theory, and optimal control theory. They are often used to describe the cost or the performance attached to some decision process with respect to a given optimal policy. The theory of idempotent measures provides a natural and rather general functional model to describe and analyze most of these optimization problems. In this context, l.s.c. functions arise as the limiting idempotent measures associated with a sequence of distributions in a logarithmic scale. The aim of this short section is to provide an introduction on the relations between large-deviation analysis and idempotent measures theory (see [205, 240]). These ideas induce a new, modern way of thinking about large-deviation principles. We first need quite a bit of preliminary notation.

We define the N-logarithmic addition/multiplication operations $(\oplus^N, \odot)$ on $\mathbb{R} \cup \{-\infty\}$ and denoted by

$$a \oplus^N b = \frac{1}{N} \log \left(e^{Na} + e^{Nb}\right) \quad \text{and} \quad a \odot b = a + b \; \left(= \frac{1}{N} \log \left(e^{Na} \cdot e^{Nb}\right)\right)$$

It is easy to see from these definitions that $-\infty$ and 0 are the neutral elements of the operations $\oplus_N$ and $\odot$ and the set $(\mathbb{R}\cup\{-\infty\},\oplus_N,\odot)$ is a semiring. It is important to note that for any sequence of numbers a_N and b_N we have

$$\limsup_{N\to\infty}(a_N \oplus^N b_N) = (\limsup_{N\to\infty} a_N) \oplus (\limsup_{N\to\infty} b_N) \quad \text{with} \quad a \oplus b = a \vee b$$

To prove this assertion, we use the fact that

$$0 \leq (a \oplus^N b - a \oplus b) \leq 1/N \tag{10.10}$$

and $\limsup_{N\to\infty}(a_N \oplus b_N) = (\limsup_{N\to\infty} a_N) \oplus (\limsup_{N\to\infty} b_N)$. Note that the display above is in general not met if we replace $\limsup_{N\to\infty}$ by $\liminf_{N\to\infty}$ but we have

$$(\liminf_{N\to\infty} a_N) \oplus (\liminf_{N\to\infty} b_N) \leq \liminf_{N\to\infty}(a_N \oplus b_N) = \liminf_{N\to\infty}(a_N \oplus^N b_N)$$

For instance for $a_N = (-1)^N = -b_N$, we have

$$a_N \oplus b_N = 1 > (\liminf_{N\to\infty} a_N) \oplus (\liminf_{N\to\infty} b_N) = -1$$

This dequantization of the semiring $(\mathbb{R}\cup\{-\infty\},\oplus^N,\odot)$ into the so-called $(\max,+)$ semiring $(\mathbb{R}\cup\{-\infty\},\oplus,\odot)$ is the cornerstone of idempotent measure and functional analysis.

Let $\mathcal{B}_b(M,\mathbb{R}\cup\{-\infty\})$ be the set of all upper-bounded measurable functions from M into $\mathbb{R}\cup\{-\infty\}$. Idempotent analysis is concerned with the study of linear operators on $\mathcal{B}_b(M,\mathbb{R}\cup\{-\infty\})$ and taking values in the semirings associated with the previously defined operations.

Definition 10.2.3 *A $(\oplus^N,\odot)$-integral operator is linear mapping L_N from the set $\mathcal{B}_b(M,\mathbb{R}\cup\{-\infty\})$ into the semiring $(\mathbb{R}\cup\{-\infty\},\oplus^N,\odot)$. In reference to the traditional theory of integration, sometimes we use the notation, for any $f \in \mathcal{B}_b(M,\mathbb{R}\cup\{-\infty\})$,*

$$L_N(f) = \int^{\oplus^N} f \odot dL_N$$

A $(\oplus,\odot)$-integral operator L is defined in the same way by replacing the logarithmic operation $\oplus^N$ by $\oplus$.

To emphasize the role of this functional framework in large-deviation analysis, we introduce the following definition.

Definition 10.2.4 *Let P^N be a sequence of probability measures on a topological space M equipped with a σ-field $\sigma(M) \supset \mathcal{B}(M)$. We associate with P^N the set function*

$$L_N : A \in \sigma(M) \longrightarrow L_N[A] = \frac{1}{N}\log P^N(A) \in [-\infty, 0]$$

with the conventions $\log 0 = -\infty$. More generally, we define the N-logarithmic integral of a measurable function $f : M \to \mathbb{R} \cup \{-\infty\}$ by setting

$$L_N(f) = \frac{1}{N} \log \int e^{Nf} \, dP^N$$

Using a simple manipulation, we check that for any $A, B \in \sigma(M)$ with $A \cap B = \emptyset$

$$L_N[A \cup B] = L_N[A] \oplus^N L_N[B]$$

Using the same arguments, we check that the resulting integral operator L_N is $(\oplus^N, \odot)$-linear in the sense that for any pair of upper-bounded measurable functions $f, g : M \to \mathbb{R} \cup \{-\infty\}$ and for any $a, b \in \mathbb{R} \cup \{-\infty\}$

$$L_N[a \odot f \oplus^N b \odot g] = a \odot L_N[f] \oplus^N b \odot L_N[g] \tag{10.11}$$

Furthermore, using (10.10), we easily prove that

$$0 \le L_N[f \oplus^N g] - L_N[f \oplus g] \le 1/N \tag{10.12}$$

We also introduce the limiting set functions $L^\star$ and $L_\star$ defined for any $A \in \sigma(M)$ by

$$L^\star[A] = \limsup_{N \to \infty} L_N[A] \quad \text{and} \quad L_\star[A] = \liminf_{N \to \infty} L_N[A]$$

From previous considerations, we readily check that $L^\star$ is an idempotent probability measure on $(M, \sigma(M))$ in the sense that

$$L^\star(\emptyset) = -\infty, \quad L^\star(M) = 0, \quad \text{and} \quad L^\star(A \cup B) = L^\star(A) \oplus L^\star(B)$$

for any $A, B \in \sigma(M)$ with $A \cap B = \emptyset$. In terms of integral operators, we also deduce from (10.12) that

$$L^\star(a \odot f \oplus b \odot g) = a \odot L^\star(f) \oplus b \odot L^\star(g)$$

for any pair of upper-bounded measurable functions $f, g : M \to \mathbb{R} \cup \{-\infty\}$ and for any pair $a, b \in \mathbb{R} \cup \{-\infty\}$. Using standard calculations, we observe that the formula above is met for any pair of sets (A, B) and we have the idempotent property $L^\star(A) = L^\star(A) \oplus L^\star(A)$. We summarize the discussion above with the following proposition.

Proposition 10.2.1 *The N-logarithmic integral operators L_N associated with a sequence of probability measures P^N on a topological space M (equipped with a σ-field $\sigma(M) \supset \mathcal{B}(M)$) are $(\oplus^N, \odot)$-integral operators. In addition, the limiting operator $L^\star = \limsup_{N \to \infty} L_N$ is a $(\oplus, \odot)$-integral operator.*

To illustrate these constructions and to keep the ideas as simple as possible, let us examine the case where the state space is finite, $M = \{1, \ldots, d\}$, and equipped with the discrete Borel σ-field $B(M) = \sigma(M)$. Also let P^N be the sequence of distributions on M defined by

$$P^N(du) = \sum_{i=1}^d e^{-Nh^N(i)} \, \delta_i(du)$$

for some $h^N(i) \geq 0$ such that $P^N(M) = 1$. In this simple situation, we readily check that

$$L_N[A] = \oplus_{i \in A}^N \, h^N(i)$$

From previous estimations, we note that for every $A \subset B(M)$ we have

$$-\inf_{i \in A} h^\star(i) \leq L_\star[A] \leq L^\star[A] = -\inf_{i \in A} h_\star(i)$$

with $\liminf_{N \to \infty} h^N(i) = h_\star(i)$ and $\limsup_{N \to \infty} h^N(i) = h^\star(i)$. In this situation, we see that $h_\star = h^\star$ if and only if P^N satisfies the LDP with rate function $h_\star$. More generally, we find that a sequence of distributions P^N on M satisfies the LDP with rate function $H : M \to [0, \infty]$ if and only if for any $A \in \sigma(M)$ we have

$$L[\mathring{A}] \leq L_\star[A] \leq L^\star[A] \leq L[\overline{A}] \qquad (10.13)$$

with the idempotent measure L on $\sigma(M)$ defined by $L[A] = -H(A) = -\inf_A H$. The $(\oplus, \odot)$-integral corresponding to an idempotent measure L of the previous form is simply given for any $f \in \mathcal{B}_b(M, \mathbb{R} \cup \{-\infty\})$ by

$$L(f) = \int^\oplus f \odot dL = \sup_{u \in M} (f(u) + L(u)) = \sup_{u \in M} (f(u) - H(u))$$

To get one step further we recall that for any $f \in \mathcal{B}_b(M, \mathbb{R} \cup \{-\infty\})$ we have $\mathring{f} \leq f \leq \overline{f}$ with the pair l.s.c./u.s.c. (upper semicontinuous) closures $(\mathring{f}, \overline{f})$ of f defined by

$$\mathring{f} = \sup \{\inf_A f \, : \, u \in A \text{ open}\}$$
$$\overline{f} = \inf \{\sup_A f \, : \, u \in A \text{ closed}\}$$

These functions are sometimes called the inferior and superior limits of f. To illustrate these new notions, let Ind_A be the $(\oplus, \odot)$ indicator function of a set $A \in \sigma(M)$ and defined by $\text{Ind}_A(u) = 0$ for $u \in A$ and $-\infty$ otherwise. By the closure definition, it is easily seen that

$$\mathring{\text{Ind}}_A = \text{Ind}_{\mathring{A}} \quad \text{and} \quad \overline{\text{Ind}_A} = \text{Ind}_{\overline{A}}$$

Consequently, (10.13) is equivalent to saying

$$L(\mathring{f}) \leq L_\star[f] \leq L^\star[f] \leq L[\bar{f}] \tag{10.14}$$

for any indicator functions $f \in \{\mathrm{Ind}_A : A \in \sigma(M)\}$. In this connection, we already mention that Varadhan's integral lemma (Lemma 10.4.1) can be reformulated in terms of idempotent measures by saying that (10.14) also holds true for any upper-bounded measurable function $f : M \to \mathbb{R} \cup \{-\infty\}$. As the initiated reader may have noticed, most of the large-deviation results can be reformulated into an equivalent statement in idempotent analysis. It is clearly out of the scope of this book to present a catalog of all of these connections. Nevertheless, we finally simply state that Fenchel transformations of l.s.c. functions correspond to characteristic functions of an idempotent probability measure.

10.2.3 Some Regularity Properties

In the further development of this section, we use the following natural idempotent measure notation. We associate with an l.s.c. function $H : M \to [0, \infty]$ the set function still denoted by H and defined for any $A \subset M$ by

$$H(A) = \inf_{m \in A} H(x)$$

with the convention $H(\emptyset) = \infty$. Note that for any $A, B \subset M$ and $a \in \mathbb{R}$ we have

$$H(A \cup B) = H(A) \wedge H(B) \quad \text{and} \quad (a + H)(A) = a + H(A)$$

as well as $H(A \cap B) \geq H(A) \vee H(B)$. We easily check the idempotent property of these set functions; namely, for any $A \subset M$ we have $H(A) = H(A) \wedge H(A)$.

Definition 10.2.5 *Let $\pi : M_1 \to M_2$ be a continuous function between a pair (M_1, M_2) of Hausdorff topological spaces. Also let $H : M_1 \to [0\infty]$ be an l.s.c. function. We will denote by $H \circ \pi^{-1} : M_2 \to [0, \infty]$ the π-image of H defined for any $u_2 \in M_2$ by*

$$H \circ \pi^{-1}(u_2) = \inf \{H(u_1) : u_1 \in M_1 \text{ s.t. } \pi(u_1) = u_2\} \in [0, \infty] \tag{10.15}$$

This function corresponds to the π-image of an idempotent measure. Indeed, for any $A \subset M_2$, we have

$$(H \circ \pi^{-1})(A) = \inf_{m_2 \in A}(H \circ \pi^{-1})(m_2) = \inf_{m_1 \in \pi^{-1}(A)} H(m_1) = H(\pi^{-1}(A)) \tag{10.16}$$

The first good news is that good rate functions are preserved under continuous transformations. This elementary result gives probably one of the simplest and most powerful ways to transfer an LDP.

Theorem 10.2.1 *Let $\pi : M_1 \to M_2$ be a continuous function between a pair (M_1, M_2) of Hausdorff topological spaces. The π-image $H \circ \pi^{-1} : M_2 \to [0, \infty]$ of a good rate function $H : M_1 \to [0, \infty]$ is a good rate function. In particular, if H governs the LDP associated with a collection of measures Q^N, then the π-image measures $Q^N \circ \pi^{-1}$ satisfy the LDP with good rate function $H \circ \pi^{-1}$.*

Proof:
Since H is a good rate function, the infimum in (10.15) is obtained at some point. Using the fact that π is continuous, we conclude that the level sets $H_\pi^{-1}([0,a]) = \pi(H^{-1}([0,a])) \subset M_2$ are compact. On the other hand, for any open (resp. closed) set $A \subset M_2$, the set $\pi^{-1}(A) \subset M_1$ is open (resp. closed). From (10.16), we easily check that $Q^N \circ \pi^{-1}$ satisfies the LDP with rate $H \circ \pi^{-1}$. ∎

We observe that a lower semicontinuous function V satisfies at every point $u \in M$

$$V(u) = \sup \{V(A) \; : \; u \in A, \; A \text{ open}\} \quad (10.17)$$

Also notice that the supremum in the display above can also be taken over any open neighborhood topological basis. In this case, by the topological regularity of M, for any $u \in M$ with $V(u) < \infty$ and for any $\varepsilon > 0$ there exists a pair of neighborhoods A, B of u such that $x \in \overline{B} \subset A$ and

$$V(u) \geq V(\overline{B}) \geq V(A) \geq V(u) - \varepsilon$$

We quote the first elementary but reassuring result of the theory of large deviations.

Proposition 10.2.2 *For any sequence P^N of probability measures on a Hausdorff and regular topological space, there is at most one rate function governing the LDP of P^N.*

Proof:
Suppose there were two such rates H_1 and H_2 and some point $u \in M$ on which $H_1(u) > H_2(u)$. From previous comments and because of the lower semicontinuity, there exists a neighborhood B of u such that $x \in \overline{B}$ and

$$H_1(u) \geq H_1(\overline{B}) \geq H_1(u) - \varepsilon$$

Since the functions H_1, H_2 govern the LDP, we also have

$$-H_1(\overline{B}) \geq -H_2(B)(\geq -H_2(u))$$

This would imply that $H_1(u) \leq H_2(u) + \varepsilon$ for any $\varepsilon > 0$, yielding a contradiction with our assumption $H_1(u) > H_2(u)$. ∎

We end this section with two interesting properties of rate functions and sequences of measures on a metric state space. The first one provides some nice topological regularity properties of good rate functions with respect to open and closed neighborhoods.

Proposition 10.2.3 *Let H be a good rate function on some metric space (M, d). For any $u \in M$ and $\varepsilon > 0$, we have*

$$\lim_{\varepsilon \to 0} H(\overline{B}(u, \varepsilon)) = H(u) = \lim_{\varepsilon \to 0} H(B(u, \varepsilon))$$

where $\overline{B}(u, \varepsilon)$ denotes the closure of the open ball $B(u, \varepsilon) = \{v \in M : d(u, v) < \varepsilon\}$.

Proof:
Because of the l.s.c., we have $H(u) = \lim_{\varepsilon \downarrow 0} H(B(u, \varepsilon))$. To check the first equality, it clearly suffices to prove that for any $\varepsilon' > 0$

$$a_{\varepsilon'} =_{\text{def.}} \varepsilon' + \lim_{\varepsilon \to 0} H(\overline{B}(u, \varepsilon)) \geq H(u) \tag{10.18}$$

When the r.h.s. term is infinite, the result is trivial. Otherwise, because the level sets of H are compact, $C_{\varepsilon'}(u, \varepsilon) = \overline{B}(u, \varepsilon) \cap H^{-1}([0, a_{\varepsilon'}])$ forms a decreasing sequence of nonempty compact sets (as $\varepsilon \downarrow 0$) and

$$\cap_{\varepsilon > 0} C_{\varepsilon'}(u, \varepsilon) = C_{\varepsilon'}(u, 0) = \{u\} \in H^{-1}([0, a_{\varepsilon'}]) \Rightarrow (10.18)$$

This ends the proof of the proposition. ∎

In the next proposition, we already present a universal way to obtain a weak LDP and identify the rate function.

Proposition 10.2.4 *Let P^N be a sequence of distributions on some metric space (M, d). Suppose that for any $u \in M$ we have*

$$\lim_{\epsilon \to 0} \liminf_{N \to \infty} \frac{1}{N} \log P^N(B(u, \epsilon)) = -H(u) = \lim_{\epsilon \to 0} \limsup_{N \to \infty} \frac{1}{N} \log P^N(\overline{B}(u, \epsilon)) \tag{10.19}$$

for some function $H : M \to [0, \infty]$. Then H is l.s.c. and it governs the weak LDP of the sequence P^N. Inversely, if a good rate function H governs the LDP of some sequence of distributions P^N, then (10.19) holds true.

Proof:
For any open set $A \subset M$ and $u \in A$, there exists some $\varepsilon_0 > 0$ such that $B(u, \epsilon) \subset A$ for any $0 \leq \varepsilon \leq \varepsilon_0$. Under our assumptions, we find that

$$\liminf_{N \to \infty} \frac{1}{N} \log P^N(B(u, \epsilon)) \leq \liminf_{N \to \infty} \frac{1}{N} \log P^N(A)$$

Letting $\varepsilon \to 0$ and taking the supremum of $(-H)(u)$ over A, we conclude that H governs the LDP lower bound. Let u_n be an approximation sequence of u such that $d(u_n, u) < 1/n$. Under our assumptions, for any $\delta > 0$ there exists some n_δ such that for any $n \geq n_\delta$

$$\liminf_{N \to \infty} \frac{1}{N} \log P^N(B(u, 1/n)) \leq -H(u) + \delta$$

On the other hand, for each n there exists some sufficiently large m_n such that $B(u_n, 1/m) \subset B(u, 1/n)$ for any $m \geq m_n$,

$$\liminf_{N \to \infty} \frac{1}{N} \log P^N(B(u_n, 1/m)) \leq -H(u) + \delta$$

Letting $m \to \infty$, we conclude that for any $n \geq n_\delta$ we find that

$$H(u_n) \geq H(u) - \delta$$

We conclude that $\liminf_{n \to \infty} H(u_n) = \sup_{\delta > 0} \inf_{n \geq n_\delta} H(u_n) \geq H(u)$ and consequently H is an l.s.c. function. For any open ε-covering of a compact set $A \subset \cup_{u \in A} B(u, \varepsilon)$, we extract a finite covering

$$A \subset \cup_{1 \leq i \leq d} B(u_i, \varepsilon) \subset \cup_{1 \leq i \leq d} \overline{B}(u_i, \varepsilon)$$

and by the union of event bounds we find that

$$\limsup_{N \to \infty} \frac{1}{N} \log P^N(A) \leq -\inf_{1 \leq i \leq d} -\limsup_{N \to \infty} \frac{1}{N} \log P^N(\overline{B}(u_i, \varepsilon))$$

Letting $\varepsilon \to 0$, we conclude that

$$\limsup_{N \to \infty} \frac{1}{N} \log P^N(A) \leq -\inf_{1 \leq i \leq d} H(u_i) \leq -H(A)$$

By Proposition 10.2.3, the last assertion is a direct consequence of the definition of the LDP. This ends the proof of the proposition. ∎

10.3 Crámer's Method

As the reader may have noticed, proving LDP lower bounds is a very challenging question. Various rough exponential upper bounds have already been analyzed in Section 7.4 in the context of particle approximation models of Feynman-Kac distributions on some measurable spaces $(E_n, \mathcal{E}_n)$. For instance, for strictly positive potential functions, from the exponential estimate (7.28) stated in Corollary 7.4.3, we find that

$$\limsup_{N \to \infty} \frac{1}{N} \log \mathbb{P}_{\eta_0}^N \left(|\eta_n^N(f_n) - \eta_n(f_n)| > \varepsilon \right) \leq -\frac{\varepsilon^2}{2b(n)^2}$$

for any measurable functions $f_n : E_n \to \mathbb{R}$ such that $\mathrm{osc}(f_n) \le 1$, and with

$$b(n) \le 2 \sum_{q=0}^{n} r_{q,n}\ \beta(P_{q,n})$$

To get one step further in our discussion, we recall that these crude estimates were essentially obtained applying the Chernov-Hoeffding exponential inequality (see Lemma 7.3.2) to the random variables $\eta_n^N(f_n)$ for each test function f_n. One natural strategy to improve these estimates and obtain more precise exponential upper bounds that are valid for any test function is to analyze a "uniform version" of the Chernov-Hoeffding inequality. This idea goes back to Crámer [62] and is not restricted to particle approximation models. Because of its importance in practice, we have chosen to present this strategy in an abstract and general framework.

The forthcoming strategy we are about to present needs to strengthen our topological conditions and to impose some additional algebraic structure to the set M. In the further development of this section, we suppose that M is a Hausdorff topological vector space (recall that these spaces are necessarily regular). In analogy with the exponential moments used in the proof of the Chernov-Hoeffding inequality, we introduce the following.

Definition 10.3.1 *The asymptotic logarithmic moment-generating function of a sequence of distribution P^N on M (with topological dual $M^\star$) is the function Λ defined as follows*

$$\Lambda : V \in M^\star \to \Lambda(V) = \lim_{N\to\infty} \frac{1}{N} \log \int_M e^{N<V,m>}\ P^N(dm) \in \mathbb{R} \cup \{\infty\}$$

Proposition 10.3.1 *The function Λ and its Legendre-Fenchel transformation $\Lambda^\star$ defined by*

$$\Lambda^\star : m \in M \to \Lambda^\star(m) = \sup_{V \in M^\star} (<V,m> -\Lambda(V)) \in [0,\infty]$$

are convex functions. In addition, $\Lambda^\star$ is l.s.c., and for any compact set $A \subset M$ we have

$$\limsup_{N\to\infty} \frac{1}{N} \log P^N(A) \le -\Lambda^\star(A) \qquad (10.20)$$

Proof:
The convexity of Λ is easily checked using the Hölder inequality, while the convexity $\Lambda^\star$ is a simple consequence of its definition. The l.s.c. of $\Lambda^\star$ is proved by recalling that the supremum of a collection of continuous functions is l.s.c. Suppose that $\Lambda^\star(m) \ge \delta > 0$ for any $m \in A$ and some $\delta > 0$; otherwise, we have $\Lambda^\star(A) = 0$ and the result is trivial. Using similar arguments and by definition of $\Lambda^\star$, we can associate with each $m \in M$ a point $V_m \in M^\star$ such that

$$<V_m, m> -\Lambda(V_m) \ge \Lambda^\star(m) - \delta$$

Since $V_m \in M^\star$, we also have $\sup_{m' \in A_m} <V_m, (m-m')> \le \delta$ for some neighborhood $A_m \in M$ of each point m. By the exponential version of Markov's inequality, we find that

$$\begin{aligned}
P^N(A_m) &= \int_{A_m} e^{-<NV_m,(m-m')>+<NV_m,(m-m')>} P^N(dm') \\
&\le e^{N\delta} \int_M e^{-N<V_m,(m-m')>} P^N(dm') \\
&= e^{N(\delta-<V_m,m>)} \int_M e^{N<V_m,m'>} P^N(dm')
\end{aligned}$$

from which we deduce that

$$\frac{1}{N} \log P^N(A_m) \le \delta - <V_m, m> + \frac{1}{N} \log \int_M e^{N<V_m,m'>} P^N(dm')$$

The set $A \subset \cup_{m \in A} A_m$ being compact, we can extract a finite cover $A \subset \cup_{1 \le i \le d} A_{m_i}$ and under our assumptions the estimate above implies that

$$\begin{aligned}
\limsup_{N \to \infty} \frac{1}{N} \log P^N(A) &\le \delta - \inf_{1 \le i \le d}(<V_{m_i}, m_i> - \Lambda(V_{m_i})) \\
&\le 2\delta - \inf_{1 \le i \le d} \Lambda^\star(m_i) \le 2\delta - \Lambda^\star(A)
\end{aligned}$$

Taking $\delta \to 0$, the end of the proof of the proposition is completed. ∎

In many interesting practical situations, the Legendre-Fenchel transform of the asymptotic logarithmic moment-generating function coincides with the rate function governing the LDP of the sequence P^N. The first objective is to extend the LDP upper bounds (10.20) to any closed subset. Intuitively speaking, this extension should be possible as soon as the probability mass of the sequence P^N is exponentially concentrated on a compact set. The right notion that makes this result precise is the following.

Definition 10.3.2 *We say that a sequence of distributions P^N is exponentially tight if for any $a > 0$ there exists a compact set A_a such that*

$$\limsup_{N \to \infty} \frac{1}{N} \log P^N(M - A_a) \le -a$$

As an aside, when P^N is exponentially tight, an LDP lower bound would imply that

$$-a \ge \limsup_{N \to \infty} \frac{1}{N} \log P^N(M - A_a) \ge -H(M - A_a)$$

In this case, this shows that all the level sets $H^{-1}([0,a]) \subset A_a$ are necessarily compact.

10. Large-Deviation Principles

As mentioned above, another important consequence of the exponential tightness property is that the proof of the LDP upper bound for closed sets reduces to that of the LDP upper bound for compact sets.

Proposition 10.3.2 *Let P^N be an exponentially tight sequence of distributions on an Hausdorff topological space M equipped with the Borel σ-field $B(M)$. A function H governs the LDP upper bounds for any closed sets as soon as it governs the LDP upper bounds for all compact sets.*

Proof:
To prove this assertion, we simply observe that for any closed set $B \in (M - H^{-1}[0, \varepsilon])$, for some $\varepsilon > 0$ (otherwise we have $H(B) = 0$ and the desired implication is trivially checked), we have

$$P^N(B) \leq P^N(A_\varepsilon \cap B) + P^N(M - A_\varepsilon)$$

When the LDP upper bound holds true for the compact set $B \cap A_\varepsilon$ we conclude that for any $H(B) > \varepsilon$

$$\limsup_{N \to \infty} \frac{1}{N} \log P^N(B) \leq (-H(A_\varepsilon \cap B)) \wedge (-\varepsilon) = -\varepsilon$$

Taking the infimum of $(-\varepsilon)$ over all $\varepsilon < H(B)$ yields the desired upper bound. ■

To illustrate this notion of exponential tightness, we come back to particle approximation models of Feynman-Kac distributions and the crude exponential upper-bound presented in Section 7.4. We further assume that E_n are Polish state spaces equipped with the Borel sigma-field $\mathcal{E}_n$ and the potential functions are strictly positive. In this situation, we recall that $\mathcal{P}(E_n)$ equipped with the weak topology is again a Polish space. By a theorem of Prohorov, a closure $\overline{A}_n$ of a set $A_n \subset \mathcal{P}(E_n)$ is compact if and only if the set A_n is tight; that is, if for each $\delta_n > 0$ there exists a compact set $C_n \subset E_n$ such that $\inf_{m_n \in A_n} m_n(C_n) \geq 1 - \delta_n$. Using the fact that any single probability measure on the Polish space is tight, for any $\delta_n > 0$ we associate with the Feynman-Kac distribution η_n a compact set $C_n(\delta_n) \subset E_n$ such that $\eta_n(C_n(\delta_n)) \geq 1 - \delta_n$. The reader will immediately notice that the sets

$$A_n(\eta_n, \delta_n) = \{m_n \in \mathcal{P}(E_n) \; : \; m_n(C_n(\delta_n/2)) - \eta_n(C_n(\delta_n/2)) \geq -\delta_n/2\}$$

are compact and by construction we have

$$\inf_{m_n \in A_n(\eta_n, \delta_n)} m_n(C_n(\delta_n/2)) \geq 1 - \delta_n$$

and

$$A_n^c(\eta_n, \delta_n) \;\subset\; \{m_n \in \mathcal{P}(E_n) \; : \; |\eta_n(C_n(\delta_n/2)) - m_n(C_n(\delta_n/2))| > \delta_n/2\}$$

Using the exponential estimates (7.21) presented in Theorem 7.4.3, by the union of event bounds, we find that

$$\mathbb{P}_{\eta_0}^N \left((\eta_p^N)_{0 \leq p \leq n} \notin \prod_{p=0}^{n} A_p(\eta_p, \delta_p) \right) \leq c(n) \; e^{-N \wedge_{p=0}^{n} \delta_p^2 / d(n)}$$

for sufficiently large N and for some finite constants $c(n)$ and $d(n) < \infty$ whose values only depend on the time parameter. This proves that the laws of the particle density profiles $(\eta_0^N, \ldots, \eta_n^N)$ are exponentially tight in $\Sigma_n(E)$ equipped with the product topology, and by Proposition 10.3.2 the analysis of the LDP upper bound reduces to that of the LDP upper bounds on compact sets.

Proving the LDP lower bounds is often a delicate problem. Several different strategies have been presented in the literature. One possible route is to follow the proof of Cramér's theorem on the LDP for independent and $\mathbb{R}^d$-valued random variables. This technique has been initiated by Gärtner and Ellis in the context of non-iid and $\mathbb{R}^d$-valued random sequences. It has been further extended by Baldi to any sequence of M-random variables. We refer the reader to [112] for more precise informations as well as for a complete list of references. This strategy gives fruitful results as soon as the asymptotic logarithmic moment-generating function has some nice regularity properties, and only if the expected rate function is convex. The proof of this result depends on several deep results on convex analysis that are not discussed here and thus it will be omitted. For more details, we refer to Corollary 4.5.27 in [112]. The next theorem is due to Dembo and Zeitouni. It is a somewhat weaker version of Baldi's theorem for Banach-valued random sequences, but it provides a precise connection between the regularity of Λ and the desired LDP lower bounds.

Theorem 10.3.1 (Baldi, Dembo-Zeitouni) *Let P^N be an exponentially tight sequence of probability measures on a Banach space M. Suppose that the asymptotic logarithmic moment-generating function $\Lambda : M^\star \to \mathbb{R}$ is finite, l.s.c. with respect to the M-topology on $M^\star$, and Gateaux differentiable; that is, $\varepsilon \to \Lambda(V_1 + \varepsilon.V_2)$ is differentiable at $\varepsilon = 0$ for any pair $V_1, V_2 \in M^\star$. Then P^N satisfies the LDP with the good rate function $\Lambda^\star$.*

10.4 Laplace-Varadhan's Integral Techniques

In Section 10.2, we saw that LDP and good rate functions are preserved under continuous transformations. This property allows transfer of LDP from a sequence of distributions P^N into another as soon as Q^N is the image measure of P^N with respect to some continuous transformation between the state spaces. This quite elementary result is often used in practice and provides a simple way to transfer LDP when the corresponding random variables are connected by some regular transformation. In this section, we examine the situation in which the structural connection above is replaced by an integral continuity property. The first outstanding result in this direction is the following theorem due to Varadhan and often called Varadhan's or the Laplace-Varadhan Lemma in the literature on LDP.

Theorem 10.4.1 (Laplace-Varadhan) *Let M be a Hausdorff and regular topological space equipped with a σ-field $\sigma(M) \supset B(M)$. Suppose that a sequence of probability measures $Q^N \in \mathcal{P}(M)$, $N \geq 1$, satisfies the LDP on M with a good rate function $H : M \to [0, \infty]$. Then, for any bounded l.s.c. function $V : M \to \mathbb{R}$ and for any open set A, we have*

$$\liminf_{N\to\infty} \frac{1}{N} \log \int_A e^{NV} \, dQ^N \geq \sup_A (V - H) \tag{10.21}$$

In addition, for any bounded u.s.c. function $V : M \to \mathbb{R}$ and for any closed set A, we have

$$\limsup_{N\to\infty} \frac{1}{N} \log \int_A e^{NV} \, dQ^N \leq \sup_A (V - H) \tag{10.22}$$

Inversely, if (10.21) and (10.22) hold true for some l.s.c. function H, then Q^N satisfies the LDP on M with a rate function $H : M \to [0, \infty]$.

Proof:
We first prove the lower bound. Since V is an l.s.c. function, we recall that $V(u) = \sup\{V(B) : u \in B, B \text{ open} \subset M\}$, for every $u \in M$. Consequently, for any $u \in A$ and $\varepsilon > 0$, there exists an open neighborhood $B \subset A$ of u such that $V(u) \leq \varepsilon + V(B)$. It follows that

$$Q^N(1_A \, e^{NV}) \geq Q^N(e^{NV} \, 1_B) \geq e^{V(u) - \varepsilon} \, Q^N(B)$$

Hence it follows that $\frac{1}{N} \log Q^N(1_A \, e^{NV}) \geq (V(u) - \varepsilon) + \frac{1}{N} \log Q^N(B)$. Taking into account the LDP lower bound on open sets B, we also have for any $u \in B$

$$\liminf_{N\to\infty} \frac{1}{N} \log Q^N(1_A \, e^{NV}) \geq (V(u) - \varepsilon) - H(B) \geq (V(u) - \varepsilon) - H(u)$$

Taking the supremum over all points $u \in A$ and letting $\varepsilon \to 0$ we find that

$$\liminf_{N\to\infty} \frac{1}{N} \log Q^N(1_A \, e^{NV}) \geq \sup_A (V - H)$$

To prove the upper bound, we fix some $\varepsilon > 0$ and an arbitrary $b < \infty$ and we cover the sets $A_b = A \cap H^{-1}([0, b])$ with a collection of sufficiently small neighborhoods $B(u)$, $u \in A_b$ such that

$$\sup_{B(u)} V \leq V(u) + \varepsilon \quad \text{and} \quad \inf_{B(u)} H \geq H(u) - \varepsilon$$

Taking into account that A is closed and the level sets are compact, the trace set A_b is also compact and we can extract a finite cover $A_b \subset \mathcal{B}_n = \cup_{i=1}^n B(x_i)$ and hence

$$\begin{aligned}Q^N(e^{NV} 1_A) &\leq Q^N(e^{NV} 1_{\mathcal{B}_n}) + Q^N(e^{NV} 1_{\mathcal{B}_n^c}) \\ &\leq \sum_{i=1}^n e^{N(V(u_i)+\varepsilon)} Q^N(B(u_i)) + e^{N\|V\|} Q^N(\mathcal{B}_n^c)\end{aligned}$$

It now follows from the LDP upper bound that

$$\begin{aligned}\limsup_{N\to\infty} \frac{1}{N} \log Q^N(e^{NV} 1_A) &\leq \vee_{i=1}^n ([V(u_i) + \varepsilon] - \\ &\quad H(\overline{B(u_i)})) \vee (\|V\| - H(\mathcal{B}_n^c)) \\ &\leq \vee_{i=1}^n ([V(u_i) - H(u_i) + 2\varepsilon]) \vee (\|V\| - b) \\ &\leq (\sup_A [V - H] + 2\varepsilon) \vee (\|V\| - b)\end{aligned}$$

We end the proof of (10.22) by letting $(\varepsilon, b) \to (0, \infty)$. The last assertion is a simple consequence of the definition of the LDP. ∎

As we already mentioned in the introduction, the Laplace-Varadhan integral lemma can also be regarded as a powerful change of reference probability technique that allows transfer of large-deviation principles from a sequence of probability measures Q^N to another P^N. The integral connection usually consists of a pair (P^N, Q^N) of absolutely continuous measures such that

$$\frac{dP^N}{dQ^N} = \exp(NV) \qquad Q^N\text{-a.e.} \qquad (10.23)$$

for some measurable mapping $V : M \to \mathbb{R}$. When V is bounded continuous, the theorem above allows transfer of an LDP on Q^N to the sequence P^N. Rephrasing Theorem 10.4.1, we can state the following corollary.

Corollary 10.4.1 *Assume that Q^N satisfies an LDP with good rate function $H : M \to [0, \infty]$. If P^N satisfies the continuity condition (10.23) for some $V \in \mathcal{C}_b(M)$, then it satisfies an LDP with good rate function $(H - V)$.*

The pair of distributions (P^N, Q^N) is frequently defined in terms of the image measures

$$P^N = \mathbb{P}^N \circ \pi_N^{-1} \quad \text{and} \quad Q^N = \mathbb{Q}^N \circ \pi_N^{-1}$$

for some probabilities $\mathbb{P}^N$ and $\mathbb{Q}^N$ on some measurable space Ω^N, which may depend on N and for some measurable mapping $\pi_N : \Omega^N \to M$. Observe that if $\mathbb{P}^N$ and $\mathbb{Q}^N$ are absolutely continuous and for $\mathbb{Q}^N$-a.e. $x \in \Omega^N$

$$\frac{d\mathbb{P}^N}{d\mathbb{Q}^N}(x) = \exp(NV(\pi_N(x))) \qquad (10.24)$$

then the probability images P^N and Q^N are absolutely continuous, their Radon-Nikodym derivative satisfies (10.23), and the Laplace-Varadhan lemma applies as soon as V is a bounded continuous mapping.

When (M,d) is a metric space, the author and T. Zajic have recently presented in [109] a new strategy to relax the analytic representation (10.24). The idea consists in replacing $\mathbb{P}^N$ and $\mathbb{Q}^N$ by a pair of sequences $\mathbb{P}^N_{\alpha,m}$ and $\mathbb{Q}^N_m$ indexed respectively by a parameter pair (α, m) with $\alpha \in \mathbb{R}$ and $m \in M$ and by a parameter $m \in M$. Instead of (10.24), we suppose that for any index pair $(\alpha, m) \in (\mathbb{R} \times M)$ we have $\mathbb{P}^N_{\alpha,m} \sim \mathbb{Q}^N_m$ and for $\mathbb{Q}^N_m$-a.e. $x \in \Omega^N$

$$\frac{d\mathbb{P}^N_{\alpha,m}}{d\mathbb{Q}^N_m}(x) = \exp\left(N[\alpha S_N(x,m) + V_\alpha(\pi_N(x), m)]\right) \qquad (10.25)$$

for some measurable functions $S_N : \Omega^N \times M \to \mathbb{R}$ and $V_\alpha : M \times M \to \mathbb{R}$.

We also assume that $\mathbb{P}^N_{1,m}$ is independent of m and denote the former by $\mathbb{P}^N_1$. For any $(\alpha, m) \in (\mathbb{R} \times M)$, we define the image measures

$$P^N_{\alpha,m} = \mathbb{P}^N_{\alpha,m} \circ \pi_N^{-1} \quad \text{and} \quad Q^N_m = \mathbb{Q}^N_m \circ \pi_N^{-1}$$

Lemma 10.4.1 *Suppose the sequence of probability measures Q^N_m satisfies an LDP with good rate function $H_m : M \to [0, \infty]$ for each $m \in M$. Also assume that the mappings $V_\alpha(.,m)$, $\alpha \in \mathbb{R}$, are continuous at each m, $V_\alpha(m,m) = 0$, and the exponential moment condition*

$$\limsup_{N\to\infty} \frac{1}{N} \log \int_{\Omega^N} \exp nN[S_N(x,m) + V_1(\pi_N(x), m)] \, d\mathbb{Q}^N_m(x) < \infty \qquad (10.26)$$

holds for some $(m,n) \in M \times (1,\infty)$. Then P^N_1 satisfies an LDP with good rate function

$$I : m \in M \to I(m) = H_m(m) \in [0, \infty]$$

As we shall see later, this integral transfer lemma is a natural tool for studying the LDP of mean field interacting particle models. It has been applied with success in [109, 110] to continuous time and McKean Vlasov type particle models. It will also be central in the proof of Theorem 10.1.1 provided in Section 10.7.

Before getting into the strategy of proof of this lemma, it is convenient to better connect this result with Theorem 10.4.1. To this end, we first observe that condition (10.26) is met as soon as the functions $V_1(.,m)$ and $V_n(.,m)$ are bounded for some $(m,n) \in M \times (1,\infty)$. To see this claim, we note that

$$\int_{\Omega^N} \exp nN[S_N(x,m) + V_1(\pi_N(x),m)] \, d\mathbb{Q}^N_m(x)$$
$$= \int_{\Omega^N} \exp N[nV_1(\pi_N(x),m) - V_n((\pi_N(x),m))] \, d\mathbb{P}^N_n(x) \leq e^{\|V_1 - V_n/n\|N}$$

Also notice that when $S_N = 0$ are the null mappings, we have for any $u \in M$ and $N \geq 1$, $\frac{dP_1^N}{dQ_m^N}(u) = \exp N[V_1(u,m)]$. If $V_1(\cdot, m)$ is continuous and (10.26) holds, then the family of probability measures $P_1^N = \mathbb{P}_1^N \circ \pi_N^{-1}$ satisfies the LDP with rate function $I = H_m - V_1(., m)$. In the case where $V_1(\cdot, m)$ is continuous and (10.26) holds for all m, since $V_1(u, u) = 0$ we conclude that $I(u) = H_u(u)$.

The following technical proposition states the exponential tightness property and two key estimates needed for proving our result.

Proposition 10.4.1 *Under the assumptions of the integral lemma above, the sequence of probability measures P_1^N on M is exponentially tight. For any Borel subset $A \subset M$ and for any $1/n + 1/n' = 1$, $1 < n, n' < \infty$, and $m \in M$, we have*

$$P_1^N(A) \leq Q_m^N(A)^{1/n'} \; P_{m,n}^N(A)^{1/n} \; \exp\left[N\delta_n(m, A)\right] \quad (10.27)$$
$$Q_m^N(A) \leq P_1^N(A)^{1/n} \; P_{\alpha(n),m}^N(A)^{1/n'} \; \exp\left[N \; \delta_{\alpha(n)}(m, A)/n\right] \quad (10.28)$$

with $\alpha(n) = -n'/n$ and for any $\alpha \neq 0$

$$\delta_\alpha(m, A) = \sup_{u \in A} |V_1(u, m) - V_\alpha(u, m)/\alpha|$$

Lemma 10.4.1 is an almost direct consequence of this proposition.

Proof of Lemma 10.4.1: If we take in (10.27) the closure of the ball of radius ϵ and center $m \in M$, that is

$$A = \overline{B}(m, \epsilon) = \{u \in M \; : \; d(u, m) \leq \epsilon\}$$

we find that for any conjugate integers $1/n + 1/n' = 1$ with $1 < n, n' < \infty$

$$P_1^N(\overline{B}(m, \epsilon)) \leq Q_m^N(\overline{B}(m, \epsilon))^{1/n'} \; \exp\left[N\delta_n(m, \overline{B}(m, \epsilon))\right]$$

Recalling that $\{Q_m^N \; ; \; N \geq 1\}$ satisfies the LDP with a good rate function H_m, this implies that

$$\limsup_{N \to \infty} \frac{1}{N} \log P_1^N(\overline{B}(m, \epsilon)) \leq -\frac{1}{n'} H_m(\overline{B}(m, \epsilon)) + \delta_n(m, \overline{B}(m, \epsilon)) \quad (10.29)$$

Since H_m is a good rate function, by Proposition 10.2.3 we find that

$$I(m) = H_m(m) = \lim_{\epsilon \to 0} H_m(\overline{B}(m, \epsilon))$$

Since each mapping $V_n(., m) : M \to \mathbb{R}$ is continuous at the point m and $V_n(m, m) = 0$, by the definition of δ_n we also have that

$$\lim_{\epsilon \to 0} \delta_n(m, \overline{B}(m, \epsilon)) = 0$$

Taking first the limit $\epsilon \downarrow 0$ and then $n' \to 1$ in (10.29), we find that

$$\lim_{\epsilon \to 0} \limsup_{N \to \infty} \frac{1}{N} \log P_1^N(\overline{B}(m, \epsilon)) \leq -I(m) \qquad (10.30)$$

Now if we take in (10.28) the open ball

$$A = B(m, \epsilon) = \{u \in M \: : \: d(u, m) < \epsilon\}$$

we get

$$Q_m^N(B(m, \epsilon)) \leq P_1^N(B(m, \epsilon))^{1/n} \; \exp\left[N \; \delta_{\alpha(n)}(m, B(m, \epsilon))/n\right]$$

Our assumptions on Q_m^N imply that

$$-H_m(m) \leq -H_m(B(m, \epsilon)) \leq \liminf_{N \to \infty} \frac{1}{N} \log Q_m^N(B(m, \epsilon))$$

Arguing as above, this implies that

$$\begin{aligned} -I(m) &\leq \liminf_{N \to \infty} \frac{1}{N} \log Q_m^N(B(m, \epsilon)) \\ &\leq \frac{1}{n}\left[\liminf_{N \to \infty} \frac{1}{N} \log P_1^N(B(m, \epsilon)) + \delta_{\alpha(n)}(m, B(m, \epsilon))\right]\end{aligned}$$

Considering the limit $\epsilon \downarrow 0$, one obtains for any $n > 1$

$$-n \; I(m) \leq \lim_{\epsilon \to 0} \liminf_{N \to \infty} \frac{1}{N} \log P_1^N(B(m, \epsilon))$$

Letting $n \to 1$, we get from (10.30)

$$\lim_{\epsilon \to 0} \limsup_{N \to \infty} \frac{1}{N} \log P_1^N(\overline{B}(m, \epsilon)) \leq -I(m) \leq \lim_{\epsilon \to 0} \liminf_{N \to \infty} \frac{1}{N} \log P_1^N(B(m, \epsilon))$$

Since we clearly have

$$\lim_{\epsilon \to 0} \liminf_{N \to \infty} \frac{1}{N} \log P_1^N(B(m, \epsilon)) \leq \lim_{\epsilon \to 0} \limsup_{N \to \infty} \frac{1}{N} \log P_1^N(\overline{B}(m, \epsilon))$$

it follows that for any $m \in M$

$$\lim_{\epsilon \to 0} \limsup_{N \to \infty} \frac{1}{N} \log P_1^N(\overline{B}(m, \epsilon)) = (-I)(m) = \lim_{\epsilon \to 0} \liminf_{N \to \infty} \frac{1}{N} \log P_1^N(B(m, \epsilon))$$

By Proposition 10.2.4, we conclude that I is an l.s.c. function and it governs the weak LDP for P_1^N. Since the sequence P_1^N is exponentially tight, we recall that the weak LDP is equivalent to the full LDP and the proof of the lemma is now completed. ∎

We now come to the proof of the technical proposition.

10.4 Laplace-Varadhan's Integral Techniques

Proof of Proposition 10.4.1: Fixing $n > 1$ so that (10.26) holds and denoting the left-hand side of (10.26) by nc_n, we have, for N large enough,

$$\int \left(\frac{d\mathbb{P}_1^N}{d\mathbb{Q}_m^N}\right)^n d\mathbb{Q}_m^N = \int_{\Omega^N} \exp nN[S_N(x,m) + V_1(\pi_N(x), m)] \, d\mathbb{Q}_m^N(x)$$
$$\leq \exp(nc_n N) \qquad (10.31)$$

Since each probability Q_m^N is a tight measure on M and the sequence Q_m^N, $N \geq 1$, satisfies a full LDP, one concludes that Q_m^N is exponentially tight. For any $a < \infty$, there exists a compact set $K(m,a) \subset M$ such that

$$\limsup_{N \to \infty} \frac{1}{N} \log P_m^N(K^c(m,a)) < -a \qquad \text{with} \quad K^c(m,a) = M - K(m,a)$$

To prove that P_1^N is exponentially tight, we first note that

$$P_1^N(K_n^c(m,a)) = \mathbb{P}_1^N(1_{K_n^c(m,a)}(\pi_N(x)))$$

with

$$\frac{1}{n} + \frac{1}{n'} = 1, \quad \text{and} \quad K_n^c(m,a) = K^c(m, n'(c_n + a))$$

Thus, using Holder's inequality, we check that

$$P_1^N(K_n^c(m,a)) = \mathbb{Q}_m^N\left(\left(1_{K_n^c(m,a)} \circ \pi_N\right) \frac{d\mathbb{P}_1^N}{d\mathbb{Q}_m^N}\right)$$
$$\leq Q_m^N(K_n^c(m,a))^{1/n'} \mathbb{Q}_m^N\left(\left(\frac{d\mathbb{P}_1^N}{d\mathbb{Q}_m^N}\right)^n\right)^{\frac{1}{n}}$$
$$\leq Q_m^N(K_n^c(m,a))^{1/n'} \exp(c_n N)$$

Recalling (10.31), the estimate above implies that

$$\limsup_{N \to \infty} \frac{1}{N} \log P_1^N(K_n^c(m,a)) < -\frac{1}{n'}[n'(c_n + a)] + c_n = -a$$

This clearly ends the proof of the exponential tightness of the sequence P_1^N. In the same way, for any Borel subset $A \subset M$ and for any $1/n + 1/n' = 1$, $1 < n, n' < \infty$, and $m \in M$, we have

$$P_1^N(A) = \mathbb{Q}_m^N\left((1_A \circ \pi_N) \frac{d\mathbb{P}_1^N}{d\mathbb{Q}_m^N}\right)$$
$$\leq Q_m^N(A)^{1/n'} \mathbb{Q}_m^N\left((1_A \circ \pi_N) \left(\frac{d\mathbb{P}_1^N}{d\mathbb{Q}_m^N}\right)^n\right)^{1/n}$$
$$\leq Q_m^N(A)^{1/n'}$$
$$\times \mathbb{Q}_m^N\left((1_A \circ \pi_N) \exp(nN[S_N(x,m) + V_1(\pi_N(x), m)])\right)^{1/n}$$

358   10. Large-Deviation Principles

Since we have
$$\mathbb{Q}_m^N\left((1_A \circ \pi_N)\exp\left(nN[S_N(.,m) + V_1(\pi_N(.),m)]\right)\right)$$
$$= \mathbb{Q}_m^N\left((1_A \circ \pi_N)\,\exp\left(N[nS_N(.,m) + V_n(\pi_N(.),m)]\right)\right.$$
$$\left.\times\,\exp\left(-N[V_n(\pi_N(.),m) + nV_1(\pi_N(.),m)]\right)\right)$$
$$\leq P_{n,m}^N(A)\,\exp\left(N\sup_{u\in A}|nV_1(u,m) - V_n(u,m)|\right)$$
$$= P_{n,m}^N(A)\,\exp\left(nN\delta_n(m,A)\right)$$

we find that $P_1^N(A) \leq \mathbb{Q}_m^N(A)^{1/n'}\,P_{n,m}^N(A)^{1/n}\,e^{(N\delta_n(m,A))}$. This establishes (10.27). To prove (10.28), we first use the decomposition

$$1_A(\pi_N(x)) = \left[1_A(\pi_N(x))\,\exp\left(\frac{N}{n}[S_N(x,m) + V_1(\pi_N(x),m)]\right)\right]$$
$$\times \left[1_A(\pi_N(x))\,\exp\left(-\frac{N}{n}[S_N(x,m) + V_1(\pi_N(x),m)]\right)\right]$$

and Holder's inequality to prove that

$$Q_m^N(A)$$
$$\leq \mathbb{Q}_m^N\left(1_A \circ \pi_N\,\exp\left(N[S_N(.,m) + V_1(\pi_N(.),m)]\right)\right)^{1/n}$$
$$\times \mathbb{Q}_m^N\left(1_A \circ \pi_N\,\exp\left(-N\frac{n'}{n}[S_N(.,m) + V_1(\pi_N(.),m)]\right)\right)^{1/n'}$$
$$= P_1^N(A)^{1/n}$$
$$\times \mathbb{Q}_m^N\left(1_A \circ \pi_N\,\exp\left(-N\frac{n'}{n}[S_N(.,m) + V_1(\pi_N(.),m)]\right)\right)^{1/n'}$$
(10.32)

We finally observe that

$$\mathbb{Q}_m^N\left(1_A \circ \pi_N\,\exp\left(-N\frac{n'}{n}[S_N(.,m) + V_1(\pi_N(.),m)]\right)\right)$$
$$= \mathbb{Q}_m^N\left(1_A \circ \pi_N\,\exp\left(N\alpha(n)[S_N(.,m) + V_1(\pi_N(.),m)]\right)\right)$$
$$= \mathbb{Q}_m^N\left(1_A \circ \pi_N\,\exp\left(N[\alpha(n)\,S_N(.,m) + V_{\alpha(n)}(\pi_N(.),m)]\right)\right.$$
$$\left.\times\,\exp\left(N[\alpha(n)V_1(\pi_N(.),m) - V_{\alpha(n)}(\pi_N(.),m)]\right)\right)$$
$$\leq P_{\alpha(n),m}^N(A)\,\times\,\exp\left(N|\alpha(n)|\,\delta_{\alpha(n)}(m,A)\right)$$

and from (10.32) we obtain

$$Q_m^N(A) \leq P_1^N(A)^{1/n} \; P_{\alpha(n),m}^N(A)^{1/n'} \times \exp\left(N\delta_{\alpha(n)}(m,A)/n\right)$$

This establishes (10.28), and the proof of the proposition is now completed. ∎

10.5 Dawson-Gärtner Projective Limits Techniques

In this section, we present another powerful and natural method of lifting LDP on finite-dimensional spaces to infinite dimensional ones. This projective limit approach to LDP is due to D. Dawson and J. Gärtner [73, 74] and it has been further developed by A. de Acosta in a series of three articles [1, 2, 3]. As we shall see, this technique can be interpreted as an extended version of the contraction principle to projective limit spaces. The idea is the following.

Definition 10.5.1 *Let M be a given set and let*

$$\mathcal{M} = \{(M_U, p_U) \; : \; U \in \mathcal{U}\}$$

be a collection of topological spaces M_U and maps $p_U : M \to M_U$ indexed by a set $\mathcal{U}$. The projective limit topology of M determined by $\mathcal{M}$ is the topology generated by the collection of open sets $\{p_U^{-1}(A) \; : \; A \text{ open } \subset M_U\}$.

By a direct application of the contraction theorem (Theorem 10.2.1), we prove the following proposition.

Proposition 10.5.1 *Let M be a topological space equipped with the projective limit topology determined by a collection of topological spaces and maps (M_U, p_U) indexed by some set $U \in \mathcal{U}$.*

- *For any $U \in \mathcal{U}$, the p_U-image $H \circ p_U^{-1}$ of a good rate function H on M is a good rate function on M_U.*

- *Assume that a sequence of distributions Q^N satisfies the LDP on M with the good rate function H. Then, for any $U \in \mathcal{U}$, the p_U-image measures $Q^N \circ p_U^{-1}$ satisfy the LDP on the topological space M_U with good rate function $H \circ p_U^{-1}$.*

Definition 10.5.2 *A directed set is a preordered set $(\mathcal{U}, \leq)$ with the following property: For any pair $(U, V) \in \mathcal{U}^2$, there exists some $W \in \mathcal{U}$ such that $U \leq W$ and $V \leq W$.*

Definition 10.5.3 *Let $(\mathcal{U}, \leq)$ be a directed set and let $\mathcal{M} = \{M_U \; : \; U \in \mathcal{U}\}$ be a family of Hausdorff topological spaces indexed by $\mathcal{U}$. For each pair* of indexes $V \leq U$, assume that there are given a collection of continuous mappings $p_{U,V} : M_U \to M_V$ with $p_{U,U} = Id$ satisfying the compatibility conditions

$$U \geq V \geq W \Longrightarrow p_{U,W} = p_{U,V} \circ p_{V,W}$$

Then the family $(M_U, p_{U,V})_{U \geq V}$ is called a projective (or inverse) spectrum of $\mathcal{U}$ with spaces M_U and connecting maps $p_{U,V}$.

The fact that the maps $p_{U,V}$ go in the opposite direction of the order is clearly mathematically irrelevant. Furthermore, since we have assumed that $p_{U,U} = Id$, "identifying" U with M_U, the set $(U, p_{U,V})_{U \geq V}$ is itself the projective (or inverse) spectrum of $\mathcal{U}$. We have done these two choices for later convenience. We denote by p_U the canonical projection from $\prod_{U \in \mathcal{U}} M_U$ into M_U.

Definition 10.5.4 *Given a projective spectrum $(M_U, p_{U,V})_{U \geq V}$ of $\mathcal{U}$, we introduce the product space $\prod_{U \in \mathcal{U}} M_U$ and, for each U, let p_U be its projection onto the U-factor M_U. The subspace*

$$\lim_{\mathcal{U}} \mathcal{M} = \{m \in \prod_{U \in \mathcal{U}} M_U \ : \ \forall U \geq V \ \ p_V(m) = p_{U,V}(p_U(m))\}$$

is called the projective (or inverse) limit space of the spectrum.

The product space $\prod_{U \in \mathcal{U}} M_U$ is as usual equipped with the product topology; that is, the weakest topology such that the projections p_U are continuous. These maps are clearly continuous, and the Hausdorff property of the spaces M_U, $U \in \mathcal{U}$, is clearly transferred to $\lim_{\mathcal{U}} \mathcal{M}$. The Hausdorff property and the choice of a directed set $(\mathcal{U}, \leq)$ are also not innocent. They ensure that $\lim_{\mathcal{U}} \mathcal{M}$ is a closed subset in $\prod_{U \in \mathcal{U}} M_U$, and the relative topology on $\lim_{\mathcal{U}} \mathcal{M}$ is generated by the sets

$$\{p_U^{-1}(A) \ : \ M_U \supset A \text{ open}, \ U \in \mathcal{U}\} \tag{10.33}$$

(see for instance Theorems 2.3 and 2.4 in Appendix 2, Section 2 in [126]). Notice that for any collection of closed sets F_U, $U \in \mathcal{U}$, such that $p_{U,V}(F_U) = F_V$, as soon as $U \geq V$, the set $F = \cap_{U \in \mathcal{U}} p_U^{-1}(F_U)$ is closed since it coincides with the projective limit of the spectrum $(F_U, p_{U,V}^F)$, where $p_{U,V}^F : F_U \to F_V$ stands for the restriction of the mapping $p_{U,V}$ to F_U. Since by Tychonoff's theorem a product of compact space is compact, we finally note that $\lim_{\mathcal{U}} \mathcal{M}$ is compact as soon as all the sets M_U, $U \in \mathcal{U}$, are compact. We shall always assume that a projective limit set $M = \lim_{\mathcal{U}} \mathcal{M}$ is equipped with a σ-field $\sigma(M)$ such that $\sigma(M) \supset \cup_{U \in \mathcal{U}} p_U^{-1}(\mathcal{B}(M_U))$. The next theorem is a slight modification of a theorem of Dawson and Gärtner.

Theorem 10.5.1 (Dawson-Gärtner) *Let $M = \lim_{\mathcal{U}} \mathcal{M}$ be the projective limit of the spectrum $(M_U, p_{U,V})_{U \geq V}$ of a directed set $\mathcal{U}$, and let H be a given function from M into $[0, \infty]$. Then H is a good rate function on M if* and only if there exists a collection of good rate functions I_U on each space M_U with $U \in \mathcal{U}$ and such that

$$H = \sup_{U \in \mathcal{U}} (I_U \circ p_U) \qquad (10.34)$$

A sequence of probability measures Q^N on M satisfies the LDP with some good rate function H if and only if the sequence of all image measures $Q^N \circ p_U^{-1}$ satisfies the LDP for some good rate functions I_U with $U \in \mathcal{U}$. In each situation, the rate functions are respectively given by (10.34) and $I_U = H \circ p_U^{-1}$.

Proof:
Suppose that H is a good rate function on $\lim_{\mathcal{U}} \mathcal{M}$. By a contraction argument, we see that the image functions $I_U = H \circ p_U^{-1}$ are good rate functions on M_U. Using (10.17) and (10.33), we readily prove that

$$\begin{aligned} H(m) &= \sup_{U \in \mathcal{U}} \sup \{H(p_U^{-1}(A)) \ : \ p_U(m) \in A, \text{ open} \in M_U\} \\ &= \sup_{U \in \mathcal{U}} I_U(p_U(m)) \qquad (10.35) \end{aligned}$$

In the reverse angle, suppose that I_U are good rate functions on M_U, and let H be the pointwise supremum of the maps $I_U \circ p_U$. Since each of these maps is l.s.c., the function H is l.s.c. To prove that H has compact level sets, again by the contraction theorem (Theorem 10.2.1), we notice that they have to satisfy the compatibility conditions

$$U \geq V \implies I_V = I_U \circ p_{U,V}^{-1} \quad \text{and} \quad I_V^{-1}([0,a]) = p_{U,V}[I_U^{-1}([0,a])]$$

for any $a \in [0, \infty)$. Since projective limits are inherited by closed sets, this implies that $H^{-1}([0,a]) = (\lim_{\mathcal{U}} \mathcal{M}) \cap \prod_{U \in \mathcal{U}} I_U^{-1}([0,a])$ is the projective limit of the compact level sets $I_U^{-1}([0,a])$, $U \in \mathcal{U}$. By Tychonov's theorem, we conclude that the level set $H^{-1}([0,a])$ is a compact subset of $\lim_{\mathcal{U}} \mathcal{M}$. This proves that H is a good rate function as soon as each I_U is a good rate function.

Suppose that the sequence of measures $Q^N \circ p_U^{-1}$ satisfies the LDP upper bound with the good rate functions I_U. Since the closure $\overline{A}$ of a measurable set $A \in \sigma(M)$ coincides with the projective limit of the closures $\overline{A}_U$ of the sets $A_U = p_U(\overline{A})$, previous considerations imply that $\overline{A} \cap H^{-1}([0,a])$ is the projective limit of the compact sets $\overline{A}_U \cap I_U^{-1}([0,a])$ for any $a \in [0, \infty)$. Now if $H(\overline{A}) = 0$, the result is trivial. Otherwise there exists some $a < \infty$ such that $a < H(\overline{A})$. Recalling that the projective limit of a collection of nonempty compact sets is nonempty, for any $a < H(\overline{A})$ there exists some $U \in \mathcal{U}$ such that $\overline{A}_U \cap I_U^{-1}([0,a]) = \emptyset$ (otherwise we would get a contraction). Since $A \subset p_U^{-1}(\overline{A}_U)$, applying the LDP upper bound for $Q^N \circ$

p_U^{-1}, we conclude that

$$\limsup_{N\to\infty} \frac{1}{N} \log Q^N(A) \leq \limsup_{N\to\infty} \frac{1}{N} \log Q^N \circ p_U^{-1}(\overline{A}_U)$$
$$\leq -I_U(\overline{A}_U) \leq -a$$

We complete the proof by taking the infimum of $(-a)$ over all $a < H(\overline{A})$. Suppose that $Q^N \circ p_U^{-1}$ satisfy the LDP lower bound with the good rate function I_U. For any $A \in \sigma(M)$ and $m \in A$, there exists an open set $B_U \in M_U$ for some $U \in \mathcal{U}$ such that $m \in p_U^{-1}(B_U) \subset \overset{\circ}{A} \subset A$. Applying the LDP lower bound for $Q^N \circ p_U^{-1}$, we find that

$$\liminf_{N\to\infty} \frac{1}{N} \log Q^N(A) \geq \liminf_{N\to\infty} \frac{1}{N} \log Q^N \circ p_U^{-1}(B_U)$$
$$\geq -I_U(B_U) \geq -I_U(m)$$

We readily conclude that Q^N satisfies the LDP on M with rate function $H(m) = \sup_{U\in\mathcal{U}} I_U(p_U(m))$. In the reverse situation, suppose that Q^N satisfies the LDP on M for some good rate function. By the contraction theorem (Theorem 10.2.1), the image measures $Q^N \circ p_U^{-1}$ satisfy the LDP on M_U with good rate function $I_U = H \circ p_U^{-1}$. From previous considerations, this implies that Q^N also satisfies the LDP on M with the rate function

$$\widetilde{H}(m) = \sup_{U\in\mathcal{U}} I_U(p_U(m))$$

From previous arguments or by the uniqueness of the rate function (see Proposition 10.2.2), we conclude that $\widetilde{H} = H$. This ends the proof of the theorem. ∎

We end this section with a more or less well-known min-max type theorem for good rate functions on projective limit spaces.

Theorem 10.5.2 *Let H be a good rate function on the projective limit $M = \lim_{\mathcal{U}} M$ of the spectrum $(M_U, p_{U,V})_{U\geq V}$ of a directed set $\mathcal{U}$. We let $I_U = H \circ p_U^{-1}$ be the p_U-image function of H on M_U. For any closed set $F \in \sigma(M)$, we have*

$$H(F) = \sup_{u\in\mathcal{U}} I_U(\overline{p_U(F)}) \tag{10.36}$$

Proof:
By (10.34), we first note that (10.36) can be rewritten as

$$H(F) = \inf_{m\in F} \sup_{u\in\mathcal{U}} I_U(p_U(m)) = \sup_{u\in\mathcal{U}} \inf_{m\in \overline{p_U(F)}} I_U(m) =_{\text{def.}} \widetilde{H}(F)$$

Since we have $I_U(p_U(m)) \geq I_U(\overline{p_U(F)})$ for any $m \in F$, we deduce that $H(F) \geq \widetilde{H}(F)$. To prove the reverse inequality, we assume that $\widetilde{H}(F) < \infty$;

otherwise the desired bound trivially holds true. In this case, for each $\varepsilon > 0$, the sets
$$p_U^{-1}(\overline{p_U(F)}) \cap \{H \leq \tilde{H}(F) + \varepsilon\}$$
are nonempty compact sets. The compactness results from the fact that $p_U^{-1}(\overline{p_U(F)})$ is a closed set into the compact level sets $\{H \leq \tilde{H}(F) + \varepsilon\}$. On the other hand, if these sets were empty, we would be able to find some $m \in p_U^{-1}(\overline{p_U(F)})$ such that
$$H(m) \geq H(p_U^{-1}(\overline{p_U(F)})) = I_U(\overline{p_U(F)}) \geq \tilde{H}(F) + \varepsilon = \sup_{V \in \mathcal{U}} I_V(\overline{p_V(F)}) + \varepsilon$$

This clearly yields a contradiction. Consequently, for each $\varepsilon > 0$, we have
$$[\cap_{U \in \mathcal{U}} p_U^{-1}(\overline{p_U(F)})] \cap \{H \leq \tilde{H}(F) + \varepsilon\} \neq \emptyset$$

Since F coincides with the projective limit of the sets $(\overline{p_U(F)})_{U \in \mathcal{U}}$, we have $F = \cap_{U \in \mathcal{U}} p_U^{-1}(\overline{p_U(F)})$. From previous considerations, for any $\varepsilon > 0$ there exists some point $m_\varepsilon \in F$ such that
$$H(F) \leq H(m_\varepsilon) \leq \tilde{H}(F) + \varepsilon$$

We end the proof of the theorem by letting $\varepsilon \to 0$. ∎

10.6 Sanov's Theorem

10.6.1 Introduction

Sanov's theorem is probably one of the main startling results of large deviations and Monte Carlo approximation theory. This result provides sharp asymptotic exponential rates for the convergence of the occupation measures associated with a collection of independent random variables towards the limiting sampling distribution.

The original proof of Sanov [281] assumes that the underlying random variables take values in $\mathbb{R}$. Since this pioneering article, several extensions have been presented. In the book of Dembo and Zeitouni [112] the reader will find at least three different ways to prove this theorem in the context of random sequences taking values in Polish state spaces. Most of these strategies consist in deriving Sanov's theorem as a consequence of more general LDP such as Theorem 10.3.1.

In this section, we take a different perspective. We simplify the analysis and, speaking somewhat loosely, we show that proving the strong version of Sanov's theorem in Hausdorff topological spaces is in fact equivalent to proving the corresponding statement in finite spaces. This original approach is conducted applying the Dawson-Gärtner contraction principle to

364   10. Large-Deviation Principles

a judicious and natural projective interpretation of the strong topology on the set of probability measures. Using these simplifications, the LDP upper bounds will be easily derived using the generalized Cramér method presented in Section 10.3. The proof of the LDP lower bounds is a little more delicate. It is conducted using an elegant approximation technique essentially due to Groeneboom, Oosterhoff and Ruyggaart [170].

The projective limit approach to LDP for iid sequences in the τ-topology can be conducted in various ways, depending on the projective interpretation of the τ-topology. Our strategy is based on a projective interpretation of set-additive and $[0,1]$-valued functions with respect to the class of finite partitions directed upwards by inclusion. In this interpretation, the LDP in the τ-topology (for Hausdorff topological spaces) is essentially obtained from Sanov's theorem on finite state spaces. We shall see that this state-space enlargement can be interpreted as a projective compactification of the set of probability measures. Another projective interpretation of the τ-topology without enlarging the distribution space can be derived using the class of finite subsets of bounded measurable functions directed upwards by inclusion. This alternative approach to LDP was developed by A. de Acosta in [1] (see also [112]).

10.6.2 Topological Preliminaries

In the further development of this section, E denotes an Hausdorff topological space equipped with a Borel sigma-field $\mathcal{E}$. We let $\mathbf{P}(E)$, be the set of additive set functions from $\mathcal{E}$ into $[0,1]$ and $\mathcal{P}(E) \subset \mathbf{P}(E)$ be the subset of all probability measures on $(E, \mathcal{E})$. We equip $\mathbf{P}(E)$ with the τ_1-topology of setwise convergence. More precisely, a sequence of set functions $(\mu_n)_{n \geq 0} \in \mathbf{P}(E)^{\mathbb{N}}$ τ_1-converges to some $\mu \in \mathbf{P}(E)$, as $n \to \infty$, if and only if
$$\lim_{n \to \infty} \mu_n(A) = \mu(A)$$
for all $A \in \mathcal{E}$. It is readily seen that the τ-topology of convergence on all Borel sets of E is the corresponding relative topology induced on $\mathcal{P}(E)$ by $\mathbf{P}(E)$.

Let $\mathcal{U}$ be the set of all finite and Borel partitions of E. Since each $U \in \mathcal{U}$ is a finite partition, the σ-algebra generated by U and denoted by $\sigma(U)$ is the finite set formed by $\emptyset$, E, and the sets that are unions of elements of U. We slightly abuse the notation and denote by $\mathcal{P}(U)$ and $\mathcal{B}_b(U)$ the set of probability measures on $(E, \sigma(U))$ and the Banach space of all bounded and $\sigma(U)$-measurable functions on E (equipped with the uniform norm).

Definition 10.6.1 *We associate with each $U \in \mathcal{U}$ the Kolmogorov-Smirnov metric d_U and the U-relative entropy $\mathrm{Ent}_U(.|.)$ on $\mathcal{P}(U)$ defined for any pair $(\mu, \nu) \in \mathcal{P}(U)^2$ by the formulae*

$$d_U(\mu,\nu) = 2^{-1} \sum_{1 \leq i \leq d} |\mu(U^i) - \nu(U^i)|$$

$$\mathrm{Ent}_U(\mu|\nu) = \sum_{i=1}^{d} \mu(U^i) \log(\mu(U^i)/\nu(U^i)) \qquad (10.37)$$

with the convention $0 \log 0 = 0 = 0 \log(0/0)$, and $\mathrm{Ent}_U(\mu|\nu) = \infty$ as soon as $\mu \not\ll \nu$.

As the reader may have noticed, $(\mathcal{P}(U), d_U)$ is a compact metric space. To be more precise, we note that for any d-finite partition $U = (U^i)_{1 \leq i \leq d} \in \mathcal{U}$ the mapping $\mu \in \mathcal{P}(U) \to (\mu(U^i))_{1 \leq i \leq d} \in \mathcal{S}(d)$ is clearly an homeomorphism between $\mathcal{P}(U)$ and the compact $(d-1)$-dimensional simplex $\mathcal{S}(d) = \{\alpha \in [0,1]^d : \sum_{i=1}^{d} \alpha(i) = 1\} \subset [0,1]^d$. Since the identity mappings

$$e_U : u \in (E, \mathcal{E}) \to e_U(u) = u \in (E, \sigma(U))$$

are measurable, the set $\mathcal{P}(U)$ can alternatively be regarded as the set formed by all e_U-images of set functions in $\mathbf{P}(E)$. More precisely, we have $\mathcal{P}(U) = q_U(\mathbf{P}(E))$ with the continuous projection operators

$$q_U : \mu \in \mathbf{P}(E) \to q_U(\mu) = \mu \circ e_U^{-1} \in \mathcal{P}(U)$$

To clarify the presentation, for any pair of set functions $(\mu, \nu) \in \mathbf{P}(E)^2$, sometimes we simplify the notation and we write $d_U(\mu, \nu)$ and $\mathrm{Ent}_U(\mu|\nu)$ instead of $d_U(q_U(\mu), q_U(\nu))$ and $\mathrm{Ent}_U(q_U(\mu)|q_U(\nu))$.

Note that $\mathrm{Ent}_U(.|\nu)$ is finite and continuous on the compact sets $\{\eta \in \mathcal{P}(U) : \eta \ll q_U(\nu)\}$ and $\mathrm{Ent}_U(\mu|.)$ is continuous for every fixed μ. We later use, let us quote a technical lemma on the Lipschitz property of the mapping $\mathrm{Ent}_U(\mu|.)$.

Lemma 10.6.1 *Let $U = (U^i)_{1 \leq i \leq d} \in \mathcal{U}$, $\lambda \in \mathcal{P}(U)$ and let $\rho > 0$ be a given parameter. Also let $\mu, \eta, \eta' \in \mathcal{P}(U)$ be such that*

$$\mu \ll \lambda, \quad \rho\, \lambda \leq \eta \ll \lambda, \quad \text{and} \quad \rho\, \lambda \leq \eta' \ll \lambda$$

Then we have the uniform d_U-Lipschitz inequality

$$|\mathrm{Ent}_U(\mu|\eta) - \mathrm{Ent}_U(\mu|\eta')| \leq \frac{2}{\rho \lambda^\star(U)}\, d_U(\eta, \eta')$$

with the positive constant $\lambda^\star(U) = \wedge_{i:\lambda(U_i)>0} \lambda(U^i) > 0$.

Proof:
Since $\lambda \ll \eta$ and $\lambda \ll \eta'$ we have $\mu \ll \eta$ and $\mu \ll \eta'$, from which we conclude that the two entropies in the display above are finite. We also note that

$$\text{Ent}_U(\mu|\eta) - \text{Ent}_U(\mu|\eta') = \sum_{i:\mu(U^i)>0} \mu(U^i) \log\left(\eta'(U^i)/\eta(U^i)\right)$$

Using the elementary inequality $|\log x - \log y| \leq |x-y|/(x \wedge y)$, which is valid for any $x, y > 0$, and recalling that $\mu \ll \lambda$ we find that

$$|\text{Ent}_U(\mu|\eta) - \text{Ent}_U(\mu|\eta')| \leq \frac{1}{\rho} \sum_{i:\mu(U^i)>0} \mu(U^i) \left|\frac{\eta(U^i)}{\lambda(U^i)} - \frac{\eta'(U^i)}{\lambda(U^i)}\right|$$

The end of the proof is now clear. ∎

Definition 10.6.2 *For any $\eta \in \mathcal{P}(E)$, we also denote by $H(\,\cdot\,|\eta)$ the relative entropy criterion on $\mathbf{P}(E)$ defined by*

$$H(\,\cdot\,|\eta) : \mu \in \mathbf{P}(E) \longrightarrow H(\mu|\eta) = \sup_{U \in \mathcal{U}} \text{Ent}_U(\mu|\eta) \in [0, \infty]$$

We say that a partition U is finer than another V and we write $U \geq V$ as soon as $\sigma(U) \supset \sigma(V)$. Note that $(U^i)_{1 \leq i \leq d} \geq (V^i)_{1 \leq i \leq r}$ if and only if there exists an r-partition $(a_i)_{i \leq i \leq r}$ of the set of indexes $\{1, \ldots, d\}$ such that $V^i = \cup_{j \in a_i} U^j$ for any $i \leq r$. Using the well-known variational formula of the relative entropy on $\mathcal{P}(U)$ we have

$$\text{Ent}_U(\nu|\eta) = \sup_{f \in \mathcal{B}_b(U)} (\nu(f) - \log \eta(\exp f)) \qquad (10.38)$$

Hence it is clear that for any $\mu, \nu \in \mathbf{P}(E)$

$$U \geq V \Rightarrow \text{Ent}_V(\mu|\nu) \leq \text{Ent}_U(\mu|\nu) \quad \text{and} \quad d_V(\mu,\nu) \leq d_U(\mu,\nu) \qquad (10.39)$$

We associate with any pair of partitions $U, V \in \mathcal{U}$ the smallest partition $(U \vee V)$ such that $\sigma(U) \vee \sigma(V) = \sigma(U \vee V)$. This partition is simply defined by setting

$$U \vee V = \{A \cap B \ : \ (A, B) \in (U \times V)\}$$

Since for any $A \in U$ we trivially have $A = A \cap E = \cup_{B \in V}(A \cap B)$, we conclude that $(U \vee V) \geq U$ and by symmetry arguments $(U \vee V) \geq V$. From these observations, we conclude that $(\mathcal{U}, \geq)$ is a directed set. A natural projective spectrum of $\mathcal{U}$ is defined as follows: For any $U \geq V$, we observe that the identity mappings

$$e_{U,V} : x \in (E, \sigma(U)) \to e_{U,V}(x) = x \in (E, \sigma(V))$$

are measurable and we have $\mathcal{P}(V) = p_{U,V}(\mathcal{P}(U))$ with the projection operators

$$p_{U,V} : \mu \in \mathcal{P}(U) \to p_{U,V}(\mu) = \mu \circ e_{U,V}^{-1} \in \mathcal{P}(V)$$

Using (10.39), we find that

$$U \geq V \Rightarrow \forall (\mu, \nu) \in \mathcal{P}(U) \quad d_V(p_{U,V}(\mu), p_{U,V}(\nu)) \leq d_U(\mu, \nu)$$

from which we conclude that the connecting mappings $p_{U,V}$ are Lipschitz-continuous.

Proposition 10.6.1 *The set $\mathcal{P} = ((\mathcal{P}(U), d_U), p_{U,V})_{U \geq V}$ forms a projective inverse spectrum of $\mathcal{U}$ with compact metric spaces $(\mathcal{P}(U), d_U)$ and connecting maps $p_{U,V}$. Let $h : \lim_{\mathcal{U}} \mathcal{P} \to \mathbf{P}(E)$ be the mapping that associates to a point $\mu = (\mu^U)_{U \in \mathcal{U}} \in \lim_{\mathcal{U}} \mathcal{P}$ the set function $h(\mu) \in \mathbf{P}(E)$ defined for any $A \in \mathcal{E}$ by*

$$h(\mu)(A) = \mu^U(A)$$

where $U \in \mathcal{U}$ is some finite partition of E such that $A \in \sigma(U)$. Then h is a homeomorphism between the compact spaces $\lim_{\mathcal{U}} \mathcal{P}$ and $\mathbf{P}(E)$. In addition, is inverse mapping h^{-1} is given for any $\mu \in \mathbf{P}(E)$ by $h^{-1}(\mu) = (q_U(\mu))_{U \in \mathcal{U}}$.

Proof:
To see that h is well-defined, we first need to check that $\mu^U(A) = \mu^V(A)$ for any pair of partitions U, V such that $A \in \sigma(U)$ and $A \in \sigma(V)$. This assertion is easily proved by noting that

$$A \in (\sigma(U) \cap \sigma(V)) \Longrightarrow A \in \sigma(U \vee V)$$

and by the compatibility conditions in the definition of $\lim_{\mathcal{U}} \mathcal{P}$ we have

$$e_{(U \vee V),U}^{-1}(A) = A = e_{(U \vee V),V}^{-1}(A) \Rightarrow \mu^{(U \vee V)}(A) = \mu^U(A) = \mu^V(A)$$

To prove that $h(\mu)$ is an additive set function on $\mathcal{E}$, we choose a pair of disjoint Borel sets A, B and a pair U, V of partitions with $A \in U$ and $B \in V$. Since A and B are disjoint, we have

$$\begin{aligned} A &= A \cap (E - B) = A \cap (\cup_{C \in V, C \neq B} C) \\ &= \cup_{C \in V, C \neq B}(A \cap C) \in \sigma(U \vee V) \end{aligned}$$

and by symmetry arguments $B \in \sigma(U \vee V)$. Since $\mu^{(U \vee V)} \in \mathcal{P}(U \vee V)$, this implies that

$$\begin{aligned} h(\mu)(A \cup B) &= \mu^{(U \vee V)}(A \cup B) = \mu^{(U \vee V)}(A) + \mu^{(U \vee V)}(B) \\ &= h(\mu)(A) + h(\mu)(B) \end{aligned}$$

Let us prove that h is an injection. Let $\mu, \nu \in \lim_{\mathcal{U}} \mathcal{P}$ be a pair of points such that $h(\mu) = h(\nu)$. By the definition of h, we find that $\mu^U = \nu^U$ for any $U \in \mathcal{U}$, from which we conclude that $\mu = \nu$. On the other hand, for any $\mu \in \mathbf{P}(E)$, we have $(q_U(\mu))_{U \in \mathcal{U}} \in \lim_{\mathcal{U}} \mathcal{P}$ and

$$h((q_U(\mu))_{U \in \mathcal{U}})(A) = q_U(\mu)(A) = \mu \circ e_U^{-1}(A) = \mu(A)$$

for any $A \in \sigma(U)$ for some $U \in \mathcal{U}$. We conclude that h is a bijective map from $\lim_{\mathcal{U}} \mathcal{P}$ into $\mathbf{P}(E)$ and $h^{-1}(\mu) = (q_U(\mu))_{U \in \mathcal{U}}$.

It remains to prove that h and h^{-1} are continuous. To prove this final step, we observe that for any sequence $\mu_n = (\mu_n^U)_{U \in \mathcal{U}}$ of points in $\lim_{\mathcal{U}} \mathcal{P}$ and $\mu = (\mu^U)_{U \in \mathcal{U}} \in \lim_{\mathcal{U}} \mathcal{P}$, we have the following series of equivalent assertions

$$\lim_{n \to \infty} \mu_n = \mu \text{ in } \lim_{\mathcal{U}} \mathcal{P} \Leftrightarrow \forall U \in \mathcal{U} \quad \lim_{n \to \infty} \mu_n^U = \mu^U \text{ in } (\mathcal{P}(U), d_U)$$
$$\Leftrightarrow \forall (A, U) \in (\mathcal{E} \times \mathcal{U}) \text{ s.t. } A \in \sigma(U)$$
$$\lim_{n \to \infty} \mu_n^U(A) = \mu^U(A)$$
$$\Leftrightarrow \forall A \in \mathcal{E} \quad \lim_{n \to \infty} h(\mu_n)(A) = h(\mu)(A)$$
$$\Leftrightarrow \lim_{n \to \infty} h(\mu_n) = h(\mu) \text{ in } \mathbf{P}(E)$$

This ends the proof of the proposition. ∎

The next lemma provides a representation of the relative entropy on $\mathcal{P}(E)$ in terms of the relative entropies on the compact sets $\mathcal{P}(U)$, $U \in \mathcal{U}$. As we shall see, this characterization is a particular case of the formula (10.34) presented in Theorem 10.5.1.

Lemma 10.6.2 *The domain $D_{H(\cdot|\eta)} = \{\mu \in \mathbf{P}(E) : H(\mu|\eta) < \infty\}$ of $H(\cdot|\eta)$, $\eta \in \mathcal{P}(E)$ is included in $\mathcal{P}(E)$, and for any $\mu \in \mathcal{P}(E)$, we have*

$$H(\mu|\eta) = \mathrm{Ent}(\mu|\eta) \tag{10.40}$$

Proof:
Formula (10.40) is well-known (see for instance Pinsker [267]). By the variational formula (10.38), it suffices to check that $\cup_{U \in \mathcal{U}} \mathcal{B}_b(U)$ is a dense subset of $\mathcal{B}_b(E)$. Note for instance that for any $f \in \mathcal{B}_b(E)$ with $0 \leq f(x) \leq 1$ we have $\|f - f_n\| \leq \frac{1}{n}$ with

$$f_n = \sum_{i=0}^{n} \frac{i}{n} 1_{f^{-1}([i/n, (i+1)/n))} \in \mathcal{B}(U_n(f))$$

and $U_n(f) = (f^{-1}([i/n, (i+1)/n)))_{0 \leq i \leq n} \in \mathcal{U}$. Extending this observation to any $f \in \mathcal{B}_b(E)$, we prove that $\cup_{U \in \mathcal{U}} \mathcal{B}_b(U)$ is a dense subset of $\mathcal{B}_b(E)$ and we conclude that

$$\sup_{U \in \mathcal{U}} \mathrm{Ent}_U(\nu|\eta) = \sup_{U \in \mathcal{U}} \sup_{f \in \mathcal{B}_b(U)} (\nu(f) - \log \eta(\exp f)) = \mathrm{Ent}(\nu|\eta)$$

This ends the proof of (10.40). To prove that $D_{H,\eta} \subset \mathcal{P}(E)$, we observe that

$$H(\mu|\eta) < \infty \iff \exists c < \infty : \forall U \in \mathcal{U} \quad \mathrm{Ent}_U(\mu|\eta) \leq c$$

To take the final step, we first check that the set of measures

$$\{\nu \in \mathcal{P}(U) : \mathrm{Ent}_U(\nu|\eta) \leq c\}$$

is uniformly absolutely continuous with respect to $q_U(\eta)$ in the sense that for any $\varepsilon > 0$ there exists some $\delta > 0$ such that for any $B \in \sigma(U)$ and ν such that $\mathrm{Ent}_U(\nu|\eta) \leq c$ we have that $\nu(B) < \varepsilon$ as soon as $\eta(B) < \delta$. To prove this claim, we use the fact that $u \log u \geq -1/e$ to check that

$$\begin{aligned}
\nu(B) &= \int_B 1_{\frac{d\nu}{dq_U(\eta)} \leq \frac{\varepsilon}{2\delta}} \frac{d\nu}{dq_U(\eta)} \, dq_U(\eta) + \int_B 1_{\frac{d\nu}{dq_U(\eta)} > \frac{\varepsilon}{2\delta}} d\nu \\
&\leq \varepsilon/2 + \frac{1}{\log\left(\frac{\varepsilon}{2\delta}\right)} \int_B 1_{\frac{d\nu}{dq_U(\eta)} > \frac{\varepsilon}{2\delta}} \log\left(\frac{d\nu}{dq_U(\eta)}\right) d\nu \\
&\leq \varepsilon/2 + (\mathrm{Ent}_U(\nu|\eta) + 1/e)/\log\left(\varepsilon/2\delta\right) \leq \varepsilon
\end{aligned}$$

as soon as $(c + 1/e)/\log\left(\varepsilon/2\delta\right) \leq \varepsilon/2$. Whenever $H(\mu|\eta) \leq c < \infty$, we have

$$q_U(\mu) \in \{\nu \in \mathcal{P}(U) : \mathrm{Ent}_U(\nu|\eta) \leq c\}$$

for any $U \in \mathcal{U}$. The result above readily implies that if $H(\mu|\eta) \leq c$, then for any $\varepsilon > 0$ there exists some $\delta > 0$ such that for any $B \in \mathcal{E}$ we have

$$\eta(B) < \delta \implies \mu(B) < \varepsilon \qquad (10.41)$$

Let $(B_n)_{n \geq 1}$ be a sequence of disjoint Borel sets. Since $\eta \in \mathcal{P}(E)$ is a σ-additive measure, for any $\delta > 0$ there exists some $p \geq 1$ such that

$$\eta(\cup_{k=p}^\infty B_k) = \sum_{k=p}^\infty \eta(B_k) \leq \delta$$

If we take $B = \cup_{k=p}^\infty B_k$ in (10.41), then we find that

$$0 \leq \mu(\cup_{k=1}^\infty B_k) - \sum_{k=1}^{p-1} \mu(B_k) = \mu(B) < \varepsilon$$

from which we conclude that μ is also σ-additive and $\mu \in \mathcal{P}(E)$. This ends the proof of the lemma. ∎

10.6.3 Sanov's Theorem in the τ-Topology

The next technical lemma provides large deviations probabilities for independent and identically distributed sequences on finite state space models.

Lemma 10.6.3 *Let S be a finite state space, and let $Y = (Y^i)_{1 \le i \le N}$ be a collection of N independent and S-valued random variables, identically distributed according to a measure $\eta \in \mathcal{P}(S)$. We denote by m the mapping from the product space S^N into $\mathcal{P}(S)$ that associates to each configuration $x = (x^i)_{1 \le i \le N} \in S^N$, the empirical measure $m(x) = \frac{1}{N} \sum_{i=1}^N \delta_{x^i}$. For any $\mu \in m(S^N)$ we have*

$$(N+1)^{-|S|} \le \exp\{N \operatorname{Ent}(\mu|\eta)\} \, \mathbb{P}_\eta(m(Y) = \mu) \le 1$$

Proof:
This result is rather well-known, see for instance lemma 2.1.9 p. 15 in [112]. Its proof is rather elementary. We first notice that, for any $y \in m^{-1}(\mu)$ with $\mu \in m(S^N)$, we have

$$\mathbb{P}_\eta(Y = y) = \prod_{u \in S} \eta(u)^{N\mu(u)} = \exp N \sum_{u \in S} \mu(u) \log \eta(u)$$

This yields that

$$\begin{aligned}\mathbb{P}_\eta(m(Y) = \mu) &= \mathbb{P}_\eta(Y \in m^{-1}(\mu)) \\ &= |m^{-1}(\mu)| \, \exp N \sum_{u \in S} \mu(u) \log \eta(u) \quad (10.42)\end{aligned}$$

and $|m^{-1}(\mu)| = N!/(\prod_{u \in S}(N\mu(u))!)$. Since $\mathbb{P}_\mu(m(Y) = \mu) \le 1$ we find that

$$|m^{-1}(\mu)| \le \exp -N \sum_{u \in S} \mu(u) \log \mu(u)$$

we conclude that $\mathbb{P}_\eta(m(Y) = \mu) \le \exp\{-N \operatorname{Ent}(\mu|\eta)\}$. On the other hand, recalling that $\operatorname{Ent}(\mu|\eta) \ge 0$, we note that

$$\mathbb{P}_\eta(m(Y) = \mu) \le \mathbb{P}_\mu(m(Y) = \mu)$$

This yields that

$$1 = \mathbb{P}_\eta(m(Y) \in m(S^N)) \le |m(S^N)| \mathbb{P}_\mu(m(Y) = \mu)$$

Using the fact that any $m(y) \in m(S^N)$ can be rewritten as $m(y) = \frac{1}{N} \sum_{u \in S} |\{i : y^i = u\}| \, \delta_u$, we find that $|m(S^N)| \le (N+1)^{|S|}$, from which we conclude that

$$|m^{-1}(\mu)| \ge (N+1)^{-|S|} \, \exp -N \sum_{u \in S} \mu(u) \log \mu(u)$$

This, together with (10.42), ends the proof of the lemma. ∎

With these preliminaries taken care of, we are now in a position to state and prove our first main result.

Theorem 10.6.1 *Let $(X^i)_{i\geq 1}$ be a sequence of independent, E-valued random variables, identically distributed according to a measure $\eta \in \mathcal{P}(E)$. For any N, we denote by Q^N the law on $\mathcal{P}(E)$ of the N-empirical measures $\frac{1}{N}\sum_{i=1}^N \delta_{X^i}$. For any $U \in \mathcal{U}$, the sequence of distributions $Q^N \circ q_U^{-1}$ satisfies an LDP on $(\mathcal{P}(U), d_U)$ with the good entropy rate function*

$$I_U = \mathrm{Ent}_U(\,\cdot\,|\eta)$$

Proof:
Let $\mathcal{M}(U)$ be the set of all signed measures on $\sigma(U)$. Since each $U = (U^i)_{i\leq d} \in \mathcal{U}$ is a finite d-partition, we first observe that the sets $\mathcal{B}_b(U)$ and $\mathcal{M}(U)$ are homeomorphic to $\mathbb{R}^d$ and the elementary duality mapping

$$(\mu, f) \in (\mathcal{M}(U) \times \mathcal{B}_b(U)) \longrightarrow <f,\mu> = \int_E f(x)\,\mu(dx)$$

determines a representation of $\mathcal{M}(U)^\star$ as $\mathcal{B}_b(U)$. Let $\Lambda_U : \mathcal{B}_b(U) \to \mathbb{R}$ be the logarithmic moment-generating function defined for any $f \in \mathcal{B}_b(U)$ by

$$\begin{aligned}\Lambda_U(f) &= \frac{1}{N}\log \int e^{N<f,\mu>}(Q^N \circ q_U^{-1})(d\mu) = \frac{1}{N}\log \mathbb{E}(e^{Nm(X)(f)}) \\ &= \log \eta(\exp f) = \log q_U(\eta)(\exp f) < \infty\end{aligned}$$

where $m(X) = \frac{1}{N}\sum_{i=1}^N \delta_{X^i}$. The LDP upper bound follows from the fact that $\mathcal{P}(U)$ is compact and $\mathrm{Ent}_U(\,\cdot\,|\eta)$ is the Fenchel-Legendre transform of the function Λ_U; that is, we have that

$$\mathrm{Ent}_U(\nu|\eta) = \sup_{f \in \mathcal{B}_b(U)} (\nu(f) - \log \eta(\exp f))$$

for any $\nu \in \mathcal{P}(U)$. To prove the LDP lower bound, let $A \subset \mathcal{P}(U)$ be an open set such that $\mathrm{Ent}_U(A|\eta) < \infty$ (otherwise the proof of the lower bound is trivial). For any $\delta > 0$, there exists a point $\mu \in A$ such that

$$\mathrm{Ent}_U(\mu|\eta) \leq \mathrm{Ent}_U(A|\eta) + \delta \tag{10.43}$$

Since $\mu \in A$ and A is open in $(\mathcal{P}(U), d_U)$, there exists some $\varepsilon > 0$ such that

$$\mathcal{V}_U(\mu, \varepsilon) = \{\nu \in \mathcal{P}(U) \,:\, d_U(\nu, \mu) < \varepsilon\} \subset A$$

Up to a change of index we suppose that $\mu(U^d) = \vee_{i=1}^d \mu(U^i) (> 0)$. We associate with μ the N-approximation distributions $\mu^N \in \mathcal{P}(U)$ defined by

$$\mu^N(U^i) = \begin{cases} [N\mu(U^i)]/N & \text{if } 1 \leq i < d \\ 1 - \sum_{i=1}^{d-1}[N\mu(U^i)]/N & \text{if } i = d \end{cases}$$

By construction we note that

$$\mu^N \ll \mu \; (\ll q_U(\eta)) \quad \text{and} \quad \mathrm{Ent}_U(\mu^N|\eta) < \infty$$

Since
$$d_U(\mu^N, \mu) = \sum_{1 \le i < d} \{N\mu(U^i)\}/N \le (d-1)/N$$
then we have $\mu^N \in \mathcal{V}_U(\mu, \varepsilon)$ for any $N > N_1 = (d-1)/\varepsilon$ and therefore
$$q_U^{-1}(\{\eta \in \mathcal{P}(U) : d_U(\eta, \mu^N) = 0\}) = \{\eta \in \mathcal{P}(E) : d_U(\eta, \mu^N) = 0\}$$
$$\subset q_U^{-1}(\mathcal{V}_U(\mu, \varepsilon))$$

By the definition of the empirical measure $m(X)$, it is also clear that
$$Q^N(q_U^{-1}(\mathcal{V}_U(\mu, \varepsilon))) \ge \mathbb{P}(d_U(m(X), \mu^N) = 0)$$
$$= \frac{N!}{(N\mu^N(U^1))! \ldots (N\mu^N(U^d))!} \prod_{i=1}^{d} \eta(U^i)^{(N\mu^N(U^i))}$$

By Lemma 10.6.3, we get
$$Q^N(q_U^{-1}(\mathcal{V}_U(\mu, \varepsilon))) \ge (N+1)^{-d} \exp\{-N \mathrm{Ent}_U(\mu^N | \eta)\}$$

The continuity of the entropy function $\mathrm{Ent}_U(.|\eta)$ on the set of measures $\{\mu \in \mathcal{P}(U) : \mu << q_U(\eta)\}$ now implies that, for any $N \ge N_2$ and some $N_2 \ge 1$,
$$\mathrm{Ent}_U(\mu^N | \eta) \le \mathrm{Ent}_U(\mu | \eta) + \delta$$
We finally conclude that for any $N \ge (N_1 \vee N_2)$
$$\frac{1}{N} \log Q^N(q_U^{-1}(A)) \ge \frac{1}{N} \log \mathbb{Q}^N(q_U^{-1}(\mathcal{V}_U(\mu, \varepsilon)))$$
$$\ge -\mathrm{Ent}_U(A|\eta) - d\frac{\log(N+1)}{N} - 2\delta$$

Letting $N \to \infty$ and then $\delta \to 0$ the end of the proof of the LDP lower bound is completed. This ends the proof of Theorem 10.6.1. ∎

It is convenient at this stage to make a couple of remarks. First we observe that Theorem 10.6.1 can also be derived using Sanov's theorem. Indeed the strong version of Sanov's theorem implies Q^N satisfy the LDP in $\mathcal{P}(E)$ equipped with the τ-topology with the convex and good rate function $\mathrm{Ent}(.|\eta)$. Since for any finite Borel partition $U \in \mathcal{U}$ the projection operator $q_U : \mathcal{P}(E) \to \mathcal{P}(U)$ is a τ-continuous mapping by the contraction principle, we conclude that the measures $Q^N \circ q_U^{-1}$ satisfy the LDP on the compact topological space $(\mathcal{P}(U), d_U)$ with the good rate function
$$I_U(\nu) = \inf\{\mathrm{Ent}(\mu|\eta) : \mu \in \mathcal{P}(E) \text{ s.t. } q_U(\mu) = \nu\}$$
$$= \mathrm{Ent}_U(\nu|\eta)$$

To prove the last equality we first use Lemma 10.6.2 to check that $I_U \geq \mathrm{Ent}_U(\nu|\eta)$. The reverse inequality is based on the fact that for any $\nu \in \mathcal{P}(U)$ we have
$$\mathrm{Ent}(\Omega_\eta(\nu)|\eta) = \mathrm{Ent}_U(\nu|\eta)$$
where Ω_η is the mapping from $\mathcal{P}(U)$ into $q_U^{-1}(\mathcal{P}(U))$ that associates with any $\nu \in \mathcal{P}(U)$ the measure $\Omega_\eta(\nu)$ defined by the formula
$$\Omega_\eta(\nu)(A \cap U^i) = \begin{cases} \eta(A \cap U^i)\,\nu(U^i)/\eta(U^i) & \text{if } \eta(U^i) > 0 \\ \nu(A \cap U^i) & \text{if } \eta(U^i) = 0 \end{cases}$$
for $A \in \mathcal{E}$ and $1 \leq i \leq d$, where $U = (U^i)_{1 \leq i \leq d}$. In the reverse angle, Sanov's theorem can also be derived using Theorem 10.6.1.

The end of this section is concerned with proving the following corollary.

Corollary 10.6.1 (Sanov) *Let $(X^i)_{i \geq 1}$ be a sequence of independent, E-valued random variables, identically distributed according to a measure $\eta \in \mathcal{P}(E)$. For any N, we denote by Q^N the law on $\mathcal{P}(E)$ of the N-empirical measures $\frac{1}{N}\sum_{i=1}^N \delta_{X^i}$. The sequence of distributions Q^N satisfies an LDP on $\mathcal{P}(E)$ equipped with the τ-topology with the good entropy rate function $\mathrm{Ent}(\,\cdot\,|\eta)$.*

To prove that Theorem 10.6.1 implies Sanov's theorem, we first extend Q^N to $\mathbf{P}(E)$ by setting $Q^N(\mathbf{P}(E) - \mathcal{P}(E)) = 0$ for any $N \geq 1$. Then we let P^N be the sequence of distributions on $\lim_\mathcal{U} \mathcal{P}$ defined by $P^N = Q^N \circ h$, where h is the homeomorphism from $\lim_\mathcal{U} \mathcal{P}$ into $\mathbf{P}(E)$ introduced in Proposition 10.6.1. Since the canonical projection p_U from $\lim_\mathcal{U} \mathcal{P}$ into $\mathcal{P}(U)$ can be rewritten as $p_U = q_U \circ h$, we find that $P^N \circ p_U^{-1} = Q^N \circ q_U^{-1}$ for any $U \in \mathcal{U}$. Combining Theorem 10.6.1 with Theorem 10.5.1, we conclude that the sequence of distributions $P^N = Q^N \circ h$ satisfies the LDP on $\lim_\mathcal{U} \mathcal{P}$ with the good rate function
$$\mu \in \lim_\mathcal{U} \mathcal{P} \to \sup_{U \in \mathcal{U}} \mathrm{Ent}_U(p_U(\mu)|\eta)$$

Recalling that $h : \lim_\mathcal{U} \mathcal{P} \to \mathbf{P}(E)$ is a homeomorphism and using the fact that $p_U \circ h^{-1} = q_U$ by the contraction theorem (Theorem 10.2.1), we find that Q^N satisfies the LDP on $\mathbf{P}(E)$ (equipped with the τ_1-topology) with the good rate function
$$H(\,\cdot\,|\eta) : \mu \in \mathbf{P}(E) \longrightarrow H(\mu|\eta) = \sup_{U \in \mathcal{U}} \mathrm{Ent}_U(\mu|\eta)$$

Recalling that $Q^N(\mathcal{P}(E)) = 1$, we also have $Q^N(A) = Q^N(A \cap \mathcal{P}(E))$ for any τ_1-open or closed subset $A \subset \mathbf{P}(E)$. By Lemma 10.6.2, we have
$$D_{H(\,\cdot\,|\eta)} \subset \mathcal{P}(E) \tag{10.44}$$
This yields that for any subset $A \subset \mathbf{P}(E)$
$$H(A|\eta) = H(A \cap \mathcal{P}(E)|\eta) = \mathrm{Ent}(A \cap \mathcal{P}(E) \mid \eta) \tag{10.45}$$

Since the relative topology induced on $\mathcal{P}(E)$ by $\mathbf{P}(E)$ is the τ-topology, the τ-open (respectively τ-closed) sets are of the form $A\cap\mathcal{P}(E)$ with $A\subset\mathbf{P}(E)$ τ_1-open (respectively τ_1-closed). Using the fact that Q^N satisfies the LDP on $\mathbf{P}(E)$ with rate function $H(.|\eta)$, we deduce from (10.44) and (10.45) that Q^N satisfies the LDP lower and upper bounds met on $\mathcal{P}(E)$ for any τ-open or closed subsets $A\subset\mathcal{P}(E)$. Since $H(.|\eta)$ is a good rate function on the compact set $\mathbf{P}(E)$ the level sets $H(.|\eta)^{-1}([0,a])$ are τ_1-closed and hence τ_1-compact in $\mathbf{P}(E)$. Since we have proved that subsets of $H(.|\eta)^{-1}([0,a])\subset \mathcal{P}(E)$, we conclude that these level sets are also τ-compact subsets of $\mathcal{P}(E)$. This completes the proof of Corollary 10.6.1.

10.7 Path-Space and Interacting Particle Models

To connect our McKean interpretation models and their particle approximation schemes with the integral transfer analysis presented in Section 10.4, we shall adopt hereafter a simplified system of notation. The time horizon n and the initial distribution η_0 are always fixed. We slightly abuse the notation and when no confusion can be made, suppress the corresponding superscripts. We also simplify notation and we write $(\mathbb{K}, \Omega, \Sigma)$ instead of $(\mathbb{K}_{\eta_0,n}, \Omega_n = E_{[0,n]}, \Sigma_n)$. To each $x = (x_0^i, \ldots, x_n^i)_{1\leq i\leq N} \in \Omega^N$ and $x_p = (x_p^i)_{1\leq i\leq N} \in E_p^N$, we associate the empirical measures $m(x)$ and $m(x_p)$ by setting

$$m(x) = \frac{1}{N}\sum_{i=1}^N \delta_{(x_0^i,\ldots,x_n^i)} \quad \text{and} \quad m(x_p) = \frac{1}{N}\sum_{i=1}^N \delta_{x_p^i}$$

10.7.1 Proof of Theorem 10.1.1

For a fixed distribution η on path space Ω, we denote $\mathbb{Q}_\eta^N = (\mathbb{Q}_{\Sigma(\eta)})^{\otimes N}$, the N-fold tensor product of the measure $\mathbb{Q}_{\Sigma(\eta)}$. Let $\mathbb{P}_{\alpha,\eta}^N$, $\alpha \in \mathbb{R}$, be the collection of distributions on Ω^N defined by

$$\frac{d\mathbb{P}_{\alpha,\eta}^N}{d\mathbb{Q}_\eta^N}(\omega) = \exp\left(N[\alpha S_N(\omega,\eta) + V_\alpha(\pi_N(\omega),\eta)]\right)$$

for $\mathbb{Q}_\eta^N$-a.e. $\omega \in \Omega^N$ with

$$S_N(\omega,\eta) = \sum_{p=1}^n \int_{E_{p-1}\times E_p} m(\omega_{p-1},\omega_p)(d(u_{p-1},u_p))$$

$$\times \log\left[\frac{dK_{p,m(\omega_{p-1})}(u_{p-1},\cdot)}{dK_{p,\eta_{p-1}}(u_{p-1},\cdot)}(u_p)\right]$$

and

$$V_\alpha(\mu,\eta) = -\sum_{p=1}^{n} \int_{E_{p-1}} \mu_{p-1}(du_{p-1}) \, \log\left[Z_p^{(\alpha)}(\mu_{p-1},\eta_{p-1})(u_{p-1})\right]$$
(10.46)

Under $\mathbb{P}^N_{\alpha,\eta}$, the N-IPS model $(\xi_p)_{0\le p\le n}$ is the N-interacting particle model associated with the collection of Markov transitions $K^{(\alpha)}_{p,\mu}$ from E_{p-1} into E_p defined by

$K^{(\alpha)}_{p,\mu}(u_{p-1}, du_{p-1})$

$$= \frac{1}{Z_p^{(\alpha)}(\mu,\eta_{p-1})(u_{p-1})} \left[\frac{dK_{p,\mu}(u_{p-1},\cdot)}{dK_{p,\eta_{p-1}}(u_{p-1},\cdot)}(u_p)\right]^\alpha K_{p,\eta_{p-1}}(u_{p-1}, du_p)$$

It is convenient at this stage to make a couple of remarks.

- First we observe that for $\alpha = 1$ the transitions above are not independent on η. Consequently, $\mathbb{P}^N_{1,\eta}$ do not depend on η, and the probability measure

$$\mathbb{P}^N_1 =_{\text{def.}} \mathbb{P}^N_{1,\eta}$$

coincides with the distribution of the N-particle model associated with the collection of Markov transitions $K_{p,\mu}$.

- We also observe that the parameter α measures the degree of interaction in the system. For instance, for $\alpha = 0$ we have

$$\mathbb{P}^N_{0,\eta} = \mathbb{Q}^N_\eta = (\mathbb{Q}_{\Sigma(\eta)})^{\otimes N}$$

and under $\mathbb{P}^N_{0,\eta}$ the N-particle model consists of N independent particles with elementary transitions $\mathcal{K}_{p,\eta_{p-1}}$. By Sanov's theorem, under the tensor product measure $\mathbb{Q}^N_\eta = (\mathbb{Q}_{\Sigma(\eta)})^{\otimes N}$ the laws of the empirical measures of the path particles satisfy the LDP with a good rate function H_η given by

$$H_\eta : \mu \in \mathcal{P}(\Omega) \longrightarrow H_\eta(\mu) = \text{Ent}(\mu \mid \mathbb{Q}_{\Sigma(\eta)}) \in [0,\infty]$$

When the mapping V_α satisfies the regularity conditions (10.7), the integral transfer lemma (Lemma 10.4.1) applies and we readily conclude that the law of the empirical measures of the path particles under $\mathbb{P}^N_1$ satisfies the LDP with a good rate function I given by

$$I : \mu \in \mathcal{P}(\Omega) \longrightarrow I(\mu) = H_\mu(\mu) = \text{Ent}(\mu \mid \mathbb{Q}_{\Sigma(\mu)}) \in [0,\infty]$$

This ends the proof of Theorem 10.1.1.

10.7.2 Sufficient Conditions

We end this section with some simple and easily checked conditions on the pair (G_n, M_n) under which the desired regularity conditions (10.7) are satisfied. To simplify the presentation, we only examine the situation where $K_{n,\eta}(u_{n-1}, .) = \Phi_n(\eta)$. The McKean interpretation model (8.1) can be studied along the same line of argument as the one used in the end of Section 8.5.

We recall that the one step mappings associated with the prediction flow η_n are defined for any $f_n \in \mathcal{B}_b(E_n)$ by the equation

$$\Phi_n(\eta)(f_n) = \eta(G_{n-1} M_n(f_n))/\eta(G_{n-1})$$

We suppose the pair (G_n, M_n) satisfies the regularity conditions (G) and $(M)^{\exp}$ presented in Section 3.5.2 on page 116, for some parameters $\epsilon_n(G) > 0$, some pair of functions (k_n, a_n) and some reference measures p_n. In this situation, we easily check that the mappings V_α introduced in (10.46) take the form

$$V_\alpha(\mu, \eta) = -\sum_{p=1}^{n} \log \left[\int_{E_p} \left(\frac{d\Phi_p(\mu_{p-1})}{d\Phi_p(\eta_{p-1})} \right)^\alpha d\Phi_p(\eta_{p-1}) \right]$$

Similarly, we note that the mappings (10.7) are given by

$$\int_{E_{p-1}} \mu(du_{p-1}) \log Z_n^{(\alpha)}(\mu, \eta)(u_{p-1}) = \log \left[\int_{E_p} \left(\frac{d\Phi_n(\mu)}{d\Phi_n(\eta)} \right)^\alpha d\Phi_n(\eta) \right]$$

Our objective is to prove that our regularity assumptions on the pair (G_n, M_n) ensure that for each $n \geq 1$, $\eta \in \mathcal{P}(E_{n-1})$, and $\alpha \geq 1$ the mappings

$$L_{\alpha,n}(.,\eta) : \mu \longrightarrow \int (d\Phi_n(\mu)/d\Phi_n(\eta))^\alpha \, d\Phi_n(\eta)$$

are bounded and continuous at each η. We start by noting that for any $\mu \in \mathcal{P}(E_{n-1})$ the distributions $\Phi_n(\mu)$ and p_n are mutually absolutely continuous and

$$\exp(-a_n(u_n)) \leq \frac{d\Phi_n(\mu)}{dp_n}(u_n) = \frac{\mu(G_{n-1} \, k_n(.,u_n))}{\mu(G_{n-1})} \leq \exp a_n(u_n)$$

This readily yields that $\Phi_n(\mu)$, $\mu \in \mathcal{P}(E_{n-1})$, forms a collection of mutually absolutely continuous distributions and for any pair $(\mu, \eta) \in \mathcal{P}(E_{n-1})^2$ we have

$$\exp(-2a_n) \leq d\Phi_n(\mu)/d\Phi_n(\eta) \leq \exp(2a_n)$$

Under our assumptions, the desired boundedness condition is clearly met. Let us check that $V_{\alpha,n}(.,\eta)$, $\alpha \geq 1$, is continuous at η. To this end, we observe that

$$|L_{\alpha,n}(\mu,\eta) - L_{\alpha,n}(\eta,\eta)| \leq \int \left|\left(\frac{d\Phi_n(\mu)}{d\Phi_n(\eta)}\right)^\alpha - 1\right| d\Phi_n(\eta)$$

$$\leq \alpha \int \left|\frac{d\Phi_n(\mu)}{d\Phi_n(\eta)} - 1\right| e^{2(\alpha-1)a_n} d\Phi_n(\eta)$$

Using the decomposition (7.31) presented in the proof of Corollary 7.4.4, we find that

$$\left|\frac{d\Phi_n(\mu)}{d\Phi_n(\eta)}(u_n) - 1\right| = e^{2a_n(u_n)} \left|\frac{\mu(G_{n-1})}{\eta(G_{n-1})} - 1\right| + \left|\frac{\mu(G_{n-1} k_n(.,u_n))}{\eta(G_{n-1} k_n(.,u_n))} - 1\right|$$

From our conditions on the pair (a_n, G_n), now it suffices to check the continuity of the mappings

$$\mu \longrightarrow L'_{\alpha,n}(\mu,\eta)$$
$$= \int \left|\frac{\mu(G_{n-1} k_n(.,u_n))}{\eta(G_{n-1} k_n(.,u_n))} - 1\right| e^{2(\alpha-1)a_n(u_n)} \Phi_n(\eta)(du_n)$$

at $\eta \in \mathcal{P}(E_{n-1})$. We fix n, $v_n \in E_n$, and η, and we let f_{v_n} be the function on E_{n-1} defined by

$$f_{v_n}(u_{n-1}) = G_{n-1}(u_{n-1}) k_n(u_{n-1}, v_n)/\eta(G_{n-1} k_n(.,v_n))$$

We also note that $0 \leq f_{v_n}(u_{n-1}) \leq (e^{2a_n(v_n)}/\varepsilon_{n-1}(G))$ and

$$\Phi_n(\eta)(du_n) = \frac{\eta(G_{n-1} k_n(.,u_n))}{\eta(G_{n-1})} p_n(du_n) \leq e^{a_n(u_n)} p_n(du_n)$$

In this notation, we find that

$$0 \leq L'_{\alpha,n}(\mu,\eta) \leq \int |\mu(f_{v_n}) - \eta(f_{v_n})| e^{(2\alpha-1)a_n(v_n)} p_n(dv_n)$$

Under our assumptions, the dominated convergence theorem applies. For any sequence $\mu_n \to \eta$ weakly, we find that $\lim_{n\to\infty} L'_{\alpha,n}(\mu_n,\eta) = 0$.

10.8 Particle Density Profile Models

10.8.1 Introduction

In Section 10.7, we have examined the LDP for the McKean particle measures $\mathbb{K}^N_{n,\eta_0}$ associated with a fairly general class of McKean interpretations of the measure-valued process (10.1). The proof of Theorem 10.1.1 was based on an appropriate change of probability measure so that the law of the N-particle model $(\xi_0^i, \ldots, \xi_n^i)_{1 \leq i \leq N}$ consists of a regular Laplace distribution on path product spaces $(E_0 \times \ldots \times E_n)^N$. This strategy is therefore restricted to regular McKean models such that $K_{n+1,\mu}(u_n, .) \sim K_{n+1,\eta}(u_n, .)$ for any pair of measures $(\mu, \eta) \in \mathcal{P}(E_n)$ and therefore does not apply to complete genealogical tree models. To remove this condition, we shall be dealing here with the flow of particle density profiles $(\eta_p^N)_{0 \leq p \leq n}$ associated with the simple McKean interpretation

$$K_{n,\eta}(u_{n-1}, .) = \Phi_n(\eta)$$

of the nonlinear model (10.1). Before entering into some details about the proof of Theorem 10.1.2, it is useful to examine some direct consequences of Theorem 10.1.1. We recall that the flow $(\eta_p^N)_{0 \leq p \leq n}$ and the particle McKean measure $\mathbb{K}_{n,\eta_0}^N = \frac{1}{N} \sum_{i=1}^N \delta_{(\xi_0^i, \ldots, \xi_n^i)}$ are connected by the formula

$$\Sigma_n(\mathbb{K}_{n,\eta_0}^N) = (\eta_p^N)_{0 \leq p \leq n} \tag{10.47}$$

where Σ_n is the continuous mapping that associates with a given measure on the path space $E_{[0,n]} = \prod_{p=0}^n E_p$ the flow of its time marginals (see (10.6)).

Definition 10.8.1 *For any $n \geq 0$ and $N \geq 1$, we denote by Q_n^N the law of the N-particle density profiles $(\eta_p^N)_{0 \leq p \leq n}$ on $\mathcal{P}^n(E)$.*

The following corollary of the LDP on path space presented in Theorem 10.1.1 allows us to identify the candidate rate function that governs the LDP for the particle density profiles.

Corollary 10.8.1 *In the context of the statement of Theorem 10.1.1, for any $n \in \mathbb{N}$, Q_n^N satisfy the LDP in $\mathcal{P}^n(E)$ equipped with the product weak topology and with the good rate function J_n on $\mathcal{P}^n(E)$ given by*

$$J_n((\mu_p)_{0 \leq p \leq n}) = \sum_{p=0}^n \mathrm{Ent}(\mu_p \mid \Phi_p(\mu_{p-1})) \tag{10.48}$$

Proof:
Recalling that the LDP is preserved under continuous mappings (see Theorem 10.2.1, in Section 10.2), we deduce from (10.47) that $\Sigma_n(\mathbb{K}_{n,\eta_0}^N)$ satisfies the LDP on $\mathcal{P}(\mathcal{P}^n(E))$ (equipped with the weak topology) with the good rate function J_n defined by

$J_n((\mu_p)_{p \leq n})$

$= \inf \{\mathrm{Ent}(\mu \mid \mathbb{Q}_{\Sigma_n(\mu)}) \ : \ \mu \in \mathcal{P}(E_{[0,n]}) \text{ s.t. } \Sigma_n(\mu) = (\mu_p)_{p \leq n}\}$

To prove (10.48), we first note that

$$\Sigma_n(\mu_0 \otimes \ldots \otimes \mu_n) = (\mu_p)_{0 \leq p \leq n}$$
$$\mathbb{Q}_{(\mu_p)_{0 \leq p \leq n}} = \eta_0 \otimes \Phi_1(\mu_0) \otimes \ldots \otimes \Phi_n(\mu_{n-1})$$

from which we find that

$$\mathrm{Ent}(\mu_0 \otimes \ldots \otimes \mu_n \mid \mathbb{Q}_{(\mu_p)_{0\leq p\leq n}}) = \sum_{p=0}^{n} \mathrm{Ent}(\mu_p \mid \Phi_p(\mu_{p-1}))$$

and $J_n((\mu_p)_{0\leq p\leq n}) \leq \sum_{p=0}^{n} \mathrm{Ent}(\mu_p \mid \Phi_p(\mu_{p-1}))$. To prove the reverse inequality, we recall that for any $\mu \in \mathcal{P}(E_{[0,n]})$ we have

$$\begin{aligned}
\mathrm{Ent}(\mu \mid \mathbb{Q}_{\Sigma_n(\mu)}) &= \sup_{V_n \in \mathcal{C}_b(E_{[0,n]})} (\mu(V_n) - \log \mathbb{Q}_{\Sigma_n(\mu)}(e^{V_n})) \\
&\geq \sup_{v_0 \in \mathcal{C}_b(E_0),\ldots,v_n \in \mathcal{C}_b(E_n)} \sum_{p=0}^{n} [\mu_p(v_p) - \log \Phi_p(\mu_{p-1})(e^{v_p})] \\
&= \sum_{p=0}^{n} \mathrm{Ent}(\mu_p \mid \Phi_p(\mu_{p-1}))
\end{aligned}$$

The end of the proof of the corollary is now clear. ∎

10.8.2 Strong Large-Deviation Principles

The proof of Theorem 10.1.2 is based on a projective limit interpretation of the product τ-topology in the spirit of the proof of Sanov's Theorem presented in Section 10.6. All the notation and most results presented in this section will be used in the forthcoming development. We encourage the reader to make a brief visit to this section before entering into more details.

Since we shall be working with nonhomogeneous Hausdorff topological spaces $(E_n, \mathcal{E}_n)$, we will use the subscript $(.)_n$ to denote the corresponding objects. We denote by $\mathbf{P}(E_n)$ the set of additive set functions from $\mathcal{E}_n$ into $[0,1]$ equipped with the τ_1-topology of setwise convergence. We also equip the Cartesian product $\mathbf{P}^n(E) = \prod_{p=0}^{n} \mathbf{P}(E_p)$ with the product topology. We notice that the product τ-topology on $\mathcal{P}^n(E)$ coincides with the relative topology induced on $\mathcal{P}(E_n)$ by τ_1-topology on $\mathbf{P}(E_n)$. It is also convenient to recall that $\mathbf{P}(E_n)$ is homeomorphic to a subset of the algebraic dual $\mathcal{B}_b(E_n)^*$ of $\mathcal{B}_b(E_n)$ equipped with the $\mathcal{B}_b(E_n)$-topology (see for instance Theorem C3 on p. 315 in [112]).

Without further mention, we shall assume that 1) $\mathbf{P}(E_n)$ is furnished with a σ-algebra that contains the Borel σ-field associated with the τ_1-topology on $\mathbf{P}(E_n)$ and 2) the mappings Φ_n are continuous from $\mathbf{P}(E_{n-1})$ into $\mathbf{P}(E_n)$ with $\Phi_n(\mathcal{P}(E_{n-1})) \subset \mathcal{P}(E_n)$. From previous observations, it is easy to check that this continuity condition is met for Feynman-Kac models as soon as the potential functions are strictly positive. 3) The regularity condition (H) stated on page 338 is met.

Our immediate objective is to provide a projective limit interpretation of $\mathbf{P}^n(E)$. This program is achieved hereafter using topological arguments similar to the ones we used in Section 10.6. We let $\mathcal{U}_n$ be the set of all finite and Borel partitions U_n of E_n, and we associate with each $U_n \in \mathcal{U}_n$ the σ-algebra $\sigma(U_n)$ generated by the partition U_n. We recall that a partition U_n is said to be finer than another partition $V_n \in \mathcal{U}_n$, and we write $U_n \geq V_n$ as soon as $\sigma(U_n) \supset \sigma(V_n)$. We also equip the Cartesian product $\mathcal{U}^n = \prod_{p=0}^n \mathcal{U}_p$ with the partial ordering defined by

$$(U_p)_{0 \leq p \leq n} \leq (V_p)_{0 \leq p \leq n} \iff \forall\, 0 \leq p \leq n \quad U_p \leq V_p$$

As we did for the directed set $(\mathcal{U}_n, \geq)$, it is not difficult to prove that $(\mathcal{U}^n, \geq)$ is again a directed set. We slightly abuse the notation and for any $U_n \in \mathcal{U}_n$ we denote by $\mathcal{B}_b(U_n)$ the set of bounded and $\sigma(U_n)$-measurable functions on E_n and by $\mathcal{P}(U_n)$ the set of probability measures on $(E_n, \sigma(U_n))$. We equip the set $\mathcal{P}(U_n)$, $U_n = (U_n^i)_{1 \leq i \leq d_n} \in \mathcal{U}_n$, with the metric Kolmogorov-Smirnov metric d_{U_n} (see Definition 10.6.1).

Whenever $U_n \geq V_n$, we recall that we have $\mathcal{P}(U_n) = q_{U_n}(\mathbf{P}(E_n))$ and $\mathcal{P}(V_n) = p_{U_n, V_n}(\mathcal{P}(U_n))$ with the corresponding continuous projection operators q_{U_n} and p_{U_n, V_n}.

By Proposition 10.6.1, we know that $\mathcal{P}_n = ((\mathcal{P}(U_n), d_{U_n}), p_{U_n, V_n})_{U_n \geq V_n}$ is a projective inverse spectrum of $\mathcal{U}_n$. In addition, the mapping h_n that associates with a point $\mu_n = (\mu_n^{U_n})_{U_n \in \mathcal{U}_n} \in \lim_{\mathcal{U}_n} \mathcal{P}_n$ the set function $h_n(\mu_n) \in \mathbf{P}(E_n)$ defined for any $A_n \in \mathcal{E}_n$ by

$$h_n(\mu_n)(A_n) = \mu^{U_n}(A_n) \tag{10.49}$$

for some $U_n \in \mathcal{U}_n$ with $A_n \in \sigma(U_n)$ is a homeomorphism between the compact spaces $\lim_{\mathcal{U}_n} \mathcal{P}_n$ and $\mathbf{P}(E_n)$. The projective limit interpretation of the product τ_1-topology on $\mathbf{P}^n(E)$ is notationally more consuming but can be conducted in a similar fashion.

We associate with each $U^n = (U_p)_{0 \leq p \leq n} \in \mathcal{U}^n$ the Cartesian product

$$\mathcal{P}^n(U) = \prod_{p=0}^n \mathcal{P}(U_p)$$

equipped with the product topology inherited by the metric spaces $\mathcal{P}(U_n)$. For any $U^n \geq V^n$, we denote respectively by q_{V^n} and p_{U^n, V^n} the continuous and canonical projections from $\mathbf{P}^n(E)$, and respectively from $\mathcal{P}^n(U)$, into $\mathcal{P}^n(V)$ and defined for any $\mu = (\mu_p)_{0 \leq p \leq n} \in \mathbf{P}^n(E)$ and $\nu = (\nu_p)_{0 \leq p \leq n} \in \mathcal{P}^n(U)$ by the formulae

$$q_{V^n}(\mu) = (q_{V_p}(\mu_p))_{0 \leq p \leq n} \quad \text{and} \quad p_{U^n, V^n}(\nu) = (p_{U_p, V_p}(\nu_p))_{0 \leq p \leq n}$$

By construction, the set $\mathcal{P}^n = (\mathcal{P}^n(U), p_{U^n, V^n})_{U^n \geq V^n}$ forms a projective inverse spectrum of $\mathcal{U}^n$ with compact metric spaces $\mathcal{P}^n(U)$ and connecting maps p_{U^n, V^n}. In addition, it is not difficult to check that the change of coordinate mappings

$$\pi_n : \mu = (\mu_0^{U_0}, \ldots, \mu_n^{U_n})_{U_0, \ldots, U_n} \to \pi_n(\mu) = (\pi_n^0(\mu), \ldots, \pi_n^n(\mu))$$
$$= ((\mu_0^{U_0})_{U_0}, \ldots, (\mu_n^{U_n})_{U_n}) \quad (10.50)$$

is a homeomorphism between $\lim_{\mathcal{U}^n} \mathcal{P}^n$ and $\prod_{p=0}^n \lim_{\mathcal{U}_p} \mathcal{P}_p$. Recalling that $\prod_{p=0}^n \lim_{\mathcal{U}_p} \mathcal{P}_p \simeq \prod_{p=0}^n \mathbf{P}(E_p)$, we now have all the topological machinery we need to easily prove the following.

Proposition 10.8.1 *For any $n \geq 0$ the mapping*

$$h^n : \mu \in \lim_{\mathcal{U}^n} \mathcal{P}^n \to h^n(\mu) = (h_0(\pi_n^0(\mu)), \ldots, h_n(\pi_n^n(\mu))) \in \mathbf{P}^n(E)$$

defined in terms of the collection of maps (h_p, π_n^p), $p \leq n$, introduced in (10.49) and (10.50) is a homeomorphism.

Having established that $\mathbf{P}^n(E)$ equipped with the product τ_1-topology is homeomorphic to the compact space $\lim_{\mathcal{U}^n} \mathcal{P}^n$, and before getting into more serious business on large deviations, we now define and analyze more precisely the candidate rate function presented in (10.48) in Corollary 10.8.1.

Definition 10.8.2 *For any $n \geq 0$, we denote by J_n the l.s.c. function defined by*

$$J_n : \mu = (\mu_p)_{p \leq n} \in \mathbf{P}^n(E) \longrightarrow J_n(\mu) = \sum_{p=0}^n H_p(\mu_p | \Phi_p(\mu_{p-1}))$$

with the convention $\Phi_0(\nu) = \eta_0$ for $p = 0$ and the relative entropy criterion H_p on $(\mathbf{P}(E_p) \times \mathbf{P}(E_p))$ defined by

$$H_p(\cdot | \cdot) : (\mu, \nu) \in \mathbf{P}(E_p) \times \mathbf{P}(E_p) \longrightarrow H_p(\mu | \nu) = \sup_{U_p \in \mathcal{U}_p} \mathrm{Ent}_{U_p}(\mu | \nu) \in [0, \infty]$$

Definition 10.8.3 *We associate with each $U^n \in \mathcal{U}^n$ the image function*

$$J_n^U = J_n \circ q_{U^n}^{-1} : \mathcal{P}^n(U) \to [0, \infty]$$

of J_n with respect to q_{U^n} defined for any $\mu \in \mathcal{P}^n(U)$ by the formula

$$J_n^U(\mu) = \inf\{J_n(\nu) \; : \; \nu \in \mathbf{P}^n(E) \text{ such that } q_{U^n}(\nu) = \mu\}$$

The l.s.c. property of J_n comes from the fact that it can be obtained as the supremum of τ-continuous functions

$$J_n(\mu) = \sup_{U^n \in \mathcal{U}^n} \sup_{f \in \mathcal{B}_b^n(U)} \sum_{p=0}^n [\mu_p(f_p) - \log \Phi_p(\mu_{p-1})(e^{f_p})]$$

where we have used the notation $\mathcal{B}_b^n(U) = \prod_{p=0}^n \mathcal{B}_b(U_p)$ and the traditional convention $\Phi_p(\mu_{p-1}) = \eta_0$ for $p = 0$.

Lemma 10.8.1 *For any $n \geq 0$, the domain $D_{J_n} = \{J_n < \infty\}$ of J_n is included in $\mathcal{P}^n(E)$, and for any $\mu = (\mu_p)_{p \leq n} \in \mathcal{P}^n(E)$ and $U^n \in \mathcal{U}^n$ we have*

$$J_n(\mu) = \sum_{p=0}^{n} \text{Ent}(\mu_p | \Phi_p(\mu_{p-1})) \tag{10.51}$$

and

$$J_n^U(q_{U^n}(\mu)) = \sum_{p=0}^{n} \inf_{\nu \in q_{U_{p-1}}^{-1}(\{q_{U_{p-1}}(\mu_{p-1})\})} \text{Ent}_{U_p}(q_{U_p}(\mu_p)|\Phi_p(\nu)) \tag{10.52}$$

Proof:
We prove the lemma by induction on the time parameter. For $n = 0$, the result is a direct consequence of Lemma 10.6.2. Suppose we have proved the lemma at rank $(n-1)$. Since for any $\mu = (\mu_p)_{p \leq n}$ we have

$$J_n(\mu) = J_{n-1}((\mu_p)_{p<n}) + H_n(\mu_n | \Phi_n(\mu_{n-1}))$$

by the induction hypothesis we find that

$$J_n(\mu) \iff (\mu_p)_{p<n} \in \mathcal{P}^{n-1}(E) \quad \text{and} \quad H_n(\mu_n | \Phi_n(\mu_{n-1})) < \infty$$

Invoking again Lemma 10.6.2, we conclude that

$$J_n(\mu) \iff (\mu_p)_{p<n} \in \mathcal{P}^{n-1}(E) \quad \text{and} \quad \mu_n \in \mathcal{P}(E_n)$$

The proof of (10.51) is now a straightforward consequence of (10.40), page 368. We prove (10.52) using the same line of argument as the one used on page 371. This ends the proof of the lemma. ■

The end of this section is essentially devoted to the proof of Theorem 10.1.2. Before getting into it, it is convenient to make some remarks.

By Lemma 10.8.1, we first observe that the LDP for the τ_1-topology on $\mathbf{P}^n(E)$ implies the the LDP for the τ-topology on $\mathcal{P}^n(E)$ and hence for the weak topology with always the same rate function. As we noted earlier (see p.341), in the case of Feynman-Kac models Φ_n on Polish spaces, the exponential tightness (for the weak topology) of the laws of $(\eta_p^N)_{p \leq n}$ (proved at the end of Section 10.3) combined with the LDP lower bound (for the weak topology) shows that the rate function J_n has compact level sets (for the weak topology). This shows that for Feynman-Kac type particle models the LDP for the τ_1-topology on $\mathbf{P}^n(E)$ with (good) rate function J_n implies the LDP for the weak topology on $\mathcal{P}^n(E)$ with good rate function J_n.

After this brief digression, we now come to the following proof.

Proof of Theorem 10.1.2:

To prepare the proof of the LDP upper bound, we first recall that any closed F of $\mathbf{P}^n(E)$ has the form

$$F = \cap_{U^n \in \mathcal{U}^n} q_{U^n}^{-1}(F_{U^n})$$

where F_{U^n} stands for the τ_1-closure of the set $q_{U^n}(F)$. Since $\mathbf{P}^n(E)$ is compact, the sets $q_{U^n}^{-1}(F_{U^n})$ are compact and for any $\varepsilon > 0$ one can find a finite and open ε-covering

$$q_{U^n}^{-1}(F_{U^n}) \subset \cup_{i=1}^m \mathcal{V}_n^{U^n}(\mu^i, \varepsilon)$$

with $\mu^i = (\mu_p^i)_{p \leq n} \in F$ and $\mathcal{V}_p^{U^n}(\mu^i, \varepsilon) = (\prod_{0 \leq q \leq p} \mathcal{V}_p^{U^n}(\mu_q^i, \varepsilon))$ for any $p \leq n$. Under our continuity assumptions, we can choose these open neighborhoods such that for any $\eta_p \in \mathcal{V}_p^{U^n}(\mu_p^i, \varepsilon)$

$$d_{U_p}(\eta_p, \mu_p^i) \vee d_{U_{p+1}}(\Phi_{p+1}(\eta_p), \Phi_{p+1}(\mu_p^i)) \leq \varepsilon$$

In this case, we have for each $(f_p)_{p \leq n} \in \mathcal{B}_b^n(U)$

$$\mathbb{P}((\eta_p^N)_{p \leq n} \in \mathcal{V}_n^{U^n}(\mu^i, \varepsilon))$$

$$\leq e^{\varepsilon N \|f_n\|} \, \mathbb{E}(e^{N(\eta_n^N - \mu_n^i)(f_n)} \, 1_{\mathcal{V}_{n-1}^{U^n}(\mu^i, \varepsilon)}((\eta_p^N)_{p < n}))$$

$$= e^{\varepsilon N \|f_n\|} \, \mathbb{E}(e^{-N(\mu_n^i(f_n) - \log \Phi_n(\eta_{n-1}^N)(e^{f_n}))} \, 1_{\mathcal{V}_{n-1}^{U^n}(\mu^i, \varepsilon)}((\eta_p^N)_{p < n}))$$

$$\leq e^{\varepsilon N c(f_n) - N(\mu_n^i(f_n) - \log \Phi_n(\mu_{n-1}^i)(e^{f_n}))} \, \mathbb{P}((\eta_p^N)_{p < n} \in \mathcal{V}_{n-1}^{U^n}(\mu^i, \varepsilon))$$

for some finite constant $c(f_n) < \infty$ whose values only depend on the supremum norm of f_n. A simple induction now yields that

$$\mathbb{P}((\eta_p^N)_{p \leq n} \in \mathcal{V}_n^{U^n}(\mu^i, \varepsilon)) \leq e^{\varepsilon N \sum_{p \leq n} c(f_p) - N \sum_{p \leq n} (\mu_p^i(f_p) - \log \Phi_p(\mu_{p-1}^i)(e^{f_p}))}$$

from which we conclude that

$$\limsup_{\varepsilon \to 0} \limsup_{N \to \infty} \frac{1}{N} \log \mathbb{P}((\eta_p^N)_{p \leq n} \in \mathcal{V}_n^{U^n}(\mu^i, \varepsilon))$$

$$\leq -\sum_{p \leq n}(\mu_p^i(f_p) - \log \Phi_p(\mu_{p-1}^i)(e^{f_p}))$$

Taking the infimum over all $(f_p)_{p \leq n} \in \mathcal{B}_b^n(U)$ and by (10.52), we conclude that

$$\limsup_{\varepsilon \to 0} \limsup_{N \to \infty} \frac{1}{N} \log \mathbb{P}((\eta_p^N)_{p \leq n} \in \mathcal{V}_n^{U^n}(\mu^i, \varepsilon))$$

$$\leq -\sum_{p \leq n} \mathrm{Ent}_{U_n}(\mu_p^i | \log \Phi_p(\mu_{p-1}^i)) \leq -J_n^U(q_{U^n}(\mu^i)) \leq -J_n^U(F_{U^n})$$

By the union of event bounds, we find that

$$\limsup_{N\to\infty} \frac{1}{N} \log \mathbb{P}((\eta_p^N)_{p\leq n} \in q_{U^n}^{-1}(F_{U^n})) \leq -J_n^U(F_{U^n})$$

and therefore

$$\limsup_{N\to\infty} \frac{1}{N} \log \mathbb{P}((\eta_p^N)_{p\leq n} \in F) \leq -J_n^U(F_{U^n}) \leq -J_n^U(F_{U^n})$$

We end the proof of the LDP upper bound by taking the infimum over all U^n, invoking the min-max theorem (Theorem 10.5.2) and recalling that $J_n(F \cap \mathcal{P}^n(E)) = J_n(F)$ (see Lemma 10.8.1).

Our final objective is to prove that the rate function J_n governs the LDP lower bounds on $\mathbf{P}^n(E)$. We use an inductive proof with respect to the time parameter. Note that for $n = 0$ the desired LDP results from Sanov's theorem (Theorem 10.6.1). Suppose the desired LDP is proved at time $(n-1)$. Let $A \subset \mathbf{P}^n(E)$ be a τ_1-open set such that $J_n(A) < \infty$ (otherwise the proof of the lower bound is as usually trivial). Invoking Lemma 10.8.1, for any $\delta > 0$ there exists a point $\mu \in A \cap \mathcal{P}^n(E)$ such that

$$J_n(\mu) \leq J_n(A) + \delta \tag{10.53}$$

Since A is open, we can find a collection of strictly positive numbers $(\varepsilon_p)_{0 \leq p \leq n}$ and a sequence of finite partitions $U_k = (U_p^i)_{1 \leq i \leq d_p}$ of the sets E_p such that

$$\mathcal{C}_{\varepsilon,n}^U(\mu) = \prod_{0 \leq p \leq n} B_{U_p}(\mu_p, \varepsilon_p) \subset A$$

with the open neighborhood $B_{U_p}(\mu_p, \varepsilon_p)$ of $\mu_p \in \mathcal{P}(E_p)$ given by

$$B_{U_p}(\mu_p, \varepsilon_p) = \{\nu_p \in \mathbf{P}(E_p) \,:\, d_{U_p}(\mu_p, \nu_p) < \varepsilon_p\}$$

Up to a change of index we suppose that $\mu_n(U_n^d) = \vee_{i=1}^{d_n} \mu_n(U_n^i) (> 0)$, and we associate with $\mu_n \in \mathcal{P}(E_n)$ the N-approximation distributions $\mu_n^N \in \mathcal{P}(U_n)$ defined by

$$\mu_n^N(U_n^i) = \begin{cases} [N\mu_n(U_n^i)]/N & \text{if } 1 \leq i < d_n \\ 1 - \sum_{i=1}^{d_n-1}[N\mu_n(U_n^i)]/N & \text{if } i = d_n \end{cases}$$

By construction we note that

$$\mu_n^N \ll q_{U_n}(\mu_n) \ll q_{U_n}(\Phi_n(\mu_{n-1})) \quad \text{and} \quad \text{Ent}_{U_n}(\mu_n^N | \Phi_n(\mu_{n-1})) < \infty$$

Under our assumptions, we also have for any $\nu_{n-1} \in B_{U_{n-1}}(\mu_{n-1}, \varepsilon_{n-1})$ and $1 \leq i \leq d_n$

$$\mu_n^N(U_n^i) > 0 \implies \Phi_n(\nu_{n-1})(U_n^i) > 0$$

Since $d_{U_n}(\mu_n, \mu_n^N) \leq (d_n - 1)/N$, we find that $\mu_n^N \in B_{U_n}(\mu_n, \varepsilon_n)$ as soon as $N > N_n^1 = (d_n - 1)/\varepsilon_n$ and hence

$$\mathcal{C}_{\varepsilon,n-1}^U(\mu) \times \{\nu_n \in \mathcal{P}(E_n) \ : \ d_{U_n}(\mu_n^N, \nu_n) = 0\} \subset A$$

By the definition of the N-particle model and by Lemma 10.6.3, it is also clear that

$$\mathbb{P}(d_{U_n}(\mu_n^N, \eta_n^N) = 0 \mid \eta_{n-1}^N)$$
$$= \frac{N!}{(N\mu_n^N(U_n^1))! \ldots (N\mu_n^N(U_n^{d_n}))!} \prod_{i=1}^{d_n} \Phi_n(\eta_{n-1}^N)(U_n^i)^{N\mu_n^N(U_n^i)}$$
$$\geq \exp\left[-N \operatorname{Ent}_{U_n}(\mu_n^N \mid \Phi_n(\eta_{n-1}^N)) - d \log(N+1)\right]$$

Under our assumptions, we also have for any $\nu_{n-1} \in \mathcal{P}(E_{n-1})$

$$\rho_n \ q_{U_n}(\lambda_n) \leq q_{U_n}(\Phi_n(\nu_{n-1})) \ll q_{U_n}(\lambda_n)$$

and from previous observations

$$\mu_n^N \ll q_{U_n}(\Phi_n(\mu_{n-1})) \ll q_{U_n}(\lambda_n)$$

By Lemma 10.6.1, we obtain the uniform Lipschitz estimate

$$|\operatorname{Ent}_U(\mu_n^N | \Phi_n(\nu_{n-1})) - \operatorname{Ent}_U(\mu_n^N | \Phi_n(\mu_{n-1}))|$$
$$\leq \frac{2}{\rho_n \lambda_n^\star(U_n)} \ d_{U_n}(\Phi_n(\nu_{n-1}), \Phi_n(\mu_{n-1}))$$

with the positive constant $\lambda_n^\star(U_n) = \wedge_{i:\lambda_n(U_n^i)>0} \lambda_n(U_n^i) > 0$. Since the mapping Φ_n is τ_1-continuous for every $\delta > 0$ there exists some τ_1-open neighborhood $\mathcal{O}_{\delta,n}(\mu_{n-1}) \subset \mathbf{P}(E_{n-1})$ of μ_{n-1} (which depends on U_n) such that

$$\mathcal{O}_{\delta,n}(\mu_{n-1}) \subset B_{U_{n-1}}(\mu_{n-1}, \varepsilon_{n-1})$$

and on the set of events $\{\eta_{n-1}^N \in \mathcal{O}_{\delta,n}(\mu_{n-1})\}$ we have

$$\operatorname{Ent}_{U_n}(\mu_n^N \mid \Phi_n(\eta_{n-1}^N)) \leq \operatorname{Ent}_{U_n}(\mu_n^N \mid \Phi_n(\mu_{n-1})) + \delta \ (<\infty)$$

The continuity of the function $\operatorname{Ent}_{U_n}(. | \Phi_n(\mu_{n-1}))$ on the set of measures $\{\nu_n \in \mathcal{P}(U_n) \ : \ \nu_n \ll q_{U_n}(\Phi_n(\mu_{n-1}))\}$ now implies that

$$\operatorname{Ent}_{U_n}(\mu_n^N \mid \Phi_n(\mu_{n-1})) \leq \operatorname{Ent}_{U_n}(\mu_n \mid \Phi_n(\mu_{n-1})) + \delta$$
$$\leq H(\mu_n \mid \Phi_n(\mu_{n-1})) + \delta$$

for any $N \geq N_n^2$ and some $N_n^2 \geq 1$. It is now convenient to observe that

$$Q_n^N(A) = \mathbb{P}((\eta_0^N, \ldots, \eta_n^N) \in A)$$
$$\geq \mathbb{P}(d_{U_n}(\mu_n^N, \eta_n^N) = 0 \ , \ \eta_{n-1}^N \in \mathcal{O}_{\delta,n}(\mu_{n-1})$$
$$\text{and } (\eta_k^N)_{p<n-1} \in \mathcal{C}_{\varepsilon,n-2}^U(\mu))$$

Using a simple calculation, we deduce the lower bound

$$Q_n^N(A)$$
$$\geq \inf_{\nu \in \mathcal{O}_{\delta,n}(\mu_{n-1})} \mathbb{P}\left(d_{U_n}(\mu_n^N, \eta_n^N) = 0 \mid \eta_{n-1}^N = \nu\right) \; Q_{n-1}^N\left(\mathcal{D}_{(\delta,\varepsilon),n}(\mu)\right)$$

with the τ_1-open neighborhood $\mathcal{D}_{(\delta,\varepsilon),n}(\mu) \in \mathbf{P}^{n-1}(E)$ of $(\mu_p)_{p \leq n-1}$ defined by

$$\mathcal{D}_{(\delta,\varepsilon),n}(\mu) = \mathcal{C}_{\varepsilon,n-2}^U(\mu) \times \mathcal{O}_{\delta,n}(\mu_{n-1})$$

From previous estimations, we find that for any $N \geq (N_n^1 \wedge N_n^2)$

$$\tfrac{1}{N} \log Q_n^N(A)$$
$$\geq - H_n(\mu_n \mid \Phi_n(\mu_{n-1})) - \tfrac{d \log (N+1)}{N} - 2\delta + \tfrac{1}{N} \log Q_{n-1}^N\left(\mathcal{D}_{(\delta,\varepsilon),n}(\mu)\right)$$
$$= -J_n(\mu) - \tfrac{d \log (N+1)}{N} - 2\delta + J_{n-1}((\mu_p)_{p \leq n}) + \tfrac{1}{N} \log Q_{n-1}^N\left(\mathcal{D}_{(\delta,\varepsilon),n}(\mu)\right)$$

Furthermore, by (10.53) and our induction hypothesis, we have

$$\liminf_{N \to \infty} \tfrac{1}{N} \log Q_n^N(A)$$
$$\geq -J_n(A) - 3\delta + [J_{n-1}((\mu)_{p<n}) - J_{n-1}(\mathcal{D}_{(\delta,\varepsilon),n}(\mu))] \geq -J_n(A) - 3\delta$$

for any $\delta > 0$. Letting δ tend to 0, we conclude that J_n governs the LDP lower bound on $\mathbf{P}^n(E)$. This ends the proof of Theorem 10.1.2. ∎

11
Feynman-Kac and Interacting Particle Recipes

11.1 Introduction

This chapter offers a series of Feynman-Kac and interacting particle modeling recipes that can be combined with one another and applied to every application discussed in this book. We have chosen to present this catalog of Feynman-Kac techniques for several reasons.

First of all, as soon as a particular estimation problem has been identified as that of solving a Feynman-Kac formula of the form (1.3), the particle-approximation models are dictated by the pair potentials/kernels (G_n, M_n). In this sense, this section provides a unifying description of a class of seemingly different particle algorithms currently used in applied literature.

The second reason is that some of these techniques, such as the branching strategies, the importance sampling changes of measures, the changes of potential functions, and the multilevel decompositions of the state, can be used in practice to increase the efficiency of Monte Carlo particle algorithms.

Finally, the complete mathematical analysis and the precise numerical comparisons between these particle techniques have only been started, and various open problems remain to be solved. The forthcoming descriptions together with the mathematical asymptotic analysis described in the final chapters of this book not only suggest several avenues of research but we hope will influence the reader to design new ideas in the future.

This chapter is organized as follows. In Section 11.2, we briefly review the interacting Metropolis model introduced in Section 5.4. We illustrate the impact of this particle MCMC algorithm in the context of the popular Ising ferromagnetic spin model. For applications to Bayesian spectral analysis, we refer the reader to [71].

In the next two sections, 11.3 and 11.4, we provide a brief discussion on Feynman-Kac path measures and their interacting particle and genealogical tree interpretations.

The next three sections, 11.5, 11.6, and 11.7, present some essential and additional tools such as conditional exploration techniques and excursion-valued particle models. Some of these techniques are not new. For instance, the branching excursion models presented in Section 11.7 offer very accurate stochastic and adaptive grid approximations. These ideas were applied originally as a natural heuristic scheme in tracking and global positioning systems in [43, 105, 103, 106, 107]. They were also analyzed rigorously by the author in the article [76] published in 1998. More recently, the same branching excursion models have been applied with success to generate self-avoiding random walks by Fauenkron, Causo, and Grassberger in [149] and to related protein-folding problems by Liu and Zhang [235] under the botanical names "Markovian anticipation", "lookahead strategies" and "sampling importance sampling pilot exploration resampling".

In Section 11.8, we design a new class of branching particle interpretation models. Our reasons to include this material are two folds. First we believe that the readers may benefit to have a friendly presentation on some easy simulation algorithms that can be directly implemented on a personal computer. The second reason is that the mathematical analysis of some of the branching models presented hereafter differs from the one discussed in this book. Thus, this presentation also suggests new research avenues. In this connection, we mention that the literature on genetic algorithms abounds with various classes of selection generation strategies (see for instance [22, 310] and references therein). More recently in particle filters literature, there is an interest to find the "most efficient" selection procedure. The branching rules are often presented as intuitive but heuristic schemes with no precise asymptotic results.

In Section 11.8.2 we provide a pair of practical and easy to use local $\mathbb{L}_2$-conditions on the local branching rule that ensures convergence to the desired Feynman-Kac model. In Section 11.8.3, we illustrate these conditions with Poisson, Bernoulli and other branching mechanisms such as the Baker remainder stochastic sampling. The latter procedure was introduced by J. Baker in [21, 22] in 1985 to reduce the computational efforts of genetic type models. It was rencently rediscovered and applied with success by Liu and Chen in [228] in 1998. The first well founded asymptotic analysis and related continuous time branching schemes with random populations sized can also be found in [67, 96, 97].

In Section 11.8.4, we derive an new uniform convergence theorem which is valid for any branching model with fixed population size. In Section 11.8.5, we also show that there is apparently no hope to obtain a uniform estimate for branching models with random population size. We hope that these rather elementary results will help the reader in developing new and more sophisticated asymptotic results such as propagation-of-chaos properties, fluctuations theorems, large-deviation principles, and related concentration results.

11.2 Interacting Metropolis Models

11.2.1 Introduction

One strategy to generate approximate samples according to a given distribution π is to interpret the target distribution π as the limiting distribution of a judicious Feynman-Kac distribution flow. This modeling technique is presented in full detail in Section 5.4. In the present section, we merely content ourselves in briefly reviewing this Metropolis-Feynman-Kac model. We also illustrate some consequences of the results developed in earlier sections in the context of the Ising model. In this physical ferromagnetic spin model, atoms sit at the nodes of sites s in the d-dimensional cubic lattice $S = [-M, M]^d$. At each site s, the atom has a positive or negative spin $y(s) \in \{+1, -1\}$, and each pair of atoms $y(s)$ and $y(s')$ has an interaction energy $J_{s,s'} y(s) y(s')$ that, in addition to the interaction strength parameter $J_{s,s'}$, depends on the distance between the pair of sites (s, s'). The potential or the energy function of a configuration x is often defined by the formula

$$H(y) = \sum_{s \in S} \sum_{s' \in V_s} J_{s,s'} y(s) y(s') + \sum_{s \in S} h_s y(s)$$

where for any $s \in S$, $V_s = \{s' \in S : \vee_{p \leq d} |s_p - s'_p| \leq 1\}$. One typical question arising in practice is not only to find the extremal energy configurations but also to generate random samples with the Boltzmann-Gibbs distributions on $E = \{-1, +1\}^S$ defined by

$$\pi(dy) = \frac{1}{\nu(e^{-H})} \; e^{-H(y)} \; \nu(dy) \qquad (11.1)$$

with the uniform distribution ν. In this situation, we see that the interaction function between atoms is not related to some temporal interpretation of the state but to a particular neighborhood structure. On the other hand, the uniform distribution ν can also be regarded as the limiting probability of some Markov exploration chain on the state of all configurations. We can choose for instance

$$K(y, y') = \frac{1}{|\mathcal{V}(y)|} 1_{\mathcal{V}(y)}(y') \qquad (11.2)$$

where $|\mathcal{V}(y)|$ stands for the cardinality of the neighborhood subset of y given by $\mathcal{V}(y) = \{y' \in E : \sum_{s \in S} 1_{y(s)}(y'(s)) \leq 1\}$. By symmetry arguments, we readily find that K is ν-reversible; that is, $\nu(y')K(y',y) = \nu(y)K(y,y')$. We mention that these Boltzmann-Gibbs distributions are often used as "prior models" in Bayesian image analysis [313] and pattern theory [168].

11.2.2 Feynman-Kac-Metropolis and Particle Models

We start with a pair of Markov kernels (K, L) such that

$$(\pi \times L)_2 << (\pi \times K)_1 \tag{11.3}$$

with the distributions on the transition space $(E \times E)$ defined by

$$\begin{aligned}
(\pi \times L)_2(d(y,y')) &= \pi(y')L(y',dy) \\
(\pi \times K)_1(d(y,y')) &= \pi(y)K(y,dy')
\end{aligned}$$

Then we associate with the pair (K, L) the Metropolis potential ratio on $(E \times E)$ given by

$$G = d(\pi \times L)_2/d(\pi \times K)_1$$

In the case of the Ising model (11.1) with $K = L$ and given by (11.2), we find that for any $y' \in \mathcal{V}(y)$

$$\begin{aligned}
G(y,y') &= \exp\{-(H(y') - H(y))\} \\
&= \exp\left\{(y(s) - y'(s))\left(h_s + \sum_{s' \in V_s} J_{s,s'}y(s')\right)\right\}
\end{aligned}$$

where s stands for the only possible site where y' and y may differ. The main interest in introducing the Metropolis potential ratio above comes from the following key Feynman-Kac formula (see Theorem 5.5.1, Section 5.5)

$$\mathbb{E}_\pi^L(f_n(Y_n,\ldots,Y_0) \mid Y_n = y) = \frac{\mathbb{E}_y^K(f_n(Y_0,\ldots,Y_n) \prod_{p=0}^{n-1} G(Y_p,Y_{p+1}))}{\mathbb{E}_y^K(\prod_{p=0}^{n-1} G(Y_p,Y_{p+1}))}$$
(11.4)

We recall that $\mathbb{E}_\pi^L$ stands for the expectation operator with respect to the law of a (canonical) Markov chain with initial distribution π and elementary transitions L. In the same way, $\mathbb{E}_y^K$ is the expectation operator with respect to the law $\mathbb{P}_y^K$ of a (canonical) Markov chain starting at y and with elementary transitions K. Then we observe that under $\mathbb{P}_y^K$ the random sequence of transitions defined by the pairs $X_n = (Y_n, Y_{n+1})$ forms a Markov chain with initial distribution $\eta_0 = (\delta_y \times K)_1$ and its elementary transitions are given by

$$M^K((y,y'),d(z,z')) = \delta_{y'}(dz) \, K(z,dz')$$

11.2 Interacting Metropolis Models

In this notation, we readily find the formula

$$\begin{aligned}\mathbb{E}_\pi^L(\varphi(Y_1, Y_0) \mid Y_{n+1} = y) &= \frac{\mathbb{E}_y^K(\varphi(Y_n, Y_{n+1}) \prod_{p=0}^n G(Y_p, Y_{p+1}))}{\mathbb{E}_y^K(\prod_{p=0}^n G(Y_p, Y_{p+1}))} \\ &= \frac{\mathbb{E}_y^K(\varphi(X_n) \prod_{p=0}^n G(X_p))}{\mathbb{E}_y^K(\prod_{p=0}^n G(X_p))} \\ &= \widehat{\eta}_n(\varphi)\end{aligned} \qquad (11.5)$$

where $\widehat{\eta}_n$ represents the updated Feynman-Kac model associated with the pair potential/kernel (G, M^K).

Before getting further into our discussion, by (11.5) we observe that in some sense and as $n \to \infty$ we have $\widehat{\eta}_n \longrightarrow (\pi \times L)_2$ as soon as the Markov kernel L is sufficiently mixing. For more details, we refer the reader to Section 5.5.4. To illustrate this assertion, we return to the Ising model example. In this situation, we note that any pair of configurations $(y, y') \in E$ can be joined by a K-admissible path of length $m \leq m(d) = (2M+1)^d = |S|$. More precisely, there always exists a path $y_0, y_1, \ldots, y_m \in E$ such that $y_0 = y$, $y_i \in \mathcal{V}(y_{i+1})$ for all $i < m$ and $y_m = y'$. Since we have $|\mathcal{V}(y)| = (1 + |S|)$, for each $y \in E$, this yields the rather crude lower bound

$$\varepsilon(d) = \inf_{y, y', y''} \frac{dK^{m(d)}(y, \cdot)}{dK^{m(d)}(y', \cdot)}(y'') \geq (1 + |S|)^{-m(d)}$$

We conclude that $\beta(K^{m(d)}) \leq (1 - \varepsilon(d))$, where $\beta(K^{m(d)})$ represents the Dobrushin ergodic coefficient of $K^{m(d)}$ (see for instance (4.12) on page 128). Recalling that $K = L$, we conclude that the mixing condition $(L)_m$ introduced on page 182 is met with $m = m(d)$ and $\epsilon(L) = \varepsilon(d)$. Thus, by Theorem 5.5.2, we conclude that

$$\|\widehat{\eta}_{m(d)+n+1} - (\pi \times L)_2\|_{\text{tv}} \leq 2\varepsilon(d)^{-1} \left(1 - \varepsilon(d)\right)^{\lfloor n/m(d) \rfloor} \qquad (11.6)$$

In Section 5.5, we have seen that the distribution flow $\widehat{\eta}_n$ can be interpreted as the evolution of the laws of a nonlinear (or nonhomogeneous) Feynman-Kac-Metropolis model. The display above illustrates the main property of this class of models, namely that their decay rates to the desired target distribution do not depend on the nature of the limiting distribution. The final step is now clear. The particle simulation algorithm coincides with the particle interpretation of the Feynman-Kac models described above. For the convenience of the reader, we briefly describe one rather generic type of E^N-valued particle model:

$$(Y_n^i)_{1 \leq i \leq N} \xrightarrow{\text{exploration}} (Y_n'^i)_{1 \leq i \leq N} \xrightarrow{\text{selection}} (Y_{n+1}^i)_{1 \leq i \leq N}$$

Initially, we start for instance with N particles in the same location $Y_0^i = y$. During the exploration stage, each particle Y_n^i evolves to a new location

$Y_n'^i$ randomly chosen with the distribution $K(Y_n^i, \cdot)$. During the selection stage, we sample randomly N random variables Y_{n+1}^i with the acceptance/rejection rules

$$Y_{n+1}^i = \begin{cases} Y_n'^i & \text{with proba} \quad G(Y_n^i, Y_n'^i)/(\vee_{j=1}^N G_n(Y_n^j, Y_n'^j)) \\ Z_n^i & \text{with proba} \quad 1 - G(Y_n^i, Y_n'^i)/(\vee_{j=1}^N G(Y_n^j, Y_n'^j)) \end{cases}$$

where $(Z_n^i)_{1 \leq i \leq N}$ represents a sequence of conditionally independent random variables with the discrete distribution $\sum_{i=1}^N \frac{G(Y_n^i, Y_n'^i)}{\sum_{j=1}^N G(Y_n^j, Y_n'^j)} \delta_{Y_n'^i}$. In practice, the initial choice $Y_0^i = y$ is of course far from being optimal if the only objective is to sample according to the target distribution π. Nevertheless, in this case the resulting genealogical tree occupation measure

$$\pi_{y,n}^N = \frac{1}{N} \sum_{i=1}^N \delta_{(Y_{0,n}^i, Y_{1,n}^i, \ldots, Y_{n,n}^i)}$$

is an N-particle approximation of the distribution

$$\pi_{y,n} = \mathbb{P}_\pi^L((Y_n, \ldots, Y_0) \in \cdot \mid Y_n = y)$$

For more details, we again refer the reader to Section 5.5. Using simple manipulations, we deduce that for any $x, x', x'' \in (E \times E)$ we have

$$\frac{d(M^K)^{m(d)+1}(x, \cdot)}{d(M^K)^{m(d)+1}(x', \cdot)}(x'') \geq \varepsilon(d) \quad \text{and} \quad G(x) \geq e^{-\text{osc}(H)} G(x')$$

In other words, the time homogeneous pair (G, M^K) satisfies the regularity conditions (G) and $(M)_m$ introduced in Section 3.5.2 with $m = m(d) + 1$, $\epsilon(M) \geq \varepsilon(d)$ and $\epsilon(G) \geq e^{-\text{osc}(H)}$. These elementary observations allows to apply most of the asymptotic results presented in Chapter 4 and Chapters 7 to 10 to analyze the asymptotic behavior of these Feynman-Kac models and their particle interpretations as the time parameter or as the size of the system tends to infinity.

For instance, in the case of the Ising model described above and as a direct consequence of (11.6) and (7.26) in Theorem 7.4.4, we find that for any $f \in \text{Osc}_1(E)$

$$\mathbb{E}\left(\left|\frac{1}{N}\sum_{i=1}^N f(Y_{m(d)+n+1}^i) - \pi(f)\right|^p\right)^{1/p} \leq \frac{b(p)}{\sqrt{N}} + 2\varepsilon(d)^{-1}(1-\varepsilon(d))^{\lfloor n/m(d) \rfloor}$$

with

$$b(p) = c\, m(d)\, d(p)^{1/p}\, \varepsilon(d)^{-3} \exp\{(2m(d)+1)\,\text{osc}(H)\}$$

Similarly, for any nondecreasing pair sequences $(q(N), n(N))$ such that $n(N)q^2(N) = o(N)$, we have by Corollary 8.9.2

$$\limsup_{N \to \infty} \frac{N}{n(N)q^2(N)} \|\mathbb{P}_{y,n(N)}^{(N,q(N))} - \pi_{y,n(N)}^{\otimes q(N)}\|_{\text{tv}} \leq c\,\varepsilon(d)^{-2}\, e^{2m(d)\,\text{osc}(H)}$$

where $\mathbb{P}_{y,n}^{(N,q)}$ represents the distribution of the first q-path ancestral lines from the origin up to time n.

11.2.3 Interacting Metropolis and Gibbs Samplers

The objective of this section is to better connect the interacting Metropolis model discussed above with the more traditional Gibbs sampler. More precisely, we show that the N-particle model with "simple" Gibbs mutation transitions coincides with N independent copies of the Gibbs sampler. Let $(E, \mathcal{E})$ be some measurable space and let $\pi \in \mathcal{P}(E^d)$ be the distribution of a d-dimensional random vector $U = (U^i)_{1 \leq i \leq d}$. For each index i, we let $U_i = (U^j)_{1 \leq j \leq d,\ j \neq i}$ be the $(d-1)$-dimensional vector deduced from U by simply deleting the ith coordinate and let π_i be its marginal distribution. We further require that the distribution π can be disintegrated with respect to the distribution π_i and we have the formula

$$\pi(d(u^1,\ldots,u^d)) = \pi_i(du_i)\, \pi^i(u_i, du^i)$$

for some Markov transition π^i from E^{d-1} into E, where du_i stands for an infinitesimal of the point $u_i = (u^j)_{1 \leq j \leq d,\ j \neq i}$. We finally introduce the Markov transition k_i on E^d defined by

$$k_i(f)(u) = \int_E \pi^i(u_i, dv^i) f(\theta_i(u, v^i))$$

for any $(f, u) \in (\mathcal{B}_b(E^d) \times E^d)$ and where $\theta_i(u, v^i) \in E^d$ is the d-vector defined for any $1 \leq k \leq d$ by

$$\theta_i(u, v^i)^k = v^i\, 1_{k=i} + u^i\, 1_{k \neq i}$$

By construction, we clearly have that $(\pi \times k_i)_1 = (\pi \times k_i)_2$ for any $1 \leq i \leq d$; in other words, each k_i is reversible with respect to π. This only indicates that each k_i is a candidate to construct a Markov chain Monte Carlo algorithm with limiting measure π. As the reader has certainly noticed, these kernels are usually degenerate, and the resulting chain will behave very poorly. The simple Gibbs sampler is a homogeneous Markov chain Y_n with an elementary transition of the form $(k_1 \ldots k_d)^m$ for some $m \geq 1$. For $m = 1$, the random transition consists in changing successively each coordinate of the initial state. For $m > 1$, we repeat this mechanism m times. To take the final step, we notice that $(\pi \times K)_1 = (\pi \times L)_2$ for any pair of transitions (K, L) of the form

$$K = (k_1 \ldots k_d)^m \quad \text{and}\quad L = (k_d k_{d-1} \ldots k_1)^m$$

for some $m \geq 1$. We now leave the reader to check that the N interacting Metropolis model associated with the pair (K, L) coincides with N independent samples of the Gibbs algorithm.

11.3 An Overview of some General Principles

In earlier sections, we have developed a particle methodology for solving numerically Feynman-Kac distributions of the form

$$\widehat{\mathbb{Q}}_n(d(x_0,\ldots,x_n)) = \frac{1}{\widehat{\mathcal{Z}}_n} \left\{ \prod_{q=0}^n G_q(x_q) \right\} \mathbb{P}_n(d(x_0,\ldots,x_n)) \qquad (11.7)$$

where G_n is a sequence of nonnegative potential functions on some measurable spaces $(E_n, \mathcal{E}_n)$ and $\mathbb{P}_n$ is the distribution defined by

$$\mathbb{P}_n(d(x_0,\ldots,x_n)) = \eta_0(dx_0) M_1(x_0, dx_1) \ldots M_n(x_{n-1}, dx_n)$$

where M_n is a sequence of Markov transitions from E_{n-1} into E_n and η_0 some initial distribution on E_0. The particle interpretation of these measures is not unique. It can be defined alternatively from the dynamical equation of the nth time marginal $\widehat{\eta}_n$ or from one of the prediction flows η_n given for any $f_n \in \mathcal{B}_b(E_n)$ by the equation

$$\eta_n(f_n) = \gamma_n(f_n)/\gamma_n(1) \quad \text{with} \quad \gamma_n(f_n) = \mathbb{E}\left(f_n(X_n) \prod_{p=0}^{n-1} G_p(X_p) \right)$$

In Section 2.4.3, we have seen that the updated model can also be interpreted as a prediction model associated with the pair of updated potentials/transitions $(\widehat{G}_n, \widehat{M}_n)$ defined in (2.13) and (2.14). In this sense, the analysis of the prediction model provides a unifying treatment of both situations. Its evolution equation has the form

$$\eta_{n+1} = \eta_n K_{n+1,\eta_n}$$

where $K_{n,\eta}$ is a nonunique sequence of Markov transitions. The particle interpretation model associated with a given choice of transitions consists of a sequence of Markov chains on the product spaces E_n^N with elementary transitions

$$\text{Proba}(\xi_n \in d(x_n^1,\ldots,x_n^N) \mid \xi_{n-1}) = \prod_{i=1}^N K_{n+1,m(\xi_n)}(\xi_{n-1}^i, dx_n^i)$$

Under appropriate regularity conditions on the collection $K_{n,\eta}$, the particle occupation measures $\eta_n^N = \frac{1}{N}\sum_{i=1}^N \delta_{\xi_n^i}$ converge in some sense and as $N \to \infty$ to the desired distribution η_n. In Section 2.5.3, we have presented at least six different choices of possible transitions. Given a pair of potentials/transitions (G_n, M_n), we can take for instance

$$K_{n+1,\eta_n} = S_{n,\eta_n} M_{n+1} \qquad (11.8)$$

11.3 An Overview of some General Principles

where S_{n,η_n} is the collection of selection transitions given by the formula

$$S_{n,\eta_n}(x_n, .) = \frac{G_n(x_n)}{\eta_n\text{-ess-sup}(G_n)} \delta_{x_n} + \left(1 - \frac{G_n(x_n)}{\eta_n\text{-ess-sup}(G_n)}\right) \Psi_n(\eta_n) \tag{11.9}$$

For $(0, 1]$-valued potential functions G_n, we can alternatively choose

$$S_{n,\eta_n}(x_n, .) = G_n(x_n) \, \delta_{x_n} + (1 - G_n(x_n)) \, \Psi_n(\eta_n) \tag{11.10}$$

We recall that Ψ_n is the Boltzmann-Gibbs transformation associated with the potential function G_n and defined by

$$\Psi_n(\eta_n)(dx_n) = \frac{1}{\eta_n(G_n)} \, G_n(x_n) \, \eta_n(dx_n)$$

Loosely speaking, one advantage in choosing the first selection transition is that it increases the proportion of particles that do not interact. The interacting particle model associated with the first McKean interpretation model is defined in terms of a two-step mutation/selection transition

$$\xi_n \xrightarrow{S_{n,m(\xi_n)}} \widehat{\xi}_n \xrightarrow{M_{n+1}} \xi_{n+1} \tag{11.11}$$

At time $n = 0$, the system consists of N iid random particles with common distribution η_0. During the selection stage, each particle ξ_n^i selects a new location with the discrete distribution

$$S_{n,m(\xi_n)}(\xi_n^i, .) = \frac{G_n(\xi_n^i)}{\vee_{j=1}^N G_n(\xi_n^j)} \delta_{\xi_n^i} + \left(1 - \frac{G_n(\xi_n^i)}{\vee_{j=1}^N G_n(\xi_n^j)}\right) \Psi_n(m(\xi_n))$$

During the mutation stage, we evolve randomly each selected particle according to the Markov transition M_{n+1}. An important and distinctive feature of these particle approximation models is that they can also be used to estimate recursively in time the partition functions $\gamma_n(1)$ as well as the Feynman-Kac path measures (11.7). The last assertion will be discussed in some detail in Section 11.4. The central idea to approximate the unnormalized distributions γ_n is to use the easily derived product formula

$$\gamma_n(f_n) = \eta_n(f_n) \prod_{p=0}^{n-1} \eta_p(G_p) \tag{11.12}$$

Mimicking this key representation, we construct a natural unbiased estimate simply by setting

$$\gamma_n^N(f_n) = \eta_n^N(f_n) \prod_{p=0}^{n-1} \eta_p^N(G_p)$$

For more details on the asymptotic behavior of these approximation measures, we refer the reader to Chapters 7 and 9.

11.4 Descendant and Ancestral Genealogies

The evolutionary particle model described at the end of Section 11.3 has a natural birth and death interpretation. We refer the reader to Chapter 3. Tracing back in time the genealogy of each current individual $\widehat{\xi}_n^i$, we define the ancestor lines

$$(\widehat{\xi}_{0,n}^i, \widehat{\xi}_{1,n}^i, \ldots, \widehat{\xi}_{n-1,n}^i, \widehat{\xi}_{n,n}^i)$$

This genealogical tree structure can be used to estimate the Feynman-Kac path measures $\widehat{\mathbb{Q}}_n$ defined in (11.7). In some sense, as $N \to \infty$ we have

$$\widehat{\mathbb{Q}}_n^N = \frac{1}{N} \sum_{i=1}^N \delta_{(\widehat{\xi}_{0,n}^i, \widehat{\xi}_{1,n}^i, \ldots, \widehat{\xi}_{n-1,n}^i, \widehat{\xi}_{n,n}^i)} \longrightarrow \widehat{\mathbb{Q}}_n \qquad (11.13)$$

The precise meaning of this asymptotic result is described in Chapters 7 to 10. This tree-based particle model also provides a natural way to estimate the conditional distributions of $\widehat{\mathbb{Q}}_n$ with respect to the time horizon given by the following proposition.

Proposition 11.4.1 *We fix a time horizon $n \geq 0$ and we let $(Y_0, \ldots, Y_n)$ be a random path on $(E_0 \times \ldots \times E_n)$ with distribution $\widehat{\mathbb{Q}}_n$. Also let $\widehat{\mathbb{Q}}_{p,n}$ be the marginal of $\widehat{\mathbb{Q}}_n$ w.r.t. the first $(p+1)$-coordinates. For each $p \leq n$, a version $\widehat{\mathbb{Q}}_{n|p}(x_p, d(x_{p+1}, \ldots, x_n))$ of the conditional distribution $(Y_{p+1}, \ldots, Y_n)$ given $(Y_0, \ldots, Y_p) = (x_0, \ldots, x_p)$ only depends on $Y_p = x_p$ and it is given for $\widehat{\mathbb{Q}}_{p,n}$-a.e. $(x_0, \ldots, x_p)$ by the formula*

$$\widehat{\mathbb{Q}}_{n|p}(x_p, d(x_{p+1}, \ldots, x_n)) = \frac{\prod_{q=p+1}^n G_q(x_q)}{\widehat{\mathcal{Z}}_{n|p}(x_p)} \mathbb{P}_{n|p}(x_p, d(x_{p+1}, \ldots, x_n))$$

with

$$\mathbb{P}_{n|p}(x_p, d(x_{p+1}, \ldots, x_n)) = M_{p+1}(x_p, dx_{p+1}) \ldots M_n(x_{n-1}, dx_n)$$

and the normalizing constants $\widehat{\mathcal{Z}}_{n|p}(x_p) = \mathbb{E}(\prod_{q=p+1}^n G_q(X_q) | X_p = x_p)$.

Proof:
We first notice that the marginal of $\widehat{\mathbb{Q}}_n$ with respect to the first $p+1$ coordinates $(x_0, \ldots, x_p)$ is given by

$$\widehat{\mathbb{Q}}_{p,n}(d(x_0, \ldots, x_p)) = \frac{\widehat{\mathcal{Z}}_{n|p}(x_p)}{\widehat{\mathcal{Z}}_n} \prod_{q=0}^p G_q(x_q) \, \mathbb{P}_p(d(x_0, \ldots, x_p))$$

Then we readily check that

$$\widehat{\mathbb{Q}}_n(d(x_0, \ldots, x_p)) = \widehat{\mathbb{Q}}_{p,n}(d(x_0, \ldots, x_p)) \, \widehat{\mathbb{Q}}_{n|p}(x_p, d(x_{p+1}, \ldots, x_n))$$

from which the end of the proof is straightforward. ∎

Loosely speaking, any individual in the genealogical tree, say $\widehat{\xi}^i_{p,n}$, can be regarded as the common ancestor of a particle evolution model from time p up to the current time n. The resulting tree of descendant individuals provides a particle approximation of the $\widehat{\mathbb{Q}}_n$-conditional distribution of a canonical path $(Y_p, \ldots, Y_n)$ given $Y_p = \widehat{\xi}^i_{p,n}$.

To describe more precisely the particle approximations of the conditional distributions $\widehat{\mathbb{Q}}_{n|p}$, it is convenient to count the number of descendant individual of each ancestor. To define these numbers, we first fix a time horizon $n \geq 0$ and we let $\widehat{N}_{0,n}$ be the number of ancestors at the origin 0; in other words,

$$\widehat{N}_{0,n} = \text{Card}\{1 \leq i_0 \leq N : \widehat{\xi}^{i_0}_{0,n}\}$$

We also let $\widehat{\mathcal{A}}_{0,n} = \{\widehat{\chi}^{i_0}_{0,n} : 1 \leq i_0 \leq \widehat{N}_{0,n}\}$ be the set of these initial ancestors. Each of these ancestors $\widehat{\chi}^{i_0}_{0,n}$ is the parent of $\widehat{N}^{i_0}_{1,n}$ individuals

$$\widehat{\chi}^{i_0}_{0,n} \longrightarrow \widehat{\chi}^{i_0,i_1}_{1,n}, \quad i_1 = 1, \ldots, \widehat{N}^{i_0}_{1,n}$$

at level $p = 1$. In the same way, each of these individuals $\widehat{\chi}^{i_0,i_1}_{1,n}$ is the parent of $\widehat{N}^{i_0,i_1}_{2,n}$ individuals

$$\widehat{\chi}^{i_0,i_1}_{1,n} \longrightarrow \widehat{\chi}^{i_0,i_1,i_2}_{2,n}, \quad i_2 = 1, \ldots, \widehat{N}^{i_0,i_1}_{2,n}$$

at level $p = 2$, and so on, up to the current population at level $p = n$. To clarify the presentation, we slightly abuse the notation and we write $i_{[0,p]}$ instead of $(i_0, i_1, \ldots, i_p)$. We notice that a particular individual $\widehat{\chi}^{i_{[0,p]}}_{p,n}$ at level p has had $\widehat{d}_{p,n}(i_{[0,p]})$ descendants *at level* n and we have the backward inductive formula

$$\widehat{d}_{p,n}(i_{[0,p]}) = \sum_{i_{p+1}=1}^{\widehat{N}^{i_{[0,p]}}_{p+1,n}} \widehat{d}_{p+1,n}(i_{[0,p]}, i_{p+1}) \tag{11.14}$$

with the constant terminal conditions $\widehat{d}_{n,n}(i_{0,n}) = 1$ so that

$$\widehat{d}_{n-1,n}(i_{[0,n-1]}) = \widehat{N}^{i_{[0,n-1]}}_{n,n}$$

Also notice that we have the backward integer decomposition

$$N = \sum_{i_0=1}^{\widehat{N}_{0,n}} \widehat{d}_{0,n}(i_0) = \sum_{i_0=1}^{\widehat{N}_{0,n}} \sum_{i_1=1}^{\widehat{N}^{i_0}_{1,n}} \widehat{d}_{1,n}(i_{[0,1]}) = \sum_{i_0=1}^{\widehat{N}_{0,n}} \sum_{i_1=1}^{\widehat{N}^{i_0}_{1,n}} \sum_{i_2=1}^{\widehat{N}^{i_{[0,1]}}_{2,n}} \widehat{d}_{2,n}(i_{[0,2]}) = \ldots$$

The disintegration formula of the genealogical path-particle distribution (11.13) is now given by the following proposition.

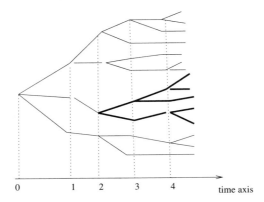

FIGURE 11.1. Genealogy of a given ancestor

Proposition 11.4.2 *We fix a time horizon $n \geq 0$ and we let $(Y_0, \ldots, Y_n)$ be a random path on $(E_0 \times \ldots \times E_n)$ with distribution $\widehat{\mathbb{Q}}_n^N$. For each $p \leq n$, a version*
$$\widehat{\mathbb{Q}}_{n|p}^N(\widehat{\chi}_{p,n}^{i_{[0,p]}}, d(x_{p+1}, \ldots, x_n))$$
of the conditional distribution $(Y_{p+1}, \ldots, Y_n)$ given $Y_p = \widehat{\chi}_{p,n}^{i_{[0,p]}}$ is defined by the empirical measure of the genealogical tree model of the ancestor $\widehat{\chi}_{p,n}^{i_{[0,p]}}$, namely

$$\widehat{\mathbb{Q}}_{n|p}^N(\widehat{\chi}_{p,n}^{i_{[0,p]}}, \cdot) = \frac{1}{\widehat{d}_{p,n}(i_{[0,p]})} \sum_{i_{p+1}=1}^{\widehat{N}_{p+1,n}^{i_{[0,p]}}} \sum_{i_{p+2}=1}^{\widehat{N}_{p+2,n}^{i_{[0,p+1]}}} \cdots \sum_{i_n=1}^{\widehat{N}_{n,n}^{i_{[0,n-1]}}} \delta_{(\widehat{\chi}_{p+1,n}^{i_{[0,p+1]}}, \ldots, \widehat{\chi}_{n,n}^{i_{[0,n]}})}$$

In Figure 11.1, we have presented with a thick line the genealogical tree model associated with a given ancestor $\widehat{\chi}_{2,5}^{i_{[0,2]}}$ with time label 2. This picture corresponds to the situation where $(p,n) = (2,5)$, and $N = 15$. In this situation, we note that the backward decomposition formula (11.14) reads

$$\widehat{d}_{2,5}(i_{[0,2]}) = 5 = 3 + 2 = (2+1) + 2$$

and we have $\widehat{N}_{3,5}^{i_{[0,2]}} = 2$, $(\widehat{N}_{4,5}^{i_{[0,2]},1}, \widehat{N}_{4,5}^{i_{[0,2]},2}) = (2,1)$, and

$$(\widehat{N}_{5,5}^{i_{[0,2]},1,1}, \widehat{N}_{5,5}^{i_{[0,2]},1,2}, \widehat{N}_{5,5}^{i_{[0,2]},2,1}) = (2,1,2)$$

We use the same disintegration technique for the prediction Feynman-Kac path-measure $\mathbb{Q}_n$ (defined as in 11.7, with taking the product of potential functions up to time $(n-1)$) and its genealogical path-particle approximation

$$\mathbb{Q}_n^N = \frac{1}{N} \sum_{i=1}^N \delta_{(\xi_{0,n}^i, \xi_{1,n}^i, \ldots, \xi_{n-1,n}^i, \xi_{n,n}^i)}$$

11.4 Descendant and Ancestral Genealogies

In the display above, $(\xi_{p,n}^i)_{0\leq p\leq n}$ represents the ancestral line of the current individual $\xi_{n,n}^i = \xi_n^i$.

We denote by $(d_{p,n}(i_{[0,p]}), N_{p+1,n}^{i_{[0,p]}}, \chi_{p,n}^{i_{[0,p]}})$ the quantities defined as

$$(\widehat{d}_{p,n}(i_{[0,p]}), \widehat{N}_{p+1,n}^{i_{[0,p]}}, \widehat{\chi}_{p,n}^{i_{[0,p]}})$$

by replacing the updated ancestral lines $\widehat{\xi}_{p,n}^i$ by the predicted lines $\xi_{p,n}^i$.

Definition 11.4.1 *For any $0 \leq p \leq n$, and any multi-index $i_{[0,p]} = (i_0,\ldots,i_p)$, with $1 \leq i_k \leq N_{k,n}^{i_{[0,k-1]}}$, $k \leq p$, the descendant genealogical tree model of the pth ancestor $\chi_{p,n}^{i_{[0,p]}} \in E_p$ at time n is the random tree starting at $\chi_{p,n}^{i_{[0,p]}}$ and given by*

$$\chi_{p,n}^{i_{[0,p]}} \longrightarrow \chi_{p+1,n}^{i_{[0,p]}, i_{p+1}} \longrightarrow \cdots \longrightarrow \chi_{n,n}^{i_{[0,p]}, i_{p+1},\ldots, i_n}$$

The descendant genealogy of a given ancestor forms a random tree valued process that evolves in accordance with the selection/mutation transitions of the particle model. Figure 11.2 gives a schematic picture of the descendant genealogical tree models from time p to time $n = p + 4$ associated with a given individual x at time p. Note that any particle $\xi_p^i = x$ in the complete genealogical tree model at time p can be regarded as the current descendant individual of an historical process; that is, we have

$$\chi_{p,p}^{i_{[0,p]}^p} = x$$

for some multi-index $i_{[0,p]}^p = (i_0^p, \ldots, i_p^p)$. When the descendant genealogy of x still survives at time $n > p$ we also have $\chi_{p,k}^{i_{[0,p]}^k} = x$, for any k from p to n, and for some multi-index $i_{[0,p]}^k = (i_0^k, \ldots, i_p^k)$. In other words, x is the common ancestor of a sequence of genealogical trees. In Figure 11.2, we have presented the descendant tree models at time $p+1$, $p+2$, $p+3$, and $n = p+4$. Note that the number of descendant individuals is not fixed but it depends on the evolution of the whole particle model. From previous arguments, we observe that the path-particle measures $\mathbb{Q}_{k|p}^N(x,\cdot)$ provide "an N-approximation" of the conditional Feynman-Kac path-measures $\mathbb{Q}_{k|p}(x,\cdot)$. We denote by $\mathbb{Q}_{[k]|p}^N(x,\cdot)$ and $\mathbb{Q}_{[k]|p}(x,\cdot)$ their marginal w.r.t. the time parameter k. For instance, in the time homogeneous situation we have for any $l \geq 0$

$$\mathbb{Q}_{[p+l]|p}(x,\cdot) = \eta_l^{(x)}$$

where $\eta_l^{(x)}$ represents the solution of the prediction Feynman-Kac flow starting at $\eta_0^{(x)} = \delta_x$ at time $l = 0$. Using the product formula (11.12) we have

$$\mathbb{E}_{p,x}\left(f(X_{p+l}) \prod_{q=p}^{p+l-1} G(X_q)\right) = \mathbb{Q}_{[p+l]|p}(f)(x) \prod_{q=p}^{p+l-1} \mathbb{Q}_{[q]|p}(G)(x)$$

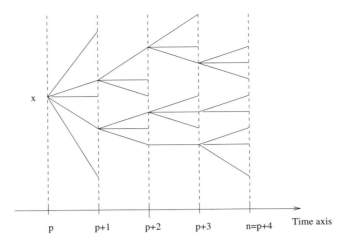

FIGURE 11.2. Complete Descendant Genealogy

These unnormalized measures can be estimated using the x-descendant genealogical measures $\mathbb{Q}^N_{[p+k]|p}(x,.)$ with the product formula

$$f \in \mathcal{B}_b(E) \longrightarrow \mathbb{Q}^N_{[p+l]|p}(f)(x) \prod_{q=p}^{p+l-1} \mathbb{Q}^N_{[q]|p}(G)(x) \qquad (11.15)$$

11.5 Conditional Explorations

In the algorithm presented in (11.11), the particles randomly explore the state space using an "a priori" mutation transition. The latter terminology has been chosen in reference to the filtering literature. It simply expresses the fact that the mutation transition does not depend on the potential functions. In the present section, we design a more accurate exploration strategy that depends on the potential functions. The key idea is simply based on the decomposition

$$M_{n+1}(x_n, dx_{n+1})\, G_{n+1}(x_{n+1}) = \widehat{G}_n(x_n)\, \widehat{M}_{n+1}(x_n, dx_{n+1})$$

with

$$\widehat{M}_{n+1}(x_n, dx_{n+1}) = \frac{M_{n+1}(x_n, dx_{n+1})\, G_{n+1}(x_{n+1})}{M_{n+1}(G_{n+1})(x_n)} \qquad (11.16)$$

and $\widehat{G}_n(x_n) = M_{n+1}(G_{n+1})(x_n)$. From the display above, we find that

$$\widehat{\mathbb{Q}}_n(d(x_0,\ldots,x_n)) = \frac{1}{\widehat{\mathcal{Z}}_n} \left\{ \prod_{q=0}^{n-1} \widehat{G}_q(x_q) \right\} \widehat{\mathbb{P}}_n(d(x_0,\ldots,x_n)) \qquad (11.17)$$

with the path distribution $\widehat{\mathbb{P}}_n$ given by

$$\widehat{\mathbb{P}}_n(d(x_0,\ldots,x_n)) = \widehat{\eta}_0(dx_0)\widehat{M}_1(x_0,dx_1)\ldots\widehat{M}_n(x_{n-1},dx_n)$$

From these formulae, one deduces that the updated Feynman-Kac flow $\widehat{\eta}_n$ satisfies the equation

$$\widehat{\eta}_{n+1} = \widehat{\eta}_n \widehat{K}_{n+1,\widehat{\eta}_n} \qquad (11.18)$$

with the collection of Markov transitions $\widehat{K}_{n+1,\widehat{\eta}_n} = \widehat{S}_{n,\widehat{\eta}_n}\widehat{M}_{n+1}$ defined as in (11.8) by replacing the pair (G_n, M_n) by $(\widehat{G}_n, \widehat{M}_n)$. The particle interpretation model associated with (11.18) is now defined as in (11.11) by replacing the pairs (G_n, M_n) by $(\widehat{G}_n, \widehat{M}_n)$. More precisely, this particle model is again defined by a selection/mutation transition

$$\xi_n \xrightarrow{\widehat{S}_{n,m(\xi_n)}} \widehat{\xi}_n \xrightarrow{\widehat{M}_{n+1}} \xi_{n+1} \qquad (11.19)$$

Note that during the selection stage each particle ξ_n^i selects a new location with the discrete distribution

$$\widehat{S}_{n,m(\xi_n)}(\xi_n^i,\cdot) = \frac{\widehat{G}_n(\xi_n^i)}{\vee_{j=1}^N \widehat{G}_n(\xi_n^j)} \delta_{\xi_n^i} + \left(1 - \frac{\widehat{G}_n(\xi_n^i)}{\vee_{j=1}^N \widehat{G}_n(\xi_n^j)}\right) \widehat{\Psi}_n(m(\xi_n))$$

where $\widehat{\Psi}_n$ is the Boltzmann-Gibbs transformation associated with the potential function $\widehat{G}_n$. During the mutation stage, we randomly evolve each selected particle according to the Markov transition $\widehat{M}_{n+1}$. Particle models with conditional mutations are particularly useful when the regularity condition is not met for the initial potential functions G_n but for the new reference functions $M_n(G_n) = \widehat{G}_n$ (see Exercise 11.9.6). Nevertheless, and as we have already mentioned in Chapter 3, page 100, sampling random transitions according to $\widehat{M}_{n+1}$ is usually time-consuming, and we need to resort to an additional level of approximation.

One natural idea is to sample the transition $\widehat{\xi}_n^i \rightsquigarrow \xi_n^i$ from a given selected particle, say $\widehat{\xi}_n^i = \xi_n^j$, and to use at each step the particle approximation

$$M_{n+1}(\xi_n^j, dx_{n+1}) \simeq M_{n+1}^{N'}(\xi_n^j, dx_{n+1}) =_{\text{def.}} \frac{1}{N'}\sum_{k=1}^{N'} \delta_{U_{n+1}^{j,k}}$$

where $(U_{n+1}^{j,k})_{1\leq k\leq N'}$ is an auxiliary collection of N' conditionally independent random variables with distribution $M_{n+1}(\xi_n^j,\cdot)$. Replacing in (11.16) the transition M_{n+1} by its N'-approximation, we have in some sense

$$\widehat{M}_{n+1}(\xi_n^j,\cdot) \simeq \widehat{M}_{n+1}^{N'}(\xi_n^j,\cdot) =_{\text{def.}} \sum_{k=1}^{N'} \frac{G_{n+1}(U_{n+1}^{j,k})}{\sum_{l=1}^{N'} G_{n+1}(U_{n+1}^{j,l})} \delta_{U_{n+1}^{j,k}}$$

and $\quad \widehat{G}_n(\xi_n^j) \simeq \widehat{G}_n^{N'}(\xi_n^j) =_{\text{def.}} \frac{1}{N'}\sum_{k=1}^{N'} G_{n+1}(U_{n+1}^{j,k})$

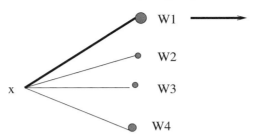

FIGURE 11.3. particle conditional exploration

In Figure 11.2, we have presented with a thick line a single approximation conditional exploration starting from x and based on $N' = 4$ exploration transitions $\xi_n^j = x \rightsquigarrow U_{n+1}^{j,k}$, $1 \leq k \leq 4$, with respective weights $G_{n+1}(U_{n+1}^{j,k}) = Wk$.

11.6 State-Space Enlargements and Path-Particle Models

Let $p = (p_n)_{n \geq 0}$ be an increasing sequence of time parameters such that $p_0 = 0$ and $p_n < p_{n+1}$ for all $n \geq 0$. For any $s < t$, we denote by

$$x_{[s,t)} = (x_q)_{s \leq q < t} \in E_{[s,t)} = \prod_{s \leq q < t} E_q$$

a generic excursion from time s up to time t (excluded). We associate with p the collection of potential functions, Markov transitions, and initial distributions $(G_n^{(p)}, M_n^{(p)}, \eta_0^{(p)})$ on excursion spaces and defined by

$$G_n^{(p)}(x_{[p_n, p_{n+1})}) = \prod_{p_n \leq q < p_{n+1}} G_q(x_q)$$

$$M_n^{(p)}(x_{[p_{n-1}, p_n)}, dx_{[p_n, p_{n+1})}) = \prod_{p_n \leq q < p_{n+1}} M_q(x_{q-1}, dx_q)$$

and

$$\eta_0^{(p)}(dx_{[0,p_1)}) = \eta_0(dx_0) \, M_1(x_0, dx_1) \ldots M_{p_1-1}(x_{p_1-2}, dx_{p_1-1})$$

Using this notation, we observe that any path from 0 up to time p_{n+1} can be decomposed with respect to the time mesh p with the excursion decomposition

$$x_{[0,p_{n+1})} = (x_{[0,p_1)}, \ldots, x_{[p_n, p_{n+1})}) \in E_{[0, p_{n+1})} = E_0^{(p)} \times \ldots \times E_n^{(p)}$$

11.6 State-Space Enlargements and Path-Particle Models

where $E_n^{(p)} = E_{[p_n, p_{n+1}]}$ for all $n \geq 0$. Let $\widehat{\mathbb{Q}}_n^{(p)}$ be the updated Feynman-Kac distribution defined as in (11.7) by replacing the triplet (G_n, M_n, η_0) by $(G_n^{(p)}, M_n^{(p)}, \eta_0^{(p)})$.

Proposition 11.6.1 *For any $n \geq 0$, we have $\widehat{\mathbb{Q}}_{p_{n+1}-1} = \widehat{\mathbb{Q}}_n^{(p)}$.*

Proof:
It suffices to observe that $\prod_{q=0}^{p_{n+1}-1} G_q(x_q) = \prod_{q=0}^{n} G_q^{(p)}(x_q)$ and $\mathbb{P}_{p_{n+1}-1} = \mathbb{P}_n^{(p)}$ with

$$\mathbb{P}_n^{(p)}(d(x_{[0,p_1]}, \ldots, x_{[p_n, p_{n+1}]}))$$
$$= \eta_0^{(p)}(dx_{[0,p_1]}) \, M_1^{(p)}(x_{[0,p_1]}, dx_{[p_1,p_2]}) \ldots M_n^{(p)}(x_{[p_{n-1},p_n]}, dx_{[p_n,p_{n+1}]})$$

∎

Using Proposition 11.6.1, we extend the particle model defined in (11.11) into an excursion-valued particle interpretation. To be more precise, let $\xi_n^{(p)}, \widehat{\xi}_n^{(p)}$ be the Markov chain defined as in (11.11) by replacing the set of parameters (E_n, G_n, M_n, η_0) by $(E_n^{(p)}, G_n^{(p)}, M_n^{(p)}, \eta_0^{(p)})$. At time $n = 0$, the system consists of N iid excursions

$$\xi_0^{(p),i} = (\zeta_0^i, \ldots, \zeta_{p_1-1}^i)$$

with common distribution $\eta_0^{(p)}$. During the selection stage, we randomly select N excursions

$$\widehat{\xi}_n^{(p),i} = \widehat{\zeta}_{[p_n,p_{n+1}]}^i = (\widehat{\zeta}_{p_n}^i, \ldots, \widehat{\zeta}_{p_{n+1}-1}^i)$$

with the discrete distribution

$$S_{n,m(\xi_n^{(p)})}^{(p)}(\xi_n^{(p),i}, \cdot)$$

$$= \frac{G_n^{(p)}(\xi_n^{(p),i})}{\vee_{j=1}^N G_n^{(p)}(\xi_n^{(p),j})} \, \delta_{\xi_n^{(p),i}} + \left(1 - \frac{G_n^{(p)}(\xi_n^{(p),i})}{\vee_{j=1}^N G_n^{(p)}(\xi_n^{(p),j})}\right) \Psi_n^{(p)}(m(\xi_n^{(p)}))$$

In the display above $\Psi_n^{(p)}$ represents the Boltzmann-Gibbs transformation on the excursion space $E_n^{(p)}$ associated with the potential function $G_n^{(p)}$. During the mutation stage, we randomly evolve each selected excursion according to the Markov transition $M_{n+1}^{(p)}$

$$\widehat{\xi}_n^{(p),i} = \widehat{\zeta}_{[p_n,p_{n+1}]}^i \xrightarrow{M_{n+1}^{(p)}} \xi_{n+1}^{(p),i} = \zeta_{[p_{n+1},p_{n+2}]}^i$$

In other words, we randomly sample N excursions of length $(p_{n+2} - p_{n+1})$ from the final selected sites

$$\widehat{\zeta}_{p_{n+1}-1}^i \xrightarrow{M_{p_{n+1}}} \zeta_{p_{n+1}}^i \xrightarrow{M_{p_{n+1}+1}} \zeta_{p_{n+1}+1}^i \xrightarrow{M_{p_{n+1}+2}} \cdots \xrightarrow{M_{p_{n+2}-1}} \zeta_{p_{n+2}-1}^i$$

This model and the one introduced in (11.11) are mathematically equivalent, but in practice the excursion particle model often gives more accurate results. For instance, in the simple filtering example discussed in the introduction of the book, the radar observations only give information on the distance between the sensor and the target. To estimate the speed and acceleration components of the signal, we clearly need to use at least three observations. Therefore, in terms of the excursion particle model presented above, it is natural to update the particle excursions using at least three likelihood functions. In this situation, it is more judicious to take $(p_{n+1} - p_n) \geq 3$.

11.7 Conditional Excursion Particle Models

We can also combine the excursion particle model described in Section 11.6 with the conditional exploration techniques presented in Section 11.5. The corresponding conditional excursion particle model is defined as the one presented in Section 11.6 by replacing $(G_n^{(p)}, M_n^{(p)}, \eta_0^{(p)})$ by the triplets $(\widehat{G}_n^{(p)}, \widehat{M}_n^{(p)}, \widehat{\eta}_0^{(p)})$ defined by

$$\widehat{M}_n^{(p)}(x_{[p_{n-1},p_n]}, dx_{[p_n,p_{n+1}]})$$

$$= \frac{M_n^{(p)}(x_{[p_{n-1},p_n]}, dx_{[p_n,p_{n+1}]})\, G_n^{(p)}(x_{[p_n,p_{n+1}]})}{M_n^{(p)}(G_n^{(p)})(x_{[p_{n-1},p_n]})}$$

and

$$\widehat{G}_{n-1}^{(p)}(x_{[p_{n-1},p_n]}) = M_n^{(p)}(G_n^{(p)})(x_{[p_{n-1},p_n]})$$
$$\widehat{\eta}_0^{(p)} = \Psi_0^{(p)}(\eta_0^{(p)}) \qquad (11.20)$$

As in Section 11.6, to sample the conditional excursion transition

$$\widehat{\xi}_n^{(p),i} = (\widehat{\zeta}_{[p_n,p_{n+1}]}^i) \rightsquigarrow \xi_{n+1}^{(p),i} = (\zeta_{[p_{n+1},p_{n+2}]}^i)$$

from a given selected excursion, say $\widehat{\xi}_n^{(p),i} = \xi_n^{(p),j}$, we can use the N'-particle approximation

$$M_{n+1}^{(p)}(\xi_n^{(p),j}, .) \simeq M_{n+1}^{(p),N'}(\xi_n^{(p),j}, .) =_{\text{def.}} \frac{1}{N'} \sum_{k=1}^{N'} \delta_{U_{[p_n,p_{n+1}]}^{j,k}}$$

where $(U_{[p_{n+1},p_{n+2}]}^{j,k})_{1 \leq k \leq N'}$ represents an auxiliary collection of N' conditionally independent random excursions with distribution $M_{n+1}^{(p)}(\xi_n^{(p),j}, .)$.

Replacing in the definition of $(\widehat{M}_n^{(p)}, \widehat{G}_n^{(p)})$ the transition $M_{n+1}^{(p)}$ by its N'-approximation $M_{n+1}^{(p),N'}$, we have

$$\widehat{M}_{n+1}^{(p)}(\xi_n^{(p),j},\,.\,) \simeq \widehat{M}_{n+1}^{(p),N'}(\xi_n^{(p),j},\,.\,)$$

$$=_{\text{def.}} \sum_{k=1}^{N'} \frac{G_{n+1}^{(p)}(U_{[p_{n+1},p_{n+2}]}^{j,k})}{\sum_{l=1}^{N'} G_{n+1}^{(p)}(U_{[p_{n+1},p_{n+2}]}^{j,l})} \delta_{\xi_{n+1}^{j,k}}$$

and $\widehat{G}_n^{(p)}(\xi_n^{(p),j}) \simeq \widehat{G}_n^{(p),N'}(\xi_n^{(p),j}) =_{\text{def.}} \frac{1}{N'} \sum_{k=1}^{N'} G_{n+1}^{(p)}(U_{[p_{n+1},p_{n+2}]}^{j,k})$

The resulting (N, N')-approximation model is essentially the same as the one we discussed at the end of Section 11.5 except that we are using here an additional excursion exploration mechanism.

11.8 Branching Selection Variants

11.8.1 Introduction

In this section, we design branching particle interpretations of the Feynman-Kac model $(\eta_n, \widehat{\eta}_n)$ associated with a pair of potential functions and Markov transitions (G_n, M_n) on some measurable state spaces $(E_n, \mathcal{E}_n)$ (see Section 11.3 for a more precise definition of these distribution flows).

The best pedagogical way to introduce these branching strategies is to interpret the interacting particle models presented in Section 11.3 as a birth and death Markov chain. Here again, to simplify the presentation we restrict ourselves to $(0, 1]$-valued potential functions G_n and McKean transitions given by

$$K_{n+1,\eta_n} = S_{n,\eta_n} M_{n+1}$$

with the selection kernel

$$S_{n,\eta_n}(x, dy) = G_n(x)\, \delta_x(dy) + (1 - G_n(x))\, \Psi_n(\eta_n)(dy) \qquad (11.21)$$

We use the same notations as in (1.16), and we denote by $|g_n| = \sum_{i=1}^N g_n^i$ the number of particles which have *accepted* to stay in the same location and we let $(\widetilde{\xi}_n^j)_{|g_n|<j\leq N}$ be a collection of $\widetilde{N}_n = (N - |g_n|)$ conditionally independent and identically distributed selected sites with law $\Psi_n(m(\xi_n))$. After the selection mechanism, the total number of particle at site ξ_n^i is clearly given by

$$b_n^i = g_n^i + \widetilde{b}_n^i \quad \text{and} \quad \widetilde{b}_n^i = \text{Card}\{|g_n| < j \leq N : \widetilde{\xi}_n^j = \xi_n^i\}$$

where

$$(\widetilde{b}_n^1, \ldots, \widetilde{b}_n^N) \stackrel{\text{def.}}{=} \text{Multinomial}(\widetilde{N}_n, W_n^1, \ldots, W_n^N) \quad \text{with} \quad W_n^i = \frac{G_n(\xi_n^i)}{Nm(\xi_n)(G_n)} \qquad (11.22)$$

406   11. Feynman-Kac and Interacting Particle Recipes

The N-valued random variables $(b_n^i)_{1 \leq i \leq N}$ can be interpreted as the random number of births of the individuals $(\xi_n^i)_{1 \leq i \leq N}$. Note that we have

$$\begin{aligned}\mathbb{E}(b_n^i|\xi_n) &= \mathbb{E}(g_n^i|\xi_n) + \mathbb{E}\Big(\sum_{j=|g_n|+1}^{N} 1_{\xi_n^i}(\widetilde{\xi}_n^j) \mid \xi_n\Big) \\ &= G_n(\xi_n^i) + (1 - m(\xi_n)(G_n))\ G_n(\xi_n^i)/m(\xi_n)(G_n) \\ &= G_n(\xi_n^i)/m(\xi_n)(G_n)\end{aligned}$$

and therefore, with some obvious abusive notation

$$\mathbb{E}\Big(\frac{1}{N}\sum_{i=1}^{N} b_n^i\ \delta_{\xi_n^i} \mid \xi_n\Big) = \Psi_n(m(\xi_n))$$

In much the same way, the local branching $\mathbb{L}_2$-mean errors are given by

$$\mathbb{E}((\textstyle\sum_{i=1}^{N} b_n^i f_n(\xi_n) - N\ m(\xi_n) S_{n,m(\xi_n)}(f_n))^2|\xi_n)$$

$$= N\ m(\xi_n) S_{n,m(\xi_n)}(f_n - S_{n,m(\xi_n)}(f_n))^2$$

$$= N\ \Psi_n(m(\xi_n))(f_n - \Psi_n(m(\xi_n))f_n)^2 - N\ m(\xi_n)(G_n^2(f_n - \Psi_n(m(\xi_n))f_n)^2)$$

$$= N\ \Psi_n(m(\xi_n))(f_n - \Psi_n(m(\xi_n))(f_n))^2$$
$$- N\ m(\xi_n)(G_n)\ \Psi_n(m(\xi_n))(G_n(f_n - \Psi_n(m(\xi_n))(f_n))^2)$$

It is also instructive to compare the local variance of this model with the one of the simple genetic model associated with the McKean selection kernel

$$S_{n,\eta_n}(x_n, dx_n) = \Psi_n(\eta_n)(dx_n) \qquad (11.23)$$

In this situation, the selection transition simply consists in sampling randomly N conditionally independent random particles $\widehat{\xi}_n^i$ with the same discrete distribution $\Psi_n(m(\xi_n))$ The number of particle have selected the i-th site

$$b_n^i = \text{Card}\{1 \leq j \leq N : \widehat{\xi}_n^j = \xi_n^i\}$$

can be interpreted as the number of births of the individual ξ_n^i and we have

$$(b_n^1, \ldots, b_n^N) \stackrel{\text{def.}}{=} \text{Multinomial}(N, W_n^1, \ldots, W_n^N) \qquad (11.24)$$

The local error variance is now given by

$$\mathbb{E}((\textstyle\sum_{i=1}^{N} b_n^i f_n(\xi_n) - N\ m(\xi_n) S_{n,m(\xi_n)}(f_n))^2|\xi_n)$$

$$= N\ \Psi_n(m(\xi_n))(f_n - \Psi_n(m(\xi_n))f_n)^2$$

We readily observe that this local variance is greater than the one of the interacting model associated with the McKean transition (11.21). In comparison to (11.23), the McKean selection transition (11.21) contains an important extra acceptance term which allow particles with high potential to stay in the same level. In this sense, the simple genetic selection contains too much randomness. For instance, in the somewhat degenerate case where G_n are constant functions $G_n = 1$, the local selection variance associated with the transition (11.21) is null while the one (11.23) is equal to $m(\xi_n))(f_n - m(\xi_n)f_n)^2$.

The question discussed above is clearly related to the Efron and the Bayesian weighted bootstrap theory (see for instance [25] and references therein). Roughly speaking, the original idea of the bootstrap is that a sample empirical measure calculated from an *observed* sample distribution is sufficiently like the unknown distribution of the observations X. As a result, we can use the bootstrapped sample to estimate some statistical quantities of the unknown distribution such as the mean or the variance.

To better connect the bootstrap ideas with the previous discussion, it is convenient to interpret $X = (X^i)_{1 \leq i \leq N}$ as N independent observations of some unknown distribution η. In this interpretation, $\Psi(m(X))$ is an observation of the unknown updated distribution $\Psi(\eta)$ and finally $\widehat{X} = (\widehat{X}^i)_{1 \leq i \leq N}$ is the corresponding bootstrapped sample. The question addressed above is now essentially equivalent to finding a bootstrap strategy with respect to the observed distribution $\Psi(m(X))$. Using some standard abusive notation, if $\overline{X} = (\overline{X}^i)_{1 \leq i \leq N}$ represents a sequence of iid random variables with common law $\Psi(\eta)$, then for any test function f and as $N \to \infty$ we have

$$\sqrt{N}(m(\overline{X}) - \Psi(\eta))(f) \to_d \mathcal{N}(0, \overline{\sigma}^2(f))$$

with $\overline{\sigma}^2(f) = \Psi(\eta)((f - \Psi(\eta)(f))^2)$. On the other hand, using the same techniques as those presented in Chapter 9 we prove that

$$\sqrt{N}(m(\widehat{X}) - \Psi(m(X)))(f) \to_d \mathcal{N}(0, \sigma(f))$$

The variances $\sigma(f)$ in the two situations examined in (11.23) and (11.21) are respectively given by $\sigma(f) = \overline{\sigma}(f)$ and by

$$\begin{aligned}\sigma^2(f) &= \eta[S_\eta(f - S_\eta(f))^2] \\ &= \overline{\sigma}^2(f) - \eta(G^2(f - \Psi(\eta)f)^2) \leq \overline{\sigma}^2(f)\end{aligned}$$

These rather elementary comparisons indicate that whenever the observed distribution have the form $\Psi(\eta) = \eta S_\eta$ we expect to construct more accurate estimates when bootstrapping in accordance to S_η.

11.8.2 Description of the Models

The idea behind the forthcoming construction is to replace the multinomial branching transition by different type of distributions such as Poisson, Bernoulli, Binomial and others. As an important rule, we keep the acceptance mechanism and we only change the distribution of the branching numbers $\widetilde{b}_n^i$. Our models will not be restricted to fixed population size but they take values in the state space

$$\mathbf{E}_n = \bigcup_{p \in \mathbb{N}} (\{p\} \times E_n^p)$$

with the convention $E_n^p = \{\Delta\}$ a cemetery point if $p = 0$. The parameter $p \in \mathbb{N}$ represents the size of the system. The initial number of particles $N_0 \in \mathbb{N}$ is a fixed non-random number which represents the precision parameter of the branching approximation model. The evolution in time of the branching model

$$(N_n, \xi_n) \longrightarrow (\widehat{N}_n, \widehat{\xi}_n) \longrightarrow (N_{n+1}, \xi_{n+1})$$

is conducted as follows. As before, the initial particle system $\xi_0 = (\xi_0^i)_{1 \le i \le N_0}$ consists of N_0 independent and identically distributed particles with common law η_0. At the time n, the particle system ξ_n consists of N_n particles. If $N_n = 0$, the system collapses, and we let $(\widehat{N}_p, \widehat{\xi}_p) = (N_{p+1}, \xi_{p+1}) = (0, \Delta)$ for any $p \ge n$. Otherwise, each particle ξ_n^i branches into a random number of offsprings b_n^i, $1 \le i \le N_n$, and the mechanism is chosen so that the following un-bias and $\mathbb{L}_2$ conditions hold. For any $f_n \in \text{Osc}(E_n)$, we have on the event $\{N_n > 0\}$

$$\mathbb{E}(\sum_{i=1}^{N_n} b_n^i f(\xi_n^i) \mid \xi_n) = N_n \Psi_n(m(\xi_n))(f_n)$$

$$\mathbb{E}([\sum_{i=1}^{N_n} b_n^i f_n(\xi_n^i) - N_n \Psi_n(m(\xi_n))(f_n)]^2 \mid \xi_n) \le c\, N_n \qquad (11.25)$$

At the end of this stage, the particle system $\widehat{\xi}_n$ consists of $\widehat{N}_n = \sum_{i=1}^{N_n} b_n^i$ particles denoted by

$$\widehat{\xi}_n^i = \xi_n^k \qquad 1 \le k \le N_n \qquad \sum_{l=1}^{k-1} b_n^l + 1 \le i \le \sum_{l=1}^{k-1} b_n^l + b_n^k \qquad (11.26)$$

The mutation transition is defined as before, except if we have $\widehat{N}_n = 0$. In this case, the system dies, and we set $(N_p, \xi_p) = (\widehat{N}_p, \widehat{\xi}_p) = (0, \Delta)$ for any $p > n$.

The N_0-approximation measures of the Feynman-Kac distributions η_n and $\widehat{\eta}_n$ are respectively defined by

$$\eta_n^N = \frac{1}{N_0} \sum_{i=1}^{N_n} \delta_{\xi_n^i} \in \mathcal{M}_+(E_n) \quad \text{and} \quad \widehat{\eta}_n^N = \frac{1}{N_0} \sum_{i=1}^{\widehat{N}_n} \delta_{\widehat{\xi}_n^i} \in \mathcal{M}_+(E_n)$$

with the convention that $\sum_{\emptyset} = 0$, the null measure on E_n. By construction, we observe that

$$\eta_n^N = 1_{N_n > 0} \frac{1}{N_0} \sum_{i=1}^{N_n} \delta_{\xi_n^i} \quad \text{and} \quad \widehat{\eta}_n^N = 1_{\widehat{N}_n > 0} \frac{1}{N_0} \sum_{i=1}^{N_n} b_n^i \, \delta_{\xi_n^i}$$

In terms of these particle measures, the conditions (11.25) are equivalent on the event $\{N_n > 0\}$ to the following ones

$$N_0 \, \mathbb{E}(\widehat{\eta}_n^N(f_n) \mid \xi_n) = N_n \Psi_n(\eta_n^N)(f_n)$$

$$\mathbb{E}([N_0 \, \widehat{\eta}_n^N(f_n) - N_n \Psi_n(\eta_n^N)(f_n)]^2 \mid \xi_n) \leq c \, N_n$$

Finally, observe that on the event $\{N_n > 0\}$ we have

$$\mathbb{E}([N_0 \widehat{\eta}_n^N(f_n) - N_0 \Psi_n(\eta_n^N)(f_n)]^2 \mid \xi_n)$$

$$= \mathbb{E}([N_0 \, \widehat{\eta}_n^N(f_n) - N_n \Psi_n(\eta_n^N)(f_n)]^2 \mid \xi_n) + [N_0 - N_n]^2 \, [\Psi_n(\eta_n^N)(f_n)]^2$$

$$\leq c \, (N_n + [N_0 - N_n]^2)$$

which in turn implies that

$$\mathbb{E}([\widehat{\eta}_n^N(f_n) - \Psi_n(\eta_n^N)(f_n)]^2 \mid \xi_n) \leq c \, (N_n/N_0^2 + [1 - N_n/N_0]^2) \quad (11.27)$$

Finally, we observe that size of the system doesn't change during the exploration phase $N_n = \widehat{N}_{n-1}$, and we have the unbias property

$$\mathbb{E}(\eta_n^N(f_n) \mid \widehat{\xi}_{n-1}) = \widehat{\eta}_{n-1}^N M_n(f_n)$$

and the local sampling $\mathbb{L}_2$-estimates

$$N_0 \, \mathbb{E}([\eta_n^N(f_n) - \widehat{\eta}_{n-1}^N M_n(f_n)]^2 \mid \widehat{\xi}_{n-1}) = \widehat{\eta}_{n-1}^N M_n(f_n - M_n(f_n))^2 \leq c \, N_n/N_0$$

11.8.3 Some Branching Selection Rules

In Section 11.8.1, we have already presented two multinomial type strategies (11.22) and (11.24). These two interacting models correspond to the two different choices of McKean selection transitions given (11.21) and (11.23). We have also seen that the first McKean interpretation (11.21) with the multinomial branching (11.22) reduces the local variance of the branching selection model.

In this section, we provide a series of branching laws which satisfy the pair condition (11.25). To clarify the presentation, we restrict our discussion to branching interpretations of the second McKean model (11.23), thus leaving aside the improvement we can get in practice by using the first McKean interpretation. Nevertheless, we emphasize that all the forthcoming branching rules can be combined without further work with the acceptance/rejection McKean selection transition (11.21).

We describe branching selections at a given time n, and it is also implicitly assume that the system at that time has not collapsed. In other words, all the forthcoming laws are defined on the event $\{N_n > 0\}$. We also simplify the notation and we set

$$N_n W_n^i = G_n(\xi_n^i)/\eta_n^N(G_n)$$

We recall that $\lfloor a \rfloor$ (respectively $\{a\} = a - \lfloor a \rfloor$) is the integer part (resp. the fractional part) of a number $a \in \mathbb{R}$.

- *Remainder stochastic sampling rules.* Each particle ξ_n^i first branches directly into a fixed and "deterministic" number of offsprings $\overline{b}_n^i = \lfloor N_n W_n^i \rfloor$ so that the intermediate population consists of $\overline{N}_n \stackrel{\text{def.}}{=} \sum_{i=1}^{N_n} \overline{b}_n^i$ particles. It can be seen that at least one particle has one offspring. Otherwise, we would have

$$\forall i \ \overline{b}_n^i = \lfloor N_n W_n^i \rfloor = 0 \Longrightarrow \forall i \ N_n W_n^i = \{N_n W_n^i\} < 1$$

which would contradict the fact that $N_n = \sum_{i=1}^{N_n} N_n W_n^i$. Therefore, using this preliminary deterministic branching rule, the particle model never collapses. Nevertheless, to ensure that the unbias and $\mathbb{L}_2$ conditions are met, we need to introduce an additional branching rule. Two strategies can be underlined.

One natural way to keep the size of the system fixed is to introduce in this population $\widetilde{N}_n = N_n - \overline{N}_n = \sum_{i=1}^{N_n} \{N_n W_n^i\}$ additional particles. To do this, we introduce an additional sequence of branching numbers

$$(\widetilde{b}_n^1, \ldots, \widetilde{b}_n^{N_n})$$

$$\stackrel{\text{def.}}{=} \text{Multinomial}(\widetilde{N}_n, \frac{\{N_n W_n^1\}}{\sum_{j=1}^{N_n} \{N_n W_n^j\}}, \ldots, \frac{\{N_n W_n^{N_n}\}}{\sum_{j=1}^{N_n} \{N_n W_n^j\}}) \qquad (11.28)$$

and we set $b_n^i = \overline{b}_n^i + \widetilde{b}_n^i$. In other words, each particle ξ_n^i again produces a number of $\widetilde{b}_n^i$ additional offsprings. Note that the multinomial random numbers (11.28) can alternatively be defined as follows

$$\widetilde{b}_n^k = \text{Card}\left\{1 \leq j \leq \widetilde{N}_n \ ; \ \widetilde{\xi}_n^j = \xi^k\right\} \qquad 1 \leq k \leq N_n$$

where $(\widetilde{\xi}_n^1, \ldots, \widetilde{\xi}_n^{\widetilde{N}_n})$ are $\widetilde{N}_n$ independent random variables with common law $\sum_{i=1}^{N_n} \frac{\{N_n W_n^i\}}{\sum_{j=1}^{N_n} \{N_n W_n^j\}} \delta_{\xi_n^i}$. It is easily checked that (11.25) are satisfied.

We can alternatively use the independent Bernoulli resampling numbers defined

$$\mathbb{P}(\tilde{b}_n^i = 1|\xi_n) = 1 - \mathbb{P}(\tilde{b}_n^i = 0|\mathcal{F}_n) = \{N_n W_n^i\} \qquad (11.29)$$

Also note that condition (11.25) is met since we have

$$\begin{aligned}\mathbb{E}(b_n^i|\xi_n) &= N_n W_n^i \\ \mathrm{Var}(b_n^i|\xi_n) &= \mathrm{Var}(\tilde{b}_n^i|\mathcal{F}_n) = \{N_n W_n^i\}(1 - \{N_n W_n^i\}) \in [0, 1/4]\end{aligned} \qquad (11.30)$$

- *Independent branching numbers* The *Poisson branching numbers* are defined as a sequence b_n^i of conditionally independent random numbers with distribution given for any $k \geq 0$ by

$$\mathbb{P}(b_n^i = k|\xi_n) = \exp\left(-N_n W_n^i\right) \frac{(N_n W_n^i)^k}{k!} \qquad (11.31)$$

Since we have $\mathbb{E}(b_n^i|\mathcal{F}_n) = N_n W_n^i = V(b_n^i|\mathcal{F}_n)$, we readily check that conditions (11.25) are met. The *binomial branching numbers* are defined as a sequence b_n^i of conditionally independent random numbers with distribution given for any $0 \leq k \leq N_n$ by

$$\mathbb{P}(b_n^i = k|\xi_n) = \frac{N_n!}{k!(N_n - k)!} (W_n^i)^k (1 - W_n^i)^{N_n - k} \qquad (11.32)$$

In this case, the pair condition (11.25) follows from the fact that

$$\mathbb{E}(b_n^i|\xi_n) = N_n W_n^i \quad \text{and} \quad \mathrm{Var}(b_n^i|\xi_n) = N_n W_n^i(1 - W_n^i)$$

11.8.4 Some $\mathbb{L}_2$-mean Error Estimates

In this section, we investigate the asymptotic behavior of the branching particle models described in Section 11.8.2. Our approach follows essentially the same lines of arguments as the one developed in Section 7.4.3. For a brief overview on the Feynman-Kac semigroups involved in the forthcoming analysis, we recommend the reader to start his/her study with Section 7.2 and Section 7.4.3. Our first task is to quantify the probability of extinction.

Lemma 11.8.1 *The total mass process N_n is a non-negative and integer valued martingale with respect to the filtration $\mathcal{F}_n = \sigma(\xi_p, p \leq n)$. In addition, we have for any $n \geq 0$*

$$N_0 \, \mathbb{E}(\sup_{0 \leq p \leq n} (1 - N_p/N_0)^2) \leq c\,n \quad \text{and} \quad N_0 \, \mathbb{P}(N_n = 0) \leq c\,n$$

Proof:

To check the martingale property, it suffices to notice that

$$\mathbb{E}(N_{n+1}|\mathcal{F}_n) = \mathbb{E}(\widehat{N}_n|\mathcal{F}_n) = \sum_{i=1}^{N_n} \mathbb{E}(b_n^i|\mathcal{F}_n) = N_n$$

By Doob's maximal inequality, the proof of the $\mathbb{L}_2$-estimate amounts to prove that $\mathbb{E}(N_n^2) \leq N_0^2 + c\, n\, N_0$. The latter is easily checked by induction, and using the fact that

$$\mathbb{E}((N_n - N_{n-1})^2|\mathcal{F}_{n-1}) = \mathbb{E}((\sum_{i=1}^{N_{n-1}} (b_{n-1}^i - N_{n-1} W_{n-1}^i))^2|\mathcal{F}_{n-1}) \leq c N_{n-1}$$

The second estimate is a simple consequence of the first one. To see this claim, we use Chebichev's inequality to check that for any $\epsilon \in]0,1[$

$$\mathbb{P}(N_n = 0) \leq \mathbb{P}(N_n \leq \epsilon N_0) \leq \mathbb{P}(N_0 - N_n \geq (1-\epsilon)N_0)$$
$$\leq \mathbb{P}(|N_0 - N_n| \geq (1-\epsilon)N_0) \leq \frac{cn}{N_0(1-\epsilon)^2}$$

We end the proof of the lemma by letting $\epsilon \to 0$.

∎

This simple lemma, combined with an elementary induction with respect to the time parameter, already gives some useful $\mathbb{L}_2$-mean error estimates for the branching approximation model. To describe this technique, we recall that $N_{n+1} = \widehat{N}_n = \sum_{i=1}^{N_n} b_n^i > 0 \implies N_n > 0$, from which we find the decomposition

$$|\eta_{n+1}^N - \eta_{n+1}|\, 1_{N_{n+1}>0}$$

$$\leq |\eta_{n+1}^N - \widehat{\eta}_n^N M_{n+1}|\, 1_{\widehat{N}_n>0} + |[\widehat{\eta}_n^N - \Psi_n(\eta_n^N)] M_{n+1}|\, 1_{N_n>0}$$

$$+ |\Phi_{n+1}(\eta_n^N) - \Phi_{n+1}(\eta_n)|\, 1_{N_n>0}$$

Now, on the event $\{N_n > 0\}$, we notice that

$$[\Phi_{n+1}(\eta_n^N) - \Phi_{n+1}(\eta_n)](f_{n+1}) = \frac{1}{\eta_n^N(\tilde{G}_n)}\, [\eta_n^N - \eta_n](\tilde{f}_{n+1})$$

with the functions $\tilde{f}_{n+1}$ and $\tilde{G}_n$ defined by

$$\tilde{f}_{n+1} = \tilde{G}_n\, (M_{n+1}(f_{n+1}) - \Phi_{n+1}(\eta_n)(f_{n+1})) \quad \text{and} \quad \tilde{G}_n = G_n/\eta_n(G_n)$$

This readily yields that

$$|[\Phi_{n+1}(\eta_n^N) - \Phi_{n+1}(\eta_n)](f_{n+1})| 1_{N_n>0} \leq c\, |[\eta_n^N - \eta_n](\tilde{f}_{n+1})|\, 1_{N_n>0}$$

On the other hand, by definition of the branching model, and using (11.27), we have that

$$\mathbb{E}(\mathbb{E}([(\eta_{n+1}^N - \widehat{\eta}_n^N M_{n+1})(f_{n+1})]^2 \mid \xi_n, \widehat{\xi}_n) 1_{\widehat{N}_n > 0} \mid \xi_n)$$

$$\leq c \, \mathbb{E}(\widehat{N}_n/N_0^2) = c \, N_n/N_0^2$$

and

$$\mathbb{E}([\widehat{\eta}_n^N(f_n) - \Psi_n(\eta_n^N)(f_n)]^2 \mid \xi_n) 1_{N_n > 0} \leq c \, (N_n/N_0^2 + [1 - N_n/N_0]^2)$$

Using Lemma 11.8.1, if we set

$$I_n^N = \sqrt{N_0} \sup_{f_n \in \mathrm{Osc}(E_n)} \mathbb{E}(([\eta_n^N - \eta_n](f_n) 1_{N_n > 0})^2)^{1/2}$$

then we find that $I_{n+1}^N \leq c \, [I_n^N + \sqrt{n}]$ from which we get the rather crude estimate

$$\sup_{N_0 \geq 1} \sup_{f_n \in \mathrm{Osc}(E_n)} \mathbb{E}(([\eta_n^N - \eta_n](f_n) 1_{N_n > 0})^2)^{1/2} \leq c^n$$

for some $c > 1$. One drawback of the elementary induction technique presented above is that it over-estimates the $\mathbb{L}_2$-mean errors. To get one step further, we first note that the branching rules performed at each stage of the algorithm can be interpreted as local perturbations of the limiting evolution equation. To improve the above rather crude estimation, it is therefore convenient to refine our analysis so that to enter the stability properties of the limiting measure-valued process. As usual, these properties are expressed in terms of the pair parameters $(r_{p,n}, \beta(P_{p,n}))$ introduced in (7.3) on page 218 (see also Section 7.4.3).

Theorem 11.8.1 *For any $n \geq 0$, and $f_n \in \mathrm{Osc}(E_n)$ we have*

$$\sqrt{N_0} \, \mathbb{E}\left(|[\eta_n^N - \eta_n](f_n) \, 1_{N_n > 0}|^2\right)^{\frac{1}{2}} \leq c \sum_{q=0}^{n} (1 + \sigma_p(N)) \, r_{p,n} \, \beta(P_{p,n})$$

with $\sigma_p^2(N) = N_0 \, \mathrm{Var}(N_{p-1}/N_0)$ and the convention $N_{-1} = N_0$ for $p = 0$.

Proof:
The proof is based on the following decomposition

$$(\eta_n^N - \eta_n) \, 1_{N_n > 0} = \sum_{q=0}^{n} [\Phi_{q,n}(\eta_q^N) - \Phi_{q,n}(\Phi_q(\eta_{q-1}^N))] \, 1_{N_n > 0} \quad (11.33)$$

Arguing as in the proof of Theorem 7.4.4, we readily prove the inequality

$$1_{N_n > 0} \, |[\Phi_{q,n}(\eta_q^N) - \Phi_{q,n}(\Phi_q(\eta_{q-1}^N))](f_n)|$$
$$\leq 1_{N_q > 0} \, r_{q,n} \, \beta(P_{q,n}) \, |[\eta_q^N - \Phi_q(\eta_{q-1}^N)](f_{q,n}^N)| \quad (11.34)$$

with the random function $f_{q,n}^N = Q_{q,n}^N(f_n)/\|Q_{q,n}^N(f_n)\|$ and the random operator $Q_{q,n}^N$ from $\mathcal{B}_b(E_n)$ into $\mathcal{B}_b(E_q)$ defined in (7.25), page 245. To take the final step, we use the estimate

$$1_{N_q>0}|[\eta_q^N - \Phi_q(\eta_{q-1}^N)](f_{q,n}^N)| \leq I_1 + I_2$$

with

$$\begin{aligned} I_1 &= 1_{\widehat{N}_{q-1}>0}|[\eta_q^N - \widehat{\eta}_{q-1}^N M_q](f_{q,n}^N)| \\ I_2 &= 1_{N_{q-1}>0}|[\widehat{\eta}_{q-1}^N - \Psi_{q-1}(\eta_{q-1}^N)](M_q f_{q,n}^N)| \end{aligned}$$

By definition of the branching particle model, we have

$$\mathbb{E}(I_1^2|\xi_{q-1},\widehat{\xi}_{q-1})$$

$$= 1_{\widehat{N}_{q-1}>0}\frac{1}{N_0^2}\sum_{i=1}^{\widehat{N}_{q-1}} M_q[f_{q,n}^N - M_q(f_{q,n}^N)]^2(\widehat{\xi}_{q-1}^i) \leq c\, N_q/N_0^2$$

and

$$\mathbb{E}(I_1^2|\xi_{q-1})$$

$$= 1_{N_{q-1}>0}\frac{1}{N_0^2}\mathbb{E}([\sum_{i=1}^{N_{q-1}} b_{q-1}^i (M_q f_{q,n}^N)(\xi_n^i)$$

$$- N_{q-1}\Psi_{q-1}(\eta_{q-1}^N)(M_q f_{q,n}^N))]^2|\xi_{q-1})$$

$$+ 1_{N_{q-1}>0}\mathbb{E}([1 - N_{q-1}/N_0]^2[\Psi_{q-1}(\eta_{q-1}^N)(M_q f_{q,n}^N))]^2|\xi_{q-1})$$

$$\leq c\,(N_{q-1}/N_0^2 + [1 - N_{q-1}/N_0]^2)$$

By Lemma 11.8.1, these almost sure estimates yield that

$$\mathbb{E}(I_1^2) \vee \mathbb{E}(I_2^2) \leq c\,(1/N_0 + \mathbb{E}([1 - N_{q-1}/N_0]^2))$$

from which we conclude that

$$\mathbb{E}(1_{N_q>0}|[\eta_q^N - \Phi_q(\eta_{q-1}^N)](f_{q,n}^N)|^2)^{1/2} \leq c\,(1/\sqrt{N_0} + \mathbb{E}([1 - N_{q-1}/N_0]^2)^{1/2})$$

The end of the proof is now a simple consequence of (11.33) and (11.34). ∎

Corollary 11.8.1 *We further assume that the regularity conditions* (G) *and* $(M)_m$ *are met for some parameters* $(m, \epsilon_n(G), \epsilon_n(M))$ *such that*

$$\epsilon(G) = \wedge_n \epsilon_n(G) > 0 \quad \text{and} \quad \epsilon(M) = \wedge \epsilon_n(M) > 0$$

Then, for any $n \geq 0$, *and* $f_n \in \mathrm{Osc}_1(E_n)$, *we have*

$$\sqrt{N_0}\,\mathbb{E}\left(|[\eta_n^N - \eta_n](f_n)\,1_{N_n>0}|^2\right)^{\frac{1}{2}}$$

$$\leq c\,m\,(1 + [N_0\,\mathrm{Var}(N_n/N_0)]^{1/2})/(\epsilon(G)^{2m-1}\epsilon^3(M))$$

In particular, we have

$$\sqrt{N_0}\, \mathbb{E}\left(|[\eta_n^N - \eta_n](f_n)\, 1_{N_n>0}|^2\right)^{\frac{1}{2}} \leq c\, m\, (1+\sqrt{n})/(\epsilon(G)^{2m-1}\epsilon^3(M))$$

and for conservative particle models (i.e. $N_n = N_0$), we have the uniform estimate

$$\sqrt{N_0}\, \sup_{n\geq 0} \mathbb{E}\left(|[\eta_n^N - \eta_n](f_n)|^2\right)^{\frac{1}{2}} \leq c\, m\, /(\epsilon(G)^{2m-1}\epsilon^3(M))$$

Proof:
Arguing as in the proof of theorem 7.4.4, we have the estimates

$$\beta(P_{p,n}) \leq (1 - \epsilon^2(M)\epsilon^{m-1}(G))^{\lfloor (n-p)/m \rfloor} \quad \text{and} \quad r_{p,n} \leq \epsilon^{-1}(M)\epsilon^{-m}(G)$$

from which we find that

$$\begin{aligned}
\sum_{q=0}^{n} r_{q,n}\, \beta(P_{q,n}) &= \epsilon^{-1}(M)\epsilon^{-m}(G) \sum_{q=0}^{n} (1 - \epsilon^2(M)\epsilon^{m-1}(G))^{\lfloor q/m \rfloor} \\
&\leq m\epsilon^{-1}(M)\epsilon^{-m}(G) \sum_{k=0}^{\lfloor n/m \rfloor} (1 - \epsilon^2(M)\epsilon^{m-1}(G))^k
\end{aligned}$$

Now, recalling that N_n is a martingale we see that $\mathbb{E}(N_n^2) = \mathbb{E}(\langle N\rangle_n)$ is an increasing sequence, and by Lemma 11.8.1, we have for any $p \leq n$

$$\mathrm{Var}(N_p/N_0) = (\mathbb{E}(\langle N/N_0\rangle_p) - 1) \leq \mathrm{Var}(N_n/N_0) \leq c\, n/N_0$$

The end of the proof of the corollary is now a simple consequence of the estimate stated in Theorem 11.8.1. ∎

Mimicking the product formula presented in Lemma 2.3.1, we adopt the following

Definition 11.8.1 *The N_0-particle approximation measures γ_n^N of the unnormalized measures γ_n are defined for any $f_n \in \mathcal{B}_b(E_n)$ by the product formula*

$$\gamma_n^N(f_n) = \eta_n^N(f_n) \prod_{p=0}^{n-1} m(\xi_p)(G_p)$$

Proposition 11.8.1 *For any $n \geq 0$ and $f_n \in \mathcal{B}_b(E_n)$, the random sequence defined by*

$$\Gamma_{p,n}(f_n) = \sqrt{N_0}\, [\gamma_p^N(Q_{p,n}(f_n)) - \gamma_p(Q_{p,n}(f_n))], \quad p \leq n$$

is an $\mathcal{F}$-martingale with increasing process given by the formula

$$N_0\, \langle \Gamma_{\cdot,n}(f_n)\rangle_p$$

$$= \sum_{k=0}^{p} 1_{N_{k-1}>0}\, [\prod_{l=0}^{k-1} m(\xi_l)(G_l)]^2$$

$$\times\, \mathbb{E}(([N_0\eta_k^N - N_{k-1}\Phi_k(\eta_{k-1}^N)](Q_{k,n}f_n))^2 \mid \mathcal{F}_{k-1})$$

Proof:
We use the decomposition
$$\gamma_n^N(f_n) - \gamma_n(f_n) = \sum_{p=0}^{n} [\gamma_p^N(Q_{p,n}(f_n)) - \gamma_{p-1}^N(Q_{p-1,n}(f_n))]$$

with the convention $\gamma_{-1}^N Q_{-1,n} = \eta_0 Q_{0,n} = \gamma_n$ for $p = 0$. Notice that

$$\gamma_{p-1}^N(Q_{p-1,n}(f_n)) = 1_{N_{p-1}>0} \frac{N_{p-1}}{N_0} \Phi_p(m(\xi_{p-1}))(Q_{p,n}) \prod_{k=0}^{p-1} m(\xi_k)(G_k)$$

On the other hand, since we have

$$1_{N_p>0} = 1_{N_{p-1}>0} + 1_{N_{p-1}>0, N_p>0} - 1_{N_{p-1}>0} = 1_{N_{p-1}>0} - 1_{N_{p-1}>0, N_p=0}$$

and $1_{N_p=0, N_{p-1}>0}\ \eta_p^N = 0$, we find that

$$\begin{aligned}\gamma_p^N(Q_{p,n}(f_n)) &= 1_{N_p>0}\ \eta_p^N(Q_{p,n}(f_n)) \prod_{k=0}^{p-1} m(\xi_k)(G_k) \\ &= 1_{N_{p-1}>0}\ \eta_p^N(Q_{p,n}(f_n)) \prod_{k=0}^{p-1} m(\xi_k)(G_k)\end{aligned}$$

Recalling that $\Phi_p(\eta_{p-1}^N) = \Phi_p(m(\xi_{p-1}))$, this readily yields that

$$\gamma_n^N(f_n) - \gamma_n(f_n) = \sum_{p=0}^{n} 1_{N_{p-1}>0} \left[\prod_{k=0}^{p-1} m(\xi_k)(G_k)\right]$$
$$\times [\eta_p^N(Q_{p,n}(f_n)) - \tfrac{N_{p-1}}{N_0} \Phi_p(\eta_{p-1}^N)(Q_{p,n}(f_n))]$$

By construction of the branching model, we have for any test function $f_p \in \mathcal{B}_b(E_p)$, and on the event $\{N_{p-1} > 0\}$

$$\mathbb{E}(\eta_p^N(f_p) \mid \xi_{p-1}, \widehat{\xi}_{p-1}) = \frac{1}{N_0} \sum_{i=1}^{N_{p-1}} b_{p-1}^i M_p(f_p)(\xi_{p-1}^i)$$

from which we conclude that

$$N_0\ \mathbb{E}(\eta_p^N(f_p) \mid \mathcal{F}_{p-1}) = N_{p-1} \Phi_p(m(\xi_{p-1}))(f_p)$$

The end of the proof is now clear. ∎

Corollary 11.8.2 *For any $n \geq 0$ and $f_n \in \mathcal{B}_b(E_n)$, we have*

$$\mathbb{E}(\gamma_n^N(f_n)) = \gamma_n(f_n) \quad \text{and} \quad \sup_{N_0 \geq 1} N_0\ \mathbb{E}([\gamma_n^N(f_n) - \gamma_n(f_n)]^2) < \infty$$

Proof:
In view of (11.25), we observe that
$$\mathbb{E}(([N_0\eta_k^N - N_{k-1}\Phi_k(\eta_{k-1}^N)](Q_{k,n}f_n))^2 \mid \mathcal{F}_{k-1}) \leq c\, N_{k-1}$$
The end of the proof is now a simple consequence of Proposition 11.8.1. ∎

11.8.5 Long Time Behavior

In Corollary 11.8.1, we have presented a pair or regularity conditions on the Feynman-Kac models which ensures uniform $\mathbb{L}_2$-estimates with respect to the time parameter. These asymptotic properties are essential in practice to calibrate the initial number of particles needed to achieve a desired precision degree. The main difficulty in the study of the long time behavior of branching models with independent branching numbers is that the total size process

$$N_n = N_0 + \sum_{p=1}^{n}(N_p - N_{p-1}) = N_0 + \sum_{p=1}^{n}\sum_{i=1}^{N_{p-1}}(b_p^i - N_{p-1}W_{p-1}^i)$$

is an martingale with increasing process

$$\langle N\rangle_n = N_0^2 + \sum_{p=0}^{n-1}\sum_{i=1}^{N_p} \mathbb{E}\left((b_p^i - N_p W_p^i)^2 | \mathcal{F}_p\right)$$

The only way to ensure a uniform convergence result is to ensure that

$$\sup_{n\geq 0} E\left(\langle N\rangle_n^2\right) = \sum_{p=1}^{\infty}\mathbb{E}\left(|N_p - N_{p-1}|^2\right) < \infty$$

Unfortunately, these processes are usually far from being uniformly integrable and we confess that we haven't find a particle model with conditionally independent population size which met this integrability property. If we consider the Poisson branching numbers, we have

$$\mathbb{E}\left((b_p^i - N_p W_p^i)^2/\mathcal{F}_p\right) = N_p W_p^i \qquad \forall 1 \leq i \leq N_{p-1}$$

from which we find that $N_0 E\left((\eta_n^N(1) - \eta_n(1))^2\right) = n \; (\to \infty$ as $n \to \infty)$. In the same vein, for the binomial branching model we have

$$E\left((b_p^i - N_p W_p^i)^2/\mathcal{F}_p\right) = N_p W_p^i(1 - W_p^i)$$

If we assume that $1/a \leq G_n(x) \leq a$ for some $a \geq 1$ and $N_0 > a^2$ then one gets $\sum_{i=1}^{N_p} N_p W_p^i(1 - W_p^i) \geq N_p - a^2$, which again implies that

$$E\left((\eta_n^N(1) - \eta_n(1))^2\right) \geq n\left(\frac{1}{N_0} - \frac{a^2}{N_0^2}\right) \xrightarrow[n\to\infty]{} \infty$$

418   11. Feynman-Kac and Interacting Particle Recipes

Although the Bernoulli particle model seems to be the most efficient one (since the independent random variables b_n^i have minimal variance), the forthcoming elementary example shows that, even in this case, one cannot expect to approximate the desired measures uniformly with respect to time. Let us assume that the state space $E_n = \{0,1\}$ and the pair potential/transitions (G_n, M_n) are homogeneous and chosen so that

$$G(1) = 3G(0) > 0 \qquad M(x, dy) = \nu(dy) = \frac{1}{2}\delta_0(dy) + \frac{1}{2}\delta_1(dy)$$

In this case, ξ_p^i are N_p independent random variables with common law ν and we find that

$$\forall \epsilon > 0 \qquad \mathbb{P}(|\frac{1}{N_p}\sum_{i=1}^{N_p} G(\xi_p^i) - \nu(G)| \geq \epsilon G(0) \mid N_p) \leq \frac{5}{\epsilon^2 N_p} \qquad (11.35)$$

Noticing that $\nu(G)/G(0) = 2 = 3\nu(G)/G(1)$ and $G(0) \leq G(1)$, on the set

$$\Omega_\epsilon = \left\{ |\frac{1}{N_p}\sum_{i=1}^{N_p} G(\xi_p^i) - \nu(G)| \leq \epsilon G(0) \right\}$$

we have that

$$\left| \frac{G(0)}{\frac{1}{N_p}\sum_{i=1}^{N_p} G(\xi_p^i)} - \frac{1}{2} \right| \leq \frac{\epsilon}{2(2-\epsilon)} \leq \frac{\epsilon}{2}$$

and

$$\left| \frac{G(1)}{\frac{1}{N_p}\sum_{i=1}^{N_p} G(\xi_p^i)} - \frac{3}{2} \right| \leq \frac{3\epsilon}{2(2/3-\epsilon)} \leq \frac{9\epsilon}{2}$$

as soon as $\epsilon \in (0, 1/9)$. This, in turns, implies that

$$\left[\frac{G(0)}{\frac{1}{N_p}\sum_{i=1}^{N_p} G(\xi_p^i)} \right] = 0 \quad \text{and} \quad \left[\frac{G(1)}{\frac{1}{N_p}\sum_{i=1}^{N_p} G(\xi_p^i)} \right] = 1$$

and

$$\xi_p^i = 0 \implies \{N_p W_p^i\} \ (1 - \{N_p W_p^i\}) \geq \frac{1}{4}(1-\epsilon)^2$$

$$\xi_p^i = 1 \implies \{N_p W_p^i\} \ (1 - \{N_p W_p^i\}) \geq \frac{1}{4}(1-9\epsilon)^2$$

It is then clear that on the set Ω_ϵ we have the lower bounds

$$\mathbb{E}((b_p^i - N_p W_p^i)^2 \mid \mathcal{F}_p) = \{N_p W_p^i\} \ (1 - \{N_p W_p^i\}) \geq \frac{1}{4}(1-9\epsilon)^2$$

This, together with (11.35), shows that

$$\mathbb{E}(\sum_{i=1}^{N_p}(b_p^i - N_p W_p^i)^2) \geq \frac{N_0}{4}(1-9\epsilon)^2$$

One concludes that $E\left((\eta_n^N(1) - \eta_n(1))^2\right) \geq \frac{n}{4N_0}(1-9\epsilon)^2$ ($\uparrow \infty$ as $n \uparrow \infty$).

11.8.6 Conditional Branching Models

In this section, we show that the interacting particle model (with multinomial branching laws) can be obtained by conditioning a Poisson branching particle model to have constant population size. For any $N_0 \geq 1$, we denote by
$$(\Omega, (\mathcal{F}_n, \widehat{\mathcal{F}}_n)_{n \geq 0}, (N_n, \xi_n, \widehat{N}_n, \widehat{\xi}_n)_{n \geq 0}, \mathbb{P}^{PB}_{N_0})$$
the canonical Markov model which realizes the Poisson branching particle model starting with N_0 particles, and by $\mathbb{P}^{MB}_{N_0}$ the distribution (on the canonical space) of the multinomial branching genetic model discussed in (11.24).

Proposition 11.8.2 *For any $A \in \vee_n(\mathcal{F}_n \vee \widehat{\mathcal{F}}_n)$, we have*
$$\mathbb{P}^{PB}_{N_0}(A | N = N_0) = \mathbb{P}^{MB}_{N_0}(A) \qquad \mathbb{P}^{PB}_{N_0} - a.s. \tag{11.36}$$

Proof:
Conditionally on the event $\{N = N_0\} = \bigcap_{n \geq 0} \{N_n = N_0\}$, we have
$$(\xi_n, \widehat{\xi}_n) \in (E_n^{N_0} \times E_n^{N_0}) \qquad \mathbb{P}^{PB}_{N_0} - a.s.$$
On the other hand, by construction of the mutation transition, we have for any $n \geq 0$, $x, z \in E^{N_0}$
$$\mathbb{P}^{PB}_{N_0}\left(\xi_n \in dz | N = N_0, \widehat{\xi}_{n-1} = x\right) = \mathbb{P}^{MB}_{N_0}\left(\xi_n \in dz | \widehat{\xi}_{n-1} = x\right)$$
Since changes in the number of particles only take place at branching selections, to prove (11.36) it suffices to check that for any $n \geq 0$ and $x, z \in E^{N_0}$
$$\mathbb{P}^{PB}_{N_0}\left(\widehat{\xi}_n \in dz | N = N_0, \xi_n = x\right) = \mathbb{P}^{MB}_{N_0}\left(\widehat{\xi}_n \in dz | \xi_n = x\right)$$
By definition of the Poisson branching model, we have for each $n \geq 0$, $x \in E^{N_0}$, and $k \in \mathbb{N}^{N_0}$,
$$\mathbb{P}^{PB}_{N_0}(b_n = k | N = N_0, \xi_n = x) = \mathbb{P}^{PB}_{N_0}(b_n = k | N_n = N_0, \xi_n = x)$$
$$= \frac{1}{Z(n, N_0)} \prod_{i=1}^{N_0} \exp(-N_0 W_n^i) \frac{(N_0 W_n^i)^{k_i}}{k_i!}$$
with $W_n^i = \frac{g_n(x^i)}{\sum_{j=1}^{N_0} g_n(x^j)}$, and the normalizing constants
$$Z(n, N_0) = \sum_{k_1 + \ldots + k_{N_0} = N_0} \prod_{i=1}^{N_0} \exp(-N_0 W_n^i) \frac{(N_0 W_n^i)^{k_i}}{k_i!} = e^{-N_0} N_0^{N_0} / N_0!$$
It is now not difficult to see that
$$\mathbb{P}^{PB}_{N_0}(b_n = k | N = N_0, \xi_n = x) = \frac{N_0!}{(k_1!) \ldots (k_{N_0}!)} (W_n^1)^{k_1} \ldots (W_n^{N_0})^{k_{N_0}}$$

and therefore $\mathbb{P}_{N_0}^{\mathrm{PB}}(b_n = k | N = N_0, \xi_n = x) = \mathbb{P}_{N_0}^{\mathrm{MB}}(b_n = k | \xi_n = x)$. The end of the proof is now clear. ∎

If we use multinomial branching laws, one still has the freedom to adapt the size parameter so that to produce a given number of offsprings. To this end, let $a = (a_n; n \geq 0)$ be the path numbers of offsprings we want to have at each stage of the algorithm (i.e. $N_0 = a_0, N_1 = a_1, \ldots, N_n = a_n, \ldots$). The corresponding branching laws are defined by replacing at each time n the law (11.24) by the multinomial distribution

$$(b_n^1, \ldots, b_n^{a_n}) = \mathrm{Multinomial}\left(a_{n+1}, W_n^1, \ldots, W_n^{a_n}\right) \qquad (11.37)$$

We let $P_{N_0}^{\mathrm{PB}(a)}$ be the distribution of the particle model with multinomial branchings corrections (11.37) (and starting with N_0 particles). Arguing as above, one proves that

Proposition 11.8.3 *For any* $A \in \vee_n(\mathcal{F}_n \vee \widehat{\mathcal{F}}_n)$ *we have*

$$\mathbb{P}_{N_0}^{PB}(A | N = a) = \mathbb{P}_{N_0}^{MB(a)}(A) \qquad \mathbb{P}_{N_0}^{PB} - a.s.$$

The continuous time version of Proposition 11.8.2 was proved by Etheridge and March in [134] in their study of the connections between critical branching superprocesses and the Fleming-Viot interacting particle systems. The continuous time version of Proposition 11.8.3 was proved by Perkins [265] in his precise study of the structural properties of Dawson-Watanabe and Fleming-Viot processes.

11.9 Exercises

Exercise 11.9.1: [Boltzmann-Gibbs and Feynman-Kac models] We recall that the Boltzmann-Gibbs transformation of a given distribution $\nu \in \mathcal{P}(E)$ with respect to some nonnegative potential function G with $\nu(G) > 0$ is the measure defined by $\Psi(\nu)(dx) = G(x)\,\nu(dx)/\nu(G)$. Suppose that E is a Cartesian product $E = (E_0 \times \ldots \times E_n)$ of some "elementary" measurable spaces and the pair potential/measure (G, ν) has the form

$$G(x_0, \ldots, x_n) = \prod_{p=0}^n G_p(x_p)$$

$$\nu(d(x_0, \ldots, dx_n)) = \eta_0(dx_0) \prod_{p=1}^n M_p(x_{p-1}, dx_p)$$

for some Markov kernels M_n from E_{n-1} into E_n and a distribution $\eta_0 \in \mathcal{P}(E_0)$ and for some sequence of potential functions G_n on E_n. Check that the Boltzmann-Gibbs distribution $\Psi(\nu)$ coincides with the Feynman-Kac distribution on path space

$$\widehat{\mathbb{Q}}_n(d(x_0,\ldots,x_n)) = \frac{1}{\widehat{\mathcal{Z}}_n} \left\{ \prod_{p=0}^n G_p(x_p) \right\} \mathbb{P}_n(d(x_0,\ldots,x_n))$$

where $\mathbb{P}_n$ is the distribution of the trajectory $(X_p)_{0\leq p \leq n}$ of a Markov chain with initial distribution η_0 and transitions M_n.

Exercise 11.9.2: [Sequential Monte Carlo integrations] Let $(E'_n, \mathcal{E}'_n)_{n\geq 0}$ be a sequence of measurable spaces, and let $\pi_n \in \mathcal{M}_+(E_n)$ be a sequence of positive and bounded measures on some Cartesian products $E_n = (E'_0 \times \ldots \times E'_n)$ with $\pi_n(1) > 0$. Suppose that we want to evaluate for any $n \geq 0$ and any bounded measurable function f_n on E_n the integrals

$$\pi_n(f_n) = \int_{E'_0 \times \ldots \times E'_n} f_n(x_0,\ldots,x_n)\, \pi_n(d(x_0,\ldots,x_n))$$

Further assume that the measures π_n can be disintegrated in the sense that

$$\pi_n(d(x_0,\ldots,x_n)) = \pi_{n-1}(d(x_0,\ldots,x_{n-1}))\, \pi_{n-1,n}((x_0,\ldots,x_{n-1}), dx_n) \quad (11.38)$$

for some collection of measurable transitions $\pi_{n,n+1}$ from E_n into E'_{n+1} with

$$G_n(x_0,\ldots,x_n) = \pi_{n,n+1}(1)(x_0,\ldots,x_n) > 0 \quad (11.39)$$

Let X'_n be the nonanticipative sequence of E'_n-valued random variables with initial distribution $\eta_0(dx_0) = \pi_0(dx_0)/\pi_0(1)$ and "elementary transitions"

$$\mathbb{P}(X'_n \in dx_n \mid X'_0 = x_0,\ldots,X'_{n-1} = x_{n-1}) = \frac{\pi_{n-1,n}((x_0,\ldots,x_{n-1}), dx_n)}{\pi_{n-1,n}(1)(x_0,\ldots,x_{n-1})}$$

Show that the random path sequence defined by $X_n = (X'_0,\ldots,X'_n)$ is an E_n-valued Markov chain and we have the Feynman-Kac representation formula

$$\pi_n(f_n) = \pi_0(1)\, \mathbb{E}_{\eta_0}\left(f_n(X_n) \prod_{p=0}^{n-1} G_p(X_p) \right)$$

Exercise 11.9.3: [Restricted Markov chain models] Let Y'_n be an E'_n-valued Markov chain with initial distribution π_0 and elementary Markov transitions P'_n. Also let $A_n \in \mathcal{E}'_n$ be a given collection of measurable sets such that $\pi_0(A_0) > 0$ and $P'_n(x_{n-1}, A_n) > 0$ for any $x_{n-1} \in A_{n-1}$. We denote by π_n the distribution of the random paths restricted to the tube

($\prod_{p=0}^n A_p$). More formally, π_n is defined for any $f_n \in \mathcal{B}_b(\prod_{p=0}^n E'_p)$ by the formula

$$\pi_n(f_n) = \mathbb{E}\left(f_n(Y'_0, \ldots, Y'_n) \, 1_{A_0 \times \ldots \times A_n}(Y'_0, \ldots, Y'_n)\right) \quad (11.40)$$

We denote by X'_n the Markov chain from A_{n-1} into A_n with initial distribution $\eta_0(dx_0) = \pi(dx_0) 1_{A_0}/\pi_0(A_0)$ and elementary transitions given for any $x_{n-1} \in A_{n-1}$ by

$$M'_n(x_{n-1}, dx_n) = \frac{P'_n(x_{n-1}, dx_n) \, 1_{A_n}}{P'_n(x_{n-1}, A_n)}$$

- Show that π_n can be rewritten in the Feynman-Kac form

$$\pi_n(f_n) = \pi_0(A_0) \, \mathbb{E}\left(f_n(X'_0, \ldots, X'_n) \prod_{p=1}^n P'_p(X'_{p-1}, A_p)\right)$$

- Prove that the multiplicative property (11.38) holds true on the sets ($\prod_{p=0}^n A_p$) and

$$\pi_{n-1,n}((x_0, \ldots, x_{n-1}), dx_n) = P'_n(x_{n-1}, A_n) M'_n(x_{n-1}, dx_n)$$

Check that the corresponding particle simulation models can be interpreted as an interacting acceptance/rejection technique.

Exercise 11.9.4: [Maxima distribution functions] Let Y'_n be a nonanticipative sequence of random variables taking values in some measurable spaces $(E'_n, \mathcal{E}'_n)$. Also let V_n be a sequence of measurable functions on E'_n. If we take in (11.40) the sets $A_n = V_n^{-1}((-\infty, l])$ for some $l \in \mathbb{R}$ and then prove that $\pi_n(1) = \mathbb{P}(\sup_{p < n} V_p(Y'_p) \leq l)$

$$\pi_n = \text{Law}(Y'_0, \ldots, Y'_n \; ; \; \sup_{p \leq n} V_p(Y'_p) \leq l)$$

Design a particle approximation model of these quantities.

Exercise 11.9.5: [Hitting time probabilities] We consider time-homogeneous state spaces $E_n = E$ and a given measurable subset $A_n = A \in \mathcal{E}$. Check that the prediction Feynman-Kac measure π_n introduced in Exercise 11.9.3 coincides with the law of a Markov path $(Y'_0, \ldots, Y'_n)$ given the fact that it has never exited the set A after n steps; that is, for any $f_n \in \mathcal{B}_b(E^{n+1})$

$$\eta_n(f_n) = \pi_n(f_n)/\pi_n(1) = \mathbb{E}(f_n(Y'_0, \ldots, Y'_n) \mid T > n)$$
$$\pi_n(1) = \mathbb{P}(T > n) = \prod_{p=0}^n \eta'_p(A)$$

where η'_n stands for the nth time marginal of η_n and T the first time Y'_n enters into $(E - A)$. The estimation of first passage probabilities arises in various engineering problems, including catastrophic failures, buffer exceedance overflows, and financial ruin processes. Check that

$$\eta'_n(A) = \mathbb{P}(T > n \mid T \geq n) \quad \text{and} \quad \mathbb{P}(T \geq n) = \prod_{p=0}^{n} \mathbb{P}(T > p \mid T \geq p)$$

Finally, prove that $\mathbb{P}(T = n) = \pi_n(1) - \pi_{n+1}(1) = \eta'_n(A) \prod_{p=0}^{n-1} \eta'_p(E - A)$ and construct a particle approximation model of these quantities.

Exercise 11.9.6: Let $E_n = \mathbb{R}^d$, and let (G_n, M_n) be defined by the formulae

$$\begin{aligned} G_n(x_n) &= g_n(y_n, x_n) \\ &\stackrel{\text{def.}}{=} \exp\left\{ -\frac{1}{2}(y_n - c_n x_n) r_n^{-1} (y_n - c_n x_n)^T \right\} \end{aligned}$$

and

$$M_n(x_{n-1}, dx_n)$$

$$= \frac{|q_n^{-1/2}|}{(2\pi)^{d/2}} \exp\left\{ -\frac{1}{2}(x_n - a_n(x_{n-1})) q_n^{-1} (x_n - a_n(x_{n-1}))^T \right\}$$

where z^T represents the transpose of a column vector z, $y_n \in \mathbb{R}^{d'}$ is a given d'-dimensional vector, q_n, r_n are $(d \times d)$ and (d', d'_n) symmetric and nonnegative matrices, c_n is a $(d' \times d)$-matrix, and finally a_n is a bounded drift function on $\mathbb{R}^d$. In this example, we see that regions with high potential correspond to regions where $c_n x_n$ is close to the fixed and given parameter y_n.

1. Check that for any fixed vector x_{n-1} the function

$$(x_n, y_n) \in \mathbb{R}^d \times \mathbb{R}^{d'} \to \frac{dM_n(x_{n-1}, \cdot)}{dx_n}(x_n) \frac{|r_n^{-1/2}|}{(2\pi)^{d'/2}} g_n(y_n, x_n)$$

is the joint density of the $(d + d')$ Gaussian vector

$$(X_n, Y_n) = (a_n(x_{n-1}) + W_n, c_n X_n + V_n)$$

where $(W_n, V_n) \in \mathbb{R}^{d+d'}$ is a pair of independent centered Gaussian vectors with covariance matrices $\mathbb{E}(W_n W_n^T) = q_n$ and $\mathbb{E}(V_n V_n^T) = r_n$.

2. Show that the conditional Markov transition $\widehat{M}_n(x_{n-1}, \cdot)$ introduced in (11.16) coincides in this situation with the Gaussian distribution

on $\mathbb{R}^d$ with mean $m_n(x_{n-1})$ and covariance matrix $s_n(x_{n-1})$ defined by

$$\begin{aligned} m_n(x_{n-1}) &= a_n(x_{n-1}) + q_n c_n^T (c_n q_n c_n^T + r_n)^{-1}(y_n - c_n a_n(x_{n-1})) \\ s_n(x_{n-1}) &= q_n - q_n c_n^T (c_n q_n c_n^T + r_n)^{-1} c_n q_n \end{aligned}$$

3. Also prove that the corresponding potential functions $\widehat{G}_n = M_n(G_n)$ have the form

$$\widehat{G}_n(x_{n-1})$$

$$\propto \exp\{-\tfrac{1}{2}(y_n - c_n a_n(x_{n-1}))(c_n q_n c_n^T + r_n)^{-1}(y_n - c_n a_n(x_{n-1}))^T\}$$

Describe the mutation/selection transitions of the particle model associated with the pair $(\widehat{G}_n, \widehat{M}_n)$.

Exercise 11.9.7: [Baker's selection [22]] We consider Baker's remainder stochastic scheme introduced in (11.28). Check that $\mathbb{E}(\tilde{M}^i \mid X) = \{NW^i\}$, and conclude that for any $f \in \mathcal{B}_b(E)$ we have the unbiased property $\mathbb{E}(m(\widehat{X})(f)|X) = \Psi(m(X))(f)$. Recalling that $W^i = G(X^i)/\sum_{j=1}^N G(X^j)$, show that

$$\sum_{i=1}^N \frac{\{NW^i\}}{\sum_{j=1}^N \{NW^j\}} \delta_{X^i} = \widetilde{\Psi}(m(X))$$

with the Boltzmann-Gibbs transformation $\widetilde{\Psi}$ defined by the formula

$$\forall (\eta, f) \in \mathcal{P}(E) \times \mathcal{B}_b(E) \qquad \widetilde{\Psi}(\eta)(f) = \eta(G_\eta f)/\eta(G_\eta)$$

with the potential function $G_\eta(x) = \{G(x)/\eta(G)\}$. From the question above, deduce the equivalence in distribution

$$N\left(m(\widehat{X}) - \Psi(m(X))\right) = \tilde{N}\left(m(\tilde{X}) - \widetilde{\Psi}(m(X))\right) 1_{\tilde{N}>0}$$

where $m(\tilde{X}) = \frac{1}{\tilde{N}} \sum_{i=1}^{\tilde{N}} \delta_{\tilde{X}^i}$ represents the empirical measure associated with a sequence of $\tilde{N}$ independent random variables with common law $\widetilde{\Psi}(m(X))$. Using Lemma 7.3.3, prove that for any $p \geq 1$ and $\mathrm{osc}(f) \leq 1$ we have

$$\sqrt{N}\,\mathbb{E}(|(m(\widehat{X}) - \Psi(m(X)))(f)|^p \mid X)^{\frac{1}{p}} \leq d(p)^{\frac{1}{p}}$$

with the sequence of constants $d(p)$ defined in (7.7).

Exercise 11.9.8: Let $(e_i)_{i\geq 1}$ be a sequence of independent random variable with common exponential distribution

$$\mathbb{P}(e_1 \in dx) = \lambda\, e^{-\lambda x}\, 1_{\mathbb{R}_+}(x)\, dx$$

We also set $T_i = \theta^i(e_1, \ldots, e_N) =_{def.} \sum_{k=1}^{i} e_i$ for each $i \leq N$. Check that $\theta = (\theta^i)_{i \leq N}$ is a diffeomorphism between $\mathbb{R}_+^N$ and the set

$$C_N = \{(t_1, \ldots, t_N) \; : \; 0 < t_1 < \ldots < t_N\}$$

Prove that

$$\theta^{-1}(t_1, \ldots, t_N) = (t_1, t_2 - t_1, \ldots, t_N - t_{N-1}) \quad \text{and} \quad \text{Jac}(\theta^{-1}) = 1$$

Conclude that

$$\mathbb{P}((T_1, \ldots, T_N) \in d(t_1, \ldots, t_N)) = 1_{C_N}(t_1, \ldots, t_N) \; \lambda^N \; e^{-Nt_N} \; dt_1 \ldots dt_N$$

If we set $T_{N+1} = T_N + e_{N+1}$, then check that

$$\begin{aligned}\mathbb{P}(T_{N+1} \in dt) &= \lambda^{N+1} \; e^{-\lambda t} \; 1_{\mathbb{R}_+}(t) \; [\int_{0 < t_1 < \ldots < t_N < t} dt_1 \ldots dt_N] \; dt \\ &= \lambda \; e^{-\lambda t} \; \frac{(\lambda t)^N}{N!} 1_{\mathbb{R}_+}(t) \; dt\end{aligned}$$

Using the decomposition

$$\mathbb{P}((T_1, \ldots, T_{N+1}) \in d(t_1, \ldots, t_{N+1}))$$
$$= (N! \; 1_{0 < t_1 < \ldots < t_N < t_{N+1}} \; t_{N+1}^{-N} \; dt_1 \ldots dt_N)$$
$$\times (\lambda \; e^{-\lambda t_{N+1}} \; \frac{(\lambda t_{N+1})^N}{N!} 1_{\mathbb{R}_+}(t_{N+1}) \; dt_{N+1})$$

prove that

$$\mathbb{P}((T_1/T_{N+1}, \ldots, T_N/T_{N+1}, T_{N+1}) \in d(u_1, \ldots, u_N, t_{N+1}))$$
$$= \mathbb{P}((T_1/T_{N+1}, \ldots, T_N/T_{N+1}) \in d(u_1, \ldots, u_N)) \times \mathbb{P}(T_{N+1} \in dt_{N+1})$$

with

$$\mathbb{P}((T_1/T_{N+1}, \ldots, T_N/T_{N+1}) \in d(u_1, \ldots, u_N))$$
$$= N! \; 1_{0 < u_1 < \ldots < u_N < 1} \; du_1 \ldots du_N$$

Conclude that $(T_1/T_{N+1}, \ldots, T_N/T_{N+1})$ is an uniform order statistic. Describe a simulation algorithm to sample the three multinomial type branching models described respectively in (11.22), (11.24) and (11.28).

Exercise 11.9.9: Let U be a uniform $[0, 1]$-valued random variable, and let $\lambda > 0$ be a given parameter. Show that the random variable $e = \lambda \log(1/U)$ is an exponential random variable with parameter λ. We let $(e_i)_{i \geq 1}$ be a sequence of independent exponential random variables with parameter

$\lambda = 1$, and we set $T_i = \sum_{k=1}^{i} e_i$. Using Exercise 11.9.8, prove that for any $w \geq 0$ we have

$$\begin{aligned}\mathbb{P}(T_{m+1} > w) - \mathbb{P}(T_m > w) &= \int_w^\infty e^{-u}\frac{u^m}{m!}du - \int_w^\infty e^{-u}\frac{u^{m-1}}{(m-1)!}du \\ &= \int_w^\infty \frac{\partial}{\partial u}\left(-e^{-u}\frac{u^m}{m!}\right) du = e^{-u}\frac{w^m}{m!}\end{aligned}$$

Conclude that
$$\mathbb{P}(T_n \leq w < t_{m+1}) = e^{-u}\frac{w^m}{m!}$$

We let b the first time $m \geq 0$ we have $T_{m+1} > w$. Prove that

$$\mathbb{P}(b = m) = e^{-u}\frac{w^m}{m!}$$

Describe a simulation algorithm to sample the Poisson branching model described in (11.31).

12
Applications

12.1 Introduction

This rather long chapter focuses on the applications of Feynman-Kac modeling strategies and their interacting particle interpretations to a variety of practical problems. The field of applications includes spectral analysis of Feynman-Kac and Schrödinger semigroups, rare event estimation, sequential analysis of probability ratio tests, Dirichlet problems with boundary conditions, directed polymer simulations, and nonlinear filtering problems. As an initiated reader will immediately notice, all these problems consist of solving a more or less complex Feynman-Kac distribution. At the risk of repetition, we have chosen to include this chapter because we felt that there is no textbook or journal article that really illustrates the potential applications of Feynman-Kac and particle models. In the opposite situation, a reader not initiated on Feynman-Kac and particle models is recommended to read Chapter 11 before entering into the former exposition. Chapter 11 leaves out theoretical issues and it guides the reader through most of the important concepts and techniques needed in applications. For a more thorough training on Feynman-Kac and particle models, it is convenient to read Chapters 2 and 3.

We do not pretend to present in each particular application the most efficient algorithm with the optimal branching selection distribution or the best choice of exploration excursions. The approach we have taken here rather emphasizes the Feynman-Kac modeling of a given estimation problem. As soon as we have developed a sufficiently generic Feynman-Kac interpretation, we roughly design a rather general but basic particle approximation model. In general, we leave aside the possible improvements we could obtain by using one or the other recipes presented in Chapter 11.

It is of course out of the scope of this chapter to provide a catalog with detailed numerical comparisons between these interacting particle approximation models and some other more traditional techniques such as the extended Kalman-Bucy filter often used in "almost" linear/Gaussian filtering problems or any other alternative estimation models. To offer a way of comparison and to better connect the particle methodology with more classical literature on each application area, sometimes we describe particular situations where explicit calculations of the desired quantities can be derived. These examples can also serve practitioners for testing numerically the accuracy of particle models. The proof of these explicit and analytical solutions is often housed in a series of exercises at the end of each section.

To avoid repetition, we will not restate in each particular application area all the convergence results we can deduce from Chapters 7 to 10 on the asymptotic analysis of particle models. Sometimes we illustrate the impact of some asymptotic theorem in a specific application. But as a general rule we prefer to give some precise reference to a specific convergence theorem.

From the applied probability viewpoint, the present chapter is certainly one of the most important chapters of the book. The interested reader can try to develop a collection of particle approximation models in each application subject and can also find and interpret a selected asymptotic convergence theorem. For a more thorough training on practical estimation problems, we provide a brief catalog on selected journal articles in the applied literature. For applications of particle methods to tracking and visual detection of objects, we recommend to the reader the chain of articles [17, 148, 164, 212, 282, 143]. Applications of particle methods to global positioning systems can be found in [16, 43]. The multisplitting particle analysis of rare events is described in the chain of articles [11, 157, 158, 159, 303, 304, 305]. We also mention applications in image analysis [187] as well as in biology with gene estimations in DNA sequences [227, 228, 229, 230, 231, 233], data assimilation and inverse problems for ocean monitoring and prediction [136, 137, 138, 139, 218], and in finance with economic time series [270]. See also the multiauthor book [125] as well as the monograph [227] and references therein.

In each application area, the same Feynman-Kac and particle models are often expressed using different language. To guide the reader and to better connect our mathematical models with the more "applied" literature on this subject, we provide hereafter a short discussion on these different terminologies. Conditional distributions and filtering problems are one of the most typical examples of Feynman-Kac models arising in various scientific disciplines. In weather and oceanography literature, this estimation problem is instead called data assimilation with reference to the huge amount of observations provided by atmospheric and/or oceanographic measurements. Here the updating and prediction transitions are respectively called the model analysis and the model forecast (see for instance [139] and references therein). In this context the particle methodology is instead used to estimate the error covariance matrices in an extended Kalman filter. Sometimes the empirical measures are called "ensembles", and the resulting particle approximation models are simplified into the so-called "ensemble Kalman filters".

In Bayesian literature, the filtering model is preferably expressed in terms of a Bayes formula relating an "a priori" model with the desired "a posteriori" distribution. The latter measures are sometimes called the "beliefs" (see for instance [212]). In this branch of applied statistics, particle approximation models have taken various names such as "sampling-importance-resampling filters," "condensation filters," or "bootstrap filters," but it seems that the natural terminology "particle filters" is nowadays adopted. We hope that these modern particle methodologies will continue to serve as a bridge between the frequentist and Bayesian viewpoints.

In Monte Carlo Markov chain methods, Feynman-Kac particle models are also called "sequential Monte Carlo methods" (often abbreviated SMCM) to emphasize probably the nonrecursive drawback of traditional Monte Carlo Markov chain methods. In this context, the abstract prediction Feynman-Kac models on path space are sometimes expressed using recursive abstract formulae that basically read

$$\mathbb{Q}_n(d(x_0,\ldots,x_n)) \propto \mathbb{Q}_{n-1}(d(x_0,\ldots,x_{n-1})) \, Q_n(x_{n-1}, dx_n)$$

for some positive kernel Q_n from E_{n-1} into E_n. Whenever the normalizing constants $Q_n(1) > 0$, if we take $G_{n-1} = Q_n(1)$ and $M_n = Q_n/Q_n(1)$, these path measures coincide with the one introduced in (1.3) on page 11 (see also Exercise 11.9.2).

Boltzmann-Gibbs or Feynman-Kac formulae also arise in statistical physics, biology, and financial mathematics, and more generally in applied probability but in these areas the terminology is rather more stable and often coincides with that adopted in this book.

12.2 Random Excursion Models

12.2.1 Introduction

This section focuses on Feynman-Kac distributions on excursion spaces. We first design a multilevel modeling technique that reduces the analysis of these rather complex functionals to "simple" discrete time Feynman-Kac models. Then we apply the particle methodology to solve these formulae numerically. To motivate this section, we illustrate the impact of these modeling techniques with a brief discussion on some different application areas.

430   12. Applications

In engineering science, these models can be used to represent the law of a random process in some rare event regime. These rare events may represent a catastrophic failure or a buffer exceedance. For instance, in modern communication networks, several packets of information are sent from a source to a target destination. During their transmission, they visit several nodes in the network. At each node, they wait until the service capacity of the buffer is sufficiently high, otherwise the packet is lost. In practice, the buffers are sufficiently large and these events are hopefully rare events. To study the performance of these networks, one is not only interested in estimating the probability of overflows but also how these events happen. In this particular situation, the corresponding Feynman-Kac path model represents the law of the queueing process in this rare event regime.

In physics, excursion models may represent the distribution of a path particle in an absorbing medium with hard and soft obstacles (see Section 12.2.5). For a more practical illustration, we can think of a radiation source model that emits neutron particles in a containment (see for instance [146]). In this context, the absorption potential depends on the nature of the shielding environment. The choice of the hard obstacle sets depends on the problem at hand. For instance, if we are interested in computing the probability that a particle escapes the containment before disintegrating in some particular region of the configuration space, then the hard obstacle set will be chosen as this portion of the configuration space. We again refer the interested reader to Section 12.2.5 for a more thorough study of particle evolution models in an absorbing medium with only hard obstacles.

Feynman-Kac excursion models also provide a natural probabilistic interpretation of the solution of Dirichlet problems with boundary conditions. This subject that is pinched up between partial differential equations, linear operators, and probability theory also arises in a variety of engineering applications. For instance, in financial mathematics, Feynman-Kac distributions are often used to model option price evolutions. In this context, the hard obstacle sets usually represents some levels at which the option becomes worthless, while the potential function is interpreted as an instantaneous interest rate (see for instance the pedagogical textbook of Lamberton and Lapeyre [213]). In applied probability, these models are also used to analyze the possible limiting behaviors of a given Markov process (see [279]) or to capture the interplay between the geometry of the domain and the behavior of a stochastic process as it approaches the boundaries (see for instance [268, 295]).

The main idea behind the forthcoming excursion modeling techniques is to decompose the state space into a judicious choice of threshold subsets related to the system evolution. This decomposition reflects the successive levels the stochastic process needs to cross before entering into the relevant rare event. A rough description of the splitting particle method is as follows.

When a particle starting at some level does not succeed in entering into the next one, it is killed, but each time it enters into a closer level of the rare set, it slips into several offsprings. Between the levels, these offsprings evolve as independent copies of the stochastic process of interest until they reach (or do not) an even closer level, and so on. Loosely speaking, the branching particles are attracted by gateway regions from which the rare event is more likely to happen. In this sense, these excursion splitting techniques make the occurrence of rare events more frequent. Thus, they can also be regarded as an alternative to traditional importance-sampling methods.

These branching evolutionary algorithms were originally discussed in physics by Kahn and Harris to estimate particle transmission events [177]. Since that time, several variations and refinements have been suggested by analysts and designers in telecommunication and computer systems. The most currently used nowadays is the RESTART algorithm introduced in one of the pioneering articles of Villen-Altamirano et al. [303, 304, 305]. These models were further developed in a series of three articles of Glasserman et al. in [157, 158, 159]; we also recommend [298] for applications to communication networks as well as the article of Garvel and Kroese [153] for some details on the computer implementation of these algorithms. Most of the algorithms presented in this literature (except [157, 158, 159], which are based on judicious Bernoulli simplified models, large deviations, and fluctuation analysis) are essentially based on heuristic schemes with no really precise mathematical analysis. Moreover it is commonly assumed that the transition probabilities between the levels, and thus the desired rare event probability are known.

The objective of this section is to design a novel adaptive particle splitting method to estimate these rare events. The central idea is to represent the distribution of the process in the rare event regime in terms of a class of Feynman-Kac measures in the space of excursion.

12.2.2 Dirichlet Problems with Boundary Conditions

In this section, we design a strategy to estimate a given Feynman-Kac measure on excursion space by a Feynman-Kac distribution flow. We illustrate the impact of this modeling technique by an original particle interpretation of Dirichlet problems with boundary condition. In Section 12.2.4 we shall examine related Dirichlet models with boundary hard obstacles.

We let X'_n be a Markov chain taking values in some measurable spaces $(E'_n, \mathcal{E}'_n)$. We let T be a finite stopping time w.r.t. the filtration generated by X'_n. We associate with a sequence of $[0,1]$-valued potential functions G'_n on $E'_{[0,n]}$ the Feynman-Kac distribution

$$\widehat{\gamma}(f) = \mathbb{E}\left(f(T, X'_{[0,T]}) \prod_{p=0}^{T} G'_p(X'_{[0,p]}) \right) \qquad (12.1)$$

where f is a bounded measurable test function on the excursion space
$$E = \cup_{n \geq 0}(\{n\} \times E'_{[0,n]})$$
and we have use the notation $X'_{[0,n]} =_{\text{def.}} (X'_p)_{0 \leq p \leq n}$ for every $n \geq 0$. Now, we consider the E-valued and stopped Markov chain
$$X_n = (T \wedge n, X'_{[0,T \wedge n]}) \in E_n =_{\text{def.}} \cup_{0 \leq p \leq n}(\{p\} \times E'_{[0,p]})$$
and the potential function G_n on E_n defined by
$$G_n(T \wedge n, X'_{[0,T \wedge n]}) = G'_{T \wedge n}(X'_{[0,T \wedge n]})^{1_{T \wedge n = n}} = \begin{cases} G'_n(X'_{[0,n]}) & \text{if } n \leq T \\ 1 & \text{if } n > T \end{cases} \quad (12.2)$$

For instance, let X' be the simple random walk defined in Example 2.2.1. Suppose $X'_0 \in (0, \infty)$ and let T be the first time X' hits 0. In this case, the stopped process $X'_{T \wedge n}$ coincides with the random walk on $\mathbb{N}$ where the origin is an absorbing barrier.

We associate with the stopped Markov chain X_n and the potential G_n the Feynman-Kac distributions on E_n defined by
$$\widehat{\gamma}_n(f) = \mathbb{E}\left(f(X_n) \prod_{p=0}^{n} G_p(X_p)\right)$$

For more details on stopped Markov processes, we refer the reader to Section 2.2.3. The next proposition allow us to interpret the Feynman-Kac measures in excursion-spaces (12.1) as the limiting measures of Feynman-Kac semigroups.

Proposition 12.2.1 *For any $n \geq 0$ and $f \in \mathcal{B}_b(E)$, with $\|f\| \leq 1$, we have*
$$|\widehat{\gamma}_n(f) - \widehat{\gamma}(f)| \leq 2\mathbb{P}(T > n) \quad (\longrightarrow 0 \text{ as } n \to \infty)$$
and
$$0 \leq \widehat{\gamma}_n(1) - \widehat{\gamma}(1) \leq \mathbb{P}(T > n)$$

Proof:
We first observe that
$$\widehat{\gamma}_n(f) = \mathbb{E}\left(f(X_n) 1_{T \leq n} \prod_{p=0}^{n} G_p(X_p)\right) + \mathbb{E}\left(f(X_n) 1_{T > n} \prod_{p=0}^{n} G_p(X_p)\right)$$
and
$$\mathbb{E}(f(X_n) 1_{T \leq n} \prod_{p=0}^{n} G_p(X_p)) = \mathbb{E}(f(T, X'_{[0,T]}) 1_{T \leq n} \prod_{p=0}^{T} G'_p(X'_{[0,p]}))$$
$$= \widehat{\gamma}(f) - \mathbb{E}(f(T, X'_{[0,T]}) \prod_{p=0}^{T} G'_p(X'_{[0,p]}) 1_{T > n})$$

Therefore, we find that

$$\widehat{\gamma}_n(f) - \widehat{\gamma}(f)$$

$$= \mathbb{E}\left(\left[f(n, X'_{[0,n]}) - f(T, X'_{[0,T]})\prod_{p=n+1}^{T} G'_p(X'_{[0,p]})\right]\right.$$

$$\left. \times \ \prod_{p=0}^{n} G'_p(X'_{[0,p]}) 1_{T>n}\right)$$

from which the end of the proof is clear. ∎

If $\widehat{\gamma}(1) > 0$, then we can defined the normalized distributions

$$\widehat{\eta}_n(f) = \widehat{\gamma}_n(f)/\widehat{\gamma}_n(1) \quad \text{and} \quad \widehat{\eta}(f) = \widehat{\gamma}(f)/\widehat{\gamma}(1)$$

and by Proposition 12.2.1, we have

$$\sup_{f:\operatorname{osc}(f)\leq 1}|\widehat{\eta}_n(f) - \widehat{\eta}(f)| \leq 2\mathbb{P}(T>n)/\widehat{\gamma}(1) \quad (\longrightarrow 0 \text{ as } n\to\infty)$$

To prove this assertion, we simply use the decomposition

$$\widehat{\eta}_n(f) - \widehat{\eta}(f) = \frac{\widehat{\gamma}_n(1)}{\widehat{\gamma}(1)} \times \widehat{\eta}_n\left(\frac{1}{\widehat{\gamma}(1)}(f - \widehat{\eta}(f))\right)$$

As an aside, we note that

$$\widehat{Q}_{n+1}(f)(t_n,(x_0,\ldots,x_{t_n}))$$

$$=_{\text{def.}} M_{n+1}(G_{n+1}f)(t_n,(x_0,\ldots,x_{t_n}))$$

$$= f(t_n,(x_0,\ldots,x_{t_n}))\ G'_{t_n}(t_n,(x_0,\ldots,x_{t_n}))^{1_{t_n=n+1}}$$

$$= f(t_n,(x_0,\ldots,x_{t_n}))$$

as soon as

$$(t_n < n) \quad \text{or} \quad (t_n = n \text{ and } (x_0,\ldots,x_n) \notin A_{n+1})$$

where M_n is the Markov transition of X_n and A_n is the set-realization of the stopping time T (see Section 2.2.3, page 53). From the observation above, we find the fixed point equations

$$\widehat{\gamma}\widehat{Q}_{n+1} = \widehat{\gamma} \quad \text{and} \quad \widehat{\eta} = \widehat{\Phi}_{n+1}(\widehat{\eta})$$

with the one-step transition $\widehat{\eta}_{n+1} = \widehat{\Phi}_{n+1}(\widehat{\eta}_n)$ associated with the updated Feynman-Kac flow. We can improve a little Proposition 12.2.1 when T is the entrance time of X' into a set of the form $(B \cup C)$, with $B \cap C = \emptyset$, and f is the indicator test function

$$f(T, X'_{[0,T]}) = 1_B(X'_T)$$

In this case, arguing as in the proof of Proposition 12.2.1, we find that

$$\widehat{\gamma}_n(f) = \widehat{\gamma}(f) - \widehat{\gamma}(f_n) \quad \text{with} \quad f_n(T, X'_{[0,T]}) = 1_{(n,\infty)}(T) \, 1_B(X'_T)$$

from which we conclude that

$$0 \leq \widehat{\gamma}(f) - \widehat{\gamma}_n(f) \leq \mathbb{P}(X'_T \in B, \ T > n)$$

If T is not almost surely finite, the above analysis remains valid on the event $(T < \infty)$. More precisely, Proposition 12.2.1 holds true if we replace $f(X_T)$ and $f(X_n)$ by $f(X_T)1_{T<\infty}$, and $f(X_n)1_{T<\infty}$, and $\mathbb{P}(n < T)$ by $\mathbb{P}(n < T < \infty)$.

Proposition 12.2.1 can be extended to bounded potential functions G'_n as soon as we have $\lambda = \sup_n \log \|G'_n\|$ and $\mathbb{E}(e^{\lambda T}) < \infty$. In this case, we check that

$$|\widehat{\gamma}_n(f) - \widehat{\gamma}(f)| \leq 2\mathbb{E}(e^{\lambda T} 1_{T>n}) \quad (\longrightarrow 0 \text{ as } n \to \infty)$$

The Feynman-Kac models in excursion space presented above provide a nice probabilistic interpretation of Dirichlet problems with boundary conditions. For instance, let X'_n be a time homogeneous Markov chain with Markov transitions M' on some measurable space $(E', \mathcal{E}')$, and let G' be a $[0,1]$-valued potential function on E'. If we let T be the exit time of X'_n from a measurable set $A \in \mathcal{E}'$, then for each $f \in \mathcal{B}(E')$ the functions

$$D(f)(x) = \mathbb{E}_x \left(f(X'_T) \prod_{p=1}^T G'(X'_p) \right)$$

satisfy the pair equations

$$\begin{cases} D(f)(x) &= f(x) & \text{if } x \in A^c \\ D(f)(x) &= M'(G'D(f))(x) & \text{if } x \in A \end{cases}$$

If we interpret D as a bounded integral operator, with some obvious abusive notations we have

$$\begin{cases} \mu D &= \mu & \text{if } \mu \in \mathcal{P}(A^c) \\ \mu D &= \mu \widehat{Q}' D & \text{if } \mu \in \mathcal{P}(A) \end{cases}$$

with the integral operator $\widehat{Q}'(f) = M'(G'f)$.

The particle approximation model associated with the Feynman-Kac model $\widehat{\eta}_n$ consists in stopped excursion-valued particles, that evolve and interact according to the potential function G_n introduced in (12.2). In the exit time case discussed above, the excursions are stopped as soon as they exit the set A. In this situation, they potential value is equal to 1. The other particles explore the state space, and interact with the whole configuration, in accordance with the absorption potential function G'. By construction, the algorithm stops as soon as all the particles have exited the set A. Also observe that for $G_n = 1$, the particle interpretation model reduces to N iid excursion-valued particles.

We end this section with an important observation. By Proposition 12.2.1, the function $D(f)(x)$ can be approximated by the unnormalized Feynman-Kac flow

$$D_n(f)(x) = \mathbb{E}_x \left(f(X'_{n \wedge T}) \prod_{p=1}^{n} G_p(p \wedge T, X'_{p \wedge T}) \right)$$

with the potential function

$$G_p(p \wedge T, X'_{p \wedge T}) = G'(X'_{p \wedge T})^{1_{p \wedge T = p}}$$

By the multiplicative formula presented in Proposition 2.3.1, we have

$$D_n(f)(x) = \eta_n^{(x)}(fG_n) \prod_{p=1}^{n-1} \eta_p^{(x)}(G_p)$$

where $\eta_n^{(x)}$ represents the prediction Feynman-Kac flow associated to the stopped process and the potential functions G_n, and starting at δ_x at time $n = 0$. One rather crude numerical approximation of $D_n(f)(x)$ will be to start at each $x \in A$ a separate particle model. One alternative and more judicious strategy is to evolve a single particle approximation model, properly initialized in A, and use the descendant genealogical tree approximation models (11.15) described in Section 11.4.

Let us make life slightly more complicated by considering only the excursions of a time-homogeneous Markov chain X' from a set A into a particular subset $B \subset A^c$. In other words, we are given a partition $A^c = B \cup C$ and the set C is regarded as an hard and absorbing obstacle. More formally, we suppose that X' starts in A and exits the set at a random time T. The law of the excursions ending in B are given by the Feynman-Kac measures

$$\widehat{\gamma}(f) = \mathbb{E}\left(f(T, X'_{[0,T]})1_B(X')\right) = \mathbb{E}\left(f(T, X'_{[0,T]}) \prod_{p=1}^{T} 1_{A \cup B}(X'_p)\right)$$

If we let R be the first time X' hits C, then we have

$$\begin{aligned}\widehat{\gamma}(1) &= \mathbb{P}(T < R) = \mathbb{P}(X'_T \in B) \\ \widehat{\eta}(f) &= \widehat{\gamma}(f)/\widehat{\gamma}(1) = \mathbb{E}(f(T, X'_{[0,T]}) \mid T < R)\end{aligned}$$

This formulation, combined with Proposition 12.2.1, allows us to estimate $\widehat{\gamma}(f)$ by the Feynman-Kac flow associated to the stopped excursion-valued process and defined by

$$\widehat{\gamma}_n(f) = \mathbb{E}(f(T \wedge n, X'_{[0, T \wedge n]}) \prod_{p=1}^{n} 1_{A \cup B}(X'_p)^{1_{T \wedge p = p}})$$

We readily observe that

$$\widehat{\gamma}_n(1) = \mathbb{P}(T \wedge n < R) \quad \text{and} \quad \widehat{\eta}_n(f) = \mathbb{E}(f(T \wedge n, X'_{[0,T\wedge n]}) \mid T \wedge n < R)$$

The particle interpretation model consists in stopped excursion-valued particles. A particle that enters into C is killed, and instantly a randomly chosen excursion in $A \cup B$ duplicates. Note that the excursions from A to B are stopped for always (since their potential value is equal to 1). The particle model is stopped as soon as all the particles exit A, or are absorbed by C.

When the obstacle set C is too large and too attractive, most of the excursions hit the set C. In this case the particle approximation model is not really efficient. Two strategies can be underlined. The first idea is to change the reference measure so that the excursions become more likely to avoid the set C. To guide guide the reader, we let $\mathbb{E}$ and $\overline{\mathbb{E}}$ be the expectations operator with respect to the law of a Markov chain X' with transitions M'_n and $\overline{M}'_n$. Suppose that $M'_n(x, .) \ll \overline{M}'_n(x, .)$ for all $x \in E'$. In this case, the Feynman-Kac measures $\widehat{\gamma}_n$ can be rewritten as follows

$$\widehat{\gamma}_n(f)$$

$$= \overline{\mathbb{E}}\left(f(T \wedge n, X'_{[0,T\wedge n]}) \prod_{p=1}^n \left(1_{A\cup B}(X'_p) \frac{dM'_p(X'_{p-1},\cdot)}{d\overline{M}'_p(X'_{p-1},\cdot)}(X'_p)\right)^{1_{T\wedge p=p}}\right)$$

The particle interpretation model is defined as before except that the particle explore the state space with the Markov transitions $\overline{M}'_n$ and they are updated using the Radon-Nikodym potential functions. The second idea is to introduce a judicious multilevel decomposition, and then freeze the particles as soon as they enter into a level from which the next excursion is more likely to enter in B. This more sophisticated strategy is described in detail from Section 12.2.3 to Section 12.2.5. In some instances, the varianace in the central limit theorem can be explicitely computed or at least estimated, and compared with more crude Monte carlo methods. The interested reader is referred to Section 12.2.7.

12.2.3 Multilevel Feynman-Kac Formulae

Let $(X_t)_{t \in I}$ be a strong Markov process taking values in some metric state space (S, d) with discrete or continuous time index $I = \mathbb{R}_+$ or $I = \mathbb{N}$. For discrete time models we recall that the strong Markov property always holds (see for instance Section 2.2.3). We suppose X is defined on the canonical filtered probability space $(\Omega = D(I, S), \mathcal{F} = (\mathcal{F}_t)_{t\in I}, (\mathbb{P}_x)_{x\in S})$ of left-continuous and right-limited paths $D(I, S)$ from I into S (for $I = \mathbb{N}$, note that $D(I, S) = S^I$). For any distribution $\eta_0 \in \mathcal{P}(S)$, we write $\mathbb{P}_{\eta_0} = \int_S \eta_0(dx) \, \mathbb{P}_x$, the distribution of X with initial distribution η_0.

We consider a nonempty measurable subset $A \subset S$ and we let T be the first time X_t exits from A. We further suppose that the complementary set

A^c is decomposed into two disjoint subsets $A^c = B \cup C$. We let R be the first time X hits the set C. By definition, we have $T \leq R$ with $T < R$ as soon as X hits B before C. Finally, we assume that T is a finite stopping time in the sense that $\mathbb{P}_x(T < \infty) = 1$ for any $x \in S$. In other words, if we let T_D the entrance time into a measurable set $D \subset S$, then we have

$$T = T_{B \cup C} = T_B \wedge T_C, \qquad R = T_C \quad \text{and} \quad (T < R) \iff (T_B < T_C)$$

In addition, suppose that T_C is a finite stopping time, let $X_t^C = X_{t \wedge T_C}$ be the stopped process associated with T_C, and let T_B^C be the first time X_t^C enters in B. Then, we clearly have the equivalent formulations

$$(T < R) \iff (T_B < T_C) \iff (T_B^C < \infty)$$

On the event $(T_B^C < \infty)$, we have $T = T_B = T_B^C$ and the random path $(X_t)_{t \in [0, T_B^C]} = (X_t)_{t \in [0,T]}$ represents the excursion from the origin up to the entrance time in B. In the opposite case (with the usual convention $\inf_\emptyset = \infty$), we have $T_B^C = \infty$ on the event $T_C \leq T_B$ and $(X_t)_{t \in [0, T_B^C]} = (X_t)_{t \in [0, T_C]}$ the excursion from the origin up to the entrance time in C.

Note that even if T is *bounded* it may take arbitrary large values and the process may be trapped in A for an arbitrary long period. We are interested in solving numerically functional expectations of the form

$$\Gamma(F) = \mathbb{E}_{\nu_0}\left(F(T, X_T) \exp\left\{-\int_0^T V_s(X_s)ds\right\} 1_{T<R}\right) \qquad \text{if } I = \mathbb{R}_+$$

$$\Gamma(F) = \mathbb{E}_{\nu_0}\left(F(T, X_T) \left\{\prod_{p=1}^T G_p(X_p)\right\} 1_{T<R}\right) \qquad \text{if } I = \mathbb{N}$$

(12.3)

where F is a bounded measurable function on $(I \times S)$ and $V : (s,x) \in (I \times S) \to V_s(x) \in \mathbb{R}_+$, $G : (s,x) \in (I \times S) \to G_s(x) \in [0,1]$ is a pair of bounded potential functions.

The empty set $C = \emptyset$ is not excluded. In this situation, we use the convention that $R = \infty$ and the resulting models reduce to the class of excursion models discussed in section 12.2.2. In this section, the set C has to be thought of as a hard obstacle set the Markov particle tries to avoid.

To fix these ideas, let us suppose that X_0 starts in some particular region $A_0 \subset A$ with an initial distribution ν_0. During its excursion from A_0 to $A^c (= B \cup C)$, the process passes through a decreasing sequence of level sets $(B_n)_{n=0,\ldots,m} \in \mathcal{S}^{m+1}$ with

$$B = B_m \subset \ldots \subset B_1 \subset B_0$$

The splitting parameter m and the choice of the level sets $(B_n)_{0 \leq n \leq m}$ depend on the problem at hand, but it is important to choose these quantities such that

$$A_0 = B_0 - B_1 \quad \text{and} \quad B_0 \cap C = \emptyset$$

We refer the reader to Section 12.2.5 for some worked-out examples of splitting levels in the context of rare event analysis. To capture the behavior of X between the different levels $(B_n)_{0 \leq n \leq m}$, we let T_n, $1 \leq n \leq m$, be the first time X hits $B_n \cup C$; that is

$$T_n = \inf\{0 \leq t \,:\, X_t \in B_n \cup C\}$$

We associate with these entrance times the discrete stochastic sequence of excursions

$$\mathcal{X}_n = (T_n, (X_t \,;\, T_{n-1} \leq t \leq T_n)) \in E = \cup_{s \leq t}\{t\} \times D([s,t], S) \quad (12.4)$$

By construction, we also notice that the random sequence of level-crossing times is increasing:

$$T_0 = 0 \leq T_1 \leq \ldots \leq T_m = T$$

By a direct inspection, we see that if $T_n < R$, then the second component of $\mathcal{X}_n$ represents the excursion of the process X between the successive levels B_{n-1} and B_n so that T_n can be alternatively defined by the inductive formulae

$$T_n \;=\; \inf\{T_{n-1} \leq t \,:\, X_t \in B_n \cup C\}$$

Under our assumptions, we also observe that these entrance times are finite and

$$(T < R) = (T_m < R) = (T_1 < R, \ldots, T_m < R)$$

By the strong Markov property, we prove the following result.

Proposition 12.2.2 *The stochastic sequence* $(\mathcal{X}_n)_{0 \leq n \leq m}$ *defined by*

$$\mathcal{X}_n = (T_n, (X_t \,;\, T_{n-1} \leq t \leq T_n)) \in E = \cup_{s \leq t}(\{t\} \times D([s,t], S))$$

forms a Markov chain taking values in the set of excursions E.

In this interpretation, it is also important to note that the level indexes $n \in \{0, \ldots, m\}$ are regarded as the time indexes of the excursion Markov model. Whenever C is nonempty, it may happen that the excursion starting at some level, say B_{n-1}, visits C before entering into the next desired level B_n. In this case, we have

$$T_n = R \Longrightarrow \forall p > n \quad T_p = R \quad \text{and} \quad X_p = X_R \in C$$

In the opposite case, it may happen that a given excursion starting at some level, say $B_{p-1} - B_p$,

$$\mathcal{X}_p = (T_p, (X_t \,;\, T_{p-1} \leq t \leq T_p)) \quad \text{with} \quad X_{T_{p-1}} \in B_{p-1}$$

enters "directly" into some level $X_{T_p} \in B_n \subset B_p$ with $n \geq p$ without visiting the set $B_p - B_n$. In this case, recalling that $B_n \subset \ldots \subset B_{p+1} \subset B_p$, we have $X_{T_p} = X_{T_{p+1}} = \ldots = X_{T_n} \in B_n$, $T_p = T_{p+1} = \ldots = T_n$, and $\mathcal{X}_p = \mathcal{X}_{p+1} = \ldots = \mathcal{X}_n$. In other words, in this case the process is *frozen* during $(n-p)$ units of time. This apparently innocent observation is in fact one of the key ingredients of the corresponding particle interpretation models. Loosely speaking, the particle excursions that successfully enter into some level will also be *frozen*. If some of the others never, succeed they will be killed and instantly a randomly chosen "frozen particle" duplicates. One way to check whether or not a random path has succeeded in reaching the desired nth level is to consider the potential functions $\mathcal{G}_n$ on E defined for each $t \in I$ and $x = (x_r)_{s \leq r \leq t} \in D([s,t], S)$ with $s \leq t$ by

$$\mathcal{G}_n(t,x) = 1_{B_n}(x_t) \exp\left\{-\int_s^t V_r(x_r) dr\right\} \quad \text{if} \quad I = \mathbb{R}_+$$

$$\mathcal{G}_n(t,x) = 1_{B_n}(x_t) \prod_{p=s+1}^t G_p(x_p) \quad \text{if} \quad I = \mathbb{N}$$

In this notation, we have for each n

$$(T_n < R) = (T_1 < R, \ldots, T_n < R) = (\mathcal{G}_1(\mathcal{X}_1) = 1 \ldots, \mathcal{G}_n(\mathcal{X}_n) = 1)$$

$$(\mathcal{X}_0, \ldots, \mathcal{X}_n)$$
$$= ((0, X_0), (T_1, (X_t \; ; \; 0 \leq t \leq T_1)), \ldots, (T_n, (X_t \; ; \; T_{n-1} \leq t \leq T_n)))$$

In the further development, we slightly abuse the notation and sometimes write $[X_t \; ; \; 0 \leq t \leq T_n]$ instead of $(\mathcal{X}_0, \ldots, \mathcal{X}_n)$, the sequence of excursions of X between the levels $B_0, \ldots, B_n$. Using elementary calculations, we prove the following proposition.

Proposition 12.2.3 (Multilevel Feynman-Kac models) *For any $n \in \mathbb{N}$ and $f_n \in \mathcal{B}_b(E^{n+1})$, we have*

$$\mathbb{E}_{\nu_0}\left(f_n(\mathcal{X}_0, \ldots, \mathcal{X}_n) \prod_{p=1}^n \mathcal{G}_p(\mathcal{X}_p)\right)$$

$$= \mathbb{E}_{\nu_0}\left(f_n([X_t \; ; \; 0 \leq t \leq T_n]) \, 1_{T_n < R} \, \exp\left\{-\int_0^{T_n} V_s(X_s) ds\right\}\right) \quad \text{if} \quad I = \mathbb{R}_+$$

$$= \mathbb{E}_{\nu_0}(f_n([X_t \; ; \; 0 \leq t \leq T_n]) \, 1_{T_n < R} \, \{\textstyle\prod_{p=1}^{T_n} G_p(X_p)\}) \quad \text{if} \quad I = \mathbb{N}$$

The particle interpretations of these discrete generation Feynman-Kac models should now be clear to the reader. For instance, the simplest particle interpretation in the context of continuous time models goes as follows. We start with N independent copies $(\chi_0^i)_{1\leq i\leq N}$ of X_0 and during the mutation stage we randomly evolve these particles up to the first time they hit the first-level set $(B_1 \cup C)$. If $(\chi_1^i(t))_{0\leq t\leq T_1^i}$ denotes the excursion of the ith particle from A_0 to $(B_1 \cup C)$, we compute the weights

$$\mathcal{G}_1(T_1^i, (\chi_1^i(t))_{0\leq t\leq T_1^i}) = 1_{B_1}(\chi_{T_1^i}^i) \exp\left\{-\int_0^{T_1^i} V_s(\chi_1^i(s))ds\right\}$$

Note that if an excursion hits C, its weight is 0; otherwise the exponential weight represents the strength of the soft obstacles it has visited during its evolution to B_1. In this sense, we can interpret these weights as the *predicted lifetime* of each particle. During the selection transition, with a probability $\exp\{-\int_0^{T_1^i} V_s(\chi_1^i(s))ds\}$, the excursion survives. Otherwise, the particle dies and instantly one of the excursions having succeeded in reaching B_1 is randomly chosen with a probability proportional to its predicted lifetime and splits into two identical copies. At the second step, we again evolve the selected excursions up to the first time they reach the set $(B_2 \cup C)$, and we update the particle configuration in accordance with their predicted lifetime between levels B_1 and B_2, and so on.

We refer the reader to Sections 11.3 and 11.4 for a more thorough discussion on these particle algorithms. We can clearly combine these multisplitting strategies with the stopped excursion valued particle approximation models described in the end of section 12.2.2. Precise excursion particle models evolving in an absorbing medium with only hard obstacles will also be described in Section 12.2.5.

12.2.4 Dirichlet Problems with Hard Boundary Conditions

In this section, we illustrate the abstract Feynman-Kac models presented in Section 12.2.3 in the context of Dirichlet problems with boundary conditions. We use the same notation and convention as there. In the homogeneous and discrete time case, by the Markov property we easily check that the function

$$h(x) = \mathbb{E}_x\left(f(X_T)\prod_{p=1}^T G(X_p)\,1_{T<R}\right) = \mathbb{E}_x\left(f(X_T)1_B(X_T)\prod_{p=1}^T G(X_p)\right)$$

is a solution of the Dirichlet problem

$$\begin{cases} M(Gh)(x) = h(x) & \text{for } x \in A \\ h(x) = f(x)1_B(x) & \text{for } x \notin A \end{cases}$$

where $M(x, dy)$ represents the Markov transitions of the chain X_n, $A \subset S$ is a given subset, and $A^c = (B \cup C)$ is a partition of the complementary

set A^c. If we choose a constant potential functions, say $G = 1$, then the function
$$h(x) = \mathbb{E}_x(f(X_T)1_{T<R}) = \mathbb{E}_x(f(X_T)1_{X_T \in B})$$
is a nonnegative harmonic function in A with the boundary values $h = f1_B$ on A^c.

To illustrate these models in continuous time settings, we let L be the second-order linear differential operator
$$L(h)(x) = \sum_{1 \leq i \leq d} a_i(x)\frac{\partial h}{\partial x^i}(x) + \frac{1}{2}\sum_{1 \leq i,j \leq d}(bb^\star)_{i,j}(x)\frac{\partial^2 h}{\partial x^i \partial x^j}(x)$$
where a_i are bounded Lipschitz functions on $S = \mathbb{R}^d$ and $bb^\star$ is symmetric and uniformly strictly positive definite (i.e., of full rank). The Dirichlet problem is now to find a function h that satisfies the equations
$$\begin{cases} L(h)(x) = V(x)h(x) & \text{for } x \in A \\ h(x) = g(x) & \text{for } x \in \partial A \end{cases}$$
where A is a bounded and open set with smooth boundary ∂A and (V, g) is a given pair of bounded functions. The probabilistic interpretation of this problem is as follows. First we observe that L is the infinitesimal generator of the d-dimensional stochastic process X_t, $t \in \mathbb{R}_+$, defined by the stochastic differential equation
$$dX_t^i = a_i(X_t)dt + \sum_{j=1}^d b_{i,j}(X_t)\, dW_t^i$$
where W_t^i, $1 \leq i \leq d$, are independent Wiener processes. Note that our regularity assumption on $b_{i,j}(x)$ ensures that $\mathbb{P}_x(T < \infty) = 1$ (in the opposite case, the "diffusion" may never succeed in reaching the boundary). In this situation, for any partition $A^c = (B \cup C)$, the probabilistic representation of the solution of the Dirichlet problem with $g = f1_B$ is given by the Feynman-Kac formula
$$h(x) = \mathbb{E}_x\left(f(X_T)\, \exp\{-\int_0^T V(X_s)ds\}\, 1_{T<R}\right)$$

In some very special cases, explicit solutions exist. For instance, if we take $(b, f) = (Id, 1)$, $(a, V) = (0, 0)$, the open annulus $A = A_{\varepsilon_1, \varepsilon_2} = \{x \in \mathbb{R}^d : \varepsilon_1 < |x| < \varepsilon_2\}$ with $0 < \varepsilon_1 < \varepsilon_2$, $C = \{|x| \geq \varepsilon_2\}$, and $B = \{|x| \leq \varepsilon_1\}$, then we have $X_t = W_t$ and for any $x \in A$
$$h(x) = \mathbb{E}_x(1_B(W_T)) = \mathbb{P}_x(W_T \leq \varepsilon_1) = \begin{cases} \frac{\log \varepsilon_2 - \log |x|}{\log \varepsilon_2 - \log \varepsilon_1} & \text{if } d = 2 \\[1em] \frac{|x|^{2-d} - \varepsilon_2^{2-d}}{\varepsilon_1^{2-d} - \varepsilon_2^{2-d}} & \text{if } d \geq 3 \end{cases}$$
(12.5)

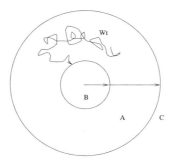

FIGURE 12.1. Brownian evolution on the annulus

In the case where $V(x) = \lambda > 0$, and the constant function $f(x) = 1$, the function $h(x) = \mathbb{E}_x(e^{-\lambda T} 1_{T<R})$ coincides with the Laplace transform of T on the event $(T < R) = (X_T \in B)$. In general, these functional Feynman-Kac models cannot be solved analytically and their numerical solution requires extensive calculations. Suppose for instance that the obstacle set C is too "large" in the sense that most of the realizations of X_t tend to end in C. In this case, we would need to sample a large number of independent copies of X_t to find at least one that reaches the desired target boundary B. This shows that a naive Monte Carlo method will need too many particles to get some reasonable statistical accuracy. As mentioned earlier, one advantage of the multisplitting technique comes from the fact that "good" particles are frozen as soon as they succeed in reaching some rare level and these leading individuals duplicates into several offsprings. As a result, these decomposition levels behave as gateways from which the particles have more chance to hit the desired regions on the boundary. One drawback of these particle models is that their accuracy is intimately related to a judicious level decomposition of the state space.

Example 12.2.1 (Absorption event) *Suppose the state $S = B \cup D$ is decomposed in two separate regions B and D. The process X evolves in the region D, which contains a collection of "soft" and "hard" obstacles represented respectively by the potential functions G_p or V_s and by a subset $C \subset D$. The particle is instantly killed as soon as it enters the "hard" obstacle set C. In this context, the quantities*

$$\mathbb{E}_{\nu_0}\left(1_{T<R} \prod_{p=1}^{T} G_p(X_p)\right) \quad \text{and} \quad \mathbb{E}_{\nu_0}\left(1_{T<R} \exp\left\{-\int_0^T V_s(X_s)ds\right\}\right)$$

represent the probability of exiting the pocket of obstacles D without being killed. More generally, the measures defined in Proposition 12.2.3 represent the distribution of the path process on this event. The sequence $(B_n)_{0 \leq n \leq m}$ represents the exit levels the process needs to reach to get out of D (before being killed). We notice that the pocket of obstacles D may be decomposed

12.2 Random Excursion Models   443

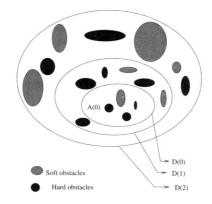

FIGURE 12.2. Soft/hard obstacles

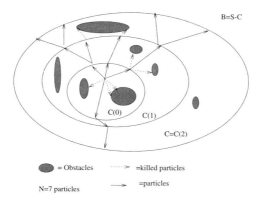

FIGURE 12.3. Genealogical model, [exit of C(2) before killing] (N=4)

into a collection of subpockets $D_0 \subset D_1 \subset \ldots \subset D_m = D$. Each D_p may intersect the set of obstacles C (see Figure 12.2). Let $B_0 = S - C$ and let B_p, $p = 0 \leq p \leq m$, be the decreasing sequence of subsets defined by $B_{p+1} = S - (C \cup D_p)$. By direct inspection, we see that this decomposition satisfies the desired properties with the target set $B_{m+1} = S - (C \cup D_m) = S - D = B$ and the initial set $A_0 = B_0 - B_1 = D_0 - D_0 \cap C$. Figure 12.2 provides a simple example of decomposition of a medium with hard and soft obstacles.

In Figure 12.3, we have illustrated the genealogical particle model associated with a killed Markov particle X evolving in a pocket of obstacles $D \subset S$. In this example, the pocket D is decomposed into three regions, $D_0 \subset D_1 \subset D_2 = D$. The decreasing sequence of exit levels is given by $B_{p+1} = S - (C \cup D_p)$ $p = 0, 1, 2$, with $B_0 = S - C$, $B = B_3 = S - D$, $A_0 = D_0 - D_0 \cap C$ (and the desired target set here is $B = B(3)$).

12.2.5 Rare Event Analysis

We use the same notation and conventions as were introduced in Section 12.2.3. For the convenience of the reader, we recall that $(X_t)_{t \in I}$ is a strong Markov process taking values in some metric state space (S, d) with discrete or continuous time index $I = \mathbb{R}_+ = [0, \infty)$ of $\mathbb{N}$. The process X starts in some Borel set $A_0 \subset S$ with a given distribution $\nu_0 \in \mathcal{P}(S)$. We also consider a pair of Borel subsets (B, C) such that $A_0 \cap C = \emptyset = B \cap C$. We associate with this pair the first time T the process hits $(B \cup C)$ and the hitting time R of the set C. Note that

$$T = \inf\{0 \leq t \,:\, X_t \in B \cup C\} \leq R = \inf\{t \geq 0 \,;\, X_t \in C\}$$

We also assume that (A_0, B, C) is chosen so that for any initial state $x \in A_0$ we have $\mathbb{P}_x(T < \infty) = 1$. As we shall see later, the Feynman-Kac splitting models developed in this section are still valid if we replace this condition by the weaker assumption that $\mathbb{P}_x(T < \infty) > 0$ for any $x \in A_0$. Nevertheless, since the branching models are built on excursion particles between successive levels, the first condition ensures that any of these excursions is finite. One would like to estimate the quantities

$$\begin{aligned} \mathbb{P}(T < R) &= \mathbb{P}(X_T \in B) \\ \mathrm{Law}(X_t \,;\, 0 \leq t \leq T \mid T < R) &= \mathrm{Law}(X_t \,;\, 0 \leq t \leq T \mid X_T \in B) \end{aligned} \tag{12.6}$$

It often happens that most of the realizations of X never reach the target set B but are "attracted" and "absorbed" by some non empty set C. These rare events are difficult to analyze numerically. One strategy to estimate these events is to consider the sequence of level-crossing excursions $\mathcal{X}_n$ associated with a splitting of the state space and defined in (12.4) on page 438, namely

$$\mathcal{X}_n = (T_n, (X_t \,;\, T_{n-1} \leq t \leq T_n)) \in E = \cup_{s \leq t}(\{t\} \times D([s, t], S))$$

with the entrance times $T_n = \inf\{0 \leq t \,:\, X_t \in B_n \cup C\}$. Following the arguments used in Section 12.2.3 to check whether or not an excursion succeeds in entering into the desired nth level, we consider the potential functions $\mathcal{G}_n$ on E defined for each $t \in I$ and $x = (x_r)_{s \leq r \leq t} \in D([s, t], S)$ with $s \leq t$ by $\mathcal{G}_n(t, x) = 1_{B_n}(x_t)$. Rephrasing proposition 12.2.3 on page 439, we obtain the following Feynman-Kac representation of the desired quantities (12.6).

Proposition 12.2.4 *For any n and for any $f_n \in \mathcal{B}_b(E^{n+1})$, we have that*

$$\mathbb{E}(f_n([X_t \,;\, 0 \leq t \leq T_n]) \,;\, T_n < R) = \mathbb{E}\left(f_n(\mathcal{X}_0, \ldots, \mathcal{X}_n) \prod_{p=0}^{n} \mathcal{G}_p(\mathcal{X}_p)\right)$$

Notice that if f_n is defined for any $s_n \leq t_n$ and $x_n \in D([s_n, t_n])$ by

$$f_n((t_0, x_0), \ldots, (t_n, x_n)) = \varphi_n(t_0, \ldots, t_n)$$

for some $\varphi_n \in \mathcal{B}_b(I^n)$, then we find that

$$\mathbb{E}(f_n([X_t \, ; \, 0 \leq t \leq T_n]) \mid T_n < R) = \mathbb{E}(\varphi_n(T_0, \ldots, T_n) \mid T_n < R)$$

We recall that the prediction Feynman-Kac model $\eta_n \in \mathcal{P}(E)$ is defined by

$$\eta_n(f) = \gamma_n(f)/\gamma_n(1) \quad \text{with} \quad \gamma_n(f) = \mathbb{E}\left(f(\mathcal{X}_n) \prod_{p=0}^{n-1} \mathcal{G}_p(\mathcal{X}_p)\right)$$

and corresponds to the conditional distributions

$$\eta_n(f) = \mathbb{E}(f(T_n, (X_t \, ; \, T_{n-1} \leq t \leq T_n)) \mid T_{n-1} < R)$$

Furthermore, it satisfies the measure-valued dynamical system

$$\eta_{n+1} = \Phi_{n+1}(\eta_n) \quad \text{with} \quad \eta_0 = \delta_0 \otimes \nu_0 \tag{12.7}$$

The mappings Φ_{n+1} from $\mathcal{P}_n(E) = \{\eta : \eta(\mathcal{G}_n) > 0\}$ into $\mathcal{P}(E)$ are defined by

$$\Phi_{n+1}(\eta) = \Psi_n(\eta)\mathcal{M}_{n+1} = \int_E \Psi_n(\eta)(du) \, \mathcal{M}_{n+1}(u, .)$$

The Markov kernels $\mathcal{M}_n(u, dv)$ represent the Markov transitions of the chain of excursions $\mathcal{X}_n$. The updating mappings Ψ_n are defined from $\mathcal{P}_n(E)$ into $\mathcal{P}_n(E)$ and for any $\eta \in \mathcal{P}_n(E)$ and $f \in \mathcal{B}_b(E)$ by the formula $\Psi_n(\eta)(f) = \eta(f\mathcal{G}_n)/\eta(\mathcal{G}_n)$.

Lemma 12.2.1 *For any $n \geq 0$, we have*

$$\mathbb{P}(T_n < R) \;=\; \widehat{\gamma}_n(1) = \gamma_{n+1}(1) = \prod_{p=0}^n \eta_p(\mathcal{G}_p)$$

In addition, we have $\mathbb{P}(T_n < R \mid T_{n-1} < R) = \eta_n(\mathcal{G}_n)$ and for any $f \in \mathcal{B}_b(E)$

$$\eta_n(f) \;=\; \mathbb{E}(f(T_n, (X_t \, ; \, T_{n-1} \leq t \leq T_n)) \mid T_{n-1} < R)$$
$$\widehat{\eta}_n(f) \;=\; \Psi_n(\eta_n)(f) = \mathbb{E}(f(T_n, (X_t \, ; \, T_{n-1} \leq t \leq T_n)) \mid T_n < R)$$

This lemma gives a Feynman-Kac interpretation of the probability of rare events. Since the potentials are indicator functions, it is more judicious to rewrite the Boltzmann-Gibbs transformations $\Psi_n(\eta) = \eta S_{n,\eta}$ in terms of the selection Markov transitions $S_{n,\eta}(u, dv)$ on E defined by

$$S_{n,\eta}(u, dv) = (1 - 1_{\mathcal{G}_n^{-1}\{1\}}(u)) \; \Psi_n(\eta)(dv) + 1_{\mathcal{G}_n^{-1}\{1\}}(u) \; \delta_u(dv)$$

Note that $\mathcal{G}_n^{-1}\{1\}$ represents the collection of excursions in S entering the nth level B_n; that is

$$\mathcal{G}_n^{-1}\{1\} = \{u = (t, (u_r)_{s \leq r \leq t}) \in (\{t\} \times D([s,t])) , \ s \leq t \text{ and } u_t \in B_n\}$$

In this notation, the equation (12.7) can be rewritten as follows

$$\eta_{n+1} = \eta_n K_{n+1,\eta_n} \quad \text{with} \quad K_{n+1,\eta_n} = S_{n,\eta_n} \mathcal{M}_{n+1} \qquad (12.8)$$

To motivate this section, we describe hereafter some more or less academic but especially instructive situations.

Example 12.2.2 (Ballistic event) *Suppose the state $S = A \cup C$ is decomposed into two separate regions A and C. The process X starts in A, and we want to estimate the probability of the entrance time into a target $B \subset A$ before exiting A. In this context, the conditional distribution (12.6) represents the law of the process in this "ballistic" regime.*

Example 12.2.3 (An elementary gambler's ruin process) *We consider a simple random walk $X_n = x + \sum_{i=1}^n \varepsilon_i$ on $E = \mathbb{Z}$, starting at some $x \in \mathbb{Z}$ where $(e_i)_{i \geq 1}$ is a sequence of independent and identically distributed random variables with common law*

$$\mathbb{P}(\varepsilon_1 = +1) = p \quad \text{and} \quad \mathbb{P}(\varepsilon_1 = -1) = q$$

with $p, q \in (0,1)$ and $p + q = 1$. If we use the convention $\sum_\emptyset = 0$, then we can interpret X_n as the amount of money won or lost by a player starting with $x \in \mathbb{Z}$ euros in a gambling game where he/she wins and loses 1 euro with respective probabilities p and q. If we let $a < x < b$ be two fixed parameters, one interesting question is to compute the probability that the player will succeed in winning $b - x$ euros, never losing more than $x - a$ euros. More formally, this question becomes that of computing the probability that the chain X_n (starting at some $x \in (a,b)$) reaches the set $B = [b, \infty)$ before entering into the set $C = (-\infty, a]$. When $p < q$ (i.e., $p < 1/2$), the random walk X_n tends to move to the left, and it becomes less and less likely that X_n will succeed in reaching the desired level B. Following Exercise 12.2.9, we check that

$$\mathbb{P}_x(R < \infty) = 1 \quad \text{and} \quad \mathbb{P}_x(T < R) = \frac{(q/p)^x - (q/p)^a}{(q/p)^b - (q/p)^a} \qquad (12.9)$$

Example 12.2.4 (Birth and death model) *We consider a simple random walk X_n on $E = \mathbb{N}$ where the state 0 is an absorbing barrier and the elementary transition probabilities are defined for any $x > 0$ by*

$$\mathbb{P}(X_n = x+1 | X_{n-1} = x) = p(x) \quad \mathbb{P}(X_n = x-1 | X_{n-1} = x) = q(x)$$

where for $x > 0$ we have $p(x), q(x) \in (0,1)$ and $p(x) + q(x) = 1$ and the absorbing condition $\mathbb{P}(X_n = 0 | X_{n-1} = 0) = 1$. We let $\mathbb{P}_x$ be the distribution of the Markov chain X_n starting at $X_0 = x$ at time $n = 0$. We can interpret X_n as the dynamical population model. Given a population size $X_n = x$, we have a birth $X_{n+1} = X_n + 1$ with a probability $p(x)$; otherwise an individual dies as $X_{n+1} = X_n - 1$ with a probability $q(x)$. In this context, one typical question is to evaluate the probability that the population size reaches some upper level $b < \infty$ before extinction. More formally, this consists in evaluating the probability that X_n (starting at some $X_0 = x > 0$) hits the level $B = [b, \infty)$ before hitting the absorbing barrier $C = \{0\}$. If $\sum_{y \geq 0} \{\prod_{z=1}^{y} \frac{q(z)}{p(z)}\} = \infty$, then in Exercise 12.2.11 we will see that, for any $x \in \mathbb{N}$, $\mathbb{P}_x(R < \infty) = 1$.

12.2.6 Asymptotic Particle Analysis of Rare Events

The N-particle model associated with a given collection of transitions $\mathcal{K}_{n,\eta}$ is described in Section 3.2. The precise description of the particle motion is a little involved, mainly because the state space is the set of excursions and the potentials are indicator functions. As a result, when all the particles miss the potential support, the algorithm is stopped and the system goes into some cemetery state. Because of its importance in practice, we provide next a detailed presentation. In the context of rare event, the particle model consists in evolving a collection of N excursion-valued particles

$$\xi_n^i = (T_{n-1}^i, (\zeta_n^i(t) \, ; \, T_{n-1}^i \leq t \leq T_n^i)) \in E \cup \{\Delta\}$$
$$\widehat{\xi}_n^i = (\widehat{T}_{n-1}^i, (\widehat{\zeta}_n^i(t) \, ; \, \widehat{T}_{n-1}^i \leq t \leq \widehat{T}_n^i)) \in E \cup \{\Delta\}$$

The auxiliary point Δ stands for a cemetery or coffin point. The random time pairs (T_{n-1}^i, T_n^i) and $(\widehat{T}_{n-1}^i, \widehat{T}_n^i)$ represent the length of the corresponding excursions. At the time $n = 0$, the initial system consists of N independent and identically distributed S-valued random variables $\xi_0^i = (0, \zeta_0^i)$ with common law $\eta_0 = \delta_0 \otimes \nu_0$. Since we have $\mathcal{G}_0(0, u) = 1$, there is no updating transition at time $n = 0$ and we set $\widehat{\xi}_0^i = (0, \zeta_0^i)$ for each $1 \leq i \leq N$. As an aside, if we use the convention $T_{-1}^i = \widehat{T}_{-1}^i = 0$, and if we set $T_0^i = \widehat{T}_0^i = 0$, then these initial variables $(\xi_0^i, \widehat{\xi}_0^i)$ can be rewritten in the excursion form

$$\xi_0^i = (0, \zeta_0^i(0)) = (T_0^i, (\zeta_0^i(t) \, ; \, T_{-1}^i \leq t \leq T_0^i))$$
$$\widehat{\xi}_0^i = (0, \widehat{\zeta}_0^i(0)) = (\widehat{T}_0^i, (\widehat{\zeta}_0^i(t) \, ; \, \widehat{T}_{-1}^i \leq t \leq \widehat{T}_0^i))$$

Mutation: The mutation stage $\widehat{\xi}_n \to \xi_{n+1}$ at time $(n+1)$ is defined as follows. If $\widehat{\xi}_n = \Delta$, we set $\xi_{n+1} = \Delta$. Otherwise, during mutation, each selected excursion

$$\widehat{\xi}_n^i = (\widehat{T}_n^i, (\widehat{\zeta}_n^i(t) \, ; \, \widehat{T}_{n-1}^i \leq t \leq \widehat{T}_n^i))$$

evolves randomly and independently of each other according to the Markov transition $\mathcal{M}_{n+1}$ of the chain $\mathcal{X}_n$. In other words

$$\xi_{n+1}^i = (T_{n+1}^i, (\zeta_{n+1}^i(t) \, ; \, T_n^i \leq t \leq T_{n+1}^i))$$

448   12. Applications

is a random variable with distribution $\mathcal{M}_{n+1}(\widehat{\xi}_n^i, .)$. More precisely, we set $T_n^i = \widehat{T}_n^i$, and the particle $\widehat{\zeta}_n^i(t)$ at time $t = T_n^i$ evolves randomly as a copy $(\zeta_{n+1}^i(s))_{s \geq T_n^i}$ of the excursion process $(X_s)_{s \geq T_n^i}$ starting at $X_{T_n^i} = \widehat{\zeta}_n^i(T_n^i)$ at time $s = T_n^i$ and up to the first time T_{n+1}^i it visits B_{n+1} or returns to C. The stopping time T_{n+1}^i represents the first time $t \geq T_n^i$ the ith excursion hits the set $B_{n+1} \cup C$.

Selection: The selection mechanisms $\xi_{n+1} \to \widehat{\xi}_{n+1}$ are defined as follows. In the mutation stage, we have sampled N excursions

$$\xi_{n+1}^i = (T_{n+1}^i, (\zeta_{n+1}^i(t) \; ; \; T_n^i \leq t \leq T_{n+1}^i))$$

Some of these particles have succeeded in reaching the desired set B_{n+1}, and the other ones have entered into C. We denote by

$$I^N(n+1) = \{i : \zeta_{n+1}^i(T_{n+1}^i) \in B_{n+1}\}$$

the labels of the particles having reached the $(n+1)$st level, and we set $m(\xi_{n+1}) = \frac{1}{N} \sum_{i=1}^{N} \delta_{\xi_{n+1}^i}$. Two situations may occur. If $I^N(n+1) = \emptyset$, then none of the particles have succeeded in hitting the desired level. In this case, we have $m(\xi_{n+1}) \notin \mathcal{P}_{n+1}(E)$. Therefore the algorithm has to be stopped and we set $\widehat{\xi}_{n+1} = \Delta$. Otherwise, the selection transition is defined as follows. Each particle

$$\widehat{\xi}_{n+1}^i = (\widehat{T}_{n+1}^i, (\widehat{\zeta}_{n+1}^i(t) \; ; \; \widehat{T}_n^i \leq t \leq \widehat{T}_{n+1}^i))$$

is sampled according to the selection distribution

$$S_{n, m(\xi_{n+1})}(\xi_{n+1}^i, dv)$$

$$= 1_{B_{n+1}}(\zeta_{n+1}^i(T_{n+1}^i)) \; \delta_{\xi_{n+1}^i}(dv) + 1_{B_{n+1}^c}(\zeta_{n+1}^i(T_{n+1}^i)) \; \Psi_n(m(\xi_{n+1}))(dv)$$

More precisely, if the ith excursion has reached the desired level (i.e., $\zeta_{n+1}^i(T_{n+1}^i) \in B_{n+1}$), then we set $\widehat{\xi}_{n+1}^i = \xi_{n+1}^i$. In the opposite case we have $\zeta_{n+1}^i(T_{n+1}^i) \notin B_{n+1}$ when the particle has not reached the $(n+1)$st level but it has visited the set C. In this case, $\widehat{\xi}_{n+1}^i$ is chosen randomly and uniformly in the set

$$\{\xi_{n+1}^j \; ; \; \zeta_{n+1}^j(T_{n+1}^j) \in B_{n+1}\} = \{\xi_{n+1}^j \; ; \; j \in I^N(n+1)\}$$

of excursions having entered into B_{n+1}. In other words, each particle that doesn't enter into the $(n+1)$st level is killed, and instantly a different particle in the B_{n+1} level splits into two offsprings. For each time $n < \tau^N = \inf\{n \geq 0 \; ; \; \forall 1 \leq i \leq N, \; \zeta_n^i(T_n^i) \in C\}$, the N-particle approximation measures $(\gamma_n^N, \eta_n^N, \widehat{\eta}_n^N)$ associated with $(\gamma_n, \eta_n, \widehat{\eta}_n)$ are defined by

$$\gamma_n^N = \left[\prod_{p=0}^{n-1} \eta_p^N(\mathcal{G}_p)\right] \times \eta_n^N \quad \text{with} \quad \eta_n^N = \frac{1}{N}\sum_{i=1}^{N} \delta_{\xi_n^i}$$

$$\widehat{\eta}_n^N = \Psi_n(\eta_n^N) = \frac{1}{\text{Card}(I^N(n))} \sum_{i \in I^N(n)} \delta_{(T_n^i, (\xi_n^i(t)\ ;\ T_{n-1}^i \leq t \leq T_n^i))}$$

We also notice that

$$\widehat{\gamma}_n^N(1) = \gamma_n^N(\mathcal{G}_n) = \prod_{p=0}^{n} \eta_p^N(\mathcal{G}_p) = N^{-n} \prod_{p=1}^{n} \text{Card}(I^N(p))$$

In other words, $\widehat{\gamma}_n^N(1)$ is the proportion product of excursions having entered levels $B_1, \ldots, B_n$. Also notice that $\widehat{\eta}_n^N$ is the occupation measure of the excursions entering the nth level. The corresponding genealogical tree model is defined in the same way by tracking back in time the whole ancestor line of current individuals. Here the path-particles at time n take values in $E_n = E^{n+1}$ and can be written as

$$\chi_n^i = (\xi_{0,n}^i, \ldots, \xi_{n,n}^i) \quad \text{and} \quad \widehat{\chi}_n^i = (\widehat{\xi}_{0,n}^i, \ldots, \widehat{\xi}_{n,n}^i) \in E_n$$

with, for each $0 \leq p \leq n$,

$$\xi_{p,n}^i = (T_{p,n}^i, (\zeta_{p,n}^i(t)\ ;\ T_{p-1,n}^i \leq t \leq T_{p,n}^i))$$
$$\widehat{\xi}_{p,n}^i = (\widehat{T}_{p,n}^i, (\widehat{\zeta}_{p,n}^i(t)\ ;\ \widehat{T}_{p-1,n}^i \leq t \leq \widehat{T}_{p,n}^i)) \in E$$

The updated particle density profiles associated with this genealogical tree-based algorithm are defined by

$$\nu_n^N = \frac{1}{\text{Card}(I^N(n))} \sum_{i \in I^N(n)} \delta_{(\xi_{0,n}^i, \ldots, \xi_{n,n}^i)}$$

The asymptotic analysis of these particle measures is discussed in Chapters 7 to 10. For instance, rephrasing Theorem 7.4.1 and Proposition 7.4.1 we check that

Theorem 12.2.1 *For any $n \geq 0$ and $N \geq 1$ we have*

$$\mathbb{P}(\tau^N \leq n) \leq a(n) \exp(-N/b(n))$$

The particle estimates are unbiased, $\mathbb{E}(\widehat{\gamma}_n^N(1) 1_{n<\tau^N}) = \mathbb{P}(T_n < R)$, and for any $p \geq 1$ and $n \geq 0$ we have

$$\sqrt{N}\ \mathbb{E}(|\widehat{\gamma}_n^N(1) 1_{n<\tau^N} - \mathbb{P}(T_n < R)|^p)^{\frac{1}{p}} \leq a(p) b(n)$$

In addition, for any $f_n \in \mathcal{B}_b(E_n)$, with $\|f_n\| \leq 1$ we have that

$$\sqrt{N}\mathbb{E}(|\nu_n^N(f_n) 1_{n<\tau^N} - \mathbb{E}(f_n([X_t,\ 0 \leq p \leq T_n]) \mid T_n < R)|^p)^{\frac{1}{p}} \leq a(p) b(n)$$

for some finite constants $a(p), b(n) < \infty$ whose values only depend respectively on the parameters p and n.

To visualize these splitting particle models, we end this section with two ballistic recursion events. In both cases, the state $S = A \cup C$ is decomposed into two disjoint Borel sets A and C. The target set B is a subset of A. The decreasing sequence $(B_n)_{0 \leq n \leq m}$ represents the physical levels the process X needs to enter before reaching the desired target B.

Example 12.2.5 *When $S = \mathbb{R}^d$ is the Euclidean space, we can think of a sequence of centered decreasing balls with radius $1/(n+1)$*

$$B_n = \mathcal{B}(0, 1/(n+1)) \subset \mathbb{R}^d \quad \text{and} \quad C = S - \mathcal{B}(0, 1+\epsilon)$$

for some $\epsilon \in (0,1)$. Further assume that the process X exits the ball of radius $(1+\epsilon)$ in finite time. In this example, $\mathbb{P}(T < R)$ is the probability that X hits the smallest ball

$$B_m = \mathcal{B}(0, 1/(m+2))$$

starting with $1/2 < |X_0| \leq 1$ and before exiting the ball of radius $(1+\epsilon)$. The distribution (12.6) represents the conditional distribution of the process X in this ballistic regime.

Example 12.2.6 *When $S = \mathbb{R}_+$, we can choose for instance the intervals*

$$B_n = [n+1, \infty) \quad \text{and} \quad C = [0, 1-\epsilon]$$

For instance, if we consider a birth and death process or a queueing network processing jobs, the level B_n represents the population size or the number of jobs in the queue. In this case, $\mathbb{P}(T < R) = \mathbb{P}(X_T > m)$ can be interpreted as the probability that the population model or the queue length reaches some critical rare levels.

In Figure 12.4, we illustrate the genealogical particle model for a particle X evolving in a set $A \subset S$ with recurrent subset $C = S - A$. To reach the desired target set B_4, the process needs to pass the sequence of levels

$$B_0 \supset B_1 \supset B_2 \supset B_3 \supset B_4$$

12.2.7 Fluctuation Results and Some Comparisons

In this short section, we briefly explain why the splitting particle methodology often increases the numerical efficiency of Monte Carlo methods. By way of comparison, we consider the naive method based on N independent copies X_t^i of the Markov process X_t. We also let T^i be the entrance time of X^i into the set $B \cup C$. The corresponding unbiased estimator of $\mathbb{P}(T < R)$

12.2 Random Excursion Models

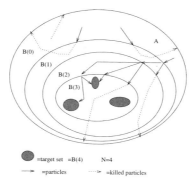

FIGURE 12.4. Genealogical model, [ballistic regime, target B(4)] ($N=4$)

is simply given by $N^{-1}\sum_{i=1}^{N} 1_{B_n}(X^i_{T_n})$. By the traditional central limit theorem for independent random variables, the random sequence

$$\overline{W}_n^N = \frac{1}{\sqrt{N}} \sum_{i=1}^{N}(1_{B_n}(X^i_{T_n}) - \mathbb{P}(T_n < R))$$

converges in law (as $N \to \infty$) to a centered Gaussian random variable $\overline{W}_n$ such that

$$\mathbb{E}(\overline{W}_n^2) = \overline{\sigma}_n^2 =_{\text{def.}} \mathbb{P}(T_n < R)(1 - \mathbb{P}(T_n < R))$$

The study of the fluctuation of the particle splitting particle approximation models is discussed in Chapter 9. For instance, using Proposition 9.4.1, and Remark 9.4.1 we have the following theorem

Theorem 12.2.2 *For any $0 \leq n \leq m+1$, the sequence of random variables*

$$W_{n+1}^N = \sqrt{N}(1_{\tau^N > n} \gamma_{n+1}^N(1) - \mathbb{P}(T_n < R))$$

converges in law (as N tends to ∞) to a Gaussian random variable W_{n+1} with mean 0 and variance

$$\sigma_n^2 = \sum_{p=0}^{n+1} \gamma_p(1)^2 \, \eta_{p-1}(K_{p,\eta_{p-1}}(Q_{p,n+1}(1) - K_{p,\eta_{p-1}}Q_{p,n+1}(1))^2)$$

(12.10)

The collection of functions $Q_{p,n+1}(1)$ on the excursion space $E = \cup_{s \leq t}(\{t\} \times D([s,t],S))$ are defined for any $x = (x(u))_{s \leq u \leq t} \in D([s,t],S)$ and $s \leq t$ by

$$Q_{p,n+1}(1)(t,x) = 1_{B_p}(x(t)) \, \mathbb{P}(T_n < R \mid T_p = t, \, X_{T_p} = x(t))$$

Explicit calculations of σ_n are in general difficult to obtain since they rely on an explicit knowledge of the semigroup $Q_{p,n}$. Next we provide an alternative formulation related to absorption times and rare event analysis. For any $p \leq q \leq n$, we set

$$\Delta^n_{p,q}(t,x) = \mathbb{P}(T_n < R \mid T_q = t,\ X_{T_q} = x)/\mathbb{P}(T_n < R \mid T_p < R)$$

After some elementary computations (see also page 305), we first prove that the variance σ_n associated with the McKean transition $K_{n,\eta}$ given in (12.8) takes the form

$$\sigma_n^2 = \mathbb{P}(T_n < R)^2\ (a_n - b_n)$$

with

$$a_n = \frac{1}{\gamma_{n+1}(1)^2} \sum_{p=0}^{n+1} \gamma_p(1)^2\ \eta_p((Q_{p,n+1}(1) - \eta_p Q_{p,n+1}(1))^2)$$

$$b_n = \frac{1}{\gamma_{n+1}(1)^2} \sum_{p=1}^{n+1} \gamma_p(1)^2\ \eta_{p-1}(\mathcal{G}_{p-1}^2(M_p Q_{p,n+1}(1) - \eta_p Q_{p,n+1}(1))^2)$$

Then we observe that $\gamma_p(1) = \mathbb{P}(T_{p-1} < R)$ and

$$\eta_p Q_{p,n+1}(1) = \gamma_{n+1}(1)/\gamma_p(1) = \mathbb{P}(T_n < R \mid T_{p-1} < R)$$

from which we conclude that

$$a_n = \sum_{p=0}^{n+1} \mathbb{E}([\Delta^n_{p-1,p}(T_p, X_{T_p}) 1_{T_p < R} - 1]^2 \mid T_{p-1} < R)$$

In much the same way, we find

$$b_n = \sum_{p=0}^{n} \mathbb{E}(1_{T_p < R}\ [\Delta^n_{p,p}(T_p, X_{T_p}) - 1]^2 \mid T_{p-1} < R)$$

$$= \sum_{p=0}^{n} \mathbb{P}(T_p < R \mid T_{p-1} < R)\ \mathbb{E}([\Delta^n_{p,p}(T_p, X_{T_p}) - 1]^2 \mid T_p < R)$$

Loosely speaking, these two formulations indicate that the renormalized variance $\sigma_n^2/\mathbb{P}(T_n < R)^2$ is the sum of the "local variances" induced by the particle-splitting transition at each stage of the algorithm. When these local errors are uniformly bounded, then we have

$$\sigma_n^2 \leq c\ (n+1)\ \mathbb{P}(T_n < R)^2$$

for some finite constant $c < \infty$ whose values do not depend on the splitting parameter n. We illustrate this result for the case of the simple random walk X_n on $S = \mathbb{Z}$ and described in Example 12.2.3 (see also Exercise 12.2.9).

To simplify the presentation, let us suppose that the chain X_n starts at the origin $X_0 = 0$ and the thresholds are given by
$$C = (-\infty, 0) \quad \text{and} \quad B_n = [n, \infty) \quad n = 1, 2, \ldots$$
In this situation, for any $0 \leq n_1 \leq n_2$, we have from (12.9)
$$\mathbb{P}(T_{n_2} < R \mid T_{n_1} < R) = \mathbb{P}(X_{T_{n_2}} = n_2 \mid X_{T_{n_1}} = n_1) = \frac{(q/p)^{n_1+1} - 1}{(q/p)^{n_2+1} - 1}$$
from which we prove that $b_n = 0$ and
$$\begin{aligned}
a_n &= 1 + \sum_{r=1}^{n+1} \{\mathbb{P}(T_r > R | T_{r-1} < R) \\
&\quad + \mathbb{P}(T_r < R | T_{r-1} < R)[\tfrac{\mathbb{P}(T_n < R \mid T_r < R)}{\mathbb{P}(T_n < R \mid T_{r-1} < R)} - 1]^2\} \\
&= 1 + \sum_{r=1}^{n+1} \{\tfrac{(q/p)^{r+1} - (q/p)^r}{(q/p)^{r+1} - 1} + \tfrac{(q/p)^r - 1}{(q/p)^{r+1} - 1}[\tfrac{(q/p)^{r+1} - (q/p)^r}{(q/p)^r - 1}]^2\} \\
&= 1 + ((q/p) - 1) \sum_{r=1}^{n+1} \tfrac{1}{1 - (p/q)^r} \\
&\leq 1 + ((q/p) - 1)(n+1) \tfrac{1}{1 - (p/q)} = 1 + (n+1)(q/p) \leq (n+1)/p
\end{aligned}$$
These calculations imply that
$$\sigma_n^2 \leq \mathbb{P}(T_n < R)^2 (n+1)/p \quad \text{and} \quad \mathbb{P}(T_n < R) = \frac{(q/p) - 1}{(q/p)^{n+1} - 1}$$
and after some manipulations, recalling that $p < q$, one finds that for any $n \geq 1$
$$\begin{aligned}
\overline{\sigma}_n^2 / \sigma_n^2 &\geq \frac{p}{n+1} \times \frac{1 - \mathbb{P}(T_n < R)}{\mathbb{P}(T_n < R)} \\
&\geq p\,[(q/p)^n - 1]/(n+1) \to \infty \quad \text{as} \quad n \to \infty
\end{aligned}$$
Related comparisons between the limiting variances for interacting and non-interacting methods in the context of filtering methods can be found in [88].

12.2.8 Exercises

Exercise 12.2.5: Let W_t be a d-dimensional Wiener process with $d \geq 2$ and let $A_{\varepsilon_1, \varepsilon_2} = \{x \in \mathbb{R}^d \mid \varepsilon_1 < |x| < \varepsilon_2\}$ with $0 < \varepsilon_1 < \varepsilon_2$. We consider the partition $A_{\varepsilon_1, \varepsilon_2}^c = B_{\varepsilon_1} \cup C_{\varepsilon_2}$ with $B_{\varepsilon_1} = \{|x| \leq \varepsilon_1\}$ and $C_{\varepsilon_2} = \{|x| \geq \varepsilon_2\}$.

- Using formula (12.5) and recalling that for any x with $0 \leq \varepsilon_1 < |x| < \varepsilon_2$ we have
$$\mathbb{P}_x(W_t \text{ hits } 0 \text{ before } C_{\varepsilon_2}) = \lim_{\varepsilon_1 \to 0} \mathbb{P}_x(W_t \text{ hits } B_{\varepsilon_1} \text{ before } C_{\varepsilon_2})$$
check that W_t never hits points.

- In much the same way, prove that for any $0 < \varepsilon_1 < |x|$

$$\mathbb{P}_x(W_t \text{ hits } B_{\varepsilon_1}) = \begin{cases} 1 & \text{for } d = 2 \\ (\varepsilon_1/|x|)^{d-2} & \text{for } d \geq 3 \end{cases}$$

- Using the Markov property, show that for $d \geq 3$

$$\mathbb{P}_x(W_t \text{ hits } B_{\varepsilon_1} \text{ after time t})$$
$$= \int \frac{1}{(2\pi t)^{d/2}} e^{-\frac{1}{2t}(x-z)^2} (\varepsilon_1/|z|)^{d-2} \, dz \quad (\to 0 \text{ as } t \to \infty)$$

- Conclude that W_t is recurrent in two dimensions and it wanders off to ∞ in dimension $d > 3$.

Exercise 12.2.6: Let $S = A \cup A^c$ be a partition of the state space S. We consider a Markov chain X_n starting at A and we denote by T the first time the chain exits the set A; that is, $T = \inf \{n \geq 0 : X_n \in A^c\}$. We further assume that there exists some $m \geq 1$ such that $\delta_m(A, A^c) =_{\text{def.}} \inf_{x \in A} \mathbb{P}_x(X_m \in A^c) > 0$.

- Show that for any $x \in A$ we have $\mathbb{P}_x(T > m) \leq (1 - \delta_m(A, A^c))$ and conclude that for any $n \geq 1$

$$\mathbb{P}_x(T > nm) = \mathbb{P}_x(T > nm \mid T > (n-1)m) \, \mathbb{P}_x(T > (n-1)m)$$
$$\leq (1 - \delta_m(A, A^c))^n$$

- Check that for any $n = pm + q$ with $p \geq 0$ and $0 \leq q < m$ we have

$$\mathbb{P}_x(T > n) \leq \mathbb{P}_x(T > pm) \leq (1 - \delta_m(A, A^c))^{-1}(1 - \delta_m(A, A^c))^{n/m}$$

and conclude that for any $x \in A$ we have $\mathbb{P}_x(T < \infty) = 1$.

Exercise 12.2.7: Let $(\varepsilon_n)_{n \geq 1}$ be a collection of independent and identically distributed random variables on the lattice $S = \mathbb{Z}^d$ with common law $\frac{1}{2d} \sum_{|e| \leq 1} \delta_e$. Let $A \subset \mathbb{Z}^d$ be a finite set. We consider the simple random walk $X_n = X_{n-1} + \varepsilon_n$ starting at some point $X_0 = x \in A$. Check that

$$m > \sup_{x \in A} |x| \implies \delta_m(A, A^c) = \inf_{x \in A} \mathbb{P}_x(X_m \in A^c) \geq (1/2d)^m$$

where $|x|$ stands for the minimal length of an admissible path joining 0 to x. Using Exercise 12.2.6, show that $\mathbb{P}_x(T < \infty) = 1$, where T represents the entrance time of the chain into A^c.

Exercise 12.2.8: Let $(\varepsilon_n)_{n\geq 1}$ be a collection of independent and identically distributed random variables with common law μ on $S = \mathbb{R}$. We further assume that $\mathbb{E}(\varepsilon_1) = 0$ and $c = \mathbb{E}(\varepsilon_1^2)^{1/2} > 0$ and we let $A = [a,b]$ be a finite interval. We consider the Markov chain $X_n = X_{n-1} + \varepsilon_n$ starting at some point $X_0 = x \in A$.

- Show that for any $m \geq 1$ we have $\mathbb{E}((\sum_{p=1}^m \varepsilon_p)^2) = mc^2$ and deduce that
$$\sqrt{m} > (|a|+b)/c \Longrightarrow \mathbb{P}\left(|\sum_{p=1}^m \varepsilon_p| > |a|+b\right) > 0$$

- Check that for any $\sqrt{m} > (|a|+b)/c$ we have $\inf_{x \in A} \mathbb{P}_x(X_m \in A^c) > 0$, and using Exercise 12.2.6 show that $\mathbb{P}_x(T < \infty) = 1$, where T represents the entrance time of the chain into A^c. Prove that the same result holds true if we replace the nonnegative variance condition by the fact that $\mathbb{E}(|\varepsilon_1|) < \infty$ and $\mathbb{P}(\varepsilon_1 = 0) < 1$ (see for instance Proposition 7.2.3 in [278]).

- Prove that if $\mathbb{E}(\varepsilon_1) < 0$, then for any $x \in \mathbb{R}$ we have $\mathbb{P}_x(R < \infty) = 1$, where R denotes the entrance time of the set $C = (-\infty, a]$.

Exercise 12.2.9: [Gambler's ruin] We consider the simple random walk model X_n starting at some $x \in [a,b]$ and described in Example 12.2.3. We further assume that $q > p$. We introduce the function
$$x \in [a, \infty) \longrightarrow \alpha(x) = \mathbb{P}_x(R < \infty) \quad \text{with} \quad R = \inf\{n \geq 0 \,:\, X_n = a\}$$
as well as the first time the chain X_n reaches one of the boundaries
$$T = \inf\{n \geq 0 \,:\, X_n \in \{a,b\}\} \ (\leq R)$$

- Check that if we have $|x - y| > n$ or $(y - x) \neq n + 2k$, for some $k \geq 1$ then $\mathbb{P}_x(X_n = y) = 0$. The case where $(y - x) = k - (n - k)$, with $0 \leq k \leq n$, corresponds to situations where the chain has moved k steps to the right and $(n - k)$ to the left. Prove that $\mathbb{P}_x(X_n = y) = C_n^k \, p^k \, q^{n-k}$.

- Show that α is the minimal solution of the equation defined for any $x > a$ by $\alpha(x) = p\alpha(x+1) + q\alpha(x-1)$ with the boundary condition $\alpha(a) = 1$.

- Whenever $p < q$, we recall that the general solution of the equation above has the form $\alpha(x) = A + B(q/p)^x$ with $\alpha(a) = 1 = A + B(q/p)^a$ so that $\alpha(x) = 1 + B\{(q/p)^x - (q/p)^a\}$. Deduce from the above that $\mathbb{P}_x(R < \infty) = 1$ for any x.

12. Applications

- Check that for any $n \geq 0$ and $\lambda > 0$ we have

$$\mathbb{P}_x(R \geq n) = \mathbb{P}_x(X_n \geq a) \leq e^{-\lambda a}\,\mathbb{E}_x(e^{\lambda X_n}) = e^{-\lambda(a-x)}\,(pe^{\lambda}+qe^{-\lambda})^n$$

- If we choose $\lambda = \frac{1}{2}\log(q/p) \in (0,\infty)$, then prove that

$$\mathbb{P}_x(R \geq n) \leq (p/q)^{(x-a)/2}\,(4pq)^{n/2}$$

- Deduce from the above that for $p \neq 1/2$

$$\mathbb{E}_x(T) \leq \mathbb{E}_x(R) = \sum_{n \geq 1} \mathbb{P}_x(R \geq n) \leq \frac{(4pq)^{1/2}}{1-(4pq)^{1/2}}\,(p/q)^{(x-a)/2}$$

- Show that for any $a < x < b$ the stochastic process $M_n = (q/p)^{X_n}$ is a $\mathbb{P}_x$-martingale with respect to the filtration $F_n = \sigma(X_0, \ldots, X_n)$ and if $p < q$, then $\mathbb{P}_x$-a.s. on the event $\{T \geq n\}$ we have that

$$\mathbb{E}_x(|M_{n+1} - M_n| \mid F_n) \leq 2(q/p)^b\,(q-p)$$

- Since we have $\mathbb{E}_x(T) < \infty$ and $\mathbb{E}_x(|M_{n+1} - M_n| \mid F_n)1_{T \geq n} < c$ for some finite constant by a well-known martingale theorem of Doob (see for instance Theorem 2 on p. 486 in [287]), prove that $\mathbb{E}_x(M_T) = \mathbb{E}_x(M_0) = (q/p)^x$, and deduce that for any $x \in [a,b]$

$$(q/p)^x = (q/p)^b \mathbb{P}_x(T < R) + (q/p)^a(1 - \mathbb{P}_x(T < R))$$

Finally conclude that for any $p \neq q$ we have

$$\mathbb{P}_x(T < R) = \frac{(q/p)^x - (q/p)^a}{(q/p)^b - (q/p)^a} \qquad (12.11)$$

- Using the strong Markov property, check that for any p,q the function $\beta(x) = \mathbb{P}_x(T < R) = \mathbb{E}_x(1_b(X_T))$ satisfies the equation

$$\beta(x) = p\beta(x+1) + q\beta(x-1)$$

for any $x \in (a,b)$, with the boundary conditions $(\beta(a), \beta(b)) = (0, 1)$. For $p \neq q$, check that the function (12.11) is the unique solution, and for $p = q = 1/2$ prove that the solution is given for any $x \in [a,b]$ by

$$\mathbb{P}_x(T < R) = (x-a)/(b-a)$$

Exercise 12.2.10: Let W_t be a standard Wiener process on $\mathbb{R}$. A Brownian motion $W_t - mt$ with negative drift (i.e., $m > 0$) can be defined as the limit of the simple random walk defined (by a slight abuse of notation) by $X_t = \sqrt{\Delta t} \sum_{i=1}^{[t/\Delta t]} \varepsilon_i$, where ε_n is a sequence of properly scaled iid Bernoulli random variables $\mathbb{P}(\varepsilon_n = 1) = 1 - \mathbb{P}(\varepsilon_n = -1) = p = \frac{1}{2}(1 - m\sqrt{\Delta t})$. If we let $\Delta t \to 0$, then check that $\mathbb{E}(X_t) = -m \, \Delta t [t/(\Delta t)] \to -mt$ and

$$\mathbb{E}((X_t - \mathbb{E}(X_t))^2) = \Delta t [t/(\Delta t)](1 - m^2 \Delta t) \to t$$

Using Exercise 12.2.9 and the fact that

$$((1-p)/p)^{1/\sqrt{\Delta t}} = (1 + m\sqrt{\Delta t})^{1/\sqrt{\Delta t}}(1 - m\sqrt{\Delta t})^{-1/\sqrt{\Delta t}} \to e^{2m}$$

check that for any $x \in [a, b]$ we have $\mathbb{P}_x(T < R) = \frac{e^{2mx} - e^{2ma}}{e^{2mb} - e^{2ma}}$, where T is the first time the process hits one of the boundaries $\{a, b\}$ and R the first time it hits a.

Exercise 12.2.11: [Birth and death model] We consider the simple random walk model X_n on $E = \mathbb{N}$ described in Example 12.2.4.

- In the homogeneous situation $(p(x), q(x)) = (p, q)$ from Exercise 12.2.9 check that for any $x \in \mathbb{N}$ we have $\alpha(x) = \mathbb{P}_x(R < \infty) = 1$ with the time of absorption $R = \inf\{n \geq 0 \, : \, X_n = 0\}$.

- More generally, prove that the function α is the minimal solution of the equation $\alpha(x) = p(x)\alpha(x+1) + q(x)\alpha(x-1)$ with the boundary condition $\alpha(0) = 1$.

- Using the fact that

$$p(x)\{\alpha(x+1) - \alpha(x)\} + q(x)\{\alpha(x-1) - \alpha(x)\} = 0$$

if we set $\Delta \alpha(x) = \alpha(x) - \alpha(x-1)$, then prove that

$$\Delta \alpha(x+1) = \frac{q(x)}{p(x)} \Delta \alpha(x) = \left\{ \prod_{y=1}^{x} \frac{q(y)}{p(y)} \right\} \Delta \alpha(1)$$

- Deduce from the above that

$$\alpha(x) = 1 + \sum_{y=1}^{x} \Delta \alpha(y) = 1 - \sum_{y=1}^{x} \left\{ \prod_{z=1}^{y-1} \frac{q(z)}{p(z)} \right\} (1 - \alpha(1))$$

- Two situations may occur. First check that if $\sum_{y \geq 0} \{\prod_{z=1}^{y} \frac{q(z)}{p(z)}\} = \infty$, then we have $\alpha(x) = 1$ for any $x \in \mathbb{N}$. In the opposite situation, we can choose any $\alpha(1)$ such that $0 \leq (1 - \alpha(1))[\sum_{y \geq 1} \{\prod_{z=1}^{y-1} \frac{q(z)}{p(z)}\}] \leq 1$.

If we take $\alpha(1) = 1 - [\sum_{y\geq 1}\{\prod_{z=1}^{y-1} \frac{q(z)}{p(z)}\}]^{-1}$, then prove that the minimal solution $\alpha(x)$ is given by

$$\alpha(x) = [\sum_{y\geq x}\{\prod_{z=1}^{y} \frac{q(z)}{p(z)}\}]/[\sum_{y\geq 0}\{\prod_{z=1}^{y} \frac{q(z)}{p(z)}\}]$$

Exercise 12.2.12: Let X'_n, (B, C) and B_n be a Markov chain, the pair target/obstacle sets and the multilevel decomposition described in the introduction and in Section 12.2.3. We fix a finite time horizon h and we set

$$Y'_n = (n, X'_n) \in E'_n = (\{n\} \times E')$$
$$\mathbf{B}_n = ([0, h) \times B_n) \quad \text{and} \quad \mathbf{C} = (\{h\} \times E')$$

Note that the final time horizon h is regarded as an hard obstacle set the process tries to avoid during its excursions between the level sets B_n. We let $\mathbf{G}_n$ be the potential functions on

$$E = \cup_{p\leq q} E'_{[p,q]} \quad \text{with} \quad E'_{[p,q]} =_{def.} (E'_p \times E'_{p+1} \times \ldots \times E'_q)$$

defined for any $p \leq q$ and $y = (t, x_t)_{p\leq t\leq q} \in E'_{[p,q]}$ by

$$\mathbf{G}_n(y) = 1_{\mathbf{B}_n}(q, x_q) = 1_{[0,h)}(q)\, 1_{B_n}(x_q)$$

We consider the excursion-valued Markov chain

$$Y_n = (Y'_t\ ;\ T_{n-1} \leq t \leq T_n) \in E$$

where $T_n = T_{B_n \cup C}$ is the entrance time of X' into the set $B_n \cup C$. Check that

$$\mathbb{E}_{\nu_0}(f_n(Y_0, \ldots, Y_n) \prod_{p=1}^{n} \mathbf{G}_p(Y_p))$$

$$= \mathbb{E}_{\nu_0}(f_n([(t, X'_t)\ ;\ 0 \leq t \leq T_n])\, 1_{X'_{T_n} \in B_n,\, T_n < h})$$

with

$$[(t, X'_t)\ ;\ 0 \leq t \leq T_n]$$
$$= ((0, X'_0), ((t, X'_t)\ ;\ 0 \leq t \leq T_1), \ldots, ((t, X'_t)\ ;\ T_{n-1} \leq t \leq T_n))$$

12.3 Change of Reference Measures

12.3.1 Introduction

In Section 2.4.2, we have seen that a given distribution on path space may have different Feynman-Kac interpretations. The objective of this section is to extend these ideas to excursion models and better connect these interpretations with importance sampling methods. In Section 12.3.3, we illustrate these changes of measure in the sequential analysis of probability ratio tests. In Section 12.3.4, we apply the multislipping particle methodology to estimate the accuracy of these statistical tests. To formalize these changes of measure suppose we are given a collection of Markov transitions $\overline{M}_n$ from E_{n-1} into E_n such that for any $x_{n-1} \in E_{n-1}$ we have $M_n(x_{n-1}, .) << \overline{M}_n(x_{n-1}, .)$. Also suppose that $\overline{\eta}_0$ is a given distribution on E_0 such that $\eta_0 << \overline{\eta}_0$ and the corresponding Radon-Nikodym derivatives are bounded positive functions. If we set

$$\overline{\mathbb{P}}_n(d(x_0, \ldots, x_n)) = \overline{\eta}_0(dx_0)\overline{M}_1(x_0, dx_1)\ldots\overline{M}_n(x_{n-1}, dx_n)$$

then we find that $\mathbb{P}_n << \overline{\mathbb{P}}_n$ and

$$\frac{d\mathbb{P}_n}{d\overline{\mathbb{P}}_n}(x_0, \ldots, x_n) = \prod_{p=0}^n \frac{dM_n(x_{p-1}, .)}{d\overline{M}_p(x_{p-1}, .)}(x_p)$$

with the conventions $M_0(x_{-1}, dx_0) = \eta_0(dx_0)$ and $\overline{M}_0(x_{-1}, dx_0) = \overline{\eta}_0(dx_0)$ for $p = 0$. By Ionescu-Tulcea's extension theorem (see for instance Theorem 2 on p. 249 in [287]), the distributions $(\mathbb{P}_n, \overline{\mathbb{P}}_n)$ can be extended into a unique pair of distributions $(\mathbb{P}, \overline{\mathbb{P}})$ on the canonical space of infinite sequences $(\Omega = \prod_{n \geq 0} E_n, \mathcal{F} = (\mathcal{F}_n)_{n \geq 0}, (X_n)_{n \geq 0})$. If we denote by $\mathbb{E}(.)$ and $\overline{\mathbb{E}}(.)$ the corresponding expectation operators, then we find that

$$\mathbb{E}(f_n(X_{[0,n]})) = \overline{\mathbb{E}}\left(f_n(X_{[0,n]}) \prod_{p=0}^n \frac{dM_p(X_{p-1}, .)}{d\overline{M}_p(X_{p-1}, .)}(X_p)\right)$$

for any bounded measurable function f_n on $E_{[0,n]} = \prod_{p=0}^n E_p$, and with $X_{[0,n]} =_{\text{def.}} (X_p)_{0 \leq p \leq n}$ for every $n \geq 0$. Note that the formulae displayed above can be interpreted as a Feynman-Kac path measure with potential functions given by the Radon-Nikodym derivatives. This elementary observation shows that one can generate path samples of a Markov process with transitions M_n, using a genealogical tree model associated with $\overline{M}_n$-exploration particles with updating mechanisms dictated by the Radon-Nikodym potential functions. More generally, we have the following proposition.

Proposition 12.3.1 *Let T be a stopping time. Then, for any bounded measurable function f on the excursion space $\cup_{n\geq 0}(\{n\} \times E_{[0,n]})$, we have*

$$\mathbb{E}(f_T(X_{[0,T]})1_{T<\infty}) = \overline{\mathbb{E}}\left(f_T(X_{[0,T]}) \prod_{p=0}^{T} \frac{dM_p(X_{p-1},\cdot)}{d\overline{M}_p(X_{p-1},\cdot)}(X_p) \, 1_{T<\infty}\right)$$

Proof:
Since $\overline{\mathbb{P}}(T < \infty) = 1$, the proof is an immediate consequence of the following equalities

$$\begin{aligned}\mathbb{E}(f_T(X_{[0,T]})1_{T<\infty}) &= \sum_{n\geq 0} \mathbb{E}(f_n(X_{[0,n]})1_{T=n}) \\ &= \sum_{n\geq 0} \overline{\mathbb{E}}\left(f_n(X_{[0,n]}) \prod_{p=0}^{n} \frac{dM_p(X_{p-1},\cdot)}{d\overline{M}_p(X_{p-1},\cdot)}(X_p) \, 1_{T=n}\right)\end{aligned}$$

■

By the definition of $\overline{\mathbb{P}}_n$, we notice that the Feynman-Kac path measure introduced in (11.7) can alternatively be rewritten as

$$\widehat{\mathbb{Q}}_n(d(x_0,\ldots,x_n)) = \frac{1}{\widehat{Z}_n}\left\{\prod_{q=0}^{n} \overline{G}_q(x_{q-1},x_q)\right\} \overline{\mathbb{P}}_n(d(x_0,\ldots,x_n)) \quad (12.12)$$

with $\overline{G}_0(x_{-1},x_0) = G_0(x_0)\frac{d\eta_0}{d\overline{\eta}_0}(x_0)$ and for any $q \geq 1$

$$\overline{G}_q(x_{q-1},x_q) = G_q(x_q) \frac{dM_q(x_{q-1},\cdot)}{d\overline{M}_q(x_{q-1},\cdot)}(x_q)$$

This observation, together with Proposition 12.3.1, yields the following Feynman-Kac model in excursion space.

Proposition 12.3.2 *Let G_n be a sequence of $[0,1]$-valued potential functions on some measurable spaces $(E_n, \mathcal{E}_n)$. For any stopping time T and any bounded measurable function f on $\cup_{n\geq 0}(\{n\} \times E_{[0,n]})$, we have*

$$\mathbb{E}\left(f_T(X_{[0,T]}) \prod_{p=0}^{T} G_p(X_p) \, 1_{T<\infty}\right)$$
$$= \overline{\mathbb{E}}\left(f_T(X_{[0,T]}) \prod_{p=0}^{T} \overline{G}_q(X_{p-1},X_p)1_{T<\infty}\right)$$

12.3.2 Importance Sampling

The changes of reference measure described in Section 12.3.1 provide a new tuning parameter in the numerical solution a particular Feynman-Kac model. These modeling strategies also offer significant benefits. For instance, let us suppose that the functions G_n are not bounded and/or not strictly positive but the new potentials $\widehat{G}_n = M_{n+1}(G_n)$ satisfy these properties. In this case, it is more judicious to use the particle model associated with a change of reference measure (see Section 2.4.3). Now suppose that a direct simulation of the chain X_n is inefficient. This happens for instance when a precise simulation technique of X_n is not known, when the oscillations of the potential functions are too large or when the chain is too slow to achieve a given rare event. One strategy to solve this problem is to use a different exploration distribution (see for instance Exercice 12.3.8). This can be performed using the change of reference probability measures presented earlier. For instance, the change of measure presented in (11.17) corresponds to the particular situation where $\overline{M}_n$ are chosen to be the transitions $\widehat{M}_n$ defined in (11.16). In this case, the particle mutations are related to the potential function, and the resulting particle is more refined in regions with high potential.

From the statistical point of view, formula (12.12) can be also interpreted as an importance sampling plan. Importance sampling is a classical technique commonly used in statistics to speed up simulations and increase the efficiency of a given Monte Carlo method. The common idea here is to modify the Feynman-Kac representation formula by replacing the reference Markov chain distribution with a new one in order to facilitate the simulations and/or to increase the accuracy of the adaptive stochastic grid. In importance sampling literature, the new distribution is also called the "twisted distribution", and the corresponding change of measure derivative is referred to as the "likelihood ratio" or "the weighted function". In this context, the choice of the pair $(\overline{G}_n, \overline{M}_n)$ is often called the sampling plan, and the strategy is to choose a judicious "twisted distribution" to increase the probability of occurrence of the desired events. This statistical interpretation leads immediately to the question of the optimal change of probability measure. The answer depends on the choice of some criteria with respect to all possible changes of reference measures. If we use the variance criterion associated with the central limit theorem developed in Chapter 9, we are led to solve a nonhomogeneous variational problem in distribution space. To our knowledge, no satisfactory general criteria and solutions have been proposed in the literature on this subject. Rather, we believe that the choice of the "twisted distribution" strongly depends on the problem at hand and on the events of interest. Loosely speaking, the counterpart of increasing the probability of occurrence of some events is to decrease the chances of the opposite. For more details, we refer the reader to the simple but illustrative Exercise 11.9.6 as well as the worked-out examples provided at the end of this section.

12.3.3 Sequential Analysis of Probability Ratio Tests

Suppose we are given a pair of Markov transitions $(M_n, \overline{M}_n)$ from some measurable space $(E, \mathcal{E})$ into itself. Also let $\mathbb{P}_y$ and $\overline{\mathbb{P}}_y$ be the distributions of a Markov chain with elementary transitions M_n and $\overline{M}_n$ and starting at $y \in E$. We suppose that these distributions are defined on a common canonical space $(\Omega, \mathcal{F} = (\mathcal{F}_n)_{n \geq 0}, (Y_n)_{n \geq 0})$ and we denote by $\mathbb{P}_{y,n}$ and $\overline{\mathbb{P}}_{y,n}$ their restrictions on $(\Omega, \mathcal{F}_n)$. We further assume that for any $y \in E$ the probability measures $M_n(y, .)$ and $\overline{M}_n(y, .)$ are absolutely continuous with each other. Under this condition the probability measures $\mathbb{P}_y$ and $\overline{\mathbb{P}}_y$ are locally absolutely continuous in the sense that for any $n \geq 0$ and $(y_p)_{0 \leq p \leq n} \in E^{n+1}$ (with $y_0 = y$) we have

$$\frac{d\mathbb{P}_{y,n}}{d\overline{\mathbb{P}}_{y,n}}(y_0, \ldots, y_n) = \prod_{p=1}^{n} \frac{dM_p(y_{p-1}, .)}{d\overline{M}_p(y_{p-1}, .)}(y_p)$$

One important question is to check whether or not the distributions $\mathbb{P}_y$ and $\overline{\mathbb{P}}_y$ are absolutely continuous. This problem arises in various situations such as in sequential analysis and particularly in the study of statistical hypothesis probability ratio tests. We refer the interested reader to any classical textbook on statistics (see for instance the monograph of D. Siegmund [289]). In this context, the answer to the question above is in general negative and $\mathbb{P}_y$ and $\overline{\mathbb{P}}_y$ are orthogonal (or singular) in the sense that there exists some $A \in \mathcal{F}_\infty = \sigma(\cup_{n \geq 0} \mathcal{F}_n)$ such that $\mathbb{P}_y(A) = \overline{\mathbb{P}}_y(A^c) = 1$. In the further development of this section, we shall work under this assumption. We refer the reader to the set of exercises provided at the end of this section. Under this condition, it is readily checked that under $\overline{\mathbb{P}}_y$ the stochastic sequence

$$Z_n(Y) = \frac{d\mathbb{P}_{y,n}}{d\overline{\mathbb{P}}_{y,n}}(Y_0, \ldots, Y_n)$$

is a martingale. Since $\overline{\mathbb{E}}_y(Z_n(Y)) = 1$, the limit $\lim_{n \to \infty} Z_n(Y) = Z_\infty(Y)$ exists $\overline{\mathbb{P}}_y$-a.e. and we have

$$\mathbb{P}_y(\lim_{n \to \infty} Z_n(Y) = \infty) = 1 = \overline{\mathbb{P}}_y(\lim_{n \to \infty} Z_n(Y) = 0)$$

To prove the first equality assertion, we combine the fact that $\mathbb{P}_y$ and $\overline{\mathbb{P}}_y$ are orthogonal with the formula

$$\mathbb{P}_y(A) = \int_A Z_\infty(Y) \, d\overline{\mathbb{P}}_y + \mathbb{P}_y(A \cap (Z_\infty(Y) = \infty))$$

which is valid for any $A \in \mathcal{F}_\infty$ (see for instance Theorem 1, p. 525 in [287]). To prove the second equality in the display above, we use the fact that $\overline{\mathbb{P}}_{y,n} \ll \mathbb{P}_{y,n}$ and by symmetry arguments $\overline{\mathbb{P}}_y(\lim_{n \to \infty} \overline{Z}_n(Y) = \infty) = 1$ with the $\mathbb{P}_y$-martingale $\overline{Z}_n(Y) = Z_n^{-1}(Y)$.

The simple hypothesis testing problem is defined as follows. Suppose we have a sample from a Markov chain $Y = (Y_n)_{n\geq 0}$ (starting at $Y_0 = y$ at time $n = 0$) and we want to know whether Y is a random sample from $\mathbb{P}_y$ (hypothesis (H)) or a sample from $\overline{\mathbb{P}}_y$ (hypothesis $(\overline{H})$). From previous considerations, if one of the hypotheses (H) or $(\overline{H})$ is true, then we respectively have that

$$\lim_{n\to\infty} Z_n(Y) = \infty \quad \text{or} \quad \lim_{n\to\infty} Z_n(Y) = 0$$

Thus, one natural way to test one of these two hypotheses is to choose a pair of parameters $0 < a < 1 < b$ and wait until the first time T the random sequence $Z_n(Y)$ exits the domain (a, b), namely

$$T = \inf\{n \geq 0 \ : \ Z_n(Y) \notin (a, b)\} \tag{12.13}$$

Then we accept or reject (H) if we have respectively $Z_T(Y) \geq b$ or $Z_T(Y) \leq a$. The region (a, b) is called the critical region. Note that in both situations T is a bounded stopping time so that the algorithm above always terminates in finite time. To get one step further, we introduce the random time

$$R = \inf\{n \geq 0 \ : \ Z_n(Y) \leq a\} \ (\geq T)$$

One rather simple way to measure the accuracy of the test is to estimate the probability of rejecting $(\overline{H})$ if $(\overline{H})$ is true. These probabilities are the so-called type I errors (or size of the test). They are defined formally by the formula

$$\begin{aligned}\overline{\mathbb{P}}_y(Z_T(Y) \geq b) &= \overline{\mathbb{P}}_y(T < R) \\ &= \mathbb{E}_y(Z_T^{-1}(Y)\, 1_{Z_T(Y)\geq b})(\leq b^{-1}\mathbb{P}_y(Z_T(Y) \geq b))\end{aligned} \tag{12.14}$$

Type II errors are defined as the probability of accepting $(\overline{H})$ when it is false or equivalently of rejecting (H) when it is true:

$$\mathbb{P}_y(Z_T(Y) \leq a) = \overline{\mathbb{P}}_y(Z_T(Y)\, 1_{Z_T(Y)\leq a})(\leq a\, \overline{\mathbb{P}}_y(Z_T(Y) \leq a)) \tag{12.15}$$

We mention that the probability $\mathbb{P}_y(Z_T(Y) > a)$ of rejecting $(\overline{H})$ when it is false is called the power of the test. Also note that in view of (12.14) it is tempting to take b very large but recalling (12.13), and in doing so we increase the probability (12.15). Thus we can only compare two tests with the same power (or with the same size).

12.3.4 A Multisplitting Particle Approach

Note that the expression (12.14) is a decreasing function of b and it has the same form as the one introduced in (12.6). As we already mentioned, these probabilities can rarely be solved explicitly. They rather belong to the class of rare event estimation problems discussed in Section 12.2.5 and can be estimated using the splitting methodology. In the present situation, the target and obstacle sets are respectively given by $B = [b, \infty)$ and $C = (-\infty, a]$. The underlying Markov process is the multiplicative chain $(X_n, Y_n) = (Z_n(Y), Y_n)$ defined by the recursion

$$X_n = X_{n-1} \times \frac{dM_n(Y_{n-1}, \cdot)}{d\overline{M}_n(Y_{n-1}, \cdot)}(Y_n) \in S = \mathbb{R}_+ \quad \text{with} \quad X_0 = 1 \quad (12.16)$$

where Y_n is a Markov chain with elementary transitions $\overline{M}_n$ and starting at $Y_0 = y \in (a, b)$. We will of course not repeat the integral description of the particle algorithm associated with this model. To give a brief illustration, let us assume for simplicity that b is an integer. In this case we can choose to split the particles whenever they hit the intervals $B_n = [n+1, \infty)$, $1 \leq n < b$. To sample the first mutation step, we sample N independent excursions $(Y_p^i)_{0 \leq p \leq T_1^i}$ with elementary transitions $\overline{M}_p$ (starting at y) up to the first time they hit $(B_1 \cup C)$

$$T_1^i = \inf\left\{p \geq 0 \;:\; \prod_{q=1}^{p} \frac{dM_q(Y_{q-1}^i, \cdot)}{d\overline{M}_q(Y_{q-1}^i, \cdot)}(Y_q^i) \in (B_1 \cup C)\right\}$$

Each excursion ending in C is killed, and instantly a particle randomly chosen among those that reach B_1 splits into two offsprings. After this updating stage, we evolve independently the particles in B_1 up to the first time the corresponding probability ratio hits $(B_2 \cup C)$. Then, we update this new configuration as before. We kill the excursions ending in C and duplicate the ones in B_2, and so on. Formula (12.14) also suggests two additional particle interpretations. If we define the potential function $G(x, y) = x/y$ on $(0, \infty)^2$, then by the definition of X_n we obtain the equations

$$\overline{\mathbb{P}}_y(T < R) = \mathbb{P}_y(Z_T^{-1}(Y) \, 1_{Z_T(Y) \geq b}) = \mathbb{E}_1\left(\prod_{p=1}^{T} G(X_{p-1}, X_p) \, 1_{T<R}\right) \tag{12.17}$$

where $\mathbb{E}_1(.)$ represents the expectation operator with respect to the law of the Markov chain defined in (12.16) under $\mathbb{P}_y$. The second expression can be approximated using the excursion particle models developed in Section 12.2. Since under $\mathbb{P}_y$ the Markov chain X_n tends to infinity, the exploration excursions are more likely to reach the upper levels. Nevertheless, the updating mechanism will favor excursions that have not entered into these levels too fast. The first expression in (12.17) is related to the more traditional importance sampling Monte Carlo method. The corresponding estimator is defined by $\frac{1}{N}\sum_{i=1}^{N} Z_{T^i}^{-1}(Y^i) \, 1_{Z_{T^i}(Y^i) \geq b}$, where $(Y^i)_{1 \leq i \leq N}$ is a sequence of N independent copies of the chain Y under $\mathbb{P}_y$. The variance of this estimate is proportional to

$$\overline{\mathbb{E}}_y(Z_T^{-1}(Y) \, 1_{Z_T(Y) \geq b}) - \overline{\mathbb{P}}_y(T < R)^2 \leq b^{-1} \, \overline{\mathbb{P}}_y(T < R) - \overline{\mathbb{P}}_y(T < R)^2$$

This clearly shows the improvements obtained in comparison with the rather crude Monte Carlo estimate $\frac{1}{N}\sum_{i=1}^{N} 1_{Z_{T^i}(Y^i) \geq b}$ based on N independent copies of the chain Y under $\mathbb{P}_y$.

12.3.5 Exercises

Exercise 12.3.3: [Importance sampling] Suppose we want to evaluate the integral $\mu(G)$ of a nonnegative and bounded potential function G with respect to some distribution μ on some measurable space $(E, \mathcal{E})$. We associate with a sequence of independent random variables $(X^i)_{i \geq 1}$ with common distribution μ the empirical measures $\mu^N = \frac{1}{N}\sum_{i=1}^{N} \delta_{X^i}$.

- Check that $\mathbb{E}(\mu^N(G)) = \mu(G)$ and
$$N\mathbb{E}((\mu^N(G) - \mu(G))^2) = \sigma_\mu(G) =_{\text{def.}} \mu((G - \mu(G))^2)$$

- For any probability measure $\overline{\mu}$ such that $\mu << \overline{\mu}$, prove that $\mu(G) = \overline{\mu}(\overline{G})$ with $\overline{G} = G\,\frac{d\mu}{d\overline{\mu}}$. We let $\overline{\mu}^N = \frac{1}{N}\sum_{i=1}^{N} \delta_{\overline{X}^i}$ be the occupation measure associated with a sequence of N independent random variables $(\overline{X}^i)_{i \geq 1}$ with common distribution $\overline{\mu}$. Prove that $\mathbb{E}(\overline{\mu}^N(\overline{G})) = \mu(G)$ and
$$N\mathbb{E}((\overline{\mu}^N(\overline{G}) - \mu(G))^2) = \sigma_{\overline{\mu}}(\overline{G}) \quad =_{\text{def.}} \overline{\mu}((\overline{G} - \mu(G))^2)$$
$$= \sigma_\mu(G) - \mu\left(G^2\left(1 - \frac{d\mu}{d\overline{\mu}}\right)\right)$$

- Roughly speaking, from the equation above, we see that a reduction of variance is obtained as soon as $\overline{\mu}$ is chosen such that $\frac{d\mu}{d\overline{\mu}} < 1$ on regions where G is more likely to take large values. In other words, it is judicious to choose a new reference distribution $\overline{\mu}$ so that the sampled particles $\overline{X}^i$ are more likely to visit regions with high potential. For instance, if $G = 1_A$ is the indicator function of some measurable set $A \in \mathcal{E}$, then prove that
$$\sigma_{\overline{\mu}}(\overline{G}) = \sigma_\mu(G) - \mu\left(1_A\left(1 - \frac{d\mu}{d\overline{\mu}}\right)\right)$$
If we choose $\overline{\mu}$ such that $\frac{d\mu}{d\overline{\mu}}(x) \leq 1 - \delta$ for any $x \in A$, then check that
$$\overline{\mu}(A) \geq \mu(A)/(1-\delta) \quad \text{and} \quad \sigma_{\overline{\mu}}(\overline{G}) + \delta\mu(A) \leq \sigma_\mu(G)$$

- Show that the optimal distribution $\overline{\mu}$ is the Boltzmann-Gibbs measure $\overline{\mu}(dx) = \Psi(\mu)(dx) = \mu(G)^{-1}G(x)\mu(dx)$ in the sense that $\sigma_{\overline{\mu}}(\overline{G}) = 0$. This optimal strategy is clearly hopeless since the normalizing constant $\mu(G)$ is precisely the constant we want to estimate!

- As a bad choice example, if we take $\overline{\mu} = \mu(G^{-2})^{-1}G^{-2}(x)\mu(dx)$, then check that $\sigma_{\overline{\mu}}(\overline{G}) \geq \mu(G^4)/\mu(G^2) - \mu(G)^2 \geq \sigma_\mu(G)$.

Exercise 12.3.4: [Simple random walk] Let $(\varepsilon_n)_{n\geq 0}$ be independent and identically distributed random variable with common law $\mathbb{P}(\varepsilon_n = 1) = 1 - \mathbb{P}(\varepsilon = -1) = p \in (0,1)$. We consider the simple random walk X_n on $E = \mathbb{Z}$ defined by $X_n = \sum_{p=0}^n \varepsilon_n$. Suppose we want to evaluate (using a Monte Carlo scheme) the probability that X_n enters into a subset $A \subset \mathbb{N} - \{0\}$. If we have $p < 1/2$, then the random walk X_n tends to move to the left. One natural way to increase the probability that the random walk visits the set A is to change p by some $\overline{p} \in (p,1)$. In this case, the random walk $\overline{X}_n$ defined as X_n by replacing p by $\overline{p}$ is more likely to move to the right, and as a result the event $(\overline{X}_n \in A)$ is more likely than $(X_n \in A)$. The expected value of $f(X_n) = 1_A(X_n)$ and the particle approximation mean using the standard Monte Carlo method are given respectively by

$$\mathbb{E}(f(X_n)) = \mathbb{P}(X_n \in A) \quad \text{and} \quad m(X_n)(f) = \frac{1}{N}\sum_{i=1}^N 1_A(X_n^i)$$

where $(X_n^i)_{i\geq 1}$ is a collection of independent copies of X_n.

- We let P_n be the distribution of the random sequence $(\varepsilon_p)_{0\leq p\leq n} \in \{-1,+1\}^{n+1}$. Check that

$$P_n(d(u_0,\ldots,u_n)) = (p(1-p))^{\frac{n+1}{2}} (p/(1-p))^{\frac{1}{2}\sum_{k=0}^n u_k}$$

- We let $\overline{P}_n$ be the distribution of the random sequence $(\overline{\varepsilon}_p)_{0\leq p\leq n}$ defined as $(\varepsilon_p)_{0\leq p\leq n}$ by replacing p by some $\overline{p} \in (0,1)$. Deduce from the first question that $P_n << \overline{P}_n$ and

$$\frac{dP_n}{d\overline{P}_n}(u_0,\ldots,u_n) = G_n\left(\sum_{k=0}^n u_k\right) \quad \text{with} \quad G_n(x) = u(p,\overline{p})^{\frac{n+1}{2}} v(p,\overline{p})^{\frac{x}{2}}$$

and $(u(p,\overline{p}), v(p,\overline{p})) = (\frac{p(1-p)}{\overline{p}(1-\overline{p})}, \frac{p(1-\overline{p})}{\overline{p}(1-p)})$.

- Check that $\mathbb{E}(f(X_n)) = \mathbb{E}(f(\overline{X}_n)G_n(\overline{X}_n))$ for any $f \in \mathcal{B}_b(\mathbb{Z})$.

- Let $(\overline{X}_n^i)_{i\geq 1}$ be a collection of independent copies of $\overline{X}_n$. By the central limit theorem, prove that the sequence of random variables

$$W_n^N(f) = \sqrt{N}(m(X_n)(f) - \mathbb{E}(f(X_n)))$$
$$\overline{W}_n^N(f) = \sqrt{N}(m(\overline{X}_n)(f_n G_n) - \mathbb{E}(f(X_n)))$$

converges in law, as $N \to \infty$, to a pair of Gaussian random variables with mean 0 and respective variance $\sigma_n(f)$ and $\overline{\sigma}_n(f)$ defined by

$$\begin{aligned}
\sigma_n^2(f) &= \mathbb{E}(f(X_n)^2) - \mathbb{E}(f(X_n))^2 \\
\overline{\sigma}_n^2(f) &= \mathbb{E}(f(\overline{X}_n)^2 G_n(\overline{X}_n)^2) - \mathbb{E}(f(X_n))^2 \\
&= \sigma_n^2(f) + \mathbb{E}(f(X_n)^2 (G_n(X_n) - 1))
\end{aligned}$$

- Prove that, for any indicator functions $f = 1_A$ with $A \subset (G_n \leq 1/a_n)$, for some $a_n \geq 1$ we have

$$\overline{\sigma}_n^2(f) \leq a_n^{-1}\, \mathbb{P}(X_n \in A) - \mathbb{P}(X_n \in A)^2 \leq \sigma_n^2(f)$$

Exercise 12.3.5: Let $(\varepsilon_n)_{n \geq 0}$ and $(\overline{\varepsilon}_n)_{n \geq 0}$ be two collections of independent and identically distributed exponential random variable with respective intensity parameter $\lambda, \overline{\lambda} > 0$. Use the same line of reasoning as in the previous exercise to evaluate the occurrence of the event $X_n = \sum_{p=0}^n \varepsilon_n \in A \subset (0, \infty)$.

Exercise 12.3.6: [Importance sampling and large deviations] Let $m(X) = \frac{1}{N} \sum_{i=1}^N \delta_{X^i}$ be the empirical measure associated with a sequence of N independent and identically distributed random variables $X = (X^i)_{1 \leq i \leq N}$ with common law μ (on some measurable space $(E, \mathcal{E})$). We consider a nonnegative and bounded test function V on E. Suppose we want to evaluate the quantity $p(x) = \mathbb{P}(m(X)(V) \geq x)$ for some $x > \mu(V) = \mathbb{E}(m(X)(V))$. We consider N' independent copies $(X_j)_{1 \leq j \leq N'}$ of the N-dimensional random vector X and we set $S_{N'} = \frac{1}{N'} \sum_{j=1}^{N'} 1_{[x,\infty)}(m(X_j)(V))$. Check that $S_{N'}$ is an unbiased estimate in the sense that $\mathbb{E}(S_{N'}) = p(x)$ and for any $N' \geq 1$ we have

$$N'\mathbb{E}([S_{N'} - \mathbb{P}_\mu(m(X)(V) \geq x)]^2) = p(x)(1 - p(x)) \tag{12.18}$$

- Prove that $p(x) \leq e^{-N H_V(x)}$ with $H_V(x) = \sup_{\lambda \geq 0}(\lambda x - \log \mu(e^{\lambda V}))$.

- Check that the supremum in the display above is attained at some $\lambda_{V,x} \in (0, \infty)$ that satisfies the equation

$$x = \Psi_{x,V}(\mu)(V) \quad \text{with} \quad \Psi_{x,V}(\mu)(du) = \frac{1}{\mu(e^{\lambda_{V,x} V})}\, e^{\lambda_{V,x} V(u)}\, \mu(du)$$

- Let $m(\overline{X}) = \frac{1}{N} \sum_{i=1}^N \delta_{\overline{X}^i}$ be the empirical measure associated with a sequence of N independent and identically distributed random variables $\overline{X} = (\overline{X}^i)_{1 \leq i \leq N}$ with common law $\Psi_{x,V}(\mu)$. Prove that $p(x) =$

468   12. Applications

$\mathbb{E}(1_{m(\overline{X})(V) \geq x} Z(\overline{X}))$ with $Z(\overline{X}) = e^{-N[\lambda_{V,x} m(\overline{X})(V) - \log \mu(e^{\lambda_{V,x} V})]}$. We consider N' independent copies $(\overline{X}_j)_{1 \leq j \leq N'}$ of the N-dimensional random vector $\overline{X}$ and we set

$$\overline{S}_{N'} = \frac{1}{N'} \sum_{j=1}^{N'} Z(\overline{X}_j) \, 1_{[x,\infty)}(m(\overline{X}_j)(V))$$

Prove that $\mathbb{E}(\overline{S}_{N'}) = \mathbb{P}(m(X)(V) \geq x)$ and for any $N' \geq 1$ we have

$$N' \mathbb{E}([\overline{S}_{N'} - \mathbb{P}_\mu(m(X)(V) \geq x)]^2)$$

$$= \mathbb{E}(1_{[x,\infty)}(m(X)(V)) \, Z(X)) - p(x)^2 \leq p(x) \, [e^{-NH(x)} - p(x)]$$

Compare this growth rate with the one obtained without changing the reference sampling measure (12.18).

Exercise 12.3.7: Let X_n be a Markov chain taking values in some measurable spaces $(E_n, \mathcal{E}_n)$, with elementary transitions M_n, and initial distribution η_0. We consider a collection of Markov transitions $\overline{M}_n$ satisfying the regularity conditions stated in Section 12.3.1. We introduce the pair potentials/transitions $(\mathbf{G}_n, \mathbf{M}_n)$ on the transition spaces $\mathbf{E}_n = (E_n, E_{n+1})$ defined by

$$\mathbf{M}_n((x,y), d(x', y')) = \delta_y(dx') \, M_{n+1}(x', dy')$$
$$\mathbf{G}_n(x, y) = \frac{dM_{n+1}(x, \cdot)}{d\overline{M}_{n+1}(x, \cdot)}(y)$$

Describe the Feynman-Kac model associated with the pair $(\mathbf{G}_n, \mathbf{M}_n)$ with initial distribution $\eta_0 \times \overline{M}_1$. Show that the corresponding genealogical approximation model can be interpreted as a particle simulation method of the Markov chain X_n.

Exercise 12.3.8: Let X_n be a random walk on $\mathbb{Z}$ starting at the origin with elementary transitions

$$M_n(x, dy) = p_n(x) \, \delta_{x+1}(dy) + q_n(x) \, \delta_{x-1}(dy)$$

where $p_n(x)$ and $q_n(x)$ are $(0,1)$-valued numbers such that $p_n(x) + q_n(x) = 1$. We let $\overline{M}_n$ be the Markov transition defined as M_n by replacing the pair $(p_n(x), q_n(x))$ by another pair of $(0,1)$-valued parameters $(\overline{p}_n(x), \overline{q}_n(x))$ such that $\overline{p}_n(x) + \overline{q}_n(x) = 1$. Prove that for any $x, y \in \mathbb{Z}$ with $(y - x) \in \{-1, +1\}$ we have

$$\frac{dM_n(x, \cdot)}{d\overline{M}_n(x, \cdot)}(y) = p_n(x)/\overline{p}_n(x) \, 1_{x+1}(y) + q_n(x)/\overline{q}_n(x) \, 1_{x-1}(y)$$
$$= (p_n(x)/\overline{p}_n(x))^{\frac{1+(y-x)}{2}} \, (q_n(x)/\overline{q}_n(x))^{\frac{1-(y-x)}{2}}$$

Deduce that for any $f_n \in \mathcal{B}_b(\mathbb{Z}^{n+1})$ we have

$$\mathbb{E}(f_n(X_{[0,n]})) = \overline{\mathbb{E}}\left(f_n(X_{[0,n]}) \prod_{k=1}^n G_k(X_{k-1}, X_k)\right)$$

with the potential functions

$$\begin{aligned}G_k(X_{k-1}, X_k) &= \left(\frac{p_n(X_{k-1})}{\overline{p}_n(X_{k-1})}\right)^{\frac{1+\Delta X_k}{2}} \left(\frac{q_n(X_{k-1})}{\overline{q}_n(X_{k-1})}\right)^{\frac{1-\Delta X_k}{2}} \\ &= \left(\frac{(p_n q_n)}{(\overline{p}_n \overline{q}_n)}(X_{k-1})\right)^{\frac{1}{2}} \left(\frac{(p_n \overline{q}_n)}{(\overline{p}_n q_n)}(X_{k-1})\right)^{\frac{\Delta X_k}{2}}\end{aligned}$$

In the above display, $\Delta X_k = (X_k - X_{k-1})$ and $\overline{\mathbb{E}}$ is the expectation w.r.t. the law of the random walk starting at the origin and evolving with the Markov transitions $\overline{M}_n$. Describe a genealogical particle simulation method of the random walk with transitions M_n using the potential functions G_n and the mutation transitions $\overline{M}_n$. If we choose $\overline{p}_n = q_n > \overline{q}_n = p_n$, then prove that

$$G_k(X_{k-1}, X_k) = (p_n(X_{k-1})/q_n(X_{k-1}))^{\Delta X_k} \geq 1 \iff \Delta X_k = -1$$

In this situation, deduce that the particles are more likely to move to the right, and the selection transition tends to favor the particles moving to the left.

12.4 Spectral Analysis of Feynman-Kac-Schrödinger Semigroups

In this section, we apply the particle methodology for the numerical solution of the Lyapunov exponent of Feynman-Kac and Schrödinger type semigroups on some classical Banach spaces. In some situations, these important spectral quantities also coincide with the principal eigenvalues of positive operators.

Except in some particular situations such as for the well-known harmonic oscillator or for the one dimensional neutron model discussed page 149 (see also Theorem 10.1 and the example on pp. 67–68 in [176]), explicit and analytic descriptions of these exponents are generally not available and we need to resort to some kind of approximation. Several strategies have been suggested in the literature. To name a few, the perturbation theory proposes in some instances asymptotic expansions of isolated eigenvalues (see for instance Kato [200]). Several characterizations of Lyapunov exponents have been suggested in mathematics literature in the beginning of

1950s, including the work of H. Wieland [312], Krein and Rutman [208], Birkhoff [32] and Harris [176]. Donsker and Varadhan have also presented in a series of papers (see for instance [117]) a theory of large deviations that expresses the well-known Raleigh-Ritz representation of the top eigenvalue in terms of a variational problem in distribution space. In some particular situations, this global optimization problem can be solved by using for instance some kind of stochastic global search algorithm or specific Hilbert projection techniques.

Our approach consists in expressing these spectral exponents and the corresponding eigenfunctions in terms of the fixed point of a nonlinear Feynman-Kac distribution flow. These key functional representations bring some new light on connections between the spectral theory of Schrödinger operators and interacting measure-valued processes. They also provide a natural microscopic particle interpretation of these spectral quantities. Furthermore, the uniform convergence analysis derived in Section 7.4.3, lays solid theoretical foundations for the asymptotic and the long time behavior of these particle approximation models. This approach, which has been presented in [102] for continuous and discrete time models, was influenced by the recent works of Burdzy, Holyst, Ingerman, and March [186, 38], Hetherington [179], Sznitman [295], and earlier joint work of the author with Guionnet [86] and Doucet [82].

12.4.1 Lyapunov Exponents and Spectral Radii

We consider a time-homogeneous Markov chain

$$\left(\Omega = E^{\mathbb{N}}, \mathcal{F} = (\mathcal{F}_n)_{n \geq 0}, X = (X_n)_{n \geq 0}, (\mathbb{P}_x)_{x \in E}\right)$$

taking values in a measurable space $(E, \mathcal{E})$ with Markov transitions M. We also let G be a measurable potential function on E that satisfies the regularity condition (G) stated on page 115, and we set $r = \sup_{x,y} \{G(x)/G(y)\} < \infty$. We associate with the pair (G, M) the integral operator on the Banach space $(\mathcal{B}_b(E), \|\cdot\|)$ defined by

$$Q(x, dy) = G(x)\, M(x, dy) \qquad (12.19)$$

Note that $\|Q\| = \sup_{\|f\|=1} \|Q(f)\| \leq \|G\|$. The semigroup Q^n on $\mathcal{B}_b(E)$ associated with Q is defined by the formula $Q^n = QQ^{n-1}$ with $Q^0 = Id$. For any $f \in \mathcal{B}_b(E)$, we observe that $Q^n(f)$ is also given by the Feynman-Kac formula

$$Q^n(f)(x) = \mathbb{E}_x\left(f(X_n) \prod_{p=0}^{n-1} G(X_p)\right)$$

Since the potential is assumed to be bounded, one finds that Q^n is a collection of bounded operators on $\mathcal{B}_b(E)$ with norm

$$\|Q^n\| = \sup_{\|f\|=1} \|Q^n(f)\| = \|Q^n(1)\|\ (\leq \|G\|^n)$$

The Lyapunov exponent or the spectral radius $\operatorname{Lyap}(Q) \in [0, +\infty]$ of the semigroup Q on the Banach space $\mathcal{B}_b(E)$ is the quantity defined by subadditive arguments as

$$\operatorname{Lyap}(Q) = \operatorname{Spr}(Q) = \lim_{n \to \infty} \|Q^n\|^{1/n} = \inf_{n \geq 1} \|Q^n\|^{1/n} \qquad (12.20)$$

12.4.2 Feynman-Kac Asymptotic Models

Next we propose a way to relate the logarithmic exponents

$$\Lambda(G) = \log \operatorname{Lyap}(Q) = \lim_{n \to \infty} \frac{1}{n} \sup_{x \in E} \log Q^n(1)(x)$$

with the long time average of the Feynman-Kac distribution flow model η_n defined for any $f \in \mathcal{B}_b(E)$ by $\eta_n(f) = \gamma_n(f)/\gamma_n(1)$ with

$$\gamma_n(f) = \eta_0 Q^n(f) = \mathbb{E}_{\eta_0}\left(f(X_n) \prod_{p=0}^{n-1} G(X_p) \right)$$

where η_0 is a given probability measure on E and $\mathbb{E}_{\eta_0}(.)$ represents the expectation operator with respect to the distribution $\mathbb{P}_{\eta_0} = \int \eta_0(dx) \mathbb{P}_x$. We recall that the distribution flow η_n satisfies the time-homogeneous and nonlinear equations

$$\eta_n = \Phi(\eta_{n-1}) = \Psi(\eta_{n-1}) M$$

with the Boltzmann-Gibbs transformation $\Psi : \mathcal{P}(E) \to \mathcal{P}(E)$ defined for any $\eta \in \mathcal{P}(E)$ and $f \in \mathcal{B}_b(E)$ by

$$\Psi(\eta)(f) = \eta(Gf)/\eta(G)$$

We denote by Φ^n, $n \geq 0$, the corresponding nonlinear evolution semigroup defined by

$$\Phi^n = \Phi^{n-1} \circ \Phi \quad \text{and} \quad \Phi^0 = Id$$

The choice of the initial distribution may vary. When $\eta_0 = \delta_x$, sometimes we write $\gamma_n^{(x)}$ and $\eta_n^{(x)}$, the corresponding measures. In this notation, we have

$$\gamma_n^{(x)}(f) = \delta_x Q^n(f) = Q^n(f)(x) \quad \text{and} \quad \eta_n^{(x)}(f) = \Phi^n(\delta_x)(f) = \frac{Q^n(f)(x)}{Q^n(1)(x)}$$

On the other hand, we observe that

$$\gamma_n(1) = \gamma_{n-1}(G) = \eta_{n-1}(G)\, \gamma_{n-1}(1) \quad \text{and} \quad \gamma_n(1) = \prod_{p=0}^{n-1} \eta_p(G)$$

472   12. Applications

This yields that

$$\Lambda_n^{(x)}(G) =_{\text{def.}} \frac{1}{n} \log Q^n(1)(x) = \frac{1}{n} \log \gamma_n^{(x)}(1) = \frac{1}{n} \sum_{p=0}^{n-1} \log \eta_p^{(x)}(G)$$

We note that the logarithmic exponents $\Lambda(G)$ can be rewritten as

$$\Lambda(G) = \lim_{n\to\infty} \sup_{x\in E} \Lambda_n^{(x)}(G)$$

Our next objective is to express the Lyapunov exponent $\text{Lyap}(Q)$ in terms of the fixed point η_∞ of the semigroup Φ. The existence and the possibility of approximating such a representation will depend on the asymptotic stability properties of Φ. These properties are discussed in some detail in Section 4.3 and 4.4. In the present section we simplify the presentation and we use the following contraction condition:

(Φ) For any $\mu_1, \mu_2 \in \mathcal{P}(E)$ and $n \geq 0$ we have

$$\|\Phi^n(\mu_1) - \Phi^n(\mu_2)\|_{\text{tv}} \leq \beta(\Phi^n) \, \|\mu_1 - \mu_2\|_{\text{tv}} \quad \text{with} \quad \beta(\Phi) = \sum_{n\geq 0} \beta(\Phi^n) < \infty$$

This contraction property is difficult to check in practice. Several sufficient conditions in terms of the pair (G, M) are derived in Chapter 4. For instance, suppose the Markov kernel M satisfies the mixing condition $(M)_m$ stated in page 139 for some $m \geq 1$ and some $\varepsilon(M) = \varepsilon \in (0, 1]$. Then, using Proposition 4.3.5, we find that

$$\beta(\Phi^n) \leq 2\epsilon^{-1} \, r^m \left(1 - \epsilon^2 \, r^{(m-1)}\right)^{[n/m]}$$

In addition, we have the following estimate, which does not depend on the potential function

$$\|\Phi^n(\mu_1) - \Phi^n(\mu_2)\|_{\text{tv}} \leq (1 - \varepsilon^2 \, r^{(m-1)})^n$$

By the Banach fixed point theorem the contraction condition (Φ) implies the existence of a unique fixed point $\eta_\infty \in \mathcal{P}(E)$. The interplay between the latter and the quantities (G, M) is described by the fixed point formula

$$\eta_\infty(f) = \Phi(\eta_\infty)(f) = \eta_\infty(GM(f))/\eta_\infty(G)$$

If we take $f = Q^n(1)$, we readily find the recursive formula

$$\eta_\infty(Q^{n+1}(1)) = \eta_\infty(Q^n(1)) \, \eta_\infty(Q(1)) = \eta_\infty(Q^n(1)) \, \eta_\infty(G)$$

Thus we have $\eta_\infty(Q^n(1)) = (\eta_\infty(G))^n$, from which we conclude that

$$\log \eta_\infty(G) = \frac{1}{n} \log \mathbb{E}_{\eta_\infty} \left(\prod_{p=0}^{n-1} G(X_p) \right)$$

12.4 Spectral Analysis of Feynman-Kac-Schrödinger Semigroups

Using the inequality $|\log x - \log y| \leq |x-y|/(x \wedge y)$, which is valid for any $x, y > 0$, we also find the estimates

$$|\Lambda_n^{(x)}(G) - \Lambda_n^{(y)}(G)| \leq \frac{1}{n} \sum_{p=0}^{n-1} |\log \eta_p^{(x)}(G) - \log \eta_p^{(y)}(G)|$$

$$\leq \frac{1}{\inf_E G} \frac{1}{n} \sum_{p=0}^{n-1} |\eta_p^{(x)}(G) - \eta_p^{(y)}(G)| \leq 2\frac{r}{n}\beta(\Phi)$$

and in much the same way

$$|\Lambda_n^{(x)}(G) - \log \eta_\infty(G)| \leq \frac{2}{n} r\beta(\Phi) \qquad (12.21)$$

We summarize the discussion above with the following proposition.

Proposition 12.4.1 *When condition (Φ) holds true, then the logarithmic Lyapunov exponent is given by the formulae*

$$\Lambda(G) = \log \eta_\infty(G) = \lim_{n \to \infty} \Lambda_n^{(x)}(G) \quad \text{with} \quad \Lambda_n^{(x)}(G) = \frac{1}{n} \sum_{p=0}^{n-1} \log \eta_p^{(x)}(G)$$

In addition, we have the uniform estimates

$$\sup_{x \in E} \|\eta_n^{(x)} - \eta_\infty\|_{\mathrm{tv}} \leq \beta(\Phi^n) \quad \text{and} \quad n \sup_{x \in E} |\Lambda_n^{(x)}(G) - \Lambda(G)| \leq 2r\beta(\Phi)$$

12.4.3 Particle Lyapunov Exponents

The choice of a particle interpretation of the Feynman-Kac semigroups introduced in Section 12.4.2 is not unique. We refer the reader to Section 2.5.3 as well as to Chapter 11 for a thorough discussion on different McKean interpretation models. To fix the ideas, we consider here the N particle approximation models introduced in (11.11), and we assume that the potential function G satisfies condition (G) for some $\varepsilon(G) = 1/r \in (0,1]$ with $1 \leq r < \infty$ and the mutation transition M satisfies the mixing condition $(M)_m$ for some $m \geq 1$ and some $\varepsilon(M) = \varepsilon > 0$.

We recall that during the mutation stage the particles ξ_n^i evolve randomly according to the Markov transitions M. During the selection, each of these particles ξ_n^i randomly selects an individual $\widehat{\xi}_n^i$ according to the distribution

$$S_{m(\xi_n)}(\xi_n^i, .) = \frac{G(\xi_n^i)}{\vee_{j=1}^N G(\xi_n^j)} \delta_{\xi_n^i} + \left(1 - \frac{G(\xi_n^i)}{\vee_{j=1}^N G(\xi_n^j)}\right) \Psi(m(\xi_n)) \qquad (12.22)$$

In other words, with a probability $\frac{G(\xi_n^i)}{\vee_{j=1}^N G(\xi_n^j)}$, it remains in the same site, and in the opposite event it jumps to a newly selected site randomly chosen with distribution $\Psi(m(\xi_n))$. For $(0, 1]$-valued potentials G, the regions of the state space where $G < 1$ can be interpreted as soft obstacles. When the particles evolve in these regions, their lifetime is decreasing and they try to escape by selecting individuals with higher potential value. In the opposite case, the particles evolving in state-space regions where $G = 1$ are not affected by the selection pressure and evolve randomly according to M elementary transitions.

We consider hereafter the N-particle approximation model of the flow $\eta_n^{(x)}$ and we denote by $\eta_n^{(N,x)}$ the N-particle approximation measures $\eta_n^{(N,x)} = \frac{1}{N} \sum_{i=1}^{N} \delta_{\xi_n^i}$. Under condition $(M)_m$, we have proved in theorem 7.4.4 the following uniform estimate: For any $f \in \mathcal{B}_b(E)$, $\|f\| \leq 1$, and $N \geq 1$

$$\sqrt{N} \sup_{x \in E,\ n \geq 0} \mathbb{E}\left(|\eta_n^{(N,x)}(f) - \eta_n^{(x)}(f)|^p\right)^{1/p} \leq a(p)\ c(m, \epsilon)$$

for some finite constants $c(m, \epsilon) < \infty$ whose values depend on the triplet (ϵ, m, G). If we apply this result to $f = G/\|G\|$, we find that

$$\sqrt{N} \sup_{x \in E,\ n \geq 0} \mathbb{E}\left(|\eta_n^{(N,x)}(G) - \eta_n^{(x)}(G)|^p\right)^{1/p} \leq a(p)\ c(m, \epsilon)\ \|G\|$$

from which we conclude that

$$\sqrt{N} \sup_{x \in E,\ n \geq 0} \mathbb{E}\left(|\Lambda_n^{(N,x)}(G) - \Lambda_n^{(x)}(G)|^p\right)^{1/p} \leq a(p)\ c(m, \epsilon)\ \|G\|$$

with the N-particle approximation of the Lyapunov exponent

$$\Lambda_n^{(N,x)}(G) = \frac{1}{n} \sum_{p=0}^{n-1} \log \eta_p^{(N,x)}(G) = \frac{1}{n} \sum_{p=0}^{n-1} \log \frac{1}{N} \sum_{i=1}^{N} G(\xi_p^i)$$

If we combine this uniform estimate with (12.21), we obtain the following.

Theorem 12.4.1 *Suppose condition $(M)_m$ is met for some $m \geq 1$ and some $\varepsilon > 0$. Then, for any $N \geq 1$, we have*

$$\sup_{x \in E,\ n \geq \sqrt{N}} \sqrt{N}\ \mathbb{E}\left(|\Lambda_n^{(N,x)}(G) - \Lambda(G)|^p\right)^{1/p} < \infty$$

and for any $p \geq 1$ and $f \in \mathcal{B}_b(E)$ with $\|f\| \leq 1$

$$\sup_{x \in E,\ n \geq \frac{1}{2\varepsilon^2} \log N} \sqrt{N}\ \mathbb{E}\left(|\eta_n^{(N,x)}(f) - \eta_\infty(f)|^p\right)^{1/p} \leq a(p)\ c(m, \epsilon)$$

for some finite constant $c(m, \varepsilon) < \infty$ that depends on the triplet (ε, m, G).

12.4.4 Hard, Soft and Repulsive Obstacles

In an earlier section, we assumed that the potential function G cannot take null values, thus excluding some interesting physical situations. Suppose that G is a $[0,1]$-valued function, and let $\widehat{E} = G^{-1}((0,1])$. In this situation, the logarithmic exponent of the semigroup $Q(f) = GM(f)$ is given by

$$\Lambda(G) = \lim_{n\to\infty} \frac{1}{n} \sup_{x\in\widehat{E}} \log \mathbb{E}_x \left(\prod_{p=0}^{n-1} G(X_p) \right) \in [-\infty, 0] \qquad (12.23)$$

A trivial example to illustrate this situation is to choose the indicator function $G = 1_A$ of some measurable set $A \in \mathcal{E}$. In this case, we clearly have for any $f \in \mathcal{B}_b(E)$

$$\forall x \in \widehat{E} = A \qquad \mathbb{E}_x \left(f(X_n) \prod_{p=0}^{n-1} G(X_p) \right) = \mathbb{E}_x(f(X_n)\, 1_{T\geq n})$$

where T is the exit time of A; that is, $T = \inf\{n \geq 0 : X_n \notin A\}$. Two interpretations can be underlined. In the first one, the set $A^c = G^{-1}(0)$ is regarded as a hard obstacle and T as the killing time of the particle when it hits A. The second dual interpretation is to interpret the set A as a trap where the particle spends some time before visiting A^c. Loosely speaking, in this particular situation, we have in some sense and for large values of n

$$\sup_{x\in A} \mathbb{P}_x(T \geq n) \simeq e^{-n\,|\Lambda(G)|}$$

The larger is $\Lambda(G)(\in [-\infty, 0])$, the smaller is the strength of the obstacle A^c or the larger is the trapping effect of A. We further assume that the triplet (η_0, G, M) satisfies the accessibility condition $(\mathcal{A})$ introduced in (2.16); that is, for any $x \in \widehat{E}$, we have $M(x, \widehat{E}) > 0$ and $\eta_0(\widehat{E}) > 0$. The main simplification due to this condition is that the prediction and updated Feynman-Kac models

$$\eta_n(f) = \gamma_n(f)/\gamma_n(1) \quad \text{and} \quad \widehat{\eta}_n(f) = \Psi(\eta_n)(f) = \gamma_n(fG)/\gamma_n(G)$$

are well-defined for any time $n \geq 0$ (for more details, we refer the reader to Section 2.5). Therefore we also have the asymptotic Feynman-Kac interpretation

$$\Lambda(G) = \lim_{n\to\infty} \sup_{x\in\widehat{E}} \frac{1}{n} \sum_{p=0}^{n-1} \log \eta_p^{(x)}(G) \qquad (12.24)$$

The particle approximation model of the Feynman-Kac prediction flow $\eta_n^{(x)}$ described in Section 12.4.3 is defined in the same way, but it may happen that at a given time τ^N all the configurations ξ_n exit the set $\widehat{E}$ and the algorithm is stopped. Also note that if a given particle $\xi_n^i \notin \widehat{E}$, then during the selection stage it jumps to a new selected individual in $\widehat{E}$. In a birth and death interpretation, a particle evolving in $E-\widehat{E}$ is killed and instantly a new, randomly selected individual in $\widehat{E}$ splits into two offsprings. In this sense, the set $E-\widehat{E}$ can be interpreted as a hard obstacle set. We again refer the reader to Section 2.5 for more details on these particle evolution models in absorbing media. In Chapter 3, we have developed several strategies to estimate the probability of the events $\{\tau^N < n\}$ and the asymptotic analysis of the corresponding particle approximation models.

Our next objective is to design an alternative particle interpretation. The key idea is to turn the hard obstacle set into a repulsive obstacle. To describe this particle algorithm, we first observe that for any $x \in \widehat{E}$ we have that

$$M(x,dy)\, G(y) = \widehat{G}(x)\, \widehat{M}(x,dy)$$

with

$$\widehat{G}(x) = M(G)(x) \quad \text{and} \quad \widehat{M}(x,dy) = \frac{M(x,dy)\, G(y)}{M(G)(x)}$$

For instance, in the case where $G = 1_{\widehat{E}}$ is the indicator function of some measurable subset $\widehat{E}$, we have that

$$\widehat{G}(x) = M(x,\widehat{E}) \quad \text{and} \quad \widehat{M}(x,dy) = \frac{M(x,dy)\, 1_{\widehat{E}}(y)}{M(x,\widehat{E})}$$

From this observation, we readily check that

$$\gamma_n(fG) = \mathbb{E}_{\eta_0}\left(f(X_n)\prod_{p=0}^{n} G_p(X_p)\right) = \eta_0(G)\, \widehat{\mathbb{E}}_{\widehat{\eta}_0}\left(f(X_n)\prod_{p=0}^{n-1} \widehat{G}_p(X_p)\right)$$

where $\widehat{\eta}_0 = \Psi(\eta_0)$ and $\widehat{\mathbb{E}}_{\widehat{\eta}_0}(.)$ represents the expectation with respect to the law of a Markov chain X_n with initial distribution $\widehat{\eta}_0$ and Markov transitions $\widehat{M}$. By (12.23), we finally conclude that

$$\Lambda(G) = \lim_{n\to\infty} \frac{1}{n+1} \sup_{x\in\widehat{E}} \log \widehat{\mathbb{E}}_x\left(\prod_{p=0}^{n-1} \widehat{G}_p(X_p)\right) \qquad (12.25)$$

This asymptotic Feynman-Kac interpretation of $\Lambda(G)$ is defined as in (12.23) by replacing (G, M) by $(\widehat{G}, \widehat{M})$. Thus, the corresponding particle approximation model is defined as the one described in Section 12.4.3, but the particle explores the state space $\widehat{E}$ with the elementary transitions $\widehat{M}$ and the selection transition is defined as in (12.22) by replacing the potential function G by $\widehat{G}$. Note that in the former model the particle will never visit the hard obstacle set $E - \widehat{E}$. In this sense, when replacing M by $\widehat{M}$, we turn the hard obstacle set into a soft obstacle set. From these observations, we see that the whole asymptotic analysis presented in Section 12.4.3 is still valid if we replace (G, M) by $(\widehat{G}, \widehat{M})$. Finally we refer the reader to the end of Section 4.4 for examples of mutation transitions $\widehat{M}$ satisfying the mixing condition $(M)_m$.

12.4.5 Related Spectral Quantities

In this section, we discuss the interplay between the Lyapunov exponent and some other related spectral quantities. In what follows, Q is a given bounded operator on $\mathcal{B}_b(E)$ such that $f \geq 0 \Longrightarrow Q(f) \geq 0$. This condition is clearly satisfied for the operator $Q(f) = G\ M(f)$. Suppose there exists a probability μ on $(E, \mathcal{E})$ and a constant $c \geq 0$ such that the following inequalities are satisfied for any $f \in \mathcal{B}_b(E)$:

$$\mu(|Q(f)|) \leq c\,\mu(|f|) \tag{12.26}$$

Under this condition, the image $Q(f)$ of a function $f \in \mathcal{B}_b(E)$ negligible with respect to μ remains negligible. Thus Q is a well-defined operator on $L_\infty(\mu)$. From the upper bound above, Q can be extended in a unique way as a bounded operator on $\mathbb{L}_1(\mu)$. We also observe that for any $f \in \mathcal{B}_b(E)$ we have

$$\mu(Q(f)^2) \leq \mu(Q(f^2)Q(1)) \leq \|Q(1)\|\,\mu(Q(f^2))$$

so that Q is also a well-defined operator on $\mathbb{L}_2(\mu)$. From this observation, we are led to consider the corresponding notion of spectral radius,

$$\mathrm{Spr}_{2,\mu}(Q) := \lim_{n \to \infty} \|Q^n\|_{2,\mu}^{1/n} = \inf_{n \geq 1} \|Q^n\|_{2,\mu}^{1/n}$$

where

$$\|Q^n\|_{2,\mu}^2 := \sup_{f \in L_2(\mu) \setminus \{0\}} \mu[(Q(f))^2]/\mu[f^2]$$

If E is finite and μ gives positive weight to any of its points, then the equivalence of norms on finite-dimensional space (in this case the algebra of $E \times E$ matrices) enables us to see that $\mathrm{Spr}(Q) = \mathrm{Spr}_{2,\mu}(Q)$, but this equality is not always satisfied. Even when E is finite, it is easy to construct an example for which $\|Q\| > \|Q\|_{2,\mu}$ with a probability μ not charging the whole set E (what is always true in this finite context is that $\|Q\|_{\infty,\mu} = \|Q\|_{2,\mu}$). Nevertheless, under a symmetry assumption, we have the following result.

Lemma 12.4.1 *If Q is self-adjoint on $\mathbb{L}_2(\mu)$, then we have $\mathrm{Spr}(Q) \geq \mathrm{Spr}_{2,\mu}(Q)$.*

Proof:
Let a function $f \in \mathbb{L}_2(\mu)$ and an integer $n \geq 1$ be given. Using the symmetry of Q^n, we obtain

$$\mu[(Q^n(f))^2] \leq \mu[Q^n(f^2)Q^n(1)]$$
$$= \mu[f^2 Q^{2n}(1)] \leq \|Q^{2n}(1)\|\,\mu[f^2]$$

Taking a supremum over $f \in \mathbb{L}_2(\mu) \setminus \{0\}$, this shows that

$$\|Q^n\|_{2,\mu}^{1/n} \leq \|Q^{2n}(1)\|^{1/(2n)}$$

thus, letting n go to infinity, we obtain the previous bound. ∎

To prove a reverse inequality, we assume that Q can be written as a density kernel with respect to μ; namely, that there exists a measurable mapping $q: E \times E \to \mathbb{R}_+$ such that for any $f \in \mathcal{B}_b(E)$, $x \in E$

$$Q(f)(x) = \int q(x,y) f(y) \, \mu(dy)$$

Lemma 12.4.2 *Under the hypothesis that $\sup_{x \in E} \int q(x,y)^2 \, \mu(dy) < +\infty$, we have $\mathrm{Spr}(Q) \leq \mathrm{Spr}_{2,\mu}(Q)$. In addition, if Q is self-adjoint (i.e., q is symmetric, $\mu \otimes \mu$-a.s.), then we have that $\mathrm{Spr}(Q) = \mathrm{Spr}_{2,\mu}(Q)$.*

Proof:
We have for any integer number $n \geq 1$ and point $x \in E$, by the Cauchy-Schwartz inequality,

$$\begin{aligned}
Q^n(1)(x) &= \int q(x,y) Q^{n-1}(1)(y) \, \mu(dy) \\
&\leq \sqrt{\int q(x,y)^2 \, \mu(dy)} \sqrt{\mu[Q^{n-1}(1)^2]} \\
&= \sqrt{\int q(x,y)^2 \, \mu(dy)} \, \|Q^{n-1}\|_{2,\mu}
\end{aligned}$$

Taking the supremum over $x \in E$ and then the nth root, and finally letting n be large, we obtain the affirmations of the lemma. ∎

Let Q' be the semigroup on $\mathcal{B}_b(E)$ defined by $Q'(f) = G^{1/2} M(f G^{1/2})$. Observe that $G^{1/2} Q'(f \, G^{-1/2}) = G \, M(f)$ for any $f \in \mathcal{B}_b(E)$. Since the mapping $f \to G^{1/2} f$ is an isomorphism from $\mathcal{B}_b(E)$ into itself, we conclude that $Q(f) = G \, M(f)$ and Q' have the same spectrum, the same eigenvalues, and the same spectral radius (notions to be understood in the Banach space $\mathcal{B}_b(E)$; see for instance [200]). The main simplifications in working with Q' is that if M is reversible with respect to a probability μ (i.e. $\mu(f_1 M(f_2)) = \mu(M(f_1) f_2)$) then the same is true for Q'. In this case, Q' can be extended as an operator auto-adjoint in $\mathbb{L}_2(\mu)$. Finally, we have the following equivalences.

Proposition 12.4.2 *If $M(x, \cdot) \sim \mu$ and $\sup_{x \in E} \|dM(x,\cdot)/d\mu\|_{\mathbb{L}_2(\mu)} < +\infty$ for any $x \in E$, then we have*

$$\mathrm{Lyap}(Q) = \mathrm{Spr}(Q) = \mathrm{Spr}(Q') = \mathrm{Spr}_{2,\mu}(Q')$$

12.4.6 Exercises

In the next series of exercises, we analyze the connections between the spectral analysis of the semi-group Q and $\widehat{Q}$ with

$$Q(x, dy) = G(x)\, M(x, dy) \quad \text{and} \quad \widehat{Q}(x, dy) = M(x, dy)\, G(y)$$

and the limiting distributions η_∞ and $\widehat{\eta}_\infty$ (whenever there exist) of the Feynman-Kac semi-group Φ and $\widehat{\Phi}$ defined by

$$\Phi(\eta) = \Psi(\eta) M \quad \text{and} \quad \widehat{\Phi}(\eta) = \Psi(\eta M) \qquad (12.27)$$

where Ψ represents the Boltzmann-Gibbs transformation associated to the potential function G. As usually, and unless otherwise is stated, we assume that the potential function G is bounded and non negative and M is a Markov transition on some measurable space $(E, \mathcal{E})$. Finally, note that any bounded positive integral operator Q such that $Q(1)(x) \in (0, \infty)$ can be written as above by setting

$$G(x) = Q(1)(x) \quad \text{and} \quad M(x, dy) = Q(x, dy)/Q(1)(x)$$

Exercise 12.4.3: Suppose there exists a positive eigenvector $h_g \in \mathcal{B}_b(E)$ such that $Q(h_g) = GM(h_g) = e^{\Lambda(G)}\, h_g$. We further assume that M is reversible with respect to some distribution μ, and we let $\mu_g \in \mathcal{P}(E)$ be defined by $\mu_g(f) = \mu(h_g M(f))/\mu(h_g)$. Check that $\Phi(\mu_g) = \mu_g$.

Exercise 12.4.4: We consider the Feynman-Kac interpretation (12.24) of the log-Lyapunov exponent $\Lambda(G)$. Using (12.25), check that

$$\Lambda(G) = \limsup_{n \to \infty} \frac{1}{n+1} \sum_{p=0}^{n} \log \widehat{\eta}_p^{(x)} M(G)$$

We further assume that the pair conditions $((G), (M)_m)$ are met for some $m \geq 1$ and some $\varepsilon = \varepsilon(M) > 0$ and $r = 1/\varepsilon(G) < \infty$. Using (4.34), show that the updated semigroup $\widehat{\Phi}^n$ satisfies condition (Φ) with

$$\beta(\widehat{\Phi}^n) \leq 2\varepsilon^{-1} r^{(m-1)} (1 - \varepsilon^2 r^{1-m})^{[n/m]}$$

Show that the (unique) fixed points $(\eta_\infty, \widehat{\eta}_\infty)$ of the mappings $(\Phi, \widehat{\Phi})$ are connected by the formulae $\eta_\infty = \widehat{\eta}_\infty M$ and $\widehat{\eta}_\infty = \Psi(\eta_\infty)$. Under the assumptions of Exercise 12.4.3, prove that $\widehat{\eta}_\infty(f) = \mu(h_g f)/\mu(h_g)$.

Exercise 12.4.5: We consider a pair of time homogeneous potential/transition (G, M) on some measurable space $(E, \mathcal{E})$ satisfying conditions (G)

and $(M)_m$ for some parameters m and $(\varepsilon(G), \varepsilon(M)) = (1/r, e)$. Check that $\widehat{\Phi}$ has a unique fixed point $\widehat{\eta}_\infty = \widehat{\Phi}(\widehat{\eta}_\infty)$. We further assume that M is reversible with respect to some measure $\mu \in \mathcal{P}(E)$. Show that $\widehat{\eta}_\infty$ and μ are absolutely continuous, and

$$h(x) =_{\text{def.}} \frac{d\widehat{\eta}_\infty}{d\mu}(x) \in [\epsilon/r^m, r^m/\epsilon]$$

Using the fixed point equation, prove that for any $g \in \mathbb{L}_1(\mu)$ we have

$$\mu(g\, Q(h)) = \widehat{\eta}_\infty M(G)\, \mu(g\, h)$$

with the integral operator $Q(f) = GM(f)$. Deduce from the above that

$$Q(h) = \lambda_G\, h \quad \mu-\text{a.s} \quad \text{with} \quad \lambda_G = \widehat{\eta}_\infty M(G) = \eta_\infty(G)$$

Inversely, suppose that we have $Q(g) = \lambda\, g \quad \mu-\text{a.s}$ for some $\lambda > 0$ and some non negative and bounded function g with $\mu(g) > 0$ (By Perron-Frobenius theorem, λ coincide with the top eigenvalue of the semi-group Q on $\mathbb{L}_1(\mu)$). Then, prove that

$$\widehat{\eta}_\infty(dx) = \frac{1}{\mu(g)}\, g(x)\, \mu(dx) \quad \text{and} \quad \lambda = \lambda_G$$

Exercise 12.4.6: Prove that the following assertions are satisfied for any $\lambda \in \mathbb{R}$ and any bounded function h

$$Q(h) = \lambda h \implies \widehat{Q}(g) = \lambda g \quad \text{with} \quad g = M(h)$$
$$\widehat{Q}(h) = \lambda h \implies Q(g) = \lambda g \quad \text{with} \quad g = hG$$

Exercise 12.4.7: We assume that M is reversible with respect to some positive measure μ. We also assume that

$$\widehat{Q}(h) = \lambda h \tag{12.28}$$

for some $\lambda > 0$, and some non negative function h. Let Ψ_g be the Boltzmann-Gibbs transformation associated to some positive potential function g such that $\mu(g) \in (0, \infty)$. Show that the following assertions are satisfied.

- If $\mu(h) \in (0, \infty)$, then the measure

$$\widehat{\eta}_\infty =_{\text{def.}} \Psi_{Gh}(\mu) \in \mathcal{P}(E)$$

is a fixed point of the mapping $\widehat{\Phi}$, and we have $\lambda = \widehat{\eta}_\infty M(G)$. In addition, this result holds true if (12.28) is only met on the set $G^{-1}((0, \infty))$, and as soon as $\mu(hG) \in (0, \infty)$.

12.4 Spectral Analysis of Feynman-Kac-Schrödinger Semigroups

- If $\mu(h) \in (0, \infty)$, then the measure

$$\eta_\infty =_{def.} \widehat{\eta}_\infty M = \Psi_h(\mu) \in \mathcal{P}(E)$$

is a fixed point of the mapping Φ, and we have $\lambda = \eta_\infty(G)$.
Inversely, suppose that Φ has a fixed point $\eta_\infty \in \mathcal{P}(E)$ with a bounded density $h = \frac{d\eta_\infty}{d\mu}$. In this case, prove that for any $g \in \mathbb{L}_1(\mu)$ we have

$$\mu(g\, \widehat{Q}(h)) = \eta_\infty(G)\, \mu(g\, h)$$

Deduce that

$$\widehat{Q}(h) = \lambda_G\, h \quad \mu - \text{a.s. with} \quad \lambda_G = \eta_\infty(G)$$

Let h' be the modification of h defined by

$$h' = h + \lambda_G^{-1}\, (\widehat{Q}(h) - \lambda_G h)\, 1_A \quad \text{with} \quad A = \{\widehat{Q}(h) \neq \lambda_G h\}$$

Check that $\widehat{Q}(h')(x) = \lambda_G\, h'(x)$, for any $x \in E$.

Exercise 12.4.8: In this exercise, we construct a Feynman-Kac semigroup having a prescribed eigenvalue and eigenvector. To this end, we suppose that M is reversible with respect to some positive measure μ on some measurable space E. We let $\lambda > 0$ and h be a fixed non negative and bounded function such that $M(h) > 0$. Check that

$$G = \lambda h / M(h) \Longrightarrow Q(h) = \lambda h$$

Let Ψ_h be the Boltzmann-Gibbs transformation associated to the potential function h. Deduce that the measure $\widehat{\eta}_\infty = \Psi_h(\mu) \in \mathcal{P}(E)$ is a fixed point of the mapping $\widehat{\Phi}$ associated to the pair (G, M) and defined in (12.27). If $\widehat{\gamma}_n$ is the un-normalized Feynman-Kac flow associated to the pair potential/transitions (G, M), then check that for any $f \in \mathcal{B}_b(E)$ we have that

$$\widehat{\gamma}_n(f) = \lambda^{n+1}\, \eta_0(h)\, \mathbb{E}_{\widehat{\eta}_0}(f_h(X_n)) \prod_{p=1}^n \frac{h(X_p)}{M(h)(X_{p-1})})$$

with $f_h = f/M(h)$, and $\widehat{\eta}_0 = \Psi_h(\eta_0)$. Finally, check that

$$\widehat{\gamma}_n(f) = \lambda^{n+1}\, \eta_0(h)\, \widehat{\eta}_0 M_h^n(f_h)$$

where M_h is the Markov transition defined by

$$M_h(x, dy) = M(x, dy) h(y) / M(h)(x)$$

Exercise 12.4.9: We assume that M is reversible with respect to some positive measure μ. We also assume that $Q(h) = \lambda h$ for some $\lambda > 0$, and some non negative function h with $\mu(h) \in (0, \infty)$. If Ψ_h be the Boltzmann-Gibbs transformation associated to h, then prove the following assertions.

- The measure $\eta_\infty =_{def.} \Psi_h(\mu)M \in \mathcal{P}(E)$ is a fixed point of the mapping Φ, and we have $\lambda = \eta_\infty(G)$.

- The measure $\widehat{\eta}_\infty =_{def.} \Psi_h(\mu) \in \mathcal{P}(E)$ is a fixed point of the mapping $\widehat{\Phi}$.

Inversely, suppose that $\widehat{\Phi}$ has a fixed point $\widehat{\eta}_\infty \in \mathcal{P}(E)$, with a bounded density $h = \frac{d\widehat{\eta}_\infty}{d\mu}$. In this case, prove that for any $g \in \mathbb{L}_1(\mu)$ we have

$$\mu(g\ \widehat{Q}(h)) = \widehat{\eta}_\infty M(G)\ \mu(g\ h)$$

Deduce that

$$\widehat{Q}(h) = \lambda_G\ h \quad \mu - \text{a.s.} \quad \text{with} \quad \lambda_G = \widehat{\eta}_\infty M(G)$$

Modify h up to a μ-null set so that to have $\widehat{Q}(h) = \lambda_G\ h$ on all the set E.

Exercise 12.4.10: Let (G, M) be a pair of potential transitions on some measurable space $(E, \mathcal{E})$. We further assume that M is reversible with respect to some positive measure μ and we have $\mu(G) \in (0, \infty)$ and $M(G)(x) \in (0, \infty)$, for any $x \in \widehat{E} =_{def.} G^{-1}((0, \infty))$. We consider the pair of updated potential/transitions $(\widehat{G}, \widehat{M})$ on $\widehat{E}$ defined in (2.13) and (2.14).

- Check that $\widehat{M}$ is reversible with respect to $\widehat{\mu} =_{def.} \Psi(\widehat{\Psi}(\mu)) \in \mathcal{P}(\widehat{E})$ where Ψ, resp. $\widehat{\Psi}$, denotes the Boltzmann-Gibbs transformation associated to the potential function G, resp. $\widehat{G}$.

- Let M be the bi-Laplace transition on $E = \mathbb{R}$ defined by

$$M(x, dy) = \frac{c}{2}\ e^{-c|y-x|}\ dy$$

for some $c > 0$. We also let $G = 1_{[0,L]}$ be the indicator potential function of the interval $[0, L]$. Prove that the updated potential/transitions $(\widehat{G}, \widehat{M})$ on $[0, L]$ defined in (2.13) and (2.14) have the following form

$$\widehat{G}(x) = M(x, [0, L]) = 1 - 2^{-1}(e^{-cx} + e^{-c(L-x)})$$
$$\widehat{M}(x, dy) = \frac{c}{2 - (e^{-cx} + e^{-c(L-x)})}\ e^{-c|y-x|}\ 1_{[0,L]}(y)\ dy$$

Also check that $\widehat{M}$ on $[0, L]$ is reversible with respect to the measure $\widehat{\mu} \in \mathcal{P}([0, L])$ defined by

$$\widehat{\mu}(dx) = c_L\ 1_{[0,L]}(x)\ (1 - 2^{-1}(e^{-cx} + e^{-c(L-x)}))\ dx \qquad (12.29)$$

where c_L represent a normalizing constant.

12.4 Spectral Analysis of Feynman-Kac-Schrödinger Semigroups

Exercise 12.4.11: We consider the indicator potential function and the bi-Laplace transition (G, M) introduced in the second part of exercise 12.4.10. For any $\beta > 0$ we set

$$h_\beta(x) = \sin(c\beta x) + \beta \cos(c\beta x)$$

- Check that $\int_0^x c\, e^{cy}\, h_\beta(y)\, dy = e^{cx} \sin(c\beta x)$ and

$$\int_x^L c\, e^{-cy}\, h_\beta(y)\, dy$$

$$= \tfrac{2 e^{-cx}}{1+\beta^2}\, h_\beta(x) - e^{-cx} \sin(c\beta x) + e^{-cL}\left(\sin(c\beta L) - \tfrac{2}{1+\beta^2}\, h_\beta(L)\right)$$

- Deduce that for any $x \in [0, L]$ we have

$$(1+\beta^2)\, \widehat{Q}(h_\beta)(x) = h_\beta(x) - \frac{e^{-c(L-x)}}{2}\left((1-\beta^2)\sin(c\beta L) + 2\beta \cos(c\beta L)\right)$$

with the bounded integral operator $\widehat{Q}$ on $[0, L]$ defined by

$$\widehat{Q}(f) = M(fG) = \widehat{G}\, \widehat{M}(f)$$

- If $cL = \pi/2$, then check that for $\beta = 1$ we have for any $x \in [0, L] (= [0, \pi/(2c)])$

$$\widehat{Q}(h_1)(x) = \frac{1}{2}\, h_1(x) > 0$$

- If $cL \in (\pi/2, \pi)$, then prove that we can choose $\beta \in (\pi/(2cL), 1)$ such that

$$\tan(cL\beta) + \frac{2\beta}{1-\beta^2} = 0$$

In this situation, verify that for any $x \in [0, L]$ we have

$$\widehat{Q}(h_\beta)(x) = \frac{1}{1+\beta^2}\, h_\beta(x)$$

Check that $h_\beta(x) > 0$ for any $x \in [0, \pi/(2c\beta)]$, and for any $x \in (\pi/(2c\beta), L]$

$$\beta + \tan(c\beta x) \leq \beta - \frac{2\beta}{1-\beta^2} = -\beta\, \frac{1+\beta^2}{1-\beta^2} < 0$$

Deduce that $h_\beta(x) > 0$ for any $x \in [0, L]$.

- In both of the situations examined above, and using Exercise 12.4.7, prove that the measure $\widehat{\eta}_\infty(dx) \propto_{def.} 1_{[0,L]}(x)\, h_\beta(x)\, dx \in \mathcal{P}([0, L])$ is a fixed point of the mapping $\widehat{\Phi}$.

Exercise 12.4.12: Returning to the Example 4.4.1, we assume that $E = \mathbb{Z}$ and M is the Markov transition defined by

$$M(x, dy) = p(-1)\,\delta_{x-1}(dy) + p(0)\,\delta_x(dy) + p(+1)\,\delta_{x+1}(dy)$$

with $p(i) \in (0,1)$ and $\sum_{|i|\leq 1} p(i) = 1$. If we take $\widehat{E} = [0,m]$ for some $m \geq 1$ and let $G = 1_{\widehat{E}}$, then prove that for any pair $(x,y) \in [0,m]$ $M^m(x,\{y\}) \geq \wedge_{|i|\leq 1} p(i)^m$ and $\widehat{G}(x) = M(x, [0,m]) \in [\wedge_{|i|\leq 1} p(i), 1]$. Deduce that the pair potential/kernel $(\widehat{G}, \widehat{M})$ satisfies the regularity conditions $((G),(M)_m)$ with $\varepsilon(\widehat{G}) \geq \wedge_{|i|\leq 1} p(i)$ and $\varepsilon(\widehat{M}) \geq \wedge_{|i|\leq 1} p(i)^m$. Show that $\widehat{M}(x, dy) = M(x, dy)$, for any $x \in (0,m)$, and on the boundary $\partial \widehat{E} = \{0, m\}$

$$\widehat{M}(0, dy) = \frac{p(0)}{p(0)+p(+1)}\delta_x(dy) + \frac{p(+1)}{p(0)+p(+1)}\delta_{x+1}(dy)$$

$$\widehat{M}(m, dy) = \frac{p(0)}{p(0)+p(-1)}\delta_x(dy) + \frac{p(-1)}{p(0)+p(-1)}\delta_{x-1}(dy)$$

In the same way, prove that for any $x \in (0, m)$

$$\widehat{G}(x) = 1, \quad \widehat{G}(0) = 1 - p(-1) \quad \text{and} \quad \widehat{G}(m) = 1 - p(+1)$$

Describe the particle approximation model associated with the pair $(\widehat{G}, \widehat{M})$.

12.5 Directed Polymers Simulation

12.5.1 Feynman-Kac and Boltzmann-Gibbs Models

In biology and industrial chemistry, flexible polymer models describe the chemical and kinetic structure of macromolecules in a given solvent. The polymer chain at time n is regarded as a sequence of random variables

$$X_n = U_{[0,n]} =_{\text{def.}} (U_0, \ldots, U_n) \in E_n = \underbrace{E \times \ldots \times E}_{(n+1) \text{ times}}$$

taking values in some metric space (E, d). The elementary states $(U_p)_{0 \leq p \leq n}$ represent the monomers of the macromolecules X_n. The length parameter n represents the degree of polymerization. The monomers are connected by chemical bonds and interact with one another as well as with the chemicals in the solvent. The energy of a polymerization sequence

$$X_0 = U_0 \longrightarrow X_1 = (U_0, U_1) \longrightarrow X_2 = (U_0, U_1, U_2) \longrightarrow \ldots \longrightarrow X_n = U_{[0,n]}$$

is defined in terms of a Boltzmann potential

$$\exp -\frac{1}{\epsilon} \sum_{p=0}^{n} V_p(U_{[0,p]}) \qquad (12.30)$$

The parameter $\epsilon \in \mathbb{R}_+$ represents the temperature of the solvent, and each potential function

$$V_n : (u_0, \ldots, u_n) \in E_n \to V_n(u_0, \ldots, u_n) \in \mathbb{R}_+$$

reflects the local intermolecular energy between the monomer $U_n = u_n$ in the polymer chain $X_{n-1} = (u_0, \ldots, u_{n-1})$ during the nth polymerization

$$X_{n-1} = (u_0, \ldots, u_{n-1}) \longrightarrow X_n = ((u_0, \ldots, u_{n-1}), u_n)$$

The potential functions V_n depend on the nature of the solvent and the physico-chemical structure of the polymer. At low temperature, $\epsilon \to 0$, the interaction between monomers may be strongly repulsive at short distances and attractive or repulsive at larger ones. For instance, the monomers may tend to avoid being closed on each other. These excluded volume effects and repulsive interactions can be modeled by choosing a potential function satisfying the following condition:

$$V_n(u_0, \ldots, u_n) = 0 \iff u_n \notin \{u_0, \ldots, u_{n-1}\} \qquad (12.31)$$

In this situation, every self-interaction is penalized by a factor $e^{-\frac{1}{\varepsilon}V_n}$ so that the energy of an elementary polymerization is minimal iff the new monomer differs from the previous ones. In this context, the inverse temperature parameter ε is sometimes called the strength of repulsion. In the opposite case, at high temperature, $\epsilon \to \infty$, the interaction forces disappear. In this situation, it is commonly assumed that X_n is an E_n-valued Markov chain with elementary transitions M_n and initial distribution η_0. By the definition of the chain $X_n = U_{[0,n]}$, this Markovian hypothesis implies that the Markov transitions M_n have the form

$$M_n(x_{n-1}, dx_n) = \delta_{x_{n-1}}(d(u_0, \ldots, u_{n-1})) \, P_n((u_0, \ldots, u_{n-1}), du_n)$$

for any $x_{n-1} \in E_{n-1}$ and $x_n = (u_0, \ldots, u_n) \in E_n$ and for some Markov transition P_n from E_n into E. Also note that whenever $(U_n)_{n \geq 0}$ is itself a Markov chain, the transitions P_n have the form

$$P_n((u_0, \ldots, u_{n-1}), du_n) = P'_n(u_{n-1}, du_n)$$

for some Markov transitions P'_n from E into E. In summary, we see that the distribution of an abstract polymer chain with polymerization degree n is defined for any $f_n \in \mathcal{B}_b(E_n)$ by the Feynman-Kac distribution on path space

$$\widehat{\eta}_n(f_n) = \widehat{\gamma}_n(f_n)/\widehat{\gamma}_n(1)$$

486   12. Applications

with

$$\widehat{\gamma}_n(f_n) = \mathbb{E}\left(f_n(X_n) \prod_{p=0}^{n} G_p(X_p)\right) = \mathbb{E}\left(f_n(U_{[0,n]}) \prod_{p=0}^{n} G_p(U_{[0,p]})\right) \quad (12.32)$$

and the potential functions

$$G_n : x_n \in E_n \mapsto G_n(x_n) = \exp\left\{-\frac{V_n(x_n)}{\varepsilon}\right\} \in [0,1]$$

Another important quantity is the Feynman-Kac prediction model η_n defined as above by taking in formula (12.32) a product up to time $(n-1)$. More precisely, we have

$$\eta_n(f_n) = \gamma_n(f_n)/\gamma_n(1) \quad \text{with} \quad \gamma_n(f_n) = \mathbb{E}\left(f_n(X_n) \prod_{p=0}^{n-1} G_p(X_p)\right)$$

It is finally important to recall that the so-called partition function $\widehat{\gamma}_n(1)$ can also be expressed in terms of the flow $(\eta_p)_{p \leq n}$ with the product formula

$$\widehat{\gamma}_n(f_n) = \widehat{\eta}_n(f_n) \prod_{p=0}^{n} \eta_p(G_p) = \eta_n(f_n G_n) \prod_{p=0}^{n-1} \eta_p(G_p) \quad (12.33)$$

This rather elementary description allows us to construct a natural unbiased and on-line particle estimation of the partition function (see for instance Section 11.3).

In the biostatistics literature, these Feynman-Kac models seem to be preferably expressed in terms of the Boltzmann-Gibbs distribution densities

$$q_n(x_n) \propto \exp\left\{-\frac{H_n(x_n)}{\varepsilon}\right\} p_n(x_n)$$

The function p_n represents the density of a random sequence $X_n = U_{[0,n]}$ taking values in a multidimensional state space $E_n = (\mathbb{R}^d)^{n+1}$. The interaction energy of a given sequence $x_n = (u_0, \ldots, u_{n-1}, u_n) = (x_{n-1}, u_n)$ is measured by the Hamiltonian function H_n defined alternatively by one of the equivalent expressions

$$H_n(x_n) = \sum_{p=0}^{n} V_p(u_0, \ldots, u_p) = H_{n-1}(x_{n-1}) + V_n(x_{n-1}, u_n)$$

In this notation, the Feynman-Kac formulae (12.32) can alternatively be rewritten in terms of the Boltzmann-Gibbs distribution

$$\widehat{\eta}_n(dx_n) = \frac{1}{\nu_n(e^{-H_n/\varepsilon})} e^{-H_n(x_n)/\varepsilon} \nu_n(dx_n) \quad (12.34)$$

where $\nu_n \in \mathcal{P}(E_n)$ represents the distribution of the path $X_n = U_{[0,n]}$.

It is of course out of the scope of this section to present a complete and precise catalog on all the polymer simulation models that can be derived using the particle methodology described in chapter 11. The reasons are twofold. First there does not exist a universally effective particle simulation algorithm that applies to all directed polymer models. Although all of these simulation techniques are built on the same particle methodology, their accuracy really depends on various tuning parameters such as the selection period, the branching rules, population size controls, the choice of the mutation transitions, the length of the exploration excursions, and others. In addition, the improvements we can get using one or another refinement strategy are strongly related to the precise nature of the chemico-physical interactions of the model at hand. On the other hand, and due to the incomplete knowledge of the intermolecular potentials in real macromolecules, there does not exist a precise chemico-physical polymer model. In order to get some insight, the literature abounds with simplified models with diverse repulsive and attractive interaction energy landscapes.

This section is organized as follows. In Section 12.5.2, we provide a brief discussion on evolutionary particle algorithms and their connection with more traditional Metropolis type models such as the slithering tortoise or the reptilian type algorithms. In the next two Sections 12.5.3 and 12.5.4, we have collected some commonly used simplified models with repulsive and attractive interactions. We also underline some natural connections with self-avoiding and reinforced random walks. In the final section, 12.5.5, we apply the genealogical-tree-based methodologies presented in Chapter 3 (see also Section 11.4) to sample polymer chains associated with a given collection of intermolecular potentials.

12.5.2 Evolutionary Particle Simulation Methods

One challenging question in biostatistics is to generate independent samples according to the Feynman-Kac or Boltzmann-Gibbs distributions on path space introduced in (12.32). Several strategies have been suggested in the literature, including traditional Monte Carlo methods such as Metropolis-Hastings type models [216]. In the context of self-avoiding random walks, two classical Monte Carlo strategies can be underlined: the Berreti-Sokal, or the slithering tortoise algorithm [30], and the reptilian or slithering snake algorithm [210, 308]. These models first consist in modifying randomly and locally the monomers of a given chain with fixed polymerization degree. Then these small deformations are accepted or rejected. The recurrent drawbacks of these Monte Carlo algorithms are the following. The potential energy function usually has too many local minima, and its oscillations tend to slow down the convergence of the algorithm. Furthermore, they are not recursive with respect to the polymerization degree parameter. In this connection, we mention that if we are only interested in generating macromolecule samples with a fixed polymerization degree n, then we can take advantage of the Boltzmann-Gibbs representation (12.34) and alternatively use the interacting Metropolis approximation models presented in Section 11.2 and in Chapter 5. Various heuristic-like and recursive strategies have been recently suggested, such as chain growth methods [26, 167, 226, 235, 115] and the pioneering Rosenbluth's pruned-enrichment technique [280]. The basic and common strategy is as follows. We start with a random sequence of macromolecules of a given polymerization degree, say p_1. We compute the Boltzmann weight of each one. Then we properly eliminate bad configurations with high interaction energy and replace them by cloning the good ones. In biostatistics literature this selection stage is often called "enrichment". Finally, we grow each selected macromolecule up to a polymerization degree $p_2(\geq p_1)$, and so on. Here again the choice of the selection/mutation transitions is not unique. In some situations, the growing mechanism is almost dictated by the problem at hand. For instance, suppose that $\widehat{\eta}_n$ represents the path distribution of a self-avoiding random walk $U_{[0,n]}$. In this case, it is tempting to choose a growing/mutation transition that chooses randomly among the "free" neighbors (see [280]). Another idea is to send out an auxiliary collection of exploration paths of a given length to test the local environment. Then we select one of them in accordance with its interaction energy (see [235]). A wide range of selection/enrichment mechanisms have also been suggested in the literature. Some of them are based on branching selection variants such as those presented in Section 11.8 (see [235]). Some authors also suggest the use of weights' thresholds to detect the "best" selection period and avoid excessive duplications (see [167, 227, 307]).

Numerical results tend to indicate the superiority of these evolutionary type algorithms, but to our knowledge no sufficient analysis has been done to justify that these natural models are well-founded. The Feynman-Kac representation model (12.32) and the particle recipes developed in Chapter 11 clearly show that each of these algorithms coincides with a particular particle interpretation of the Feynman-Kac formulae (12.32). These evolutionary particle algorithms are essentially dictated by the dynamical structure of the Feynman-Kac model (12.32). In this connection, we mention that in biostatistics literature these recursions are preferably expressed with some obvious abusive notation as follows:

$$\widehat{\eta}_n(dx_n) \propto \widehat{\eta}_{n-1}(dx_{n-1}) \ M_n(x_{n-1}, dx_n) \ \exp\left\{-[H_n(x_n) - H_{n-1}(x_{n-1})]/\varepsilon\right\}$$

12.5.3 Repulsive Interaction and Self-Avoiding Markov Chains

At low temperature, $\varepsilon \to 0$, and under appropriate regularity conditions, the Feynman-Kac measures (12.32) with repulsive interaction potentials

(12.31) converge in some sense to the distributions $\widehat{\gamma}$ defined by

$$\widehat{\gamma}_n(f_n) = \mathbb{E}\left(f_n(X_n)\prod_{p=0}^{n}G_p(X_p)\right) \quad (12.35)$$

with the indicator potential functions

$$G_n = 1_{\widehat{E}_n} \text{ and } \widehat{E}_n = \{(u_0,\ldots,u_n)\in E_n \,;\, u_n \notin \{u_0,\ldots,u_{n-1}\}\} \quad (12.36)$$

When the underlying stochastic sequence X_n is such that

$$\begin{aligned}\widehat{\gamma}_n(1) &= \mathbb{P}(X_1 \in \widehat{E}_1,\ldots,X_n \in \widehat{E}_n)\\ &= \mathbb{P}(\forall\, 0 \leq p < q \leq n,\; U_p \neq U_q) > 0\end{aligned}$$

then the distribution flow $\widehat{\eta}_n$ is well-defined and we have

$$\begin{aligned}\widehat{\eta}_n &= \mathrm{Law}(X_n \mid X_1 \in \widehat{E}_1,\ldots,X_n \in \widehat{E}_n)\\ &= \mathrm{Law}(U_{[0,n]} \mid \forall 0 \leq p < q \leq n,\; U_p \neq U_q) \quad (12.37)\end{aligned}$$

An Excursion Feynman-Kac Model

The simplified directed polymer model described in (12.37) can also be regarded as a path-particle evolution model in an absorbing medium with hard obstacles $E_n - \widehat{E}_n$. More precisely, if we let $T = \inf\{n \geq 0 \,:\, X_n \notin \widehat{E}_n\}$ be the first time X_n exits the set $\widehat{E}_n$, then $\widehat{\eta}_n$ represents the law of the path particle $X_n = U_{[0,n]}$ given the fact that it has not been absorbed at time n. In other words, in this notation we have that $\widehat{\eta}_n = \mathrm{Law}(X_n \mid T > n)$.

Self-Avoiding Random Walks

One of the simplest mathematical models with self-repulsive interaction is the self-avoiding random walk model (abbreviated SAW). This rather elementary probabilistic model is often used in practice mainly because various authors seem to agree that it captures some qualitative features of polymer conformations. An SAW of length n is defined as a realization of the path of a simple random walk on the d-dimensional lattice $E = \mathbb{Z}^d$ that visits points no more than once. In more precise language, an SAW of length n is a random path $X_n = U_{[0,n]}$ distributed according to the Feynman-Kac distribution $\widehat{\eta}_n$ introduced in (12.37). Recalling that the Markov transitions of U_n are defined by $P(u,.) = (2d)^{-1}\sum_{|e|=1} 1_{u+e}$ and assuming that it starts at the origin $U_0 = 0$, we readily check that the partition functions are given by

$$\widehat{\gamma}_n(1) = \mathbb{P}(\forall 0 \leq p < q \leq n\; U_p \neq U_q) = |\mathbf{S}_n|/(2d)^n \quad (12.38)$$

where $\mathbf{S}_n$ is the set of self-avoiding random walks of length n and starting at 0. In the same way, we check that $\widehat{\eta}_n$ is the uniform distribution on $\mathbf{S}_n$

and

$$\eta_{n+1}(u_0,\ldots,u_n,u_{n+1}) = \frac{1}{|\mathbf{S}_n|}\, 1_{\mathbf{S}_n}(u_0,\ldots,u_n)\, \frac{1}{2d}\sum_{|e|=1} 1_{u_n+e}(u_{n+1})$$

Note that in this particular case we also have $\eta_n(G_n) = |\mathbf{S}_n|/(|\mathbf{S}_{n-1}|(2d))$.

Related Repulsive Interaction Models

Repulsive interactions can also be modeled by Boltzmann type potentials (12.30) with $V_n = 1_{\widehat{E}_n^c}$ or $V_n(u_0,\ldots,u_n) = \sum_{p=0}^{n-1} 1_{\{u_p\}}(u_n)$. The latter potential corresponds to the Edward model and will be discussed in Exercise 12.5.3. To model repulsive interactions at larger distances we can use for instance the excluded-volume potentials functions $G_n(u_0,\ldots,u_n) = 1_{\mathcal{V}_{n-1}(u_0,\ldots,u_{n-1})^c}(u_n)$, where $\mathcal{V}_{n-1}(x_{n-1})$ is a given neighborhood of $x_{n-1} = (u_0,\ldots,u_{n-1})$. Note that in this situation we have

$$\widehat{\eta}_n = \mathrm{Law}(U_{[0,n]}|\ \forall 1 \leq p \leq n\ \ U_p \notin \mathcal{V}_{p-1}(U_{[0,p-1]}))$$

as soon as $\widehat{\gamma}_n(1) = \mathbb{P}(\forall 1 \leq p \leq n\ \ U_p \notin \mathcal{V}_{p-1}(U_{[0,p-1]})) > 0$.

12.5.4 Attractive Interaction and Reinforced Markov Chains

The attractive interaction situation is closely related to self-interacting and reinforced Markov chains. For instance, if we choose the potential functions

$$G_n(u_0,\ldots,u_n) = \sum_{p=0}^{n-1} 1_{u_p}(u_n)$$

then during polymerizations the monomers are attracted to each other. In addition, when U_n is a homogeneous Markov chain with transition P' on a countable set E, then the Markov transitions $\widehat{M}_{n+1}$ defined in (11.16) are now given for any $x_n = (u_0,\ldots,u_n)$, $y_n = (v_0,\ldots,v_n) \in E^{n+1}$, and $v \in E$ by

$$\widehat{M}_{n+1}(x_n,(y_n,v)) = 1_{x_n}(y_n) \sum_{p=0}^{n} \frac{P'(v_n,v_p)}{\sum_{q=0}^{n} P'(v_n,v_q)}\, 1_{v_p}(v)$$

as soon as $\sum_{q=0}^{n} P'(v_n,v_q) > 0$.

12.5.5 Particle Polymerization Techniques

We can clearly combine the Feynman-Kac modeling techniques presented in this section with the particle recipes described in Chapter 11 and Section 12.2 to design a collection of particle approximation and simulation models. In this context, the corresponding genealogical-tree-based algorithms can also be interpreted as particle polymerization models. For instance the simple genetic N-particle model associated with the Feynman-Kac model (12.32) consists of N polymer chains with degree n

$$\xi_n^i = (\zeta_{0,n}^i, \ldots, \zeta_{n,n}^i) \quad \text{and} \quad \widehat{\xi}_n^i = (\widehat{\zeta}_{0,n}^i, \ldots, \widehat{\zeta}_{n,n}^i) \in E_n$$

During the selection stage, we randomly choose N polymer chains $\widehat{\xi}_n^i$ with common law

$$\sum_{i=1}^N \frac{e^{-\frac{1}{\varepsilon} V_n(\zeta_{0,n}^i, \ldots, \zeta_{n,n}^i)}}{\sum_{j=1}^N e^{-\frac{1}{\varepsilon} V_n(\zeta_{0,n}^j, \ldots, \zeta_{n,n}^j)}} \delta_{(\zeta_{0,n}^i, \ldots, \zeta_{n,n}^i)} \qquad (12.39)$$

This mechanism is intended to favor minimal energy polymerizations. For instance, in the case of repulsive interaction (12.31), a given polymer with degree n, say $(\zeta_{0,n}^i, \ldots, \zeta_{n,n}^i)$, has more chance of being selected if the last monomer $\zeta_{n,n}^i$ added during the nth sampled polymerization differs from the previous ones; that is, if $\zeta_{n,n}^i \notin \{\zeta_{0,n}^i, \ldots, \zeta_{n-1,n}^i\}$. During the mutation transition, each selected polymer $\widehat{\xi}_n^i$ evolves randomly according to the transition M_{n+1} of the path chain

$$X_n = U_{[0,n]} \longrightarrow X_{n+1} = U_{[0,n+1]} = (U_{[0,n]}, U_{n+1}) = (X_n, U_{n+1})$$

at time $n+1$; that is

$$\begin{aligned}\xi_{n+1}^i &= ((\zeta_{0,n+1}^i, \ldots, \zeta_{n,n+1}^i), \zeta_{n+1,n+1}^i) \\ &= ((\widehat{\zeta}_{0,n}^i, \ldots \ldots, \widehat{\zeta}_{n,n}^i), \zeta_{n+1,n+1}^i) \in E_{n+1} = (E_n \times E)\end{aligned} \qquad (12.40)$$

where $\zeta_{n+1,n+1}^i$ is a random variable with distribution $P_n(\widehat{\xi}_n^i, .)$. Various asymptotic estimates can be derived from Chapters 7 to 10. For instance, if we let $\mathbb{P}_n^{(N,q)}$ be the distribution of the first q-path particles of polymerization degree n

$$\mathbb{P}_n^{(N,q)} = \text{Law}[(\zeta_{0,n}^1, \ldots, \zeta_{n,n}^1), \ldots, (\zeta_{0,n}^q, \ldots, \zeta_{n,n}^q)]$$

then using Theorem 8.3.3 we have the following proposition.

Proposition 12.5.1 *For any $q \leq N$ and $n \geq 1$, we have the strong propagation-of-chaos estimates*

$$\|\mathbb{P}_n^{(N,q)} - \eta_n^{\otimes q}\|_{\text{tv}} \leq c(\varepsilon, n) \, q^2/N$$

for some finite constant $c(\varepsilon, n)$ whose value only depends on the pair time and cooling parameter (n, ε).

492   12. Applications

Loosely speaking, this result shows that particle models produce asymptotically independent blocks of random variables with common law η_n. In this sense, we can say that particle interpretations are particle simulation techniques for sampling polymers with a given Boltzmann-Gibbs measure. Moreover, mimicking the product formula (12.33), we construct a natural particle approximation of the partition functions $\widehat{\gamma}_n(1)$ by setting

$$\widehat{\gamma}_n^N(1) = \prod_{p=0}^{n} \eta_p^N(\exp(-V_p/\varepsilon)) \qquad (12.41)$$

where $\eta_n^N = \frac{1}{N}\sum_{i=1}^{N} \delta_{(\zeta_{0,n}^i, \zeta_{0,n}^i, \ldots, \zeta_{n,n}^i)}$ stands for the N-approximation measures of the prediction Feynman-Kac flow η_n. Precise asymptotic properties of these unbiased estimators can be found in Chapters 7 to 10, including central limit theorems and exponential estimates.

Conditional Mutations

An alternative particle polymerization technique consists in using the simple genetic N-particle model of the distribution flow $\widehat{\eta}_n$ defined in (12.32). This particle simulation strategy is again defined by a genetic selection/mutation mechanism. During the selection stage, we choose randomly N polymers $\widehat{\xi}_n^i = (\widehat{\zeta}_{0,n}^i, \ldots, \widehat{\zeta}_{n,n}^i)$ with common law

$$\sum_{i=1}^{N} \frac{\widehat{G}_n(\zeta_{0,n}^i, \ldots, \zeta_{n,n}^i)}{\sum_{j=1}^{N}\widehat{G}_n(\zeta_{0,n}^j, \ldots, \zeta_{n,n}^j)} \delta_{(\zeta_{0,n}^i, \ldots, \zeta_{n,n}^i)}$$

with the potential functions $\widehat{G}_n$ given for any $x_n \in E_n$ by the formula

$$\widehat{G}_n(x_n) = M_{n+1}(e^{-\frac{V_{n+1}}{\varepsilon}})(x_n) = \int_E e^{-\frac{V_{n+1}(x_n, u)}{\varepsilon}} P_{n+1}(x_n, du)$$

During the mutation transition, each selected particle $\widehat{\xi}_n^i = (\widehat{\zeta}_{p,n}^i)_{0 \leq p \leq n}$ evolves randomly according to the transition $\widehat{M}_{n+1}$ defined in (11.16). That is, we have that

$$\begin{aligned}
\xi_{n+1}^i &= ((\zeta_{0,n+1}^i, \ldots, \zeta_{n,n+1}^i), \zeta_{n+1,n+1}^i) \\
&= ((\widehat{\zeta}_{0,n}^i, \ldots, \widehat{\zeta}_{n,n}^i), \zeta_{n+1,n+1}^i) \in E_{n+1} = E_n \times E
\end{aligned}$$

where $\zeta_{n+1,n+1}^i$ is a random variable with distribution

$$\widehat{P}_{n+1}(\widehat{\xi}_n^i, du) \propto P_{n+1}(\widehat{\xi}_n^i, du) \exp\left\{-\frac{V_{n+1}(\widehat{\xi}_n^i, u)}{\varepsilon}\right\}$$

Self-Avoiding Particle Models

At low temperature, $\epsilon \to 0$, the Feynman-Kac polymer measures (12.32) with repulsive interaction take the form (12.35) and the discrete selection distributions (12.39) tends to the uniform measure

$$|I^N(n)|^{-1} \sum_{i \in I^N(n)} \delta_{(\zeta^i_{0,n},\ldots,\zeta^i_{n,n})}$$

with $I^N(n) = \{i\ ;\ (\zeta^i_{0,n},\ldots,\zeta^i_{n,n}) \in \widehat{E}_n\}$. In this situation, it may happen that all the N polymerizations have intersected and $I^N(n) = \emptyset$. In Section 7.4.1, we have seen that this event has an exponentially small probability. When the potentials are indicator functions, it is more judicious to use the particle algorithm associated with the McKean interpretation model (11.10). In this situation, the selection transition consists in sampling each polymerization $\widehat{\xi}^i_n = (\widehat{\zeta}^i_{0,n},\ldots,\widehat{\zeta}^i_{n,n})$ according to the distribution

$$\mathcal{S}_{n,m(\xi_n)}(\xi^i_n,.) = 1_{\widehat{E}_n}(\xi^i_n)\ \delta_{\xi^i_n} + 1_{\widehat{E}^c_n}(\xi^i_n)\ \frac{1}{|I^N(n)|} \sum_{j \in I^N(n)} \delta_{\xi^j_n}$$

The polymers $\xi^i_n = (\zeta^i_{0,n},\ldots,\zeta^i_{n,n}) \in \widehat{E}_n$ without self-intersections are not affected by the selection stage, and we set $\widehat{\xi}^i_n = \xi^i_n$. In the opposite case, the polymer chains $\xi^i_n = (\zeta^i_{0,n},\ldots,\zeta^i_{n,n}) \notin \widehat{E}_n$ with self-intersections are killed and replaced by a collection of polymers $\widehat{\xi}^i_n = (\widehat{\zeta}^i_{0,n},\ldots,\widehat{\zeta}^i_{n,n})$ randomly and uniformly chosen in the set $\{(\zeta^j_{0,n},\ldots,\zeta^j_{n,n})\ ;\ j \in I^N(n)\}$. Arguing as in (12.41), we construct a particle approximation of the partition functions $\widehat{\gamma}_n(1)$ by setting

$$\widehat{\gamma}^N_n(1) = 1_{\tau^N > n} \prod_{p=0}^n (|I^N(p)|/N) \qquad (12.42)$$

where $\tau^N = \inf\{n \geq 0\ :\ I^N(n) = 0\}$ represents the first time the particle algorithm is stopped. We again refer the reader to Chapters 7 to 10 for precise asymptotic properties of these particle estimates.

In Figure 12.5, we have presented a self-avoiding polymerization model associated with $N = 7$ particles. The dotted lines stands for killed self-intersecting lines and the thick lines represent the branching evolution of self-avoiding pomylers.

Related Particle Models

The particle simulation models described above can be refined in various ways. For instance, we can improve the accuracy of the exploration grid using the conditional branching excursion strategies described in Section 11.7. The resulting particle model is again decomposed into a mutation/selection

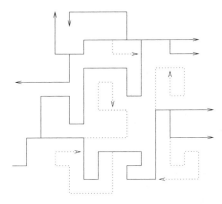

FIGURE 12.5. Particle polymerizations

transition, but in the former the particle mutation also depends on the potential functions. The precise and formal description of this genealogical-tree-based simulation algorithm is notationally time-consuming and it is better understood using the abstract and general models presented in Section 11.7. Roughly speaking, it is described as follows. Initially, we sample N independent copies $(U_0^i)_{1 \le i \le N}$ of U_0. From each one, we evolve N' exploration paths $(U_{[0,p_1)}^{i,j})_{1 \le j \le N'}$ of a given length, say p_1 (≥ 1), and we choose randomly one of these auxiliary excursions with a probability proportional to its Boltzmann weight

$$G_0^{(p)}(U_{[0,p_1)}^{i,j}) = \prod_{0 \le k < p_1} G_k(U_{[0,k]}^{i,j}) = \exp\left(-\frac{1}{\epsilon} \sum_{0 \le k < p_1} V_k(U_{[0,k]}^{i,j})\right)$$

with

$$G_k(u_0, \ldots, u_k) = \exp\left\{-\frac{1}{\varepsilon} V_k(u_0, \ldots, u_k)\right\}$$

The resulting sequence of excursions $\xi_0^{(p),i} = \zeta_{[0,p_1)}^i$ can be regarded as N-approximation samples from the conditional distribution

$$\widehat{\eta}_0^{(p)}(d(u_0, \ldots, u_{p_1-1})) \propto e^{-\frac{1}{\epsilon} \sum_{0 \le k < p_1} V_k(u_0, \ldots, u_k)} \, \eta_0^{(p)}(d(u_0, \ldots, u_{p_1-1}))$$

where $\eta_0^{(p)}$ represents the distribution of the random sequence $X_0^{(p)} = (U_k)_{0 \le k < p_1}$. To define the next two-step selection/mutation transitions, we again evolve from each $\zeta_{[0,p_1)}^i$ a sequence of N' independent excursions $(U_{[p_1,p_2)}^{i,j})_{1 \le j \le N'}$ of the chain $U_{[p_1,p_2)}$ starting at $U_{p_1-1} = \zeta_{p_1-1}^i$ (for some length $(p_2 - p_1)$ (≥ 1)). With some obvious abusive notation, we denote by $M_1^{(p)}$ the Markov transition from $U_{[0,p_1)} \rightsquigarrow U_{[0,p_2)} = (U_{[0,p_1)}, U_{[p_1,p_2)})$

$$M_1^{(p)}(u, d(v, w)) = \delta_u(dv) \, \mathbb{P}(U_{[0,p_1)} \in dw \mid U_{[0,p_1)} = v)$$

12.5 Directed Polymers Simulation   495

By construction, we have the estimate

$$M_1^{(p)}(\zeta^i_{[0,p_1)}, \cdot) \overset{N'\uparrow\infty}{\simeq} M_1^{(p),N'}(\zeta^i_{[0,p_1)}, \cdot) =_{\text{def.}} \frac{1}{N'}\sum_{j=1}^{N'} \delta_{U^{i,j}_{[p_1,p_2)}}$$

The first selection transition

$$\xi_0^{(p)} = (\zeta^k_{[0,p_1)})_{1\leq k\leq N} \longrightarrow \widehat{\xi}_0^{(p)} = (\widehat{\zeta}^k_{[0,p_1)})_{1\leq k\leq N}$$

consists in randomly choosing N paths, say $\widehat{\zeta}^k_{[0,p_1)} = \zeta^{i_k}_{[0,p_1)}$, with a probability proportional to

$$\begin{aligned}
M_1^{(p),N'}(G_1^{(p)})(\zeta^{i_k}_{[0,p_1)}) &= \frac{1}{N'}\sum_{j=1}^{N'} \exp\left(-\frac{1}{\epsilon}\sum_{p_1\leq q<p_2} V_q(\zeta^{i_k}_{[0,p_1)}, U^{i_k,j}_{[p_1,q]})\right) \\
&\overset{N'\uparrow\infty}{\simeq} M_1^{(p)}(G_1^{(p)})(\zeta^{i_k}_{[0,p_1)})
\end{aligned}$$

with

$$G_1^{(p)}(u_0,\ldots,u_{p_2-1}) = \prod_{p_1\leq k<p_2} G_k(u_0,\ldots,u_k)$$

The conditional mutation

$$\widehat{\xi}_0^{(p),k} = \widehat{\zeta}^k_{[0,p_1)} \longrightarrow \xi_1^{(p),k} = (\widehat{\zeta}^k_{[0,p_1)}, \zeta^k_{[p_1,p_2)})$$

consists in extending each selected path $\widehat{\zeta}^k_{[0,p_1)} = \zeta^{i_k}_{[0,p_1)}$ with one of the auxiliary excursions, say $\zeta^k_{[p_1,p_2)} = U^{i_k,j}_{[p_1,p_2)}$, randomly chosen with a probability proportional to

$$\exp\left(-\frac{1}{\epsilon}\sum_{p_1\leq q<p_2} V_q(\zeta^{i_k}_{[0,p_1)}, U^{i_k,j}_{[p_1,q]})\right) = G_1^{(p)}(\zeta^{i_k}_{[0,p_1)}, U^{i_k,j}_{[p_1,p_2)})$$

To define the next selection/mutation stages, we again evolve from each path $\xi_1^{(p),i}$ a sequence of N' independent excursions $(U^{i,j}_{[p_2,p_3)})_{1\leq j\leq N'}$ of the chain $U_{[p_2,p_3)}$ starting at $U_{p_2-1} = \zeta^i_{p_2-1}$, and so on. Note that the N'-sequences $U^{i,j}_{[p_n,p_{n+1}]}$ are used to generate approximate polymerizations with initial distribution $\widehat{\eta}_0^{(p)}$ and the "conditional" mutation transitions $\widehat{M}_n^{(p)}$. For a more formal and precise presentation of these branching strategies, we refer the reader to Section 11.7 (see also Sections 11.5 and 11.6).

12.5.6 Exercises

Exercise 12.5.2: [Rosenbluth's pruned-enrichment model [280]] We consider the SAW model described on page 489 on the square lattice $d = 2$.

Check that in this case the transitions M_n of the path-valued Markov chain $X_n = U_{[0,n]}$ are given for any $x_n = (u_0, \ldots, u_n)$, $y_n = (v_0, \ldots, v_n) \in (\mathbb{Z}^2)^n$ and $v \in \mathbb{Z}^2$ by

$$M_{n+1}(x_n, (y_n, v)) = 1_{x_n}(y_n) \frac{1}{4} \sum_{i=1}^{4} 1_{u_n + e_i}(v)$$

with $e_1 = (1, 0)$, $e_2 = (0, 1)$, $e_3 = (-1, 0)$, and $e_4 = (-1, -1)$. Given a sequence $x_n = (u_0, \ldots, u_n)$ of polymerization degree n, we denote by

$$\mathcal{C}(x_n) = \{e_i \,; 1 \leq i \leq 4, \ u_n + e_i \notin \{u_0, \ldots, u_n\}\}$$

the set of indexes of available directions for placing the next monomer without intersecting x_n. Note that a given SAW x_n may be trapped in the sense that it cannot be extended to a new SAW; i.e., $\mathcal{C}(x_n) = \emptyset$. Prove that whenever $|\mathcal{C}(x_n)| \neq 0$ the "conditional" transitions $\widehat{M}_{n+1}$ defined in (11.16) are defined for any $y_n = (v_0, \ldots, v_n) \in (\mathbb{Z}^d)^n$ and $v \in \mathbb{Z}^d$ by the formula

$$\widehat{M}_{n+1}(x_n, (y_n, v)) = 1_{x_n}(y_n) \, |\mathcal{C}(x_n)|^{-1} \sum_{e \in \mathcal{C}(x_n)} 1_{u_n + e}(v)$$

A Markov chain with these transitions, is sometimes called a myopic SAW or a "true" SAW (see [6]). Finally, check that the mutation stage in the particle model associated with the pair $(\widehat{G}_n, \widehat{M}_n)$ consists in extending each path, avoiding the occupied neighbors, and the potential function counts the proportion of occupied neighbors around the last visited site; i.e., $\widehat{G}_n(x_n) = |\mathcal{C}(x_n)|/4$.

Exercise 12.5.3: [Edwards' model [130]] Suppose U_n is the simple random walk on $E = \mathbb{Z}^d$, and the potential functions G_n are given by

$$G_n(u_0, \ldots, u_n) = \exp\left[-\frac{1}{\epsilon} \sum_{p=0}^{n-1} 1_{\{u_p\}}(u_n)\right]$$

Show that in this case the unnormalized Feynman-Kac distributions introduced in (12.32) are given by the formulae

$$\widehat{\gamma}_n(f_n) = \mathbb{E}\left(f_n(U_{[0,n]}) \exp\left\{-\frac{1}{\epsilon} \sum_{0 \leq p < q \leq n} 1_{U_p = U_q}\right\}\right)$$

These measures are sometimes called the weakly SAW or the Domb-Joyce model. There exist various conjectures related to this polymer model. For instance, the order of magnitude of distance $|U_n|$ between the endpoints of the polymer $X_n = U_{[0,n]}$ is not known in dimensions $d = 2,3,4$. For $d \geq 5$, Hara and Slade have proved in [173] that $|U_n| \simeq \sqrt{n}$, and for $d = 1$ Greven and den Hollander have checked in [169] that $|U_n| \simeq n$. Construct a genealogical tree model to generate approximate samples of the Edwards-Domb-Joyce model.

Exercise 12.5.4: [Lawler [214]] We consider the SAW model described on page 489. Using the fact that any $(p+q)$ SAW is a concatenation on p and q steps prove that $|\mathbf{S}_{p+q}| \leq |\mathbf{S}_p| \, |\mathbf{S}_q|$. Recalling that an SAW cannot return to the most recently visited site, check that $d^n \leq |\mathbf{S}_n| \leq 2d \, (2d-1)^{n-1}$. By subadditivity arguments, prove that the connective constant $c(d)$ defined by $c(d) = \lim_{n \to \infty} |\mathbf{S}_n|^{1/n}$ exists and $c(d) \in [d, (2d-1)]$. The exact values of $c(d)$ are unknown. Using (12.42) and (12.38), propose a particle estimation of these connective constants.

12.6 Filtering/Smoothing and Path estimation

12.6.1 Introduction

Feynman-Kac distributions and their particle approximation models play a major role in the theory of nonlinear filtering. We recall that the filtering problem consists in computing the conditional distributions of a state signal X given a sequence of observations Y. To understand the motivation behind this problem, we can think of the signal X as being the Markovian model for the time evolution of a target in tracking problems. The observation process Y represents the noisy and partial information delivered by some sensors such as RADAR (Radio Detection and Ranging) or SONAR (Sound Navigation and Ranging). Of course, the exact values of the signal X and the values of the various disturbance sources are not known but it is reasonable to assume that we know their statistical structure.

Filtering problems arise in various application areas, including applied probability, engineering science, and particularly in advanced signal processing, as well as in financial mathematics and biology. They provide a natural prediction/updating probabilistic model for the on-line estimation of some quantity evolving in some sensor environment. Each applied scientific discipline tends to use a different language to express and analyze the same filtering problems. For the convenience of the reader and to better connect these application areas, we have collected four different ways to introduce a nonlinear filtering problem.

In the first probabilistic interpretation, the signal/observation pair is regarded as a two-component Markov chain. In engineering literature, we instead start with a Markov signal process given by a dynamical physical equation, and the observation sequence is instead given by a sensor equation. Another abstract way to introduce the filtering problem consists in introducing a new reference probability measure. The last interpretation comes from the Bayesian literature.

A Markov chain filtering model

Let $(X, Y) = \{(X_n, Y_n) \; ; \; n \geq 0\}$ be a Markov chain taking values in some product spaces $\{(E_n \times F_n) \; ; \; n \geq 0\}$. Here $\{(F_n, \mathcal{F}_n) \; ; \; n \geq 0\}$ is an auxiliary sequence of measurable spaces. We further assume that the initial distribution ν_0 and the Markov transitions T_n of (X, Y) have the form

$$\nu_0(d(x_0, y_0)) = g_0(x_0, y_0) \, \eta_0(dx_0) \, q_0(dy_0) \qquad (12.43)$$
$$T_n((x_{n-1}, y_{n-1}), d(x_n, y_n)) = g_n(x_n, y_n) \, M_n(x_{n-1}, dx_n) \, q_n(dy_n) \qquad (12.44)$$

where, for each $n \in \mathbb{N}$, $g_n : E_n \times F_n \to (0, \infty)$ is a strictly positive function, $q_n \in \mathcal{P}(F_n)$, $\eta_0 \in \mathcal{P}(E_0)$ and M_n are Markov transitions from E_{n-1} into E_n.

Engineering presentation

In engineering and advanced signal processing literature, an alternative and more classical way to define the pair (signal/observation) Markov process (X, Y) is as follows. The signal X_n is a Markov chain with transition probability kernels M_n and taking values at each time n in some measurable space $(E_n, \mathcal{E}_n)$. In some instances, X_n is described by a dynamical equation

$$X_n = F_n(X_{n-1}, W_n) \qquad (12.45)$$

where W_n represents a sequence of independent random variables taking values in some measurable space $(S_n^w, \mathcal{S}_n^w)$ and $F_n : E_{n-1} \times S_n^w \to E_n$ is a given measurable drift function. In this case, we readily check that

$$M_n(f_n)(x_{n-1}) = \mathbb{E}(f_n(F_n(x_{n-1}, W_n)))$$

The observation process is defined for each $n \geq 0$ by a sensor equation

$$Y_n = H_n(X_n, V_n) \qquad (12.46)$$

The sequence V_n is independent of X and represents the noise sources. It consists of a collection of independent random variables taking values in some auxiliary measurable spaces $(S_n^v, \mathcal{S}_n^v)$. For each $n \geq 0$, the random variable V_n is distributed according to a probability measure $q_n \in \mathcal{P}(S_n^v)$. The collection of measurable functions $H_n : E_n \times S_n^v \to F_n$ is chosen so that

$$\mathbb{P}(H_n(x_n, V_n) \in dy_n) = g_n(x_n, y_n) \, q_n(dy_n) \qquad (12.47)$$

for each $x_n \in E_n$. In other words, the laws of $H_n(x_n, V_n)$ and V_n are absolutely continuous and $g_n(x_n, .)$ is the corresponding density.

A change of reference probability model

This technique is particularly useful in modeling continuous time nonlinear filtering problems. In the discrete time case, the idea is to consider the canonical process associated with the chain (X, Y) with initial distribution ν_0

$$(\Omega = \prod_{n \geq 0}(E_n \times F_n), \ \mathcal{G} = (\mathcal{G}_n)_{n \in \mathbb{N}}, \ (X, Y) = (X_n, Y_n)_{n \in \mathbb{N}}, \ \mathbb{P})$$

Let $\overline{\mathbb{P}}$ be the probability measure on $(\Omega, \mathcal{F})$ defined by its restrictions $\overline{\mathbb{P}}_n$ to $\Omega_n = \prod_{p=0}^{n}(E_p \times F_p)$

$$\overline{\mathbb{P}}_n(d((x_0, y_0), \ldots, d(x_n, y_n))) = \overline{\mathbb{P}}_n^X(d(x_0, \ldots, dx_n)) \ \overline{\mathbb{P}}_n^Y(d(y_0, \ldots, dy_n))$$

with

$$\begin{aligned}\overline{\mathbb{P}}_n^X(d(x_0, \ldots, dx_n)) &= \eta_0(dx_0) M_1(x_0, dx_1) \ldots M_n(x_{n-1}, dx_n) \\ \overline{\mathbb{P}}_n^Y(d(y_0, \ldots, dy_n)) &= q_0(dy_0) \ q_1(dy_1) \ldots q_n(dy_n)\end{aligned}$$

In other words, under this new reference measure $\overline{\mathbb{P}}$, X is again a Markov chain with initial distribution η_0, and $Y = (Y_n)_{n \geq 0}$ is a sequence of random variables independent of X and independent with respective distributions $q = (q_n)_{n \geq 0}$. Let $\mathbb{P}_n$ be the restriction of $\mathbb{P}$ to Ω_n. By the definition of the Markov kernel (12.44) of the chain (X_n, Y_n) under $\mathbb{P}$, the distributions $\mathbb{P}_n$ and $\overline{\mathbb{P}}_n$ are absolutely continuous with one another. Their Radon-Nikodym derivatives are defined for $\overline{\mathbb{P}}_n$-a.e. $((x_0, y_0), \ldots, (x_n, y_n)) \in \Omega_n$ by the formula

$$\frac{d\mathbb{P}_n}{d\overline{\mathbb{P}}_n}(((x_0, y_0), \ldots, (x_n, y_n))) = \prod_{p=0}^{n} g_p(x_p, y_p)$$

Using one of these interpretations, we find that

$$\mathbb{P}_n((X_0, \ldots, X_n) \in d(x_0, \ldots, x_n), (Y_0, \ldots, Y_n) \in d(y_0, \ldots, y_n))$$

$$= \left[\prod_{p=0}^{n} g_p(x_p, y_p)\right] \times [\eta_0(dx_0) M_1(x_0, dx_1) \ldots, M_n(x_{n-1}, dx_n)]$$

A Bayesian filtering presentation

In the Bayesian literature, the authors sometimes abuse the notation and adopt a simplified and intuitive presentation of a filtering problem. In this notation, the conditional distributions of Y_n given X_n are instead denoted by

$$\begin{aligned}p_n^{Y/X}(y_n | x_n) \ dy_n &= \mathbb{P}(Y_n \in dy_n \mid X_n = x_n) \\ &= \mathbb{P}(H_n(x_n, V_n) \in dy_n)\end{aligned}$$

500   12. Applications

The quantity dy_n has to be understood as a given probability measure on $(F_n, \mathcal{F}_n)$. To connect this notation with (12.47) we have

$$p_n^{Y/X}(y_n|x_n) \, dy_n = g_n(x_n, y_n) \, q_n(dy_n)$$

In other words, dy_n stands for $q_n(dy_n)$, and $p_n^{Y/X}$ represents the likelihood potential function

$$p_n^{Y/X}(y_n|x_n) = g_n(x_n, y_n)$$

In the same line of ideas, the elementary transitions of X_n are instead written in this field as

$$p_n^X(x_n|x_{n-1}) \, dx_n = \mathbb{P}(X_n \in dx_n \mid X_{n-1} = x_{n-1})$$

The quantity dx_n is more difficult to connect appropriately to our abstract Markov kernels. The notation above must to be thought of as

$$p_n^X(x_n|x_{n-1}) \, dx_n = M_n(x_{n-1}, dx_n) \quad \text{and} \quad p_0^X(x_0) \, dx_0 = \eta_0(dx_0)$$

Some authors also suppress the superscripts $(.)^{Y|X}$ and $(.)^X$ and the time index. In this simplified notation, we have

$$\mathbb{P}((X_0, \ldots, X_n) \in d(x_0, \ldots, x_n))$$
$$= p(x_0) \, p(x_1|x_0) \ldots p(x_n|x_{n-1}) \, dx_0 \ldots dx_n$$

and

$$\mathbb{P}((Y_0, \ldots, Y_n) \in d(y_0, \ldots, y_n) \mid (X_0, \ldots, X_n) = (x_0, \ldots, x_n))$$
$$= p(y_0|x_0) \, \ldots p(y_n|x_n) \, dy_0 \ldots dy_n$$

12.6.2   Motivating Examples

The literature on Bayesian statistics, sequential Monte Carlo methods, and other engineering sciences abounds with applications of particle algorithms to filtering problems. It is clearly out of the scope of this section to present a precise catalog on all of these applications. We rather refer the interested reader to the list of referenced articles. To illustrate this section and better connect the particle methodology developed in this book with the existing applied literature, we provide a brief discussion on some typical filtering problems currently studied in engineering literature (see another complementary series of examples provided in Section 12.6.2).

Positioning and Tracking Problems

One typical estimation problem arising in engineering literature is to estimate the dynamics of a moving object evolving in some sensor environment.

12.6 Filtering/Smoothing and Path estimation

For instance, in classical tracking problems, we estimate a target motion using radar or sonar observations. The physical measurements are often related to some signal arrival time delays or Doppler effects. In the context of global positioning systems (GPS), the electronic device delivers position estimates by measuring arrival times of a series of signals emitted by a satellite [43, 202]. In mobile robot localization problems [148, 111, 212, 282], the measurement data are collected from the robot's observations, such as its distance to a wall. In people tracking problems, we first need to design a simplified human body model. Then the observation process is as usual related to some image/audio sensors [184, 187, 188]. In navigational positioning problems, the ships are equipped with devices that measure their relative range with respect to some reference point [292].

To illustrate this rather general class of models, we present a simple positioning problem in wireless communication networks. This example is taken from [259]. The signal process is a simple Markov chain that represents the random evolution of a vehicle. The components of the state vector $X_n = (X_n^1, X_n^2, X_n^3)$ represent respectively the position, velocity, and acceleration coordinates. The location components depend on the network of streets and roads on which the vehicle travels. For the pair speed/acceleration components, we can use the physical model described in the introductory Section 1.1. The vehicle X_n evolves in a wireless radio environment. At each time, we receive radio measurements from several base stations on the position of the vehicle. Assuming that these stations are located at some fixed sites, say B^i, $i \in I$, a generic model of multisensor measurements is

$$Y_n^i = d(X_n^1, B^i) + V_n^i, \quad i \in I$$

where $d(.,.)$ represents some pseudo-distance criterion and V_n^i a collection of independent sensor perturbations. In these wireless network positioning problems, the vehicle process X_n often uses sharp turns, and its random dynamics are strongly nonlinear. As a result, this filtering problem is far from being linear/Gaussian, and an extended Kalman-Bucy filter often offers poor estimation results. Notice that this elementary model can be extended in various ways. For instance, we can consider moving base radio stations or multiple vehicle trackings or consider position tracking of microcell and mobile phones. The latter application area has recently received much interest. More details, as well as precise comparisons with the traditional Kalman filter approach, can be found in the referenced articles.

Multiple Models Estimation

Let X_n^1 be a Markov chain taking values in some measurable space $E^{(1)}$ (equipped with a σ-algebra $\mathcal{E}^{(1)}$) with initial distribution $\eta_0^{(1)}$ and elementary transitions $M_n^{(1)}$. Given this chain we suppose the pair process (X_n^2, Y_n)

is a given $\mathbb{R}^{p+q}$-valued Markov chain defined by the recursive relations

$$\begin{cases} X_n^2 = A_n(X_n^1)\, X_{n-1}^2 + a_n(X_n^1) + B_n(X_n^1)\, W_n\,, & n \geq 1 \\ Y_n = C_n(X_n^1)\, X_n^2 + c_n(X_n^1) + V_n\,, & n \geq 0 \end{cases} \quad (12.48)$$

for some measurable mappings (A_n, B_n, C_n) from $E^{(1)}$ into the sets of matrices and some drift functions (a_n, c_n) with appropriate dimensions. As traditionally, the sequences of random variables W_n and V_n are independent and independent of X_0^2 and X^1. They take values in $\mathbb{R}^{d_w}$ and $\mathbb{R}^{d_v}$ and are distributed according to a centered Gaussian distribution with covariance matrices

$$R_n^v = \mathbb{E}(V_n\, V_n') \quad \text{and} \quad R_n^w = \mathbb{E}(W_n\, W_n')$$

Given X^1, the initial random variable X_0^2 is a Gaussian random variable in $\mathbb{R}^p$ with a mean and covariance matrix that only depend on X_0^1 and are denoted by

$$\begin{aligned} \text{Mean}_0(X_0^1) &= \mathbb{E}(X_0^2|X_0^1) \\ \text{Cov}_0(X_0^1) &= \mathbb{E}((X_0^2 - \mathbb{E}(X_0^2|X_0^1))\,(X_0^2 - \mathbb{E}(X_0^2|X_0^1))'|X_0^1) \end{aligned}$$

These linear/Gaussian models arise in various application areas such as in multimodel estimation. In this context, the process X^1 represents the possible values of the system parameters as well as the different noise levels. For instance, in the space shuttle orbiter entry model proposed by Ewell in [142], when the acceleration enters below some level, the shuttle dynamics switch to some cruise navigation. Related switching models associated with judicious thresholds can be found in [236] and [292].

These multimodel filtering problems are often solved numerically by using a judicious hypothesis-testing technique on a collection of likely linear-optimal filters associated with each possible value of the system parameters. The only interaction occurs when we combine these models appropriately to obtain the output estimate. These rather well-known techniques go back to a pioneering article of Magill [238] on system identification and published in 1965; see also Bar-Shalom and Fortmann [24] for applications to missile-tracking models with different types of maneuvers. These ideas are also related to model-fusion strategies. In the latter, the multiple Kalman prediction models are regarded as measurements delivered by a virtual sensor.

These hypothesis-testing algorithms provide quite accurate results when we have a small number of likely hypotheses (see for instance [292] and references therein). In more general instances, the structure of the set of hypotheses is more complex and may also vary in time. As mentioned by Stengel in [292], page 405-406, one natural idea is to refine the filter adaptation by dropping the filters associated with less likely hypotheses and duplicating the ones associated with the most probable set of parameters.

The engineering literature on tracking maneuvering targets or on failure detections abounds on heuristic-like algorithms based on these evolutionary ideas. We refer the interested reader to the filter-spawning method presented by Fisher in [145] in the context of the VISTA F-16 actuator failure estimation or the switching algorithm [224] as well as the interacting multiple model algorithm (IMM) of Blom and Bar-Shalom [34]; see also [221, 222, 283, 284, 314] for precise application models. Related interesting schemes can be found in [58, 224, 225, 241]. In Section 12.6.7, we will show that the Feynman-Kac modeling and the particle methodology described in this book provide a natural and firm theoretical treatment on multiple fusion estimation models.

Stochastic Volatility Estimation

The extended Black-Scholes model describing the dynamics of the price of a given risky asset is defined by the stochastic equation

$$dY_t = Y_t(rdt + X_t dV_t) \qquad (12.49)$$

where r is an instantaneous interest rate and V_t is a standard Wiener motion. Assume that the observed volatility process X_t satisfies the equation

$$dX_t = -a\ (X_t - X_0)dt + bdW_t$$

for some fixed parameters $a, b \geq 0$ and a standard Wiener motion, independent of V_t. If we discretize the time using the Euler approximation with a fixed mesh Δ, then we obtain with some obvious abusive notation the discrete time filtering model

$$\begin{aligned} X_n &= X_{n-1} - a(X_{n-1} - X_0)\Delta + b\sqrt{\Delta}\ W_n \\ Y_n &= Y_{n-1}(1 + r\Delta + X_n\sqrt{\Delta}\ V_n) \end{aligned}$$

where V_n and W_n are independent sequences of iid standard Gaussian variables. Note that, using the explicit solution of (12.49), we can alternatively use the discrete observation model

$$Y_n = Y_{n-1} \exp\left([r - X_n^2/2]\Delta + X_n\sqrt{\Delta}\ V_n\right)$$

We notice that, using a classical state-space enlargement technique, we can include the parameters (a, b) in the state space. In a different but related context, Viens has adapted in [302] a general particle-filtering method of a joint work of the author with Jacod and Protter [90] in order to address the question of stochastic volatility filtering in financial math. He has used this method for solving a stochastic portfolio optimization problem under a partially observed stochastic volatility model, using elements of stochastic control, and providing a Monte Carlo method that solves the filtering and the stochastic control problem in unison. Stochastic volatility estimation has been proposed using particle methods for several years. The most popular method consists of invoking filtering by an ARCH/GARCH model, as proposed by Nelson in [255]. Related models and numerical methods can be found in the chain of articles [35, 37, 53, 131, 147, 155, 257].

Hidden Markov Models

Hidden Markov chains (HMM) are particular examples of filtering problems for which the signal/observation model has a fixed and deterministic component. We assume that the unknown component θ belongs to some measurable space $(S, \mathcal{S})$, and we associate with each θ the pair signal/observation model defined as in (12.44) by the formulae

$$\nu_{\theta,0}(d(x_0, y_0)) = g_{\theta,0}(x_0, y_0)\, \eta_{\theta,0}(dx_0)\, q_0(dy_0) \quad (12.50)$$
$$T_{\theta,n}((x_{n-1}, y_{n-1}), d(x_n, y_n)) = g_{\theta,n}(x_n, y_n)\, M_{\theta,n}(x_{n-1}, dx_n)\, q_n(dy_n) \quad (12.51)$$

In the display above, $g_{\theta,n}$, $\eta_{\theta,0}$ and $M_{\theta,n}$ are collections of positive functions, measures, and Markov transitions on appropriate state spaces (see page 498) and indexed by θ. The HMM problem is as follows. We observe a series of measurements Y_p, $p \leq n$, corresponding to some unknown value of the parameter say $\theta^\star$. These HMM and related stochastic autoregressive models occur in various application areas, including in speech recognition [190], biology [56], neurosciences [150], and economics [171, 172, 52].

The numerical estimation techniques fall into two categories, namely the maximum likelihood and the Bayesian estimators. These two approaches are discussed below.

In the Bayesian approach, we suppose that the unknown parameter $\theta^\star$ is a realization of some random variable θ with distribution $r \in \mathcal{P}(S)$. In this situation, if we take $\mathcal{X}_n = (X_n, \theta)$, then we see that the pair sequence $(\mathcal{X}_n, Y_n)$ is again a Markov chain of the same form as the one described in (12.44). These ideas can be extended in a natural way by considering the unknown parameter θ as a realization of the initial condition θ_0 of an auxiliary Markov chain θ_n. This Bayesian methodology proposes a way to reduce the HMM problem to a classical filtering problem.

The maximum likelihood estimators are defined as the sequence of parameters $\widehat{\theta}_n$ that maximize the conditional log-likelihood functions defined by

$$\Lambda_n(\theta, \theta^\star)$$

$$= \log \int_{E_{[0,n]}} \left\{ \prod_{p=0}^n g_{\theta,p}(x_p, Y_p) \right\} \eta_{\theta,0}(dx_0) \prod_{p=1}^n M_{\theta,p}(x_{p-1}, dx_p) \quad (12.52)$$

where $(Y_n)_{n \geq 0}$ represents a series of observations of the parameter $\theta^\star$.

12.6.3 Feynman-Kac Representations

To simplify the presentation, we fix the sequence of observations $Y = y$. A version of the conditional distributions of the signal states given their noisy observations is expressed in terms of Feynman-Kac formulae of the same type as the ones discussed above. More precisely, let G_n be the non-homogeneous function on E_n defined for any $x_n \in E_n$ by

$$G_n(x_n) = g_n(x_n, y_n) \qquad (12.53)$$

Note that G_n depends on the observation value y_n at time n. In this notation, the conditional distributions of the path $X_{[0,n]} =_{def.} (X_0, \ldots, X_n)$ given the sequence of observations $Y_{[0,n]} =_{def.} (Y_0, \ldots, Y_n)$ from the origin up to time n are given by the path Feynman-Kac measures

$$\widehat{\mathbb{Q}}_n(d(x_0, \ldots, x_n))$$
$$= \mathbb{P}_n(X_{[0,n]} \in d(x_0, \ldots, x_n) \mid Y_{[0,n]} = (y_0, \ldots, y_n))$$
$$= \frac{1}{\widehat{\mathcal{Z}}_n} \left[\prod_{p=0}^{n} G_p(x_p) \right] \times [\eta_0(dx_0) M_1(x_0, dx_1) \ldots, M_n(x_{n-1}, dx_n)]$$

with the normalizing constants

$$\widehat{\mathcal{Z}}_n = \int_{E_{[0,n]}} \left[\prod_{p=0}^{n} G_p(x_p) \right] \times [\eta_0(dx_0) M_1(x_0, dx_1) \ldots, M_n(x_{n-1}, dx_n)] \qquad (12.54)$$

The prediction and updated marginal distributions are defined for any test function $f_n \in \mathcal{B}_b(E_n)$ by

$$\eta_n(f_n) = \gamma_n(f_n)/\gamma_n(1) \quad \text{and} \quad \widehat{\eta}_n(f_n) = \widehat{\gamma}_n(f_n)/\widehat{\gamma}_n(1)$$

with the unnormalized distributions

$$\gamma_n(f_n) = \mathbb{E}\left(f_n(X_n) \prod_{p=0}^{n-1} G_p(X_p) \right) \quad \text{and} \quad \widehat{\gamma}_n(f_n) = \gamma_n(G_n f_n)$$

Due to the choice of potential functions (12.53), the distributions η_n and $\widehat{\eta}_n$ coincide respectively with the one-step predictor and the optimal filter

$$\eta_n = \text{Law}(X_n \mid Y_{[0,n-1]} = (y_0, \ldots, y_{n-1}))$$
$$\widehat{\eta}_n(f) = \eta_n(fG_n)/\eta_n(G_n) = \text{Law}(X_n \mid Y_{[0,n]} = (y_0, \ldots, y_n))$$

with $Y_{[0,n]} = (Y_0, \ldots, Y_n)$. Notice that the normalizing constants $\widehat{\mathcal{Z}}_n$ introduced in (12.54) coincide with the quantities $\widehat{\gamma}_n(1) = \gamma_n(G_n)$ and they can be expressed in terms of the prediction flow η_n with the product formula

$$\widehat{\mathcal{Z}}_n = \widehat{\gamma}_n(1) = \prod_{p=0}^{n} \eta_p(G_p) \qquad (12.55)$$

Taking the logarithm, we obtain the so-called conditional log-likelihood functions

$$\Lambda_n = \frac{1}{n+1} \log \widehat{\mathcal{Z}}_n = \frac{1}{n+1} \sum_{p=0}^{n} \log \eta_p(G_p) \qquad (12.56)$$

It is also interesting to examine the situation where X is a path-space model; namely, suppose we have that

$$X_n = X'_{[0,n]} =_{\text{def.}} (X'_0, \ldots, X'_n) \in E_n = E'_{[0,n]} =_{\text{def.}} (E'_0 \times \cdots \times E'_n)$$

where X'_n is a Markov chain taking values in some measurable spaces $(E'_n, \mathcal{E}'_n)$ with initial distribution η_0 and transitions M'_n. In this situation, the observation sequence (12.46) takes the form

$$Y_n = H_n(X'_{[0,n]}), V_n)$$

This means that the information delivered by sensors at each time n depends on the whole path of the signal X' back from the origin and up to time n. Note that in this case the function $g_n((x'_0, \ldots, x'_n), y_n)$ depends on the current observation $Y_n = y_n$ and on the whole path-coordinates $(x'_0, \ldots, x'_n)$. This type of sensor is in fact much more general than those arising in practice. In classical filtering problems, the observation sequence is instead defined by

$$Y_n = H'_n(X'_n, V_n)$$

for some appropriate function $H'_n : E'_n \times S_n \to F_n$ and the resulting function $g_n(., y_n)$ only depends on the endpoint coordinate x'_n of the path $(x'_0, \ldots, x'_n)$ that is

$$g_n((x'_0, \ldots, x'_n), y_n) = g'_n(x'_n, y_n)$$

for some strictly positive function $g'_n : E'_n \to]0, \infty[$. We emphasize that in this particular situation the pair process (X'_n, Y_n) has the same form as before. It is a Markov chain taking values in the measurable spaces $(E'_n \times F_n)$. The initial distribution and the Markov transitions of (X', Y) are defined as in (12.43) and (12.44) by replacing (g_n, M_n) by (g'_n, M'_n). From these observations, one concludes that

$$\begin{aligned}\widehat{\eta}'_n &= \text{Law}(X'_n \mid Y_{[0,n]} = (y_0, \ldots, y_n)) \\ \widehat{\eta}_n &= \text{Law}(X'_{[0,n]} \mid Y_{[0,n]} = (y_0, \ldots, y_n))\end{aligned} \qquad (12.57)$$

12.6 Filtering/Smoothing and Path estimation

In connection with the engineering presentation of a filtering problem given in (12.45) and (12.46), we observe that the random sequence

$$W_{[0,n]} =_{\text{def.}} (X_0, W_1, \ldots, W_n) \in E_0 \times S_1^w \times \ldots \times S_n^w$$

forms a Markov chain and versions of the conditional distributions

$$\widetilde{\eta}_n = \text{Law}(W_{[0,n]} \mid Y_{[0,n]} = (y_0, \ldots, y_n))$$

are also given by the Feynman-Kac path measures defined by

$$\widetilde{\eta}_n(f_n) = \widetilde{\gamma}_n(f_n)/\widetilde{\gamma}_n(1) \quad \text{with} \quad \widetilde{\gamma}_n(f_n) = \mathbb{E}\left(f_n(W_{[0,n]}) \prod_{p=0}^{n} G_p(X_n)\right)$$

The functional representations of the conditional distributions presented above clearly belong to the same class of Feynman-Kac distribution flow models discussed in this book. In filtering literature, the nonlinear evolution equations of these models are usually called the nonlinear filtering equations. The two major problems concern the study of the stability properties and the long time behavior of these equations and then their numerical solution. The first question is related to the fact that the initial condition of the signal is usually unknown and any filter, even the optimal Kalman-Bucy filter, in the linear/Gaussian situation is initialized using erroneous parameters. The second question is more recurrent in applied literature; namely, how to solve the filtering equation. Except in some very particular situations, the optimal filter equation is a nonlinear equation in an infinite-dimensional state space, and it is known that there does not exist any finite realization (see for instance [50]). We emphasize that the two questions above are intimately related. For instance, the stability properties of the filtering equations ensure that local numerical errors do not propagate. We recall that these robustness properties allow us to derive several uniform convergence estimates with respect to the time parameter (see Section 7.4.3).

The stochastic analysis and the particle methodology described in this book give some partial answers to both of these problems. For instance, the stability properties of the filtering equations can be derived using the Feynman-Kac contraction properties discussed in Chapter 4, Section 4.3. On the other hand the numerical solution of these measure-valued processes can be conducted using the particle methodology developed in Chapter 3. To avoid unnecessary repetition, it is of course out of the scope of this section to review all the consequences of these results. Because of their importance in practice and for the convenience of the reader, we provide in the next two sections two rather short discussions on these two problems with some precise references to chapters and sections on these subjects.

12.6.4 Stability Properties of the Filtering Equations

The long time behavior of the filtering equation can be studied using the contraction properties of the nonlinear Feynman-Kac semigroups derived in Chapter 4. This chapter presents several functional contraction inequalities for Feynman-Kac semigroups with respect to various entropy-like criteria. In filtering settings, these semigroups have natural interpretations in terms of conditional distributions. To be more precise, it is first convenient to introduce a simplified system of notation. For any $k, l \geq 0$, we set $Y_k^{k+l} = (Y_k, \ldots, Y_{k+l})$. We also slightly abuse the notation and, for any $n \leq k$ and $l \geq 0$, we write

$$p(y_k^{k+l} \mid X_n = x_n)$$
$$= \int_{E_{n+1} \times \ldots \times E_{k+l}} \left\{ \prod_{p=k}^{k+l} g_p(x_p, y_p) \right\} \prod_{p=n+1}^{k+l} M_p(x_{p-1}, dx_p) \qquad (12.58)$$

the conditional density of the random vector Y_k^{k+l} given $X_n = x_n$ (with respect to the $(l+1)$-tensor distribution $(q_k \otimes \ldots \otimes q_{k+l})$) and evaluated at y_k^{k+l}. This abusive and complex system of notation is currently used in Bayesian statistics as well as in engineering literature. Notice that whenever the observation sequence is fixed, the quantities (12.58) only depend on the parameter $x_n \in E_n$. Recalling that $G_n(x_n) = g_n(x_n, y_n)$, the densities (12.58) are better expressed in terms of the pair of Feynman-Kac semigroups $(Q_{p,n}, \widehat{Q}_{p,n})$ introduced in Section 2.7 and defined by

$$Q_{p,n}(f_n)(x_p) = \mathbb{E}_{p,x_p} \left(f(X_n) \prod_{k=p}^{n-1} G_k(X_k) \right)$$

$$\widehat{Q}_{p,n}(f_n)(x_p) = \mathbb{E}_{p,x_p} \left(f(X_n) \prod_{k=p+1}^{n} G_k(X_k) \right)$$

Indeed, an elementary manipulation yields that

$$G_{p,n}(x_p) =_{\text{def.}} Q_{p,n}(1)(x_p) = p(y_p^{n-1} \mid X_p = x_p)$$
$$\widehat{G}_{p,n}(x_p) =_{\text{def.}} \widehat{Q}_{p,n}(1)(x_p) = p(y_{p+1}^{n} \mid X_p = x_p)$$

From previous observations, it is also not difficult to check that the normalized Feynman-Kac semigroups $(P_{p,n}, \widehat{P}_{p,n})$ associated with $(Q_{p,n}, \widehat{Q}_{p,n})$ have the following interpretation:

$$P_{p,n}(x_p, dx_n) = \mathbb{P}(X_n \in dx_n \mid Y_{p+1}^{n-1} = y_{p+1}^{n-1}, X_p = x_p)$$
$$\widehat{P}_{p,n}(x_p, dx_n) = \mathbb{P}(X_n \in dx_n \mid Y_{p+1}^{n} = y_{p+1}^{n}, X_p = x_p)$$

To get to the final step in our discussion, we recall that the nonlinear semigroups $(\widehat{\Phi}_{p,n}, \Phi_{p,n})$ of the conditional distribution flows $(\widehat{\eta}_n, \eta_n)$ are expressed in terms of the Markov transitions $(P_{p,n}, \widehat{P}_{p,n})$ and the potential functions $(G_{p,n}, \widehat{G}_{p,n})$ with the formulae

$$\Phi_{p,n}(\mu) = \Psi_{p,n}(\mu) P_{p,n} \quad \text{and} \quad \widehat{\Phi}_{p,n}(\mu) = \widehat{\Psi}_{p,n}(\mu) \widehat{P}_{p,n} \qquad (12.59)$$

where $(\Psi_{p,n}, \widehat{\Psi}_{p,n})$ are the Boltzmann-Gibbs transformations associated with the pair $(G_{p,n}, \widehat{G}_{p,n})$ and defined by

$$\Psi_{p,n}(\mu)(f_n) = \mu(f_n G_{p,n})/\mu(G_{p,n}) \quad \text{and} \quad \widehat{\Psi}_{p,n}(\mu)(f_n) = \mu(f_n \widehat{G}_{p,n})/\mu(\widehat{G}_{p,n})$$

We finally recall that these conditional distributions can be regarded as the transitions of a nonhomogeneous Markov chain. This observation combined with the Boltzmann-Gibbs representations (12.59) of the semigroups $(\widehat{\Phi}_{p,n}, \Phi_{p,n})$ is one of the key points of our approach to the stability of Feynman-Kac semigroups developed in Section 4.3. More precisely, for any $p \leq q \leq n$, we have the decompositions

$$P_{p,n} = R_{p,q}^{(n)} P_{q,n} \quad \text{and} \quad \widehat{P}_{p,n} = \widehat{R}_{p,q}^{(n)} \widehat{P}_{q,n}$$

with the Markov transitions

$$R_{p,q}^{(n)}(x_p, dx_q) = \mathbb{P}(X_q \in dx_q \mid Y_{p+1}^{n-1} = y_{p+1}^{n-1}, X_p = x_p)$$
$$\widehat{R}_{p,q}^{(n)}(x_p, dx_q) = \mathbb{P}(X_q \in dx_q \mid Y_{p+1}^{n} = y_{p+1}^{n}, X_p = x_p)$$

To give a flavor of the stability properties that can be deduced from Chapter 4, let us assume that the signal transitions M_n satisfy the mixing condition $(M)_m$ for $m = 1$ and some sequence of numbers $\varepsilon_n(M)$ with $\varepsilon = \inf_n \varepsilon_n(M) > 0$ (see on page 116). Note that this condition ensures that the Markov transitions T_n of the pair Markov chain (X_n, Y_n) given in (12.44) have the same mixing property; that is,

$$T_n((x_{n-1}, y_{n-1}), d(x_n, y_n)) \geq \varepsilon\, T_n((x'_{n-1}, y'_{n-1}), d(x'_n, y'_n))$$

Rephrasing Proposition 4.4.2, we find that

$$\beta(\widehat{R}_{p,p+1}^{(n)}) \leq (1 - \varepsilon^2) \quad \text{and} \quad \widehat{G}_{p,n}(x_p) \geq \varepsilon\, \widehat{G}_{p,n}(x'_p)$$

for any pair $(x_n, x'_n) \in E_n^2$. We conclude that the semigroup $\widehat{\Phi}_{p,n}$ of the optimal filter is exponentially asymptotically stable in the sense that

$$\begin{aligned}
\beta(\widehat{P}_{p,p+n}) &= \sup_{\mu_1, \mu_2} \|\widehat{\Phi}_{p,p+n}(\mu_1) - \widehat{\Phi}_{p,p+n}(\mu_2)\|_{\text{tv}} \\
&= \sup_{x_p, x'_p} \|\mathbb{P}(X_{p+n} \in \cdot \mid Y_{p+1}^{p+n} = y_{p+1}^{p+n}, X_p = x_p) \\
&\qquad - \mathbb{P}(X_{p+n} \in \cdot \mid Y_{p+1}^{p+n} = y_{p+1}^{p+n}, X_p = x'_p)\|_{\text{tv}} \\
&\leq (1 - \varepsilon^2)^n
\end{aligned}$$

If we take $(\mu_1, \mu_2) = (\eta_0 M_1 \ldots M_p, \widehat{\eta}_p)$, then we find that

$$\widehat{\Phi}_{p,p+n}(\mu_1) = \mathbb{P}(X_{p+n} \in \cdot \mid Y_{p+1}^{p+n} = y_{p+1}^{p+n})$$
$$\widehat{\Phi}_{p,p+n}(\mu_2) = \mathbb{P}(X_{p+n} \in \cdot \mid Y_0^{p+n} = y_0^{p+n})$$

From previous inequalities, we readily deduce the uniform estimate for the approximation and finite m-memory filters

$$\|\mathbb{P}(X_{p+m} \in \cdot \mid Y_{p+1}^{p+m} = y_{p+1}^{p+m}) - \mathbb{P}(X_{p+m} \in \cdot \mid Y_0^{p+m} = y_0^{p+m})\|_{\mathrm{tv}} \leq (1-\varepsilon^2)^m$$

12.6.5 Asymptotic Properties of Log-likelihood Functions

The uniform and strong exponential stability estimates provided in Section 12.6.4 also appear to be useful in the asymptotic analysis of the conditional log-likelihood functions introduced in (12.52). Suppose the pair of signals/observations (X_n, Y_n) forms a time-homogeneous Markov chain with initial distribution ν_0 and elementary transitions $T_{\theta,n} = T_\theta$ given in (12.50) and (12.51). From the product formula (12.55), we first find that

$$\Lambda_n(\theta, \theta^\star) = \frac{1}{n+1} \sum_{p=0}^{n} \log \eta_{\theta, Y_0^{p-1}, p}(G_{\theta, Y_p})$$
$$= \frac{1}{n+1} \log \eta_{\theta, 0}(G_{\theta, Y_0}) + \frac{1}{n+1} \sum_{p=0}^{n-1} \log \widehat{\eta}_{\theta, Y_0^p, p} M_\theta(G_{\theta, Y_{p+1}})$$

where Y_0^n represents a series of observations or the parameter $\theta^\star$ and with some obvious and usual abusive notation

$$\widehat{\eta}_{\theta, y_0^n, n}(dx_n) = \mathbb{P}_\theta(X_n \in dx_n \mid Y_0^n = y_0^n)$$
$$\eta_{\theta, y_0^{n-1}, n}(dx_n) = \mathbb{P}_\theta(X_n \in dx_n \mid Y_0^{n-1} = y_0^{n-1})$$
$$G_{\theta, y_n}(x_n) = g_\theta(x_n, y_n)$$

for any realization sequence $y_0^n \in (F_0 \times \ldots \times F_n)$. With some obvious abusive notation, we have

$$p_\theta(y_{n+1} \mid y_0^n)$$

$$= \widehat{\eta}_{\theta, y_0^n, n} M_\theta(G_{\theta, y_{n+1}})$$

$$= \int_{E_{n+1}} \mathbb{P}_\theta(X_n \in dx_n \mid Y_0^n = y_0^n) \, M_\theta(x_n, dx_{n+1}) \, g_\theta(x_{n+1}, y_{n+1})$$

Suppose that the Markov transitions M_θ satisfy the mixing condition $(M)_m$ for $m = 1$ and for some sequence of numbers $\varepsilon_\theta(M)$ such that

$$\varepsilon = \inf_\theta \varepsilon_\theta(M) > 0$$

Under this rather strong uniform mixing condition, all the estimates derived in Section 12.6.4 remain valid. For instance, we have

$$\|\mathbb{P}_\theta(X_{p+m} \in \cdot \mid Y_{p+1}^{p+m} = y_{p+1}^{p+m}) - \mathbb{P}_\theta(X_{p+m} \in \cdot \mid Y_0^{p+m} = y_0^{p+m})\|_{\mathrm{tv}}$$
$$\leq (1-\varepsilon^2)^m$$
(12.60)

Let $\widehat{\eta}'_{\theta,Y_0^n,n}$ be the solution of the filtering equation starting at some erroneous initial condition η'_0, and let $\Lambda'_n(\theta,\theta^\star)$ be the corresponding log-likelihood function. Arguing as in Section 12.4 and assuming for simplicity that $\sup_\theta \|g_\theta\| < \infty$, we conclude that

$$|\log \widehat{\eta}_{\theta,y_0^n,n} M_\theta(G_{\theta,y_{n+1}}) - \log \widehat{\eta}'_{\theta,y_0^n,n} M_\theta(G_{\theta,y_{n+1}})| \leq 2\varepsilon^{-1}\,(1-\varepsilon^2)^n$$

and therefore $n\|\Lambda'_n - \Lambda_n\| \leq c(\varepsilon)$, $\mathbb{P}_{\theta^\star}$-a.s., for some constant whose values only depend on ε. This shows that the conditional log-likelihood function does not depend asymptotically on the initial condition of the filter. In much the same way, if we denote by $\Lambda_n^{(m)}$ the log-likelihood function associated with the finite m-memory filter, then by (12.60) we readily prove that

$$\|\Lambda_n^{(m)} - \Lambda_n\| \leq 2\varepsilon^{-1}(1-\varepsilon^2)^m\,(1-m/n)^+$$

with $a^+ = \max(a,0)$. We fix the memory length m and we denote by $(U_n)_{n \geq m-1}$ the $(E \times F)^{m+1}$-valued Markov process

$$U_n = ((X_{n-m+1}, Y_{n-m+1}), \ldots, (X_{n+1}, Y_{n+1}))$$

Under our assumptions, we have for any $u_n, u'_n \in (E \times F)^{m+1}$

$$\mathcal{T}_\theta^{m+1}(u_n, \cdot) \geq \varepsilon^{m+1}\,\mathcal{T}_\theta^{m+1}(u'_n, \cdot)$$

where $\mathcal{T}_\theta$ represents the Markov transition of U_n. This shows that U_n is exponentially asymptotically stable and it has a unique invariant measure ν_θ. We readily deduce the almost sure convergence

$$\lim_{n \to \infty} \Lambda_n^{(m)}(\theta,\theta^\star) = \Lambda^{(m)}(\theta,\theta^\star) =_{\mathrm{def.}} \mathbb{E}_{\nu_{\theta^\star}}(\log \widehat{\eta}_{\theta,m-1,Y_0^{m-1}}^{(m)} M_\theta(G_{\theta,Y_m}))$$

This result can be proven using for instance the Poisson equation and classical martingale convergence theorems. Recalling that

$$\widehat{\eta}_{\theta,m-1,y_0^{m-1}}^{(m)} M_\theta(G_{\theta,y_m}) = p_\theta(y_m \mid y_0^{m-1})$$

the limit criterion $\Lambda^{(m)}$ is often rewritten as follows

$$\Lambda^{(m)}(\theta,\theta^\star) = \mathbb{E}_{\nu_{\theta^\star}}(\log p_\theta(y_m \mid y_0^{m-1}))$$

Using related arguments, we can prove that $\Lambda_n(\theta,\theta^\star)$ converge almost surely with respect to the law of the stationary process (X_n,Y_n) and as $n \to \infty$ to some deterministic function $\Lambda(\theta,\theta^\star)$. These results combined with some appropriate regularity conditions on the function $\theta \to (g_\theta, M_\theta)$ imply that $\Lambda(\theta,\theta^\star) \leq \Lambda(\theta^\star,\theta^\star)$ with the equality if and only if $\theta = \theta^\star$ (see for instance [121] and references therein).

12.6.6 Particle Approximation Measures

We first recall that the flows η_n and $\widehat{\eta}_n$ are solutions of nonlinear equations with various McKean interpretations (see Section 11.3 and Section 2.5.3). Each McKean interpretation is attached to a different evolutionary particle approximation model (see Chapter 3). To give a numerical sound to this section, we roughly describe the evolution of the simple genetic approximation model. In the latter and between two observations, the particles evolve as independent copies of the signal. When an observation arrives, we select randomly better-fitted individuals with respect to their likelihoods. This simple algorithm can be refined in various ways using the particle recipes presented in Chapter 11. For instance, we can change the sampling distribution and use conditional excursion type mutations to refine the precision of the stochastic grid (see for instance Section 11.7). Notice that in this situation the pair potentials/transitions $(\widehat{G}_n, \widehat{M}_n)$ introduced in (11.16) have the form

$$\widehat{\eta}_0 = \mathbb{P}(X_0 \in dx_0 \mid Y_0 = y_0)$$
$$\widehat{G}_n(x_{n-1}) = p(y_n \mid x_{n-1})$$
$$\widehat{M}_n(x_{n-1}, dx_n) = \mathbb{P}(X_n \in dx_n \mid X_{n-1} = x_{n-1}, Y_n = y_n)$$

More generally, the triplets $(\widehat{\eta}_0^{(p)}, \widehat{G}_n^{(p)}, \widehat{M}_n^{(p)})$ corresponding to the extended excursion model and defined in (11.20) take the form

$$\widehat{\eta}_0^{(p)} = \mathbb{P}(X_{[0,p_1)} \in dx_{[0,p_1)} \mid Y_{[0,p_1)} = y_{[0,p_1)})$$
$$\widehat{G}_{n-1}^{(p)}(x_{[p_{n-1},p_n)}) = p(y_{[p_n,p_{n+1})} \mid X_{p_n-1} = x_{p_n-1})$$

and

$$\widehat{M}_n^{(p)}(x_{[p_{n-1},p_n)}, dx_{[p_n,p_{n+1})})$$
$$= \mathbb{P}(X_{[p_n,p_{n+1})} \in dx_{[p_n,p_{n+1})} \mid X_{p_n-1} = x_{p_n-1}, Y_{[p_n,p_{n+1})} = y_{[p_n,p_{n+1})})$$

The accuracy and the computational cost of the selection stage can also be improved using one of the branching rules proposed in Section 11.8.

Each branching particle model has a natural birth and death interpretation that induces the important notions of the ancestral line of each current individual and the corresponding genealogical trees. A review of these path-space models is provided in Section 11.4. The occupation measures of these path-particle historical processes provide a natural approximation of the laws (12.57) of the path of a signal given a series of observations. More precise models can be found in Section 3.4 and their asymptotic analysis is described in full detail in Chapters 7, 8, 9, and 10. We also mention that, mimicking the product formula (12.55), we construct an on-line and unbiased particle estimation of the normalizing constants $\widehat{\mathcal{Z}}_n$ introduced in (12.54). We again refer to Section 11.3 for some details on these particle approximation models.

12.6.7 A Partially Linear/Gaussian Filtering Model

Quenched and Annealed Feynman-Kac models

In this section, we examine the nonlinear filtering model with a linear/Gaussian component described in (12.48) on page 502. To simplify the presentation we restrict the presentation to homogeneous measurable mappings $(A_n, B_n, C_n) = (A, B, C)$ and null drift functions $(a_n, c_n) = 0$. The extension to the general case is straighforward (see Section 2.5.4). We use the modeling techniques presented in Section 2.6 to introduce a quenched Kalman-Bucy equation. In this context, the quenched flow is Gaussian and can be solved explicitly for any realization of the randomness. The annealed distributions are difficult to solve in practice. In the filtering context, they represent the conditional distributions of the "nonlinear part" of the signal given the observations. We connect the annealed and quenched models in terms of the Feynman-Kac model in distribution space introduced in Section 2.6. Let us denote by $\mathcal{N}(m, P)$ a Gaussian distribution on $\mathbb{R}^p$ with mean vector $m \in \mathbb{R}^p$ and covariance matrix $P \in \mathbb{R}^{p \times p}$

$$\mathcal{N}(m, P)(dx) = \frac{1}{(2\pi)^{p/2} \sqrt{|P|}} \exp\left[-\frac{1}{2}(x-m) P^{-1} (x-m)'\right] dx$$

By direct inspection, we see that the pair signal $X_n = (X_n^1, X_n^2) \in E_n = E^{(1)} \times \mathbb{R}^p$ forms a Markov chain. Its Markov transitions take the form

$$M_n((x_{n-1}, z_{n-1}), d(x_n, z_n)) = M_n^{(1)}(x_{n-1}, dx_n) \, M_{x_n, n}^{(2)}(z_{n-1}, dz_n)$$

with the Gaussian transition on $\mathbb{R}^p$

$$M_{x_n, n}^{(2)}(z_{n-1}, dz_n) = \mathbb{P}((A(x_n) \, z_{n-1} + B(x_n) \, W_n) \in dz_n)$$

Similarly, the initial distribution η_0 of the pair (X_0^1, X_0^2) is given by

$$\eta_0(d(x_0, z_0)) = \eta_0^{(1)}(dx_0) \, \eta_{x_0, 0}^{(2)}(dz_0)$$

with $\eta_{x_0,0}^{(2)} = \mathcal{N}(\mathrm{Mean}_0(x_0), \mathrm{Cov}_0(x_0))$. This pair signal model is clearly of the same form as the one discussed in Section 2.6. More precisely, the distribution of the path $(X_0^1, \ldots, X_n^1)$ is defined by

$$\mathbb{P}_{\eta_0^{(1)}, n}^{(1)}(d(x_0, \ldots, x_n)) = \eta_0^{(1)}(dx_0) \, M_1^{(1)}(x_0, dx_1) \ldots M_n^{(1)}(x_{n-1}, dx_n)$$

and, given a realization $X^1 = x = (x_n)_{n \geq 0} \in (E^{(1)})^{\mathbb{N}}$, the second component $X^2 = (X_n^2)_{n \geq 0}$ forms an $\mathbb{R}^p$-valued Markov chain and the conditional distribution of the path $X_{[0,n]}^2 =_{\text{def.}} (X_0^2, \ldots, X_n^2)$ is defined by the formula

$$\mathbb{P}_{[x], n}^{(2)}(d(z_0, \ldots, z_n)) = \eta_{x_0, 0}^{(2)}(dz_0) \, M_{x_1, 1}^{(2)}(z_0, dz_1) \ldots M_{x_n, n}^{(2)}(z_{n-1}, dz_n)$$

514   12. Applications

From this expression, we notice that $\mathbb{P}^{(2)}_{[x],n}$ only depends on $(x_0, \ldots, x_n)$.

Let $G_n : E^{(1)} \times \mathbb{R}^p \to (0, \infty)$ be the likelihood functions defined by

$$G_n(x_n, z_n) = g_n((x_n, z_n), y_n) = \frac{d\mathcal{N}(C(x_n)\,z_n, R^v_n)}{d\mathcal{N}(0, R^v_n)}(y_n)$$

From the considerations above we find that

$$\mathbb{P}_{\eta_0, n}(X_{[0,n]} \in d((x_0, z_0), \ldots, (x_n, z_n)) \,|\, Y_{[0,n]} = (y_0, \ldots, = y_n))$$

$$= \widehat{\mathcal{Z}}_n^{-1} \left\{ \prod_{p=0}^n G_p(x_p, z_p) \right\} \; \mathbb{P}^{(1)}_{\eta_0^{(1)}, n}(d(x_0, \ldots, x_n)) \; \mathbb{P}^{(2)}_{[x],n}(d(z_0, \ldots, z_n))$$

with $X_{[0,n]} =_{\mathrm{def.}} (X_0, \ldots, X_n)$, $Y_{[0,n]} =_{\mathrm{def.}} (Y_0, \ldots, Y_n)$, and the normalizing constant $\widehat{\mathcal{Z}}_n > 0$. If we write $X^i_{[0,n]} =_{\mathrm{def.}} (X^i_0, \ldots, X^i_n)$ for $i = 1, 2$ then we find that

$$\mathbb{P}_{\eta_0, n}(X^2_{[0,n]} \in d(z_0, \ldots, z_n) \,|\, X^1_{[0,n]} = (x_0, \ldots, x_n),\, Y_{[0,n]} = (y_0, \ldots, y_n))$$

$$= \widehat{\mathcal{Z}}_{[x],n}^{-1} \left\{ \prod_{p=0}^n G_p(x_p, z_p) \right\} \; \mathbb{P}^{(2)}_{[x],n}(d(z_0, \ldots, z_n))$$

with the normalizing constant $\widehat{\mathcal{Z}}_{[x],n} > 0$. The marginal distributions are defined for any $f \in \mathcal{B}_b(\mathbb{R}^p)$ by the Feynman-Kac formulae

$$\eta^{(2)}_{[x],n}(f) = \gamma^{(2)}_{[x],n}(f)/\gamma^{(2)}_{[x],n}(1)$$

with

$$\gamma^{(2)}_{[x],n}(f) = \mathbb{E}^{(2)}_{[x,n]}\left(f(X^2_n) \prod_{p=0}^{n-1} G_{x_p, p}(X^2_p) \right) \qquad (12.61)$$

with the "random" potential functions

$$G_{x_n, n} : z \in \mathbb{R}^p \longrightarrow G_{x_n, n}(z) = G_n(x_n, z) \in (0, \infty)$$

It is also convenient to consider their updated versions

$$\widehat{\eta}^{(2)}_{[x],n}(f) = \widehat{\gamma}^{(2)}_{[x],n}(f)/\widehat{\gamma}^{(2)}_{[x],n}(1) \quad \text{with} \quad \widehat{\gamma}^{(2)}_{[x],n}(f) = \gamma^{(2)}_{[x],n}(f G_{x_n, n}) \qquad (12.62)$$

The annealed marginal distributions on $E^{(1)}$ are defined for any $f_n \in \mathcal{B}_b(E^{(1)})$ by the Feynman-Kac formula

$$\eta^{(1)}_n(f) = \gamma^{(1)}_n(f)/\gamma^{(1)}_n(1) \quad \text{with} \quad \gamma^{(1)}_n(f) = \mathbb{E}_{\eta_0}\left(f(X^1_n) \prod_{p=0}^{n-1} G_p(X_p) \right)$$

$$\widehat{\eta}^{(1)}_n(f) = \widehat{\gamma}^{(1)}_n(f)/\widehat{\gamma}^{(1)}_n(1) \quad \text{with} \quad \widehat{\gamma}^{(1)}_n(f) = \mathbb{E}_{\eta_0}\left(f(X^1_n) \prod_{p=0}^{n} G_p(X_p) \right)$$

12.6 Filtering/Smoothing and Path estimation

In our context, these Feynman-Kac flows represent the one-step predictors and the optimal filters

$$\eta^{(2)}_{[x],n} = \text{Law}(X^2_n \mid Y_{[0,n-1]} = (y_0,\ldots,y_{n-1}),\ X^1_{[0,n]} = (x_0,\ldots,x_n))$$
$$\widehat{\eta}^{(2)}_{[x],n} = \text{Law}(X^2_n \mid Y_{[0,n]} = (y_0,\ldots,y_n),\ X^1_{[0,n]} = (x_0,\ldots,x_n))$$

and

$$\eta^{(1)}_n = \text{Law}(X^1_n \mid Y_{[0,n-1]} = (y_0,\ldots,y_{n-1}))$$
$$\widehat{\eta}^{(1)}_n = \text{Law}(X^1_n \mid Y_{[0,n]} = (y_0,\ldots,y_n))$$

Quenched Kalman-Bucy Filters

The quenched marginal distributions can be solved using the traditional Kalman-Bucy filter (see Section 2.5.4). More precisely, for any realization of the chain $X^1 = x$ and for any sequence of observations $Y = y$, the one-step predictor and the optimal filter are Gaussian distributions

$$\eta^{(2)}_{[x],n} = \mathcal{N}(\widehat{X}^{(2)\,-}_{(x,y),n}, P^-_{x,n}) \quad \text{and} \quad \widehat{\eta}^{(2)}_{[x],n} = \mathcal{N}(\widehat{X}^{(2)}_{(x,y),n}, P_{x,n})$$

As traditionally we slightly abuse the notation and suppress the dependence on the observation sequence. In this notation, we write $\widehat{X}^{(2)\,-}_{x,n}$ and $\widehat{X}^{(2)}_{x,n}$ instead of $\widehat{X}^{(2)\,-}_{(x,y),n}$ and $\widehat{X}^{(2)}_{(x,y),n}$. The synthesis of the conditional mean and covariance matrices is carried out using the traditional Kalman-Bucy recursive equations (see Section 2.5.4). For $n = 0$, we recall that the initial conditions of the latter are given by

$$\widehat{X}^{(2)\,-}_{x,0} = \text{Mean}_0(x_0) \quad \text{and} \quad P^-_{x,0} = \text{Cov}_0(x_0)$$

The filter equation is decomposed into the traditional two step updating/prediction transitions

$$(\widehat{X}^{(2)\,-}_{x,n}, P^-_{x,n}) \xrightarrow{\text{Updating}} (\widehat{X}^{(2)}_{x,n}, P_{x,n}) \xrightarrow{\text{Prediction}} (\widehat{X}^{(2)\,-}_{x,n+1}, P^-_{x,n+1})$$

These two mechanisms are defined as follows

- Updating: This transition depends on the current observation $Y_n = y_n$ and it is defined by the relations

$$\begin{cases} \widehat{X}^{(2)}_{x,n} = \widehat{X}^{(2)\,-}_{x,n} + \mathbf{G}_{x,n}\,(y_n - C(x_n)\,\widehat{X}^{(2)\,-}_{x,n}) \\ P_{x,n} = (I - \mathbf{G}_{x,n}\,C(x_n))\,P^-_{x,n} \end{cases}$$

with the gain matrix

$$\mathbf{G}_{x,n} = P^-_{x,n}\,C(x_n)'\,[C(x_n)\,P^-_{x,n}\,C(x_n)' + R^v_n]^{-1}$$

516   12. Applications

- Prediction: This transition does not depend on the observation and it is given by the simple relations

$$\begin{cases} \widehat{X}^{(2)\,-}_{x,n+1} &= A(x_{n+1})\,\widehat{X}^{(2)}_{x,n} \\ P^{-}_{x,n+1} &= A(x_{n+1})\,P_{x,n}\,A(x_{n+1})' + B(x_{n+1})\,R^{w}_{n+1}\,B(x_{n+1})' \end{cases}$$

A Feynman-Kac Model in Distribution Space

In our context the Feynman-Kac model in distribution space presented in Section 2.6.2 is defined in terms of the Markov chain

$$X'_n = (X^1_n, \eta^{(2)}_{[X^1],n}) \in E' = (E^{(1)} \times \mathcal{P}(\mathbb{R}^p))$$

From previous considerations, the second component is a random Gaussian distribution

$$\eta^{(2)}_{[x],n} = \mathcal{N}(\widehat{X}^{(2)\,-}_{x,n}, P^{-}_{x,n})$$

It corresponds to the one step predictor associated with a realization of the chain X^1. Its evolution in time is given by the Kalman-Bucy equation, which can be written in terms of a measure-valued process

$$\eta^{(2)}_{[X^1],n} = \Phi^{(2)}_n((X^1_{n-1}, X^1_n), \eta^{(2)}_{[X^1],n-1})$$

with initial condition

$$\eta^{(2)}_{[X^1],0} = \mathcal{N}(\widehat{X}^{(2)\,-}_{x,0}, P^{-}_{x,0}) = \mathcal{N}(\text{Mean}_0(X^1_0), \text{Cov}_0(X^1_0))$$

The nonlinear nature of the filtering problem leads to a collection of mappings

$$\Phi^{(2)}_{n+1}((u,v), .)\,, \quad u,v \in E^{(1)}$$

that preserve the subset $\text{Gauss}(\mathbb{R}^p) \subset \mathcal{P}(\mathbb{R}^p)$ of Gaussian distributions on $\mathbb{R}^p$. From Kalman-Bucy recursions, we find that

$$\Phi^{(2)}_{n+1}((u,v), \mathcal{N}(m,P)) = \mathcal{N}(\text{Mean}_{n+1}((u,v),(m,P)),\ \text{Cov}_{n+1}(u,v)) \quad (12.63)$$

The quantities $\text{Mean}_{n+1}((u,v),(m,P))$ and $\text{Cov}_{n+1}((u,v),(m,P))$ are computed using the following updating/prediction rules.

$$\begin{cases} \text{Mean}_{n+1}((u,v),(m,P)) &= A(v)\,\widehat{m}(u) \\ \text{Cov}_{n+1}((u,v),(m,P)) &= A(v)\,\widehat{P}(u)\,A(v)' + B(v)\,R^{w}_{n+1}\,B(v)' \end{cases}$$

with the updated pair $(\widehat{m}(u), \widehat{P}(u))$ defined by

$$\begin{cases} \widehat{m}(u) &= m + \mathbf{G}_n(u)\,(y_n - C(u)\,m) \\ \widehat{P}(u) &= (I - \mathbf{G}_n(u)\,C(u))\,P \end{cases}$$

with the gain matrix $\mathbf{G}_n(u) = P\, C(u)'\, [C(u)\, P\, C(u)' + R_n^v]^{-1}$. For more details we refer the reader to Section 2.5.4.

The Markov chain X_n' has transitions defined for any $f' \in \mathcal{B}_b(E')$ and $(u, \eta) \in E'$ by

$$M_n'(f_n')(u, \eta) = \int_{E_n^{(1)}} M_n^{(1)}(u, dv)\, f_n'(v, \Phi_n^{(2)}((u, v), \eta))$$

We also see that the elementary transition of the distribution component $\eta_{[X^1],n}^{(2)}$ is deterministic given the first one X_n^1. This can be summarized by the synthetic formula

$$(X_{n-1}^1, \eta_{[X^1],n-1}^{(2)}) \longrightarrow (X_n^1, \eta_{[X^1],n}^{(2)}) = (X_n^1, \Phi_n^{(2)}((X_{n-1}^1, X_n^1), \eta_{[X^1],n-1}^{(2)}))$$

We consider the annealed potential functions

$$G_n' : (x, \mu) \in E' \longrightarrow G_n'(x, \mu) = \int_{\mathbb{R}^p} \mu(dz)\, G_n(x, z) \in (0, \infty) \quad (12.64)$$

Since we have

$$\begin{aligned} G_n'(u, \mathcal{N}(m, P)) &= \mathcal{N}(m, P)(G_{u,n}) \\ &= \frac{d\mathcal{N}(C(u)\, m, C(u)\, P\, C(u)' + R_n^v)}{d\mathcal{N}(0, R_n^v)}(y_n) \end{aligned} \quad (12.65)$$

we conclude that

$$G_n'(x_n, \eta_{[x],n}^{(2)}) = \frac{d\mathcal{N}(C(x_n)\, \widehat{X}_{x,n}^{(2)-}, C(x_n) P_{x,n}^- C(x_n)' + R_n^v)}{d\mathcal{N}(0, R_n^v)}(y_n)$$

We finally associate with the pair (X_n', G_n') the distribution flows $(\eta_n', \widehat{\eta}_n')$ on E' defined for any $f' \in \mathcal{B}_b(E')$ by

$$\eta_n'(f') = \gamma_n'(f')/\gamma_n'(1) \quad \text{and} \quad \widehat{\eta}_n'(f') = \widehat{\gamma}_n'(f')/\widehat{\gamma}_n'(1) \quad (12.66)$$

with

$$\gamma_n'(f') = \mathbb{E}_{\eta_0}\left(f'(X_n') \prod_{p=0}^{n-1} G_p'(X_p')\right) \quad \text{and} \quad \widehat{\gamma}_n'(f') = \gamma_n'(f' G_n')$$

By Proposition 2.6.4, if we choose $f'(x, \eta) = f(x)$ for some $f \in \mathcal{B}_b(E^{(1)})$, then we find that

$$\begin{aligned} \eta_n'(f') &= \eta_n^{(1)}(f) = \mathbb{E}_{\eta_0}(f(X_n^1)\, |Y_{[0,n-1]} = (y_0, \ldots, y_{n-1})) \\ \widehat{\eta}_n'(f') &= \widehat{\eta}_n^{(1)}(f) = \mathbb{E}_{\eta_0}(f(X_n^1)\, |Y_{[0,n]} = (y_0, \ldots, y_n)) \end{aligned}$$

In the same way for any $f \in \mathcal{B}_b(\mathbb{R}^p)$ we find that

$$\eta_n'(f') = \mathbb{E}_{\eta_0}(f(X_n^2)\, |\, Y_{[0,n-1]} = (y_0, \ldots, y_{n-1}))$$

as soon as $f'(x, \eta) = \eta(f)$, and

$$\eta'_n(f') = \mathbb{E}_{\eta_0}(f(X_n^2) \mid Y_{[0,n]} = (y_0, \ldots, y_n))$$

as soon as $f'(x, \eta) = \eta(G_{x,n}f)/\eta(G_{x,n})$

Much more is true. If we consider the signal/observation filtering model

$$\begin{cases} X'_n &= (X_n^1, \eta^{(2)}_{[X^1],n}) \\ Y_n &= C(X_n^1) \, \widehat{X}^{(2)-}_{X^1,n} + \widehat{V}_{X^1,n} \end{cases} \quad (12.67)$$

with the quenched innovation sequence

$$\widehat{V}_{X^1,n} = Y_n - \mathbb{E}(Y_n \mid Y_{[0,n-1]}, X^1_{[0,n]}) = Y_n - C(X_n^1)\widehat{X}^{(2)-}_{X^1,n}$$

then we find that

$$\begin{aligned} \eta'_n &= \operatorname{Law}(X_n^1, \eta^{(2)}_{[X^1],n} \mid Y_{[0,n-1]} = (y_0, \ldots, y_{n-1})) \\ \widehat{\eta}'_n &= \operatorname{Law}(X_n^1, \eta^{(2)}_{[X^1],n} \mid Y_{[0,n]} = (y_0, \ldots, y_n)) \end{aligned}$$

Speaking somewhat loosely in this interpretation we see that the potential function

$$G'_n(x_n, \eta^{(2)}_{[x],n})$$

$$\propto \exp\left[-\tfrac{1}{2}(y_n - C(x_n) \, \widehat{X}^{(2)-}_{x,n}) R_n^v(x_n)^{-1}(y_n - C(x_n) \, \widehat{X}^{(2)-}_{x,n})'\right]$$

with $R_n^v(x_n) = (C(x_n)P_{x,n}^- C(x_n)' + R_n^v)$, represents the probability that the observation $Y_n = y_n$ would be made given $Y_{[0,n-1]}$ and the value $X_n^1 = x_n$. The observation model in (12.67) is sometimes called a "virtual sensor" in the literature on multimodel estimation. For static models $X_n^1 = X_0^1$ taking values in a finite set, the filtering equations associated with (12.67) coincide with the so-called multiple hypothesis testing algorithm (MHT).

Interacting Kalman-Bucy Filters

In this section, we briefly discuss the simple genetic model associated with the Feynman-Kac distribution flow η'_n. By construction, we first notice that the algorithm consists here of N (state, measure)-valued particles

$$\xi_n^i = (\zeta_n^i, \mu_n^i) \quad \text{and} \quad \widehat{\xi}_n^i = (\widehat{\zeta}_n^i, \widehat{\mu}_n^i) \in E' = E^{(1)} \times \operatorname{Gauss}(\mathbb{R}^p)$$

The initial configuration $\xi_0^i = (\zeta_0^i, \mu_0^i)$ is defined by N independent random variables ζ_0^i with common distributions $\eta_0^{(1)}$. The N measure components are simply given by $\mu_0^i = \mathcal{N}(m_0^i, P_0^i)$ with initial mean and covariance matrix

$$m_0^i = \operatorname{Mean}_0(\zeta_0^i) \quad \text{and} \quad P_0^i = \operatorname{Cov}_0(\zeta_0^i)$$

12.6 Filtering/Smoothing and Path estimation

The selection transition consists in randomly choosing N particles $\widehat{\xi}_n^i = (\widehat{\zeta}_n^i, \widehat{\mu}_n^i)$ with common law

$$\sum_{i=1}^{N} \frac{G_n'(\zeta_n^i, \mu_n^i)}{\sum_{j=1}^{N} G_n'(\zeta_n^j, \mu_n^j)} \delta_{(\zeta_n^i, \mu_n^i)}$$

If we set $\mu_n^i = \mathcal{N}(m_n^i, P_n^i) \in \text{Gauss}(\mathbb{R}^p)$, then by (12.65) the weights are given at each step by the Radon-Nikodym derivatives

$$G_n'(\zeta_n^i, \mu_n^i) = \mu_n^i(G_{\zeta_n^i, n}) = \frac{d\mathcal{N}(C(\zeta_n^i) m_n^i, C(\zeta_n^i) P_n^i C(\zeta_n^i)' + R_n^v)}{d\mathcal{N}(0, R_n^v)}(y_n)$$

During the mutation transition, the evolution of the selected (path, measure) particles

$$\widehat{\xi}_n^i = (\widehat{\zeta}_n^i, \widehat{\mu}_n^i) \longrightarrow \xi_{n+1}^i = (\zeta_{n+1}^i, \mu_{n+1}^i)$$

is defined as follows. First, each selected particle $\widehat{\zeta}_n^i$ evolves according to the transition $M_{n+1}^{(1)}$ so that ζ_{n+1}^i are conditionally independent random variables with respective distributions $M_{n+1}^{(1)}(\widehat{\zeta}_n^i, .)$. Then, given the selected states $\widehat{\zeta}_n^i$ and ζ_{n+1}^i, the measure component μ_{n+1}^i is defined by the deterministic transition

$$\widehat{\mu}_n^i \to \mu_{n+1}^i = \Phi_{n+1}^{(2)}((\widehat{\zeta}_n^i, \zeta_{n+1}^i), \widehat{\mu}_n^i)$$

From (12.63), we find that

$$\widehat{\mu}_n^i = \mathcal{N}(\widehat{m}_n^i, \widehat{P}_n^i) \in \text{Gauss}(\mathbb{R}^p) \Rightarrow \mu_{n+1}^i = \mathcal{N}(m_{n+1}^i, P_{n+1}^i) \in \text{Gauss}(\mathbb{R}^p)$$

with

$$\begin{aligned} m_{n+1}^i &= \text{Mean}_{n+1}((\widehat{\zeta}_n^i, \zeta_{n+1}^i), (\widehat{m}_n^i, \widehat{P}_n^i)) \\ P_{n+1}^i &= \text{Cov}_{n+1}((\widehat{\zeta}_n^i, \zeta_{n+1}^i), (\widehat{m}_n^i, \widehat{P}_n^i)) \end{aligned}$$

Let $\eta_n'^N$ be the particle density profiles associated with this N-interacting Kalman-Bucy filter

$$\eta_n'^N = \frac{1}{N} \sum_{i=1}^{N} \delta_{(\zeta_n^i, \mu_n^i)}$$

The asymptotic behavior of these empirical measures is discussed from Chapter 7 to Chapter 10. To illustrate the impact of these results, here we give next a simple $\mathbb{L}_p$ mean error estimate presented in Section 7.4.

Proposition 12.6.1 *For each $p \geq 1$, and any $f \in \mathcal{B}_b(E^{(1)} \times \mathcal{P}(\mathbb{R}^p))$, we have for each $N \geq 1$*

$$\sqrt{N} \, \mathbb{E}(|\eta_n'^N(f) - \eta_n'(f)|^p)^{\frac{1}{p}} \leq a(p) b(n) \, \|f\|$$

Note that the potential functions G'_n do not satisfy the regularity condition (G) stated on page 115. Nevertheless, we can prove Proposition 12.6.1 combining the arguments developed in Section 7.4 for general non negative potential functions with some traditional cut-off arguments.

As traditionally, if instead of the Markov chain $X'_n = (X^1_n, \eta^{(2)}_{X^1,n})$ we consider the Markov chain in path space

$$\mathcal{X}'_n = (X^1_{[0,n]}, \eta^{(2)}_{X^1,n}) \quad \text{with} \quad X^1_{[0,n]} = (X^1_0, \ldots, X^1_n)$$

then the empirical measures associated with the resulting N genealogical-tree-based algorithm

$$\eta'^N_n = \frac{1}{N} \sum_{i=1}^{N} \delta_{((\zeta^i_{0,n}, \ldots, \zeta^i_{n,n}), \mu^i_n)}$$

converge as $N \to \infty$ to the distributions in path space

$$\eta'_n = \mathrm{Law}(X^1_{[0,n]}, \eta^{(2)}_{[X^1],n} \mid Y_{[0,n-1]})$$

and the same $\mathbb{L}_p$-estimates hold.

12.6.8 Exercises

Exercise 12.6.2: Let $V = (V_n)_{n \geq 1}$ be a sequence of independent random variables such that $\mathbb{E}(V_n) = 0$ and $\sigma_n = \mathbb{E}(V_n^2)^{1/2} < \infty$. We consider a sequence of observations $Y_n = X + V_n$ of a single random variable X that we assumed to be independent of V and such that $\sigma = \mathbb{E}(X^2)^{1/2} < \infty$.

- If $\widehat{X}^n = \frac{1}{n} \sum_{p=1}^{n} Y_n$, then check that $\mathbb{E}((X - \widehat{X}^n)^2) = n^{-2} \sum_{p=1}^{n} \sigma_n^2$ and conclude that

$$\lim_{n \to \infty} n^{-1} \sum_{p=1}^{n} \sigma_n^2 < \infty \implies \lim_{n \to \infty} \mathbb{E}(X \mid Y_{[0,n]}) = X \quad \text{in } \mathbb{L}^2(\mathbb{P})$$

- We further assume that X and V_n are Gaussian random variables. and we set $\widehat{X}_n = \mathbb{E}(X \mid Y_{[0,n]})$. Using for instance the Kalman recursions provided on page 515 check that

$$\widehat{X}_{n+1} = \widehat{X}_n + g_{n+1}(Y_{n+1} - \widehat{X}_n)$$

with the gain term $g_{n+1} = q_n/(q_n + \sigma^2_{n+1})$, where $q_n = \mathbb{E}((X - \widehat{X}_n)^2)$. Using the fact that $q_n = (1 - g_n)q_{n-1}$, show that

$$q_{n+1}/q_n = 1 - g_{n+1} = q_n^{-1}/(q_n^{-1} + \sigma^{-2}_{n+1}) \quad \text{and} \quad q^{-1}_{n+1} = q^{-1}_n + \sigma^{-2}_{n+1}$$

Conclude that $\widehat{X}_{n+1} = (q_{n+1}/q_n) \widehat{X}_n + (1 - q_{n+1}/q_n) Y_{n+1}$ and

$$\lim_{n \to \infty} \mathbb{E}((X - \widehat{X}_n)^2) = 0 \iff \sum_{n \geq 1} \sigma_n^{-2} = \infty$$

Exercise 12.6.3: This exercise is taken from [121]. We extend the HMM model presented in Section 12.6.5 to the time index $\mathbb{Z}$. We fix some parameter $\theta \in S$ and an observation sequence $y = (y_n)_{n \in \mathbb{Z}}$. For any $p, q, n \in \mathbb{Z}$ with $p \leq q \leq n$, we slightly abuse the notation and we let

$$\widehat{\eta}^{(x)}_{\theta, y^n_{p+1}, n} = \widehat{P}_{p,n}(x, .) = \mathbb{P}_\alpha(X_n \in . \mid Y^n_{p+1} = y^n_{p+1}, X_p = x)$$

be the solution of the filtering equation starting at $\widehat{\eta}_{\theta, p} = \delta_x$, with $x \in E$, at time $p \in \mathbb{Z}$. In the same abusive notation, check that

$$\widehat{\eta}^{(x)}_{\theta, Y^n_{p+1}, n} M_\theta(G_{\theta, y_{n+1}}) = \widehat{P}_{p,n} M_\theta(G_{\theta, y_{n+1}})(x) = p_\theta(y_{n+1} \mid y^n_{p+1}, X_p = x)$$

- Derive a Feynman-Kac representation of $\widehat{P}_{p,n}(x_p, .)$ and prove that

$$\widehat{P}_{p,n} = \widehat{R}^{(n)}_{p,q} \widehat{P}_{q,n}$$
$$\widehat{R}^{(n)}_{p,q}(x, dx') = \mathbb{P}_\alpha(X_q \in dx' \mid Y^n_{p+1} = y^n_{p+1}, X_p = x)$$

- Conclude that for any $(x, x') \in E^2$ and $(p, p') \in \mathbb{Z}^2$ we have the uniform estimates

$$\|\widehat{P}_{p,n}(x, .) - \widehat{P}_{p',n}(x', .)\|_{\text{tv}} \leq \beta(\widehat{P}_{p \vee p', n}) \leq (1 - \varepsilon^2)^{n - (p \vee p')}$$

and

$$|\log \widehat{\eta}^{(x)}_{\theta, y^n_{p+1}, n} M_\theta(G_{\theta, y_{n+1}}) - \log \widehat{\eta}^{(x')}_{\theta, y^n_{p'+1}, n} M_\theta(G_{\theta, y_{n+1}})|$$

$$\leq 2\varepsilon^{-1}(1 - \varepsilon^2)^{n - (p \vee p')}$$

- Let $\overline{\mathbb{P}}_\theta$ be the distribution of the stationary Markov chain $(X_n, Y_n)_{n \in \mathbb{Z}}$ with time index $\mathbb{Z}$. Deduce that $(\log \widehat{\eta}^{(x)}_{\theta, Y^n_{-m+1}, n} M_\theta(G_{\theta, Y_{n+1}}))_{m \geq 0}$ is a uniform Cauchy sequence that converges $\overline{\mathbb{P}}_{\theta^\star}$-a.s. to some $\Lambda_{\infty, n}(\theta, \theta^\star) \in \mathbb{L}_1(\overline{\mathbb{P}}_{\theta^\star})$ whose values do not depend on x.

- For $p' = 0$ and $p = -m$, $m \in \mathbb{N}$, show that

$$\|\widehat{P}_{0,n}(x, .) - \widehat{P}_{-m,n}(x', .)\|_{\text{tv}} \leq (1 - \varepsilon^2)^n$$

and

$$|\log \widehat{\eta}^{(x')}_{\theta, y^n_1, n} M_\theta(G_{\theta, y_{n+1}}) - \log \widehat{\eta}^{(x)}_{\theta, y^n_{-m+1}, n} M_\theta(G_{\theta, y_{n+1}}) -| \leq 2\varepsilon^{-1}(1 - \varepsilon^2)^n$$

Deduce the uniform $\overline{\mathbb{P}}_{\theta^\star}$-a.s. estimate

$$|\log \widehat{\eta}^{(x')}_{\theta, Y^n_1, n} M_\theta(G_{\theta, Y_{n+1}}) - \Lambda_{\infty, n}(\theta, \theta^\star)| \leq 2\varepsilon^{-1}(1 - \varepsilon^2)^n$$

Since $(\Lambda_{\infty, n}(\theta, \theta^\star))_{n \geq 0}$ forms a $\overline{\mathbb{P}}_{\theta^\star}$-stationary sequence prove that

$$\lim_{n \to \infty} \Lambda_n(\theta, \theta^\star) = \overline{\mathbb{E}}_\theta(\Lambda_{\infty, 0}(\theta, \theta^\star)) \quad (\overline{\mathbb{P}}_{\theta^\star}\text{-a.e. and in } \mathbb{L}_1(\overline{\mathbb{P}}_{\theta^\star}))$$

Exercise 12.6.4: Let $(X^{i,N})_{1\leq i\leq N}$ be an exchangeable sequence of random variables, taking values in some measurable space $(E, \mathcal{E})$. Also let $d\mu \propto e^{-V} d\lambda$ be the Boltzmann-Gibbs measure associated to a reference measure λ, and to a nonnegative potential function with $\lambda(e^{-V}) > 0$. Suppose we have, for any $q \leq N$, the following propagation-of-chaos estimate

$$\|\text{Law}(X^{1,N}, \ldots, X^{q,N}) - \mu^{\otimes q}\|_{\text{tv}} \leq \epsilon_N(q)$$

where $\lim_{N\to\infty} \epsilon_N(q(N)) = 0$, for some $\lim_{N\to\infty} q(N) = \infty$.

- Let V_λ be the λ-essential infimum of V. Prove that for any $\delta > 0$, we have

$$\mathbb{P}\left(\wedge_{1\leq i\leq q(N)} V(X^{i,N}) > V_\lambda + \delta\right) \leq \epsilon_N(q(N)) + (1-\mu(V\leq V_\lambda+\delta))^{q(N)}$$

- We consider the elementary 1-dimensional filtering model defined by the pair Markov chain

$$\begin{cases} X_n = a_n(X_{n-1}) + W_n, & X_0 = W_0 \\ Y_n = b_n(X_n) + V_n \end{cases}$$

where W_n, V_n are iid Gaussian variables with common distribution $\mathcal{N}(0, 2/\beta)$, with $\beta > 0$. Check that a version of the conditional distribution of $W_{[0,n]} = (W_0, \ldots, W_n)$ given $Y_{[0,n]} = (Y_0, \ldots, Y_n) = y_{[0,n]}$ is given by the formula

$$d\mu_n = d\mathbb{P}\left(W_{[0,n]} \in \cdot \mid Y_{[0,n]} = y_{[0,n]}\right) \propto e^{-\beta V_{n,y_{[0,n]}}} d\lambda_n \quad (12.68)$$

where λ_n stands for the Lebesgue measure on $\mathbb{R}^{n+1}$, and the potential function $V_{n,y_{[0,n]}}$ is defined by

$$V_{n,y_{[0,n]}}(w_{[0,n]}) = \sum_{p=0}^{n} w_p^2 + \sum_{p=0}^{n}(y_p - b_p(x_p^w))^2$$

In the above display, x_n^w represents the solution of the controlled system $x_n^w = a_n(x_{n-1}^w) + w_n$, starting at $x_0 = w_0$. Let $U_n^i = (W_{p,n}^i)_{0\leq p\leq n}$, $1 \leq i \leq N$, be the genealogical tree model associated to the Feynman-Kac distribution (12.68). Using Theorem 8.3.3, check that

$$\|\text{Law}(U_n^1, \ldots, U_n^q) - \mu_n^{\otimes q}\|_{\text{tv}} \leq \frac{q^2}{N} b(n)$$

Prove that for any $q(N) = o(1/\sqrt{N})$ we have the convergence in probability

$$\lim_{N\to\infty} \wedge_{1\leq i\leq q(N)} V_{n,y_{[0,n]}}(U_n^i) = \inf V_{n,y_{[0,n]}}$$

References

[1] A. de Acosta. On large deviations of empirical measures in the τ-topology. *J. Appl. Probab.*, 31A:41–47, 1994. Studies in applied probability.

[2] A. de Acosta. Projective systems in large deviation theory. II. Some applications. In *Probability in Banach spaces, 9 (Sandjberg, 1993)*, volume 35 of *Progr. Probab.*, pages 241–250. Birkhäuser, Boston, 1994.

[3] A. de Acosta. Exponential tightness and projective systems in large deviation theory. In *Festschrift for Lucien Le Cam*, pages 143–156. Springer, New York, 1997.

[4] D. Aldous and U. Vazirani, Go With the Winners Algorithms. In *Proc. 35th Symp. Foundations of Computer Sci.*, pages 492-501, 1994.

[5] J.-M. Alliot, D. Delahaye, J.-L. Farges, and M. Schoenauer. Genetic algorithms for automatic regrouping of air traffic control sectors. In J.R. McDonnell, R.G. Reynolds, and D.B. Fogel, editors, *Proceedings of the 4th Annual Conference on Evolutionary Programming*, pages 657–672. MIT Press, Cambridge, 1995.

[6] D.J. Amit, G. Parisi, and L. Peliti. Asymptotic behavior of the "true" self avoiding walk. *Phys. Rev. B*, 27:1635–1645, 1983.

[7] C. Andrieu and A. Doucet. Optimal estimation of amplitude and phase modulated signals. *Monte Carlo Methods Appl.*, 7(1–2):1–14, 2001.

[8] C. Andrieu, A. Doucet, and W.J. Fitzgerald. An introduction to Monte Carlo methods for Bayesian data analysis. In *Nonlinear Dynamics and Statistics (Cambridge, 1998)*, pages 169–217. Birkhäuser, Boston, 2001.

[9] C. Andrieu, A. Doucet, W.J. Fitzgerald, and J.-M. Pérez. Bayesian computational approaches to model selection. In *Nonlinear and Nonstationary Signal Processing (Cambridge, 1998)*, pages 1–41. Cambridge Univiversity Press, Cambridge, 2000.

[10] C. Andrieu, A. Doucet, and E. Punskaya. Sequential Monte Carlo methods for optimal filtering. In *Sequential Monte Carlo Methods in Practice*, Statistics for Engineering and Information Science. Sci., pages 79–95. Springer, New York, 2001.

[11] S. Asmussen and R.Y. Rubinstein. Steady state rare events simulations in queueing models and its complexity properties. *Advances in queueing, Probab. Stochastics Ser.*, pages 429–461, CRC, Boca Raton, FL, 1995.

[12] R. Assaraf, M. Caffarel et A. Khelif, Diffusion Monte Carlo methods with a fixed number of walkers, *Phys. Rev. E*, vol. 61, no. 4, pp. 4566-4575, 2000.

[13] R. Atar, F. Viens, and O. Zeitouni. Robustness of zakai's equation via Feynman-Kac representations. In Q. Zhang. W.M. McEneaney, G. Yin, editors, *Stochastic Analysis, Control, Optimization and Applications*, pages 339–352. Birkhauser, Boston, 1999.

[14] K. B. Athreya and P. Jagers, editors. *Classical and Modern Branching Processes*, volume 84 of The IMA Volumes in Mathematics and Its Applications. Papers from the IMA Workshop held at the University of Minnesota, Minneapolis, MN, June 13–17, 1994, Springer-Verlag, New York, 1997..

[15] R. Azencott. Grandes déviations et applications. In P.L. Hennequin, editor, *'Ecole d'ÉtÉ de Saint Flour VIII*, Lecture Notes in Mathematics 774, pages 1–176. Springer-Verlag, Berlin, 1980.

[16] B. Azimi-Sadaji and P.S. Krishnaprasad. Approximate nonlinear filtering and its applications for gps. *Proceedings of 39th IEEE Conference on Decision and Control*, 1579-84, Sydney, Australia, Dec. 2000.

[17] B. Azimi-Sadaji and P.S. Krishnaprasad. Change detection for non linear systems, a particle filtering approach. *Proceedings of 2002 American Control Conference*, ACC2002.

[18] D.A. Bader, J.J., and R. Chellappa. Scalable data parallel algorithms for texture synthesis and compression using Gibbs random fields. Technical Report CS-TR-3123 and UMIACS-TR-93-80, UMIACS and Electrical Engineering, University of Maryland, College Park, MD, 1993.

[19] R.R. Bahadur and R. Ranga Rao. On deviations of the sample mean. *Ann. Math. Stat.*, 31:1015–1027, 1960.

[20] R.R. Bahadur and S.L. Zabell. Large deviations of the sample mean in general vector spaces. *Ann. Probab.*, 7:587–621, 1979.

[21] J. Baker. Adaptive selection methods for genetic algorithms. In J. Grefenstette, editor, *Proceedings of the International Conference on Genetic Algorithms and Their Applications*. L. Erlbaum Associates, Hillsdale, NJ, 1985.

[22] J. Baker. Reducing bias and inefficiency in the selection algorithm. In J. Grefenstette, editor, *Proceedings of the Second International Conference on Genetic Algorithms and Their Applications*. L. Erlbaum Associates, Hillsdale, NJ, 1987.

[23] A. Bakirtzis, S. Kazarlis, and V. Petridis. A genetic algorithm solution to the economic dispatch problem. *IEE Proceedings-C*. Vol. 141, No. 4, pp. 377-382, July 1994.

[24] Y. Bar-Shalom and T.E. Fortmann. *Tracking and Data Associations*. Academic Press, New York, 1988.

[25] P. Barbe and P. Bertail. *The Weighted Bootstrap*. Lecture Notes in Statistics 98. Springer-Verlag, Berlin, 1995.

[26] U. Bastolla, H. Frauenkron, E. Gerstner, W. Nadler and P. Grassberger. Testing a new Monte Carlo algorithm for protein folding, *Proteins: Structure, Function and Genetics* 32, 52-66 (1998).

[27] N. Bellomo and M. Pulvirenti. Generalized kinetic models in applied sciences. In *Modeling in Applied Sciences*. Modeling and Simulation in Science, Engineering, and Technology, 1–19. Birkhäuser, Boston, 2000.

[28] N. Bellomo and M. Pulvirenti, editors. *Modeling in Applied Sciences*. Modeling and Simulation in Science, Engineering, and Technology. Birkhäuser, Boston, 2000.

[29] B. Berge, I.D. Chueshov, and P.A. Vuillermot. Solutions to certain parabolic SPDE's driven by Wiener processes. *Stochastic Process. Appl.*, 92:237–263, 2001.

[30] A. Berreti and A.D. Sokal. *J. Stat. Phys.*, 40(485), 1985.

[31] L. Bertini and G. Giacomin. On the long-time behavior of the stochastic heat equation. *Probab. Theory Related Fields*, 114(3):279–289, 1999.

[32] G. Birkhoff. Positivity and criticality *PSAM*, vol. 11, 111–126, (1957).

[33] R. Bleck, C. Rooth, D. Hu, and L.T. Smith. Salinity-driven thermohaline transients in a wind and thermohaline forced isopycnic coordinate model of the north atlantic. *J. of Phys. Oceanogr.*, 22:1486–1515, 1992.

[34] H.A.P. Blom and Y. Bar-Shalom. The interacting multiple model algorithm for systems with Markovian switching coefficients. *IEEE Trans. on Autom. Control*, 38(3):780–783, 1998.

[35] T. Bollerslev and P.E. Rossi. In P.E. Rossi, editor, *Introduction to Modelling Stock Market Volatility. Bridging the Gap to Continuous Time.* Academic Press, New York, 1996.

[36] A.A. Borovkov. Boundary-value problems for random walks and large deviations in function spaces. *Theory Probab. Appl.*, 12:575–595, 1967.

[37] D. Brigo and B. Hanzon. On some filtering problems arising in mathematical finance. The interplay between insurance, finance, and control. *Insurance Math. Econ.*, 22(1):53–64, 1998.

[38] K. Burdzy, R. Holyst, and P. March. A Fleming-Viot particle representation of Dirichlet Laplacian. *Commun. Math. Phys.*, 214:679–703, 2000.

[39] B.P. Carlin, N.G. Polson, and D.S. Stoffer. A Monte-Carlo approach to nonnormal and nonlinear state-space modeling. *J. Am. Stat. Assoc.*, 87(418):493–500, 1992.

[40] R.A. Carmona and S.A. Molchanov. Parabolic Anderson model and intermittency. *Mem. Am. Math. Soc. 108*, no. 518, (1994).

[41] R.A. Carmona, S.A. Molchanov, and F.G. Viens. Sharp upper bound on exponential behavior of a stochastic partial differential equation. *Random Operators Stochastic Equations*, 4(1):43–49, 1996.

[42] R.A. Carmona and F. Viens. Almost-sure exponential behavior of a stochastic Anderson model with continuous space parameter. *Stochastics Stochastics Rep.*, 62(3-4), 251–273, 1998.

[43] H. Carvalho, P. Del Moral, A. Monin, and G. Salut. Optimal nonlinear filtering in GPS/INS integration. *IEEE Trans. Aerosp. Electron. Syst.*, 33(3):835–850, 1997.

[44] O. Catoni. Rough large deviations estimates for simulated annealing: Application to exponential schedules. *Ann. Probab.*, 20:1109–1146, 1992.

[45] R. Cerf. Asymptotic convergence of a genetic algorithm. *C. R. Acad. Sci. Paris Sér. I Math.*, 319(3):271–276, 1994.

[46] R. Cerf. A new genetic algorithm. *C. R. Acad. Sci. Paris Sér. I Math.*, 319(9):999–1004, 1994.

[47] R. Cerf. A new genetic algorithm. *Ann. Appl. Probab.*, 6(3):778–817, 1996.

[48] R. Cerf. Asymptotic convergence of genetic algorithms. *Adv. Appl. Probab.*, 30(2):521–550, 1998.

[49] F. Cérou, P. Del Moral F. LeGland, and P. Lezaud. *Genetic genealogical models in rare event analysis*. Publications du Laboratoire de Statistiques et Probabilités, Toulouse III, 2002.

[50] M. Chaleyat-Maurel and D. Michel. Des résultats de non existence de filtres de dimension finie. *C. R. Acad. Sc. de Paris Série I Math.*, 296, no. 22, 933–936, 1983.

[51] R. Chen, J.S. Liu, and W.H. Wong. Rejection control and sequential importance sampling. *J. Am. Stat. Assoc.*, 93(443):1022–1031, 1998.

[52] S. Chib, S. Kim, and S. Shephard. Stochastic volatility: likelihood inference and comparison with ARCH models. *Rev. Econ. Stud.*, 65:361–394, 1998.

[53] S. Chib, F. Nardari, and N. Shephard. Markov chain Monte-Carlo methods for generalized stochastic volatility models. *J. of Econ.* 108, 281-316, 1998.

[54] Y.S. Chow and H. Teichter. *Probability Theory, Independence, Interchangeability and Martingales*, 2nd ed., Springer Texts in Statistics, Springer-Verlag, New York, 1988.

[55] K.L. Chung. *A Course in Probability Theory*. A Series of Monographs and Textbook, 2nd Ed., Probability and Mathematical Statistics, vol. 21, Academic Press, New York, 1974.

[56] G.A. Churchill. Stochastic models for heterogeneous DNA sequences. *Bull. Math. Biol.*, 51:79–94, 1989.

[57] T.C. Clapp and S.J. Godsill. Fix lag smoothing using sequential importance sampling. In A.P. Dawid, J.M. Bernardo, J.O. Berger, and A.F.M. Smith, editors, *Bayesian Statistics*, pages 743–752. Oxford University Press, Oxford, 1999.

[58] C.S. Clark. Multiple model adaptive estimation and control redistribution performance on the VISTA F-16 during partial actuator impairments. MS Thesis, School of Engineering, Air Force Institute of Technology, Wright-Patterson AFB, OH, 1997.

[59] J.M.C. Clark, D.L. Ocone, and C. Coumarbatch. Relative entropy and error bounds for filtering of Markov processes. *Math. Control Signal Syst.*, 12(4):346–360, 1999.

[60] J.E. Cohen, Y. Iwasa, G. Rautu, M.B. Ruskai, E. Seneta, and G. Zbaganu. Relative entropy under mappings by stochastic matrices. *Linear Algebra Appl.*, 179:211–235, 1993.

[61] F. Comets. Large deviations for a conditional probability distribution. Applications to random interacting Gibbs measures. *Probab. Theory Related Fields*, 80:407–432, 1989.

[62] H. Crámer. Sur un nouveau théorème limite de la théorie des probabilités. *Act. Sci. et ind.*, 3:5–23, 1938.

[63] D. Crisan, J. Gaines, and T.J. Lyons. A particle approximation of the solution of the Kushner-Stratonovitch equation. *SIAM J. Appl. Math.*, 58(5):1568–1590, 1998.

[64] D. Crisan and T.J. Lyons. Nonlinear filtering and measure valued processes. *Probab. Theory Related Fields*, 109:217–244, 1997.

[65] D. Crisan and T.J. Lyons. A particle approximation of the solution of the Kushner-Stratonovitch equation. *Probab. Theory Related Fields*, 115(4):549–578, 1999.

[66] D. Crisan, P. Del Moral, and T.J. Lyons. Interacting particle systems approximations of the Kushner-Stratonovitch equation. *Adv. in Appl. Probab.*, 31(3):819–838, 1999.

[67] D. Crisan, P. Del Moral, and T.J. Lyons. Non linear filtering using branching and interacting particle systems. *Markov Processes Related Fields*, 5(3):293–319, 1999.

[68] I. Csiszár. Eine informationstheoretische Ungleichung und ihre Anwendung auf den Beweis der Ergodizität von Markoffschen Ketten. *Magyar Tud. Akad. Mat. Kutató Int. Közl.*, 8:85–108, 1963.

[69] I. Csiszar. Sanov property, generalized i-projection and a conditional limit theorem. *Ann. Probab.*, 12(3):768–793, 1984.

[70] D. Dacunha-Castelle. Formule de Chernov pour une suite de variables réelles. In *Grandes déviations et Applications Statistiques*, pages 19–24. Astérisque 68, Paris, 1979.

[71] M. Davy, P. Del Moral, and A. Doucet. Methodes Monte-Carlo sequentielles pour l'analyse spectrale bayesienne. In *Proceedings of the GRETSI Conference*, Paris 2003.

[72] D. Dawson. Measure-valued Markov processes. In P.L. Hennequin, editor, *Lectures on Probability Theory. Ecole d'Eté de Probabilités de Saint-Flour XXI-1991*, Lecture Notes in Mathematics 1541. Springer-Verlag, Berlin, 1993.

[73] D. Dawson and J. Gärtner. Large deviations from the McKean Vlasov limit for weakly interacting diffusions. *Stochastics*, 20:247–308, 1987.

[74] D. Dawson and J. Gärtner. Analytic aspects of multilevel large deviations. In *Asymptotic Methods in Probability and Statistics (Ottawa, ON, 1997)*, pages 401–440. North-Holland, Amsterdam, 1998.

[75] P. Del Moral. Non-linear filtering: interacting particle resolution. *Markov Processes Related Fields*, 2(4):555–581, 1996.

[76] P. Del Moral. Measure valued processes and interacting particle systems. Application to nonlinear filtering problems. *Ann. Appl. Probab.*, 8(2):438–495, 1998.

[77] P. Del Moral and M. Doisy. Maslov idempotent probability calculus. Part I. *Theory Probab. Appl.*, 43(4):735–751, 1998.

[78] P. Del Moral and M. Doisy. Maslov idempotent probability calculus. Part II. *Theory Probab. Appl.*, 44(2):384–400, 1999.

[79] P. Del Moral and M. Doisy. On the applications of Maslov optimization theory. *Math. Notes*, 69(2):232–244, 2001.

[80] P. Del Moral and A. Doucet. On a class of genealogical and interacting metropolis models. J. Azéma, M. Emery, M. Ledoux, and M. Yor, editors, *Séminaire de Probabilités XXXVII*, Lecture Notes in Mathematics no. 1832, pp. 415–446. Springer-Verlag, Berlin, 2004.

[81] P. Del Moral, A. Doucet, and G. Peters. Sequential Monte Carlo samplers, *Technical Report, Cambridge University*, CUED/F-INFENG/TR 443, Dec. 2002.

[82] P. Del Moral and A. Doucet. Particle motions in absorbing medium with hard and soft obstacles. To appear in *Stochastic Analysis and Applications*, 2004.

[83] P. Del Moral and A. Guionnet. Large deviations for interacting particle systems. Applications to nonlinear filtering problems. *Stochastic Processes Appl.*, 78:69–95, 1998.

[84] P. Del Moral and A. Guionnet. A central limit theorem for nonlinear filtering using interacting particle systems. *Ann. Appl. Probab.*, 9(2):275–297, 1999.

[85] P. Del Moral and A. Guionnet. On the stability of measure valued processes with applications to filtering. *C. R. Acad. Sc. de Paris Série I Math.*, 329(5):429–434 (1999).

[86] P. Del Moral and A. Guionnet. On the stability of interacting processes with applications to filtering and genetic algorithms. *Ann. Inst. Henri Poincaré*, 37(2):155–194, 2001.

[87] P. Del Moral and J. Jacod. Interacting particle filtering with discrete observations. In N.J. Gordon, A. Doucet, and J.F.G. de Freitas, editors, *Sequential Monte-Carlo Methods in Practice*. Springer-Verlag, New York, 2001.

[88] P. Del Moral and J. Jacod. Interacting particle filtering with discrete-time observations: asymptotic behaviour in the Gaussian case. In *Stochastics in Finite and Infinite Dimensions*, Trends in Mathematics, pages 101–122. Birkhäuser, Boston, 2001.

[89] P. Del Moral and J. Jacod. The Monte-Carlo method for filtering with discrete-time observations: Central limit theorems. In *Numerical Methods and stochastics (Toronto, ON, 1999)*, volume 34 of Fields Inst. Commun., pages 29–53. American Mathematical Society, Providence, RI, 2002.

[90] P. Del Moral, J. Jacod, and P. Protter. The Monte Carlo method for filtering with discrete time observations. *Probab. Theory Related Fields*, 120:346–368, 2001.

[91] P. Del Moral, M.A. Kouritzin, and L. Miclo. On a class of discrete generation interacting particle systems. *Electron. J. Probab.*, 6(16):1–26, 2001.

[92] P. Del Moral, L. Kallel, and J. Rowe. Modeling genetic algorithms with interacting particle systems. *Rev. Mat., Teoria apl.*, 8(2):19-78, 2001.

[93] P. Del Moral, M. Ledoux, and L. Miclo. On contraction properties of Markov kernels. *Probab. Theory Related Fields*, 126:395–420, 2003.

[94] P. Del Moral and L. Miclo. On the convergence and the applications of the generalized simulated annealing. *SIAM J. Control Optim.*, 37(4):1222–1250, 1999.

[95] P. Del Moral and L. Miclo. About the strong propagation of chaos for interacting particle approximations of Feynman-Kac formulae. Publications du Laboratoire de Statistique et Probabilités, no. 08-00, Université Paul Sabatier, Toulouse, France, 2000.

[96] P. Del Moral and L. Miclo. Asymptotic results for genetic algorithms with applications to nonlinear estimation. In L. Kallel and B. Naudts, editors, *Proceedings of the Second EvoNet Summer School on Theoretical Aspects of Evolutionary Computing*, Natural Computing Series. Springer-Verlag, New York, 2000.

[97] P. Del Moral and L. Miclo. Branching and interacting particle systems approximations of Feynman-Kac formulae with applications to nonlinear filtering. In J. Azéma, M. Emery, M. Ledoux, and M. Yor, editors, *Séminaire de Probabilités XXXIV*, Lecture Notes in Mathematics 1729, pages 1–145. Springer-Verlag, Berlin, 2000.

[98] P. Del Moral and L. Miclo. A Moran particle system approximation of Feynman-Kac formulae. *Stochastic Processes Appl.*, 86:193–216, 2000.

[99] P. Del Moral and L. Miclo. Genealogies and increasing propagation of chaos for Feynman-Kac and genetic models. *Ann. Appl. Probab.*, 11(4):1166–1198, 2001.

[100] P. Del Moral and L. Miclo. On the stability of non linear Feynman-Kac semi-groups. *Annales de la Faculté des Sciences de Toulouse*, 11(2):135–175, 2002.

[101] P. Del Moral and L. Miclo. Annealed Feynman-Kac models. *Commun. Math. Phys.*, 235(2):191–214, 2003.

[102] P. Del Moral and L. Miclo. Particle approximations of Lyapunov exponents connected to Schrödinger operators and Feynman-Kac semigroups. *ESAIM: Probability and Statistics*, no. 7, pp. 171–208, 2003.

[103] P. Del Moral, J.C. Noyer, G. Rigal, and G. Salut. Traitement particulaire du signal radar, détection, estimation et reconnaissance de cibles aériennes. Technical report, LAAS/CNRS, Toulouse, 1992.

[104] P. Del Moral, J.C. Noyer, and G. Salut. Résolution particulaire et traitement non-linéaire du signal : application radar/sonar. In *Traitement du signal*, (12):4, 287-301, 1995.

[105] P. Del Moral, G. Rigal, and G. Salut. Estimation et commande optimale non linéaire. Technical Report 2, LAAS/CNRS, Toulouse, March 1992. Contract DRET-DIGILOG.

[106] P. Del Moral, G. Rigal, and G. Salut. Estimation et commande optimale non-linéaire : un cadre unifié pour la résolution particulaire. Technical report, LAAS/CNRS, Toulouse, 1992. Contract DRET-DIGILOG-LAAS/CNRS.

[107] P. Del Moral, G. Rigal, and G. Salut. Filtrage non-linéaire nongaussien appliqué au recalage de plates-formes inertielles. Technical report, LAAS/CNRS, Toulouse, 1992. STCAN/DIGILOG-LAAS/CNRS contract no. A.91.77.013.

[108] P. Del Moral and G. Salut. Random particle methods in (max,+) optimization problems. In J. Gunawardena, editor, *Idempotency*, Publications of the Newton Institute, pages 383–392. Cambridge University Press, Cambridge, 1998.

[109] P. Del Moral and T. Zajic. On Laplace-Varadhan's integral lemma. *C. R. Acad. Sci. Paris Série I Math.*, 334(8):693–698, 2002.

[110] P. Del Moral and T. Zajic. A note on the Laplace-Varadhan integral lemma. *Bernoulli*, 9(1):49–65, 2003.

[111] F. Dellaert, D. Fox, W. Burgard, and S. Thrun. Monte-Carlo localization for mobile robots. *IEEE International Conference on Robotics and Automation, ICRA99*, IEEE, New York, 1999.

[112] A. Dembo and O. Zeitouni. *Large Deviations Techniques and Application*. Jones and Bartlett Publishers, Boston, 1993.

[113] H. Derin. The use of Gibbs distributions in image processing. In Blake and H. V. Poor, editors, Communications and Networks, pages 266–298. Springer-Verlag, New York, 1986.

[114] J.-D. Deuschel and D.W. Stroock. *Large Deviations*. Pure and Applied Mathematics 137. Academic Press, New York, 1989.

[115] K.A. Dill T.C. Beutler. *Protein Sci.*, 5(2037), 1996.

[116] R.L. Dobrushin. Central limit theorem for nonstationnary Markov chains, i,ii. *Theory of Probability and its Applications*, 1(1 and 4):66–80 and 330–385, 1956.

[117] M.D. Donsker and R.S. Varadhan. Asymptotic evaluation of certain wiener integrals for large time. *Functional integration and its applications (Proc. Internat. Conf., London, 1974)*, pp. 15–33. Clarendon Press, Oxford, 1975.

[118] M.D. Donsker and S.R.S. Varadhan. Asymptotic evaluation of certain Markov process expectations for large time, i. *Commun. Pure Appl. Math.*, 28:1–47, 1975.

[119] M.D. Donsker and S.R.S. Varadhan. Asymptotic evaluation of certain Markov process expectations for large time, ii. *Commun. Pure Appl. Math.*, 28:279–301, 1975.

[120] M.D. Donsker and S.R.S. Varadhan. Asymptotic evaluation of certain Markov process expectations for large time, iii. *Commun. Pure Appl. Math.*, 29:389–461, 1976.

[121] R. Douc, E. Moulines, and T. Ryden. Asymptotic properties of the maximum likelihood estimator in autoregressive models with Markov regime. Preprint ENST, Paris 2003.

[122] A. Doucet and C. Andrieu. On sequential Monte Carlo sampling methods for Bayesian filtering, *Statistics and Computing*, vol. 10, no. 3, pp. 197-208, 2000.

[123] A. Doucet and C. Andrieu. Particle filtering for partially observed Gaussian state space models. *J. R. Stat. Soc. Ser. B, Stat. Methodol.*, 64(4):827–836, 2002.

[124] A. Doucet, N. de Freitas, and N. Gordon. An introduction to sequential Monte Carlo methods. In *Sequential Monte Carlo Methods in Practice*, Statistics for Engineering and Information Science, pages 3–14. Springer, New York, 2001.

[125] A. Doucet, N. de Freitas, and N. Gordon, editors. *Sequential Monte Carlo Methods in Pratice*. Statistics for engineering and Information Science. Springer, New York, 2001.

[126] J. Dugundji. *Topology*. Prentice-Hall of India, New Delhi, 1975.

[127] P. Dupuis and R.S. Ellis. *A Weak Convergence Approach to the Theory of Large Deviations*. Vol. 18, Wiley Series in Probability and Statistics, John Wiley & Sons, Chichester 2000.

[128] E.B. Dynkin. *An Introduction to Branching Measure-Valued Processes*, vol. 6 of CRM Monograph Series. American Mathematical Society, Providence, RI, 1994.

[129] E.B. Dynkin and A. Mandelbaum. Symmetric statistics, Poisson processes and multiple Wiener integrals. *Ann. Stat.*, 11:739–745, 1983.

[130] S.F. Edwards. The statistical mechanics of polymers with excluded volume. *Proc. Phys. Sci.*, 85:613–624, 1965.

[131] R.J. Elliott and J. van der Hoek. An application of hidden markov models to asset allocation problems. *Finance Stochastics*, 1:229–238, 1997.

[132] R.J. Elliott, L. Aggoun, and J.B. Moore. *Hidden Markov models*. Vol. 29, Applications of Mathematics, Springer-Verlag, New York, 1995.

[133] R.S. Ellis. *Large Deviations and Statistical Mechanics*. Springer-Verlag, New York, 1985.

[134] A. Etheridge and P. March. A note on superprocesses. *Prob. Th. Rel. Fields*, 89:141–147, 1991.

[135] G. Evensen. Sequential data assimilation with a nonlinear quasi-geotrophic model using Monte-Carlo methods to forecast error statistics. *J. Geophys. Res.*, 99:143–162, 1994.

[136] G. Evensen. Application of ensemble integrations for predictability studies and data assimilation. Monte-Carlo simulations in oceanography. Aha Huliko'a Hawaiian Winter Workshop, University of Hawaii at Manoa, 1997.

[137] G. Evensen. Sequential data assimilation for nonlinear dynamics: The ensemble Kalman filter. Oceanographic Forecasting: Conceptual Basis and Applications, N. Pinardi and J.D. Woods, editors. Springer-Verlag, Berlin, Heidelberg, 2002.

[138] G. Evensen and P.J. Van Leeuwen. Assimilation of geostat altimeter data for the agulhas current using an ensemble Kalman filter with a quasi-geotrophic model. *Mon. Weather Rev.*, 124:85–96, 1996.

[139] G. Evensen. The ensemble Kalman filter: Theoretical formulation and practical implementation. To appear in *Ocean Dyn.*, 2003.

[140] G. Evensen and V.E. Haugen. Assimilation of SLA and SST data into an OGCM for the Indian Ocean. *Ocean Dyn.*, 52:133–151, 2002.

[141] G. Evensen and V.E. Haugen. Indian Ocean circulation: An integrated model and remote sensor study. *J. Geophys. Res.*, 107:11–23, 2002.

[142] J.J. Ewell. Space shuttle orbiter entry through land navigation. *IEEE International Conference on Intelligent Robots and Systems*, pages 627–632, IEEE, New York, 1988.

[143] C.M. Ewing N.J. Gordon and D.J. Salmond. Bayesian state estimation for tracking and guidance using the bootstrap filter. *AIAA J. Guidance, Control Dyn.*, 18:1434–1443, 1995.

[144] D. Fox, F. Dellaert, W. Burgard, and S. Thrun. Using the condensation algorithm for robust, vision-based mobile robot localization. In *Proceedings of the IEEE International Conference on Computer Vision and Pattern Recognition, Fort Collins, CO*, IEEE, New York, 1999.

[145] K.A. Fisher and P.S. Maybeck. Multiple model adaptive estimation with filter spawning. *IEEE Trans. Aerosp. Electron. Syst.*, 38(3):755–769, 2002.

[146] G.S. Fishman. *Monte-Carlo. concepts, algorithms and applications.* Springer Series in Operations Research. Springer-Verlag, New York, 1996.

[147] F. Fornari and A. Mele. *Stochastic Volatility in Financial Markets - Crossing the Bridge to Continuous Time.* Kluwer, Dordrecht, 2000.

[148] D. Fox, S. Thrun, F. Dellaert, and W. Burgard. Particle filters for robot localization. In *Sequential Monte-Carlo Methods in Practice*, A. Doucet, N. de Freitas, and N. Gordon, editors, Springer-Verlag, New York, 2000.

[149] H. Frauenkron, M.S. Causo, and P. Grassberger. Two-dimensional self-avoiding walks on a cylinder *Phys. Rev.*, E 59, R16-R19 (1999).

[150] D.R. Fredkin and J.A. Rice. Correlation functions of a finite-state process with application to channel kinetics. *Math. Biosci.*, 87:161–172, 1987.

[151] J.F.G. de Freitas, M. Niranjan, A.H. Gee, and A. Doucet. Sequential Monte-Carlo methods to train neural networks models. *Neural Comput.*, 12(4):955–993, 2000.

[152] J. Gartner, W. Konig, and S.A. Molchanov. Almost sure asymptotics for the continuous parabolic Anderson model. *Probab. Theory Related Fields*, 118(4):547–573, 2000.

[153] M.J.J. Garvel and D.P. Kroese. A comparison of restart implementations. *Proceedings of the 1998 Winter Simulation Conference*, pages 601–608, IEEE Computer society Press, Piscataway, New-Jersey, 1998.

[154] S. Geman and D. Geman. Stochastic relaxation, Gibbs distributions, and the bayesian restoration of images. *IEEE Trans. on Pattern Anal. Mach. Intelligence*, 6:721–741, 1984.

[155] E. Ghysels, A.C. Harvey, and E. Renault. *Stochastic Volatility. Statistical Methods in Finance.* Handbook of Statistics 14, North-Holland, Amsterdam, 1996.

[156] F. Le Gland, C. Musso, and N. Oudjane. An analysis of regularized interacting particle methods for nonlinear filtering. In *Proceedings of the 3rd IEEE European Workshop on Computer-Intensive Methods in Control and Signal Processing*, I. Rojicek, M. Valeckova, M. Karny, and K. Warwick, editors, pp. 167-174, Prague, 1998.

[157] P. Glasserman, P. Heidelberger, P. Shahabuddin, and T. Zajic. Splitting for rare event simulation: analysis of simple cases. *Proceedings of the 1996 Winter Simulation Conference*, pages 302–308, IEEE Computer society Press, Piscataway, 1996.

[158] P. Glasserman, P. Heidelberger, P. Shahabuddin, and T. Zajic. A large deviations perspective on the efficiency of multilevel slipping. *IEEE Trans. on Autom. Control*, 43(12):1666–1679, 1998.

[159] P. Glasserman, P. Heidelberger, P. Shahabuddin, and T. Zajic. Multilevel splitting for estimating rare event probabilities. *Oper. Res.*, 47(4):585–600, 1999.

[160] S. Godsill, A. Doucet, and M. West. Maximum a posteriori sequence estimation using Monte Carlo particle filters. *Ann. Inst. Stat.. Math.*, 53(1):82–96, 2001.

[161] D.E. Goldberg. Genetic algorithms and rule learning in dynamic control systems. In *Proceedings of the First International Conference on Genetic Algorithms*, pages 8–15. L. Erlbaum Associates, Hillsdale, NJ, 1985.

[162] D.E. Goldberg. *Genetic Algorithms in Search, Optimization and Machine Learning.* Addison-Wesley, Reading, MA, 1989.

[163] N.J. Gordon, D.J. Salmon, and C. Ewing. Bayesian state estimation for tracking and guidance using the bootstrap filter. *J. Guidance Control Dyn.*, 18(6):1434–1443, 1995.

[164] N.J. Gordon, D.J. Salmon, and A.F.M. Smith. Novel approach to nonlinear/non-Gaussian Bayesian state estimation. *IEE Proc. F*, 140:107–113, 1993.

[165] C. Graham and S. Méléard. Stochastic particle approximations for generalized Boltzmann models and convergence estimates. *Ann. Probab.*, 25(1):115–132, 1997.

[166] C. Graham and S. Méléard. Probabilistic tools and Monte-Carlo approximations for some Boltzmann equations. In *CEMRACS 1999 (Orsay)*, volume 10 of ESAIM Proceedings, pages 77–126 (electronic). Société de Mathématiques Appliquées et Industrielles, Paris, 1999.

[167] P. Grassberger. Advanced sequential Monte-Carlo methods in physics. In Ed. H. Rollnik and D. Wolk, editors. NIC Series, *Proceedings NIC Symposium*, vol. 9, pages 1–12, John von Neuman Institute for Computing, Jülich, 2002.

[168] U. Grenander. *Elements of pattern theory*. The Johns Hopkins University Press. Baltimore and London, 1996.

[169] A. Greven and F. den Hollander. A variational characterization of the speed of a one dimensional self-repellent random walk. *Ann. Appl. Probab.*, 3:1067–1099, 1993.

[170] P. Groeneboom, J. Oosterhoff, and F.H. Ruymgaart. Large deviation theorems for empirical probability measures. *Ann. Probab.*, 7(4):553–586, 1979.

[171] J.D. Hamilton. A new approach to the economic analysis of nonstationary time series and the business cycle. *Econometrica*, 57:357–384, 1989.

[172] J.D. Hamilton. Analysis of time series subject to changes in regime. *J. Economet.*, 45:39–70, 1990.

[173] T. Hara and G. Slade. The lace expansion for self avoiding walk in five or more dimensions. *Rev. Math. Phys.*, 4:235–327, 1992.

[174] G.H. Hardy, J.E. Littlewood, and G. Pólya. *Inequalities*. Cambridge Mathematical Library. Cambridge University Press, Cambridge, 1988. Reprint of the 1952 edition.

[175] T.E. Harris. Some mathematical models for branching processes. In *Proceedings of the Second Berkeley Symposium on Mathematical Statistics and Probability, 1950*, pages 305–328. University of California Press, Berkeley and Los Angeles, 1951.

[176] T.E. Harris. *The Theory of Branching Processes*. Die Grundlehren der Mathematischen Wissenschaften, Bd. 119. Springer-Verlag, Berlin, 1963.

[177] T.E. Harris and H. Kahn. Estimation of particle transmission by random sampling. *Natl. Bur. Stand. Appl. Math. Ser.*, 12:27–30, 1951.

[178] W.K. Hastings. Monte-Carlo sampling methods using Markov chains and their applications. *Biometrika*, 57:97–109, 1970.

[179] J.H. Hetherington, Observations on the statistical iteration of matrices, *Physical Review A*, vol. 30, no. 5, pp. 2713–2719, 1984.

[180] C. C. Heyde, editor. *Branching Processes*, volume 99 of Lecture Notes in Statistics. Springer-Verlag, New York, 1995.

[181] T. Higushi. Monte-Carlo filter using the genetic algorithm operators. *J. Stat. Comput. Simulation*, 59(1):1–23, 1997.

[182] T. Higushi. Self-organizing time series model. In N.J. Gordon A. Doucet, J.F.G. de Freitas, editors, *Sequential Monte-Carlo Methods in Practice*, pages 428–444. Springer-Verlag, New York, 2001.

[183] J.H. Holland. *Adaptation in Natural and Artificial Systems*. University of Michigan Press, Ann Arbor, 1975.

[184] C. Hue, J.P. Le Cadre, and P. Pérez. A particle filter to track multiple objects. *IEEE Trans. Aerosp. and Electron. Syst.*, 38(3):791-812, 2002.

[185] M. Hurzeler and H.R. Künch. Monte-Carlo approximations for general state space models. *J. Comput. Graphical Stat.*, 7(2):175–193, 1998.

[186] D. Ingerman K. Burdzy, R. Holyst and P. March. Configurational transition in a fleming-viot-type model and probabilistic interpretation of laplacian eigenfunctions. *J. Phys.*, A 29:2633–2642, 1996.

[187] M. Isard and A. Blake. Contour tracking by stochastic propagation of conditional densities. *Computer Vision, ECCV'96*, B. Buxton and R. Cipolla, editors. Springer-Verlag, New York, 1996.

[188] M. Isard and A. Blake. Condensation-conditional density propagation for visual tracking. *Int. J. Comput. Vision*, 29(1):5–28, 1998.

[189] J. Jacod and A.N. Shiryaev. *Limit Theorems for Stochastic Processes*. Series of Comprehensive Studies in Mathematics 288. Springer-Verlag, New York, 1987.

[190] B.H. Juang and L.R. Rabiner. Hidden Markov models for speech recognition. *Technometrics*, 33:251–272, 1991.

[191] J.M. Johnson and Y. Rahmat-Samii. Genetic algorithms in electromagnetics. In *IEEE Antennas and Propagation Society International Symposium Digest*, volume 2, pages 1480–1483. IEEE, New York, 1996.

[192] F. Jouve, L. Kallel, and M. Schoenauer. Mechanical inclusions identification by evolutionary computation. *Eur. J. Finite Elements*, 5(5-6):619–648, 1996.

[193] F. Jouve, L. Kallel, and M. Schoenauer. Identification of mechanical inclusions. In D. Dagsgupta and Z. Michalewicz, editors, *Evolutionary Computation in Engineering*, pages 477–494. Springer-Verlag, New York, 1997.

[194] M. Kac. On distributions of certain wiener functionals. *Trans. Am. Math. Soc.*, 65:1–13, 1949.

[195] G. Kallianpur and C. Striebel. Stochastic differential equations occurring in the estimation of continuous parameter stochastic processes. Tech. Rep. 103, Department of Statistics, University of Minnesota, Minneapolis, 1967.

[196] R.E. Kalman. A new approach to linear filtering and prediction problems. *ASME Trans., J. Basic Engineering*, 82(D):35–50, 1960.

[197] R.E. Kalman and R.S. Bucy. New results in linear filtering and prediction. *ASME Trans., J. Basic Engineering*, 83(D):95–108, 1961.

[198] A. Kaneko and J.H. Park. Assimilation of coastal acoustic tomography data into a barotropic ocean model. *Geophys. Res. Lett.*, 27:3373–3376, 2000.

[199] K. Karplus, C. Barrett, and R. Hughey. Hidden Markov models for detecting remote protein homologies. *Bioinformatics*, 14(10):846-856, 1998

[200] T. Kato. *Perturbation Theory for Linear Operators*. Classics in Mathematics. Springer-Verlag, Berlin, Heidelberg, New York, 1980.

[201] A.I. Khintchin. Über einen neuen grenzwertstatz der wahrscheinlichkeitsrechnung. *Math. Ann.*, 101:745–752, 1929.

[202] S.J. Kim and R.A. Iltis. Performance comparison or particle and extended Kalman filters algorithms for GPS c/a code tracking and interference rejection. *Conference on Information Sciences and Systems. Princeton University*, 2002.

[203] M. Kimmel and D.E. Axelrod. *Branching Processes in Biology*, volume 19 of Interdisciplinary Applied Mathematics. Springer-Verlag, New York, 2002.

[204] G. Kitagawa. Monte-Carlo filter and smoother for non-Gaussian nonlinear state space models. *J. Comput. and Graphical Stat.*, 5(1):1–25, 1996.

[205] V.N. Kolokoltsov and V.P. Maslov. *Idempotent Analysis and Its Applications*, volume 401 of Mathematics and its Applications. Kluwer Academic Publishers Group, Dordrecht, 1997. Translation of *Idempotent Analysis and Its Application in optimal control* (Russian), Nauka Moscow, 1994, with an appendix by P. Del Moral.

[206] T. Koski. *Hidden Markov Models for Bioinformatics*, volume 2 of Computational Biology Series. Kluwer Academic Publishers, Dordrecht, 2001.

[207] J.H. Kotecha and P.M. Djuric. Sequential Monte-Carlo sampling detector for Rayleigh fast-fading channels. *Proceedings of the IEEE International Conference on Acoustics, Speech and Signal Processing, Istanbul, Turkey*, Springer-Verlag, New York, 2000.

[208] M.G. Krein and M.A. Rutman. Linear operators leaving invariant a cone in a Banach space. *American Mathematical Society Translation*, no. 26, 1950.

[209] K. Kremer and K. Binder. Monte carlo simulation of lattice models for macromolecules. *Comput. Phys. Rep.*, 1988.

[210] A.K. Kron, O.B. Ptitsyn, A.M. Skvortsov, and A.K. Fedorov. *Molec. Biol.*, 1(487), 1967.

[211] S. Kullback and R.A. Leibler. On information and sufficiency. *Ann. Math. Stat.*, (22):79–86, 1951.

[212] C. Kwok, D. Fox, and M. Meila. Adapatative real time particle filters for robot localization. *Proceedings of the 2003 IEEE International Conference on Robotics Automation Taipei*, Taiwan, 2003.

[213] D. Lamberton and B. Lapeyre. *Introduction to Stochastic Calculus Applied to Finance*. Chapman and Hall, London, 1996.

[214] G. Lawler. *Intersections of Random Walks. Probability and Its Applications*. Birkhaüser, Boston, 1991.

[215] C.E. Lawrence, S.F. Altschul, M.S. Bogouski, J.S. Liu, A.F. Neuwald, and J.C. Wooten. Detecting subtle sequence signals: A Gibbs sampling strategy for multiple alignment. *Science*, 262:208–214, 1993.

[216] A.R. Leach. *Molecular Modeling, Principles and Applications*. Longman-Harlow, London, 1996.

[217] M. Ledoux and M. Talagrand. *Probability in Banach spaces*. Springer-Verlag, New York, 1991.

[218] P.J. Van Leeuwen and G. Evensen. Data assimilation and inverse methods in terms of a probabilistic formulation. *Mon. Weather Rev.*, 124:2898–2913, 1996.

[219] F. LeGland and N. Oudjane. Stability and uniform approximation of nonlinear filters using the Hilbert metric, and application to particle filters. to appear in The Annals of Applied Probability (2004).

[220] F. LeGland and N. Oudjane. A robustification approach to stability and to uniform particle approximation of nonlinear filters: The example of pseudo-mixing signals. *Stochastic Processes Appl.*, 106(2):279–316, 2003.

[221] A.J. Leigh and V. Krishnamurthy. An improvement to the interacting multiple model algorithm. *IEEE Trans. on Signal Processing*, 49(12):2909–2923, 2001.

[222] D. Lerro and Y. Bar-Shalom. Interacting multiple model tracking with target amplitude feature. *IEEE Trans. Aerosp. Electron. Syst.*, 29(2):494–508, 1993.

[223] J. Li and R.M. Gray. *Image Segmentation and Compression Using Hidden Markov Models*. Kluwer Academic Publishers, Dordrecht, 2000.

[224] X.R. Li. Multiple model with variable structure: Model group switching algorithm. *Proceedings of the 36th Conference on Decision and Control, San Diego, CA*, pages 3114–3119. 1997.

[225] X.R. Li and Y. Bar-Shalom. Multiple model estimation with variable structure. *IEEE Trans. Autom. Control*, 41(4):479–493, 1996.

[226] F. Liang and W.H. Wong. Evolutionary Monte Carlo for protein folding simulations. *J. Chem. Phys.*, 115 (7), pp. 3374–3380, 2001.

[227] J.S. Liu. *Monte-Carlo Strategies in Scientific Computing*. Springer Series in Statistics, Springer, New York, 2001.

[228] J.S. Liu and R. Chen. Sequential Monte-Carlo methods for dynamic systems. *J. Am. Stat. Assoc.*, 93(443):1032–1044, 1998.

[229] J.S. Liu and S. Jensen. Computational discovery of gene regulatory binding motifs: A bayesian perspective. Tech. Rep. Department. of Statistics, Harvard University, Cambridge, 2003.

[230] J.S. Liu, A. Kong, and W.H. Wong. Sequential imputation method and Bayesian missing data problems. *J. Am. Stat. Assoc.*, 89:278–288, 1994.

[231] J.S. Liu, S. Kou, and S. Xie. Bayesian analysis of single molecule experiments. Tech. Rep., Department. of Statistics, Harvard University, Cambridge, 2003.

[232] J.S. Liu and C.E. Lawrence. Bayesian inference on biopolymer models. *Bioinformatics*, 15:38–52, 1999.

[233] J.S. Liu and T. Logvinenko. Bayesian methods in biological sequence analysis. Handbook of Statistical Genetics, 2nd ed. D.J. Balding, M. Bishop, and C. Cannings, editors. Wiley, Chichester, 2003.

[234] J.S. Liu, A.F. Neuwald, and C.E. Lawrence. Bayesian models for multiple local sequence alignment and Gibbs sampling strategies. *J. Am. Stat. Assoc.*, 90(432):1156–1170, 1995.

[235] J.S. Liu and J.Z. Zhang. A new sequential importance sampling method and its applications to the 2-dimensional hydrophobic-hydrophilic model. *J. Chem. Phys.*, 117(7), pp. 3492-3498, 2002.

[236] L. Ljung. *System Identification, Theory for the User.* Prentice Hall Information and System Sciences Series. Prentice Hall, Englewood Cliffs, NJ, 1987.

[237] I. L. MacDonald and W. Zucchii. *Hidden Markov and other Models for Discrete-Valued Time Series.* Chapman and Hall, London, 1997.

[238] D.D. Magill. Optimal adaptive estimation of sampled stochastic processes. *IEEE Trans. on Autom. Control*, 10(4):434–439, 1965.

[239] A.D. Marrs, N.J. Gordon, and D.J. Salmon. Sequential analysis of nonlinear dynamic systems using particles and mixtures. In P.C. Young, W.J. Fitzgerald, A. Walden, and R.L. Smith, editors, *Nonlinear and Nonstationary Signal Processing.* Cambridge University Press, Cambridge, 2001.

[240] V.P. Maslov. *Méthodes opératorielles.* Edition Mir, Moscow, 1987.

[241] P.S. Maybeck and R.I. Suizu. Adaptive tracker field of view variation via multiple model filtering. *IEEE Trans. Aerosp. Electron. Syst.*, 21(4):529–537, 1985.

[242] V. Melik-Alaverdian and M.P. Nightingale, Quantum Monte Carlo methods in statistical mechanics, *Internat. J. of Modern Phys. C*, vol. 10, no. 8, pp. 1409-1418, 1999.

References   543

[243] H.P. McKean, Jr. A class of Markov processes associated with nonlinear parabolic equations. *Proc. Natl. Acad. Sci. U.S.A.*, 56:1907–1911, 1966.

[244] H.P. McKean, Jr. Propagation of chaos for a class of non-linear parabolic equations. In *Stochastic Differential Equations (Lecture Series in Differential Equations, Session 7, Catholic University, 1967)*, pages 41–57. Air Force Office of Scientific Research, Arlington, VA, 1967.

[245] S. Méléard. Asymptotic behaviour of some interacting particle systems; McKean-Vlasov and Boltzmann models. In D. Talay and L. Tubaro, editors, *Probabilistic Models for Nonlinear Partial Differential Equations, Montecatini Terme, 1995*, Lecture Notes in Mathematics 1627. Springer-Verlag, Berlin, 1996.

[246] S. Méléard. Convergence of the fluctuations for interacting diffusions with jumps associated with Boltzmann equations. *Stochastics Stochastics Rep.*, 63(3–4):195–225, 1998.

[247] S. Méléard. Probabilistic interpretation and approximations of some Boltzmann equations. In *Stochastic models (Spanish) (Guanajuato, 1998)*, volume 14 of *Aportaciones Mat. Investig.*, pages 1–64. Soc. Mat. Mexicana, México, 1998.

[248] S. Méléard. Stochastic approximations of the solution of a full Boltzmann equation with small initial data. *ESAIM Probab. Stat.*, 2:23–40, 1998.

[249] N. Metropolis, A.W. Rosenbluth, M.N. Rosenbluth, E.Teller, A.H. Teller. Equation of state calculations by fast computing machines. *J. Chem. Phys.*, 90:233–241, 1953.

[250] S.P. Meyn and R.L. Tweedie. *Markov Chains and Stochastic Stability*. Communications and Control Engineering Series, Springer-Verlag London Ltd., London, 1993.

[251] J.E. Moyal. The general theory of stochastic population processes. *Acta Math.*, 108:1–31, 1962.

[252] J.E. Moyal. Multiplicative population chains. *Proc. R. Soc. Ser. A*, 266:518–526, 1962.

[253] C. Musso and N. Oudjane. Regularized particle schemes applied to the tracking problem. In *International Radar Symposium, Munich, Proceedings*, September 1998.

[254] M. Nagasawa. *Stochastic Processes in Quantum Physics*. Monographs in Mathematics, vol. 94. Birkhäuser-Verlag, Boston, 1991.

[255] D.B. Nelson. Arch models as diffusion approximations. *J. Economet.*, 45(1-2):7–38, 1990.

[256] A.F. Neuwald, J.S. Liu, and C.E. Lawrence. Gibbs motif sampling: Detection of bacterial outer membrane repeats. *Protein Sci.*, 4:1618–1632, 1995.

[257] J.N. Nielsen and M. Vestergaard. Estimation in continuous time stochastic volatility models using nonlinear filters. *Int. J. Theor. Appl. Finance*, 3(2):279–308, 2000.

[258] A. Nix and M.D. Vose. Modelling genetic algorithms with Markov chains. *Ann. Math. Artificial Intelligence*, 5:79–88, 1991.

[259] P-J. Nordlund, F. Gunnarsson and F. Gustafsson. Particle filters for positioning in wireless networks. *Proceedings of EUSIPCO*, Toulouse, France, 2002.

[260] D. Ocone. Entropy inequalities and entropy dynamics in nonlinear filtering of diffusion processes. In *Stochastic Analysis, Control, Optimization and Applications*, Systems Control Foundations and Applications, pages 477–496. Birkhäuser, Boston, 1999.

[261] P. Shahabuddin, P. Glasserman, P. Heidelberger, and T. Zajic. Multilevel splitting for estimating rare event probabilities. *Oper. Res.*, 47(4):585–600, 1999.

[262] E. Pardoux. Filtrage non linéaire et équations aux dérivés partielles stochastiques associées. In P.L. Hennequin, editor, *Ecole d'Eté de Probabilités de Saint-Flour XIX-1989*, Lecture Notes in Mathematics 1464. Springer-Verlag, Berlin, 1991.

[263] M. Peinado. Go with the winners algorithms for cliques in random graphs. *Algorithms Comput.*, 2223:525–536, 2001.

[264] M. Peinado and T. Lengauer. Go with the winners generators with applications to molecular modeling. *Random. Approx. Tech. Comput. Sci.*, 1269:135–149, 1997.

[265] E.A. Perkins. Conditional Dawson-Watanabe processes and Fleming-Viot processes. *Seminar in Stochastic Processes*, pages 142–155, 1991.

[266] D.T. Pham. Stochastic methods for sequential data assimilation in strongly nonlinear systems. *Mon. Weather Rev.*, 129:1194–1207, 1992.

[267] M.S. Pinsker. *Information and Information Stability of Random Variables and Processes*. Holden Day, San Francisco, 1964.

[268] R.G. Pinsky. *Positive Harmonic Functions and Diffusions. an Integrated and Analytic Approach.* Cambridge Studies in Advanced Mathematics, 45. Cambridge University Press, Cambridge, 1995.

[269] M.K. Pitt and N. Shephard. Filtering via simulation: Auxiliary particle filters. *J. Am. Stat. Assoc.*, 93(443):1022–1031, 1998.

[270] M.K. Pitt and N. Sheppard. Filtering via simulation: auxiliary particle filters. *J. Am. Stat. Assoc.*, 94, 590-599, 1999.

[271] D. Pollard. *Convergence of Stochastic Processes.* Springer Verlag, New York, 1984.

[272] A. Puhalskii. On functional principle of large deviations. *New Trends Probab. Stat.*, 1:198–219, 1991.

[273] M. Pulvirenti. Kinetic limits for stochastic particle systems. In *Probabilistic Models for Nonlinear Partial Differential Equations (Montecatini Terme, 1995)*, volume 1627 of Lecture Notes in Mathematics, pages 96–126. Springer, Berlin, 1996.

[274] E. Punskaya, A. Doucet, and W.J. Fitzgerald. Particle Filtering for Joint Symbol and Code Delay Estimation in DS Spread Spectrum Systems in Multipath Environment. To appear in *J. Applied Signal Processing*, 2004.

[275] L.R. Rabiner A tutorial on hidden Markov models and selected applications in speech recognition. *Proc. IEEE*, 77(2):257–285, 1989.

[276] S.T. Rachev. *Probability Metrics and the Stability of Stochastic Models.* Wiley, New York, 1991.

[277] M. Reed and B. Simon. *Methods of Modern Mathematical Physics, II, Fourier Analysis, Self Adjointness.* Academic Press, New York, 1975.

[278] S. I. Resnick. *Adventures in Stochastic Processes.* Birkhäuser, Boston, 1994.

[279] D. Revuz. *Markov Chains.* North Holland, Amsterdam, 1984.

[280] M.N. Rosenbluth and A.W. Rosenbluth. Monte-carlo calculations of the average extension of macromolecular chains. *J. Chem. Phys.*, 23:356–359, 1955.

[281] I.N. Sanov. On the probability of large deviations of random variables. Select. Transl. Math. Statist. and Probability, Vol. 1 pp. 213–244, Inst. Math. Statist. and Amer. Math. Soc., Providence, R.I., 1961.

[282] D. Schultz, W. Burgard, D. Fox, and A.B. Cremers. People tracking with a mobile robot using sample based joint probabilistic data association filters. *Int. J. Robotics Res.*, (22)2, 2003.

[283] E.A. Semerdjiev and L.S. Mihaylova. Adaptative IMM algorithm for manouevring ship tracking. *Proceedings of the first International Conference on Multisource-Multisensor Information Fusion (FUSION'98), Las Vegas, Nevada*, volume 2, pages 974–979, C.S.R.E.A. Press, Athens, Georgia, 1998.

[284] E.A. Semerdjiev, L.S. Mihaylova, and Tz. Semerdjiev. Manouevring ship model identification and imm tracking algorithm design. *Proceedings of the first International Conference on Multisource-Multisensor Information Fusion (FUSION'98), Las Vegas, Nevada*, volume 2, pages 968–973, C.S.R.E.A. Press, Athens, Georgia, 1998.

[285] J. Shapcott. Index tracking: Genetic algorithms for investment portfolio selection. Technical Report SS92-24, EPCC, Edinburgh, September 1992.

[286] T. Shiga and H. Tanaka. Central limit theorem for a system of Markovian particles with mean field interaction. *Z. Wahrschein. Verwandte Gebiete*, 69:439–459, 1985.

[287] A.N. Shiryaev. *Probability*, second edition, Volume 95 in Graduate Texts in Mathematics. Springer-Verlag, New-York, 1996.

[288] G.R. Shorack. *Probability for Statisticians*. Springer Texts in Statistics, Springer, New York, 2000.

[289] D. Siegmund. *Sequential Analysis: Tests and confidence intervals.* Springer Verlag, New York, 1985.

[290] N. Smirnoff. Über wahrscheinlichkeiten grosser abweichungen. *Rec. Soc. Math. Moscow*, 40:441–455, 1933.

[291] D.J. Spielgelhalter, W.R. Gilks, and S. Richardson. *Monte Carlo Markov Chain in Practice.* Chapman and Hall, London, 1996.

[292] R.F. Stengel. *Optimal Control and Estimation.* Dover Publications Inc., New York, 1986.

[293] D.W. Stroock. *An Introduction to the Theory of Large Deviations.* Springer-Verlag, Berlin, 1984.

[294] A.S. Sznitman. Topics in propagation of chaos. In P.L. Hennequin, editor, *Ecole d'Eté de Probabilités de Saint-Flour XIX-1989*, Lecture Notes in Mathematics 1464. Springer-Verlag, Berlin, 1991.

[295] A.S. Sznitman. *Brownian Motion Obstacles and Random Media*. Springer-Verlag, Monographs in Mathematics, New York, 1998.

[296] S. Tindel and F. Viens. Convergence of a branching particle system to the solution of a parabolic SPDE on the circle. *Random Oper. Stochastic Equations, (to appear)*, 2003.

[297] P. Torma and Cs. Szepesvri. Towards facial pose tracking. *In Proc. First Hungarian Computer Graphics and Geometry Conference Budapest*, Hungary, pp. 10-16, 2002.

[298] J. K. Townsend, Z. Haraszti, J. A. Freebersyser, and M. Devetsikiotis, Simulation of rare events in communication networks. *IEEE Commun. Mag.*, Vol. 36, No. 8, pages 36–41, 1998.

[299] D. Treyer, D.S. Weile, and E. Michielsen. The application of novel genetic algorithms to electromagnetic problems. In *Applied Computational Electromagnetics, Symposium Digest*, volume 2, pages 1382–1386, Monterey, CA, March 1997.

[300] S.R.S. Varadhan. Asymptotic probabilities and differential equations. *Commun. Pure Appl. Math.*, 19:261–286, 1966.

[301] A.D. Ventcel and M.I. Freidlin. On small perturbations of dynamical systems. *Russian Math. Surveys*, 25:1–55, 1970.

[302] F. Viens. Portfolio optimization under partially observed stochastic volatility. In *COM- CON 8. The 8th International Conference on Advances in Communication and Control*. W. Wells, editor, pages 1-12. Optim. Soft., Inc., 2002.

[303] M. Villen-Altamirano, A. Martinez-Marron, J. Gamo, and F. Fernandez-Questa. Enhancements of the accelerated simulation method restart by considering multiple thresholds. In *Proceedings of the 14th International Teletraffic Congress. The Fundamental Role of Teletraffic in the Evolution of the Telecommunication Networks*. J. Labetoulle and J.W. Roberts, editors. Elsevier Science Publishers, Amsterdam, pages 797–810, 1994.

[304] M. Villen-Altamirano and J. Villen-Altamirano. Restart: a method for accelerating rare event simulation. In *Proceedings of the 13th International Teletraffic Congress. In Queueing Performance and Control in ATM*, J.W. Cohen and C.D. Pack, editors. Elsevier Science Publishers, Amsterdam, pages 71–76, 1991.

[305] M. Villen-Altamirano and J. Villen-Altamirano. Restart: a straightforward method for fast simulation of rare events. *Proceedings of the 1994 Winter Simulation Conference*, pages 282–289. IEEE Computer Society Press, Piscataway, NJ, 1994.

[306] M. D. Vose. *The Simple Genetic Algorithm, Foundations and Theory.* The MIT Press Books, Cambridge, 1999.

[307] F.T. Wall and J.J. Erpenbeck. *J. Chem. Phys.*, 30:634–637, 1959.

[308] F.T. Wall and F. Mandel. Macromolecular dimensions obtained by an efficient Monte Carlo method without sample attrition. *J. Chem. Phys.*, Vol 63(11) pp. 4592-4595, 1975.

[309] D. Whitley. A genetic algorithms tutorial. *Statistics and Computing*, (4):65-85, 1994.

[310] D. Whitley. An Overview of Evolutionary Algorithms, *J. Information and Software Technology*, 43:817-831, 2001

[311] A.N. Van der Vaart and J.A. Wellner. *Weak Convergence and Empirical Processes with Applications to Statistics.* Springer Series in Statistics. Springer, New York, 1996.

[312] H. Wieland. Unzerlegbare, nicht negative Matrizen . *Math. Z.*, vol. 52, 642-648, 1950.

[313] G. Wrinkler. *Image Analysis, Random Fields and Markov Chain Monte Carlo Methods, a Mathematical Introduction* 2nd edition. Applications of mathematics Series 27, Springer-Verlag, New York, 2003.

[314] Y.M. Zhang and X.R. Li. Detection and diagnostic of sensor and actuator failures using IMM estimation. *IEEE Trans. Aerosp. Electron. Syst.*, 34(4):1295–1312, 1998.

Index

$(E, \mathcal{E})$, 7
$(E_n, \mathcal{E}_n)$, 48
$(\mathbb{K}_n^N)^{\odot q}$, 256
$(\mathbb{K}_n^N)^{\otimes q}$, 256
$(\oplus^N, \odot)$-integral operator, 341
$(n)_p$, 222
E_n^c, 71
$E_{(p,n]}$, 10
$E_{[0,n]}$, 459
$E_{[p,n]}$, 10
$G_{p,n}^{(q)}$, 258
$G_{p,n}$, 89
$H(\mu, \nu)$, 122
$K_{n,\eta}$, 30
$L_{t,\eta}^{(i)}$, 35
$L_{t,\eta}$, 35
$M(f)$, 9
M^n, 10
$M_1 M_2$, 9
$M_{p,n}$, 88
$P_{p,n}^{(q)}$, 258
$P_{p,n}$, 89
$Q_{p,n}^{(q)}$, 258
Q_n, 14
$Q_{p,n}$, 88
$R_{p,q}^{(n)}$, 89
$S_{n,\eta}$, 31, 73
V^+, 35
V^-, 35
$X_{[0,n]}$, 459
$\mathcal{B}_b^n(U)$, 381
$\mathcal{B}_b(E)$, 7
$\mathcal{B}_b(U)$, 364
$\mathcal{B}_b(U_n)$, 380
$\mathbb{E}_{[x]}(.)$, 83
$\mathbb{E}_\mu(.)$, 58
$\mathbb{E}_{p,\mu_p}$, 88
$\mathbb{E}_x(.)$, 58
$\mathcal{F}, \mathcal{F}_n$, 227
$\mathcal{F}_\infty$, 51
$\mathbb{K}_{\eta,n}$, 74
$\mathbb{K}_\eta$, 74
$\mathcal{L}_t^N$, 35
$\mathbb{P}_{\eta_0,[n]}^{(N,q)}$, 114, 255
$\mathbb{P}_{\eta_0,n}^{(N,q)}$, 114, 255
$\mathbb{P}_\mu(.)$, 58
$\mathbb{P}_{p,x_p}$, 87
$\mathbb{P}_x(.)$, 58

$\mathcal{P}(U)$, 364
$\mathcal{P}(U_n)$, 380
$\mathcal{P}^n(E)$, 336
$\mathcal{P}^n(U)$, 380
$\mathcal{P}_n(E_n)$, 60, 70
$\mathbf{P}(E_n)$, 379
$\mathbf{P}^n(E)$, 379
Φ_n, 70
$\Phi_{p,n}$, 89
Ψ_n, 13, 31, 61
$\Psi_{p,n}$, 133
$\mathbb{Q}_n$, 11
$\mathbb{Q}_t$, 11
Θ_n^q, 255
$\Theta_{p,n}^q$, 255
$\mathcal{Z}_n$, 11, 58, 63
$\mathcal{Z}_t$, 11
$\mathcal{B}_b(E)$, 7
$\beta(M)$, 127
$\beta(\Phi)$, 132
$\beta_H(\Phi)$, 132
$\epsilon_{p,n}(G)$, 139
η_n^N, 31, 111
η_t^N, 36
$\eta_n'^N$, 111
η_n', 63
η_n, 12, 30, 59, 88
η_t, 12
γ_n^N, 31, 111
γ_t^N, 36
$\gamma_n'^N$, 111
γ_n', 63
γ_n, 12, 30, 59, 61, 88
γ_t, 12
$\lambda(V)$, 24
$\langle q, N \rangle$, 256, 267
$\langle q \rangle$, 256
$\mathrm{Ent}_U(.,.)$, 365
$\mathrm{Osc}_1(E)$, 8
$\mathrm{osc}(f)$, 8
μM, 9
$\overline{a}_H$, 138
$\mathcal{P}(E)$, 7
τ-topology, 337
τ_1-topology, 379

$\tilde{Q}_{p,n}^{\mu_p}$, 144
$\widehat{E}_n$, 66, 68, 92
$\widehat{G}_n$, 65, 68, 70
$\widehat{G}_{p,n}$, 92
$\widehat{M}_n$, 65, 68, 70
$\widehat{P}_{p,n}$, 92
$\widehat{Q}_{p,n}$, 92
$\widehat{R}_{p,q}^{(n)}$, 92
$\widehat{\Phi}_n$, 70
$\widehat{\Phi}_{p,n}$, 92
$\widehat{\Psi}_n$, 70
$\widehat{\eta}_n$, 60, 91
$\widehat{\gamma}_n$, 60, 61, 91
ξ_n^i, 31, 97
ξ_t^i, 35
$\xi_n^{(N)}$, 31, 97
$\xi_t^{(N)}$, 35
$\xi_n^{(N,i)}$, 31, 97
$\xi_t^{(N,i)}$, 35
$\xi_{p,n}'^i$, 105
ξ_n, 31, 97
ξ_t, 35
$d_U(.,.)$, 365
$m(X)$, 221, 271
$m(\xi_n)$, 97
$m(x)$, 221, 267
$m(x)^{\odot q}$, 267
$m(x)^{\otimes q}$, 267
$\mathbf{Q}_{p,n}$, 91
$\mathcal{M}(E)$, 7
$\mathcal{M}_+(E)$, 7
$\mathcal{M}_0(E)$, 7
Ent(.), 8
(max,+)-semiring, 341

Absorbed particle, 68, 72
Absorbing condition, 446
Absorbing medium, 22
Absorption events, 71, 442
Acceptance/rejection, 41
Accessibility condition, 66, 67, 92
Adaptive dynamic, 41
Adaptive stochastic search, 40

Index

Additive set functions, 379
Ancestor, 33, 397
Ancestral line, 33, 36, 105
Angle bracket, 236, 307
Approximation measures, 111
Asymptotic stability, 122
Auto-regressive model, 56

Baker's selection, 388, 424
Baldi and Dembo-Zeitouni theorem, 351
Ballistic events, 446, 450
Bayesian
    prior and posterior, 21, 499
Birth and death process, 33, 36, 55, 446, 450, 457
Boltzmann
    operator, 69
    rarefied gas models, 108
Boltzmann entropy, 122
Boltzmann-Gibbs
    asymptotic properties, 271
    distribution, 32
    transformation, 13, 60
Bootstrap filters, 41
Branching and interacting particle systems, 41, 405
Branching excursions model, 404
Branching selections, 41, 388
Buffer overflows, 430

Canonical
    chain, 51
    space, 51
Cemetery state, 68, 71, 447
Central limit theorem, 291
    particle density profiles, 300, 301
    path space models, 322
    triangular arrays, 291, 294, 295
Chain growth methods, 488
Change of reference probability, 63, 459, 499
Chemical bonds, 484

Coffin state, 68
Colliding molecules, 40
Combinatorial transport equation, 267
Communication networks, 431, 501
Compatibility condition, 107
    continuous time, 23
    discrete time, 76
Condensation filters, 41, 429
Conditional explorations, 400, 404, 423, 492, 493
Conditions
    (G), 115
    $(I)_m$, 139
    $(L)_m$, 182
    $(M)_m$, 116
    $(Q)_m$, 139
    $(\mathcal{A})$, 67, 220
    $(\mathcal{B})$, 220
    $(M)^{(p)}$, 116
    $(M)^{\exp}$, 116
    (Φ), 472
    $(\widehat{G})$, 147
    $(\widehat{M})_m$, 147
    $(\widehat{Q})_m$, 147
    $(\mathcal{H}_a)$, 135
Connecting maps, 360
Connective constant, 497
Continuous mapping theorem, 299
Contraction coefficients, 132, 138, 472, 508
Coordinate method, 50
Covering numbers, 227
Crámer technique, 333
Creation and killing, 40
Csiszar divergence, 123
Cylinder set, 50

Data assimilation, 21
Dawson-Gärtner
    projective methods, 333, 359
Delta method, 291, 299
Descendant genealogy, 397
Directed polymer, 427

Dirichlet problems, 54, 427, 430, 431, 440
Disintegration, 396, 397
DNA sequences, 428
Dobrushin ergodic coefficient, 127
Domb-Joyce model, 496
Donsker's theorem, 292, 318
    particle models, 318
Dynkin-Mandelbaum theorem, 323, 326

Economical time series, 428
Edwards' model, 496
Elementary transition, 49
Energy function, 78
Ensemble Kalman filters, 41, 429
Entropy integral, 228
Evolutionary mathematics, 40
Exchangeable measure, 262
Excursion particles, 431
Excursion-space models, 52
Exploration, 71
Exponential tightness, 349
    particle models, 351
Extended Black-Scholes model, 503
Extended Kalman-Bucy filter, 428
Extinction Probabilities, 231

Feynman-Kac measures
    annealed models, 83
    conditional, 396
    continuous time models, 11, 12
    discrete time models, 11, 12, 47
    distribution flows, 58, 68
    distribution space models, 85
    excursion-space models, 431
    normalizing constant, 12
    path space models, 34, 62, 110
    prediction models, 60, 88, 110
    quenched models, 83
    random medium models, 81
    time marginals, 34
    unnormalized models, 60, 110
    updated models, 60, 88, 110
Feynman-Kac semigroups
    contraction properties, 132
    functional inequalities, 134, 137
    McKean models, 277
    oscillations, 133
    prediction models, 88
    stochastic models, 152
    updated models, 91
    weak regularity properties, 144
Feynman-Kac-Metropolis models, 164, 166
Financial mathematics, 41
Fluctuations, 113, 450

Galton-Watson model, 26
Gateaux differentiability, 351
Genealogical tree, 25, 396, 450
    descendant and ancestral genealogies, 396
    interacting particle models, 95
    models, 33, 36, 103
Genetic
    algorithms, 41, 77
    particle model, 40
    population, 25
Gibbs sampling, 41, 393
Glivenko-Cantelli theorem, 241
Global optimization, 41
Global positioning system, 15, 428, 500
Go with the winner, 41, 102

h-relative entropy, 122
    variational representation, 136
Hahn-Jordan decomposition, 125
Hamiltonian function, 26, 486
Hausdorff topological space, 339
Havrda-Charvat entropy, 122
Hellinger integrals, 122
Hidden Markov models, 504, 521
Hilbert-Schmidt operator, 323
Historical process, 52, 64, 103

Idempotent analysis, 332, 340

Idempotent probability measures, 332
Image processing, 41
Importance sampling, 460, 465
Inequalities
    Bürkholder, 223
    Bernstein, 222
    Berry-Esseen, 292, 306
        martingale sequences, 306, 309, 310
        particle models, 311
    Chernov-Hoeffding, 223
    Csiszar, 262
    Khinchine, 223
    Marcinkiewicz-Zygmund, 223
Infinitesimal neighborhood, 49
Integral operator, 9
Interacting jump, 75
Interacting Kalman-Bucy filters, 518
Interacting Metropolis models, 29, 41, 389
Interacting particle systems, 95, 394
Interacting process interpretation, 73, 394
Invariant measures, 157, 472
    existence and uniqueness, 160
Ionescu-Tulcea theorem, 459
Ising model, 389

Jump generator, 24

Kakutani-Hellinger integrals, 122
Kallianpur-Striebel formula, 16
Kalman-Bucy filters, 79
Killing, 22, 443
    annealed properties, 198
    interpretation, 71
    transition, 71
Kolmogorov-Smirnov metric, 365
Kushner-Stratonovitch equation, 17

Laplace-Varadhan lemma, 333, 352, 354
    extended version, 354
Large-deviation principles, 113, 331
    definition, 335
    lower bound, 340
    McKean models, 337, 374
    upper bound, 339
    weak principles, 340
Lebesgue decomposition, 123
Legendre-Fenchel transformation, 348
Levy's convergence theorem, 291
Lifetime, 23
Likelihood
    asymptotic properties, 510, 521
    functions, 504, 506
    ratio, 461
Lindeberg condition, 297
Logarithmic addition/multiplication, 340
Logarithmic moment-generating function, 349
Lower semicontinuity, 340
Lyapunov exponent, 469

Macromolecules, 484
Markov chain, 48
    canonical model, 50
    excursion-space models, 52
    nonhomogeneous, 58
    path-space models, 51, 52
    stopped models, 52
Markov kernel
    definition, 9
    operator, 9
McKean
    interpretations, 75, 77, 394
    measures, 68, 74, 111
    models, 76
Mean field particle process, 74
Metropolis-Hastings models, 41, 164, 488
Micro-statistical mechanics, 23, 40
Mixing conditions, 139
Moment-generating function
    independent sequences, 224

interacting models, 247
Monomers, 484
Multiple Hypothesis Testing algorithm, 518
Multiple models estimation, 501
Multiple Wiener integrals, 324
Multisplitting method, 428, 429, 439, 451, 463
Mutation, 32, 98
Myopic self-avoiding walks, 496

Nanbu particle model, 108
Natural evolution models, 40
Nonlinear filtering, 427, 497
    conditional distribution, 16
    definition, 15
    discrete time formulation, 17
    discrete time observations, 17
    observation process, 15
    partially linear/Gaussian, 513
    robust equation, 18
    signal process, 15
    speech separation, 20
    stability properties, 153, 508
    stochastic volatility, 20
    tracking problems, 19

Obstacles, 68, 440
    hard and soft, 22, 72, 475
    repulsive, 73, 475
Occupation measure, 33
Ocean prediction, 428
Offsprings, 39
One-dimensional neutron model, 149, 469
One-step predictor, 154, 505
Optimal control, 522
Optimal filter, 154, 505

Parabolic Anderson model, 48
Particle approximation measure, 109
Particle filters, 41, 429, 512
Particle genealogy, 103
Particle Lyapunov exponents, 473

Particle regulation, 522
Particle simulation, 41
Particles, 30
Path particles, 111
Path-space models, 51
Perturbation sequence, 15
Perturbation theory, 469
Poisson problem, 55
Polish space, 333
Polymers, 40, 484
    degree of polymerization, 25, 484
    directed polymers, 25
    intermolecular interaction, 25
    nonintersecting chains, 25
    simulation models, 487, 490
    solvent, 25
Positive operators, 469
Potential
    creation and killing, 23
Prediction, 70
Preordered set, 359
Projective limit space, 360
Projective limit topology, 359
Projective spectrum, 360
Propagation of chaos, 113
    entropy estimates, 259
    strong chaoticity, 257
    total variation estimates, 260
    weak chaoticity, 253
Protein-folding problems, 388
Prune enrichments, 41, 488, 495

Quadratic characteristic, 307
Quantum physics, 22
Quenched Kalman-Bucy filters, 515
Queueing model, 56

Radar processing, 15, 497
Raleigh-Ritz principle, 470
Random excursion models, 429, 448, 459, 489, 493
Random medium, 81
Rare events, 427, 430, 444, 463
Ratio tests, 462

Index 555

Reconfiguration, 41
Regular topological space, 339
Reinforced random walks, 487, 490
Rejuvenation, 41
Relaxation time, 144, 152, 191
Remainder stochastic sampling, 388
Reptilian algorithms, 487
Repulsive/attractive interaction, 485, 488
Resampling, 41
Restart method, 41, 431
Restricted Markov chains, 421
Ruin process, 446, 455

Sampling-importance-resampling, 41, 429
Sanov theorem, 363, 373
Satellite constellation, 501
Schrödinger
    equations, 23
    operator, 23, 24
        top eigenvalue/vector, 24
    semigroups, 427, 469
Selection, 32, 98
Self-avoiding random walks, 388, 489, 493, 495
Sequential Monte Carlo methods, 37, 421, 429
Shannon-Kullback information, 122
Shiga-Tanaka formula, 323
Simple random walk, 55
Skorohod theorem, 299
Slithering tortoise algorithms, 487
Slutsky's technique, 291, 298
Spawning, 41
Spectral analysis, 469, 477
Spectral radius, 470, 477
Statistical hypothesis, 462, 502
Stein lemma, 307
Stein's technique, 307
Stochastic linearization, 39
Stochastic volatility, 503
Stopped process, 12
Storage and dam model, 55
Strong contraction estimates, 142

Strong law of large numbers, 231
Sub-Markov property, 68
Switching, 41
Switching models, 502

Telecommunication analysis, 41
Time uniform estimate, 244
Toeplitz-Kronecker lemma, 194
Topological space, 339
Topology, 339
Total variation distance, 124
Trace class operator, 324
Tracking problems, 428, 497, 500
Transport problem, 103
Trapping analysis, 22
Trapping interpretation, 68
Tychonoff's theorem, 360
Type I/II errors, 463

unbiased estimate, 112
Updating, 70
Upper semi-continuity, 340
Upper semicontinuity, 340
Urn model, 56

Vague topology, 335
Variational entropy formula, 366

Weak topology, 335
Weak-⋆ topology, 336
Weighted bootstrap, 407

Zolotarev seminorm, 227